Connecting the Concepts

We find that when students understand the connections among the real zeros of a function, the solutions of its associated equation, and the first coordinates of the x-intercepts of its graph, a door opens to a new level of mathematical comprehension that increases the probability of success in this course. As a result, we emphasize zeros, solutions, and x-intercepts throughout the text. A summary of the connections among the concepts is shown graphically and algebraically below.

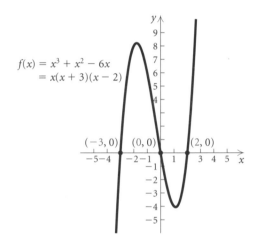

The **x-intercepts** of the graph of $f(x) = x^3 + x^2 - 6x$:

$$(-3, 0), \quad (0, 0), \quad \text{and} \quad (2, 0)$$

The **zeros** of the function $f(x) = x^3 + x^2 - 6x$:

$$-3, \quad 0, \quad \text{and} \quad 2$$

The **solutions** of the equation $x^3 + x^2 - 6x = 0$:

$$-3, \quad 0, \quad \text{and} \quad 2$$

- The **zeros** of the *function* $f(x) = x^3 + x^2 - 6x$ are the **x-coordinates of the x-intercepts** of the graph of $f(x)$.
- The **solutions** of the *equation* $x^3 + x^2 - 6x = 0$ are the **x-coordinates of the x-intercepts** of the graph of $f(x) = x^3 + x^2 - 6x$.
- The **zeros** of the *function* $f(x) = x^3 + x^2 - 6x$ are the **solutions** of the *equation* $x^3 + x^2 - 6x = 0$.

College Algebra

SECOND EDITION

College Algebra

SECOND EDITION

Judith A. Beecher

Indiana University Purdue University Indianapolis

Judith A. Penna

Indiana University Purdue University Indianapolis

Marvin L. Bittinger

Indiana University Purdue University Indianapolis

PEARSON

Addison
Wesley

Boston   San Francisco   New York
London   Toronto   Sydney   Tokyo   Singapore   Madrid
Mexico City   Munich   Paris   Cape Town   Hong Kong   Montreal

Publisher	Greg Tobin
Senior Acquisitions Editor	Anne Kelly
Executive Project Manager	Kari Heen
Assistant Editor	Katie Nopper
Editorial Assistant	Cecilia Fleming
Senior Production Supervisor	Karen Wernholm
Editorial and Production Services	Martha K. Morong/Quadrata, Inc.
Art Editor	Geri Davis/The Davis Group, Inc.
Senior Marketing Manager	Becky Anderson
Marketing Coordinator	Carolyn Buddeke
Associate Media Producer	Lynne Blaszak
Software Development	Mary Dougherty and David Malone
Senior Manufacturing Buyer	Evelyn Beaton
Cover Design	Dennis Schaefer
Text Design	Geri Davis/The Davis Group, Inc.
Composition	Beacon Publishing Services
Illustrations	Network Graphics and William Melvin

For permission to use copyrighted material, grateful acknowledgment is made to the copyright holders on page A-56, which is hereby made part of this copyright page.

Library of Congress Cataloging-in-Publication Data

Beecher, Judith A.

    College algebra/Judith A. Beecher, Judith A. Penna, Marvin L. Bittinger.--
2nd ed.

        p. cm.

    Includes index.

    ISBN 0-321-15934-9

    1. Algebra. I. Penna, Judith A. II. Bittinger, Marvin L. III. Title.

QA152.3.B44 2004

512.9--dc22   2003060898

4 5 6 7 8 9 10—VH—070605

For Our Mothers

Mary Jane Swank

Jean Melvin

Ersle Gosnell

Contents

Preface

Our challenge, and our goal, when writing this textbook was to do everything possible to help you learn the concepts and skills contained between its covers. Every feature we have included was put here with this in mind. We realize that your time is both valuable and limited, so we communicate in a highly visual way that allows you to focus easily and learn quickly and efficiently. Take advantage of the side-by-side algebraic solutions and their visualizations, the Visualizing the Graph and the Connecting the Concepts features, the vocabulary reviews, the "classify the function" exercises, and the Study Tips. We included them to enable you to make the most of your study time and to be successful in this course.

Best wishes for a positive learning experience,

Judy Beecher
Judy Penna
Marv Bittinger

CONTENT FEATURES

College Algebra, Second Edition, covers college-level algebra and is appropriate for a one-term course in precalculus mathematics. We introduce functions earlier than most college algebra texts, and our approach is more visual as well. Although a course in intermediate algebra is a prerequisite for using this text, Chapter R, "Basic Concepts of Algebra," provides sufficient review to unify the diverse mathematical backgrounds of most students.

In this second edition, we have responded to reviewers' suggestions by reorganizing and expanding the content and increasing the number of exercises. In particular, Section 1.5, More on Functions, has been divided into two sections, 1.5 and 1.6. Composition of functions has been moved from Section 4.1 to Section 1.6, where it appears with the material on the algebra of functions, and graphing of polynomial functions has been added and emphasized in Section 3.1. In addition, the second edition has nearly 1200 more exercises than the first edition. This is a 28% increase in the number of exercises.

- **Function Emphasis** Functions are the core of this course and should be presented as a thread that runs throughout the course rather than as an isolated topic. We introduce functions in Chapter 1, whereas most traditional college algebra textbooks cover equation-solving in Chapter 1. (See pp. 69, 77–79, 174, and 191.) Our approach introduces students to a relatively new concept at the beginning of the course rather than requiring them to begin with a review of what was previously covered in intermediate algebra. The concept of a function can be challenging for students. By repeatedly exposing them to the language, notation, and use of functions, demonstrating visually how functions relate to equations and graphs, and also showing how functions can be used to model real data, we hope that students will not only become comfortable with functions but also come to understand and appreciate them.

- **Visual Emphasis** Our early introduction of functions allows graphs to be used to provide a visual aspect to solving equations and inequalities. For example, we are able to show the students both algebraically and visually that the solutions of a quadratic equation $ax^2 + bx + c = 0$ are the zeros of the quadratic function $f(x) = ax^2 + bx + c$ as well as the x-intercepts of the graph of that function. This makes it possible for students, particularly visual learners, to gain a quick understanding of these concepts. (See pp. 200 and 305.)

New and Improved!

- **Emphasis on Graphing** To further enhance the visual aspect of the material, there is an increased emphasis on graphs and graphing in this edition, again reminding students how important and informative visualization can be in the discipline of mathematics. In particular, graphing of polynomial functions by hand has been added and emphasized in Section 3.1. (See pp. 74, 207–212, 252–258, 344, and 365.)

New!

- **Visualizing the Graph** We have added a new feature, Visualizing the Graph. This provides students with an opportunity to match an equation with its graph by focusing on the characteristics of the equation and the corresponding attributes of the graph. This feature appears at least once in every chapter. (See pp. 216, 300, and 368.)

- **Zeros, Solutions, and *x*-Intercepts Theme** We find that when students understand the connections among the real zeros of a function,

the solutions of its associated equation, and the first coordinates of the
x-intercepts of its graph, a door opens to a new level of mathematical
comprehension that increases the probability of success in this course. We
emphasize zeros, solutions, and x-intercepts throughout the text by using
consistent, precise terminology and including exceptional graphics. Seeing
this theme repeated in different contexts leads to a better understanding
and retention of these concepts. (See pp. 193 and 203.)

New!
- **Classifying the Function**   With a focus on conceptual understanding, students are asked periodically to identify a number of functions by their type (for example, linear, quadratic, rational, and so on). As students progress through the text, the variety of functions they know increases and these exercises become more challenging. The "classifying the function" exercises appear with the review exercises in the Skill Maintenance portion of an exercise set. (See pp. 285 and 446.)

- **Side-by-Side Feature**   Many examples are presented in a side-by-side, two-column format in which the algebraic solution of an equation appears in the left column and a graphical interpretation of the solution appears in the right column. (See pp. 193 and 386.) This enables students to visualize and comprehend the connections among the solutions of an equation, the zeros of a function, and the x-intercepts of the graph of a function.

New!
- **Vocabulary Review**   This new feature checks and reviews students' understanding of the vocabulary introduced throughout the text. It appears once in every chapter, in the Skill Maintenance portion of an exercise set, and is intended to provide a continuing review of the terms that students must know in order to be able to communicate effectively in the language of mathematics. (See pp. 124, 236, 304, and 404.)

- **Optional Review Chapter**   Chapter R, "Basic Concepts of Algebra," provides an optional review of intermediate algebra. Some or all of the topics in this chapter can be taught at the beginning of the course, or it can be used as a convenient source of information throughout the term for students who need a quick review of particular topics.

New!
- **Appendix Covering Basic Concepts from Geometry**   New to this edition is an appendix covering basic concepts from geometry that students need to know in order to work with topics pertaining to angles and triangles. Formulas from geometry are also listed near the back of the book.

- **Connecting the Concepts**   This feature highlights the importance of connecting concepts. When students are presented with concepts in a visual form—using graphs, an outline, or a chart—rather than merely in paragraphs of text, comprehension is streamlined and retention is maximized. The visual aspect of this feature invites students to stop and check their understanding of how concepts work together in one section or in several sections. This concept check in turn enhances student performance on homework assignments and exams. (See pp. 176, 203, and 346.)

- **Real-Data Applications** We encourage students to see and interpret the mathematics that appears every day in the world around them. Throughout the writing process, we conducted an energetic search for real-data applications, and the result is a variety of examples and exercises that connect the mathematical content with everyday life. Most of these applications feature source lines and frequently include charts and graphs. Many are drawn from the fields of health, business and economics, life and physical sciences, social science, and areas of general interest such as sports and travel. (See pp. 90, 343, and 354.)

- **Study Tips** Appearing in the text margin, the Study Tips provide helpful study hints throughout the text and also briefly remind students to use the electronic and print supplements that accompany the text. (See pp. 114, 136, and 271.)

- **Review Icons** Placed next to the concept that a student is currently studying, a review icon references a section of the text in which the student can find and review the topics on which the current concept is built. (See pp. 287 and 330.)

- **Technology Connections** This feature appears throughout the text to demonstrate how a graphing calculator can be used to solve problems. The technology is set apart from the traditional exposition, so that it does not intrude if no technology is desired. Although students might not be using graphing calculators, the graphing calculator windows that appear in the Technology Connection features enhance the visual element of the text, providing graphical interpretations of solutions of equations, zeros of functions, and x-intercepts of graphs of functions. (See pp. 168, 200, and 383.) Exercises that are designed to be worked using a graphing calculator are grouped together in the exercise sets under the heading Technology Connection so that they can be easily identified. A graphing calculator manual, providing keystroke-level instruction for six models of graphing calculators, is also available. (See Supplements for the Student.)

PEDAGOGICAL FEATURES

- **Chapter Openers** Each chapter opens with an application relevant to the content of the chapter. Also included is a table of contents for the chapter, listing section titles. (See pp. 51 and 329.)

- **Section Objectives** Content objectives are listed at the beginning of each section. Together with subheadings throughout the section, these objectives provide a useful outline of the section for both instructors and students. (See pp. 69 and 191.)

- **Annotated Examples** In the second edition, we have included over 570 examples designed to prepare the student fully to do the exercises. This is an increase of approximately 10% over the number of examples in the previous edition. Learning is carefully guided with the use of numerous color-coded art pieces and step-by-step annotations. Substitutions and annotations are highlighted in red for emphasis. (See pp. 196 and 376.)

- **Use of Color**   The text uses full color in an extremely functional way, as seen in the design elements and numerous pieces of art. The use of color has been carefully thought out so that it carries a consistent meaning that enhances students' ability to read and comprehend the exposition. (See pp. 150 and 330–331.)

- **Art Package**   The text contains over 1060 art pieces including photographs, situational art, and statistical graphs that not only highlight the abundance of real-world applications but also help students visualize the mathematics being discussed. (See pp. 65, 121, and 297.)

- **Five-Step Problem-Solving Process**   The basis for problem solving is a distinctive five-step process established early in the text (Section 2.1) to help students learn strategic ways to approach and solve applied problems. This process is then used throughout the text to give students a consistent framework for problem solving. (See pp. 168, 169, 172, and 201.)

- **Variety of Exercises**   There are over 5270 exercises in this text. The exercise sets are enhanced with real-data applications and source lines, detailed art pieces, tables, graphs, and photographs. In addition to the exercises that provide students practice with the concepts presented in the section, the exercise sets feature the following elements.

    Collaborative Discussion and Writing Exercises can be used in small groups or by the class as a whole to encourage students to talk and write about the key mathematical concepts in each section. (See pp. 206 and 371.)

New and Improved!

    Skill Maintenance Exercises provide an ongoing review of concepts previously presented in the course, enhancing students' retention of these concepts. New to this edition are Vocabulary Review (see pp. 124, 236, 304, and 404) and Classifying the Function (see pp. 285 and 446). Answers to *all* Skill Maintenance exercises appear in the answer section at the back of the book along with a section reference that directs students quickly and efficiently to the appropriate section of the text if they need help with an exercise. (See pp. 156, 285, and 404.)

    Synthesis Exercises appear at the end of each exercise set and encourage critical thinking by requiring students to synthesize concepts from several sections or to take a concept a step further than in the general exercises. (See pp. 275 and 304.)

    Technology Exercises are to be done using a graphing calculator. They are set apart from the non-calculator exercises, making it convenient for the instructor to avoid assigning them if graphing calculators are not being used in the course. (See pp. 83 and 389.)

- **Highlighted Information**   Important definitions, properties, and rules are displayed in screened boxes, and summaries and procedures are listed in boxes outlined in black. This organization and presentation provides for efficient learning and review. (See pp. 278 and 358.)

- **Summary and Review**   The Summary and Review at the end of each chapter contains an extensive set of review exercises along with a list of important properties and formulas covered in that chapter. This feature provides excellent preparation for chapter tests and the final examination. Answers to *all* review exercises appear in the answer section at the back of the book, along with corresponding section references. (See pp. 237–240 and 406–409.)

- **Chapter Tests**   The test at the end of each chapter allows students to test themselves and target areas that need further study before taking the in-class test. Answers to *all* Chapter Test questions appear in the answer section at the back of the book, along with corresponding section references. (See pp. 163 and 410.)

SUPPLEMENTS FOR THE INSTRUCTOR

New!

Annotated Instructor's Edition
(ISBN 0–321–23690–4)

The *Annotated Instructor's Edition* includes all the answers to the exercise sets, usually right on the page where the exercises appear. These readily accessible answers will help both new and experienced instructors prepare for class efficiently.

Instructor's Solutions Manual
(ISBN 0–321–23695–5)

The *Instructor's Solutions Manual* by Judith A. Penna contains worked-out solutions to all exercises in the exercise sets, including the Collaborative Discussion and Writing exercises.

Printed Test Bank/Instructor's Resource Guide
(ISBN 0–321–23693–9)

This supplement, prepared by Laurie Hurley, contains the following:

- Four free-response test forms for each chapter following the format of and with the same level of difficulty as the tests in the main text.
- Two multiple-choice test forms for each chapter.
- Six forms of the final examination, four with free-response questions and two with multiple-choice questions.
- Black-line masters of grids and number lines for transparency masters or test preparation.

- A videotape index.

- A correlation guide from the First Edition to the Second Edition.

MyMathLab MyMathLab is a complete, online course for Addison-Wesley mathematics textbooks that integrates interactive, multimedia instruction correlated to the textbook content. MyMathLab can be easily customized to suit the needs of students and instructors and provides a comprehensive and efficient online course-management system that allows for diagnosis, assessment, and tracking of students' progress.

Features of MyMathLab:

- Fully interactive multimedia textbooks are built in CourseCompass, a version of Blackboard™ designed specifically for Addison-Wesley.

- Chapter and section folders from the textbook contain a wide range of instructional content: videos, software tools, audio clips, animations, and electronic supplements.

- Hyperlinks take the user directly to online homework, testing, diagnosis, tutorials, and gradebooks in MathXL, Addison-Wesley's homework, tutorial, and testing system for mathematics and statistics.

- Instructors can create, copy, edit, assign, and track all tests and homework for their course as well as track student performance.

- With push-button ease, instructors can remove, hide, or annotate Addison-Wesley preloaded content, add their own course documents, or change the order in which material is presented.

- Using the communication tools found in MyMathLab, instructors can hold online office hours, host a discussion board, create communication groups within their class, send e-mails, and maintain a course calendar.

- Print supplements are available online, side by side with the textbook.

For more information, visit our Web site at www.mymathlab.com or contact your local publisher's representative for a live demonstration.

TestGen with QuizMaster
(ISBN 0–321–23692–0)

TestGen enables instructors to build, edit, print, and administer tests using a computerized bank of questions developed to cover all the objectives of the text. Instructors can modify test-bank questions or add new questions by using the built-in question editor, which allows users to create graphs, import graphics, insert math notation, and insert variable numbers or text. Tests can be printed or administered online via the Web or other network. TestGen comes packaged with QuizMaster, which allows students to take tests on a local area network. The software is available on a dual-platform Windows/Macintosh CD-ROM.

Powerpoint Lecture Presentation
(ISBN 0–321–23694–7)

This Powerpoint presentation provides lecture outlines geared specifically to the sequence of this textbook and is available within MyMathLab or at www.aw-bc.com/suppscentral.

MathXL   MathXL is an online homework, tutorial, and assessment system that uses algorithmically generated exercises correlated at the objective level to the textbook. Instructors can assign tests and homework provided by Addison-Wesley or create and customize their own tests and homework assignments. Instructors can also track their students' results and tutorial work in an online gradebook. Students can take chapter tests and receive personalized study plans that will diagnose weaknesses and link them to areas they need to study and retest. Students can also work unlimited practice problems and receive tutorial instruction for areas in which they need improvement. Selected exercises link directly to video clips and animations. MathXL can be packaged with new copies of *College Algebra*, Second Edition.

SUPPLEMENTS FOR THE STUDENT

Graphing Calculator Manual
(ISBN 0–321–23697–1)

The *Graphing Calculator Manual* by Judith A. Penna, with the assistance of Daphne Bell, contains keystroke level instruction for the Texas Instruments TI-83, TI-83 Plus, TI-84 Plus, TI-85, TI-86, and TI-89.

The *Graphing Calculator Manual* uses actual examples and exercises from *College Algebra*, Second Edition, to help teach students to use a graphing calculator. The order of topics in the *Graphing Calculator Manual* mirrors that of the text, providing a just-in-time mode of instruction.

Student's Solutions Manual
(ISBN 0–321–23698–x)

The *Student's Solutions Manual*, by Judith A. Penna, contains completely worked-out solutions with step-by-step annotations for all the odd-numbered exercises in the exercise sets in the text, with the exception of the Collaborative Discussion and Writing exercises.

MyMathLab   MyMathLab is a complete, online course for Addison-Wesley mathematics textbooks that integrates interactive, multimedia instruction correlated to the textbook content. MyMathLab can be easily customized to suit the needs of students and instructors and provides a comprehensive and efficient online course-management system that allows for diagnosis, assessment, and tracking of students' progress.

Features of MyMathLab:

- Fully interactive multimedia textbooks are built in CourseCompass, a version of Blackboard™ designed specifically for Addison-Wesley.
- Chapter and section folders from the textbook contain a wide range of instructional content: videos, software tools, audio clips, animations, and electronic supplements.
- Hyperlinks take the user directly to online homework, testing, diagnosis, tutorials, and gradebooks in MathXL, Addison-Wesley's homework, tutorial, and assessment system for mathematics and statistics.
- Print supplements are available online, side by side with the textbook.

For more information, visit our Web site at www.mymathlab.com.

InterActMath® Tutorial Web site
www.interactmath.com

This interactive tutorial Web site provides algorithmically generated practice exercises that correlate directly to the exercises in the text. A detailed worked-out example and guided solution accompany each practice exercise. The Web site recognizes student errors and provides feedback.

MathXL Tutorials on CD®
(ISBN 0–321–26767–2)

Available on CD-ROM, this interactive tutorial software provides algorithmically generated practice exercises that are correlated at the objective level to the odd-numbered exercises in the text. Every exercise in the program is accompanied by an example and a guided solution designed to involve students in the solution process. Selected problems also include a video clip to help students visualize concepts. The software tracks student activity and scores and can generate printed summaries of students' progress.

MathXL   MathXL is an online homework, tutorial, and assessment system that uses algorithmically generated exercises correlated at the objective level to the textbook. Students can take chapter tests and receive personalized study plans that will diagnose weaknesses and link them to areas they need to study and retest. Students can also work unlimited practice problems and receive tutorial instruction for areas in which they need improvement. Selected problems link directly to video clips and animations for extra instruction.

Addison-Wesley Math Tutor Center   The Addison-Wesley Math Tutor Center is staffed by qualified college mathematics instructors who tutor students on examples and exercises from the textbook. Tutoring is provided via toll-free telephone, toll-free fax, e-mail, and the Internet. Interactive web-based technology allows students and tutors to view and listen to live instruction in real-time over the Internet. The Math Tutor Center is accessed through a registration number that can be packaged with a new textbook or purchased separately. (*Note*: MyMathLab students obtain access to the Tutor Center through their MyMathLab course.)

Videotape Series
(ISBN 0–321–23691–2)

This series of videotapes, created specifically for *College Algebra*, Second Edition, features an engaging team of lecturers who provide comprehensive lessons on every objective in the text. Many of the video segments are staged as an office hour during which a student poses questions and works through problems with the instructor.

Digital Video Tutor
(ISBN 0–321–23696–3)

This supplement provides the entire set of videotapes for the text in digital format on CD-ROM, making it easy and convenient for students to watch video segments from a computer, either at home or on campus. Available for purchase with the text at minimal cost, the Digital Video Tutor is ideal for distance learning and supplemental instruction.

ACKNOWLEDGMENTS

We wish to express our heartfelt thanks to a number of people who have contributed in special ways to the development of this textbook. Our editor, Anne Kelly, encouraged and supported our vision. Kari Heen, our executive project manager, deserves special recognition for overseeing every phase of the project and keeping it moving. We are very appreciative of the marketing insight provided by Becky Anderson, our marketing manager, and of the support that we received from the entire Addison-Wesley Higher Education Group, including Karen Wernholm, senior production supervisor, Katie Nopper, assistant editor, and Ceci Fleming, editorial assistant. We also thank Lynne Blaszak, associate media producer, for her creative work on the media products that accompany this text. And we are immensely grateful to Martha Morong, for her editorial and production services, and to Geri Davis, for her text design and art editing, and for the endless hours of hard work they have done to make this a book of which we are proud. We also thank Barbara Johnson for her meticulous accuracy checking of the text.

The following reviewers made invaluable contributions to the development of this text and we thank them for that:

Sherry S. Biggers, *Clemson University Department of Mathematical Sciences*

Christine Bush, *Palm Beach Community College, Lake Worth*

Walter Czarnec, *Framingham State College*

Douglas Dunbar, *Okaloosa-Walton Community College*

Wayne Ferguson, *Northwest Mississippi Community College*

Dauhrice K. Gibson, (retired), *Gulf Coast Community College*

Jim Graziose, *Palm Beach Community College*

Joseph Lloyd Harris, *Gulf Coast Community College*

Susan K. Hitchcock, *Palm Beach Community College*

Sharon S. Hudson, *Gulf Coast Community College*

Jennifer Jameson, *Coconino Community College, Flagstaff*

Bernard F. Mathon, *Miami-Dade College*

Barry J. Monk, *Macon State College*

Darla Ottman, *Elizabethtown Community College*

Vicki Partin, *Lexington Community College*

Kathy V. Rodgers, *University of Southern Indiana*

Abdelrida Saleh, *Miami-Dade College*

Rajalakshmi Sriram, *Okaloosa-Walton Community College*

J.A.B.
J.A.P.
M.L.B.

Basic Concepts of Algebra

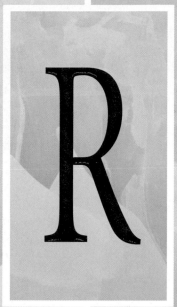

APPLICATION

Gordon deposits $100 in a retirement account each month beginning at age 25. If the investment earns 4% interest, compounded monthly, how much will have accumulated in the account when Gordon retires at age 65?

This problem appears as Exercise 90 in Section R.2.

1

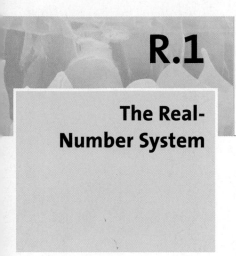

R.1

The Real-Number System

- *Identify various kinds of real numbers.*
- *Use interval notation to write a set of numbers.*
- *Identify the properties of real numbers.*
- *Find the absolute value of a real number.*

Real Numbers

In applications of algebraic concepts, we use real numbers to represent quantities such as distance, time, speed, area, profit, loss, and temperature. Some frequently used sets of real numbers and the relationships among them are shown below.

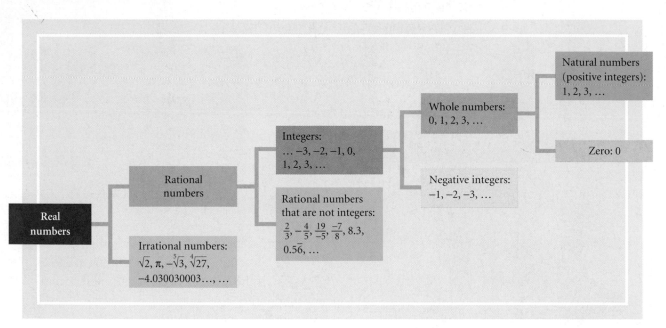

Numbers that can be expressed in the form p/q, where p and q are integers and $q \neq 0$, are **rational numbers.** Decimal notation for rational numbers either *terminates* (ends) or *repeats*. Each of the following is a rational number.

a) 0 $\qquad\qquad$ $0 = \dfrac{0}{a}$ for any nonzero integer a

b) -7 $\qquad\qquad$ $-7 = \dfrac{-7}{1}$, or $\dfrac{7}{-1}$

c) $\dfrac{1}{4} = 0.25$ $\qquad$ **Terminating decimal**

d) $-\dfrac{5}{11} = -0.\overline{45}$ $\qquad$ **Repeating decimal**

The real numbers that are not rational are **irrational numbers.** Decimal notation for irrational numbers neither terminates nor repeats. Each of the following is an irrational number.

a) $\pi = 3.1415926535\ldots$ **There is no repeating block of digits.**

$\left(\frac{22}{7}\right.$ and 3.14 are rational *approximations* of the irrational number π.$\left.\right)$

b) $\sqrt{2} = 1.414213562\ldots$ **There is no repeating block of digits.**

c) $-6.12122122212222\ldots$ **Although there is a pattern, there is no repeating block of digits.**

The set of all rational numbers combined with the set of all irrational numbers gives us the set of **real numbers.** The real numbers are *modeled* using a **number line,** as shown below.

Each point on the line represents a real number and every real number is represented by a point on the line.

$$-2.9 \qquad -\tfrac{3}{5} \quad \sqrt{3} \quad \pi \quad \tfrac{17}{4}$$

The order of the real numbers can be determined from the number line. If a number a is to the left of a number b, then a **is less than** b ($a < b$). Similarly, a **is greater than** b ($a > b$), if a is to the right of b on the number line. For example, we see from the number line above that $-2.9 < -\frac{3}{5}$, because -2.9 is to the left of $-\frac{3}{5}$. Also, $\frac{17}{4} > \sqrt{3}$, because $\frac{17}{4}$ is to the right of $\sqrt{3}$.

The statement $a \leq b$, read "a is less than or equal to b," is true if either $a < b$ is true or $a = b$ is true.

The symbol $\in$ is used to indicate that a member, or **element,** belongs to a set. Thus if we let $\mathbb{Q}$ represent the set of rational numbers, we can see from the diagram on page 2 that $0.5\overline{6} \in \mathbb{Q}$. We can also write $\sqrt{2} \notin \mathbb{Q}$ to indicate that $\sqrt{2}$ is *not* an element of the set of rational numbers.

When *all* the elements of one set are elements of a second set, we say that the first set is a **subset** of the second set. The symbol $\subseteq$ is used to denote this. For instance, if we let $\mathbb{R}$ represent the set of real numbers, we can see from the diagram that $\mathbb{Q} \subseteq \mathbb{R}$ (read "$\mathbb{Q}$ is a subset of $\mathbb{R}$").

Interval Notation

Sets of real numbers can be expressed using **interval notation.** For example, for real numbers a and b such that $a < b$, the **open interval** (a, b) is the set of real numbers between, but not including, a and b. That is,

$$(a, b) = \{x \mid a < x < b\}.$$

The points a and b are **endpoints** of the interval. The parentheses indicate that the endpoints are not included in the interval.

Some intervals extend without bound in one or both directions. The interval $[a, \infty)$, for example, begins at a and extends to the right without bound. That is,

$$[a, \infty) = \{x \mid x \geq a\}.$$

The bracket indicates that a is included in the interval.

The various types of intervals are listed below.

Intervals: Types, Notation, and Graphs

TYPE	INTERVAL NOTATION	SET NOTATION	GRAPH
Open	(a, b)	$\{x \mid a < x < b\}$	
Closed	$[a, b]$	$\{x \mid a \le x \le b\}$	
Half-open	$[a, b)$	$\{x \mid a \le x < b\}$	
Half-open	$(a, b]$	$\{x \mid a < x \le b\}$	
Open	(a, ∞)	$\{x \mid x > a\}$	
Half-open	$[a, \infty)$	$\{x \mid x \ge a\}$	
Open	$(-\infty, b)$	$\{x \mid x < b\}$	
Half-open	$(-\infty, b]$	$\{x \mid x \le b\}$	

The interval $(-\infty, \infty)$, graphed below, names the set of all real numbers, $\mathbb{R}$.

EXAMPLE 1 Write interval notation for each set and graph the set.

a) $\{x \mid -4 < x < 5\}$ **b)** $\{x \mid x \ge 1.7\}$

c) $\{x \mid -5 < x \le -2\}$ **d)** $\{x \mid x < \sqrt{5}\}$

Solution

a) $\{x \mid -4 < x < 5\} = (-4, 5);$

b) $\{x \mid x \ge 1.7\} = [1.7, \infty);$

c) $\{x \mid -5 < x \le -2\} = (-5, -2];$

d) $\{x | x < \sqrt{5}\} = (-\infty, \sqrt{5})$;

$$\xleftarrow{\quad\quad\quad\quad\quad\quad\quad\quad\quad} \overset{)}{\xrightarrow{\quad\quad\quad}}$$
$$-5 \;-4 \;-3 \;-2 \;-1 \;\;0 \;\;1 \;\;2 \;\;3 \;\;4 \;\;5$$

Properties of the Real Numbers

The following properties can be used to manipulate algebraic expressions as well as real numbers.

Properties of the Real Numbers

For any real numbers a, b, and c:

$a + b = b + a$ and $ab = ba$	Commutative properties of addition and multiplication
$a + (b + c) = (a + b) + c$ and $a(bc) = (ab)c$	Associative properties of addition and multiplication
$a + 0 = 0 + a = a$	Additive identity property
$-a + a = a + (-a) = 0$	Additive inverse property
$a \cdot 1 = 1 \cdot a = a$	Multiplicative identity property
$a \cdot \dfrac{1}{a} = \dfrac{1}{a} \cdot a = 1 \; (a \neq 0)$	Multiplicative inverse property
$a(b + c) = ab + ac$	Distributive property

Note that the distributive property is also true for subtraction since $a(b - c) = a[b + (-c)] = ab + a(-c) = ab - ac$.

EXAMPLE 2 State the property being illustrated in each sentence.

a) $8 \cdot 5 = 5 \cdot 8$ 　　　　　　　　**b)** $5 + (m + n) = (5 + m) + n$

c) $14 + (-14) = 0$ 　　　　　　　**d)** $6 \cdot 1 = 1 \cdot 6 = 6$

e) $2(a - b) = 2a - 2b$

Solution

SENTENCE	PROPERTY
a) $8 \cdot 5 = 5 \cdot 8$	Commutative property of multiplication: $ab = ba$
b) $5 + (m + n) = (5 + m) + n$	Associative property of addition: $a + (b + c) = (a + b) + c$
c) $14 + (-14) = 0$	Additive inverse property: $a + (-a) = 0$
d) $6 \cdot 1 = 1 \cdot 6 = 6$	Multiplicative identity property: $a \cdot 1 = 1 \cdot a = a$
e) $2(a - b) = 2a - 2b$	Distributive property: $a(b + c) = ab + ac$

Technology Connection

We can use the absolute-value operation on a graphing calculator to find the distance between two points. On many graphing calculators, absolute value is denoted "abs" and is found in the MATH NUM menu and also in the CATALOG.

```
abs (−2−3)
                    5
abs (3−(−2))
                    5
```

Absolute Value

The number line can be used to provide a geometric interpretation of *absolute value*. The **absolute value** of a number a, denoted $|a|$, is its distance from 0 on the number line. For example, $|-5| = 5$, because the distance of -5 from 0 is 5. Similarly, $\left|\frac{3}{4}\right| = \frac{3}{4}$, because the distance of $\frac{3}{4}$ from 0 is $\frac{3}{4}$.

Absolute Value

For any real number a,

$$|a| = \begin{cases} a, & \text{if } a \geq 0, \\ -a, & \text{if } a < 0. \end{cases}$$

When a is nonnegative, the absolute value of a is a. When a is negative, the absolute value of a is the opposite, or additive inverse, of a. Thus, $|a|$ is never negative; that is, for any real number a, $|a| \geq 0$.

Absolute value can be used to find the distance between two points on the number line.

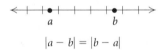

$$|a - b| = |b - a|$$

Distance Between Two Points on the Number Line

For any real numbers a and b, the **distance between a and b** is $|a - b|$, or, equivalently, $|b - a|$.

EXAMPLE 3 Find the distance between -2 and 3.

Solution The distance is

$$|-2 - 3| = |-5| = 5, \quad \text{or, equivalently,}$$
$$|3 - (-2)| = |3 + 2| = |5| = 5.$$

R.1

Exercise Set

In Exercises 1– 10, consider the numbers $-12, \sqrt{7}, 5.\overline{3}, -\frac{7}{3}, \sqrt[3]{8}, 0, 5.242242224\ldots, -\sqrt{14}, \sqrt[5]{5}, -1.96, 9, 4\frac{2}{3}, \sqrt{25}, \sqrt[3]{4}, \frac{5}{7}.$

1. Which are whole numbers?

2. Which are integers?

3. Which are irrational numbers?

4. Which are natural numbers?

5. Which are rational numbers?

6. Which are real numbers?

7. Which are rational numbers but not integers?

8. Which are integers but not whole numbers?

9. Which are integers but not natural numbers?

10. Which are real numbers but not integers?

Write interval notation. Then graph the interval.

11. $\{x \mid -3 \leq x \leq 3\}$ **12.** $\{x \mid -4 < x < 4\}$

13. $\{x \mid -4 \leq x < -1\}$ **14.** $\{x \mid 1 < x \leq 6\}$

15. $\{x \mid x \leq -2\}$ **16.** $\{x \mid x > -5\}$

17. $\{x \mid x > 3.8\}$ **18.** $\{x \mid x \geq \sqrt{3}\}$

19. $\{x \mid 7 < x\}$ **20.** $\{x \mid -3 > x\}$

Write interval notation for the graph.

21.

22.

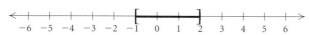

23.

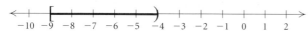

24.

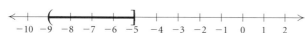

25.

26.

27.

28.

In Exercises 29–46, the following notation is used:
$\mathbb{N}$ = *the set of natural numbers,* $\mathbb{W}$ = *the set of whole numbers,* $\mathbb{Z}$ = *the set of integers,* $\mathbb{Q}$ = *the set of rational numbers,* $\mathbb{I}$ = *the set of irrational numbers, and*

$\mathbb{R}$ = *the set of real numbers. Classify the statement as true or false.*

29. $6 \in \mathbb{N}$ **30.** $0 \notin \mathbb{N}$

31. $3.2 \in \mathbb{Z}$ **32.** $-10.\overline{1} \in \mathbb{R}$

33. $-\dfrac{11}{5} \in \mathbb{Q}$ **34.** $-\sqrt{6} \in \mathbb{Q}$

35. $\sqrt{11} \notin \mathbb{R}$ **36.** $-1 \in \mathbb{W}$

37. $24 \notin \mathbb{W}$ **38.** $1 \in \mathbb{Z}$

39. $1.089 \notin \mathbb{I}$ **40.** $\mathbb{N} \subseteq \mathbb{W}$

41. $\mathbb{W} \subseteq \mathbb{Z}$ **42.** $\mathbb{Z} \subseteq \mathbb{N}$

43. $\mathbb{Q} \subseteq \mathbb{R}$ **44.** $\mathbb{Z} \subseteq \mathbb{Q}$

45. $\mathbb{R} \subseteq \mathbb{Z}$ **46.** $\mathbb{Q} \subseteq \mathbb{I}$

Name the property illustrated by the sentence.

47. $6x = x6$

48. $3 + (x + y) = (3 + x) + y$

49. $-3 \cdot 1 = -3$ **50.** $x + 4 = 4 + x$

51. $5(ab) = (5a)b$ **52.** $4(y - z) = 4y - 4z$

53. $2(a + b) = (a + b)2$ **54.** $-7 + 7 = 0$

55. $-6(m + n) = -6(n + m)$

56. $t + 0 = t$

57. $8 \cdot \dfrac{1}{8} = 1$

58. $9x + 9y = 9(x + y)$

Simplify.

59. $|-7.1|$ **60.** $|-86.2|$

61. $|347|$ **62.** $|-54|$

63. $\left| -\sqrt{97} \right|$ **64.** $\left| \dfrac{12}{19} \right|$

65. $|0|$ **66.** $|15|$

67. $\left| \dfrac{5}{4} \right|$ **68.** $\left| -\sqrt{3} \right|$

Find the distance between the given pair of points on the number line.

69. $-5,\ 6$ **70.** $-2.5,\ 0$

71. $-8,\ -2$ **72.** $\dfrac{15}{8},\ \dfrac{23}{12}$

73. 6.7, 12.1

74. -14, -3

75. $-\dfrac{3}{4}$, $\dfrac{15}{8}$

76. -3.4, 10.2

77. -7, 0

78. 3, 19

Collaborative Discussion and Writing

To the student and the instructor: The Collaborative Discussion and Writing exercises are meant to be answered with one or more sentences. These exercises can also be discussed and answered collaboratively by the entire class or by small groups. Because of their open-ended nature, the answers to these exercises do not appear at the back of the book. They are denoted by the words "Discussion and Writing."

79. How would you convince a classmate that division is not associative?

80. Under what circumstances is $\sqrt{a}$ a rational number?

Synthesis

To the student and the instructor: The Synthesis exercises found at the end of every exercise set challenge students to combine concepts or skills studied in that section or in preceding parts of the text.

Between any two (different) real numbers there are many other real numbers. Find each of the following. Answers may vary.

81. An irrational number between 0.124 and 0.125

82. A rational number between $-\sqrt{2.01}$ and $-\sqrt{2}$

83. A rational number between $-\dfrac{1}{101}$ and $-\dfrac{1}{100}$

84. An irrational number between $\sqrt{5.99}$ and $\sqrt{6}$

85. The hypotenuse of an isosceles right triangle with legs of length 1 unit can be used to "measure" a value for $\sqrt{2}$ by using the Pythagorean theorem, as shown.

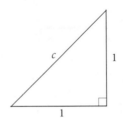

$$c^2 = 1^2 + 1^2$$
$$c^2 = 2$$
$$c = \sqrt{2}$$

Draw a right triangle that could be used to "measure" $\sqrt{10}$ units.

R.2

- *Simplify expressions with integer exponents.*
- *Solve problems using scientific notation.*
- *Use the rules for order of operations.*

Integer Exponents, Scientific Notation, and Order of Operations

Integers as Exponents

When a positive integer is used as an *exponent*, it indicates the number of times a factor appears in a product. For example, 7^3 means $7 \cdot 7 \cdot 7$ and 5^1 means 5.

For any positive integer n,

$$a^n = \underbrace{a \cdot a \cdot a \cdots a,}_{n \text{ factors}}$$

where a is the **base** and n is the **exponent.**

Zero and negative-integer exponents are defined as follows.

For any nonzero real number a and any integer m,

$$a^0 = 1 \quad \text{and} \quad a^{-m} = \frac{1}{a^m}.$$

EXAMPLE 1   Simplify each of the following.

a) 6^0 **b)** $(-3.4)^0$

Solution

a) $6^0 = 1$ **b)** $(-3.4)^0 = 1$

EXAMPLE 2   Write each of the following with positive exponents.

a) 4^{-5} **b)** $\dfrac{1}{(0.82)^{-7}}$ **c)** $\dfrac{x^{-3}}{y^{-8}}$

Solution

a) $4^{-5} = \dfrac{1}{4^5}$

b) $\dfrac{1}{(0.82)^{-7}} = (0.82)^{-(-7)} = (0.82)^7$

c) $\dfrac{x^{-3}}{y^{-8}} = x^{-3} \cdot \dfrac{1}{y^{-8}} = \dfrac{1}{x^3} \cdot y^8 = \dfrac{y^8}{x^3}$

The results in Example 2 can be generalized as follows.

For any nonzero numbers a and b and any integers m and n,

$$\frac{a^{-m}}{b^{-n}} = \frac{b^n}{a^m}.$$

(A factor can be moved to the other side of the fraction bar if the sign of the exponent is changed.)

EXAMPLE 3 Write an equivalent expression without negative exponents:

$$\frac{x^{-3}y^{-8}}{z^{-10}}.$$

Solution We move each factor to the other side of the fraction bar and change the sign of each exponent:

$$\frac{x^{-3}y^{-8}}{z^{-10}} = \frac{z^{10}}{x^3y^8}.$$

The following properties of exponents can be used to simplify expressions.

Properties of Exponents

For any real numbers a and b and any integers m and n, assuming 0 is not raised to a nonpositive power:

$$a^m \cdot a^n = a^{m+n} \qquad \text{Product rule}$$

$$\frac{a^m}{a^n} = a^{m-n} \ (a \neq 0) \qquad \text{Quotient rule}$$

$$(a^m)^n = a^{mn} \qquad \text{Power rule}$$

$$(ab)^m = a^m b^m \qquad \text{Raising a product to a power}$$

$$\left(\frac{a}{b}\right)^m = \frac{a^m}{b^m} \ (b \neq 0) \qquad \text{Raising a quotient to a power}$$

EXAMPLE 4 Simplify each of the following.

a) $y^{-5} \cdot y^3$

b) $\dfrac{48x^{12}}{16x^4}$

c) $(t^{-3})^5$

d) $(2s^{-2})^5$

e) $\left(\dfrac{45x^{-4}y^2}{9z^{-8}}\right)^{-3}$

Solution

a) $y^{-5} \cdot y^3 = y^{-5+3} = y^{-2}$, or $\dfrac{1}{y^2}$

b) $\dfrac{48x^{12}}{16x^4} = \dfrac{48}{16}x^{12-4} = 3x^8$

c) $(t^{-3})^5 = t^{-3 \cdot 5} = t^{-15}$, or $\dfrac{1}{t^{15}}$

d) $(2s^{-2})^5 = 2^5(s^{-2})^5 = 32s^{-10}$, or $\dfrac{32}{s^{10}}$

e) $\left(\dfrac{45x^{-4}y^2}{9z^{-8}}\right)^{-3} = \left(\dfrac{5x^{-4}y^2}{z^{-8}}\right)^{-3}$

$$= \dfrac{5^{-3}x^{12}y^{-6}}{z^{24}} = \dfrac{x^{12}}{5^3y^6z^{24}}, \text{ or } \dfrac{x^{12}}{125y^6z^{24}}$$

Scientific Notation

We can use scientific notation to name very large and very small positive numbers and to perform computations.

> **Scientific Notation**
>
> **Scientific notation** for a number is an expression of the type
>
> $$N \times 10^m,$$
>
> where $1 \le N < 10$, N is in decimal notation, and m is an integer.

Keep in mind that in scientific notation, positive exponents are used for numbers greater than or equal to 10 and negative exponents for numbers between 0 and 1.

EXAMPLE 5 *Undergraduate Enrollment.* In a recent year, there were 16,539,000 undergraduate students enrolled in post-secondary institutions in the United States (*Source*: U.S. National Center for Education Statistics). Convert this number to scientific notation.

Solution We want the decimal point to be positioned between the 1 and the 6, so we move it 7 places to the left. Since the number to be converted is greater than 10, the exponent must be positive.

$$16{,}539{,}000 = 1.6539 \times 10^7$$

EXAMPLE 6 *Mass of a Neutron.* The mass of a neutron is about 0.00000000000000000000000000167 kg. Convert this number to scientific notation.

Solution We want the decimal point to be positioned between the 1 and the 6, so we move it 27 places to the right. Since the number to be converted is between 0 and 1, the exponent must be negative.

$$0.00000000000000000000000000167 = 1.67 \times 10^{-27}$$

EXAMPLE 7 Convert each of the following to decimal notation.

a) 7.632×10^{-4}　　　　　　　　　　**b)** 9.4×10^5

Solution

a) The exponent is negative, so the number is between 0 and 1. We move the decimal point 4 places to the left.

$$7.632 \times 10^{-4} = 0.0007632$$

b) The exponent is positive, so the number is greater than 10. We move the decimal point 5 places to the right.

$$9.4 \times 10^5 = 940,000$$

EXAMPLE 8 *Distance to a Star.* Alpha Centauri is about 4.3 light-years from Earth. One **light-year** is the distance that light travels in one year and is about 5.88×10^{12} miles. How many miles is it from Earth to Alpha Centauri? Express your answer in scientific notation.

Solution

$$4.3 \times (5.88 \times 10^{12}) = (4.3 \times 5.88) \times 10^{12}$$
$$= 25.284 \times 10^{12} \qquad \text{This is not scientific notation because } 25.284 > 10.$$
$$= (2.5284 \times 10^1) \times 10^{12}$$
$$= 2.5284 \times (10^1 \times 10^{12})$$
$$= 2.5284 \times 10^{13} \text{ miles} \qquad \text{Writing scientific notation}$$

Order of Operations

Recall that to simplify the expression $3 + 4 \cdot 5$, we multiply 4 and 5 first to get 20 and then add 3 to get 23. Mathematicians have agreed on the following procedure, or rules for order of operations.

Rules for Order of Operations

1. Do all calculations within grouping symbols before operations outside. When nested grouping symbols are present, work from the inside out.
2. Evaluate all exponential expressions.
3. Do all multiplications and divisions in order from left to right.
4. Do all additions and subtractions in order from left to right.

EXAMPLE 9 Calculate each of the following.

a) $8(5 - 3)^3 - 20$

b) $\dfrac{10 \div (8 - 6) + 9 \cdot 4}{2^5 + 3^2}$

Solution

a)
$$8(5 - 3)^3 - 20 = 8 \cdot 2^3 - 20 \qquad \text{Doing the calculation within parentheses}$$
$$= 8 \cdot 8 - 20 \qquad \text{Evaluating the exponential expression}$$
$$= 64 - 20 \qquad \text{Multiplying}$$
$$= 44 \qquad \text{Subtracting}$$

Enter the computations in
Example 9 on a graphing
calculator as shown below.

```
8(5−3)^3−20
                        44
(10/(8−6)+9*4)/(2^5+3²)
                         1
```

To confirm that the
parentheses around the
numerator and around the
denominator are essential in
Example 9(b), enter the
computation without using
these parentheses. What is
the result?

b) $\dfrac{10 \div (8 - 6) + 9 \cdot 4}{2^5 + 3^2} = \dfrac{10 \div 2 + 9 \cdot 4}{32 + 9}$

$$= \dfrac{5 + 36}{41} = \dfrac{41}{41} = 1$$

Note that fraction bars act as grouping symbols. That is, the given expression is equivalent to $[10 \div (8 - 6) + 9 \cdot 4] \div (2^5 + 3^2)$.   ▬

EXAMPLE 10   *Compound Interest.*   If a principal P is invested at an interest rate r, compounded n times per year, in t years it will grow to an amount A given by

$$A = P\left(1 + \frac{r}{n}\right)^{nt}.$$

Suppose that \$1250 is invested at 4.6% interest, compounded quarterly. How much is in the account at the end of 8 years?

Solution   We have $P = 1250$, $r = 4.6\%$, or 0.046, $n = 4$, and $t = 8$. Substituting, we find that the amount in the account at the end of 8 years is given by

$$A = 1250\left(1 + \frac{0.046}{4}\right)^{4 \cdot 8}.$$

Next, we evaluate this expression:

$$
\begin{aligned}
A &= 1250(1 + 0.0115)^{4 \cdot 8} &&\text{Dividing} \\
&= 1250(1.0115)^{4 \cdot 8} &&\text{Adding} \\
&= 1250(1.0115)^{32} &&\text{Multiplying in the exponent} \\
&\approx 1250(1.441811175) &&\text{Evaluating the exponential expression} \\
&\approx 1802.263969 &&\text{Multiplying} \\
&\approx 1802.26. &&\text{Rounding to the nearest cent}
\end{aligned}
$$

The amount in the account at the end of 8 years is \$1802.26.   ▬

R.2

Exercise Set

Simplify.

1. 18^0

2. $\left(-\frac{4}{3}\right)^0$

3. $x^9 \cdot x^0$

4. $a^0 \cdot a^4$

5. $5^8 \cdot 5^{-6}$

6. $6^2 \cdot 6^{-7}$

7. $m^{-5} \cdot m^5$

8. $n^9 \cdot n^{-9}$

9. $y^3 \cdot y^{-7}$

10. $b^{-4} \cdot b^{12}$

11. $7^3 \cdot 7^{-5} \cdot 7$

12. $3^6 \cdot 3^{-5} \cdot 3^4$

13. $2x^3 \cdot 3x^2$

14. $3y^4 \cdot 4y^3$

15. $(-3a^{-5})(5a^{-7})$

16. $(-6b^{-4})(2b^{-7})$

17. $(5a^2b)(3a^{-3}b^4)$

18. $(4xy^2)(3x^{-4}y^5)$

19. $(6x^{-3}y^5)(-7x^2y^{-9})$

20. $(8ab^7)(-7a^{-5}b^2)$

21. $(2x)^3(3x)^2$

22. $(4y)^2(3y)^3$

23. $(-2n)^3(5n)^2$

24. $(2x)^5(3x)^2$

25. $\dfrac{b^{40}}{b^{37}}$

26. $\dfrac{a^{39}}{a^{32}}$

27. $\dfrac{x^{-5}}{x^{16}}$

28. $\dfrac{y^{-24}}{y^{-21}}$

29. $\dfrac{x^2y^{-2}}{x^{-1}y}$

30. $\dfrac{x^3y^{-3}}{x^{-1}y^2}$

31. $\dfrac{32x^{-4}y^3}{4x^{-5}y^8}$

32. $\dfrac{20a^5b^{-2}}{5a^7b^{-3}}$

33. $(2ab^2)^3$

34. $(4xy^3)^2$

35. $(-2x^3)^5$

36. $(-3x^2)^4$

37. $(-5c^{-1}d^{-2})^{-2}$

38. $(-4x^{-5}z^{-2})^{-3}$

39. $(3m^4)^3(2m^{-5})^4$

40. $(4n^{-1})^2(2n^3)^3$

41. $\left(\dfrac{2x^{-3}y^7}{z^{-1}}\right)^3$

42. $\left(\dfrac{3x^5y^{-8}}{z^{-2}}\right)^4$

43. $\left(\dfrac{24a^{10}b^{-8}c^7}{12a^6b^{-3}c^5}\right)^{-5}$

44. $\left(\dfrac{125p^{12}q^{-14}r^{22}}{25p^8q^6r^{-15}}\right)^{-4}$

Convert to scientific notation.

45. 405,000

46. 1,670,000

47. 0.00000039

48. 0.00092

49. 234,600,000,000

50. 8,904,000,000

51. 0.00104

52. 0.00000000514

53. One cubic inch is approximately equal to 0.000016 m^3.

54. The United States government collected $1,137,000,000,000 in individual income taxes in a recent year (*Source:* U.S. Internal Revenue Service).

Convert to decimal notation.

55. 8.3×10^{-5}

56. 4.1×10^6

57. 2.07×10^7

58. 3.15×10^{-6}

59. 3.496×10^{10}

60. 8.409×10^{11}

61. 5.41×10^{-8}

62. 6.27×10^{-10}

63. The amount of solid waste generated in the United States in a recent year was 2.319×10^8 tons (*Source:* Franklin Associates, Ltd.).

64. The mass of a proton is about 1.67×10^{-24} g.

Compute. Write the answer using scientific notation.

65. $(3.1 \times 10^5)(4.5 \times 10^{-3})$

66. $(9.1 \times 10^{-17})(8.2 \times 10^3)$

67. $(2.6 \times 10^{-18})(8.5 \times 10^7)$

68. $(6.4 \times 10^{12})(3.7 \times 10^{-5})$

69. $\dfrac{6.4 \times 10^{-7}}{8.0 \times 10^6}$

70. $\dfrac{1.1 \times 10^{-40}}{2.0 \times 10^{-71}}$

71. $\dfrac{1.8 \times 10^{-3}}{7.2 \times 10^{-9}}$

72. $\dfrac{1.3 \times 10^4}{5.2 \times 10^{10}}$

Solve. Write the answer using scientific notation.

73. *Chesapeake Bay Bridge-Tunnel.* The 17.6-mile-long Chesapeake Bay Bridge-Tunnel was completed in 1964. Construction costs were $210 million. Find the average cost per mile.

74. *Personal Space in Hong Kong.* The area of Hong Kong is 412 square miles. It is estimated that the population of Hong Kong will be 9,600,000 in 2050. Find the number of square miles of land per person in 2050.

75. *Nuclear Disintegration.* One gram of radium produces 37 billion disintegrations per second. How many disintegrations are produced in 1 hr?

76. *Length of Earth's Orbit.* The average distance from the earth to the sun is 93 million mi. About how far does the earth travel in a yearly orbit? (Assume a circular orbit.)

Calculate.

77. $3 \cdot 2 - 4 \cdot 2^2 + 6(3 - 1)$

78. $3[(2 + 4 \cdot 2^2) - 6(3 - 1)]$

79. $16 \div 4 \cdot 4 \div 2 \cdot 256$

80. $2^6 \cdot 2^{-3} \div 2^{10} \div 2^{-8}$

81. $\dfrac{4(8 - 6)^2 - 4 \cdot 3 + 2 \cdot 8}{3^1 + 19^0}$

82. $\dfrac{[4(8 - 6)^2 + 4](3 - 2 \cdot 8)}{2^2(2^3 + 5)}$

Compound Interest. *Use the compound interest formula from Example 10 in Exercises 83–86. Round to the nearest cent.*

83. Suppose that $2125 is invested at 6.2%, compounded semiannually. How much is in the account at the end of 5 yr?

84. Suppose that $9550 is invested at 5.4%, compounded semiannually. How much is in the account at the end of 7 yr?

85. Suppose that $6700 is invested at 4.5%, compounded quarterly. How much is in the account at the end of 6 yr?

86. Suppose that $4875 is invested at 5.8%, compounded quarterly. How much is in the account at the end of 9 yr?

Collaborative Discussion and Writing

87. Are the parentheses necessary in the expression $4 \cdot 25 \div (10 - 5)$? Why or why not?

88. Is $x^{-2} < x^{-1}$ for any negative value(s) of x? Why or why not?

Synthesis

Savings Plan. *The formula*

$$S = P \left[\frac{\left(1 + \dfrac{r}{12}\right)^{12 \cdot t} - 1}{\dfrac{r}{12}} \right]$$

gives the amount S accumulated in a savings plan when a deposit of P dollars is made each month for t years in an account with interest rate r, compounded monthly. Use this formula for Exercises 89–92.

89. Marisol deposits $250 in a retirement account each month beginning at age 40. If the investment earns 5% interest, compounded monthly, how much will have accumulated in the account when she retires 27 yr later?

90. Gordon deposits $100 in a retirement account each month beginning at age 25. If the investment earns 4% interest, compounded monthly, how much will have accumulated in the account when Gordon retires at age 65?

91. Gina wants to establish a college fund for her newborn daughter that will have accumulated $120,000 at the end of 18 yr. If she can count on an interest rate of 6%, compounded monthly, how much should she deposit each month to accomplish this?

92. Liam wants to have $200,000 accumulated in a retirement account by age 70. If he starts making monthly deposits to the plan at age 30 and can count on an interest rate of 4.5%, compounded monthly, how much should he deposit each month in order to accomplish this?

Simplify. Assume that all exponents are integers, all denominators are nonzero, and zero is not raised to a nonpositive power.

93. $(x^t \cdot x^{3t})^2$

94. $(x^y \cdot x^{-y})^3$

95. $(t^{a+x} \cdot t^{x-a})^4$

96. $(m^{x-b} \cdot n^{x+b})^x (m^b n^{-b})^x$

97. $\left[\dfrac{(3x^a y^b)^3}{(-3x^a y^b)^2} \right]^2$

98. $\left[\left(\dfrac{x^r}{y^t} \right)^2 \left(\dfrac{x^{2r}}{y^{4t}} \right)^{-2} \right]^{-3}$

R.3

Addition, Subtraction, and Multiplication of Polynomials

- *Identify the terms, coefficients, and degree of a polynomial.*
- *Add, subtract, and multiply polynomials.*

Polynomials

Polynomials are a type of algebraic expression that you will often encounter in your study of algebra. Some examples of polynomials are

$$3x - 4y, \quad 5y^3 - \tfrac{7}{3}y^2 + 3y - 2, \quad -2.3a^4, \quad \text{and} \quad z^6 - \sqrt{5}.$$

All but the first are polynomials in one variable.

> ### Polynomials in One Variable
>
> A **polynomial in one variable** is any expression of the type
>
> $$a_n x^n + a_{n-1}x^{n-1} + \cdots + a_2 x^2 + a_1 x + a_0,$$
>
> where n is a nonnegative integer, $a_n, \ldots, a_0$ are real numbers, called **coefficients,** and $a_n \neq 0$. The parts of a polynomial separated by plus signs are called **terms.** The **degree** of the polynomial is n, the **leading coefficient** is a_n, and the **constant term** is a_0. The polynomial is said to be written in **descending order,** because the exponents decrease from left to right.

EXAMPLE 1 Identify the terms of the polynomial

$$2x^4 - 7.5x^3 + x - 12.$$

Solution Writing plus signs between the terms, we have

$$2x^4 - 7.5x^3 + x - 12 = 2x^4 + (-7.5x^3) + x + (-12),$$

so the terms are

$$2x^4, \quad -7.5x^3, \quad x, \quad \text{and} \quad -12. \qquad \blacksquare$$

A polynomial, like 23, consisting of only a nonzero constant term has degree 0. It is agreed that the polynomial consisting only of 0 has *no* degree.

EXAMPLE 2 Find the degree of each polynomial.

a) $2x^3 - 9$
b) $y^2 - \tfrac{3}{2} + 5y^4$
c) 7

Solution

POLYNOMIAL	DEGREE
a) $2x^3 - 9$	3
b) $y^2 - \frac{3}{2} + 5y^4 = 5y^4 + y^2 - \frac{3}{2}$	4
c) $7 = 7x^0$	0

Algebraic expressions like $3ab^3 - 8$ and $5x^4y^2 - 3x^3y^8 + 7xy^2 + 6$ are **polynomials in several variables.** The **degree of a term** is the sum of the exponents of the variables in that term. The **degree of a polynomial** is the degree of the term of highest degree.

EXAMPLE 3   Find the degree of the polynomial

$$7ab^3 - 11a^2b^4 + 8.$$

Solution   The degrees of the terms of $7ab^3 - 11a^2b^4 + 8$ are 4, 6, and 0, respectively, so the degree of the polynomial is 6.

A polynomial with just one term, like $-9y^6$, is a **monomial.** If a polynomial has two terms, like $x^2 + 4$, it is a **binomial.** A polynomial with three terms, like $4x^2 - 4xy + 1$, is a **trinomial.**

Expressions like

$$2x^2 - 5x + \frac{3}{x}, \qquad 9 - \sqrt{x}, \quad \text{and} \quad \frac{x+1}{x^4+5}$$

are not polynomials, because they cannot be written in the form $a_nx^n + a_{n-1}x^{n-1} + \cdots + a_1x + a_0$, where the exponents are all nonnegative integers and the coefficients are all real numbers.

Addition and Subtraction

If two terms of an expression have the same variables raised to the same powers, they are called **like terms,** or **similar terms.** We can **combine,** or **collect, like terms** using the distributive property. For example, $3y^2$ and $5y^2$ are like terms and

$$3y^2 + 5y^2 = (3 + 5)y^2$$
$$= 8y^2.$$

We add or subtract polynomials by combining like terms.

EXAMPLE 4   Add or subtract each of the following.

a) $(-5x^3 + 3x^2 - x) + (12x^3 - 7x^2 + 3)$
b) $(6x^2y^3 - 9xy) - (5x^2y^3 - 4xy)$

Solution

a) $(-5x^3 + 3x^2 - x) + (12x^3 - 7x^2 + 3)$

$= (-5x^3 + 12x^3) + (3x^2 - 7x^2) - x + 3$ **Rearranging using the commutative and associative properties**

$= (-5 + 12)x^3 + (3 - 7)x^2 - x + 3$ **Using the distributive property**

$= 7x^3 - 4x^2 - x + 3$

b) We can subtract by adding an opposite:

$(6x^2y^3 - 9xy) - (5x^2y^3 - 4xy)$

$= (6x^2y^3 - 9xy) + (-5x^2y^3 + 4xy)$ **Adding the opposite of $5x^2y^3 - 4xy$**

$= 6x^2y^3 - 9xy - 5x^2y^3 + 4xy$

$= x^2y^3 - 5xy.$ **Combining like terms**

Multiplication

Multiplication of polynomials is based on the distributive property—for example,

$(x + 4)(x + 3) = x(x + 3) + 4(x + 3)$ **Using the distributive property**

$= x^2 + 3x + 4x + 12$ **Using the distributive property two more times**

$= x^2 + 7x + 12.$ **Combining like terms**

In general, to multiply two polynomials, we multiply each term of one by each term of the other and add the products.

EXAMPLE 5 Multiply: $(4x^4y - 7x^2y + 3y)(2y - 3x^2y)$.

Solution We have

$(4x^4y - 7x^2y + 3y)(2y - 3x^2y)$

$= 4x^4y(2y - 3x^2y) - 7x^2y(2y - 3x^2y) + 3y(2y - 3x^2y)$
     Using the distributive property

$= 8x^4y^2 - 12x^6y^2 - 14x^2y^2 + 21x^4y^2 + 6y^2 - 9x^2y^2$
     Using the distributive property three more times

$= 29x^4y^2 - 12x^6y^2 - 23x^2y^2 + 6y^2.$ **Collecting like terms**

We can also use columns to organize our work, aligning like terms under each other in the products.

$$
\begin{array}{r}
4x^4y - 7x^2y + 3y \\
2y - 3x^2y \\
\hline
-12x^6y^2 + 21x^4y^2 - 9x^2y^2 \\
8x^4y^2 - 14x^2y^2 + 6y^2 \\
\hline
-12x^6y^2 + 29x^4y^2 - 23x^2y^2 + 6y^2
\end{array}
$$

     Multiplying by $-3x^2y$
     Multiplying by $2y$
     Adding

We can find the product of two binomials by multiplying the **F**irst terms, then the **O**uter terms, then the **I**nner terms, then the **L**ast terms. Then we collect like terms, if possible. This procedure is sometimes called **FOIL**.

EXAMPLE 6 Multiply: $(2x - 7)(3x + 4)$.

Solution We have

$$(2x - 7)(3x + 4) = 6x^2 + 8x - 21x - 28$$
$$= 6x^2 - 13x - 28$$

We can use FOIL to find some special products.

Special Products of Binomials

$(A + B)^2 = A^2 + 2AB + B^2$ Square of a sum

$(A - B)^2 = A^2 - 2AB + B^2$ Square of a difference

$(A + B)(A - B) = A^2 - B^2$ Product of a sum and a difference

EXAMPLE 7 Multiply each of the following.

a) $(4x + 1)^2$ b) $(3y^2 - 2)^2$ c) $(x^2 + 3y)(x^2 - 3y)$

Solution

a) $(4x + 1)^2 = (4x)^2 + 2 \cdot 4x \cdot 1 + 1^2 = 16x^2 + 8x + 1$

b) $(3y^2 - 2)^2 = (3y^2)^2 - 2 \cdot 3y^2 \cdot 2 + 2^2 = 9y^4 - 12y^2 + 4$

c) $(x^2 + 3y)(x^2 - 3y) = (x^2)^2 - (3y)^2 = x^4 - 9y^2$

R.3

Exercise Set

Determine the terms and the degree of the polynomial.

1. $-5y^4 + 3y^3 + 7y^2 - y - 4$

2. $2m^3 - m^2 - 4m + 11$

3. $3a^4b - 7a^3b^3 + 5ab - 2$

4. $6p^3q^2 - p^2q^4 - 3pq^2 + 5$

Perform the operations indicated.

5. $(5x^2y - 2xy^2 + 3xy - 5) + (-2x^2y - 3xy^2 + 4xy + 7)$

6. $(6x^2y - 3xy^2 + 5xy - 3) + (-4x^2y - 4xy^2 + 3xy + 8)$

7. $(2x + 3y + z - 7) + (4x - 2y - z + 8) + (-3x + y - 2z - 4)$

8. $(2x^2 + 12xy - 11) + (6x^2 - 2x + 4) +$
$(-x^2 - y - 2)$

9. $(3x^2 - 2x - x^3 + 2) - (5x^2 - 8x - x^3 + 4)$

10. $(5x^2 + 4xy - 3y^2 + 2) - (9x^2 - 4xy + 2y^2 - 1)$

11. $(x^4 - 3x^2 + 4x) - (3x^3 + x^2 - 5x + 3)$

12. $(2x^4 - 3x^2 + 7x) - (5x^3 + 2x^2 - 3x + 5)$

13. $(a - b)(2a^3 - ab + 3b^2)$

14. $(n + 1)(n^2 - 6n - 4)$

15. $(x + 5)(x - 3)$

16. $(y - 4)(y + 1)$

17. $(x + 6)(x + 4)$

18. $(n - 5)(n - 8)$

19. $(2a + 3)(a + 5)$

20. $(3b + 1)(b - 2)$

21. $(2x + 3y)(2x + y)$

22. $(2a - 3b)(2a - b)$

23. $(y + 5)^2$

24. $(y + 7)^2$

25. $(x - 4)^2$

26. $(a - 6)^2$

27. $(5x - 3)^2$

28. $(3x - 2)^2$

29. $(2x + 3y)^2$

30. $(5x + 2y)^2$

31. $(2x^2 - 3y)^2$

32. $(4x^2 - 5y)^2$

33. $(a + 3)(a - 3)$

34. $(b + 4)(b - 4)$

35. $(2x - 5)(2x + 5)$

36. $(4y - 1)(4y + 1)$

37. $(3x - 2y)(3x + 2y)$

38. $(3x + 5y)(3x - 5y)$

39. $(2x + 3y + 4)(2x + 3y - 4)$

40. $(5x + 2y + 3)(5x + 2y - 3)$

41. $(x + 1)(x - 1)(x^2 + 1)$

42. $(y - 2)(y + 2)(y^2 + 4)$

Collaborative Discussion and Writing

43. Is the sum of two polynomials of degree n always a polynomial of degree n? Why or why not?

44. Explain how you would convince a classmate that $(A + B)^2 \neq A^2 + B^2$.

Synthesis

Multiply. Assume that all exponents are natural numbers.

45. $(a^n + b^n)(a^n - b^n)$

46. $(t^a + 4)(t^a - 7)$

47. $(a^n + b^n)^2$

48. $(x^{3m} - t^{5n})^2$

49. $(x - 1)(x^2 + x + 1)(x^3 + 1)$

50. $[(2x - 1)^2 - 1]^2$

51. $(x^{a-b})^{a+b}$

52. $(t^{m+n})^{m+n} \cdot (t^{m-n})^{m-n}$

53. $(a + b + c)^2$

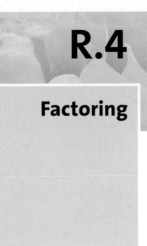

R.4

Factoring

- Factor polynomials by removing a common factor.
- Factor polynomials by grouping.
- Factor trinomials of the type $x^2 + bx + c$.
- Factor trinomials of the type $ax^2 + bx + c$, $a \neq 1$, using the FOIL method and the grouping method.
- Factor special products of polynomials.

To factor a polynomial, we do the reverse of multiplying; that is, we find an equivalent expression that is written as a product.

Terms with Common Factors

When a polynomial is to be factored, we should always look first to factor out a factor that is common to all the terms using the distributive property. We usually look for the constant common factor with the largest absolute value and for variables with the largest exponent common to all the terms. In this sense, we factor out the "largest" common factor.

EXAMPLE 1 Factor each of the following.

a) $15 + 10x - 5x^2$ **b)** $12x^2y^2 - 20x^3y$

Solution

a) $15 + 10x - 5x^2 = 5 \cdot 3 + 5 \cdot 2x - 5 \cdot x^2 = 5(3 + 2x - x^2)$

We can always check a factorization by multiplying:

$$5(3 + 2x - x^2) = 15 + 10x - 5x^2.$$

b) There are several factors common to the terms of $12x^2y^2 - 20x^3y$, but $4x^2y$ is the "largest" of these.

$$12x^2y^2 - 20x^3y = 4x^2y \cdot 3y - 4x^2y \cdot 5x$$
$$= 4x^2y(3y - 5x)$$

Factoring by Grouping

In some polynomials, pairs of terms have a common binomial factor that can be removed in a process called **factoring by grouping.**

EXAMPLE 2 Factor: $x^3 + 3x^2 - 5x - 15$.

Solution We have

$$x^3 + 3x^2 - 5x - 15 = (x^3 + 3x^2) + (-5x - 15)$$ Grouping; each group of terms has a common factor.

$$= x^2(x + 3) - 5(x + 3)$$ Factoring a common factor out of each group

$$= (x + 3)(x^2 - 5).$$ Factoring out the common binomial factor

Trinomials of the Type $x^2 + bx + c$

Some trinomials can be factored into the product of two binomials. To factor a trinomial of the form $x^2 + bx + c$, we look for binomial factors of the form

$$(x + p)(x + q),$$

where $p \cdot q = c$ and $p + q = b$. That is, we look for two numbers p and q whose sum is the coefficient of the middle term of the polynomial, b, and whose product is the constant term, c.

When we factor any polynomial, we should always check first to determine whether there is a factor common to all the terms. If there is, we factor it out first.

EXAMPLE 3   Factor: $x^2 + 5x + 6$.

Solution   First, we look for a common factor. There is none. Next, we look for two numbers whose product is 6 and whose sum is 5. Since the constant term, 6, and the coefficient of the middle term, 5, are both positive, we look for a factorization of 6 in which both factors are positive.

PAIRS OF FACTORS	SUMS OF FACTORS
1, 6	7
2, 3	5 ←

The numbers we need are 2 and 3.

The factorization is $(x + 2)(x + 3)$. We have

$$x^2 + 5x + 6 = (x + 2)(x + 3).$$

We can check this by multiplying:

$$(x + 2)(x + 3) = x^2 + 3x + 2x + 6 = x^2 + 5x + 6.$$

EXAMPLE 4   Factor: $2y^2 - 14y + 24$.

Solution   First, we look for a common factor. Each term has a factor of 2, so we factor it out first:

$$2y^2 - 14y + 24 = 2(y^2 - 7y + 12).$$

Now we consider the trinomial $y^2 - 7y + 12$. We look for two numbers whose product is 12 and whose sum is -7. Since the constant term, 12, is positive and the coefficient of the middle term, -7, is negative, we look for a factorization of 12 in which both factors are negative.

PAIRS OF FACTORS	SUMS OF FACTORS
-1, -12	-13
-2, -6	-8
-3, -4	-7 ←

The numbers we need are -3 and -4.

The factorization of $y^2 - 7y + 12$ is $(y-3)(y-4)$. We must also include the common factor that we factored out earlier. Thus we have

$$2y^2 - 14y + 24 = 2(y-3)(y-4).$$

EXAMPLE 5 Factor: $x^4 - 2x^3 - 8x^2$.

Solution First, we look for a common factor. Each term has a factor of x^2, so we factor it out first:

$$x^4 - 2x^3 - 8x^2 = x^2(x^2 - 2x - 8).$$

Now we consider the trinomial $x^2 - 2x - 8$. We look for two numbers whose product is -8 and whose sum is -2. Since the constant term, -8, is negative, one factor will be positive and the other will be negative.

Pairs of Factors	Sums of Factors
$-1,\ 8$	7
$1,\ -8$	-7
$-2,\ 4$	2
$2,\ -4$	-2

The numbers we need are 2 and −4.

We might have observed at the outset that since the sum of the factors is -2, a negative number, we need consider only pairs of factors for which the negative factor has the greater absolute value. Thus only the pairs 1, -8 and 2, -4 need have been considered.

Using the pair of factors 2 and -4, we see that the factorization of $x^2 - 2x - 8$ is $(x+2)(x-4)$. Including the common factor, we have

$$x^4 - 2x^3 - 8x^2 = x^2(x+2)(x-4).$$

Trinomials of the Type $ax^2 + bx + c, a \neq 1$

We consider two methods for factoring trinomials of the type $ax^2 + bx + c$, $a \neq 1$.

The FOIL Method

We first consider the **FOIL method** for factoring trinomials of the type $ax^2 + bx + c, a \neq 1$. Consider the following multiplication.

$$(3x+2)(4x+5) = 12x^2 + \underbrace{15x + 8x}_{} + 10$$
$$= 12x^2 + 23x + 10$$

To factor $12x^2 + 23x + 10$, we must reverse what we just did. We look for two binomials whose product is this trinomial. The product of the First terms must be $12x^2$. The product of the Outside terms plus the product of the Inside terms must be $23x$. The product of the Last terms must be 10. We know from the preceding discussion that the answer is $(3x + 2)(4x + 5)$. In general, however, finding such an answer involves trial and error. We use the following method.

To factor trinomials of the type $ax^2 + bx + c$, $a \neq 1$, using the **FOIL method:**

1. Factor out the largest common factor.

2. Find two First terms whose product is ax^2:

$$(\blacksquare\, x +\ \ )(\blacksquare\, x +\ \ ) = ax^2 + bx + c.$$

   FOIL

3. Find two Last terms whose product is c:

$$(\ x + \blacksquare\,)(\ x + \blacksquare\,) = ax^2 + bx + c.$$

   FOIL

4. Repeat steps (2) and (3) until a combination is found for which the sum of the Outside and Inside products is bx:

$$(\blacksquare\, x + \blacksquare\,)(\blacksquare\, x + \blacksquare\,) = ax^2 + bx + c.$$

   I

   O

   FOIL

EXAMPLE 6 Factor: $3x^2 - 10x - 8$.

Solution

1. There is no common factor (other than 1 or -1).

2. Factor the first term, $3x^2$. The only possibility (with positive coefficients) is $3x \cdot x$. The factorization, if it exists, must be of the form $(3x + \blacksquare\,)(x + \blacksquare\,)$.

3. Next, factor the constant term, -8. The possibilities are $(-8)(1)$, $8(-1)$, $-2(4)$, and $2(-4)$. The factors can be written in the opposite order as well: $1(-8)$, $-1(8)$, $4(-2)$, and $-4(2)$.

4. Find a pair of factors for which the sum of the outside and the inside products is the middle term, $-10x$. Each possibility should be checked by multiplying. Some trials show that the desired factorization is $(3x + 2)(x - 4)$. ▬

The Grouping Method

The second method for factoring trinomials of the type $ax^2 + bx + c$, $a \neq 1$, is known as the **grouping method,** or the *ac*-**method.**

> To factor $ax^2 + bx + c, a \neq 1$, using the **grouping method:**
>   1. Factor out the largest common factor.
>   2. Multiply the leading coefficient a and the constant c.
>   3. Try to factor the product ac so that the sum of the factors is b. That is, find integers p and q such that $pq = ac$ and $p + q = b$.
>   4. Split the middle term. That is, write it as a sum using the factors found in step (3).
>   5. Factor by grouping.

EXAMPLE 7 Factor: $12x^3 + 10x^2 - 8x$.

Solution

1. Factor out the largest common factor, $2x$:

    $$12x^3 + 10x^2 - 8x = 2x(6x^2 + 5x - 4).$$

2. Now consider $6x^2 + 5x - 4$. Multiply the leading coefficient, 6, and the constant, -4: $6(-4) = -24$.

3. Try to factor -24 so that the sum of the factors is the coefficient of the middle term, 5.

PAIRS OF FACTORS	SUMS OF FACTORS
1, −24	−23
−1, 24	23
2, −12	−10
−2, 12	10
3, −8	−5
−3, 8	5 ← ———— $-3 \cdot 8 = -24; -3 + 8 = 5$
4, −6	−2
−4, 6	2

4. Split the middle term using the numbers found in step (3):

    $$5x = -3x + 8x.$$

5. Finally, factor by grouping:

    $$6x^2 + 5x - 4 = 6x^2 - 3x + 8x - 4$$
    $$= 3x(2x - 1) + 4(2x - 1)$$
    $$= (2x - 1)(3x + 4).$$

   Be sure to include the common factor to get the complete factorization of the original trinomial:

    $$12x^3 + 10x^2 - 8x = 2x(2x - 1)(3x + 4).$$

Special Factorizations

We reverse the equation $(A + B)(A - B) = A^2 - B^2$ to factor a **difference of squares.**

$$A^2 - B^2 = (A + B)(A - B)$$

EXAMPLE 8   Factor each of the following.

a) $x^2 - 16$ **b)** $9a^2 - 25$ **c)** $6x^4 - 6y^4$

Solution

a) $x^2 - 16 = x^2 - 4^2 = (x + 4)(x - 4)$

b) $9a^2 - 25 = (3a)^2 - 5^2 = (3a + 5)(3a - 5)$

c) $6x^4 - 6y^4 = 6(x^4 - y^4)$
$$= 6[(x^2)^2 - (y^2)^2]$$
$$= 6(x^2 + y^2)(x^2 - y^2) \qquad \text{$x^2 - y^2$ can be factored further.}$$
$$= 6(x^2 + y^2)(x + y)(x - y) \qquad \text{Because none of these factors can be factored further, we have \textit{factored completely}.}$$

The rules for squaring binomials can be reversed to factor trinomials that are squares of binomials:

$$A^2 + 2AB + B^2 = (A + B)^2;$$
$$A^2 - 2AB + B^2 = (A - B)^2.$$

EXAMPLE 9   Factor each of the following.

a) $x^2 + 8x + 16$

b) $25y^2 - 30y + 9$

Solution

$$A^2 + 2 \cdot A \cdot B + B^2 = (A + B)^2$$

a) $x^2 + 8x + 16 = x^2 + 2 \cdot x \cdot 4 + 4^2 = (x + 4)^2$

$$A^2 - 2 \cdot A \cdot B + B^2 = (A - B)^2$$

b) $25y^2 - 30y + 9 = (5y)^2 - 2 \cdot 5y \cdot 3 + 3^2 - (5y - 3)^2$

We can use the following rules to factor a **sum** or a **difference of cubes:**

$$A^3 + B^3 = (A + B)(A^2 - AB + B^2);$$
$$A^3 - B^3 = (A - B)(A^2 + AB + B^2).$$

These rules can be verified by multiplying.

EXAMPLE 10   Factor each of the following.

a) $x^3 + 27$

b) $16y^3 - 250$

Solution

a) $x^3 + 27 = x^3 + 3^3$
$$= (x + 3)(x^2 - 3x + 9)$$

b) $16y^3 - 250 = 2(8y^3 - 125)$
$$= 2[(2y)^3 - 5^3]$$
$$= 2(2y - 5)(4y^2 + 10y + 25)$$

Not all polynomials can be factored into polynomials with integer coefficients. An example is $x^2 - x + 7$. There are no real factors of 7 whose sum is -1. In such a case we say that the polynomial is "not factorable," or **prime.**

CONNECTING THE CONCEPTS

A STRATEGY FOR FACTORING

A. Always factor out the largest common factor first.

B. Look at the number of terms.

Two terms: Try factoring as a difference of squares first. Next, try factoring as a sum or a difference of cubes. Do *not* try to factor a *sum* of squares.

Three terms: Try factoring as the square of a binomial. Next, try using the FOIL method or the grouping method for factoring a trinomial.

Four or more terms: Try factoring by grouping and factoring out a common binomial factor.

C. Always *factor completely.* If a factor with more than one term can itself be factored further, do so.

R.4

Exercise Set

Factor out a common factor.

1. $2x - 10$

2. $7y + 42$

3. $3x^4 - 9x^2$

4. $20y^2 - 5y^5$

5. $4a^2 - 12a + 16$

6. $6n^2 + 24n - 18$

7. $a(b - 2) + c(b - 2)$

8. $a(x^2 - 3) - 2(x^2 - 3)$

Factor by grouping.

9. $x^3 + 3x^2 + 6x + 18$

10. $3x^3 - x^2 + 18x - 6$

11. $y^3 - y^2 + 3y - 3$

12. $y^3 - y^2 + 2y - 2$

13. $24x^3 - 36x^2 + 72x - 108$

14. $5a^3 - 10a^2 + 25a - 50$

15. $a^3 - 3a^2 - 2a + 6$

16. $t^3 + 6t^2 - 2t - 12$

17. $x^3 - x^2 - 5x + 5$

18. $x^3 - x^2 - 6x + 6$

Factor the trinomial.

19. $p^2 + 6p + 8$

20. $w^2 - 7w + 10$

21. $x^2 + 8x + 12$

22. $x^2 + 6x + 5$

23. $t^2 + 8t + 15$

24. $y^2 + 12y + 27$

25. $x^2 - 6xy - 27y^2$

26. $t^2 - 2t - 15$

27. $2n^2 - 20n - 48$

28. $2a^2 - 2ab - 24b^2$

29. $y^4 - 4y^2 - 21$

30. $m^4 - m^2 - 90$

31. $2n^2 + 9n - 56$

32. $3y^2 + 7y - 20$

33. $12x^2 + 11x + 2$

34. $6x^2 - 7x - 20$

35. $4x^2 + 15x + 9$

36. $2y^2 + 7y + 6$

37. $2y^2 + y - 6$

38. $20p^2 - 23p + 6$

39. $6a^2 - 29ab + 28b^2$

40. $10m^2 + 7mn - 12n^2$

41. $12a^2 - 4a - 16$

42. $12a^2 - 14a - 20$

Factor the difference of squares.

43. $m^2 - 4$

44. $z^2 - 81$

45. $9x^2 - 25$

46. $16x^2 - 9$

47. $6x^2 - 6y^2$

48. $8a^2 - 8b^2$

49. $4xy^4 - 4xz^2$

50. $5x^2y - 5yz^4$

51. $7pq^4 - 7py^4$

52. $25ab^4 - 25az^4$

Factor the square of a binomial.

53. $y^2 - 6y + 9$

54. $x^2 + 8x + 16$

55. $4z^2 + 12z + 9$

56. $9z^2 - 12z + 4$

57. $1 - 8x + 16x^2$

58. $1 + 10x + 25x^2$

59. $a^3 + 24a^2 + 144a$

60. $y^3 - 18y^2 + 81y$

61. $4p^2 - 8pq + 4q^2$

62. $5a^2 - 10ab + 5b^2$

Factor the sum or difference of cubes.

63. $x^3 + 8$

64. $y^3 - 64$

65. $m^3 - 1$

66. $n^3 + 216$

67. $2y^3 - 128$

68. $8t^3 - 8$

69. $3a^5 - 24a^2$

70. $250z^4 - 2z$

71. $t^6 + 1$

72. $27x^6 - 8$

Factor completely.

73. $18a^2b - 15ab^2$

74. $4x^2y + 12xy^2$

75. $x^3 - 4x^2 + 5x - 20$

76. $z^3 + 3z^2 - 3z - 9$

77. $8x^2 - 32$

78. $6y^2 - 6$

79. $4y^2 - 5$

80. $16x^2 - 7$

81. $m^2 - 9n^2$

82. $25t^2 - 16$

83. $x^2 + 9x + 20$

84. $y^2 + y - 6$

85. $y^2 - 6y + 5$

86. $x^2 - 4x - 21$

87. $2a^2 + 9a + 4$

88. $3b^2 - b - 2$

89. $6x^2 + 7x - 3$

90. $8x^2 + 2x - 15$

91. $y^2 - 18y + 81$

92. $n^2 + 2n + 1$

93. $9z^2 - 24z + 16$

94. $4z^2 + 20z + 25$

95. $x^2y^2 - 14xy + 49$

96. $x^2y^2 - 16xy + 64$

97. $4ax^2 + 20ax - 56a$

98. $21x^2y + 2xy - 8y$

99. $3z^3 - 24$

100. $4t^3 + 108$

101. $16a^7b + 54ab^7$

102. $24a^2x^4 - 375a^8x$

103. $y^3 - 3y^2 - 4y + 12$

104. $p^3 - 2p^2 - 9p + 18$

105. $x^3 - x^2 + x - 1$

106. $x^3 - x^2 - x + 1$

107. $5m^4 - 20$

108. $2x^2 - 288$

109. $2x^3 + 6x^2 - 8x - 24$

110. $3x^3 + 6x^2 - 27x - 54$

111. $4c^2 - 4cd + d^2$

112. $9a^2 - 6ab + b^2$

113. $m^6 + 8m^3 - 20$

114. $x^4 - 37x^2 + 36$

115. $p - 64p^4$

116. $125a - 8a^4$

Collaborative Discussion and Writing

117. Under what circumstances can $A^2 + B^2$ be factored?

118. Explain how the rule for factoring a sum of cubes can be used to factor a difference of cubes.

Synthesis

Factor.

119. $y^4 - 84 + 5y^2$

120. $11x^2 + x^4 - 80$

121. $y^2 - \frac{8}{49} + \frac{2}{7}y$

122. $t^2 - \frac{27}{100} + \frac{3}{5}t$

123. $x^2 + 3x + \frac{9}{4}$

124. $x^2 - 5x + \frac{25}{4}$

125. $x^2 - x + \frac{1}{4}$

126. $x^2 - \frac{2}{3}x + \frac{1}{9}$

127. $(x + h)^3 - x^3$

128. $(x + 0.01)^2 - x^2$

129. $(y - 4)^2 + 5(y - 4) - 24$

130. $6(2p + q)^2 - 5(2p + q) - 25$

Factor. Assume that variables in exponents represent natural numbers.

131. $x^{2n} + 5x^n - 24$

132. $4x^{2n} - 4x^n - 3$

133. $x^2 + ax + bx + ab$

134. $bdy^2 + ady + bcy + ac$

135. $25y^{2m} - (x^{2n} - 2x^n + 1)$

136. $x^{6a} - t^{3b}$

137. $(y - 1)^4 - (y - 1)^2$

138. $x^6 - 2x^5 + x^4 - x^2 + 2x - 1$

R.5

Rational Expressions

- *Determine the domain of a rational expression.*
- *Simplify rational expressions.*
- *Multiply, divide, add, and subtract rational expressions.*
- *Simplify complex rational expressions.*

A **rational expression** is the quotient of two polynomials. For example,

$$\frac{3}{5}, \quad \frac{2}{x - 3}, \quad \text{and} \quad \frac{x^2 - 4}{x^2 - 4x - 5}$$

are rational expressions.

The Domain of a Rational Expression

The **domain** of an algebraic expression is the set of all real numbers for which the expression is defined. Since division by zero is not defined, any number that makes the denominator zero is not in the domain of a rational expression.

EXAMPLE 1   Find the domain of each of the following.

a) $\dfrac{2}{x - 3}$ **b)** $\dfrac{x^2 - 4}{x^2 - 4x - 5}$

Solution

a) Since $x - 3$ is 0 when $x = 3$, the domain of $2/(x - 3)$ is the set of all real numbers except 3.

b) To determine the domain of $(x^2 - 4)/(x^2 - 4x - 5)$, we first factor the denominator:

$$\frac{x^2 - 4}{x^2 - 4x - 5} = \frac{x^2 - 4}{(x + 1)(x - 5)}.$$

The factor $x + 1$ is 0 when $x = -1$, and the factor $x - 5$ is 0 when $x = 5$. Since $(x + 1)(x - 5) = 0$ when $x = -1$ or $x = 5$, the domain is the set of all real numbers except -1 and 5.   ▬

We can describe the domains found in Example 1 using *set-builder notation*. For example, we write "The set of all real numbers x such that x is not equal to 3" as

$$\{x \,|\, x \text{ is a real number } and\ x \neq 3\}.$$

Similarly, we write "The set of all real numbers x such that x is not equal to -1 and x is not equal to 5" as

$$\{x \,|\, x \text{ is a real number } and\ x \neq -1 \ and\ x \neq 5\}.$$

Simplifying, Multiplying, and Dividing Rational Expressions

To simplify rational expressions, we use the fact that

$$\frac{a \cdot c}{b \cdot c} = \frac{a}{b} \cdot \frac{c}{c} = \frac{a}{b} \cdot 1 = \frac{a}{b}.$$

EXAMPLE 2 Simplify: $\dfrac{9x^2 + 6x - 3}{12x^2 - 12}$.

Solution

$$\dfrac{9x^2 + 6x - 3}{12x^2 - 12} = \dfrac{3(3x^2 + 2x - 1)}{12(x^2 - 1)} \Bigg\}$$ Factoring the numerator and the denominator

$$= \dfrac{3(x + 1)(3x - 1)}{3 \cdot 4(x + 1)(x - 1)}$$

$$= \dfrac{3(x + 1)}{3(x + 1)} \cdot \dfrac{3x - 1}{4(x - 1)}$$ Factoring the rational expression

$$= 1 \cdot \dfrac{3x - 1}{4(x - 1)} \qquad\qquad \dfrac{3(x + 1)}{3(x + 1)} = 1$$

$$= \dfrac{3x - 1}{4(x - 1)} \qquad\qquad \text{Removing a factor of 1}$$

Canceling is a shortcut that is often used to remove a factor of 1.

EXAMPLE 3 Simplify each of the following.

a) $\dfrac{4x^3 + 16x^2}{2x^3 + 6x^2 - 8x}$ b) $\dfrac{2 - x}{x^2 + x - 6}$

Solution

a) $\dfrac{4x^3 + 16x^2}{2x^3 + 6x^2 - 8x} = \dfrac{2 \cdot 2 \cdot x \cdot x(x + 4)}{2 \cdot x(x + 4)(x - 1)}$ Factoring the numerator and the denominator

$$= \dfrac{\cancel{2} \cdot 2 \cdot \cancel{x} \cdot x(\cancel{x + 4})}{\cancel{2} \cdot \cancel{x}(\cancel{x + 4})(x - 1)} \qquad \begin{array}{l}\text{Removing a factor of 1:}\\ \dfrac{2x(x + 4)}{2x(x + 4)} = 1\end{array}$$

$$= \dfrac{2x}{x - 1}$$

b) $\dfrac{2 - x}{x^2 + x - 6} = \dfrac{2 - x}{(x + 3)(x - 2)}$ Factoring the denominator

$$= \dfrac{-1(x - 2)}{(x + 3)(x - 2)} \qquad 2 - x = -1(x - 2)$$

$$= \dfrac{-1(\cancel{x - 2})}{(x + 3)(\cancel{x - 2})} \qquad \text{Removing a factor of 1: } \dfrac{x - 2}{x - 2} = 1$$

$$= \dfrac{-1}{x + 3}, \text{ or } -\dfrac{1}{x + 3}$$

In Example 3(b), we saw that

$$\dfrac{2 - x}{x^2 + x - 6} \quad \text{and} \quad -\dfrac{1}{x + 3}$$

are **equivalent expressions.** This means that they have the same value for all numbers that are in *both* domains. Note that -3 is not in the domain of *either* expression, whereas 2 is in the domain of $-1/(x + 3)$ but not in the domain of $(2 - x)/(x^2 + x - 6)$ and thus is not in the domain of *both* expressions.

To multiply rational expressions, we multiply numerators and multiply denominators and, if possible, simplify the result. To divide rational expressions, we multiply the dividend by the reciprocal of the divisor and, if possible, simplify the result—that is,

$$\frac{a}{b} \cdot \frac{c}{d} = \frac{ac}{bd} \quad \text{and} \quad \frac{a}{b} \div \frac{c}{d} = \frac{a}{b} \cdot \frac{d}{c} = \frac{ad}{bc}.$$

EXAMPLE 4 Multiply or divide and simplify each of the following.

a) $\dfrac{x + 4}{x - 3} \cdot \dfrac{x^2 - 9}{x^2 - x - 2}$

b) $\dfrac{y^3 - 1}{y^2 - 1} \div \dfrac{y^2 + y + 1}{y^2 + 2y + 1}$

Solution

a) $\dfrac{x + 4}{x - 3} \cdot \dfrac{x^2 - 9}{x^2 - x - 2} = \dfrac{(x + 4)(x^2 - 9)}{(x - 3)(x^2 - x - 2)}$

Multiplying the numerators and the denominators

$$= \frac{(x + 4)(x + 3)(x - 3)}{(x - 3)(x - 2)(x + 1)}$$

Factoring and removing a factor of 1: $\dfrac{x - 3}{x - 3} = 1$

$$= \frac{(x + 4)(x + 3)}{(x - 2)(x + 1)}$$

b) $\dfrac{y^3 - 1}{y^2 - 1} \div \dfrac{y^2 + y + 1}{y^2 + 2y + 1} = \dfrac{y^3 - 1}{y^2 - 1} \cdot \dfrac{y^2 + 2y + 1}{y^2 + y + 1}$

Multiplying by the reciprocal of the divisor

$$= \frac{(y^3 - 1)(y^2 + 2y + 1)}{(y^2 - 1)(y^2 + y + 1)}$$

$$= \frac{(y - 1)(y^2 + y + 1)(y + 1)(y + 1)}{(y + 1)(y - 1)(y^2 + y + 1)}$$

Factoring and removing a factor of 1

$$= y + 1$$

Adding and Subtracting Rational Expressions

When rational expressions have the same denominator, we can add or subtract by adding or subtracting the numerators and retaining the common denominator. If the denominators differ, we must find equivalent rational expressions that have a common denominator. In general, it is most efficient to find the **least common denominator (LCD)** of the expressions.

> To find the least common denominator of rational expressions, factor each denominator and form the product that uses each factor the greatest number of times it occurs in any factorization.

EXAMPLE 5 Add or subtract and simplify each of the following.

a) $\dfrac{x^2 - 4x + 4}{2x^2 - 3x + 1} + \dfrac{x + 4}{2x - 2}$

b) $\dfrac{x}{x^2 + 11x + 30} - \dfrac{5}{x^2 + 9x + 20}$

Solution

a) $\dfrac{x^2 - 4x + 4}{2x^2 - 3x + 1} + \dfrac{x + 4}{2x - 2}$

$= \dfrac{x^2 - 4x + 4}{(2x - 1)(x - 1)} + \dfrac{x + 4}{2(x - 1)}$ **Factoring the denominators**

The LCD is $(2x - 1)(x - 1)(2)$, or $2(2x - 1)(x - 1)$.

$= \dfrac{x^2 - 4x + 4}{(2x - 1)(x - 1)} \cdot \dfrac{2}{2} + \dfrac{x + 4}{2(x - 1)} \cdot \dfrac{2x - 1}{2x - 1}$ **Multiplying each term by 1 to get the LCD**

$= \dfrac{2x^2 - 8x + 8}{(2x - 1)(x - 1)(2)} + \dfrac{2x^2 + 7x - 4}{2(x - 1)(2x - 1)}$

$= \dfrac{4x^2 - x + 4}{2(2x - 1)(x - 1)}$ **Adding the numerators**

b) $\dfrac{x}{x^2 + 11x + 30} - \dfrac{5}{x^2 + 9x + 20}$

$= \dfrac{x}{(x + 5)(x + 6)} - \dfrac{5}{(x + 5)(x + 4)}$ **Factoring the denominators**

The LCD is $(x + 5)(x + 6)(x + 4)$.

$= \dfrac{x}{(x + 5)(x + 6)} \cdot \dfrac{x + 4}{x + 4} - \dfrac{5}{(x + 5)(x + 4)} \cdot \dfrac{x + 6}{x + 6}$

Multiplying each term by 1 to get the LCD

$= \dfrac{x^2 + 4x}{(x + 5)(x + 6)(x + 4)} - \dfrac{5x + 30}{(x + 5)(x + 4)(x + 6)}$

$= \dfrac{x^2 + 4x - 5x - 30}{(x + 5)(x + 6)(x + 4)}$ **Be sure to change the sign of *every* term in the numerator of the expression being subtracted:** $-(5x + 30) = -5x - 30$

$= \dfrac{x^2 - x - 30}{(x + 5)(x + 6)(x + 4)}$

$= \dfrac{(x + 5)(x - 6)}{(x + 5)(x + 6)(x + 4)}$ **Factoring and removing a factor of 1:** $\dfrac{x + 5}{x + 5} = 1$

$= \dfrac{x - 6}{(x + 6)(x + 4)}$

Complex Rational Expressions

A **complex rational expression** has rational expressions in its numerator or its denominator or both.

> To simplify a complex rational expression:
>
> *Method 1.* Find the LCD of all the denominators within the complex rational expression. Then multiply by 1 using the LCD as the numerator and the denominator of the expression for 1.
>
> *Method 2.* First add or subtract, if necessary, to get a single rational expression in the numerator and in the denominator. Then divide by multiplying by the reciprocal of the denominator.

EXAMPLE 6   Simplify: $\dfrac{\dfrac{1}{a} + \dfrac{1}{b}}{\dfrac{1}{a^3} + \dfrac{1}{b^3}}$.

Solution

Method 1. The LCD of the four rational expressions in the numerator and the denominator is $a^3 b^3$.

$$\frac{\dfrac{1}{a} + \dfrac{1}{b}}{\dfrac{1}{a^3} + \dfrac{1}{b^3}} = \frac{\dfrac{1}{a} + \dfrac{1}{b}}{\dfrac{1}{a^3} + \dfrac{1}{b^3}} \cdot \frac{a^3 b^3}{a^3 b^3} \qquad \text{Multiplying by 1}$$
$$\text{using } \frac{a^3 b^3}{a^3 b^3}$$

$$= \frac{\left(\dfrac{1}{a} + \dfrac{1}{b}\right)(a^3 b^3)}{\left(\dfrac{1}{a^3} + \dfrac{1}{b^3}\right)(a^3 b^3)}$$

$$= \frac{a^2 b^3 + a^3 b^2}{b^3 + a^3}$$

$$= \frac{a^2 b^2 (\cancel{b + a})}{(\cancel{b + a})(b^2 - ba + a^2)} \qquad \text{Factoring and removing a}$$
$$\text{factor of 1: } \frac{b + a}{b + a} = 1$$

$$= \frac{a^2 b^2}{b^2 - ba + a^2}$$

Method 2. We add in the numerator and in the denominator.

$$\frac{\dfrac{1}{a} + \dfrac{1}{b}}{\dfrac{1}{a^3} + \dfrac{1}{b^3}} = \frac{\dfrac{1}{a} \cdot \dfrac{b}{b} + \dfrac{1}{b} \cdot \dfrac{a}{a}}{\dfrac{1}{a^3} \cdot \dfrac{b^3}{b^3} + \dfrac{1}{b^3} \cdot \dfrac{a^3}{a^3}} \longleftarrow \text{The LCD is } ab.$$

$\longleftarrow$ The LCD is a^3b^3.

$$= \frac{\dfrac{b}{ab} + \dfrac{a}{ab}}{\dfrac{b^3}{a^3b^3} + \dfrac{a^3}{a^3b^3}}$$

$$= \frac{\dfrac{b + a}{ab}}{\dfrac{b^3 + a^3}{a^3b^3}} \qquad \text{We have a single rational expression in both the numerator and the denominator.}$$

$$= \frac{b + a}{ab} \cdot \frac{a^3b^3}{b^3 + a^3} \qquad \text{Multiplying by the reciprocal of the denominator}$$

$$= \frac{(b + a)(a)(b)(a^2b^2)}{(a)(b)(b + a)(b^2 - ba + a^2)}$$

$$= \frac{a^2b^2}{b^2 - ba + a^2}$$

R.5

Exercise Set

Find the domain of the rational expression.

1. $-\dfrac{3}{4}$

2. $\dfrac{5}{8 - x}$

3. $\dfrac{3x - 3}{x(x - 1)}$

4. $\dfrac{15x - 10}{2x(3x - 2)}$

5. $\dfrac{x + 5}{x^2 + 4x - 5}$

6. $\dfrac{(x^2 - 4)(x + 1)}{(x + 2)(x^2 - 1)}$

7. $\dfrac{7x^2 - 28x + 28}{(x^2 - 4)(x^2 + 3x - 10)}$

8. $\dfrac{7x^2 + 11x - 6}{x(x^2 - x - 6)}$

Multiply or divide and, if possible, simplify.

9. $\dfrac{x^2 - y^2}{(x - y)^2} \cdot \dfrac{1}{x + y}$

10. $\dfrac{r - s}{r + s} \cdot \dfrac{r^2 - s^2}{(r - s)^2}$

11. $\dfrac{x^2 - 2x - 35}{2x^3 - 3x^2} \cdot \dfrac{4x^3 - 9x}{7x - 49}$

12. $\dfrac{x^2 + 2x - 35}{3x^3 - 2x^2} \cdot \dfrac{9x^3 - 4x}{7x + 49}$

13. $\dfrac{a^2 - a - 6}{a^2 - 7a + 12} \cdot \dfrac{a^2 - 2a - 8}{a^2 - 3a - 10}$

14. $\dfrac{a^2 - a - 12}{a^2 - 6a + 8} \cdot \dfrac{a^2 + a - 6}{a^2 - 2a - 24}$

15. $\dfrac{m^2 - n^2}{r + s} \div \dfrac{m - n}{r + s}$

16. $\dfrac{a^2 - b^2}{x - y} \div \dfrac{a + b}{x - y}$

17. $\dfrac{3x + 12}{2x - 8} \div \dfrac{(x + 4)^2}{(x - 4)^2}$

18. $\dfrac{a^2 - a - 2}{a^2 - a - 6} \div \dfrac{a^2 - 2a}{2a + a^2}$

19. $\dfrac{x^2 - y^2}{x^3 - y^3} \cdot \dfrac{x^2 + xy + y^2}{x^2 + 2xy + y^2}$

20. $\dfrac{c^3 + 8}{c^2 - 4} \div \dfrac{c^2 - 2c + 4}{c^2 - 4c + 4}$

21. $\dfrac{(x - y)^2 - z^2}{(x + y)^2 - z^2} \div \dfrac{x - y + z}{x + y - z}$

22. $\dfrac{(a + b)^2 - 9}{(a - b)^2 - 9} \cdot \dfrac{a - b - 3}{a + b + 3}$

Add or subtract and, if possible, simplify.

23. $\dfrac{5}{2x} + \dfrac{1}{2x}$ **24.** $\dfrac{10}{9y} - \dfrac{4}{9y}$

25. $\dfrac{3}{2a + 3} + \dfrac{2a}{2a + 3}$ **26.** $\dfrac{a - 3b}{a + b} + \dfrac{a + 5b}{a + b}$

27. $\dfrac{5}{4z} - \dfrac{3}{8z}$ **28.** $\dfrac{12}{x^2 y} + \dfrac{5}{xy^2}$

29. $\dfrac{3}{x + 2} + \dfrac{2}{x^2 - 4}$

30. $\dfrac{5}{a - 3} - \dfrac{2}{a^2 - 9}$

31. $\dfrac{y}{y^2 - y - 20} - \dfrac{2}{y + 4}$

32. $\dfrac{6}{y^2 + 6y + 9} - \dfrac{5}{y + 3}$

33. $\dfrac{3}{x + y} + \dfrac{x - 5y}{x^2 - y^2}$

34. $\dfrac{a^2 + 1}{a^2 - 1} - \dfrac{a - 1}{a + 1}$

35. $\dfrac{y}{y - 1} + \dfrac{2}{1 - y}$
(*Note:* $1 - y = -1(y - 1)$.)

36. $\dfrac{a}{a - b} + \dfrac{b}{b - a}$
(*Note:* $b - a = -1(a - b)$.)

37. $\dfrac{x}{2x - 3y} - \dfrac{y}{3y - 2x}$

38. $\dfrac{3a}{3a - 2b} - \dfrac{2a}{2b - 3a}$

39. $\dfrac{9x + 2}{3x^2 - 2x - 8} + \dfrac{7}{3x^2 + x - 4}$

40. $\dfrac{3y}{y^2 - 7y + 10} - \dfrac{2y}{y^2 - 8y + 15}$

41. $\dfrac{5a}{a - b} + \dfrac{ab}{a^2 - b^2} + \dfrac{4b}{a + b}$

42. $\dfrac{6a}{a - b} - \dfrac{3b}{b - a} + \dfrac{5}{a^2 - b^2}$

43. $\dfrac{7}{x + 2} - \dfrac{x + 8}{4 - x^2} + \dfrac{3x - 2}{4 - 4x + x^2}$

44. $\dfrac{6}{x + 3} - \dfrac{x + 4}{9 - x^2} + \dfrac{2x - 3}{9 - 6x + x^2}$

45. $\dfrac{1}{x + 1} + \dfrac{x}{2 - x} + \dfrac{x^2 + 2}{x^2 - x - 2}$

46. $\dfrac{x - 1}{x - 2} - \dfrac{x + 1}{x + 2} - \dfrac{x - 6}{4 - x^2}$

Simplify.

47. $\dfrac{\dfrac{x^2 - y^2}{xy}}{\dfrac{x - y}{y}}$ **48.** $\dfrac{\dfrac{a - b}{b}}{\dfrac{a^2 - b^2}{ab}}$

49. $\dfrac{\dfrac{x}{y} - \dfrac{y}{x}}{\dfrac{1}{y} + \dfrac{1}{x}}$ **50.** $\dfrac{\dfrac{a}{b} - \dfrac{b}{a}}{\dfrac{1}{a} - \dfrac{1}{b}}$

51. $\dfrac{c + \dfrac{8}{c^2}}{1 + \dfrac{2}{c}}$ **52.** $\dfrac{a - \dfrac{a}{b}}{b - \dfrac{b}{a}}$

53. $\dfrac{\dfrac{x^2 + xy + y^2}{x^2}}{\dfrac{x^2 - y^2}{y} \quad \dfrac{}{x}}$ **54.** $\dfrac{\dfrac{a^2}{b} + \dfrac{b^2}{a}}{a^2 - ab + b^2}$

55. $\dfrac{a - a^{-1}}{a + a^{-1}}$

56. $\dfrac{x^{-1} + y^{-1}}{x^{-3} + y^{-3}}$

57. $\dfrac{\dfrac{1}{x - 3} + \dfrac{2}{x + 3}}{\dfrac{3}{x - 1} - \dfrac{4}{x + 2}}$

58. $\dfrac{\dfrac{5}{x + 1} - \dfrac{3}{x - 2}}{\dfrac{1}{x - 5} + \dfrac{2}{x + 2}}$

59. $\dfrac{\dfrac{a}{1 - a} + \dfrac{1 + a}{a}}{\dfrac{1 - a}{a} + \dfrac{a}{1 + a}}$

60. $\dfrac{\dfrac{1 - x}{x} + \dfrac{x}{1 + x}}{\dfrac{1 + x}{x} + \dfrac{x}{1 - x}}$

61. $\dfrac{\dfrac{1}{a^2} + \dfrac{2}{ab} + \dfrac{1}{b^2}}{\dfrac{1}{a^2} - \dfrac{1}{b^2}}$

62. $\dfrac{\dfrac{1}{x^2} - \dfrac{1}{y^2}}{\dfrac{1}{x^2} - \dfrac{2}{xy} + \dfrac{1}{y^2}}$

Collaborative Discussion and Writing

63. When adding or subtracting rational expressions, we can always find a common denominator by forming the product of all the denominators. Explain why it is usually preferable to find the least common denominator.

64. How would you determine which method to use for simplifying a particular complex rational expression?

Synthesis

Simplify.

65. $\dfrac{(x + h)^2 - x^2}{h}$

66. $\dfrac{\dfrac{1}{x + h} - \dfrac{1}{x}}{h}$

67. $\dfrac{(x + h)^3 - x^3}{h}$

68. $\dfrac{\dfrac{1}{(x + h)^2} - \dfrac{1}{x^2}}{h}$

69. $\left[\dfrac{\dfrac{x + 1}{x - 1} + 1}{\dfrac{x + 1}{x - 1} - 1} \right]^5$

70. $1 + \dfrac{1}{1 + \dfrac{1}{1 + \dfrac{1}{1 + \dfrac{1}{x}}}}$

Perform the indicated operations and, if possible, simplify.

71. $\dfrac{n(n + 1)(n + 2)}{2 \cdot 3} + \dfrac{(n + 1)(n + 2)}{2}$

72. $\dfrac{n(n + 1)(n + 2)(n + 3)}{2 \cdot 3 \cdot 4} + \dfrac{(n + 1)(n + 2)(n + 3)}{2 \cdot 3}$

73. $\dfrac{x^2 - 9}{x^3 + 27} \cdot \dfrac{5x^2 - 15x + 45}{x^2 - 2x - 3} + \dfrac{x^2 + x}{4 + 2x}$

74. $\dfrac{x^2 + 2x - 3}{x^2 - x - 12} \div \dfrac{x^2 - 1}{x^2 - 16} - \dfrac{2x + 1}{x^2 + 2x + 1}$

R.6

Radical Notation and Rational Exponents

- *Simplify radical expressions.*
- *Rationalize denominators or numerators in rational expressions.*
- *Convert between exponential and radical notation.*
- *Simplify expressions with rational exponents.*

A number c is said to be a **square root** of a if $c^2 = a$. Thus, 3 is a square root of 9, because $3^2 = 9$, and -3 is also a square root of 9, because $(-3)^2 = 9$. Similarly, 5 is a third root (called a **cube root**) of 125, because $5^3 = 125$. The number 125 has no other real-number cube root.

> ### *nth* Root
> A number c is said to be an ***nth* root** of a if $c^n = a$.

The symbol $\sqrt{a}$ denotes the nonnegative square root of a, and the symbol $\sqrt[3]{a}$ denotes the real-number cube root of a. The symbol $\sqrt[n]{a}$ denotes the *n*th root of a, that is, a number whose *n*th power is a. The symbol $\sqrt[n]{}$ is called a **radical**, and the expression under the radical is called the **radicand**. The number n (which is omitted when it is 2) is called the **index**. Examples of roots for $n = 3$, 4, and 2, respectively, are

$$\sqrt[3]{125}, \qquad \sqrt[4]{16}, \quad \text{and} \quad \sqrt{3600}.$$

Any real number has only one real-number odd root. Any positive number has two square roots, one positive and one negative. The same is true for fourth roots or roots with any even index. The positive root is called the **principal root.** When an expression such as $\sqrt{4}$ or $\sqrt[6]{23}$ is used, it is understood to represent the principal (nonnegative) root. To denote a negative root, we use $-\sqrt{4}$, $-\sqrt[6]{23}$, and so on.

EXAMPLE 1 Simplify each of the following.

a) $\sqrt{36}$ **b)** $-\sqrt{36}$

c) $\sqrt[5]{\dfrac{32}{243}}$ **d)** $\sqrt[3]{-8}$

e) $\sqrt[4]{-16}$

Solution

a) $\sqrt{36} = 6$, because $6^2 = 36$.

b) $-\sqrt{36} = -6$, because $6^2 = 36$ and $-\left(\sqrt{36}\right) = -(6) = -6$.

c) $\sqrt[5]{\dfrac{32}{243}} = \dfrac{2}{3}$, because $\left(\dfrac{2}{3}\right)^5 = \dfrac{2^5}{3^5} = \dfrac{32}{243}$.

d) $\sqrt[3]{-8} = -2$, because $(-2)^3 = -8$.

e) $\sqrt[4]{-16}$ is not a real number, because we cannot find a real number that can be raised to the fourth power to get -16.　　　　　　　━

We can generalize Example 1(e) and say that when a is negative and n is even, $\sqrt[n]{a}$ is not a real number. For example, $\sqrt{-4}$ and $\sqrt[4]{-81}$ are not real numbers.

Technology Connection

We can find $\sqrt{36}$ and $-\sqrt{36}$ in Example 1 using the square-root feature on the keypad of a graphing calculator, and we can use the cube-root feature to find $\sqrt[3]{-8}$. We can use the xth-root feature to find higher roots.

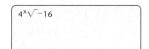 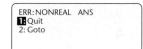

$\sqrt{(36)}$	
	6
$-\sqrt{(36)}$	
	-6

$\sqrt[3]{(-8)}$	
	-2
$5^{x}\sqrt{(32/243)}$ ▸ Frac	
	2/3

When we try to find $\sqrt[4]{-16}$ on a graphing calculator set in REAL mode, we get an error message indicating that the answer is nonreal.

$4^{x}\sqrt{-16}$

ERR:NONREAL ANS
1: Quit
2: Goto

Simplifying Radical Expressions

Consider the expression $\sqrt{(-3)^2}$. This is equivalent to $\sqrt{9}$, or 3. Similarly, $\sqrt{3^2} = \sqrt{9} = 3$. This illustrates the first of several properties of radicals, listed below.

Properties of Radicals

Let a and b be any real numbers or expressions for which the given roots exist. For any natural numbers m and n ($n \neq 1$):

1. If n is even, $\sqrt[n]{a^n} = |a|$.

2. If n is odd, $\sqrt[n]{a^n} = a$.

3. $\sqrt[n]{a} \cdot \sqrt[n]{b} = \sqrt[n]{ab}$.

4. $\sqrt[n]{\dfrac{a}{b}} = \dfrac{\sqrt[n]{a}}{\sqrt[n]{b}}$ ($b \neq 0$).

5. $\sqrt[n]{a^m} = \left(\sqrt[n]{a}\right)^m$.

EXAMPLE 2 Simplify each of the following.

a) $\sqrt{(-5)^2}$ b) $\sqrt[3]{(-5)^3}$ c) $\sqrt[4]{4} \cdot \sqrt[4]{5}$ d) $\sqrt{50}$

e) $\dfrac{\sqrt{72}}{\sqrt{6}}$ f) $\sqrt[3]{8^5}$ g) $\sqrt{216x^5y^3}$ h) $\sqrt{\dfrac{x^2}{16}}$

Solution

a) $\sqrt{(-5)^2} = |-5| = 5$ Using Property 1

b) $\sqrt[3]{(-5)^3} = -5$ Using Property 2

c) $\sqrt[4]{4} \cdot \sqrt[4]{5} = \sqrt[4]{4 \cdot 5} = \sqrt[4]{20}$ Using Property 3

d) $\sqrt{50} = \sqrt{25 \cdot 2} = \sqrt{25} \cdot \sqrt{2} = 5\sqrt{2}$ Using Property 3

e) $\dfrac{\sqrt{72}}{\sqrt{6}} = \sqrt{\dfrac{72}{6}}$ Using Property 4

$= \sqrt{12} = \sqrt{4 \cdot 3} = \sqrt{4} \cdot \sqrt{3}$ Using Property 3

$= 2\sqrt{3}$

f) $\sqrt[3]{8^5} = \left(\sqrt[3]{8}\right)^5$ Using Property 5

$= 2^5 = 32$

g) $\sqrt{216x^5y^3} = \sqrt{36 \cdot 6 \cdot x^4 \cdot x \cdot y^2 \cdot y}$

$= \sqrt{36x^4y^2}\sqrt{6xy}$ Using Property 3

$= |6x^2y|\sqrt{6xy}$ Using Property 1

$= 6x^2|y|\sqrt{6xy}$ $6x^2$ cannot be negative, so absolute-value signs are not needed for it.

h) $\sqrt{\dfrac{x^2}{16}} = \dfrac{\sqrt{x^2}}{\sqrt{16}}$ Using Property 4

$= \dfrac{|x|}{4}$ Using Property 1

In many situations, radicands are never formed by raising negative quantities to even powers. In such cases, absolute-value notation is not required. For this reason, **we will henceforth assume that no radicands are formed by raising negative quantities to even powers.** For example, we will write $\sqrt{x^2} = x$ and $\sqrt[4]{a^5b} = a\sqrt[4]{ab}$.

Radical expressions with the same index and the same radicand can be added or subtracted.

EXAMPLE 3 Perform the operations indicated.

a) $3\sqrt{8x^2} - 5\sqrt{2x^2}$

b) $\left(4\sqrt{3} + \sqrt{2}\right)\left(\sqrt{3} - 5\sqrt{2}\right)$

Solution

a) $3\sqrt{8x^2} - 5\sqrt{2x^2} = 3\sqrt{4x^2 \cdot 2} - 5\sqrt{x^2 \cdot 2}$

$= 3 \cdot 2x\sqrt{2} - 5x\sqrt{2}$

$= 6x\sqrt{2} - 5x\sqrt{2}$

$= (6x - 5x)\sqrt{2}$ Using the distributive property

$= x\sqrt{2}$

b) $\left(4\sqrt{3} + \sqrt{2}\right)\left(\sqrt{3} - 5\sqrt{2}\right) = 4\left(\sqrt{3}\right)^2 - 20\sqrt{6} + \sqrt{6} - 5\left(\sqrt{2}\right)^2$

                                Multiplying

$$= 4 \cdot 3 + (-20 + 1)\sqrt{6} - 5 \cdot 2$$
$$= 12 - 19\sqrt{6} - 10$$
$$= 2 - 19\sqrt{6}$$

An Application

The Pythagorean theorem relates the lengths of the sides of a right triangle. The side opposite the triangle's right angle is called the **hypotenuse.** The other sides are the **legs.**

The Pythagorean Theorem

The sum of the squares of the lengths of the legs of a right triangle is equal to the square of the length of the hypotenuse:

$$a^2 + b^2 = c^2.$$

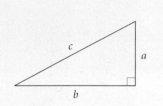

EXAMPLE 4 *Surveying.* A surveyor places poles at points A, B, and C in order to measure the distance across a pond. The distances AC and BC are measured as shown. Find the distance AB across the pond.

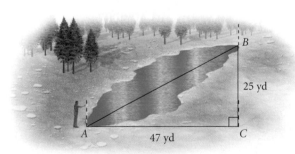

Solution We see that the lengths of the legs of a right triangle are given. Thus we use the Pythagorean theorem to find the length of the hypotenuse:

$$c^2 = a^2 + b^2$$
$$c = \sqrt{a^2 + b^2} \quad \textbf{Solving for } c$$
$$= \sqrt{25^2 + 47^2}$$
$$= \sqrt{625 + 2209}$$
$$= \sqrt{2834}$$
$$\approx 53.2.$$

The distance across the pond is about 53.2 yd.

Technology Connection

We can perform the computation in Example 4 on a calculator.

$\sqrt{(25^2 + 47^2)}$

      53.23532662

Rationalizing Denominators or Numerators

There are times when we need to remove the radicals in a denominator or a numerator. This is called **rationalizing the denominator** or **rationalizing the numerator.** It is done by multiplying by 1 in such a way as to obtain a perfect nth power.

EXAMPLE 5　Rationalize the denominator of each of the following.

a) $\sqrt{\dfrac{3}{2}}$　　　　　　　　　　b) $\dfrac{\sqrt[3]{7}}{\sqrt[3]{9}}$

Solution

a) $\sqrt{\dfrac{3}{2}} = \sqrt{\dfrac{3}{2} \cdot \dfrac{2}{2}} = \sqrt{\dfrac{6}{4}} = \dfrac{\sqrt{6}}{\sqrt{4}} = \dfrac{\sqrt{6}}{2}$

b) $\dfrac{\sqrt[3]{7}}{\sqrt[3]{9}} = \dfrac{\sqrt[3]{7}}{\sqrt[3]{9}} \cdot \dfrac{\sqrt[3]{3}}{\sqrt[3]{3}} = \dfrac{\sqrt[3]{21}}{\sqrt[3]{27}} = \dfrac{\sqrt[3]{21}}{3}$

　　　　Pairs of expressions of the form $a\sqrt{b} + c\sqrt{d}$ and $a\sqrt{b} - c\sqrt{d}$ are called **conjugates.** The product of such a pair contains no radicals and can be used to rationalize a denominator or a numerator.

EXAMPLE 6　Rationalize the numerator: $\dfrac{\sqrt{x} - \sqrt{y}}{5}$.

Solution

$$\dfrac{\sqrt{x} - \sqrt{y}}{5} = \dfrac{\sqrt{x} - \sqrt{y}}{5} \cdot \dfrac{\sqrt{x} + \sqrt{y}}{\sqrt{x} + \sqrt{y}} \qquad \text{The conjugate of } \sqrt{x} - \sqrt{y} \text{ is } \sqrt{x} + \sqrt{y}.$$

$$= \dfrac{\left(\sqrt{x}\right)^2 - \left(\sqrt{y}\right)^2}{5\sqrt{x} + 5\sqrt{y}} \qquad (a + b)(a - b) = a^2 - b^2$$

$$= \dfrac{x - y}{5\sqrt{x} + 5\sqrt{y}}$$

Rational Exponents

We are motivated to define *rational exponents* so that the properties for integer exponents hold for them as well. For example, we must have

$$a^{1/2} \cdot a^{1/2} = a^{1/2 + 1/2} = a^1 = a.$$

Thus we are led to define $a^{1/2}$ to mean $\sqrt{a}$. Similarly, $a^{1/n}$ would mean $\sqrt[n]{a}$. Again, if the laws of exponents are to hold, we must have

$$(a^{1/n})^m = (a^m)^{1/n} = a^{m/n}.$$

Thus we are led to define $a^{m/n}$ to mean $\sqrt[n]{a^m}$, or, equivalently, $\left(\sqrt[n]{a}\right)^m$.

> **Rational Exponents**
>
> For any real number a and any natural numbers m and n for which $\sqrt[n]{a}$ exists,
>
> $$a^{1/n} = \sqrt[n]{a},$$
> $$a^{m/n} = \sqrt[n]{a^m} = \left(\sqrt[n]{a}\right)^m, \quad \text{and}$$
> $$a^{-m/n} = \frac{1}{a^{m/n}}.$$

We can use the definition of rational exponents to convert between radical and exponential notation.

EXAMPLE 7 Convert to radical notation and, if possible, simplify each of the following.

a) $7^{3/4}$ b) $8^{-5/3}$

c) $m^{1/6}$ d) $(-32)^{2/5}$

Solution

a) $7^{3/4} = \sqrt[4]{7^3}$, or $\left(\sqrt[4]{7}\right)^3$

b) $8^{-5/3} = \dfrac{1}{8^{5/3}} = \dfrac{1}{\left(\sqrt[3]{8}\right)^5} = \dfrac{1}{2^5} = \dfrac{1}{32}$

c) $m^{1/6} = \sqrt[6]{m}$

d) $(-32)^{2/5} = \sqrt[5]{(-32)^2} = \sqrt[5]{1024} = 4$, or
$(-32)^{2/5} = \left(\sqrt[5]{-32}\right)^2 = (-2)^2 = 4$

EXAMPLE 8 Convert each of the following to exponential notation.

a) $\left(\sqrt[4]{7xy}\right)^5$ b) $\sqrt[6]{x^3}$

Solution

a) $\left(\sqrt[4]{7xy}\right)^5 = (7xy)^{5/4}$

b) $\sqrt[6]{x^3} = x^{3/6} = x^{1/2}$

We can use the laws of exponents to simplify exponential and radical expressions.

EXAMPLE 9 Simplify and then, if appropriate, write radical notation for each of the following.

a) $x^{5/6} \cdot x^{2/3}$ b) $(x + 3)^{5/2}(x + 3)^{-1/2}$ c) $\sqrt[3]{\sqrt{7}}$

Solution

a) $x^{5/6} \cdot x^{2/3} = x^{5/6 + 2/3} = x^{9/6} = x^{3/2} = \sqrt{x^3} = \sqrt{x^2}\sqrt{x} = x\sqrt{x}$

b) $(x + 3)^{5/2}(x + 3)^{-1/2} = (x + 3)^{5/2 - 1/2} = (x + 3)^2$

c) $\sqrt[3]{\sqrt{7}} = \sqrt[3]{7^{1/2}} = (7^{1/2})^{1/3} = 7^{1/6} = \sqrt[6]{7}$

EXAMPLE 10   Write an expression containing a single radical: $a^{1/2}b^{5/6}$.

Solution

$$a^{1/2}b^{5/6} = a^{3/6}b^{5/6} = (a^3b^5)^{1/6} = \sqrt[6]{a^3b^5}$$

R.6

Exercise Set

Simplify. Assume that variables can represent any real number.

1. $\sqrt{(-11)^2}$

2. $\sqrt{(-1)^2}$

3. $\sqrt{16y^2}$

4. $\sqrt{36t^2}$

5. $\sqrt{(b+1)^2}$

6. $\sqrt{(2c-3)^2}$

7. $\sqrt[3]{-27x^3}$

8. $\sqrt[3]{-8y^3}$

9. $\sqrt[4]{81x^8}$

10. $\sqrt[4]{16z^{12}}$

11. $\sqrt[5]{32}$

12. $\sqrt[5]{-32}$

13. $\sqrt{180}$

14. $\sqrt{48}$

15. $\sqrt{72}$

16. $\sqrt{250}$

17. $\sqrt[3]{54}$

18. $\sqrt[3]{135}$

19. $\sqrt{128c^2d^4}$

20. $\sqrt{162c^4d^6}$

21. $\sqrt[4]{48x^6y^4}$

22. $\sqrt[4]{243m^5n^{10}}$

23. $\sqrt{x^2-4x+4}$

24. $\sqrt{x^2+16x+64}$

Simplify. Assume that no radicands were formed by raising negative quantities to even powers.

25. $\sqrt{10}\,\sqrt{30}$

26. $\sqrt{28}\,\sqrt{14}$

27. $\sqrt{12}\,\sqrt{33}$

28. $\sqrt{15}\,\sqrt{35}$

29. $\sqrt{2x^3y}\,\sqrt{12xy}$

30. $\sqrt{3y^4z}\,\sqrt{20z}$

31. $\sqrt[3]{3x^2y}\,\sqrt[3]{36x}$

32. $\sqrt[5]{8x^3y^4}\,\sqrt[5]{4x^4y}$

33. $\sqrt[3]{2(x+4)}\,\sqrt[3]{4(x+4)^4}$

34. $\sqrt[3]{4(x+1)^2}\,\sqrt[3]{18(x+1)^2}$

35. $\sqrt[6]{\dfrac{m^{12}n^{24}}{64}}$

36. $\sqrt[8]{\dfrac{m^{16}n^{24}}{2^8}}$

37. $\dfrac{\sqrt[3]{40m}}{\sqrt[3]{5m}}$

38. $\dfrac{\sqrt{40xy}}{\sqrt{8x}}$

39. $\dfrac{\sqrt[3]{3x^2}}{\sqrt[3]{24x^5}}$

40. $\dfrac{\sqrt{128a^2b^4}}{\sqrt{16ab}}$

41. $\sqrt[3]{\dfrac{64a^4}{27b^3}}$

42. $\sqrt{\dfrac{9x^7}{16y^8}}$

43. $\sqrt{\dfrac{7x^3}{36y^6}}$

44. $\sqrt[3]{\dfrac{2yz}{250z^4}}$

45. $9\sqrt{50}+6\sqrt{2}$

46. $11\sqrt{27}-4\sqrt{3}$

47. $6\sqrt{20}-4\sqrt{45}+\sqrt{80}$

48. $2\sqrt{32}+3\sqrt{8}-4\sqrt{18}$

49. $8\sqrt{2x^2}-6\sqrt{20x}-5\sqrt{8x^2}$

50. $2\sqrt[3]{8x^2}+5\sqrt[3]{27x^2}-3\sqrt{x^3}$

51. $(\sqrt{3}-\sqrt{2})(\sqrt{3}+\sqrt{2})$

52. $(\sqrt{8}+2\sqrt{5})(\sqrt{8}-2\sqrt{5})$

53. $(2\sqrt{3}+\sqrt{5})(\sqrt{3}-3\sqrt{5})$

54. $(\sqrt{6}-4\sqrt{7})(3\sqrt{6}+2\sqrt{7})$

55. $(1+\sqrt{3})^2$

56. $(\sqrt{2}-5)^2$

57. $(\sqrt{5}-\sqrt{6})^2$

58. $(\sqrt{3}+\sqrt{2})^2$

59. *Distance from Airport.* An airplane is flying at an altitude of 3700 ft. The slanted distance directly to the airport is 14,200 ft. How far horizontally is the airplane from the airport?

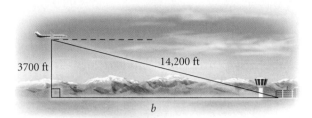

60. *Bridge Expansion.* During a summer heat wave, a 2-mi bridge expands 2 ft in length. Assuming that the bulge occurs straight up the middle, estimate the height of the bulge. (In reality, bridges are built with expansion joints to control such buckling.)

61. An *equilateral triangle* is shown below.

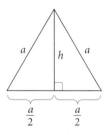

a) Find an expression for its height h in terms of a.
b) Find an expression for its area A in terms of a.

62. An isosceles right triangle has legs of length s. Find an expression for the length of the hypotenuse in terms of s.

63. The diagonal of a square has length $8\sqrt{2}$. Find the length of a side of the square.

64. The area of square $PQRS$ is 100 ft^2, and A, B, C, and D are the midpoints of the sides. Find the area of square $ABCD$.

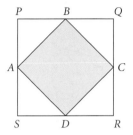

Rationalize the denominator.

65. $\sqrt{\dfrac{2}{3}}$

66. $\sqrt{\dfrac{3}{7}}$

67. $\dfrac{\sqrt[3]{5}}{\sqrt[3]{4}}$

68. $\dfrac{\sqrt[3]{7}}{\sqrt[3]{25}}$

69. $\sqrt[3]{\dfrac{16}{9}}$

70. $\sqrt[3]{\dfrac{3}{5}}$

71. $\dfrac{6}{3+\sqrt{5}}$

72. $\dfrac{2}{\sqrt{3}-1}$

73. $\dfrac{1-\sqrt{2}}{2\sqrt{3}-\sqrt{6}}$

74. $\dfrac{\sqrt{5}+4}{\sqrt{2}+3\sqrt{7}}$

75. $\dfrac{6}{\sqrt{m}-\sqrt{n}}$

76. $\dfrac{3}{\sqrt{v}+\sqrt{w}}$

Rationalize the numerator.

77. $\dfrac{\sqrt{12}}{5}$

78. $\dfrac{\sqrt{50}}{3}$

79. $\sqrt[3]{\dfrac{7}{2}}$

80. $\sqrt[3]{\dfrac{2}{5}}$

81. $\dfrac{\sqrt{11}}{\sqrt{3}}$

82. $\dfrac{\sqrt{5}}{\sqrt{2}}$

83. $\dfrac{9-\sqrt{5}}{3-\sqrt{3}}$

84. $\dfrac{8-\sqrt{6}}{5-\sqrt{2}}$

85. $\dfrac{\sqrt{a}+\sqrt{b}}{3a}$

86. $\dfrac{\sqrt{p}-\sqrt{q}}{1+\sqrt{q}}$

Convert to radical notation and simplify.

87. $x^{3/4}$

88. $y^{2/5}$

89. $16^{3/4}$

90. $4^{7/2}$

91. $125^{-1/3}$

92. $32^{-4/5}$

93. $a^{5/4}b^{-3/4}$

94. $x^{2/5}y^{-1/5}$

95. $m^{5/3}n^{7/3}$

96. $p^{7/6}q^{11/6}$

Convert to exponential notation.

97. $\left(\sqrt[4]{13}\right)^5$

98. $\sqrt[5]{17^3}$

99. $\sqrt[3]{20^2}$

100. $\left(\sqrt[5]{12}\right)^4$

101. $\sqrt[3]{\sqrt{11}}$

102. $\sqrt[3]{\sqrt[4]{7}}$

103. $\sqrt{5}\,\sqrt[3]{5}$

104. $\sqrt[3]{2}\,\sqrt{2}$

105. $\sqrt[5]{32^2}$

106. $\sqrt[3]{64^2}$

Simplify and then, if appropriate, write radical notation.

107. $(2a^{3/2})(4a^{1/2})$

108. $(3a^{5/6})(8a^{2/3})$

109. $\left(\dfrac{x^6}{9b^{-4}}\right)^{1/2}$

110. $\left(\dfrac{x^{2/3}}{4y^{-2}}\right)^{1/2}$

111. $\dfrac{x^{2/3}y^{5/6}}{x^{-1/3}y^{1/2}}$

112. $\dfrac{a^{1/2}b^{5/8}}{a^{1/4}b^{3/8}}$

113. $(m^{1/2}n^{5/2})^{2/3}$

114. $(x^{5/3}y^{1/3}z^{2/3})^{3/5}$

115. $a^{3/4}(a^{2/3}+a^{4/3})$

116. $m^{2/3}(m^{7/4}-m^{5/4})$

Write an expression containing a single radical and simplify.

117. $\sqrt[3]{6}\,\sqrt{2}$

118. $\sqrt{2}\,\sqrt[4]{8}$

119. $\sqrt[4]{xy}\,\sqrt[3]{x^2y}$

120. $\sqrt[3]{ab^2}\,\sqrt{ab}$

121. $\sqrt[3]{a^4\sqrt{a^3}}$

122. $\sqrt{a^3\sqrt[3]{a^2}}$

123. $\dfrac{\sqrt{(a+x)^3}\,\sqrt[3]{(a+x)^2}}{\sqrt[4]{a+x}}$

124. $\dfrac{\sqrt[4]{(x+y)^2}\,\sqrt[3]{x+y}}{\sqrt{(x+y)^3}}$

Collaborative Discussion and Writing

125. Explain how you would convince a classmate that $\sqrt{a+b}$ is not equivalent to $\sqrt{a}+\sqrt{b}$, for positive real numbers a and b.

126. Explain how you would determine whether $10\sqrt{26}-50$ is positive or negative without carrying out the actual computation.

Synthesis

Simplify.

127. $\sqrt{1+x^2}+\dfrac{1}{\sqrt{1+x^2}}$

128. $\sqrt{1-x^2}-\dfrac{x^2}{2\sqrt{1-x^2}}$

129. $\left(\sqrt{a^{\sqrt{a}}}\right)^{\sqrt{a}}$

130. $(2a^3b^{5/4}c^{1/7})^4 \div (54a^{-2}b^{2/3}c^{6/5})^{-1/3}$

Chapter R Summary and Review

Important Properties and Formulas

Properties of the Real Numbers

Commutative:	$a + b = b + a;$ $ab = ba$
Associative:	$a + (b + c) =$ $(a + b) + c;$ $a(bc) = (ab)c$
Additive Identity:	$a + 0 = 0 + a = a$
Additive Inverse:	$-a + a =$ $a + (-a) = 0$
Multiplicative Identity:	$a \cdot 1 = 1 \cdot a = a$
Multiplicative Inverse:	$a \cdot \dfrac{1}{a} = \dfrac{1}{a} \cdot a = 1$ $(a \neq 0)$
Distributive:	$a(b + c) = ab + ac$

Absolute Value

For any real number a,

$$|a| = \begin{cases} a, & \text{if } a \geq 0, \\ -a, & \text{if } a < 0. \end{cases}$$

Properties of Exponents

For any real numbers a and b and any integers m and n, assuming 0 is not raised to a nonpositive power:

The Product Rule: $\quad a^m \cdot a^n = a^{m+n}$

The Quotient Rule: $\quad \dfrac{a^m}{a^n} = a^{m-n} \quad (a \neq 0)$

The Power Rule: $\quad (a^m)^n = a^{mn}$

Raising a Product to a Power: $\quad (ab)^m = a^m b^m$

Raising a Quotient to a Power:

$$\left(\frac{a}{b}\right)^m = \frac{a^m}{b^m} \quad (b \neq 0)$$

Compound Interest Formula

$$A = P\left(1 + \frac{i}{n}\right)^{nt}$$

Special Products of Binomials

$$(A + B)^2 = A^2 + 2AB + B^2$$
$$(A - B)^2 = A^2 - 2AB + B^2$$
$$(A + B)(A - B) = A^2 - B^2$$

Sum or Difference of Cubes

$$A^3 + B^3 = (A + B)(A^2 - AB + B^2)$$
$$A^3 - B^3 = (A - B)(A^2 + AB + B^2)$$

Properties of Radicals

Let a and b be any real numbers or expressions for which the given roots exist. For any natural numbers m and n ($n \neq 1$):

If n is even, $\sqrt[n]{a^n} = |a|$.

If n is odd, $\sqrt[n]{a^n} = a$.

$\sqrt[n]{a} \cdot \sqrt[n]{b} = \sqrt[n]{ab}$.

$\sqrt[n]{\dfrac{a}{b}} = \dfrac{\sqrt[n]{a}}{\sqrt[n]{b}} \quad (b \neq 0)$.

$\sqrt[n]{a^m} = \left(\sqrt[n]{a}\right)^m$.

Rational Exponents

For any real number a and any natural numbers m and n for which $\sqrt[n]{a}$ exists,

$$a^{1/n} = \sqrt[n]{a},$$
$$a^{m/n} = \sqrt[n]{a^m} = \left(\sqrt[n]{a}\right)^m, \quad \text{and}$$
$$a^{-m/n} = \frac{1}{a^{m/n}}.$$

Pythagorean Theorem

$$a^2 + b^2 = c^2$$

Review Exercises

Answers for all of the review exercises appear in the answer section at the back of the book. If you get an incorrect answer, return to the section of the text indicated in the answer.

In Exercises 1–6, consider the numbers -43.89, 12, -3, $-\frac{1}{5}$, $\sqrt{7}$, $\sqrt[3]{10}$, -1, $-\frac{4}{3}$, $7\frac{2}{3}$, -19, 31, 0.

1. Which are integers?

2. Which are natural numbers?

3. Which are rational numbers?

4. Which are real numbers?

5. Which are irrational numbers?

6. Which are whole numbers?

7. Write interval notation for $\{x \mid -3 \le x < 5\}$.

Simplify.

8. $|-3.5|$

9. $|16|$

10. Find the distance between -7 and 3 on the number line.

Calculate.

11. $5^3 - [2(4^2 - 3^2 - 6)]^3$

12. $\dfrac{3^4 - (6 - 7)^4}{2^3 - 2^4}$

Convert to decimal notation.

13. 3.261×10^6

14. 4.1×10^{-4}

Convert to scientific notation.

15. 0.01432

16. $43{,}210$

Calculate. Write the answer using scientific notation.

17. $\dfrac{2.5 \times 10^{-8}}{3.2 \times 10^{13}}$

18. $(8.4 \times 10^{-17})(6.5 \times 10^{-16})$

Simplify.

19. $(7a^2b^4)(-2a^{-4}b^3)$

20. $\dfrac{54x^6y^{-4}z^2}{9x^{-3}y^2z^{-4}}$

21. $\sqrt[4]{81}$

22. $\sqrt[5]{-32}$

23. $\dfrac{b - a^{-1}}{a - b^{-1}}$

24. $\dfrac{\dfrac{x^2}{y} + \dfrac{y^2}{x}}{y^2 - xy + x^2}$

25. $\left(\sqrt{3} - \sqrt{7}\right)\left(\sqrt{3} + \sqrt{7}\right)$

26. $\left(5x^2 - \sqrt{2}\right)^2$

27. $8\sqrt{5} + \dfrac{25}{\sqrt{5}}$

28. $(x + t)(x^2 - xt + t^2)$

29. $(5a + 4b)(2a - 3b)$

30. $(5xy^4 - 7xy^2 + 4x^2 - 3) - (-3xy^4 + 2xy^2 - 2y + 4)$

Factor.

31. $x^3 + 2x^2 - 3x - 6$

32. $12a^3 - 27ab^4$

33. $24x + 144 + x^2$

34. $9x^3 + 35x^2 - 4x$

35. $8x^3 - 1$

36. $27x^6 + 125y^6$

37. $6x^3 + 48$

38. $4x^3 - 4x^2 - 9x + 9$

39. $9x^2 - 30x + 25$

40. $18x^2 - 3x + 6$

41. $a^2b^2 - ab - 6$

42. Divide and simplify:
$$\frac{3x^2 - 12}{x^2 + 4x + 4} \div \frac{x - 2}{x + 2}.$$

43. Subtract and simplify:
$$\frac{x}{x^2 + 9x + 20} - \frac{4}{x^2 + 7x + 12}.$$

Write an expression containing a single radical.

44. $\sqrt{y^5} \, \sqrt[3]{y^2}$

45. $\dfrac{\sqrt{(a + b)^3} \, \sqrt[3]{a + b}}{\sqrt[6]{(a + b)^7}}$

46. Convert to radical notation: $b^{7/5}$.

47. Convert to exponential notation:
$$\sqrt[8]{\frac{m^{32}n^{16}}{3^8}}.$$

48. Rationalize the denominator:
$$\frac{\sqrt{x} - \sqrt{y}}{\sqrt{x} + \sqrt{y}}.$$

49. How long is a guy wire that reaches from the top of a 17-ft pole to a point on the ground 8 ft from the bottom of the pole?

Collaborative Discussion and Writing

50. Anya says that $15 - 6 \div 3 \cdot 4$ is 12. What mistake is she probably making?

51. A calculator indicates that $4^{21} = 4.398046511 \times 10^{12}$. How can you tell that this is an approximation?

Synthesis

Mortgage Payments. The formula

$$M = P\left[\frac{\dfrac{r}{12}\left(1 + \dfrac{r}{12}\right)^n}{\left(1 + \dfrac{r}{12}\right)^n - 1}\right]$$

gives the monthly mortgage payment M on a home loan of P dollars at interest rate r, where n is the total number of payments (12 times the number of years). Use this formula in Exercises 52–55.

52. The cost of a house is $98,000. The down payment is $16,000, the interest rate is $6\frac{1}{2}$ %, and the loan period is 25 yr. What is the monthly mortgage payment?

53. The cost of a house is $124,000. The down payment is $20,000, the interest rate is $5\frac{3}{4}$ %, and the loan period is 30 yr. What is the monthly mortgage payment?

54. The cost of a house is $135,000. The down payment is $18,000, the interest rate is $7\frac{1}{2}$ %, and the loan period is 20 yr. What is the monthly mortgage payment?

55. The cost of a house is $151,000. The down payment is $21,000, the interest rate is $6\frac{1}{4}$ %, and the loan period is 25 yr. What is the monthly mortgage payment?

Multiply. Assume that all exponents are integers.

56. $(x^n + 10)(x^n - 4)$

57. $(t^a + t^{-a})^2$

58. $(y^b - z^c)(y^b + z^c)$

59. $(a^n - b^n)^3$

Factor.

60. $y^{2n} + 16y^n + 64$

61. $x^{2t} - 3x^t - 28$

62. $m^{6n} - m^{3n}$

Chapter R Test

1. Consider the numbers
$$-8, \ \tfrac{11}{3}, \ \sqrt{15}, \ 0, \ -5.49, \ 36, \ \sqrt[3]{7}, \ 10\tfrac{1}{6}.$$
   a) Which are integers?
   b) Which are rational numbers?
   c) Which are rational numbers but not integers?
   d) Which are integers but not natural numbers?

Simplify.

2. $\left| -\dfrac{14}{5} \right|$

3. $|19.4|$

4. $|-1.2xy|$

5. Write interval notation for $\{x \mid -3 < x \le 6\}$. Then graph the interval.

6. Find the distance between -7 and 5 on the number line.

7. Calculate: $32 \div 2^3 - 12 \div 4 \cdot 3$.

8. Convert to scientific notation: 0.0000367.

9. Convert to decimal notation: 4.51×10^6.

10. Compute and write scientific notation for the answer:
$$\frac{2.7 \times 10^4}{3.6 \times 10^{-3}}.$$

Simplify.

11. $x^{-8} \cdot x^5$

12. $(2y^2)^3 (3y^4)^2$

13. $(-3a^5 b^{-4})(5a^{-1} b^3)$

14. $(3x^4 - 2x^2 + 6x) - (5x^3 - 3x^2 + x)$

15. $(x + 3)(2x - 5)$

16. $(2y - 1)^2$

17. $\dfrac{\dfrac{x}{y} - \dfrac{y}{x}}{x + y}$

18. $\sqrt{54}$

19. $\sqrt[3]{40}$

20. $3\sqrt{75} + 2\sqrt{27}$

21. $\sqrt{18} \ \sqrt{10}$

22. $\left(2 + \sqrt{3}\right)\left(5 - 2\sqrt{3}\right)$

Factor.

23. $y^2 - 3y - 18$

24. $x^3 + 10x^2 + 25x$

25. $2n^2 + 5n - 12$

26. $8x^2 - 18$

27. $m^3 - 8$

28. Multiply and simplify:
$$\frac{x^2 + x - 6}{x^2 + 8x + 15} \cdot \frac{x^2 - 25}{x^2 - 4x + 4}.$$

29. Subtract and simplify: $\dfrac{x}{x^2 - 1} - \dfrac{3}{x^2 + 4x - 5}.$

30. Rationalize the denominator: $\dfrac{5}{7 - \sqrt{3}}.$

31. Convert to radical notation: $t^{5/7}$.

32. Convert to exponential notation: $\left(\sqrt[5]{7}\right)^3$.

33. How long is a guy wire that reaches from the top of a 12-ft pole to a point on the ground 5 ft from the bottom of the pole?

Synthesis

34. Multiply: $(x - y - 1)^2$.

Graphs, Functions, and Models

APPLICATION

*I*n 1992, 210 children from mainland China were adopted by U.S. parents. The number of Chinese children adopted ten years later, in 2002, was 6062. Find the average rate of change in the number of Chinese adoptions from 1992 to 2002. (*Source*: Bureau of Citizenship and Immigration Services)

This problem appears as Exercise 34 in Section 1.3.

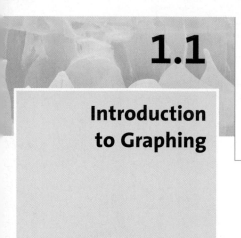

1.1

Introduction to Graphing

- *Plot points.*
- *Determine whether an ordered pair is a solution of an equation.*
- *Graph equations.*
- *Find the distance between two points in the plane and find the midpoint of a segment.*
- *Find an equation of a circle with a given center and radius, and given an equation of a circle, find the center and the radius.*
- *Graph equations of circles.*

Graphs

Graphs provide a means of displaying, interpreting, and analyzing data in a visual format. It is not uncommon to open a newspaper or magazine and encounter graphs. Examples of bar, circle, and line graphs are shown below.

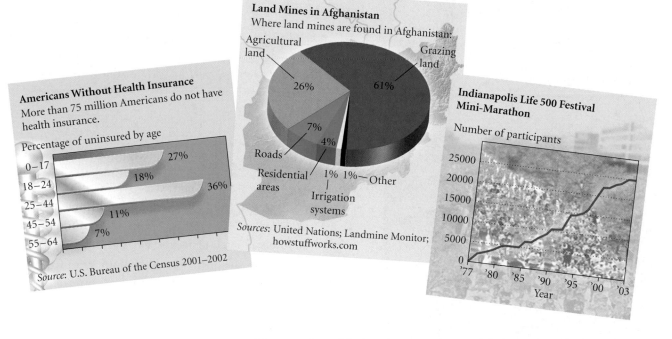

Americans Without Health Insurance
More than 75 million Americans do not have health insurance.

Percentage of uninsured by age

- 0–17: 27%
- 18–24: 18%
- 25–44: 36%
- 45–54: 11%
- 55–64: 7%

Source: U.S. Bureau of the Census 2001–2002

Land Mines in Afghanistan
Where land mines are found in Afghanistan:

- Agricultural land: 26%
- Grazing land: 61%
- Roads: 7%
- Irrigation systems: 4%
- Residential areas: 1%
- Other: 1%

Sources: United Nations; Landmine Monitor; howstuffworks.com

Indianapolis Life 500 Festival Mini-Marathon

Number of participants

Year: '77 '80 '85 '90 '95 '00 '03

Many real-world situations can be modeled, or described mathematically, using equations in which two variables appear. We use a plane to graph a pair of numbers. To locate points on a plane, we use two perpendicular number lines, called **axes,** which intersect at $(0, 0)$. We call this point the **origin.** The horizontal axis is called the ***x*-axis,** and the vertical axis is called the ***y*-axis.** (Other variables, such as a and b, can also be used.) The axes divide the plane into four regions, called **quadrants,** denoted by Roman numerals and numbered counterclockwise from the upper right. Arrows show the positive direction of each axis.

Each point (x, y) in the plane is called an **ordered pair.** The first number, x, indicates the point's horizontal location with respect to the y-axis, and the second number, y, indicates the point's vertical location with respect to

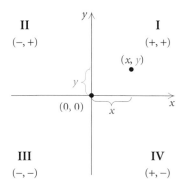

II
$(-, +)$

I
$(+, +)$

III
$(-, -)$

IV
$(+, -)$

the x-axis. We call x the **first coordinate**, **x-coordinate,** or **abscissa.** We call y the **second coordinate**, **y-coordinate,** or **ordinate.** Such a representation is called the **Cartesian coordinate system** in honor of the French mathematician and philosopher René Descartes (1596–1650).

In the first quadrant, both coordinates of a point are positive. In the second quadrant, the first coordinate is negative and the second is positive. In the third quadrant, both coordinates are negative, and in the fourth quadrant, the first coordinate is positive and the second is negative.

EXAMPLE 1 Graph and label the points $(-3, 5)$, $(4, 3)$, $(3, 4)$, $(-4, -2)$, $(3, -4)$, $(0, 4)$, $(-3, 0)$, and $(0, 0)$.

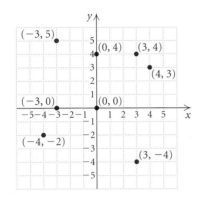

Solution To graph or **plot** $(-3, 5)$, we note that the x-coordinate, -3, tells us to move from the origin 3 units to the left of the y-axis. Then we move 5 units up from the x-axis.* To graph the other points, we proceed in a similar manner. (See the graph at left.) Note that the point $(4, 3)$ is different from the point $(3, 4)$.

Solutions of Equations

Equations in two variables, like $2x + 3y = 18$, have solutions (x, y) that are ordered pairs such that when the first coordinate is substituted for x and the second coordinate is substituted for y, the result is a true equation. The first coordinate in an ordered pair generally represents the variable that occurs first alphabetically.

EXAMPLE 2 Determine whether each ordered pair is a solution of $2x + 3y = 18$.

a) $(-5, 7)$ b) $(3, 4)$

Solution We substitute the ordered pair into the equation and determine whether the resulting equation is true.

a)
$$2x + 3y = 18$$
$$\begin{array}{c|c} 2(-5) + 3(7) \; ? \; 18 & \\ -10 + 21 & \\ 11 & 18 \quad \text{FALSE} \end{array}$$
We substitute -5 for x and 7 for y (alphabetical order).

The equation $11 = 18$ is false, so $(-5, 7)$ is not a solution.

b)
$$2x + 3y = 18$$
$$\begin{array}{c|c} 2(3) + 3(4) \; ? \; 18 & \\ 6 + 12 & \\ 18 & 18 \quad \text{TRUE} \end{array}$$
We substitute 3 for x and 4 for y.

The equation $18 = 18$ is true, so $(3, 4)$ is a solution.

---

*We first saw notation such as $(-3, 5)$ in Section R.1. There the notation represented an open interval. Here the notation represents an ordered pair. The context in which the notation appears usually makes the meaning clear.

| **Suggestions for Drawing Graphs** |

1. Calculate solutions and list the ordered pairs in a table.

2. Use graph paper.

3. Draw axes and label them with the variables.

4. Use arrows on the axes to indicate positive directions.

5. Scale the axes; that is, label the tick marks on the axes. Consider the ordered pairs found in part (1) above when choosing the scale.

6. Plot the ordered pairs, look for patterns, and complete the graph. Label the graph with the equation being graphed.

Graphs of Equations

The equation considered in Example 2 actually has an infinite number of solutions. Since we cannot list all the solutions, we will make a drawing, called a **graph,** that represents them. Shown at left are some suggestions for drawing graphs.

To Graph an Equation

To **graph an equation** is to make a drawing that represents the solutions of that equation.

Many equations of the type $Ax + By = C$ can be graphed conveniently using intercepts. The **x-intercept** of the graph of an equation is the point at which the graph crosses the x-axis. The **y-intercept** is the point at which the graph crosses the y-axis. We know from geometry that only one line can be drawn through two given points. Thus, if we know the intercepts, we can graph the line. To ensure that a computational error has not been made, it is a good idea to calculate and plot a third point as a check.

x- and y-Intercepts

An **x-intercept** is a point $(a, 0)$. To find a, let $y = 0$ and solve for x.

A **y-intercept** is a point $(0, b)$. To find b, let $x = 0$ and solve for y.

EXAMPLE 3   Graph: $2x + 3y = 18$.

Solution   To find ordered pairs that are solutions of this equation, we can replace either x or y with any number and then solve for the other variable. In this case, it is convenient to find the intercepts of the graph. For instance, if x is replaced with 0, then

$$2 \cdot 0 + 3y = 18$$
$$3y = 18$$
$$y = 6. \qquad \text{Dividing by 3}$$

Thus, $(0, 6)$ is a solution. It is the y-intercept of the graph. If y is replaced with 0, then

$$2x + 3 \cdot 0 = 18$$
$$2x = 18$$
$$x = 9. \qquad \text{Dividing by 2}$$

Thus, $(9, 0)$ is a solution. It is the x-intercept of the graph. We find a third solution as a check. If x is replaced with 5, then

$$2 \cdot 5 + 3y = 18$$
$$10 + 3y = 18$$
$$3y = 8 \qquad \text{Subtracting 10}$$
$$y = \tfrac{8}{3}. \qquad \text{Dividing by 3}$$

Thus, $\left(5, \tfrac{8}{3}\right)$ is a solution.

We list the solutions in a table and then plot the points. Note that the points appear to lie on a straight line.

x	y	(x, y)
0	6	$(0, 6)$
9	0	$(9, 0)$
5	$\tfrac{8}{3}$	$\left(5, \tfrac{8}{3}\right)$

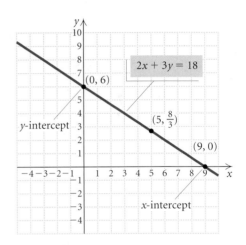

Were we to graph additional solutions of $2x + 3y = 18$, they would be on the same straight line. Thus, to complete the graph, we use a straight-edge to draw a line as shown in the figure. This line represents all solutions of the equation.

When graphing some equations, it is easier to first solve for y and then find ordered pairs. We can use the addition and multiplication principles to solve for y.

EXAMPLE 4   Graph: $3x - 5y = -10$.

Solution   We first solve for y:

$$3x - 5y = -10$$
$$-5y = -3x - 10 \qquad \text{Subtracting } 3x \text{ on both sides}$$
$$y = \tfrac{3}{5}x + 2. \qquad \text{Multiplying by } -\tfrac{1}{5} \text{ on both sides}$$

By choosing multiples of 5 for x, we can avoid fraction values when calculating y. For example, if we choose -5 for x, we get

$$y = \tfrac{3}{5}x + 2 = \tfrac{3}{5}(-5) + 2 = -3 + 2 = -1.$$

The table below lists a few more points. We plot the points and draw the graph.

x	y	(x, y)
-5	-1	$(-5, -1)$
0	2	$(0, 2)$
5	5	$(5, 5)$

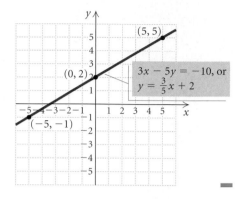

In the equation $y = \frac{3}{5}x + 2$, the value of y *depends* on the value chosen for x, so x is said to be the **independent variable** and y the **dependent variable.**

Technology Connection

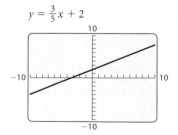

We can graph an equation on a graphing calculator. Many calculators require an equation to be entered in the form "$y =$." In such a case, if the equation is not initially given in this form, it must be solved for y before it is entered in the calculator. For the equation $3x - 5y = -10$ in Example 4, we enter $y = \frac{3}{5}x + 2$ on the equation-editor, or "$y =$", screen in the form $y = (3/5)x + 2$.

Next, we determine the portion of the xy-plane that will appear on the calculator's screen. That portion of the plane is called the **viewing window.**

The notation used in this text to denote a window setting consists of four numbers $[\text{L}, \text{R}, \text{B}, \text{T}]$, which represent the **L**eft and **R**ight endpoints of the x-axis and the **B**ottom and **T**op endpoints of the y-axis, respectively. The window with the settings $[-10, 10, -10, 10]$ is the **standard viewing window.** These settings are shown in the third window at left. On some graphing calculators, the standard window can be selected quickly using the ZSTANDARD feature from the ZOOM menu.

Xmin and Xmax are used to set the left and right endpoints of the x-axis, respectively; Ymin and Ymax are used to set the bottom and top endpoints of the y-axis. The settings Xscl and Yscl give the scales for the axes. For example, Xscl = 1 and Yscl = 1 means that there is 1 unit between tick marks on each of the axes. In this text, scaling factors other than 1 will be listed by the window unless they are readily apparent.

After entering the equation and choosing a viewing window, we can then draw the graph.

Technology Connection

A graphing calculator can be used to create a table of ordered pairs that are solutions of an equation. For the equation in Example 5, $y = x^2 - 9x - 12$, we first enter the equation on the equation-editor screen. Then we set up a table in AUTO mode by designating a value for TBLSTART and a value for ΔTBL. The calculator will produce a table starting with the value of TBLSTART and continuing by adding ΔTBL to supply succeeding x-values. For the equation $y = x^2 - 9x - 12$, we let TBLSTART $= -3$ and ΔTBL $= 1$. We can scroll up and down in the table to find values other than those shown here.

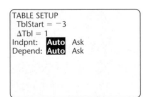

```
TABLE SETUP
  TblStart = -3
  ΔTbl = 1
Indpnt:  Auto  Ask
Depend:  Auto  Ask
```

X	Y₁
-3	24
-2	10
-1	-2
0	-12
1	-20
2	-26
3	-30

X = -3

EXAMPLE 5   Graph: $y = x^2 - 9x - 12$.

Solution   We make a table of values, plot enough points to obtain an idea of the shape of the curve, and connect them with a smooth curve. It is important to scale the axes to include most of the ordered pairs listed in the table. Here it is appropriate to use a larger scale on the y-axis than on the x-axis.

x	y	(x, y)
-3	24	$(-3, 24)$
-1	-2	$(-1, -2)$
0	-12	$(0, -12)$
2	-26	$(2, -26)$
4	-32	$(4, -32)$
5	-32	$(5, -32)$
10	-2	$(10, -2)$
12	24	$(12, 24)$

① **Select values for x.**

② **Compute values for y.**

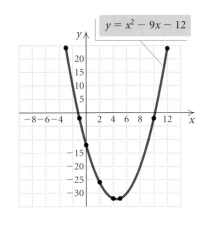

$y = x^2 - 9x - 12$

The Distance Formula

Suppose that a conservationist needs to determine the distance across an irregularly shaped pond. One way in which he or she might proceed is to measure two legs of a right triangle that is situated as shown below. The Pythagorean theorem, $a^2 + b^2 = c^2$, can then be used to find the length of the hypotenuse, which is the distance across the pond.

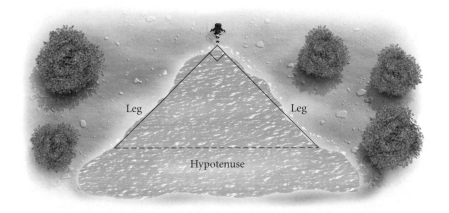

Leg   Leg

Hypotenuse

A similar strategy is used to find the distance between two points in a plane. For two points (x_1, y_1) and (x_2, y_2), we can draw a right triangle in which the legs have lengths $|x_2 - x_1|$ and $|y_2 - y_1|$.

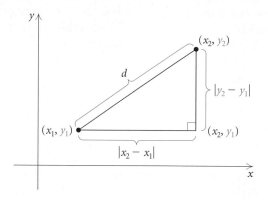

Using the Pythagorean theorem, we have

$$d^2 = |x_2 - x_1|^2 + |y_2 - y_1|^2.$$

Because we are squaring, parentheses can replace the absolute-value symbols:

$$d^2 = (x_2 - x_1)^2 + (y_2 - y_1)^2.$$

Taking the principal square root, we obtain the distance formula.

The Distance Formula

The **distance** d between any two points (x_1, y_1) and (x_2, y_2) is given by

$$d = \sqrt{(x_2 - x_1)^2 + (y_2 - y_1)^2}.$$

The subtraction of the x-coordinates can be done in any order, as can the subtraction of the y-coordinates. Although we derived the distance formula by considering two points not on a horizontal or a vertical line, the distance formula holds for *any* two points.

EXAMPLE 6   Find the distance between the pair of points.

a) $(-2, 2)$ and $(3, -6)$ **b)** $(-1, -5)$ and $(-1, 2)$

Solution   We substitute into the distance formula.

a) $d = \sqrt{[3 - (-2)]^2 + (-6 - 2)^2}$

$= \sqrt{5^2 + (-8)^2} = \sqrt{25 + 64}$

$= \sqrt{89} \approx 9.4$

b) $d = \sqrt{[-1 - (-1)]^2 + (-5 - 2)^2}$

$= \sqrt{0^2 + (-7)^2} = \sqrt{0 + 49}$

$= \sqrt{49} = 7$

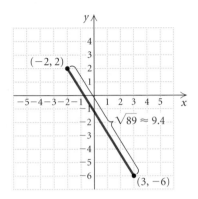

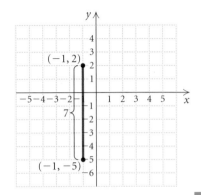

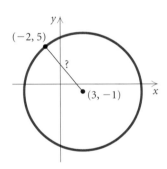

EXAMPLE 7   The point $(-2, 5)$ is on a circle that has $(3, -1)$ as its center. Find the length of the radius of the circle.

Solution   Since the length of the radius is the distance from the center to a point on the circle, we substitute into the distance formula:

$$r = \sqrt{[3 - (-2)]^2 + (-1 - 5)^2}$$ **Either point can serve as (x_1, y_1).**

$$= \sqrt{5^2 + (-6)^2} = \sqrt{61} \approx 7.8.$$ **Rounded to the nearest tenth**

The radius of the circle is approximately 7.8.

Midpoints of Segments

The distance formula can be used to develop a way of determining the *midpoint* of a segment when the endpoints are known. We state the formula and leave its proof to the exercises.

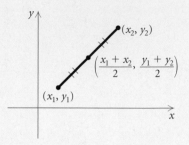

The Midpoint Formula

If the endpoints of a segment are (x_1, y_1) and (x_2, y_2), then the coordinates of the **midpoint** are

$$\left(\frac{x_1 + x_2}{2}, \frac{y_1 + y_2}{2} \right).$$

Note that we obtain the coordinates of the midpoint by averaging the coordinates of the endpoints. This is a good way to remember the midpoint formula.

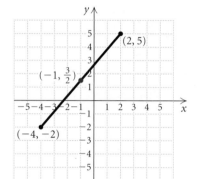

EXAMPLE 8 Find the midpoint of the segment whose endpoints are $(-4, -2)$ and $(2, 5)$.

Solution   Using the midpoint formula, we obtain

$$\left(\frac{-4 + 2}{2}, \frac{-2 + 5}{2} \right) = \left(\frac{-2}{2}, \frac{3}{2} \right)$$

$$= \left(-1, \frac{3}{2} \right).$$

EXAMPLE 9 The diameter of a circle connects two points $(2, -3)$ and $(6, 4)$ on the circle. Find the coordinates of the center of the circle.

Solution   Since the center of the circle is the midpoint of the diameter, we use the midpoint formula:

$$\left(\frac{2 + 6}{2}, \frac{-3 + 4}{2} \right), \quad \text{or} \quad \left(\frac{8}{2}, \frac{1}{2} \right), \quad \text{or} \quad \left(4, \frac{1}{2} \right).$$

The coordinates of the center are $\left(4, \frac{1}{2} \right)$.

Circles

A **circle** is the set of all points in a plane that are a fixed distance r from a *center* (h, k). Thus if a point (x, y) is to be r units from the center, we must have

$$r = \sqrt{(x - h)^2 + (y - k)^2}. \qquad \textbf{Using the distance formula}$$

Squaring both sides gives an equation of a circle. The distance r is the length of a *radius* of the circle.

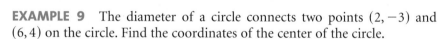

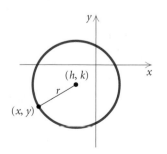

The Equation of a Circle

The equation of a circle with center (h, k) and radius r, in standard form, is

$$(x - h)^2 + (y - k)^2 = r^2.$$

EXAMPLE 10   Find an equation of the circle having radius 5 and center $(3, -7)$.

Solution   Using the standard form, we have

$$[x - 3]^2 + [y - (-7)]^2 = 5^2 \qquad \textbf{Substituting}$$
$$(x - 3)^2 + (y + 7)^2 = 25.$$

EXAMPLE 11   Graph the circle $(x + 5)^2 + (y - 2)^2 = 16$.

Solution   We write the equation in standard form to determine the center and the radius:

$$[x - (-5)]^2 + [y - 2]^2 = 4^2.$$

The center is $(-5, 2)$ and the radius is 4. We locate the center and draw the circle using a compass.

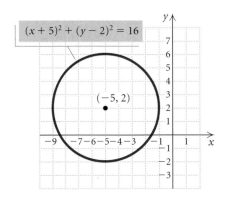

If we square $x + 5$ and $y - 2$ in the equation of the circle in Example 11, we obtain another form for the equation:

$$(x + 5)^2 + (y - 2)^2 = 16$$
$$x^2 + 10x + 25 + y^2 - 4y + 4 = 16 \qquad \textbf{Squaring}$$
$$x^2 + y^2 + 10x - 4y + 13 = 0. \qquad \textbf{Simplifying}$$

This form is the general form of the equation of the circle. With equations in general form, we can complete the square to find the center and the radius of the circle.

EXAMPLE 12 Find the center and the radius of the circle and graph it:

$$x^2 + y^2 - 8x + 2y + 13 = 0.$$

Solution   First, we regroup the terms in order to **complete the square** twice, once with $x^2 - 8x$ and once with $y^2 + 2y$:

$$x^2 + y^2 - 8x + 2y + 13 = 0$$
$$(x^2 - 8x) + (y^2 + 2y) + 13 = 0. \qquad \text{Regrouping}$$

Next, we complete the square inside each set of parentheses. We want to add something to both $x^2 - 8x$ and $y^2 + 2y$ so that each becomes the square of a binomial. For $x^2 - 8x$, we take half the x-coefficient, $\frac{1}{2}(-8) = -4$, and square it, $(-4)^2 = 16$. Then we add 0, or $16 - 16$, inside the parentheses. For $y^2 + 2y$, we have $\frac{1}{2} \cdot 2 = 1$ and $1^2 = 1$, so we add $1 - 1$ inside the parentheses.

$$(x^2 - 8x + 0) + (y^2 + 2y + 0) + 13 = 0 \qquad \text{Adding 0}$$
$$(x^2 - 8x + 16 - 16) + (y^2 + 2y + 1 - 1) + 13 = 0$$
$$(x^2 - 8x + 16) + (y^2 + 2y + 1) - 16 - 1 + 13 = 0 \qquad \text{Regrouping}$$
$$(x - 4)^2 + (y + 1)^2 - 4 = 0 \qquad \begin{array}{l}\text{Factoring and}\\\text{simplifying}\end{array}$$
$$(x - 4)^2 + (y + 1)^2 = 4 \qquad \text{Adding 4}$$
$$(x - 4)^2 + [y - (-1)]^2 = 2^2 \qquad \begin{array}{l}\text{Writing in}\\\text{standard form}\end{array}$$

The center is $(4, -1)$ and the radius is 2.

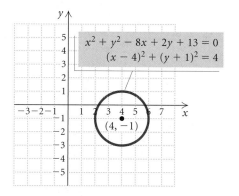

Technology Connection

When we graph a circle, we select a viewing window in which the distance between units is visually the same on both axes. This procedure is called **squaring the viewing window.** We do this so that the graph will not be distorted. A graph of the circle $x^2 + y^2 = 36$ in a nonsquared window is shown in Fig. 1.

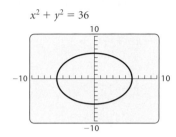

FIGURE 1　　　　　　FIGURE 2

On many graphing calculators, the ratio of the height to the width of the viewing screen is $\frac{2}{3}$. When we choose a window in which Xscl = Yscl and the length of the y-axis is $\frac{2}{3}$ the length of the x-axis, the window will be squared. The windows with dimensions $[-6, 6, -4, 4]$, $[-9, 9, -6, 6]$, and $[-12, 12, -8, 8]$ are examples of squared windows. A graph of the circle $x^2 + y^2 = 36$ in a squared window is shown in Fig. 2. Many graphing calculators have an option on the ZOOM menu that squares the window automatically.

To graph a circle, we select the CIRCLE feature from the DRAW menu and enter the coordinates of the center and the length of the radius. The graph of the circle $(x - 2)^2 + (y + 1)^2 = 16$ is shown here.

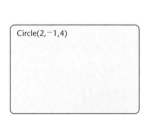

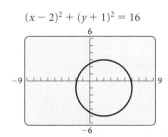

Another method for graphing circles will be discussed in Section 6.2.

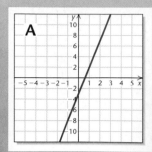

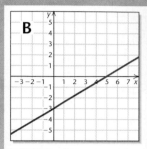

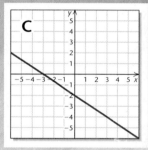

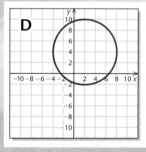

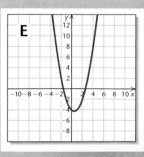

Visualizing the Graph

Match the equation with its graph.

1. $y = -x^2 + 5x - 3$

2. $3x - 5y = 15$

3. $(x - 2)^2 + (y - 4)^2 = 36$

4. $y - 5x = -3$

5. $x^2 + y^2 = \dfrac{25}{4}$

6. $15y - 6x = 90$

7. $y = -\dfrac{2}{3}x - 2$

8. $x^2 + y^2 + 6x - 2y - 6 = 0$

9. $3x + 5y = 15$

10. $y = x^2 - x - 4$

Answers on page A-3

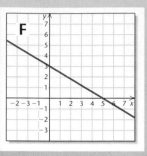

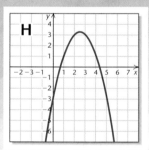

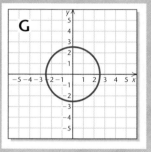

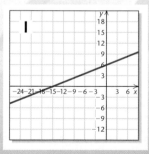

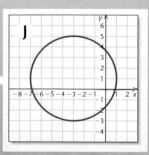

1.1

Exercise Set

Graph and label the given points.

1. $(4,0), (-3,-5), (-1,4), (0,2), (2,-2)$

2. $(1,4), (-4,-2), (-5,0), (2,-4), (4,0)$

3. $(-5,1), (5,1), (2,3), (2,-1), (0,1)$

4. $(4,0), (4,-3), (-5,2), (-5,0), (-1,-5)$

Express the data pictured in the graph as ordered pairs, letting the first coordinate represent the year and the second coordinate the amount.

5.

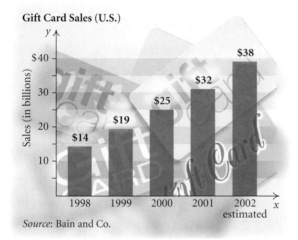

Gift Card Sales (U.S.)

Source: Bain and Co.

6.

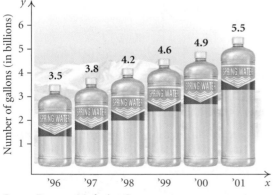

Consumption of Bottled Water (U.S.)

Source: Beverage Marketing Corp.

Use substitution to determine whether the given ordered pairs are solutions of the given equation.

7. $(1,-1), (0,3); y = 2x - 3$

8. $(2,5), (-2,-5); y = 3x - 1$

9. $\left(\frac{2}{3},\frac{3}{4}\right), \left(1,\frac{3}{2}\right); 6x - 4y = 1$

10. $(1.5,2.6), (-3,0); x^2 + y^2 = 9$

11. $\left(-\frac{1}{2},-\frac{4}{5}\right), \left(0,\frac{3}{5}\right); 2a + 5b = 3$

12. $\left(0,\frac{3}{2}\right), \left(\frac{2}{3},1\right); 3m + 4n = 6$

13. $(-0.75,2.75), (2,-1); x^2 - y^2 = 3$

14. $(2,-4), (4,-5); 5x + 2y^2 = 70$

Graph the equation.

15. $y = 3x + 5$ **16.** $y = -2x - 1$

17. $x - y = 3$ **18.** $x + y = 4$

19. $2x + y = 4$ **20.** $3x - y = 6$

21. $y = -\frac{3}{4}x + 3$ **22.** $3y - 2x = 3$

23. $5x - 2y = 8$ **24.** $y = 2 - \frac{4}{3}x$

25. $x - 4y = 5$ **26.** $6x - y = 4$

27. $3x - 4y = 12$ **28.** $2x + 3y = -6$

29. $2x + 5y = -10$ **30.** $4x - 3y = 12$

31. $y = -x^2$ **32.** $y = x^2$

33. $y = x^2 - 3$ **34.** $y = 4 - x^2$

35. $y = -x^2 + 2x + 3$ **36.** $y = x^2 + 2x - 1$

Find the distance between the pair of points. Give an exact answer and, where appropriate, an approximation to three decimal places.

37. $(4,6)$ and $(5,9)$ **38.** $(-3,7)$ and $(2,11)$

39. $(6,-1)$ and $(9,5)$ **40.** $(-4,-7)$ and $(-1,3)$

41. $(-4.2,3)$ and $(2.1,-6.4)$

42. $\left(-\frac{3}{5},-4\right)$ and $\left(-\frac{3}{5},\frac{2}{3}\right)$

43. $\left(-\frac{1}{2},4\right)$ and $\left(\frac{5}{2},4\right)$

44. $(0.6, -1.5)$ and $(-8.1, -1.5)$

45. $\left(\sqrt{3}, -\sqrt{5}\right)$ and $\left(-\sqrt{6}, 0\right)$

46. $\left(-\sqrt{2}, 1\right)$ and $\left(0, \sqrt{7}\right)$

47. $(0, 0)$ and (a, b)

48. (r, s) and $(-r, -s)$

49. The points $(-3, -1)$ and $(9, 4)$ are the endpoints of the diameter of a circle. Find the length of the radius of the circle.

50. The point $(0, 1)$ is on a circle that has center $(-3, 5)$. Find the length of the diameter of the circle.

The converse of the Pythagorean theorem is also a true statement: If the sum of the squares of the lengths of two sides of a triangle is equal to the square of the length of the third side, then the triangle is a right triangle. Use the distance formula and the Pythagorean theorem to determine whether the set of points could be vertices of a right triangle.

51. $(-4, 5)$, $(6, 1)$, and $(-8, -5)$

52. $(-3, 1)$, $(2, -1)$, and $(6, 9)$

53. $(-4, 3)$, $(0, 5)$, and $(3, -4)$

54. The points $(-3, 4)$, $(2, -1)$, $(5, 2)$, and $(0, 7)$ are vertices of a quadrilateral. Show that the quadrilateral is a rectangle. (*Hint:* Show that the quadrilateral's opposite sides are the same length and that the two diagonals are the same length.)

Find the midpoint of the segment having the given endpoints.

55. $(4, -9)$ and $(-12, -3)$

56. $(7, -2)$ and $(9, 5)$

57. $(6.1, -3.8)$ and $(3.8, -6.1)$

58. $(-0.5, -2.7)$ and $(4.8, -0.3)$

59. $(-6, 5)$ and $(-6, 8)$

60. $(1, -2)$ and $(-1, 2)$

61. $\left(-\frac{1}{6}, -\frac{3}{5}\right)$ and $\left(-\frac{2}{3}, \frac{5}{4}\right)$

62. $\left(\frac{2}{9}, \frac{1}{3}\right)$ and $\left(-\frac{2}{5}, \frac{4}{5}\right)$

63. $\left(\sqrt{3}, -1\right)$ and $\left(3\sqrt{3}, 4\right)$

64. $\left(-\sqrt{5}, 2\right)$ and $\left(\sqrt{5}, \sqrt{7}\right)$

65. Graph the rectangle described in Exercise 54. Then determine the coordinates of the midpoint of each of the four sides. Are the midpoints vertices of a rectangle?

66. Graph the square with vertices $(-5, -1)$, $(7, -6)$, $(12, 6)$, and $(0, 11)$. Then determine the midpoint of each of the four sides. Are the midpoints vertices of a square?

67. The points $\left(\sqrt{7}, -4\right)$ and $\left(\sqrt{2}, 3\right)$ are endpoints of the diameter of a circle. Determine the center of the circle.

68. The points $\left(-3, \sqrt{5}\right)$ and $\left(1, \sqrt{2}\right)$ are endpoints of the diagonal of a square. Determine the center of the square.

Find an equation for a circle satisfying the given conditions.

69. Center $(2, 3)$, radius of length $\frac{5}{3}$

70. Center $(4, 5)$, diameter of length 8.2

71. Center $(-1, 4)$, passes through $(3, 7)$

72. Center $(6, -5)$, passes through $(1, 7)$

73. The points $(7, 13)$ and $(-3, -11)$ are at the ends of a diameter.

74. The points $(-9, 4)$, $(-2, 5)$, $(-8, -3)$, and $(-1, -2)$ are vertices of an inscribed square.

75. Center $(-2, 3)$, tangent (touching at one point) to the *y*-axis

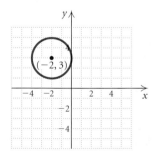

76. Center $(4, -5)$, tangent to the *x*-axis

Find the center and the radius of the circle. Then graph the circle.

77. $x^2 + y^2 = 4$

78. $x^2 + y^2 = 81$

79. $x^2 + (y - 3)^2 = 16$

80. $(x + 2)^2 + y^2 = 100$

81. $(x - 1)^2 + (y - 5)^2 = 36$

82. $(x - 7)^2 + (y + 2)^2 = 25$

83. $(x + 4)^2 + (y + 5)^2 = 9$

84. $(x + 1)^2 + (y - 2)^2 = 64$

85. $x^2 + y^2 - 6x - 2y - 6 = 0$

86. $x^2 + y^2 + 4x - 8y + 19 = 0$

87. $x^2 + y^2 + 2x + 2y = 7$

88. $x^2 + y^2 - 10x + 6y = -30$

Find the equation of the circle. Express the equation in standard form.

89.

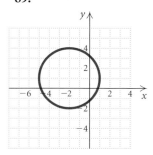

90.

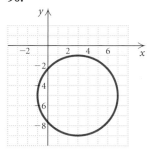

91.

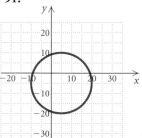

92.

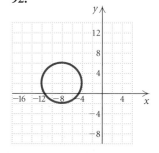

Technology Connection

Use a graphing calculator to graph the equation in the standard window.

93. $4x + y = 7$

94. $5x + y = -8$

95. $y = \frac{1}{3}x + 2$

96. $y = \frac{3}{2}x - 4$

97. $y = x^2 + 6$

98. $y = 5 - x^2$

99. $y = 2 - x^2$

100. $y = x^2 - 5x + 3$

Graph the equation in the standard window and in the given window. Determine which window better shows the shape of the graph and the x- and y-intercepts.

101. $y = 3x^2 - 6$
  $[-4, 4, -4, 4]$

102. $y = -2x + 24$
  $[-15, 15, -10, 30]$, with Xscl = 3 and Yscl = 5

103. $y = -\frac{1}{6}x^2 + \frac{1}{12}$
  $[-1, 1, -0.3, 0.3]$, with Xscl = 0.1 and Yscl = 0.1

104. $y = 8 - x^2$
  $[-3, 3, -3, 3]$

In Exercises 105 and 106, how would you change the window so that the circle is not distorted? Answers may vary.

105.

$(x + 3)^2 + (y - 2)^2 = 36$

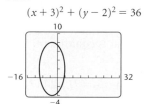

106.

$(x - 4)^2 + (y + 5)^2 = 49$

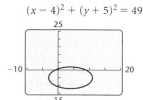

107.–112. Using a graphing calculator, graph each of the circles given in Exercises 77, 80, 81, 84, 85, and 88, respectively.

Collaborative Discussion and Writing

To the student and the instructor: The Collaborative Discussion and Writing exercises are meant to be answered with one or more sentences. They can be discussed and answered collaboratively by the entire class or by small groups. Because of their open-ended nature, the answers to these exercises do not appear at the back of the book. They are denoted by the words "Discussion and Writing."

113. Explain how the Pythagorean theorem is used to develop the equation of a circle in standard form.

114. Explain how you could find the coordinates of a point $\frac{7}{8}$ of the way from point A to point B.

Synthesis

To the student and the instructor: The Synthesis exercises found at the end of every exercise set challenge students to combine concepts or skills studied in that section or in preceding parts of the text.

115. If the point (p, q) is in the fourth quadrant, in which quadrant is the point $(q, -p)$?

Find the distance between the pair of points and find the midpoint of the segment having the given points as endpoints.

116. $\left(a, \dfrac{1}{a}\right)$ and $\left(a + h, \dfrac{1}{a + h}\right)$

117. $\left(a, \sqrt{a}\right)$ and $\left(a + h, \sqrt{a + h}\right)$

Find an equation of a circle satisfying the given conditions.

118. Center $(-5, 8)$ with a circumference of 10π units

119. Center $(2, -7)$ with an area of 36π square units

120. Find the point on the x-axis that is equidistant from the points $(-4, -3)$ and $(-1, 5)$.

121. Find the point on the y-axis that is equidistant from the points $(-2, 0)$ and $(4, 6)$.

122. Determine whether $(-1, -3)$, $(-4, -9)$, and $(2, 3)$ are collinear.

123. *Swimming Pool.* A swimming pool is being constructed in the corner of a yard, as shown. Before installation, the contractor needs to know measurements a_1 and a_2. Find them.

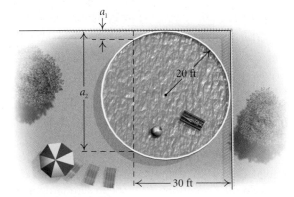

124. *An Arch of a Circle in Carpentry.* Ace Carpentry needs to cut an arch for the top of an entranceway.

The arch needs to be 8 ft wide and 2 ft high. To draw the arch, the carpenters will use a stretched string with chalk attached at an end as a compass.

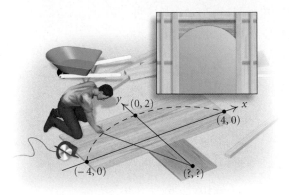

a) Using a coordinate system, locate the center of the circle.

b) What radius should the carpenters use to draw the arch?

*Determine whether each of the following points lies on the **unit circle**, $x^2 + y^2 = 1$.*

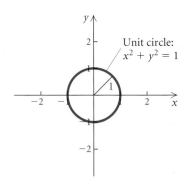

125. $\left(\dfrac{\sqrt{3}}{2}, -\dfrac{1}{2}\right)$ **126.** $(0, -1)$

127. $\left(-\dfrac{\sqrt{2}}{2}, \dfrac{\sqrt{2}}{2}\right)$ **128.** $\left(\dfrac{1}{2}, -\dfrac{\sqrt{3}}{2}\right)$

129. Prove the midpoint formula by showing that:

a) $\left(\dfrac{x_1 + x_2}{2}, \dfrac{y_1 + y_2}{2}\right)$ is equidistant from the points (x_1, y_1) and (x_2, y_2); and

b) the distance from (x_1, y_1) to the midpoint plus the distance from (x_2, y_2) to the midpoint equals the distance from (x_1, y_1) to (x_2, y_2).

130. Consider any right triangle with base b and height h, situated as shown. Show that the midpoint of the hypotenuse P is equidistant from the three vertices of the triangle.

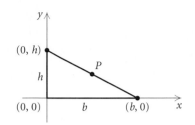

1.2

Functions and Graphs

- *Determine whether a correspondence or a relation is a function.*
- *Find function values, or outputs, using a formula.*
- *Find the domain and the range of a function.*
- *Determine whether a graph is that of a function.*
- *Solve applied problems using functions.*

We now focus our attention on a concept that is fundamental to many areas of mathematics—the idea of a *function*.

Functions

We first consider an application.

Thunder Time Related to Lightning Distance. During a thunderstorm, it is possible to calculate how far away, y (in miles), lightning is when the sound of thunder arrives x seconds after the lightning has been sighted. It is known that the distance, in miles, is $\frac{1}{5}$ of the time, in seconds. If we hear the sound of thunder 15 seconds after we've seen the lightning, we know that the lightning is $\frac{1}{5} \cdot 15$, or 3 miles away. Similarly, 5 sec corresponds to 1 mi, 2 sec to $\frac{2}{5}$ mi, and so on. We can express this relationship with a set of ordered pairs, a graph, and an equation.

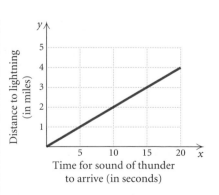

x	y	ORDERED PAIRS: (x, y)	CORRESPONDENCE
0	0	$(0, 0)$	$0 \longrightarrow 0$
1	$\frac{1}{5}$	$\left(1, \frac{1}{5}\right)$	$1 \longrightarrow \frac{1}{5}$
2	$\frac{2}{5}$	$\left(2, \frac{2}{5}\right)$	$2 \longrightarrow \frac{2}{5}$
5	1	$(5, 1)$	$5 \longrightarrow 1$
10	2	$(10, 2)$	$10 \longrightarrow 2$
15	3	$(15, 3)$	$15 \longrightarrow 3$

The ordered pairs express a relationship, or correspondence, between the first and second coordinates. We can see this relationship in the graph as well. The equation that describes the correspondence is

$$y = \frac{1}{5}x.$$

This is an example of a *function*. In this case, distance y is a function of time x; that is, y is a function of x, where x is the **independent variable** and y is the **dependent variable.**

Let's consider some other correspondences before giving the definition of a function.

DOMAIN		RANGE
To each registered student	there corresponds	an I. D. number.
To each mountain bike sold	there corresponds	its price.
To each number between -3 and 3	there corresponds	the square of that number.

In each correspondence, the first set is called the **domain** and the second set is called the **range.** For each member, or **element,** in the domain, there is *exactly one* member in the range to which it corresponds. Thus each registered student has exactly *one* I. D. number, each mountain bike has exactly *one* price, and each number between -3 and 3 has exactly *one* square. Each correspondence is a *function.*

Function

A **function** is a correspondence between a first set, called the **domain,** and a second set, called the **range,** such that each member of the domain corresponds to *exactly one* member of the range.

It is important to note that not every correspondence between two sets is a function.

EXAMPLE 1 Determine whether each of the following correspondences is a function.

a)
$$
\begin{array}{l}
-6 \searrow \\
6 \nearrow\!\!\!\!\searrow 36 \\
-3 \searrow \\
3 \nearrow\!\!\!\!\searrow 9 \\
0 \longrightarrow 0
\end{array}
$$

b) Sandra Bullock   Air Force One
George Clooney   Miss Congeniality
Harrison Ford   O Brother Where Art Thou
Julia Roberts   The Pelican Brief
Denzel Washington   Pretty Woman
  Remember the Titans

Solution

a) This correspondence *is* a function because each member of the domain corresponds to exactly one member of the range. Note that the definition of a function allows more than one member of the domain to correspond to the same member of the range.

b) This correspondence *is not* a function because there are members of the domain (Julia Roberts and Denzel Washington) that are paired with more than one member of the range (Julia with *The Pelican Brief* and *Pretty Woman* and Denzel with *The Pelican Brief* and *Remember the Titans*).

EXAMPLE 2   Determine whether each of the following correspondences is a function.

DOMAIN	CORRESPONDENCE	RANGE
a) Years in which a presidential election occurs	The person elected	A set of presidents
b) The integers	Each integer's cube root	A subset of the real numbers
c) All states in the United States	A senator from that state	The set of all U.S. senators
d) The set of all U.S. senators	The state a senator represents	All states in the the United States

Solution

a) This correspondence *is* a function, because in each presidential election *exactly one* president is elected.

b) This correspondence *is* a function, because each integer has *exactly one* cube root.

c) This correspondence *is not* a function, because each state can be paired with *two* different senators.

d) This correspondence *is* a function, because each senator represents only one state.

When a correspondence between two sets is not a function, it is still an example of a **relation.**

> ### Relation
>
> A **relation** is a correspondence between a first set, called the **domain,** and a second set, called the **range,** such that each member of the domain corresponds to *at least one* member of the range.

All the correspondences in Examples 1 and 2 are relations, but, as we have seen, not all are functions. Relations are sometimes written as sets of ordered pairs (as we saw earlier in the example on lightning) in which elements of the domain are the first coordinates of the ordered pairs and elements of the range are the second coordinates. For example, instead of writing $-3 \longrightarrow 9$, as we did in Example 1(a), we could write the ordered pair $(-3, 9)$.

EXAMPLE 3   Determine whether each of the following relations is a function. Identify the domain and the range.

a) $\{(9, -5), (9, 5), (2, 4)\}$

b) $\{(-2, 5), (5, 7), (0, 1), (4, -2)\}$

c) $\{(-5, 3), (0, 3), (6, 3)\}$

Solution

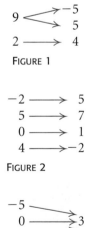

FIGURE 1

FIGURE 2

FIGURE 3

a) The relation *is not* a function because the ordered pairs $(9, -5)$ and $(9, 5)$ have the same first coordinate and different second coordinates (see Fig. 1).

The domain is the set of all first coordinates: $\{9, 2\}$.

The range is the set of all second coordinates: $\{-5, 5, 4\}$.

b) The relation *is* a function because *no* two ordered pairs have the same first coordinate and different second coordinates (see Fig. 2).

The domain is the set of all first coordinates: $\{-2, 5, 0, 4\}$.

The range is the set of all second coordinates: $\{5, 7, 1, -2\}$.

c) The relation *is* a function because *no* two ordered pairs have the same first coordinate and different second coordinates (see Fig. 3).

The domain is $\{-5, 0, 6\}$.

The range is $\{3\}$.

Notation for Functions

Functions used in mathematics are often given by equations. They generally require that certain calculations be performed in order to determine which member of the range is paired with each member of the domain.

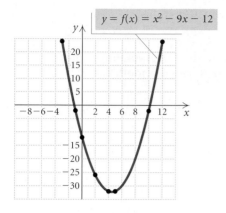

$y = f(x) = x^2 - 9x - 12$

For example, in Section 1.1 we graphed the function $y = x^2 - 9x - 12$ by doing calculations like the following:

$$\text{for } x = -2,\ y = (-2)^2 - 9(-2) - 12 = 10,$$
$$\text{for } x = 0,\ y = 0^2 - 9 \cdot 0 - 12 = -12,\quad \text{and}$$
$$\text{for } x = 1,\ y = 1^2 - 9 \cdot 1 - 12 = -20.$$

A more concise notation is often used. For $y = x^2 - 9x - 12$, the **inputs** (members of the domain) are values of x substituted into the equation. The **outputs** (members of the range) are the resulting values of y. If we call the function f, we can use x to represent an arbitrary *input* and $f(x)$—read "f of x," or "f at x," or "the value of f at x"—to represent the corresponding *output*. In this notation, the function given by $y = x^2 - 9x - 12$ is written as $f(x) = x^2 - 9x - 12$ and the above calculations would be

$$f(-2) = (-2)^2 - 9(-2) - 12 = 10,$$
$$f(0) = 0^2 - 9 \cdot 0 - 12 = -12,$$
$$f(1) = 1^2 - 9 \cdot 1 - 12 = -20.$$

> **Keep in mind that $f(x)$ *does not* mean $f \cdot x$.**

Thus, instead of writing "when $x = -2$, the value of y is 10," we can simply write "$f(-2) = 10$," which can be read as "f of -2 is 10" or "for the input -2, the output of f is 10." The letters g and h are also often used to name functions.

EXAMPLE 4 A function f is given by $f(x) = 2x^2 - x + 3$. Find each of the following.

a) $f(0)$ **b)** $f(-7)$
c) $f(5a)$ **d)** $f(a - 4)$

Solution We can think of this formula as follows:

$$f(\ \ ) = 2(\ \ )^2 - (\ \ ) + 3.$$

Then to find an output for a given input we think: "Whatever goes in the blank on the left goes in the blank(s) on the right." This gives us a "recipe" for finding outputs.

a) $f(0) = 2(0)^2 - 0 + 3 = 0 - 0 + 3 = 3$
b) $f(-7) = 2(-7)^2 - (-7) + 3 = 2 \cdot 49 + 7 + 3 = 108$
c) $f(5a) = 2(5a)^2 - 5a + 3 = 2 \cdot 25a^2 - 5a + 3 = 50a^2 - 5a + 3$
d) $f(a - 4) = 2(a - 4)^2 - (a - 4) + 3$
$$= 2(a^2 - 8a + 16) - a + 4 + 3$$
$$= 2a^2 - 16a + 32 - a + 4 + 3$$
$$= 2a^2 - 17a + 39$$

Technology Connection

We can find function values with a graphing calculator. Below, we illustrate finding $f(-7)$ from Example 4(b), first with the TABLE feature set in ASK mode and then with the VALUE feature from the CALC menu. In both screens, we see that $f(-7) = 108$.

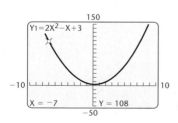

X	Y1	
-7	108	

X =

Y1=2X²−X+3

150

−10 _____ 10

X = -7   Y = 108

−50

Graphs of Functions

We graph functions the same way we graph equations. We find ordered pairs (x, y), or $(x, f(x))$, plot points, and complete the graph.

EXAMPLE 5 Graph each of the following functions.

a) $f(x) = x^2 - 5$ **b)** $f(x) = x^3 - x$ **c)** $f(x) = \sqrt{x + 4}$

Solution   We select values for x and find the corresponding values of $f(x)$. Then we plot the points and connect them with a smooth curve.

a) $f(x) = x^2 - 5$

x	$f(x)$	$(x, f(x))$
-3	4	$(-3, 4)$
-2	-1	$(-2, -1)$
-1	-4	$(-1, -4)$
0	-5	$(0, -5)$
1	-4	$(1, -4)$
2	-1	$(2, -1)$
3	4	$(3, 4)$

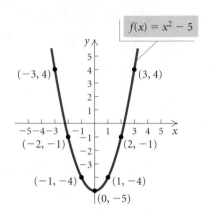

b) $f(x) = x^3 - x$ **c)** $f(x) = \sqrt{x + 4}$

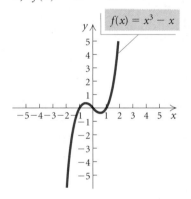 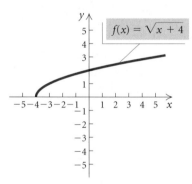

To find a function value, like $f(3)$, from a graph, we locate the input 3 on the horizontal axis, move vertically to the graph of the function, and then move horizontally to find the output on the vertical axis. For the function $f(x) = x^2 - 5$, we see that $f(3) = 4$.

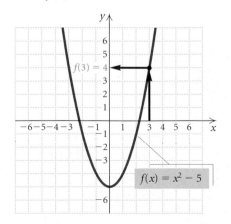

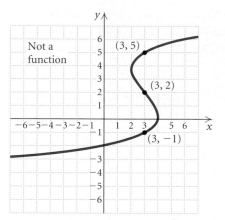

Not a function

Since 3 is paired with more than one member of the range, the graph does not represent a function.

We know that when one member of the domain is paired with two or more different members of the range, the correspondence *is not* a function. Thus, when a graph contains two or more different points with the same first coordinate, the graph cannot represent a function (see the graph at left; note that 3 is paired with −1, 2, and 5). Points sharing a common first coordinate are vertically above or below each other. This leads us to the *vertical-line test*.

The Vertical-Line Test

If it is possible for a vertical line to cross a graph more than once, then the graph is *not* the graph of a function.

To apply the vertical-line test, we try to find a vertical line that crosses the graph more than once. If we succeed, then the graph is not that of a function. If we do not, then the graph is that of a function.

EXAMPLE 6 Which of graphs (a) through (f) (in red) are graphs of functions? In graph (f), the solid dot shows that $(-1, 1)$ belongs to the graph. The open circle shows that $(-1, -2)$ does *not* belong to the graph.

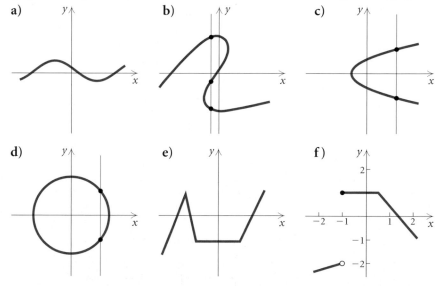

Solution Graphs (a), (e), and (f) are graphs of functions because we cannot find a vertical line that crosses any of them more than once. In (b), the vertical line crosses the graph at three points, so graph (b) is not that of a function. Also, in (c) and (d), we can find a vertical line that crosses the graph more than once, so these are not graphs of functions.

Finding Domains of Functions

When a function f, whose inputs and outputs are real numbers, is given by a formula, the *domain* is understood to be the set of all inputs for which the

INTERVAL NOTATION

REVIEW SECTION R.1.

Technology Connection

When we use a graphing calculator to find function values and a function value does not exist, the calculator indicates this with an ERROR message.

In the tables below, we see in Example 7 that $f(3)$ for $f(x) = 1/(x - 3)$ and $g(-7)$ for $g(x) = \sqrt{x} + 5$ do not exist. Thus, 3 and -7 are *not* in the domains of the corresponding functions.

$y = 1/(x - 3)$

X	Y1
1	−.5
3	ERROR

X =

$y = \sqrt{x} + 5$

X	Y1
16	9
−7	ERROR

X =

expression is defined as a real number. When a substitution results in an expression that is not defined as a real number, we say that the function value *does not exist* and that the number being substituted *is not* in the domain of the function.

EXAMPLE 7 Find the indicated function values and determine whether the given values are in the domain of the function.

a) $f(1)$ and $f(3)$, for $f(x) = \dfrac{1}{x - 3}$

b) $g(16)$ and $g(-7)$, for $g(x) = \sqrt{x} + 5$

Solution

a) $f(1) = \dfrac{1}{1 - 3} = \dfrac{1}{-2} = -\dfrac{1}{2}$;

Since $f(1)$ is defined, 1 is in the domain of f.

$f(3) = \dfrac{1}{3 - 3} = \dfrac{1}{0}$

Since division by 0 is not defined, the number 3 is not in the domain of f.

b) $g(16) = \sqrt{16} + 5 = 4 + 5 = 9$;

Since $g(16)$ is defined, 16 is in the domain of g.

$g(-7) = \sqrt{-7} + 5$

Since $\sqrt{-7}$ is not defined as a real number, the number -7 is not in the domain of g. ▬

Inputs that make a denominator 0 or that yield a negative radicand in an even root are not in the domain of a function.

EXAMPLE 8 Find the domain of each of the following functions.

a) $f(x) = \dfrac{1}{x - 3}$ b) $g(x) = \sqrt{3 - x} + 5$

c) $h(x) = \dfrac{3x^2 - x + 7}{x^2 + 2x - 3}$ d) $f(x) = x^3 + |x|$

Solution

a) The input 3 results in a denominator of 0. The domain is $\{x \mid x \neq 3\}$. We can also write the solution using interval notation and the symbol $\cup$ for the **union** or inclusion of both sets: $(-\infty, 3) \cup (3, \infty)$.

b) We can substitute any number for which the radicand is nonnegative, that is, for which $3 - x \geq 0$, or $x \leq 3$. Thus the domain is $\{x \mid x \leq 3\}$, or $(-\infty, 3]$.

c) We can substitute any real number in the numerator, but we must avoid inputs that make the denominator 0. To find those inputs, we solve $x^2 + 2x - 3 = 0$, or $(x + 3)(x - 1) = 0$. Since $x^2 + 2x - 3$ is 0 for -3 and 1, the domain consists of the set of all real numbers except -3 and 1, or $\{x \,|\, x \neq -3 \text{ and } x \neq 1\}$, or $(-\infty, -3) \cup (-3, 1) \cup (1, \infty)$.

d) We can substitute any real number for x. The domain is the set of all real numbers, $\mathbb{R}$, or $(-\infty, \infty)$. ▬

Visualizing Domain and Range

Keep the following in mind regarding the *graph* of a function:

Domain = the set of a function's inputs, found on the horizontal axis;

Range = the set of a function's outputs, found on the vertical axis.

By carefully examining the graph of a function, we may be able to determine the function's domain as well as its range. Consider the graph of $f(x) = \sqrt{4 - (x - 1)^2}$, shown at left. We look for the inputs on the x-axis that correspond to a point on the graph. We see that they extend from -1 to 3, inclusive. Thus the domain is $\{x \,|\, -1 \leq x \leq 3\}$, or $[-1, 3]$.

To find the range, we look for the outputs on the y-axis. We see that they extend from 0 to 2, inclusive. Thus the range of this function is $\{y \,|\, 0 \leq y \leq 2\}$, or $[0, 2]$.

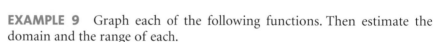

EXAMPLE 9 Graph each of the following functions. Then estimate the domain and the range of each.

a) $f(x) = \sqrt{x + 4}$

b) $f(x) = x^3 - x$

c) $f(x) = \dfrac{1}{x - 2}$

d) $f(x) = x^4 - 2x^2 - 3$

Solution

a)

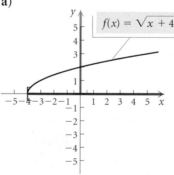

Domain = $[-4, \infty)$;
range = $[0, \infty)$

b)

Domain = all real numbers, $(-\infty, \infty)$; range = all real numbers, $(-\infty, \infty)$

c)

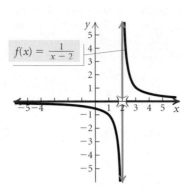

d)

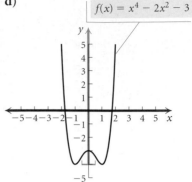

Since the graph does not touch or cross both the vertical line $x = 2$ and the x-axis $y = 0$, 2 is excluded from the domain and 0 is excluded from the range.
Domain $= (-\infty, 2) \cup (2, \infty)$;
range $= (-\infty, 0) \cup (0, \infty)$

Domain $=$ all real numbers, $(-\infty, \infty)$; range $= [-4, \infty)$

Always consider adding the reasoning of Example 8 to a graphical analysis. Think, "What can I substitute?" to find the domain. Think, "What do I get out?" to find the range. Thus, in Example 9(d), it might not appear as though the domain is all real numbers because the graph rises steeply, but by examining the equation we see that we can indeed substitute any real number for x.

Applications of Functions

EXAMPLE 10   *Speed of Sound in Air.*   The speed S of sound in air is a function of the temperature t, in degrees Fahrenheit, and is given by

$$S(t) = 1087.7 \sqrt{\frac{5t + 2457}{2457}},$$

where S is in feet per second. Find the speed of sound in air when the temperature is $0°$, $32°$, $70°$, and $-10°$ Fahrenheit.

Solution   Using a calculator, we compute the function values. We find that

$$S(0) = 1087.7 \text{ ft/sec},$$
$$S(32) \approx 1122.6 \text{ ft/sec},$$
$$S(70) \approx 1162.6 \text{ ft/sec}, \quad \text{and}$$
$$S(-10) \approx 1076.6 \text{ ft/sec}.$$

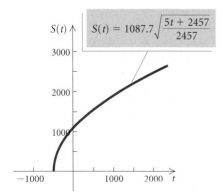

We can visualize these values using the graph at left.

CONNECTING THE CONCEPTS

FUNCTION CONCEPTS

Formula for f: $f(x) = 5 + 2x^2 - x^4$.

For every input, there is exactly one output.

$(1, 6)$ is on the graph.

For the input 1, the output is 6.

$f(1) = 6$

Domain: set of all inputs $= (-\infty, \infty)$

Range: set of all outputs $= (-\infty, 6]$

GRAPH

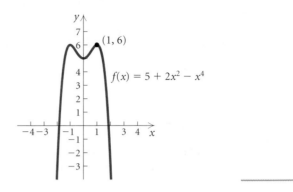

1.2

Exercise Set

In Exercises 1–14, determine whether the correspondence is a function.

1. $a \longrightarrow w$
$b \longrightarrow y$
$c \longrightarrow z$

2. $m \longrightarrow q$
$n r$
$o s$

3. $-6 \longrightarrow 36$
$-2 \longrightarrow 4$
2

4. $-3 \longrightarrow 2$
$1 \longrightarrow 4$
$5 \longrightarrow 6$
$9 \longrightarrow 8$

5. $m \longrightarrow A$
$n B$
$r C$
$s D$

6. $a \longrightarrow r$
$b s$
$c t$
d

7. WORLD'S TEN LARGEST EARTHQUAKES (1900–1999)

LOCATION AND DATE	MAGNITUDE
Chile (May 22, 1960)	9.5
Prince William Sound, Alaska (March 28, 1964)	9.2
Andreanof Islands, Aleutian Islands (March 9, 1957)	9.1
Kamchatka (November 4, 1952)	9.0
Off the coast of Ecuador (January 31, 1906)	8.8
Rat Islands, Aleutian Islands (February 4, 1965)	8.7
India–China border (August 15, 1950)	8.6
Kamchatka (February 3, 1923)	8.5
Banda Sea, Indonesia (February 1, 1938)	
Kuril Islands (October 13, 1963	

(*Source*: National Earthquake Information Center, U.S. Geological Survey)

8.

ANIMAL	SPEED ON THE GROUND (IN MILES PER HOUR)
Cheetah	70
Lion	50
Reindeer	32
Giraffe	
Elephant	25
Giant tortoise	0.17

(*Source: Time Almanac*, 1999, p. 581)

DOMAIN	CORRESPONDENCE	RANGE
9. A set of cars in a parking lot	Each car's license number	A set of letters and numbers
10. A set of people in a town	A doctor a person uses	A set of doctors
11. A set of members of a family	Each person's eye color	A set of colors
12. A set of members of a rock band	An instrument each person plays	A set of instruments
13. A set of students in a class	A student sitting in a neighboring seat	A set of students
14. A set of bags of chips on a shelf	Each bag's weight	A set of weights

Determine whether the relation is a function. Identify the domain and the range.

15. $\{(2, 10), (3, 15), (4, 20)\}$

16. $\{(3, 1), (5, 1), (7, 1)\}$

17. $\{(-7, 3), (-2, 1), (-2, 4), (0, 7)\}$

18. $\{(1, 3), (1, 5), (1, 7), (1, 9)\}$

19. $\{(-2, 1), (0, 1), (2, 1), (4, 1), (-3, 1)\}$

20. $\{(5, 0), (3, -1), (0, 0), (5, -1), (3, -2)\}$

A graph of a function is shown. Find the indicated function values.

21. $h(1)$, $h(3)$, and $h(4)$

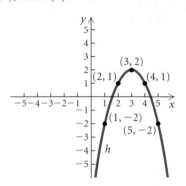

22. $t(-4)$, $t(0)$, and $t(3)$

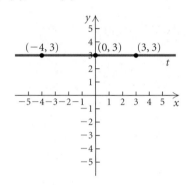

23. $s(-4)$, $s(-2)$, and $s(0)$

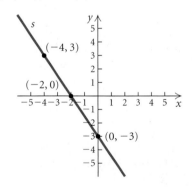

24. $g(-4)$, $g(-1)$, and $g(0)$

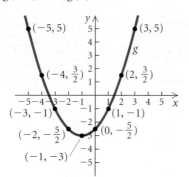

25. $f(-1)$, $f(0)$, and $f(1)$

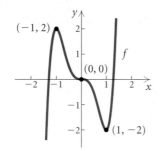

26. $g(-2)$, $g(0)$, and $g(2.4)$

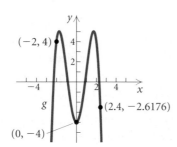

27. Given that $g(x) = 3x^2 - 2x + 1$, find each of the following.

a) $g(0)$ **b)** $g(-1)$
c) $g(3)$ **d)** $g(-x)$
e) $g(1 - t)$

28. Given that $f(x) = 5x^2 + 4x$, find each of the following.

a) $f(0)$ **b)** $f(-1)$
c) $f(3)$ **d)** $f(t)$
e) $f(t - 1)$

29. Given that $g(x) = x^3$, find each of the following.

a) $g(2)$ **b)** $g(-2)$
c) $g(-x)$ **d)** $g(3y)$
e) $g(2 + h)$

30. Given that $f(x) = 2|x| + 3x$, find each of the following.

a) $f(1)$ **b)** $f(-2)$
c) $f(-x)$ **d)** $f(2y)$
e) $f(2 - h)$

31. Given that $g(x) = \dfrac{x - 4}{x + 3}$, find each of the following.

a) $g(5)$ **b)** $g(4)$
c) $g(-3)$ **d)** $g(-16.25)$
e) $g(x + h)$

32. Given that $f(x) = \dfrac{x}{2 - x}$, find each of the following.

a) $f(2)$ **b)** $f(1)$
c) $f(-16)$ **d)** $f(-x)$
e) $f\left(-\dfrac{2}{3}\right)$

33. Find $g(0)$, $g(-1)$, $g(5)$, and $g\left(\frac{1}{2}\right)$ for

$$g(x) = \frac{x}{\sqrt{1 - x^2}}.$$

34. Find $h(0)$, $h(2)$, and $h(-x)$ for
$$h(x) = x + \sqrt{x^2 - 1}.$$

Find the domain of the function.

35. $f(x) = 7x + 4$

36. $f(x) = |3x - 2|$

37. $f(x) = 4 - \dfrac{2}{x}$

38. $f(x) = \dfrac{1}{x^4}$

39. $f(x) = \dfrac{x + 5}{2 - x}$

40. $f(x) = \dfrac{8}{x + 4}$

41. $f(x) = \dfrac{1}{x^2 - 4x - 5}$

42. $f(x) = \dfrac{x^4 - 2x^3 + 7}{3x^2 - 10x - 8}$

43. $f(x) = \sqrt{8 - x}$

44. $f(x) = x^2 - 2x$

45. $f(x) = \frac{1}{10}|x|$

46. $f(x) = \dfrac{\sqrt{x + 1}}{x}$

Graph the function. Then visually estimate the domain and the range.

47. $f(x) = |x|$ 　　　　　　 **48.** $f(x) = |x| - 2$

49. $f(x) = \sqrt{9 - x^2}$ 　　　 **50.** $f(x) = -\sqrt{25 - x^2}$

51. $f(x) = (x - 1)^3 + 2$ 　　 **52.** $f(x) = (x - 2)^4 + 1$

53. $f(x) = \sqrt{7 - x}$ 　　　 **54.** $f(x) = \sqrt{x + 8}$

55. $f(x) = -x^2 + 4x - 1$ 　 **56.** $f(x) = 2x^2 - x^4 + 5$

In Exercises 57–64, determine whether the graph is that of a function. An open dot indicates that the point does not belong to the graph.

57.

58.
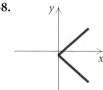

59.
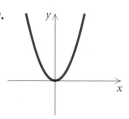

60.
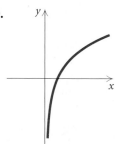

61.
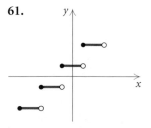

62.
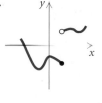

63.
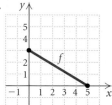

64.
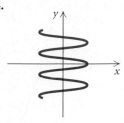

In Exercises 65–68, determine the domain and the range of each function.

65.
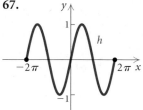

66.
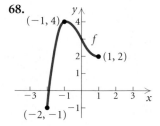

67.
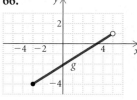

68.
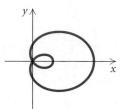

69. *Boiling Point and Elevation.*   The elevation E, in meters, above sea level at which the boiling point of water is t degrees Celsius is given by the function
$$E(t) = 1000(100 - t) + 580(100 - t)^2.$$
At what elevation is the boiling point $99.5°$? $100°$?

70. *Average Price of a Movie Ticket.*   The average price of a movie ticket, in dollars, can be estimated by the function P given by

$$P(x) = 0.1522x - 298.592$$

where x is the year. Thus, $P(2003)$ is the average price of a movie ticket in 2003. The price is lower than what might be expected due to lower prices for matinees, senior citizens' discounts, and other special prices.

a) Use the function to predict the average price in 2005 and 2010.

b) When will the average price be $8.00?

71. *Territorial Area of an Animal.* The territorial area of an animal is defined to be its defended, or exclusive, region. For example, a lion has a certain region over which it is considered ruler. It has been shown that the territorial area T, in acres, of predatory animals is a function of body weight w, in pounds, and is given by the function

$$T(w) = w^{1.31}.$$

Find the territorial area of animals whose body weights are 0.5 lb, 10 lb, 20 lb, 100 lb, and 200 lb.

Technology Connection

In Exercises 72–74, use a graphing calculator and the TABLE *feature set in* ASK *mode.*

72. Given that

$$h(x) = 3x^4 - 10x^3 + 5x^2 - x + 6,$$

find $h(-11)$, $h(7)$, and $h(15)$.

73. Given that

$$g(x) = 0.06x^3 - 5.2x^2 - 0.8x,$$

find $g(-2.1)$, $g(5.08)$, and $g(10.003)$. Round answers to the nearest tenth.

74. Find the indicated function values if they exist.

a) $f(-4)$ and $f(-6)$, for $f(x) = \dfrac{4}{(x-4)(x+6)}$

b) $g(-5)$ and $g(1)$, for $g(x) = \sqrt{x-1} + 3$

Collaborative Discussion and Writing

75. Explain in your own words what a function is.

76. Explain in your own words the difference between the domain of a function and the range of a function.

Skill Maintenance

To the student and the instructor: The Skill Maintenance exercises review skills covered previously in the text. You can expect such exercises in every exercise set. They provide excellent review for a final examination. Answers to all skill maintenance exercises, along with section references, appear in the answer section at the back of the book.

Use substitution to determine whether the given ordered pairs are solutions of the given equation.

77. $(0, -7), (8, 11);\ y = 0.5x + 7$

78. $\left(\frac{4}{5}, -2\right), \left(\frac{11}{5}, \frac{1}{10}\right);\ 15x - 10y = 32$

Graph the equation.

79. $y = (x - 1)^2$

80. $y = \frac{1}{3}x - 6$

81. $-2x - 5y = 10$

82. $(x - 3)^2 + y^2 = 4$

Synthesis

83. Give an example of two different functions that have the same domain and the same range, but have no pairs in common. Answers may vary.

84. Draw a graph of a function for which the domain is $[-4, 4]$ and the range is $[1, 2] \cup [3, 5]$. Answers may vary.

85. Draw a graph of a function for which the domain is $[-3, -1] \cup [1, 5]$ and the range is $\{1, 2, 3, 4\}$. Answers may vary.

86. Suppose that for some function f, $f(x - 1) = 5x$. Find $f(6)$.

87. Suppose that for some function g, $g(x + 3) = 2x + 1$. Find $g(-1)$.

88. Suppose $f(x) = |x + 3| - |x - 4|$. Write $f(x)$ without using absolute-value notation if x is in each of the following intervals.

a) $(-\infty, -3)$
b) $[-3, 4)$
c) $[4, \infty)$

89. Suppose $g(x) = |x| + |x - 1|$. Write $g(x)$ without using absolute-value notation if x is in each of the following intervals.

a) $(-\infty, 0)$
b) $[0, 1)$
c) $[1, \infty)$

1.3

Linear Functions, Slope, and Applications

- *Determine the slope of a line given two points on the line.*
- *Solve applied problems involving slope and linear functions.*

In real-life situations, we often need to make decisions on the basis of limited information. When the given information is used to formulate an equation or inequality that at least approximates the situation mathematically, we have created a **model.** One of the most frequently used mathematical models is *linear*—the graph of a linear model is a straight line.

Linear Functions

Let's begin to examine the connections among equations, functions, and graphs that are straight lines. Compare the graphs of linear and nonlinear functions shown here.

LINEAR FUNCTIONS

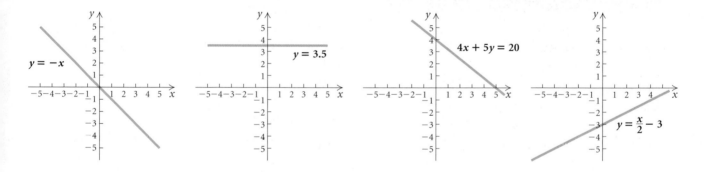

NONLINEAR FUNCTIONS

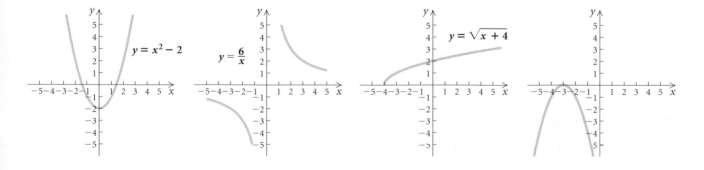

We have the following results and related terminology.

Linear Functions

A function f is a **linear function** if it can be written as

$$f(x) = mx + b,$$

where m and b are constants.

(If $m = 0$, the function is a **constant function** $f(x) = b$. If $m = 1$ and $b = 0$, the function is the **identity function** $f(x) = x$.)

Linear function:
$y = mx + b$

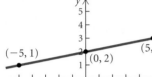
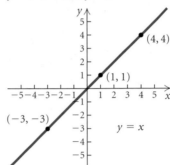

Identity function:
$y = 1 \cdot x + 0$, or $y = x$

Constant function:
$y = 0 \cdot x + b$, or $y = b$ (Horizontal line)

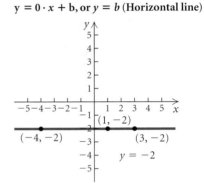

Vertical line: $x = a$
(*not* a function)

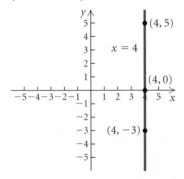

Horizontal and Vertical Lines

Horizontal lines are given by equations of the type $y = b$ or $f(x) = b$. (They are functions.)

Vertical lines are given by equations of the type $x = a$. (They are *not* functions.)

The Linear Function $f(x) = mx + b$ and Slope

To attach meaning to the constant m in the equation $f(x) = mx + b$, we first consider an application. Suppose FaxMax is an office machine business that currently has two stores in locations A and B in the same city. Their total costs for the same time period are given by two functions shown in the tables and graphs that follow. The variable x represents time, in months. The variable y represents total costs, in thousands of dollars, over that amount of time. Look for a pattern.

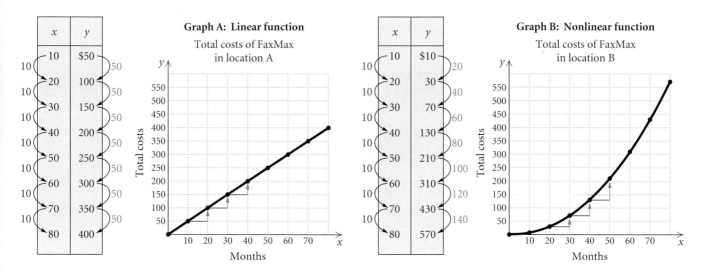

x	y
10	$50
20	100
30	150
40	200
50	250
60	300
70	350
80	400

Graph A: Linear function

Total costs of FaxMax in location A

x	y
10	$10
20	30
30	70
40	130
50	210
60	310
70	430
80	570

Graph B: Nonlinear function

Total costs of FaxMax in location B

We see in graph A that *every* change of 10 months results in a $50 thousand change in total costs. But in graph B, changes of 10 months do *not* result in constant changes in total costs. This is a way to distinguish linear from nonlinear functions. The rate at which a linear function changes, or the steepness of its graph, is constant.

Mathematically, we define the steepness, or **slope,** of a line as the ratio of its vertical change (rise) to the corresponding horizontal change (run). Slope represents the **rate of change** of y with respect to x.

Slope

The **slope m** of a line containing points (x_1, y_1) and (x_2, y_2) is given by

$$m = \frac{\text{rise}}{\text{run}}$$

$$= \frac{\text{the change in } y}{\text{the change in } x}$$

$$= \frac{y_2 - y_1}{x_2 - x_1} = \frac{y_1 - y_2}{x_1 - x_2}.$$

EXAMPLE 1 Graph the function $f(x) = -\frac{2}{3}x + 1$ and determine its slope.

Solution Since the equation for f is in the form $f(x) = mx + b$, we know it is a linear function. We can graph it by connecting two points on the graph with a straight line. We calculate two ordered pairs, plot the points, graph the function, and determine the slope:

$$f(3) = -\frac{2}{3} \cdot 3 + 1 = -2 + 1 = -1;$$

$$f(9) = -\frac{2}{3} \cdot 9 + 1 = -6 + 1 = -5;$$

Pairs: $(3, -1), (9, -5);$

$$\text{Slope} = m = \frac{y_2 - y_1}{x_2 - x_1}$$

$$= \frac{-5 - (-1)}{9 - 3} = \frac{-4}{6} = -\frac{2}{3}.$$

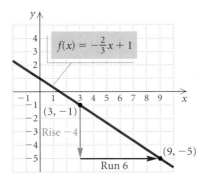

The slope is the same for any two points on a line. Thus, to check our work, note that $f(6) = -\frac{2}{3} \cdot 6 + 1 = -4 + 1 = -3$. Using the points $(6, -3)$ and $(3, -1)$, we have

$$m = \frac{-1 - (-3)}{3 - 6} = \frac{2}{-3} = -\frac{2}{3}.$$

We can also use the points in the opposite order when computing slope, so long as we are consistent:

$$m = \frac{-3 - (-1)}{6 - 3} = \frac{-2}{3} = -\frac{2}{3}.$$

Note also that the slope of the line is the number m in the equation for the function $f(x) = -\frac{2}{3}x + 1$.

The *slope* of the line given by $f(x) = mx + b$ is m.

If a line slants up from left to right, the change in x and the change in y have the same sign, so the line has a positive slope. The larger the slope, the steeper the line, as shown in Fig. 1. If a line slants down from left to right, the change in x and the change in y are of opposite signs, so the line has a negative slope. The larger the absolute value of the slope, the steeper the line, as shown in Fig. 2. When $m = 0$, $y = 0x$, or $y = 0$. Note that this horizontal line is the x-axis, as shown in Fig. 3.

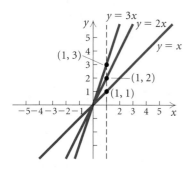

FIGURE 1

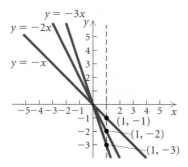

FIGURE 2

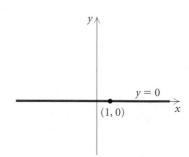

FIGURE 3

Horizontal and Vertical Lines

If a line is horizontal, the change in y for any two points is 0 and the change in x is nonzero. Thus a horizontal line has slope 0 (see Fig. 4).

If a line is vertical, the change in y for any two points is nonzero and the change in x is 0. Thus the slope is *not defined* because we cannot divide by 0 (see Fig. 5).

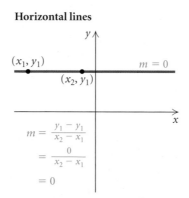

FIGURE 4

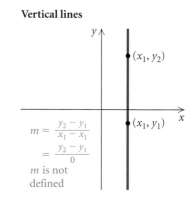

FIGURE 5

Note that zero slope and an undefined slope are two very different concepts.

EXAMPLE 2 Graph each linear equation and determine its slope.

a) $x = -2$ b) $y = \dfrac{5}{2}$

Solution

a) Since y is missing in $x = -2$, any value for y will do.

x	y
-2	0
-2	3
-2	-4

Choose any number for y; x must be -2.

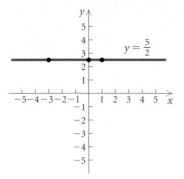

FIGURE 6

The graph (see Fig. 6) is a *vertical line* 2 units to the left of the y-axis. The slope is not defined. The graph is *not* the graph of a function.

b) Since x is missing in $y = \frac{5}{2}$, any value for x will do.

x	y
0	$\frac{5}{2}$
-3	$\frac{5}{2}$
1	$\frac{5}{2}$

Choose any number for x; y must be $\frac{5}{2}$.

FIGURE 7

The graph (see Fig. 7) is a *horizontal line* $\frac{5}{2}$, or $2\frac{1}{2}$, units above the x-axis. The slope is 0. The graph is the graph of a constant function.

Applications of Slope

Slope has many real-world applications. Numbers like 2%, 4%, and 7% are often used to represent the **grade** of a road. Such a number is meant to tell how steep a road is on a hill or mountain. For example, a 4% grade means that the road rises 4 ft for every horizontal distance of 100 ft.

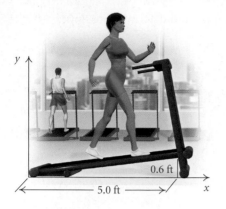

The concept of grade is also used with a treadmill. During a treadmill test, a cardiologist might change the slope, or grade, of the treadmill to measure its effect on heart rate. Another example occurs in hydrology. The strength or force of a river depends on how far the river falls vertically compared to how far it flows horizontally.

EXAMPLE 3 *Ramp for the Handicapped.* Construction laws regarding access ramps for the handicapped state that every vertical rise of 1 ft requires a horizontal run of 12 ft. What is the grade, or slope, of such a ramp?

Solution The grade, or slope, is given by $m = \frac{1}{12} \approx 0.083 \approx 8.3\%$.

Slope can also be considered as an **average rate of change.** To find the average rate of change between any two data points on a graph, we determine the slope of the line that passes through the two points.

EXAMPLE 4 *News Consumption.* In recent years, news consumption from newspapers has decreased while consumption from online sources has increased. The graphs below illustrate both trends.

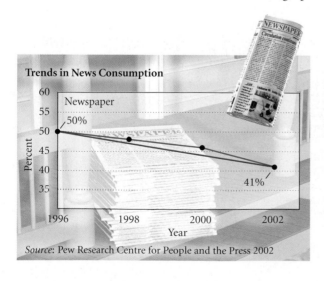

Trends in News Consumption

Newspaper

50%

41%

Source: Pew Research Centre for People and the Press 2002

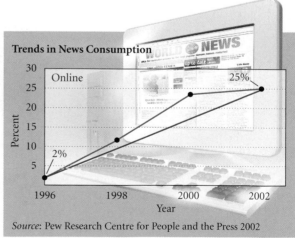

Trends in News Consumption

Online

25%

2%

Source: Pew Research Centre for People and the Press 2002

a) Find the average rate of change in news consumption from newspapers from 1996 to 2002.

b) Find the average rate of change in news consumption from online sources from 1996 to 2002.

Solution

a) We determine the coordinates of two points on the graph. In this case, we use $(1996, 50\%)$ and $(2002, 41\%)$. Then we compute the slope, or average rate of change, as follows:

$$\text{Slope} = \text{Average rate of change} = \frac{\text{change in } y}{\text{change in } x}$$

$$= \frac{41 - 50}{2002 - 1996} = \frac{-9}{6} = -\frac{3}{2} = -1\frac{1}{2}.$$

This result tells us that, on average, each year from 1996 to 2002 news consumption from newspapers decreased $1\frac{1}{2}\%$. The average rate of change over the 6-year period was a decrease of $1\frac{1}{2}\%$ per year.

b) We determine the coordinates of two points on the graph. In this case, we use $(1996, 2\%)$ and $(2002, 25\%)$. Then we compute the slope, or average rate of change, as follows:

$$\text{Slope} = \frac{25 - 2}{2002 - 1996} = \frac{23}{6} = 3\frac{5}{6}.$$

On average, each year from 1996 to 2002 news consumption from on-line sources increased $3\frac{5}{6}\%$. The average rate of change over the 6-year period was an increase of $3\frac{5}{6}\%$ per year.

Applications of Linear Functions

We now consider an application of linear functions.

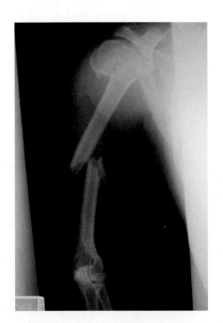

EXAMPLE 5 *Height Estimates.* An anthropologist can use linear functions to estimate the height of a male or a female, given the length of the humerus, the bone from the elbow to the shoulder. The height, in centimeters, of an adult male with a humerus of length x, in centimeters, is given by the function

$$M(x) = 2.89x + 70.64.$$

The height, in centimeters, of an adult female with a humerus of length x is given by the function

$$F(x) = 2.75x + 71.48.$$

A 26-cm humerus was uncovered in a ruins. Assuming it was from a female, how tall was she? What is the domain of this function?

Solution

We substitute into the function:

$$F(26) = 2.75(26) + 71.48 = 142.98.$$

Thus the female was 142.98 cm tall.

Theoretically, the domain of the function is the set of all real numbers. However, the context of the problem dictates a different domain. One could not find a bone with a length of 0 or less. Thus the domain consists of positive real numbers, that is, the interval $(0, \infty)$. A more realistic domain might be 20 cm to 60 cm, the interval $[20, 60]$.

Technology Connection

In Example 5, we can graph $F(x) = 2.75x + 71.48$ and use the VALUE feature to find $F(26)$.

$$y = 2.75x + 71.48$$

The female was 142.98 cm tall.

1.3

Exercise Set

In Exercises 1–4, the table of data contains input–output values for a function. Answer the following questions for each table.

a) *Is the change in the inputs x the same?*
b) *Is the change in the outputs y the same?*
c) *Is the function linear?*

1.

x	y
-3	7
-2	10
-1	13
0	16
1	19
2	22
3	25

2.

x	y
20	12.4
30	24.8
40	49.6
50	99.2
60	198.4
70	396.8
80	793.6

3.

x	y
11	3.2
26	5.7
41	8.2
56	9.3
71	11.3
86	13.7
101	19.1

4.

x	y
2	−8
4	−12
6	−16
8	−20
10	−24
12	−28
14	−32

Find the slope of the line containing the given points.

5.
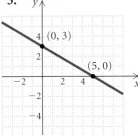

6.
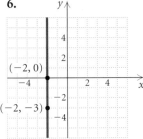

7.
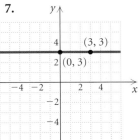

8.
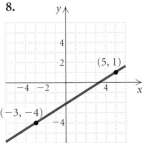

9. $(9, 4)$ and $(−1, 2)$

10. $(−3, 7)$ and $(5, −1)$

11. $(4, −9)$ and $(−5, 6)$

12. $(−6, −1)$ and $(2, −13)$

13. $(0.7, −0.1)$ and $(−0.3, −0.4)$

14. $\left(−\frac{3}{4}, −\frac{1}{4}\right)$ and $\left(\frac{2}{7}, −\frac{5}{7}\right)$

15. $(2, −2)$ and $(4, −2)$

16. $(−9, 8)$ and $(7, −6)$

17. $\left(\frac{1}{2}, −\frac{3}{5}\right)$ and $\left(−\frac{1}{2}, \frac{3}{5}\right)$

18. $(−8.26, 4.04)$ and $(3.14, −2.16)$

19. $(16, −13)$ and $(−8, −5)$

20. $(−10, −7)$ and $(−10, 7)$

21. $(\pi, −3)$ and $(\pi, 2)$

22. $\left(\sqrt{2}, −4\right)$ and $(0.56, −4)$

Graph the linear equation and determine its slope, if it exists.

23. $y = −\frac{1}{2}x + 3$ **24.** $y = \frac{3}{2}x − 4$

25. $2y − 3x = −6$ **26.** $x + 2y = 1$

27. $5x + 2y = 10$ **28.** $2y − x = 8$

29. $y = −\frac{2}{3}$ **30.** $x = 3$

31. *Credit Card Debt.*   In the past decade, the average household's credit card balance rose 173%. Find the average rate of change in the credit card balance from 1992 to 2002.

Credit Card Debt

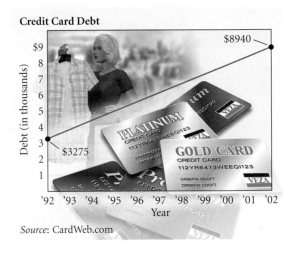

Source: CardWeb.com

32. *Unwanted Commercial e-mail.*   Between 1999 and 2002, the number of unwanted commercial e-mail messages, also called spam, sent daily rose 460%. Find the average rate of change in delivery of spam from 1999 to 2002.

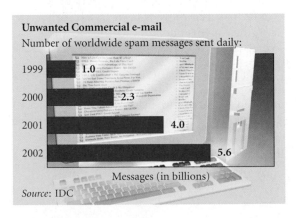

Unwanted Commercial e-mail

Number of worldwide spam messages sent daily:

1999	1.0
2000	2.3
2001	4.0
2002	5.6

Messages (in billions)

Source: IDC

33. *Overseas Adoptions.* Overseas adoptions by U.S. parents have increased by more than 200% from 1992 to 2002. In 1992, 6472 visas were issued to orphans from other countries. In 2002, 20,099 visas were issued. Find the average rate of change in the number of overseas adoptions over the 10-year period. (*Source*: National Adoption Information Clearinghouse)

34. *Chinese Adoptions.* In 1992, 210 children from mainland China were adopted by U.S. parents. The number of Chinese children adopted ten years later, in 2002, was 6062. Find the average rate of change in the number of Chinese adoptions from 1992 to 2002. (*Source*: Bureau of Citizenship and Immigration Services)

35. *Running Rate.* An Olympic marathoner passes the 10-km point of a race after 50 min and arrives at the 25-km point $1\frac{1}{2}$ hr later. Find the speed (average rate of change) of the runner.

36. *Work Rate.* As a typist resumes work on a project, $\frac{1}{6}$ of the paper has already been typed. Six hours later, the paper is $\frac{3}{4}$ done. Calculate the worker's typing rate.

37. *Ideal Minimum Weight.* One way to estimate the ideal minimum weight of a woman in pounds is to multiply her height in inches by 4 and subtract 130. Let W = the ideal minimum weight and h = height.
a) Express W as a linear function of h.
b) Find the ideal minimum weight of a woman whose height is 62 in.
c) Find the domain of the function.

38. *Pressure at Sea Depth.* The function P, given by
$$P(d) = \tfrac{1}{33}d + 1,$$
gives the pressure, in atmospheres (atm), at a depth d, in feet, under the sea.
a) Find $P(0)$, $P(5)$, $P(10)$, $P(33)$, and $P(200)$.
b) Find the domain of the function.

39. *Stopping Distance on Glare Ice.* The stopping distance (at some fixed speed) of regular tires on glare ice is a function of the air temperature F, in degrees Fahrenheit. This function is estimated by
$$D(F) = 2F + 115,$$
where $D(F)$ is the stopping distance, in feet, when the air temperature is F, in degrees Fahrenheit.
a) Find $D(0°)$, $D(-20°)$, $D(10°)$, and $D(32°)$.
b) Explain why the domain should be restricted to $[-57.5°, 32°]$.

40. *Anthropology Estimates.* Consider Example 5 and the function
$$M(x) = 2.89x + 70.64$$
for estimating the height of a male.
a) If a 26-cm humerus from a male is found in an archeological dig, estimate the height of the male.
b) What is the domain of M?

41. *Reaction Time.* Suppose that while driving a car, you suddenly see a school crossing guard standing in the road. Your brain registers the information and sends a signal to your foot to hit the brake. The car travels a distance D, in feet, during this time, where D is a function of the speed r, in miles per hour, of the car when you see the crossing guard. That reaction distance is a linear function given by
$$D(r) = \frac{11}{10}r + \frac{1}{2}.$$

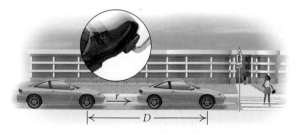

a) Find the slope of this line and interpret its meaning in this application.
b) Find $D(5)$, $D(10)$, $D(20)$, $D(50)$, and $D(65)$.
c) What is the domain of this function? Explain.

42. *Straight-Line Depreciation.* A marketing firm buys a new color printer for $5200 to print banners for a sales campaign. The printer is purchased on January 1 and is expected to last 8 yr, at the end of which time its *trade-in*, or *salvage value*, will be $1100. If the company figures the decline or depreciation in value to be the same each year, then the salvage value V, after t years, is given by the linear function

$$V(t) = \$5200 - \$512.50t, \quad \text{for } 0 \le t \le 8.$$

a) Find $V(0)$, $V(1)$, $V(2)$, $V(3)$, and $V(8)$.
b) Find the domain and the range of this function.

43. *Total Cost.* The Cellular Connection charges $60 for a phone and $29 per month under its economy plan. Write an equation that can be used to determine the total cost, $C(t)$, of operating a Cellular Connection phone for t months. Then find the total cost for 6 months.

44. *Total Cost.* Superior Cable Television charges a $65 installation fee and $80 per month for "deluxe" service. Write an equation that can be used to determine the total cost, $C(t)$, for t months of deluxe cable television service. Then find the total cost for 8 months of service.

*In Exercises 45 and 46, the term **fixed costs** refers to the start-up costs of operating a business. This includes machinery and building costs. The term **variable costs** refers to what it costs a business to produce or service one item.*

45. Kara's Custom Tees experienced fixed costs of $800 and variable costs of $3 per shirt. Write an equation that can be used to determine the total costs encountered by Kara's Custom Tees. Then determine the total cost of producing 75 shirts.

46. It's My Racquet experienced fixed costs of $950 and variable costs of $18 for each tennis racquet that is restrung. Write an equation that can be used to determine the total costs encountered by It's My Racquet. Then determine the total cost of restringing 150 tennis racquets.

Technology Connection

47. Use the VALUE feature in the CALC menu to find the function values in Exercises 38(a), 39(a), 41(b), and 42(a).

Collaborative Discussion and Writing

48. Explain as you would to a fellow student how the numerical value of slope can be used to describe the slant and the steepness of a line.

49. Discuss why the graph of a vertical line $x = a$ cannot represent a function.

Skill Maintenance

If $f(x) = x^2 - 3x$, find each of the following.

50. $f(5)$

51. $f(-5)$

52. $f(-a)$

53. $f(a + h)$

Synthesis

54. *Grade of Treadmills.* A treadmill is 5 ft long and is set at an 8% grade. How high is the end of the treadmill?

Find the slope of the line containing the given points.

55. $(-c, -d)$ and $(9c, -2d)$

56. $(r, s + t)$ and (r, s)

57. $(z + q, z)$ and $(z - q, z)$

58. $(-a - b, p + q)$ and $(a + b, p - q)$

59. (a, a^2) and $(a + h, (a + h)^2)$

60. $(a, 3a + 1)$ and $(a + h, 3(a + h) + 1)$

Suppose that f is a linear function. Determine whether each of the following is true or false.

61. $f(cd) = f(c)f(d)$

62. $f(c + d) = f(c) + f(d)$

63. $f(c - d) = f(c) - f(d)$

64. $f(kx) = kf(x)$

Let $f(x) = mx + b$. Find a formula for $f(x)$ given each of the following.

65. $f(x + 2) = f(x) + 2$

66. $f(3x) = 3f(x)$

1.4

Equations of Lines and Modeling

- *Find the slope and the y-intercept of a line given the equation $y = mx + b$, or $f(x) = mx + b$.*
- *Graph a linear equation using the slope and the y-intercept.*
- *Determine equations of lines.*
- *Given the equations of two lines, determine whether their graphs are parallel or whether they are perpendicular.*
- *Model a set of data with a linear function.*

Slope–Intercept Equations of Lines

Let's explore the effect of the constant b in linear equations of the type $f(x) = mx + b$. Compare the graphs of the equations

$$y = 3x$$

and

$$y = 3x - 2.$$

Note that the graph of $y = 3x - 2$ is a shift down of the graph of $y = 3x$, and that $y = 3x - 2$ has y-intercept $(0, -2)$. That is, the graph is parallel to $y = 3x$ and it crosses the y-axis at $(0, -2)$. The point $(0, -2)$ is the **y-intercept** of the graph.

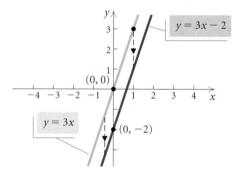

---

y-INTERCEPT

REVIEW SECTION 1.1.

Technology Connection

We can explore the effect of the constant b in linear equations of the type $f(x) = mx + b$ with a graphing calculator. Begin with the graph of $y = x$. Now graph the lines $y = x + 3$ and $y = x - 4$ in the same viewing window. Try entering these equations as $y = x + \{0, 3, -4\}$ and compare the graphs. How do the last two lines differ from $y = x$? What do you think the line $y = x - 6$ will look like?

   Try graphing $y = -0.5x$, $y = -0.5x - 4$, and $y = -0.5x + 3$ in the same viewing window. Describe what happens to the graph of $y = -0.5x$ when a number b is added.

The Slope–Intercept Equation

The linear function f given by

$$f(x) = mx + b$$

has a graph that is a straight line parallel to $y = mx$. The constant m is called the slope, and the y-intercept is $(0, b)$.

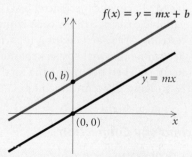

We can read the slope m and the y-intercept $(0, b)$ directly from the equation of a line written in the form $y = mx + b$.

EXAMPLE 1 Find the slope and the y-intercept of the line with equation $y = -0.25x - 3.8$.

Solution

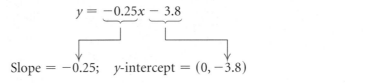

Slope $= -0.25$; y-intercept $= (0, -3.8)$ ▬

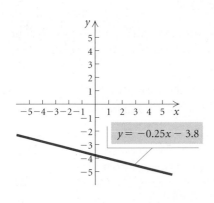

Any equation whose graph is a straight line is a **linear equation.** To find the slope and the y-intercept of the graph of a linear equation, we can solve for y, and then read the information from the equation.

EXAMPLE 2 Find the slope and the y-intercept of the line with equation $3x - 6y - 7 = 0$.

Solution   We solve for y:

$$3x - 6y - 7 = 0$$
$$-6y = -3x + 7 \qquad \text{Adding } -3x \text{ and 7 on both sides}$$
$$-\tfrac{1}{6}(-6y) = -\tfrac{1}{6}(-3x + 7) \qquad \text{Multiplying by } -\tfrac{1}{6}$$
$$y = \tfrac{1}{2}x - \tfrac{7}{6}.$$

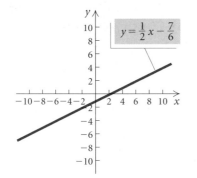

Thus the slope is $\frac{1}{2}$, and the y-intercept is $\left(0, -\frac{7}{6}\right)$.   ▬

EXAMPLE 3 A line has slope $-\frac{7}{9}$ and y-intercept $(0, 16)$. Find an equation of the line.

Solution   We use the slope–intercept equation and substitute $-\frac{7}{9}$ for m and 16 for b:

$$y = mx + b$$
$$y = -\tfrac{7}{9}x + 16.$$ ▬

EXAMPLE 4 A line has slope $-\frac{2}{3}$ and contains the point $(-3, 6)$. Find an equation of the line.

Solution   We use the slope–intercept equation, $y = mx + b$, and substitute $-\frac{2}{3}$ for m: $y = -\frac{2}{3}x + b$. Using the point $(-3, 6)$, we substitute -3 for x and 6 for y in $y = -\frac{2}{3}x + b$. Then we solve for b:

$$y = mx + b$$
$$y = -\tfrac{2}{3}x + b \qquad \text{Substituting } -\tfrac{2}{3} \text{ for } m$$
$$6 = -\tfrac{2}{3}(-3) + b \qquad \text{Substituting } -3 \text{ for } x \text{ and 6 for } y$$
$$6 = 2 + b$$
$$4 = b. \qquad \text{Solving for } b, \text{ the } y\text{-intercept}$$

The equation of the line is $y = -\frac{2}{3}x + 4$.   ▬

We can also graph a linear equation using its slope and y-intercept.

EXAMPLE 5   Graph: $y = -\frac{2}{3}x + 4$.

Solution   This equation is in slope–intercept form, $y = mx + b$. The y-intercept is $(0, 4)$. Thus we plot $(0, 4)$. We can think of the slope $\left(m = -\frac{2}{3}\right)$ as $\frac{-2}{3}$.

$$m = \frac{\text{rise}}{\text{run}} = \frac{\text{change in } y}{\text{change in } x} = \frac{-2}{3} \begin{array}{l} \leftarrow \textbf{Move 2 units down.} \\ \leftarrow \textbf{Move 3 units to the right.} \end{array}$$

Starting at the y-intercept and using the slope, we find another point by moving 2 units down and 3 units to the right. We get a new point $(3, 2)$. In a similar manner, we can move from $(3, 2)$ to find another point $(6, 0)$.

We could also think of the slope $\left(m = -\frac{2}{3}\right)$ as $\frac{2}{-3}$. Then we can start at $(0, 4)$ and move 2 units up and 3 units to the left. We get to another point on the graph, $(-3, 6)$. We now plot the points and draw the line. Note that we need only the y-intercept and one other point in order to graph the line, but it's a good idea to find a third point as a check that the second point is correct.

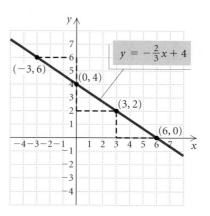

Point–Slope Equations of Lines

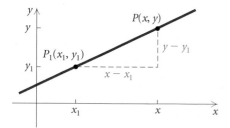

Another formula that can be used to determine an equation of a line is the *point–slope equation*. Suppose that we have a nonvertical line and that the coordinates of point P_1 on the line are (x_1, y_1). We can think of P_1 as fixed and imagine another point P on the line with coordinates (x, y). Thus the slope is given by

$$\frac{y - y_1}{x - x_1} = m.$$

Multiplying both sides by $x - x_1$, we get the *point–slope equation* of the line:

$$(x - x_1) \cdot \frac{y - y_1}{x - x_1} = m \cdot (x - x_1)$$

$$y - y_1 = m(x - x_1).$$

> ### Point–Slope Equation
>
> The **point–slope equation** of the line with slope m passing through (x_1, y_1) is
>
> $$y - y_1 = m(x - x_1).$$

If we know the slope of a line and the coordinates of one point on the line, we can find an equation of the line using either the point–slope equation,

$$y - y_1 = m(x - x_1),$$

or the slope–intercept equation,

$$y = mx + b.$$

EXAMPLE 6   Find an equation of the line containing the points $(2, 3)$ and $(1, -4)$.

Solution   We first determine the slope:

$$m = \frac{-4 - 3}{1 - 2}$$

$$= \frac{-7}{-1}$$

$$= 7.$$

Using the Point–Slope Equation: We substitute 7 for m and either of the points $(2, 3)$ or $(1, -4)$ for (x_1, y_1) in the point–slope equation. In this case, we use $(2, 3)$.

$$
\begin{array}{ll}
y - y_1 = m(x - x_1) & \textbf{Point–slope equation} \\
y - 3 = 7(x - 2) & \textbf{Substituting} \\
y - 3 = 7x - 14 & \\
y = 7x - 11 &
\end{array}
$$

Using the Slope–Intercept Equation: We substitute 7 for m and either of the points $(2, 3)$ or $(1, -4)$ for (x, y) in the slope–intercept equation and solve for b. Here we use $(1, -4)$.

$$
\begin{array}{ll}
y = mx + b & \textbf{Slope–intercept equation} \\
-4 = 7 \cdot 1 + b & \textbf{Substituting} \\
-4 = 7 + b & \\
-11 = b & \textbf{Solving for } b
\end{array}
$$

Finally, we substitute 7 for m and -11 for b in $y = mx + b$ to get

$$y = 7x - 11.$$

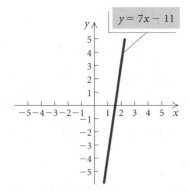

$y = 7x - 11$

Parallel Lines

Can we determine whether the graphs of two linear equations are parallel without graphing them? Let's observe three pairs of equations and their graphs.

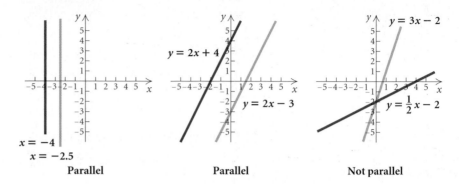

| Parallel | Parallel | Not parallel |

If two different lines, such as $x = -4$ and $x = -2.5$, are vertical, then they are parallel. Thus two equations such as $x = a_1$, $x = a_2$, where $a_1 \neq a_2$, have graphs that are *parallel lines*. Two nonvertical lines, such as $y = 2x + 4$ and $y = 2x - 3$, or, in general, $y = mx + b_1$, $y = mx + b_2$, where the slopes are the *same* and $b_1 \neq b_2$, also have graphs that are *parallel lines*.

> ### *Parallel Lines*
>
> Vertical lines are **parallel.** Nonvertical lines are **parallel** if and only if they have the same slope and different y-intercepts.

Perpendicular Lines

Can we examine a pair of equations to determine whether their graphs are perpendicular without graphing the equations? Let's observe the following pairs of equations and their graphs.

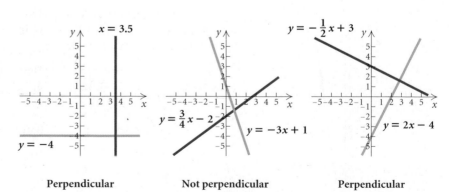

| Perpendicular | Not perpendicular | Perpendicular |

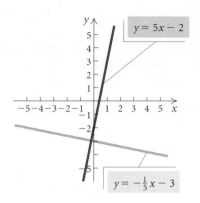

FIGURE 1

Perpendicular Lines

Two lines with slopes m_1 and m_2 are **perpendicular** if and only if the product of their slopes is -1:

$$m_1 m_2 = -1.$$

Lines are also **perpendicular** if one is vertical ($x = a$) and the other is horizontal ($y = b$).

If a line has slope m_1, the slope m_2 of a line perpendicular to it is $-1/m_1$ (the slope of one line is the *opposite of the reciprocal* of the other).

EXAMPLE 7   Determine whether each of the following pairs of lines is parallel, perpendicular, or neither.

a) $y + 2 = 5x$, $5y + x = -15$
b) $2y + 4x = 8$, $5 + 2x = -y$
c) $2x + 1 = y$, $y + 3x = 4$

Solution   We use the slopes of the lines to determine whether the lines are parallel or perpendicular.

a) We solve each equation for y:

$$y = 5x - 2, \qquad y = -\tfrac{1}{5}x - 3.$$

The slopes are 5 and $-\tfrac{1}{5}$. Their product is -1, so the lines are perpendicular (see Fig. 1).

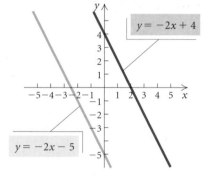

FIGURE 2

b) Solving each equation for y, we get

$$y = -2x + 4, \qquad y = -2x - 5.$$

We see that $m_1 = -2$ and $m_2 = -2$. Since the slopes are the same and the y-intercepts, $(0, 4)$ and $(0, -5)$, are different, the lines are parallel (see Fig. 2).

c) Solving each equation for y, we get

$$y = 2x + 1, \qquad y = -3x + 4.$$

Since $m_1 = 2$ and $m_2 = -3$, we know that $m_1 \neq m_2$ and that $m_1 m_2 \neq -1$. It follows that the lines are neither parallel nor perpendicular (see Fig. 3).

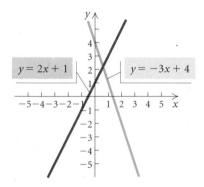

FIGURE 3

EXAMPLE 8   Write equations of the lines **(a)** parallel and **(b)** perpendicular to the graph of the line $4y - x = 20$ and containing the point $(2, -3)$.

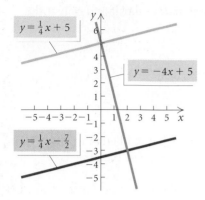

Solution We first solve $4y - x = 20$ for y to get $y = \frac{1}{4}x + 5$. Thus the slope of the given line is $\frac{1}{4}$.

a) The line parallel to the given line will have slope $\frac{1}{4}$. We use either the slope–intercept equation or the point–slope equation for a line with slope $\frac{1}{4}$ and containing the point $(2, -3)$. Here we use the point–slope equation:

$$y - y_1 = m(x - x_1)$$
$$y - (-3) = \tfrac{1}{4}(x - 2)$$
$$y + 3 = \tfrac{1}{4}x - \tfrac{1}{2}$$
$$y = \tfrac{1}{4}x - \tfrac{7}{2}.$$

b) The slope of the perpendicular line is the opposite of the reciprocal of $\frac{1}{4}$, or -4. Again we use the point–slope equation to write an equation for a line with slope -4 and containing the point $(2, -3)$:

$$y - y_1 = m(x - x_1)$$
$$y - (-3) = -4(x - 2)$$
$$y + 3 = -4x + 8$$
$$y = -4x + 5.$$

Summary of Terminology About Lines

TERMINOLOGY	MATHEMATICAL INTERPRETATION
Slope	$m = \dfrac{y_2 - y_1}{x_2 - x_1}$, or $\dfrac{y_1 - y_2}{x_1 - x_2}$
Slope–intercept equation	$y = mx + b$
Point–slope equation	$y - y_1 = m(x - x_1)$
Horizontal line	$y = b$
Vertical line	$x = a$
Parallel lines	$m_1 = m_2,\ b_1 \neq b_2$
Perpendicular lines	$m_1 m_2 = -1$

Mathematical Models

When a real-world problem can be described in mathematical language, we have a **mathematical model.** For example, the natural numbers constitute a mathematical model for situations in which counting is essential. Situations in which algebra can be brought to bear often require the use of functions as models.

Mathematical models are abstracted from real-world situations. The mathematical model gives results that allow one to predict what will happen in that real-world situation. If the predictions are inaccurate or the results of experimentation do not conform to the model, the model needs to be changed or discarded.

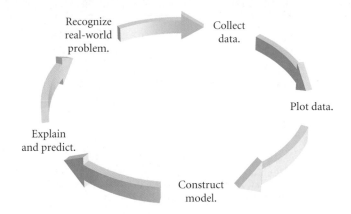

Mathematical modeling can be an ongoing process. For example, finding a mathematical model that will enable an accurate prediction of population growth is not a simple problem. Any population model that one might devise would need to be reshaped as further information is acquired.

Curve Fitting

We will develop and use many kinds of mathematical models in this text. In this chapter, we have used *linear* functions as models. Other types of functions, such as quadratic, cubic, or exponential functions, can also model data. These functions are *nonlinear*. Modeling with quadratic and cubic functions will be discussed in Chapter 3. Modeling with exponential functions will be discussed in Chapter 4.

Quadratic function:
$y = ax^2 + bx + c, a > 0$

Cubic function:
$y = ax^3 + bx^2 + cx + d, a > 0$

Exponential function:
$y = ab^x, a, b > 0$

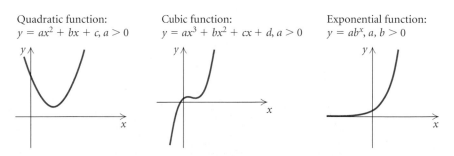

In general, we try to find a function that fits, as well as possible, observations (data), theoretical reasoning, and common sense. We call this **curve fitting;** it is one aspect of mathematical modeling.

Let's look at some data and related graphs or **scatterplots** and determine whether a linear function seems to fit the set of data.

Number of U.S. Travelers to Foreign Countries

YEAR, x	NUMBER OF U.S. TRAVELERS TO FOREIGN COUNTRIES, y (IN MILLIONS)	SCATTERPLOT	
1994, 0	4.65		It appears that the data points can be represented or modeled by a linear function. The graph is **linear.**
1995, 1	5.08		
1996, 2	5.23		
1997, 3	5.29		
1998, 4	5.63		
1999, 5	5.75		
2000, 6	6.08		

Sources: U.S. Department of Commerce; International Trade Administration; Tourism Industries

Number of Cellular Phone Subscribers

YEAR, x	NUMBER OF CELLULAR PHONE SUBSCRIBERS, y (IN MILLIONS)	SCATTERPLOT	
1990, 0	5.3		It appears that the data points cannot be modeled by a linear function. The graph is **nonlinear.**
1991, 1	7.6		
1992, 2	11.0		
1993, 3	16.0		
1994, 4	24.1		
1995, 5	33.8		
1996, 6	44.0		
1997, 7	55.3		
1998, 8	71.2		
1999, 9	86.0		
2000, 10	109.5		
2001, 11	128.4		

Source: Cellular Telecommunications Industry Association

Looking at the scatterplots, we see that the data on U.S. foreign travelers seem to be rising in a manner to suggest that a *linear function* might fit, although a "perfect" straight line cannot be drawn through the data points. A linear function does not seem to fit the data on cellular phones.

EXAMPLE 9 *Foreign Travel.*   Model the data given in the table on foreign travel on the preceding page with two different linear functions. Then with each function, predict the number of U.S. travelers to foreign countries in 2005. Of the two models, which appears to be the better fit?

Solution   We can choose any two of the data points to determine an equation. Let's use $(1, 5.08)$ and $(5, 5.75)$ for our first model.

We first determine the slope of the line:

$$m = \frac{5.75 - 5.08}{5 - 1} = 0.1675.$$

Then we substitute 0.1675 for m and either of the points $(1, 5.08)$ and $(5, 5.75)$ for (x_1, y_1) in the point–slope equation. In this case, we use $(1, 5.08)$. We get

$$y - 5.08 = 0.1675(x - 1),$$

which simplifies to

$$y = 0.1675x + 4.9125, \qquad \textbf{Model I}$$

where x is the number of years after 1994 and y is in millions.

For the second model, let's use $(0, 4.65)$ and $(6, 6.08)$. Again we first determine the slope:

$$m = \frac{6.08 - 4.65}{6 - 0} \approx 0.2383.$$

We can use the slope–intercept equation. The y-intercept, $(0, 4.65)$, is one of the data points. Substituting 0.2383 for m and 4.65 for b, we get

$$y = 0.2383x + 4.65, \qquad \textbf{Model II}$$

where x is the number of years after 1994 and y is in millions.

Now we can predict the number of U.S. foreign travelers in 2005 by substituting 11 for x $(2005 - 1994 = 11)$ in each model:

Model I: $y = 0.1675x + 4.9125$
$y = 0.1675(11) + 4.9125$
$= 6.755;$

Model II: $y = 0.2383x + 4.65$
$y = 0.2383(11) + 4.65$
$= 7.2713.$

Thus, using model I, we predict that there will be about 6.76 million U.S. foreign travelers in 2005, and using model II, we predict about 7.27 million.

Since it appears from the graphs above that model II fits the data more closely, we would choose model II over model I.

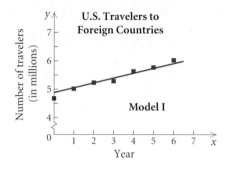

U.S. Travelers to Foreign Countries

Number of travelers (in millions)

Model I

Year

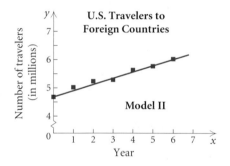

U.S. Travelers to Foreign Countries

Number of travelers (in millions)

Model II

Year

Technology Connection

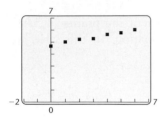

L1	L2	L3
0	4.65	-----
1	5.08	
2	5.23	
3	5.29	
4	5.63	
5	5.75	
6	6.08	

L2(7) = 6.08

FIGURE 1

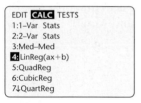

FIGURE 2

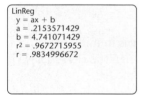

EDIT **CALC** TESTS
1:1–Var Stats
2:2–Var Stats
3:Med–Med
4:LinReg(ax+b)
5:QuadReg
6:CubicReg
7↓QuartReg

FIGURE 3

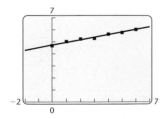

LinReg
y = ax + b
a = .2153571429
b = 4.741071429
r² = .9672715955
r = .9834996672

FIGURE 4

FIGURE 5

The Regression Line

We now consider **linear regression,** a procedure that can be used to model a set of data using a linear function. Although discussion leading to a complete understanding of this method belongs in a statistics course, we present the procedure here because we can carry it out easily using technology. The graphing calculator gives us the powerful capability to find linear models and to make predictions using them.

Consider the data presented before Example 9 on the number of U.S. travelers to foreign countries. We can fit a regression line of the form $y = mx + b$ to the data using the LINEAR REGRESSION feature on a graphing calculator.

First, we enter the data in lists on the calculator. We enter the values of the independent variable x in list L1 and the corresponding values of the dependent variable y in L2 (see Fig. 1). The graphing calculator can then create a scatterplot of the data, as shown in Fig. 2.

When we select the LINEAR REGRESSION feature from the STAT CALC menu, we find the linear equation that best models the data. It is

$$y = 0.2153571429 + 4.741071429 \qquad \textbf{Regression line}$$

(see Figs. 3 and 4). We can then graph the regression line on the same graph as the scatterplot, as shown in Fig. 5.

To predict the number of U.S. foreign travelers in 2005, we substitute 11 for x in the regression equation.

Using this model, we see that the number of U.S. travelers to foreign countries in 2005 is predicted to be about 7.11 million. Note that 7.11 is closer to the value found using model II than the one found using model I in Example 9.

Y1(11)

7.11

The Correlation Coefficient

On some graphing calculators with the DIAGNOSTIC feature turned on, a constant r between -1 and 1, called the **coefficient of linear correlation,** appears with the equation of the regression line. Though we cannot develop a formula for calculating r in this text, keep in mind that it is used to describe the strength of the linear relationship between x and y. The closer $|r|$ is to 1, the better the correlation. A positive value of r also indicates that the regression line has a positive slope, and a negative value of r indicates that the regression line has a negative slope. For the foreign travel data just discussed,

$$r = 0.9834996672,$$

which indicates a very good linear correlation.

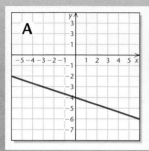

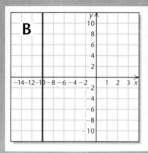

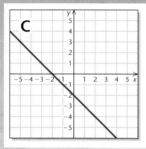

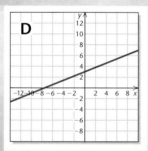

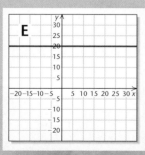

Visualizing the Graph

Match the equation with its graph.

1. $y = 20$

2. $5y = 2x + 15$

3. $y = -\dfrac{1}{3}x - 4$

4. $x = \dfrac{5}{3}$

5. $y = -x - 2$

6. $y = 2x$

7. $y = -3$

8. $3y = -4x$

9. $x = -10$

10. $y = x + \dfrac{7}{2}$

Answers on page A-7

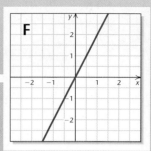

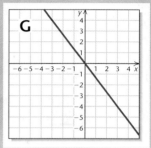

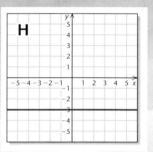

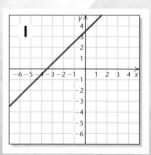

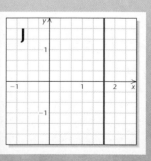

1.4

Exercise Set

Find the slope and the y-intercept of the equation.

1. $y = \frac{3}{5}x - 7$ **2.** $f(x) = -2x + 3$

3. $x = -\frac{2}{5}$ **4.** $y = \frac{4}{7}$

5. $f(x) = 5 - \frac{1}{2}x$ **6.** $y = 2 + \frac{3}{7}x$

7. $3x + 2y = 10$ **8.** $2x - 3y = 12$

9. $y = -6$ **10.** $x = 10$

11. $5y - 4x = 8$ **12.** $5x - 2y + 9 = 0$

Find the slope and the y-intercept of the graph of the linear equation. Then write the equation of the line in slope–intercept form.

13.
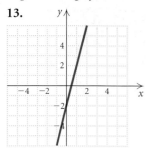

14.
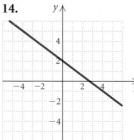

15.
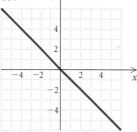

16.
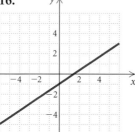

17.
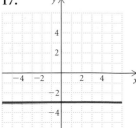

18.
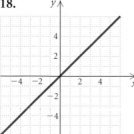

Graph the equation using the slope and the y-intercept.

19. $y = -\frac{1}{2}x - 3$ **20.** $y = \frac{3}{2}x + 1$

21. $f(x) = 3x - 1$ **22.** $f(x) = -2x + 5$

23. $3x - 4y = 20$ **24.** $2x + 3y = 15$

25. $x + 3y = 18$ **26.** $5y - 2x = -20$

Write a slope–intercept equation for a line with the given characteristics.

27. $m = \frac{2}{9}$, y-intercept $(0, 4)$

28. $m = -\frac{3}{8}$, y-intercept $(0, 5)$

29. $m = -4$, y-intercept $(0, -7)$

30. $m = \frac{2}{7}$, y-intercept $(0, -6)$

31. $m = -4.2$, y-intercept $\left(0, \frac{3}{4}\right)$

32. $m = -4$, y-intercept $\left(0, -\frac{3}{2}\right)$

33. $m = \frac{2}{9}$, passes through $(3, 7)$

34. $m = -\frac{3}{8}$, passes through $(5, 6)$

35. $m = 3$, passes through $(1, -2)$

36. $m = -2$, passes through $(-5, 1)$

37. $m = -\frac{3}{5}$, passes through $(-4, -1)$

38. $m = \frac{2}{3}$, passes through $(-4, -5)$

39. Passes through $(-1, 5)$ and $(2, -4)$

40. Passes through $(2, -1)$ and $(7, -11)$

41. Passes through $(7, 0)$ and $(-1, 4)$

42. Passes through $(-3, 7)$ and $(-1, -5)$

43. Passes through $(0, -6)$ and $(3, -4)$

44. Passes through $(-5, 0)$ and $\left(0, \frac{4}{5}\right)$

Write equations of the horizontal and the vertical lines that pass through the given point.

45. $(0, -3)$ **46.** $\left(-\frac{1}{4}, 7\right)$

47. $\left(\frac{2}{11}, -1\right)$ **48.** $(0.03, 0)$

Determine whether the pair of lines is parallel, perpendicular, or neither.

49. $y = \frac{26}{3}x - 11,$
$y = -\frac{3}{26}x - 11$

50. $y = -3x + 1,$
$y = -\frac{1}{3}x + 1$

51. $y = \frac{2}{5}x - 4,$
$y = -\frac{2}{5}x + 4$

52. $y = \frac{3}{2}x - 8,$
$y = 8 + 1.5x$

53. $x + 2y = 5,$
$2x + 4y = 8$

54. $2x - 5y = -3,$
$2x + 5y = 4$

55. $y = 4x - 5,$
$4y = 8 - x$

56. $y = 7 - x,$
$y = x + 3$

Write a slope–intercept equation for a line passing through the given point that is parallel to the given line. Then write a second equation for a line passing through the given point that is perpendicular to the given line.

57. $(3, 5),\ y = \frac{2}{7}x + 1$

58. $(-1, 6),\ f(x) = 2x + 9$

59. $(-7, 0),\ y = -0.3x + 4.3$

60. $(-4, -5),\ 2x + y = -4$

61. $(3, -2),\ 3x + 4y = 5$

62. $(8, -2),\ y = 4.2(x - 3) + 1$

63. $(3, -3),\ x = -1$

64. $(4, -5),\ y = -1$

65. *Hospital Stay.* The average length of time for a patient's stay in the hospital continues to decline. The graph below illustrates data for certain years.

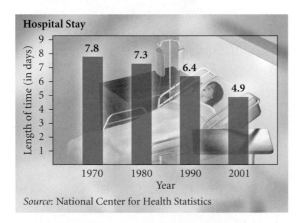

Hospital Stay

Source: National Center for Health Statistics

a) Model the data with two linear functions. Let the independent variable represent the number of years after 1970—that is, the data points are

(0, 7.8), (10, 7.3), and so on. Answers will vary depending on the data points used.

b) With each function found in part (a), predict the average length of a hospital stay in 2005.

c) Which of the two models appears to fit the data more closely (that is, provides the more realistic prediction)?

66. *Motorcycle Sales.* The following graph illustrates the annual total sales of motorcycles for certain years.

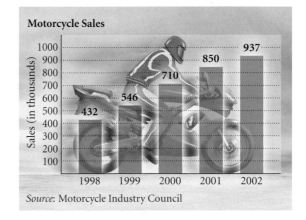

Motorcycle Sales

Source: Motorcycle Industry Council

a) Model the data with two linear functions. Let the independent variable represent the years after 1998—that is, the data points are (0, 432), (1, 546), and so on. Answers will vary depending on the data points used.

b) With each function found in part (a), predict the total motorcycle sales in 2006.

c) Which of the two models appears to fit the data more closely (that is, provides the more realistic prediction)?

67. *The Cost of Tuition at State Universities.* Model the data in the chart below with a linear function and predict the costs of college tuition in 2004–2005, in 2006–2007, and in 2010–2011. Answers will vary depending on the data points used.

COLLEGE YEAR, x	ESTIMATED TUITION, y
1997–1998, 0	$ 9,838
1998–1999, 1	10,424
1999–2000, 2	11,054
2000–2001, 3	11,717
2001–2002, 4	12,420

Sources: College Board; Senate Labor Committee

68. *Foreign Travelers to the United States.* Model the data in the chart below with a linear function and predict the number of foreign travelers visiting the United States in 2005 and in 2008. Answers will vary depending on the data points used.

YEAR, x	FOREIGN TRAVELERS TO U.S., y (IN MILLIONS)
1994, 0	4.48
1995, 1	4.33
1996, 2	4.65
1997, 3	4.78
1998, 4	4.64
1999, 5	4.85
2000, 6	5.09

Sources: U.S. Department of Commerce; International Trade Administration; Tourism Industries

69. *Egg Production.* Model the data in the chart below with a linear function and predict egg production in 2004, in 2008, and in 2012. Answers will vary depending on the data points used.

YEAR, x	EGG PRODUCTION, y (IN BILLIONS)
1994, 0	73.9
1995, 1	74.8
1996, 2	76.4
1997, 3	77.5
1998, 4	79.8
1999, 5	82.7
2000, 6	84.4
2001, 7	85.7

Sources: U.S. Department of Agriculture; National Agricultural Statistics Service

70. *Graying Work Force.* The percentage of Americans over 65 who are continuing to work is increasing. Model the data in the chart below with a linear function and predict the percentage of Americans over 65 who will be working in 2010 and in 2014. Answers will vary depending on the data points used.

YEAR, x	PERCENTAGE OVER 65 WHO ARE WORKING, y
1986, 0	10.8%
1991, 5	11.4
1996, 10	12
2001, 15	13

Source: U.S. Bureau of Labor Statistics

Technology Connection

71. *Maximum Heart Rate.* A person who is exercising should not exceed his or her maximum heart rate, which is determined on the basis of that person's sex, age, and resting heart rate. The following table relates resting heart rate and maximum heart rate for a 20-year-old man.

Resting Heart Rate, H (in beats per minute)	Maximum Heart Rate, M (in beats per minute)
50	166
60	168
70	170
80	172

Source: American Heart Association

a) Use a graphing calculator to model the data with a linear function.

b) Estimate the maximum heart rate if the resting heart rate is 40, 65, 76, and 84.

c) What is the correlation coefficient? How confident are you about using the regression line as an estimate?

72. *Study Time versus Grades.* A math instructor asked her students to keep track of how much time each spent studying a chapter on functions in her algebra–trigonometry course. She collected the information together with test scores from that chapter's test. The data are listed in the following table.

Study Time, x (in hours)	Test Grade, y (in percent)
23	81%
15	85
17	80
9	75
21	86
13	80
16	85
11	93

a) Use a graphing calculator to model the data with a linear function.

b) Predict a student's score if he or she studies 24 hr, 6 hr, and 18 hr.

c) What is the correlation coefficient? How confident are you about using the regression line as a predictor?

73. *Easter Candy Sales.* From 1999 to 2002, sales of Easter candy remained relatively constant, as shown in the following figure.

Easter Candy Sales (in billions)

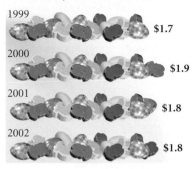

Source: National Confectioners Association

a) Use a graphing calculator to model the data with a linear function.

b) Estimate the sales of Easter candy in 2005 and in 2010.

c) Find the correlation coefficient for the regression line and determine whether the line fits the data closely.

74. a) Use a graphing calculator to fit a regression line to the data in Exercise 66.

b) Predict motorcycle sales in 2006 and compare the value with the result found in Exercise 66.

c) Find the correlation coefficient for the regression line and determine whether the line fits the data closely.

75. a) Use a graphing calculator to fit a regression line to the data in Exercise 65.

b) Predict the average length of a hospital stay in 2005 and compare the result with the estimate found with the model in Exercise 65.

c) Find the correlation coefficient for the regression line and determine whether the line fits the data closely.

Collaborative Discussion and Writing

76. A company offers its new employees a starting salary with a guaranteed 5% increase each year. Can a linear function be used to express the yearly salary as a function of the number of years an employee has worked? Why or why not?

77. If one line has a slope of $-\frac{3}{4}$ and another has a slope of $\frac{2}{5}$, which line is steeper? Why?

Skill Maintenance

Find the slope of the line containing the given points.

78. $(5, 7)$ and $(5, -7)$ **79.** $(2, -8)$ and $(-5, -1)$

Find an equation for a circle satisfying the given conditions.

80. Center $(0, 3)$, diameter of length 5

81. Center $(-7, -1)$, radius of length $\frac{9}{5}$

Synthesis

82. *Road Grade.* Using the following figure, find the road grade and an equation giving the height y as a function of the horizontal distance x.

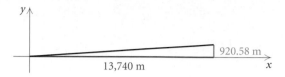

83. Find k so that the line containing the points $(-3, k)$ and $(4, 8)$ is parallel to the line containing the points $(5, 3)$ and $(1, -6)$.

84. Find an equation of the line passing through the point $(4, 5)$ and perpendicular to the line passing through the points $(-1, 3)$ and $(2, 9)$.

1.5

More on Functions

- *Graph functions, looking for intervals on which the function is increasing, decreasing, or constant, and estimate relative maxima and minima.*
- *Given an application, find a function that models the application; find the domain of the function and function values, and then graph the function.*
- *Graph functions defined piecewise.*

Because functions occur in so many real-world situations, it is important to be able to analyze them carefully.

Increasing, Decreasing, and Constant Functions

On a given interval, if the graph of a function rises from left to right, it is said to be **increasing** on that interval. If the graph drops from left to right, it is said to be **decreasing.** If the function values stay the same from left to right, the function is said to be **constant.**

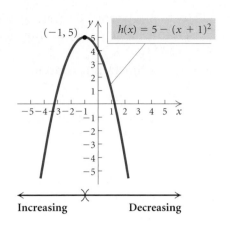

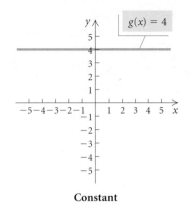

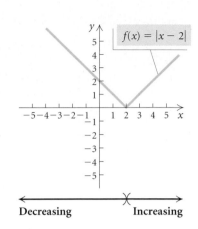

We are led to the following definitions.

Increasing, Decreasing, and Constant Functions

A function f is said to be **increasing** on an *open* interval I, if for all a and b in that interval, $a < b$ implies $f(a) < f(b)$ (see Fig. 1).

A function f is said to be **decreasing** on an *open* interval I, if for all a and b in that interval, $a < b$ implies $f(a) > f(b)$ (see Fig. 2).

A function f is said to be **constant** on an *open* interval I, if for all a and b in that interval, $f(a) = f(b)$ (see Fig. 3).

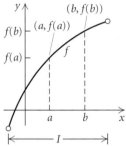

For $a < b$ in I, $f(a) < f(b)$; f is **increasing** on I.

FIGURE 1

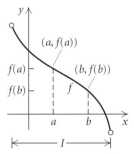

For $a < b$ in I, $f(a) > f(b)$; f is **decreasing** on I.

FIGURE 2

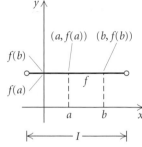

For all a and b in I, $f(a) = f(b)$; f is **constant** on I.

FIGURE 3

EXAMPLE 1 Determine the intervals on which the function in the figure at left is **(a)** increasing; **(b)** decreasing; **(c)** constant.

Solution When expressing interval(s) on which a function is increasing, decreasing, or constant, we consider only values in the *domain* of the function.

a) For x-values (that is, values in the domain) from $x = 3$ to $x = 5$, the y-values (that is, values in the range) increase from -2 to 2. Thus the function is increasing on the interval $(3, 5)$.

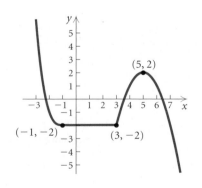

b) For all x-values from negative infinity to -1, y-values decrease; y-values also decrease for x-values from 5 to positive infinity. Thus the function is decreasing on the intervals $(-\infty, -1)$ and $(5, \infty)$.

c) For x-values from -1 to 3, y is -2. The function is constant on the interval $(-1, 3)$.

In calculus, the slope of a line tangent to the graph of a function at a particular point is used to determine whether the function is increasing, decreasing, or constant at that point. If the slope is positive, the function is increasing; if the slope is negative, the function is decreasing. Since slope cannot be both positive and negative at the same point, a function cannot be both increasing and decreasing at a specific point. For this reason, increasing, decreasing, and constant intervals are expressed in *open interval* notation. In Example 1, if $[3, 5]$ had been used for the increasing interval and $[5, \infty)$ for a decreasing interval, the function would be both increasing and decreasing at $x = 5$. This is not possible.

Relative Maximum and Minimum Values

Consider the graph shown below. Note the "peaks" and "valleys" at the x-values c_1, c_2, and c_3. The function value $f(c_2)$ is called a **relative maximum** (plural, **maxima**). Each of the function values $f(c_1)$ and $f(c_3)$ is called a **relative minimum** (plural, **minima**).

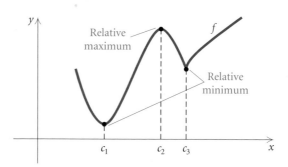

Relative Maxima and Minima

Suppose that f is a function for which $f(c)$ exists for some c in the domain of f. Then:

$f(c)$ is a **relative maximum** if there exists an *open* interval I containing c such that $f(c) > f(x)$, for all x in I where $x \neq c$; and

$f(c)$ is a **relative minimum** if there exists an *open* interval I containing c such that $f(c) < f(x)$, for all x in I where $x \neq c$.

Simply stated, $f(c)$ is a *relative* maximum if $f(c)$ is the highest point in some *open* interval, and $f(c)$ is a *relative* minimum if $f(c)$ is the lowest point in some *open* interval.

If you take a calculus course, you will learn a method for determining exact values of relative maxima and minima. In Section 2.4, we will find exact maximum and minimum values of quadratic functions algebraically. MAXIMUM and MINIMUM features on a graphing calculator can be used to approximate relative maxima and minima. In this section, we will read the values shown on a graph.

EXAMPLE 2 Using the graph shown below, determine any relative maxima or minima of the function $f(x) = 0.1x^3 - 0.6x^2 - 0.1x + 2$ and the intervals on which the function is increasing or decreasing.

Technology Connection

We can approximate relative maximum and minimum values with the MAXIMUM and MINIMUM features from the CALC menu on a graphing calculator. See the *Graphing Calculator Manual* for more information on this procedure.

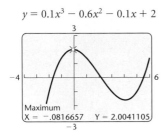

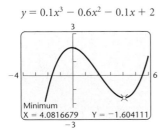

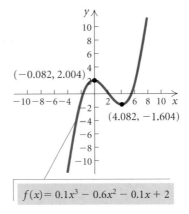

Solution We see that the *relative maximum* value of the function is 2.004. It occurs when $x = -0.082$. We also see the *relative minimum*: -1.604 at $x = 4.082$.

We note that the graph starts rising, or increasing, from the left and stops increasing at the relative maximum. From this point, the graph decreases to the relative minimum and then begins to rise again. The function is *increasing* on the intervals

$$(-\infty, -0.082) \quad \text{and} \quad (4.082, \infty)$$

and *decreasing* on the interval

$$(-0.082, 4.082).$$

Applications of Functions

Many real-world situations can be modeled by functions.

EXAMPLE 3 *Car Distance.* Emma and Brandt drive away from a campground at right angles to each other. Emma's speed is 65 mph and Brandt's is 55 mph.

a) Express the distance between the cars as a function of time.

b) Find the domain of the function.

Solution

a) Suppose 1 hr goes by. At that time, Emma has traveled 65 mi and Brandt has traveled 55 mi. We can use the Pythagorean theorem then to find the distance between them. This distance would be the length of the hypotenuse of a triangle with legs measuring 65 mi and 55 mi. After 2 hr, the triangle's legs would measure 130 mi and 110 mi. Observing that the distances will always be changing, we make a drawing and let $t =$ the time, in hours that Emma and Brandt have been driving since leaving the campground.

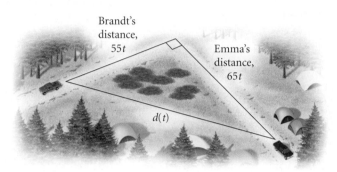

Brandt's distance, $55t$

Emma's distance, $65t$

$d(t)$

After t hours, Emma has traveled $65t$ miles and Brandt $55t$ miles. We can use the Pythagorean theorem:

$$[d(t)]^2 = (65t)^2 + (55t)^2.$$

Because distance must be nonnegative, we need consider only the positive square root when solving for $d(t)$:

$$d(t) = \sqrt{(65t)^2 + (55t)^2}$$
$$= \sqrt{4225t^2 + 3025t^2}$$
$$= \sqrt{7250t^2}$$
$$= \sqrt{7250}\,\sqrt{t^2}$$
$$\approx 85.15|t| \qquad \textbf{Approximating the root to two decimal places}$$
$$\approx 85.15t. \qquad \textbf{Since } t \geq 0, |t| = t.$$

Thus, $d(t) = 85.15t, \; t \geq 0$.

b) Since the time traveled, t, must be nonnegative, the domain is the set of nonnegative real numbers $[0, \infty)$. ▬

EXAMPLE 4   *Storage Area.*   The Sound Shop has 20 ft of dividers with which to set off a rectangular area for the storage of overstock. If a corner of the store is used for the storage area, the partition need only form two sides of a rectangle.

a) Express the floor area of the storage space as a function of the length of the partition.

b) Find the domain of the function.

c) Using the graph shown on the following page, determine the dimensions that maximize the area of the floor.

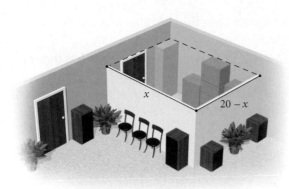

Solution

a) Note that the dividers will form two sides of a rectangle. If, for example, 14 ft of dividers are used for the length of the rectangle, that would leave $20 - 14$, or 6 ft of dividers for the width. Thus if $x =$ the length, in feet, of the rectangle, then $20 - x =$ the width. We represent this information in a sketch, as shown at left.

The area, $A(x)$, is given by

$$A(x) = x(20 - x) \qquad \text{Area} = \text{length} \cdot \text{width}$$
$$= 20x - x^2.$$

The function $A(x) = 20x - x^2$ can be used to express the rectangle's area as a function of the length.

b) Because the rectangle's length must be positive and only 20 ft of dividers is available, we restrict the domain of A to $\{x \mid 0 < x < 20\}$, that is, the interval $(0, 20)$.

c) On the graph of the function, the maximum value of the area function on the interval $(0, 20)$ appears to be 100 when $x = 10$. Thus the dimensions that maximize the area are

$$\text{Length} = x = 10 \text{ ft} \quad \text{and}$$
$$\text{Width} = 20 - x = 20 - 10 = 10 \text{ ft.}$$

Functions Defined Piecewise

Sometimes functions are defined **piecewise** using different output formulas for different parts of the domain.

EXAMPLE 5 Graph the function defined as

$$f(x) = \begin{cases} 4, & \text{for } x \leq 0, \\ 4 - x^2, & \text{for } 0 < x \leq 2, \\ 2x - 6, & \text{for } x > 2. \end{cases}$$

Solution We create the graph in three parts, as shown and described below. We list some ordered pairs in tables shown at left.

TABLE 1

x $(x \leq 0)$	$f(x) = 4$
-5	4
-2	4
0	4

TABLE 2

x $(0 < x \leq 2)$	$f(x) = 4 - x^2$
$\frac{1}{2}$	$3\frac{3}{4}$
1	3
2	0

TABLE 3

x $(x > 2)$	$f(x) = 2x - 6$
$2\frac{1}{2}$	-1
3	0
5	4

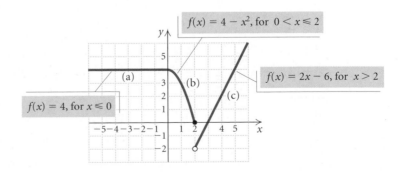

a) We graph $f(x) = 4$ *only* for inputs x less than or equal to 0, that is, on the interval $(-\infty, 0]$. (See Table 1.)

b) We graph $f(x) = 4 - x^2$ *only* for inputs x greater than 0 and less than or equal to 2, that is, on the interval $(0, 2]$. (See Table 2.)

c) We graph $f(x) = 2x - 6$ *only* for inputs x greater than 2, that is, on the interval $(2, \infty)$. (See Table 3.)

EXAMPLE 6 Graph the function defined as

$$f(x) = \begin{cases} \dfrac{x^2 - 4}{x + 2}, & \text{for } x \neq -2, \\ 3, & \text{for } x = -2. \end{cases}$$

Solution When $x \neq -2$, the denominator of $(x^2 - 4)/(x + 2)$ is nonzero, so we can simplify:

$$\frac{x^2 - 4}{x + 2} = \frac{(x + 2)(x - 2)}{x + 2} = x - 2.$$

Thus

$$f(x) = x - 2, \quad \text{for } x \neq -2.$$

The graph of this part of the function consists of a line with a "hole" at the point $(-2, -4)$, indicated by the open dot. The hole is necessary since the function is not defined for -2. At $x = -2$, we have $f(-2) = 3$ (by the definition of the function), so the point $(-2, 3)$ is plotted above the open dot.

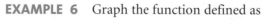

A piecewise function with importance in calculus and computer programming is the **greatest integer function,** f, denoted as $f(x) = [\![x]\!]$, or $\text{int}(x)$.

Greatest Integer Function

$f(x) = [\![x]\!] = $ the greatest integer *less than or equal to x.*

The greatest integer function pairs each input with the greatest integer *less than or equal to* that input. Thus, x-values 1, $1\frac{1}{2}$, and 1.8 are all paired with the y-value 1. Other pairings are shown below.

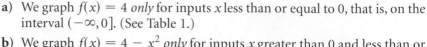

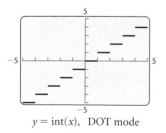

EXAMPLE 7   Graph $f(x) = [\![x]\!]$ and determine its domain and range.

Solution   The greatest integer function can also be defined as a piecewise function with an infinite number of statements.

$$
f(x) = [\![x]\!] = \begin{cases}
\vdots \\
-3, & \text{for } -3 \le x < -2, \\
-2, & \text{for } -2 \le x < -1, \\
-1, & \text{for } -1 \le x < 0, \\
0, & \text{for } 0 \le x < 1, \\
1, & \text{for } 1 \le x < 2, \\
2, & \text{for } 2 \le x < 3, \\
3, & \text{for } 3 \le x < 4, \\
\vdots
\end{cases}
$$

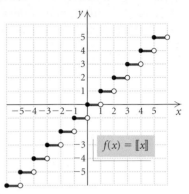

We see that the domain of $f(x)$ is the set of all real numbers, $(-\infty, \infty)$, and the range is the set of all integers, $\{\ldots, -3, -2, -1, 0, 1, 2, 3, \ldots\}$.

1.5

Exercise Set

|-12

Determine intervals on which the function is **(a)** *increasing,* **(b)** *decreasing, and* **(c)** *constant.*

1.
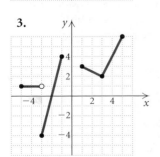

2.
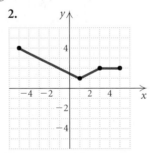

3.

4.
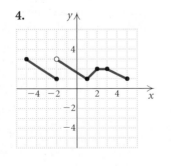

5.
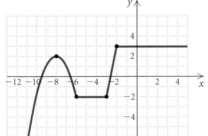

6.
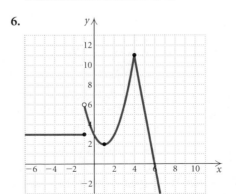

7.–12. Determine the domain and the range of each of the functions graphed in Exercises 1–6.

Using the graph, determine any relative maxima or minima of the function and the intervals on which the function is increasing or decreasing.

13. $f(x) = -x^2 + 5x - 3$

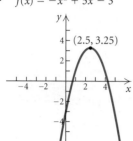

14. $f(x) = x^2 - 2x + 3$

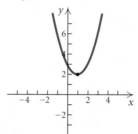

15.

$f(x) = \frac{1}{4}x^3 - \frac{1}{2}x^2 - x + 2$

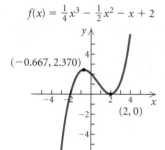

16.

$f(x) = -0.09x^3 + 0.5x^2 - 0.1x + 1$

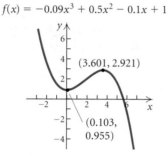

Graph the function. Estimate the intervals on which the function is increasing or decreasing and any relative maxima or minima.

17. $f(x) = x^2$

18. $f(x) = 4 - x^2$

19. $f(x) = 5 - |x|$

20. $f(x) = |x + 3| - 5$

21. $f(x) = x^2 - 6x + 10$

22. $f(x) = -x^2 - 8x - 9$

23. *Garden Area.* Creative Landscaping has 60 yd of fencing with which to enclose a rectangular flower garden. If the garden is x yards long, express the garden's area as a function of the length.

24. *Triangular Flag.* A scout troop is designing a triangular flag so that the length of its base, in inches, is 7 less than twice the height, h. Express the area of the flag as a function of the height.

25. *Rising Balloon.* A hot-air balloon rises straight up from the ground at a rate of 120 ft/min. The balloon is tracked from a rangefinder on the ground at point P, which is 400 ft from the release point Q of the balloon. Let d = the distance from the balloon to the rangefinder at point P and t = the time, in minutes, since the balloon was released. Express d as a function of t.

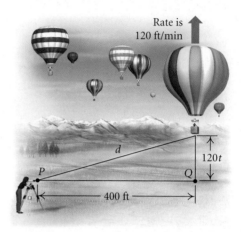

26. *Airplane Distance.* An airplane is flying at an altitude of 3700 ft. The slanted distance directly to the airport is d feet. Express the horizontal distance h as a function of d.

27. *Inscribed Rhombus.* A rhombus is inscribed in a rectangle that is w meters wide with a perimeter of 40 m. Each vertex of the rhombus is a midpoint of a side of the rectangle. Express the area of the rhombus as a function of the rectangle's width.

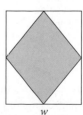

28. *Tablecloth Area.* A tailor uses 16 ft of lace to trim the edges of a rectangular tablecloth. If the tablecloth is *w* feet wide, express its area as a function of the width.

29. *Golf Distance Finder.* A device used in golf to estimate the distance *d*, in yards, to a hole measures the size *s*, in inches, that the 7-ft pin appears to be in a viewfinder. Express the distance *d* as a function of *s*.

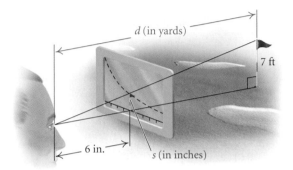

30. *Gas Tank Volume.* A gas tank has ends that are hemispheres of radius *r* feet. The cylindrical midsection is 6 ft long. Express the volume of the tank as a function of *r*.

31. *Play Space.* A daycare center has 30 ft of dividers with which to enclose a rectangular play space in a corner of a large room. The sides against the wall require no partition. Suppose the play space is *x* feet long.

a) Express the area of the play space as a function of *x*.
b) Find the domain of the function.
c) Using the graph shown below, determine the dimensions that yield the maximum area.

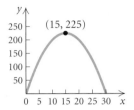

32. *Corral Design.* A rancher has 360 yd of fencing with which to enclose two adjacent rectangular corrals, one for sheep and one for cattle. A river forms one side of the corrals. Suppose the width of each corral is *x* yards.

a) Express the total area of the two corrals as a function of *x*.
b) Find the domain of the function.
c) Using the graph shown below, determine the dimensions that yield the maximum area.

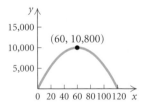

33. *Volume of a Box.* From a 12-cm by 12-cm piece of cardboard, square corners are cut out so that the sides can be folded up to make a box.

a) Express the volume of the box as a function of the length x, in centimeters, of a cut-out square.
b) Find the domain of the function.
c) Using the graph shown below, determine the dimensions that yield the maximum volume.

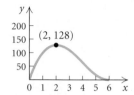

34. *Molding Plastics.* Plastics Unlimited plans to produce a one-component vertical file by bending the long side of an 8-in. by 14-in. sheet of plastic along two lines to form a **U** shape.

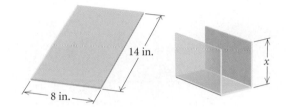

a) Express the volume of the file as a function of the height x, in inches, of the file.
b) Find the domain of the function.
c) Using the graph shown below, determine how tall the file should be in order to maximize the volume the file can hold.

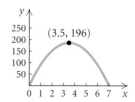

For each piecewise function, find the specified function values.

35. $g(x) = \begin{cases} x + 4, & \text{for } x \le 1, \\ 8 - x, & \text{for } x > 1 \end{cases}$
$g(-4)$, $g(0)$, $g(1)$, and $g(3)$

36. $f(x) = \begin{cases} 3, & \text{for } x \le -2, \\ \frac{1}{2}x + 6, & \text{for } x > -2 \end{cases}$
$f(-5)$, $f(-2)$, $f(0)$, and $f(2)$

37. $h(x) = \begin{cases} -3x - 18, & \text{for } x < -5, \\ 1, & \text{for } -5 \le x < 1, \\ x + 2, & \text{for } x \ge 1 \end{cases}$
$h(-5)$, $h(0)$, $h(1)$, and $h(4)$

38. $f(x) = \begin{cases} -5x - 8, & \text{for } x < -2, \\ \frac{1}{2}x + 5, & \text{for } -2 \le x \le 4, \\ 10 - 2x, & \text{for } x > 4 \end{cases}$
$f(-4)$, $f(-2)$, $f(4)$, and $f(6)$

Graph each of the following.

39. $f(x) = \begin{cases} \frac{1}{2}x, & \text{for } x < 0, \\ x + 3, & \text{for } x \ge 0 \end{cases}$

40. $f(x) = \begin{cases} -\frac{1}{3}x + 2, & \text{for } x \le 0, \\ x - 5, & \text{for } x > 0 \end{cases}$

41. $f(x) = \begin{cases} -\frac{3}{4}x + 2, & \text{for } x < 4, \\ -1, & \text{for } x \ge 4 \end{cases}$

42. $f(x) = \begin{cases} 4, & \text{for } x < -2, \\ x + 1, & \text{for } -2 < x < 3, \\ -x, & \text{for } x \ge 3 \end{cases}$

43. $f(x) = \begin{cases} x + 1, & \text{for } x \le -3, \\ -1, & \text{for } -3 < x < 4, \\ \frac{1}{2}x, & \text{for } x \ge 4 \end{cases}$

44. $f(x) = \begin{cases} \dfrac{x^2 - 9}{x + 3}, & \text{for } x \ne -3, \\ 5, & \text{for } x = -3 \end{cases}$

45. $f(x) = \begin{cases} 2, & \text{for } x = 5, \\ \dfrac{x^2 - 25}{x - 5}, & \text{for } x \ne 5 \end{cases}$

46. $f(x) = \begin{cases} \dfrac{x^2 + 3x + 2}{x + 1}, & \text{for } x \ne -1, \\ 7, & \text{for } x = -1 \end{cases}$

47. $f(x) = [\![x]\!]$

48. $f(x) = 2[\![x]\!]$

49. $g(x) = 1 + [\![x]\!]$

50. $h(x) = \frac{1}{2}[\![x]\!] - 2$

51.–56. Find the domain and the range of each of the functions defined in Exercises 39–44.

Determine the domain and the range of the piecewise function. Then write an equation for the function.

57.

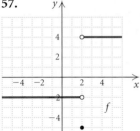

58.

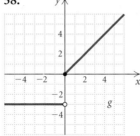

59.

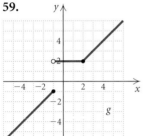

60.

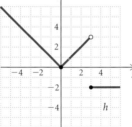

61.

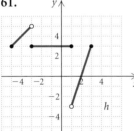

62.

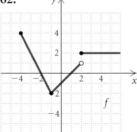

Technology Connection

63. *Temperature During an Illness.* The temperature of a patient during an illness is given by the function

$$T(t) = -0.1t^2 + 1.2t + 98.6, \quad 0 \le t \le 12,$$

where T is the temperature, in degrees Fahrenheit, at time t, in days, after the onset of the illness.

a) Graph the function using a graphing calculator.
b) Use the MAXIMUM feature to determine at what time the patient's temperature was the highest. What was the highest temperature?

64. *Advertising Effect.* A software firm estimates that it will sell N units of a new CD-ROM video game after spending a dollars on advertising, where

$$N(a) = -a^2 + 300a + 6, \quad 0 \le a \le 300,$$

and a is measured in thousands of dollars.

a) Graph the function using a graphing calculator.
b) Use the MAXIMUM feature to find the relative maximum.
c) For what advertising expenditure will the greatest number of games be sold? How many games will be sold for that amount?

Use a graphing calculator to find the intervals on which the function is increasing or decreasing. Consider the entire set of real numbers if no domain is given.

65. $f(x) = \dfrac{8x}{x^2 + 1}$

66. $f(x) = \dfrac{-4}{x^2 + 1}$

67. $f(x) = x\sqrt{4 - x^2}, \quad \text{for } -2 \le x \le 2$

68. $f(x) = -0.8x\sqrt{9 - x^2}, \quad \text{for } -3 \le x \le 3$

69. *Area of an Inscribed Rectangle.* A rectangle that is x feet wide is inscribed in a circle of radius 8 ft.

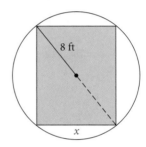

a) Express the area of the rectangle as a function of x.
b) Find the domain of the function.
c) Graph the function with a graphing calculator.
d) What dimensions maximize the area of the rectangle?

70. *Cost of Material.* A rectangular box with volume 320 ft^3 is built with a square base and top. The cost is $\$1.50/\text{ft}^2$ for the bottom, $\$2.50/\text{ft}^2$ for the sides, and $\$1/\text{ft}^2$ for the top. Let $x =$ the length of the base, in feet.

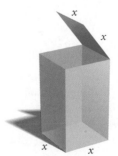

a) Express the cost of the box as a function of x.
b) Find the domain of the function.
c) Graph the function with a graphing calculator.
d) What dimensions minimize the cost of the box?

Collaborative Discussion and Writing

71. Describe a real-world situation that could be modeled by a function that is, in turn, increasing, then constant, and finally decreasing.

72. Simply stated, a *continuous function* is a function whose graph can be drawn without lifting the pencil from the paper. Examine several functions in this exercise set to see if they are continuous. Then explore the continuous functions to estimate the relative maxima and minima. For continuous functions, how can you connect the ideas of increasing and decreasing on an interval to relative maxima and minima?

Skill Maintenance

In each of Exercises 73–76, fill in the blank(s) with the correct term(s). Some of the given choices will not be used; others will be used more than once.

    constant
    function
    any
    midpoint formula
    y-intercept
    range
    domain
    distance formula
    exactly one
    identity
    x-intercept

73. A _____ is a correspondence between a first set, called the _____, and a second set called the _____, such that each member of the _____ corresponds to _____ member of the _____ .

74. The _____ is $\left(\dfrac{x_1 + x_2}{2}, \dfrac{y_1 + y_2}{2} \right)$.

75. A(n) _____ is a point $(a, 0)$.

76. A function f is a linear function if it can be written as $f(x) = mx + b$, where m and b are constants. If $m = 0$, the function is a _____ function $f(x) = b$. If $m = 1$ and $b = 0$, the function is the _____ function $f(x) = x$.

Synthesis

77. If $([\![x]\!])^2 = 25$, what are the possible inputs for x?

78. If $[\![x + 2]\!] = -3$, what are the possible inputs for x?

79. *Volume of an Inscribed Cylinder.* A right circular cylinder of height h and radius r is inscribed in a right circular cone with a height of 10 ft and a base with radius 6 ft.

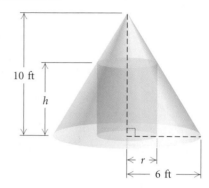

a) Express the height h of the cylinder as a function of r.
b) Express the volume V of the cylinder as a function of r.
c) Express the volume V of the cylinder as a function of h.

80. *Minimizing Power Line Costs.* A power line is constructed from a power station at point A to an island at point C, which is 1 mi directly out in the water from a point B on the shore. Point B is 4 mi downshore from the power station at A. It costs $5000 per mile to lay the power line under water and $3000 per mile to lay the power line under ground. The line comes to the shore at point S downshore from A. Let $x = $ the distance from B to S.

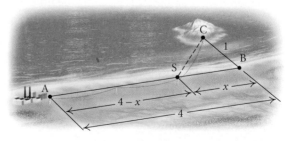

a) Express the cost C of laying the line as a function of x.
b) At what distance x from point B should the line come to shore in order to minimize cost?

1.6

The Algebra of Functions

• *Find the sum, the difference, the product, and the quotient of two functions, and determine the domains of the resulting functions.*
• *Find the composition of two functions and the domain of the composition; decompose a function as a composition of two functions.*

The Algebra of Functions: Sums, Differences, Products, and Quotients

We now use addition, subtraction, multiplication, and division to combine functions and obtain new functions.

Consider the following two functions f and g:

$$f(x) = x + 2 \quad \text{and} \quad g(x) = x^2 + 1.$$

Since $f(3) = 3 + 2 = 5$ and $g(3) = 3^2 + 1 = 10$, we have

$$f(3) + g(3) = 5 + 10 = 15,$$
$$f(3) - g(3) = 5 - 10 = -5,$$
$$f(3) \cdot g(3) = 5 \cdot 10 = 50, \quad \text{and}$$
$$\frac{f(3)}{g(3)} = \frac{5}{10} = \frac{1}{2}.$$

In fact, so long as x is in the domain of *both* f and g, we can easily compute $f(x) + g(x)$, $f(x) - g(x)$, $f(x) \cdot g(x)$, and, assuming $g(x) \neq 0$, $f(x)/g(x)$. Notation has been developed to facilitate this work.

> ### Sums, Differences, Products, and Quotients of Functions
>
> If f and g are functions and x is in the domain of each function, then
>
> $$(f + g)(x) = f(x) + g(x),$$
> $$(f - g)(x) = f(x) - g(x),$$
> $$(fg)(x) = f(x) \cdot g(x),$$
> $$(f/g)(x) = f(x)/g(x), \text{ provided } g(x) \neq 0.$$

EXAMPLE 1 Given that $f(x) = x + 1$ and $g(x) = \sqrt{x + 3}$, find each of the following.

a) $(f + g)(x)$ **b)** $(f + g)(6)$ **c)** $(f + g)(-4)$

Solution

a) $(f + g)(x) = f(x) + g(x)$
$$= x + 1 + \sqrt{x + 3} \qquad \text{This cannot be simplified.}$$

b) We can find $(f + g)(6)$ provided 6 is in the domain of *each* function. The *domain* of f is all real numbers. The *domain* of g is all real numbers x for which $x + 3 \geq 0$, or $x \geq -3$. This is the interval $[-3, \infty)$. Thus, 6 is in both domains, so we have

$$f(6) = 6 + 1 = 7, \qquad g(6) = \sqrt{6 + 3} = \sqrt{9} = 3,$$
$$(f + g)(6) = f(6) + g(6) = 7 + 3 = 10.$$

Another method is to use the formula found in part (a):

$$(f + g)(6) = 6 + 1 + \sqrt{6 + 3} = 7 + \sqrt{9} = 7 + 3 = 10.$$

c) To find $(f + g)(-4)$, we must first determine whether -4 is in the domain of each function. We note that -4 is not in the domain of g, $[-3, \infty)$. That is, $\sqrt{-4 + 3}$ is not a real number. Thus, $(f + g)(-4)$ does not exist. ▬

It is useful to view the concept of the sum of two functions graphically. In the graph below, we see the graphs of two functions f and g and their sum, $f + g$. Consider finding $(f + g)(4) = f(4) + g(4)$. We can locate $g(4)$ on the graph of g and measure it. Then we add that length on top of $f(4)$ on the graph of f. The sum gives us $(f + g)(4)$.

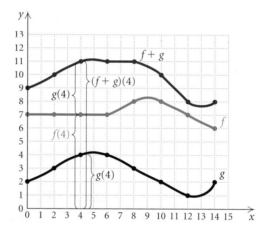

With this in mind, let's view Example 1 from a graphical perspective. Let's look at the graphs of

$$f(x) = x + 1, \qquad g(x) = \sqrt{x + 3}, \quad \text{and}$$
$$(f + g)(x) = x + 1 + \sqrt{x + 3}.$$

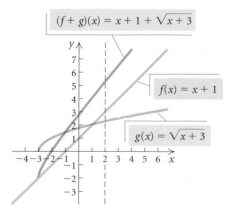

See the graph at left. Note that the domain of f is the set of all real numbers. The domain of g is $[-3, \infty)$. The domain of $f + g$ is the set of numbers in the intersection of the domains, that is, the set of numbers in both domains.

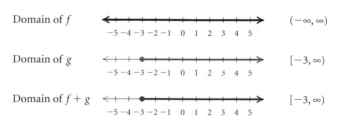

Domain of f $(-\infty, \infty)$

Domain of g $[-3, \infty)$

Domain of $f + g$ $[-3, \infty)$

Thus the domain of $f + g$ is $[-3, \infty)$.

We can confirm that the y-coordinates of the graph of $(f + g)(x)$ are the sums of the corresponding y-coordinates of the graphs of $f(x)$ and $g(x)$. Here we confirm it for $x = 2$.

$$f(x) = x + 1 \qquad\qquad g(x) = \sqrt{x + 3}$$
$$f(2) = 2 + 1 = 3; \qquad g(2) = \sqrt{2 + 3} = \sqrt{5};$$
$$(f + g)(x) = x + 1 + \sqrt{x + 3}$$
$$(f + g)(2) = 2 + 1 + \sqrt{2 + 3}$$
$$= 3 + \sqrt{5} = f(2) + g(2).$$

Let's also look at $f - g$, fg, and f/g for the functions $f(x) = x + 1$ and $g(x) = \sqrt{x + 3}$ of Example 1. The domains of $f - g$ and fg are the same as the domain of $f + g$, $[-3, \infty)$, because numbers in this interval are in the domains of *both* functions. For f/g, $g(x)$ cannot be 0. Since $\sqrt{x + 3} = 0$ when $x = -3$, -3 must be excluded and the domain of f/g is $(-3, \infty)$.

EXAMPLE 2 Given that $f(x) = x^2 - 4$ and $g(x) = x + 2$, find each of the following.

a) The domain of $f + g$, $f - g$, fg, and f/g

b) $(f + g)(x)$ c) $(f - g)(x)$

d) $(fg)(x)$ e) $(f/g)(x)$

f) $(gg)(x)$

Solution

a) The domain of f is the set of all real numbers. The domain of g is also the set of all real numbers. The domains of $f + g$, $f - g$, and fg are the set of numbers in the intersection of the domains—that is, the set of numbers in both domains, which is again the set of real numbers. For f/g, we must exclude -2, since $g(-2) = 0$. Thus the domain of f/g is the set of real numbers excluding -2, or $(-\infty, -2) \cup (-2, \infty)$.

b) $(f + g)(x) = f(x) + g(x) = (x^2 - 4) + (x + 2) = x^2 + x - 2$

c) $(f - g)(x) = f(x) - g(x) = (x^2 - 4) - (x + 2) = x^2 - x - 6$

d) $(fg)(x) = f(x) \cdot g(x) = (x^2 - 4)(x + 2) = x^3 + 2x^2 - 4x - 8$

e) $(f/g)(x) = \dfrac{f(x)}{g(x)}$

$$= \dfrac{x^2 - 4}{x + 2} \qquad\qquad \text{Note that } g(x) = 0 \text{ when } x = -2, \text{ so } (f/g)(x) \text{ is not defined when } x = -2.$$

$$= \dfrac{(x + 2)(x - 2)}{x + 2} \qquad\qquad \text{Factoring}$$

$$= x - 2 \qquad\qquad \text{Removing a factor of 1: } \dfrac{x + 2}{x + 2} = 1$$

Thus, $(f/g)(x) = x - 2$ with the added stipulation that $x \neq -2$ since -2 is not in the domain of $(f/g)(x)$.

f) $(gg)(x) = g(x) \cdot g(x) = [g(x)]^2 = (x + 2)^2 = x^2 + 4x + 4$

Study Tip

This text is accompanied by a complete set of videotapes featuring instructors presenting material and concepts from every section of the text. These videos are available in your campus math lab or on CD-ROM. When missing a class is unavoidable, you should view the video lecture(s) covering the material you missed.

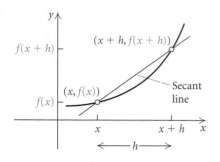

Difference Quotients

In Section 1.3, we learned that the slope of a line can be considered as an average *rate of change*. Here let's consider a nonlinear function f and draw a line through two points $(x, f(x))$ and $(x + h, f(x + h))$ as shown at left.

The slope of the line, called a **secant line,** is

$$\frac{f(x + h) - f(x)}{x + h - x},$$

which simplifies to

$$\frac{f(x + h) - f(x)}{h}. \qquad \text{Difference quotient}$$

This ratio is called the **difference quotient,** or the **average rate of change.** It is important in calculus to be able to find and simplify difference quotients.

EXAMPLE 3 For the function f given by $f(x) = 2x^2 - x - 3$, find the difference quotient

$$\frac{f(x + h) - f(x)}{h}.$$

Solution We first find $f(x + h)$:

$$\begin{aligned}
f(x + h) &= 2(x + h)^2 - (x + h) - 3 \\
&= 2[x^2 + 2xh + h^2] - x - h - 3 \\
&= 2x^2 + 4xh + 2h^2 - x - h - 3.
\end{aligned}$$

Then

$$\begin{aligned}
\frac{f(x + h) - f(x)}{h} \\
= \frac{[2x^2 + 4xh + 2h^2 - x - h - 3] - [2x^2 - x - 3]}{h} \\
= \frac{2x^2 + 4xh + 2h^2 - x - h - 3 - 2x^2 + x + 3}{h} \\
= \frac{4xh + 2h^2 - h}{h} \\
= \frac{h(4x + 2h - 1)}{h \cdot 1} = \frac{h}{h} \cdot \frac{4x + 2h - 1}{1} = 4x + 2h - 1.
\end{aligned}$$

The Composition of Functions

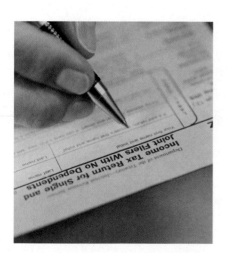

In real-world situations, it is not uncommon for the output of a function to depend on some input that is itself an output of another function. For instance, the amount a person pays as state income tax usually depends on the amount of adjusted gross income on the person's federal tax return, which, in turn, depends on his or her annual earnings. Such functions are called **composite functions.**

To illustrate how composite functions work, suppose a chemistry student needs a formula to convert Fahrenheit temperatures to Kelvin units. The formula

$$c(t) = \tfrac{5}{9}(t - 32)$$

gives the Celsius temperature $c(t)$ that corresponds to the Fahrenheit temperature t. The formula

$$k(c) = c + 273$$

gives the Kelvin temperature $k(c)$ that corresponds to the Celsius temperature c. Thus, 50° Fahrenheit corresponds to

$$c(50) = \tfrac{5}{9}(50 - 32) = \tfrac{5}{9}(18) = 10° \text{ Celsius}$$

and 10° Celsius corresponds to

$$k(10) = 10 + 273 = 283° \text{ Kelvin.}$$

We see that 50° Fahrenheit is the same as 283° Kelvin. This two-step procedure can be used to convert any Fahrenheit temperature to Kelvin units.

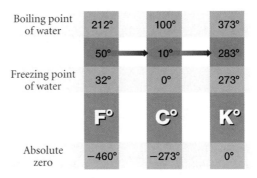

A student making numerous conversions might look for a formula that converts directly from Fahrenheit to Kelvin. Such a formula can be found by substitution:

$$k(c(t)) = c(t) + 273$$

$$= \frac{5}{9}(t - 32) + 273 \qquad \textbf{Substituting}$$

$$= \frac{5t + 2297}{9}. \qquad \textbf{Simplifying}$$

Since the last equation expresses the Kelvin temperature as a new function, K, of the Fahrenheit temperature, t, we can write

$$K(t) = \frac{5t + 2297}{9},$$

where $K(t)$ is the Kelvin temperature corresponding to the Fahrenheit temperature, t. Here we have $K(t) = k(c(t))$. The new function K is called the **composition** of k and c and can be denoted $k \circ c$ (read "k composed with c," "the composition of k and c," or "k circle c").

> ### Composition of Functions
>
> The **composite function** $f \circ g$, the **composition** of f and g, is defined as
>
> $$(f \circ g)(x) = f(g(x)),$$
>
> where x is in the domain of g and $g(x)$ is in the domain of f.

EXAMPLE 4 Given that $f(x) = 2x - 5$ and $g(x) = x^2 - 3x + 8$, find each of the following.

a) $(f \circ g)(x)$ and $(g \circ f)(x)$ **b)** $(f \circ g)(7)$ and $(g \circ f)(7)$

Solution Consider each function separately:

$f(x) = 2x - 5$ **This function multiplies each input by 2 and subtracts 5.**

and

$g(x) = x^2 - 3x + 8.$ **This function squares an input, subtracts 3 times the input from the result, and then adds 8.**

a) To find $(f \circ g)(x)$, we substitute $g(x)$ for x in the equation for $f(x)$:

$$(f \circ g)(x) = f(g(x)) = f(x^2 - 3x + 8)$$
  $x^2 - 3x + 8$ **is the input for** f**.**

$$= 2(x^2 - 3x + 8) - 5$$
  f **multiplies the input by 2 and subtracts 5.**

$$= 2x^2 - 6x + 16 - 5$$

$$= 2x^2 - 6x + 11.$$

To find $(g \circ f)(x)$, we substitute $f(x)$ for x in the equation for $g(x)$:

$$(g \circ f)(x) = g(f(x)) = g(2x - 5)$$
  $2x - 5$ **is the input for** g**.**

$$= (2x - 5)^2 - 3(2x - 5) + 8$$
  g **squares the input, subtracts three times the input, and adds 8.**

$$= 4x^2 - 20x + 25 - 6x + 15 + 8$$

$$= 4x^2 - 26x + 48.$$

b) To find $(f \circ g)(7)$, we first find $g(7)$. Then we use $g(7)$ as an input for f:

$$(f \circ g)(7) = f(g(7)) = f(7^2 - 3 \cdot 7 + 8)$$

$$= f(36) = 2 \cdot 36 - 5$$

$$= 67.$$

To find $(g \circ f)(7)$, we first find $f(7)$. Then we use $f(7)$ as an input for g:

$$(g \circ f)(7) = g(f(7)) = g(2 \cdot 7 \quad 5)$$

$$= g(9) = 9^2 - 3 \cdot 9 + 8$$

$$= 62.$$

Technology Connection

We can check our work in Example 4(b) using a graphing calculator. We enter the following on the equation-editor screen:

$$y_1 = 2x - 5$$

and

$$y_2 = x^2 - 3x + 8.$$

Then, on the home screen, we find $(f \circ g)(7)$ and $(g \circ f)(7)$ using the function notations Y1(Y2(7)) and Y2(Y1(7)), respectively.

$y_1 = 2x - 5, \quad y_2 = x^2 - 3x + 8$

Y1(Y2(7))	
	67
Y2(Y1(7))	
	62

We could also find $(f \circ g)(7)$ and $(g \circ f)(7)$ by substituting 7 for x in the equations that we found in part (a):

$$(f \circ g)(x) = 2x^2 - 6x + 11$$
$$(f \circ g)(7) = 2 \cdot 7^2 - 6 \cdot 7 + 11 = 67;$$

$$(g \circ f)(x) = 4x^2 - 26x + 48$$
$$(g \circ f)(7) = 4 \cdot 7^2 - 26 \cdot 7 + 48 = 62.$$

Note in Example 4 that, as a rule, $(f \circ g)(x) \neq (g \circ f)(x)$. We can see this graphically, as shown in the graphs at left.

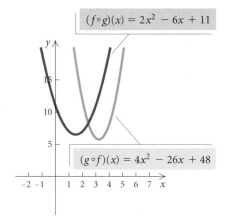

$(f \circ g)(x) = 2x^2 - 6x + 11$

$(g \circ f)(x) = 4x^2 - 26x + 48$

EXAMPLE 5   Given that $f(x) = \sqrt{x}$ and $g(x) = x - 3$:

a) Find $f \circ g$ and $g \circ f$.

b) Find the domains of $f \circ g$ and $g \circ f$.

Solution

a) $(f \circ g)(x) = f(g(x)) = f(x - 3) = \sqrt{x - 3}$

$(g \circ f)(x) = g(f(x)) = g(\sqrt{x}) = \sqrt{x} - 3$

b) The domain of f is $\{x \mid x \geq 0\}$, or the interval $[0, \infty)$. The domain of g is $(-\infty, \infty)$.

To find the domain of $f \circ g$, we first consider the domain of g, $(-\infty, \infty)$. Next, we must consider that the outputs for g will serve as inputs for f. Since the inputs for f cannot be negative, we must have

$$g(x) \geq 0, \quad \text{or } x - 3 \geq 0, \text{ or } x \geq 3.$$

Thus the domain of $f \circ g = \{x \mid x \geq 3\}$, or the interval $[3, \infty)$, as the graph in Fig. 1 below confirms.

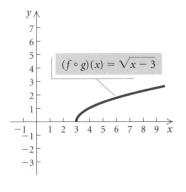

$(f \circ g)(x) = \sqrt{x - 3}$

FIGURE 1

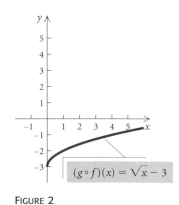

$(g \circ f)(x) = \sqrt{x} - 3$

FIGURE 2

To find the domain of $g \circ f$, we first consider the domain of f, $\{x \mid x \geq 0\}$. Next, we must consider that the outputs for f will serve as inputs for g. Since g can accept *any* real number as an input, any output from f is acceptable so the entire domain of f is the domain of $g \circ f$. That is, the domain of $g \circ f = \{x \mid x \geq 0\}$ or the interval $[0, \infty)$, as the graph in Fig. 2 above confirms.

EXAMPLE 6   Given that $f(x) = \dfrac{1}{x-2}$ and $g(x) = \dfrac{5}{x}$, find $f \circ g$ and $g \circ f$ and the domain of each.

Solution   We have

$$(f \circ g)(x) = f(g(x)) = f\left(\frac{5}{x}\right) = \frac{1}{\dfrac{5}{x} - 2} = \frac{1}{\dfrac{5-2x}{x}} = \frac{x}{5-2x};$$

$$(g \circ f)(x) = g(f(x)) = g\left(\frac{1}{x-2}\right) = \frac{5}{\dfrac{1}{x-2}} = 5(x-2).$$

The domain of $f = \{x \mid x \neq 2\}$, and the domain of $g = \{x \mid x \neq 0\}$. To find the domain of $f \circ g$, we note that since the domain of g is $\{x \mid x \neq 0\}$, 0 is *not* in the domain of $f \circ g$. Since the domain of f is $\{x \mid x \neq 2\}$, no output $g(x)$ can be 2. We solve $5/x = 2$ for x:

$$\frac{5}{x} = 2$$

$$5 = 2x$$

$$\frac{5}{2} = x.$$

We find that $\frac{5}{2}$ is also *not* in the domain of $f \circ g$. Then the domain of $f \circ g = \{x \mid x \neq 0 \ and \ x \neq \frac{5}{2}\}$, or $(-\infty, 0) \cup \left(0, \frac{5}{2}\right) \cup \left(\frac{5}{2}, \infty\right)$.

To find the domain of $g \circ f$, we recall the domain of f is $\{x \mid x \neq 2\}$. Thus, 2 is *not* in the domain of $g \circ f$. Since the domain of g is $\{x \mid x \neq 0\}$ and $f(x) = 1/(x-2)$ is never 0, there are no additional restrictions on the domain of $g \circ f$. Thus the domain of $g \circ f = \{x \mid x \neq 2\}$, or $(-\infty, 2) \cup (2, \infty)$.

Decomposing a Function as a Composition

In calculus, one often needs to recognize how a function can be expressed as the composition of two functions. In this way, we are "decomposing" the function.

EXAMPLE 7   If $h(x) = (2x - 3)^5$, find $f(x)$ and $g(x)$ such that $h(x) = (f \circ g)(x)$.

Solution   The function $h(x)$ raises $(2x - 3)$ to the 5th power. Two functions that can be used for the composition are

$$f(x) = x^5 \quad \text{and} \quad g(x) = 2x - 3.$$

We can check by forming the composition:

$$h(x) = (f \circ g)(x) = f(g(x)) = f(2x - 3) = (2x - 3)^5.$$

This is the most "obvious" answer to the question. There can be other less obvious answers. For example, if

$$f(x) = (x + 7)^5 \quad \text{and} \quad g(x) = 2x - 10,$$

then

$$h(x) = (f \circ g)(x) = f(g(x))$$
$$= f(2x - 10)$$
$$= [(2x - 10) + 7]^5 = (2x - 3)^5.$$

1.6

Exercise Set

Given that $f(x) = x^2 - 3$ and $g(x) = 2x + 1$, find each of the following, if it exists.

1. $(f + g)(5)$ **2.** $(fg)(0)$

3. $(f - g)(-1)$ **4.** $(fg)(2)$

5. $(f/g)\left(-\frac{1}{2}\right)$ **6.** $(f - g)(0)$

7. $(fg)\left(-\frac{1}{2}\right)$ **8.** $(f/g)\left(-\sqrt{3}\right)$

9. $(g - f)(-1)$ **10.** $(g/f)\left(-\frac{1}{2}\right)$

Given that $h(x) = x + 4$ and $g(x) = \sqrt{x - 1}$, find each of the following, if it exists.

11. $(h - g)(-4)$ **12.** $(gh)(10)$

13. $(g/h)(1)$ **14.** $(h/g)(1)$

15. $(g + h)(1)$ **16.** $(hg)(3)$

For each pair of functions in Exercises 17–32:

a) *Find the domain of f, g, $f + g$, $f - g$, fg, ff, f/g, and g/f.*

b) *Find $(f + g)(x)$, $(f - g)(x)$, $(fg)(x)$, $(ff)(x)$, $(f/g)(x)$, and $(g/f)(x)$.*

17. $f(x) = 2x + 3$, $g(x) = 3 - 5x$

18. $f(x) = -x + 1$, $g(x) = 4x - 2$

19. $f(x) = x - 3$, $g(x) = \sqrt{x + 4}$

20. $f(x) = x + 2$, $g(x) = \sqrt{x - 1}$

21. $f(x) = 2x - 1$, $g(x) = -2x^2$

22. $f(x) = x^2 - 1$, $g(x) = 2x + 5$

23. $f(x) = \sqrt{x - 3}$, $g(x) = \sqrt{x + 3}$

24. $f(x) = \sqrt{x}$, $g(x) = \sqrt{2 - x}$

25. $f(x) = x + 1$, $g(x) = |x|$

26. $f(x) = 4|x|$, $g(x) = 1 - x$

27. $f(x) = x^3$, $g(x) = 2x^2 + 5x - 3$

28. $f(x) = x^2 - 4$, $g(x) = x^3$

29. $f(x) = \dfrac{4}{x + 1}$, $g(x) = \dfrac{1}{6 - x}$

30. $f(x) = 2x^2$, $g(x) = \dfrac{2}{x - 5}$

31. $f(x) = \dfrac{1}{x}$, $g(x) = x - 3$

32. $f(x) = \sqrt{x + 6}$, $g(x) = \dfrac{1}{x}$

In Exercises 33–38, consider the functions F and G as shown in the following graph.

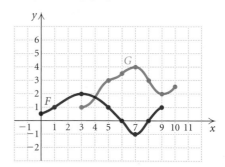

33. Find the domain of F, the domain of G, and the domain of $F + G$.

34. Find the domain of $F - G$, FG, and F/G.

35. Find the domain of G/F.

36. Graph $F + G$.

37. Graph $G - F$.

38. Graph $F - G$.

39. *Total Cost, Revenue, and Profit.* In economics, functions that involve revenue, cost, and profit are used. For example, suppose that $R(x)$ and $C(x)$ denote the total revenue and the total cost, respectively, of producing a new kind of tool for King Hardware Wholesalers. Then the difference

$$P(x) = R(x) - C(x)$$

represents the total profit for producing x tools. Given

$$R(x) = 60x - 0.4x^2 \quad \text{and} \quad C(x) = 3x + 13,$$

find each of the following.

a) $P(x)$
b) $R(100)$, $C(100)$, and $P(100)$

40. *Total Cost, Revenue, and Profit.* Given that

$$R(x) = 200x - x^2 \quad \text{and} \quad C(x) = 5000 + 8x,$$

for a new radio produced by Clear Communication, find each of the following. (See Exercise 39.)

a) $P(x)$
b) $R(175)$, $C(175)$, and $P(175)$

For each function f, construct and simplify the difference quotient

$$\frac{f(x + h) - f(x)}{h}.$$

41. $f(x) = x^2 + 1$ **42.** $f(x) = 2 - x^2$

43. $f(x) = 3x - 5$ **44.** $f(x) = -\frac{1}{2}x + 7$

45. $f(x) = 3x^2 - 2x + 1$ **46.** $f(x) = 5x^2 + 4x$

47. $f(x) = 4 + 5|x|$ **48.** $f(x) = 2|x| + 3x$

49. $f(x) = x^3$ **50.** $f(x) = x^3 - 2x$

51. $f(x) = \dfrac{x - 4}{x + 3}$ **52.** $f(x) = \dfrac{x}{2 - x}$

Given that $f(x) = 3x + 1$, $g(x) = x^2 - 2x - 6$, and $h(x) = x^3$, find each of the following.

53. $(f \circ g)(-1)$ **54.** $(g \circ f)(-2)$

55. $(h \circ f)(1)$ **56.** $(g \circ h)\left(\frac{1}{2}\right)$

57. $(g \circ f)(5)$ **58.** $(f \circ g)\left(\frac{1}{3}\right)$

59. $(f \circ h)(-3)$ **60.** $(h \circ g)(3)$

Find $(f \circ g)(x)$ and $(g \circ f)(x)$ and the domain of each.

61. $f(x) = x + 3$, $g(x) = x - 3$

62. $f(x) = \frac{4}{5}x$, $g(x) = \frac{5}{4}x$

63. $f(x) = \dfrac{4}{1 - 5x}$, $g(x) = \dfrac{1}{x}$

64. $f(x) = \dfrac{6}{x}$, $g(x) = \dfrac{1}{2x + 1}$

65. $f(x) = 3x - 7$, $g(x) = \dfrac{x + 7}{3}$

66. $f(x) = \frac{2}{3}x - \frac{4}{5}$, $g(x) = 1.5x + 1.2$

67. $f(x) = 2x + 1$, $g(x) = \sqrt{x}$

68. $f(x) = \sqrt{x - 4}$, $g(x) = \dfrac{2}{x}$

69. $f(x) = 20$, $g(x) = 0.05$

70. $f(x) = x^4$, $g(x) = \sqrt[4]{x}$

71. $f(x) = \sqrt{x + 5}$, $g(x) = x^2 - 5$

72. $f(x) = x^5 - 2$, $g(x) = \sqrt[5]{x + 2}$

73. $f(x) = x^2 + 2$, $g(x) = \sqrt{3 - x}$

74. $f(x) = 1 - x^2$, $g(x) = \sqrt{x^2 - 25}$

75. $f(x) = \dfrac{1 - x}{x}$, $g(x) = \dfrac{1}{1 + x}$

76. $f(x) = \dfrac{x^2 - 1}{x^2 + 1}$, $g(x) = \dfrac{3x - 4}{5x - 2}$

77. $f(x) = x^3 - 5x^2 + 3x + 7$, $g(x) = x + 1$

78. $f(x) = x - 1$, $g(x) = x^3 + 2x^2 - 3x - 9$

Find $f(x)$ and $g(x)$ such that $h(x) = (f \circ g)(x)$. Answers may vary.

79. $h(x) = (4 + 3x)^5$

80. $h(x) = \sqrt[3]{x^2 - 8}$

81. $h(x) = \dfrac{1}{(x - 2)^4}$

82. $h(x) = \dfrac{1}{\sqrt{3x + 7}}$

83. $h(x) = \dfrac{x^3 - 1}{x^3 + 1}$

84. $h(x) = |9x^2 - 4|$

85. $h(x) = \left(\dfrac{2 + x^3}{2 - x^3}\right)^6$

86. $h(x) = \left(\sqrt{x} - 3\right)^4$

87. $h(x) = \sqrt{\dfrac{x-5}{x+2}}$

88. $h(x) = \sqrt{1 + \sqrt{1+x}}$

89. $h(x) = (x+2)^3 - 5(x+2)^2 + 3(x+2) - 1$

90. $h(x) = 2(x-1)^{5/3} + 5(x-1)^{2/3}$

91. *Dress Sizes.*   A dress that is size x in France is size $s(x)$ in the United States, where $s(x) = x - 32$. A dress that is size x in the United States is size $y(x)$ in Italy, where $y(x) = 2(x + 12)$. Find a function that will convert French dress sizes to Italian dress sizes.

92. *Ripple Spread.*   A stone is thrown into a pond. A circular ripple is spreading over the pond in such a way that the radius is increasing at the rate of 3 ft/sec.

a) Find a function $r(t)$ for the radius in terms of t.
b) Find a function $A(r)$ for the area of the ripple in terms of the radius r.
c) Find $(A \circ r)(t)$. Explain the meaning of this function.

Technology Connection

93. Using a graphing calculator, graph the three functions in Exercise 39 in the viewing window $[0, 160, 0, 3000]$.

94. Using a graphing calculator, graph the three functions in Exercise 40 in the viewing window $[0, 200, 0, 10{,}000]$.

Collaborative Discussion and Writing

95. If $g(x) = b$, where b is a positive constant, describe how the graphs of $y = h(x)$ and $y = (h - g)(x)$ will differ.

96. Explain which values of x must be excluded from the domain of $(f \circ g)(x)$ and the domain of $(g \circ f)(x)$.

Skill Maintenance

Consider the following linear equations. Without graphing them, answer the questions below.

a) $y = x$
b) $y = -5x + 4$
c) $y = \frac{2}{3}x + 1$
d) $y = -0.1x + 6$
e) $y = 3x - 5$
f) $y = -x - 1$
g) $2x - 3y = 6$
h) $6x + 3y = 9$

97. Which, if any, have y-intercept $(0, 1)$?

98. Which, if any, have the same y-intercept?

99. Which slope down from left to right?

100. Which has the steepest slope?

101. Which pass(es) through the origin?

102. Which, if any, have the same slope?

103. Which, if any, are parallel?

104. Which, if any, are perpendicular?

Synthesis

105. Write equations of two functions f and g such that $f \circ g = g \circ f = x$. (In Section 4.1, we will study inverse functions. If $f \circ g = g \circ f = x$, functions f and g are *inverses* of each other.)

106. Write equations for two functions f and g such that the domain of $f - g$ is
$$\{x \mid x \neq -7 \text{ and } x \neq 3\}.$$

107. Find the domain of $(h/g)(x)$ given that
$$h(x) = \frac{5x}{3x - 7} \quad \text{and} \quad g(x) = \frac{x^4 - 1}{5x - 15}.$$

108. For functions h and f, find the domain of $h + f$, $h - f$, hf, and h/f if
$$h = \left\{(-4, 13), (-1, 7), (0, 5), \left(\tfrac{5}{2}, 0\right), (3, -5)\right\}, \quad \text{and}$$
$$f = \{(-4, -7), (-2, -5), (0, -3), (3, 0), (5, 2), (9, 6)\}.$$

1.7

Symmetry and Transformations

- *Determine whether a graph is symmetric with respect to the x-axis, the y-axis, and the origin.*
- *Determine whether a function is even, odd, or neither even nor odd.*
- *Given the graph of a function, graph its transformation under translations, reflections, stretchings, and shrinkings.*

Symmetry

Symmetry occurs often in nature and in art. For example, when viewed from the front, the bodies of most animals are at least approximately symmetric. This means that each eye is the same distance from the center of the bridge of the nose, each shoulder is the same distance from the center of the chest, and so on. Architects have used symmetry for thousands of years to enhance the beauty of buildings.

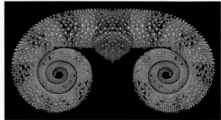

A knowledge of symmetry in mathematics helps us graph and analyze equations and functions.

Consider the points $(4, 2)$ and $(4, -2)$ that appear in the graph of $x = y^2$ (see Fig. 1). Points like these have the same x-value but opposite y-values and are **reflections** of each other across the x-axis. If, for any point (x, y) on a graph, the point $(x, -y)$ is also on the graph, then the graph is said to be **symmetric with respect to the x-axis.** If we fold the graph on the x-axis, the parts above and below the x-axis will coincide.

Consider the points $(3, 4)$ and $(-3, 4)$ that appear in the graph of $y = x^2 - 5$ (see Fig. 2). Points like these have the same y-value but opposite x-values and are **reflections** of each other across the y-axis. If, for any point (x, y) on a graph, the point $(-x, y)$ is also on the graph, then the graph is said to be **symmetric with respect to the y-axis.** If we fold the graph on the y-axis, the parts to the left and right of the y-axis will coincide.

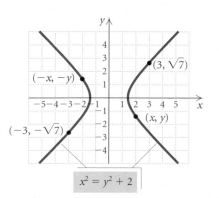

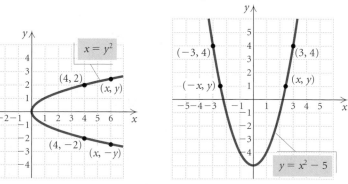

FIGURE 1

FIGURE 2

FIGURE 3

Consider the points $(3, \sqrt{7})$ and $(-3, -\sqrt{7})$ that appear in the graph of $x^2 = y^2 + 2$ (see Fig. 3). Note that if we take the opposites of the coordinates of one pair, we get the other pair. If for any point (x, y) on a graph the point $(-x, -y)$ is also on the graph, then the graph is said to be **symmetric with respect to the origin.** Visually, if we rotate the graph $180°$ about the origin, the resulting figure coincides with the original.

> ### *Algebraic Tests of Symmetry*
>
> *x-axis*: If replacing y with $-y$ produces an equivalent equation, then the graph is *symmetric with respect to the x-axis.*
>
> *y-axis*: If replacing x with $-x$ produces an equivalent equation, then the graph is *symmetric with respect to the y-axis.*
>
> *Origin*: If replacing x with $-x$ and y with $-y$ produces an equivalent equation, then the graph is *symmetric with respect to the origin.*

EXAMPLE 1 Test $y = x^2 + 2$ for symmetry with respect to the x-axis, the y-axis, and the origin.

Algebraic Solution

x-AXIS

We replace y with $-y$:

$$y = x^2 + 2$$
$$-y = x^2 + 2$$
$$y = -x^2 - 2. \qquad \text{Multiplying by } -1 \text{ on both sides}$$

The resulting equation *is not* equivalent to the original equation, so the graph *is not* symmetric with respect to the x-axis.

y-AXIS

We replace x with $-x$:

$$y = x^2 + 2$$
$$y = (-x)^2 + 2$$
$$y = x^2 + 2. \qquad \text{Simplifying}$$

The resulting equation *is* equivalent to the original equation, so the graph *is* symmetric with respect to the y-axis.

ORIGIN

We replace x with $-x$ and y with $-y$:

$$y = x^2 + 2$$
$$-y = (-x)^2 + 2$$
$$-y = x^2 + 2 \qquad \text{Simplifying}$$
$$y = -x^2 - 2.$$

The resulting equation *is not* equivalent to the original equation, so the graph *is not* symmetric with respect to the origin.

Visualizing the Solution

Let's look at the graph of $y = x^2 + 2$.

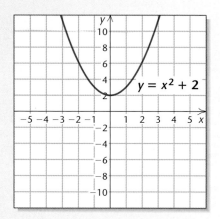

Note that if the graph were folded on the x-axis, the parts above and below the x-axis would not coincide. If it were folded on the y-axis, the parts to the left and right of the y-axis would coincide. If we rotated it $180°$ around the origin, the resulting graph would not coincide with the original graph.

Thus we see that the graph *is not* symmetric with respect to the x-axis or the origin. The graph *is* symmetric with respect to the y-axis.

EXAMPLE 2 Test $x^2 + y^4 = 5$ for symmetry with respect to the x-axis, the y-axis, and the origin.

Algebraic Solution

x-AXIS

We replace y with $-y$:

$$x^2 + y^4 = 5$$
$$x^2 + (-y)^4 = 5$$
$$x^2 + y^4 = 5.$$

The resulting equation *is* equivalent to the original equation. Thus the graph *is* symmetric with respect to the x-axis.

y-AXIS

We replace x with $-x$:

$$x^2 + y^4 = 5$$
$$(-x)^2 + y^4 = 5$$
$$x^2 + y^4 = 5.$$

The resulting equation *is* equivalent to the original equation, so the graph *is* symmetric with respect to the y-axis.

ORIGIN

We replace x with $-x$ and y with $-y$:

$$x^2 + y^4 = 5$$
$$(-x)^2 + (-y)^4 = 5$$
$$x^2 + y^4 = 5.$$

The resulting equation is equivalent to the original equation, so the graph is symmetric with respect to the origin.

Visualizing the Solution

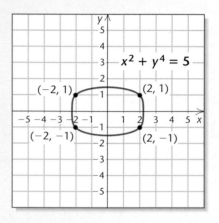

From the graph of the equation, we see symmetry with respect to both axes and with respect to the origin.

Even and Odd Functions

Algebraic Procedure for Determining Even and Odd Functions

Given the function $f(x)$:

1. Find $f(-x)$ and simplify. If $f(x) = f(-x)$, then f is even.
2. Find $-f(x)$, simplify, and compare with $f(-x)$ from step (1). If $f(-x) = -f(x)$, then f is odd.

Except for the function $f(x) = 0$, a function cannot be *both* even and odd. Thus if $f(x) \neq 0$ and we see in step (1) that $f(x) = f(-x)$ (that is, f is even), we need not continue.

Even and Odd Functions

Now we relate symmetry to graphs of functions.

Even and Odd Functions

If the graph of a function f is symmetric with respect to the y-axis, we say that it is an **even function.** That is, for each x in the domain of f, $f(x) = f(-x)$.

If the graph of a function f is symmetric with respect to the origin, we say that it is an **odd function.** That is, for each x in the domain of f, $f(-x) = -f(x)$.

An algebraic procedure for determining even and odd functions is shown at left.

EXAMPLE 3 Determine whether each of the following functions is even, odd, or neither.

a) $f(x) = 5x^7 - 6x^3 - 2x$ b) $h(x) = 5x^6 - 3x^2 - 7$

a) Algebraic Solution

$f(x) = 5x^7 - 6x^3 - 2x$

1. $f(-x) = 5(-x)^7 - 6(-x)^3 - 2(-x)$
$= 5(-x^7) - 6(-x^3) + 2x$
$(-x)^7 = (-1 \cdot x)^7 = (-1)^7 x^7 = -x^7$
$= -5x^7 + 6x^3 + 2x$

We see that $f(x) \neq f(-x)$. Thus, f is *not* even.

2. $-f(x) = -(5x^7 - 6x^3 - 2x)$
$= -5x^7 + 6x^3 + 2x$

We see that $f(-x) = -f(x)$. Thus, f is odd.

Visualizing the Solution

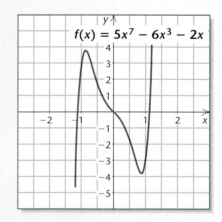

We see that the graph appears to be symmetric with respect to the origin. The function is odd.

b) **Algebraic Solution**

$h(x) = 5x^6 - 3x^2 - 7$

1. $h(-x) = 5(-x)^6 - 3(-x)^2 - 7$
        $= 5x^6 - 3x^2 - 7$

    We see that $h(x) = h(-x)$. Thus the function is even.

Visualizing the Solution

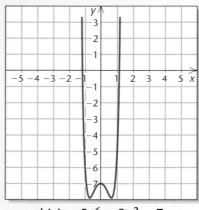

$h(x) = 5x^6 - 3x^2 - 7$

We see that the graph appears to be symmetric with respect to the y-axis. The function is even.

Transformations of Functions

The graphs of some basic functions are shown below and on the following page. Others can be seen on the inside back cover.

Identity function:
$y = x$

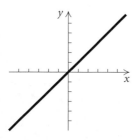

Squaring function:
$y = x^2$

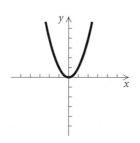

Square root function:
$y = \sqrt{x}$

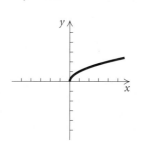

Cubing function:
$y = x^3$

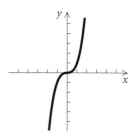

Cube root function:
$y = \sqrt[3]{x}$

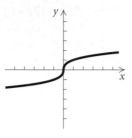

Reciprocal function:
$y = \dfrac{1}{x}$

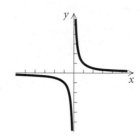

Absolute-value function:
$y = |x|$

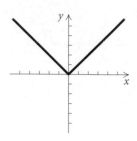

These functions can be considered building blocks for many other functions. We can create graphs of new functions by shifting them horizontally or vertically, stretching or shrinking them, and reflecting them across an axis. We now consider these **transformations.**

Vertical and Horizontal Translations

Suppose that we have a function given by $y = f(x)$. Let's explore the graphs of the new functions $y = f(x) + b$ and $y = f(x) - b$, for $b > 0$.

Consider the functions $y = \frac{1}{5}x^4$, $y = \frac{1}{5}x^4 + 5$, and $y = \frac{1}{5}x^4 - 3$ and compare their graphs. What pattern do you see? Test it with some other graphs.

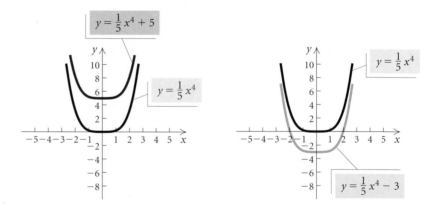

The effect of adding a constant to or subtracting a constant from $f(x)$ in $y = f(x)$ is a shift of the graph of $f(x)$ up or down. Such a shift is called a **vertical translation.**

Vertical Translation

For $b > 0$,

the graph of $y = f(x) + b$ is the graph of $y = f(x)$ shifted *up* b units;

the graph of $y = f(x) - b$ is the graph of $y = f(x)$ shifted *down* b units.

Suppose that we have a function given by $y = f(x)$. Let's explore the graphs of the new functions $y = f(x - d)$ and $y = f(x + d)$, for $d > 0$.

Consider the functions $y = \frac{1}{5}x^4$, $y = \frac{1}{5}(x - 3)^4$, and $y = \frac{1}{5}(x + 7)^4$ and compare their graphs. What pattern do you observe? Test it with some other graphs.

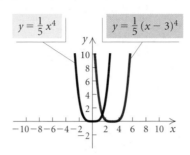
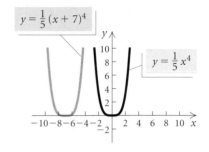

The effect of subtracting a constant from the x-value or adding a constant to the x-value in $y = f(x)$ is a shift of the graph of $f(x)$ to the right or left. Such a shift is called a **horizontal translation.**

Horizontal Translation

For $d > 0$:

> the graph of $y = f(x - d)$ is the graph of $y = f(x)$ shifted *right* d units;
>
> the graph of $y = f(x + d)$ is the graph of $y = f(x)$ shifted *left* d units.

EXAMPLE 4 Graph each of the following. Before doing so, describe how each graph can be obtained from one of the basic graphs shown on the preceding pages.

a) $g(x) = x^2 - 6$ **b)** $g(x) = |x - 4|$
c) $g(x) = \sqrt{x} + 2$ **d)** $h(x) = \sqrt{x + 2} - 3$

Solution

a) To graph $g(x) = x^2 - 6$, think of the graph of $f(x) = x^2$. Since $g(x) = f(x) - 6$, the graph of $g(x) = x^2 - 6$ is the graph of $f(x) = x^2$, shifted, or translated, *down* 6 units (see Fig. 1).

Let's compare some points on the graphs of f and g.

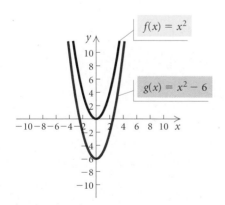

FIGURE 1

x	$f(x) = x^2$	$(x, f(x))$	$g(x) = x^2 - 6$	$(x, g(x))$
0	0	$(0, 0)$	-6	$(0, -6)$
2	4	$(2, 4)$	-2	$(2, -2)$
-2	4	$(-2, 4)$	-2	$(-2, -2)$
4	16	$(4, 16)$	10	$(4, 10)$
-4	16	$(-4, 16)$	10	$(-4, 10)$

We observe that the y-coordinate of a point on the graph of g is 6 less than the corresponding y-coordinate on the graph of f.

b) To graph $g(x) = |x - 4|$, think of the graph of $f(x) = |x|$. Since $g(x) = f(x - 4)$, the graph of $g(x) = |x - 4|$ is the graph of $f(x) = |x|$ shifted *right* 4 units (see Fig. 2).

Let's again compare points on the two graphs.

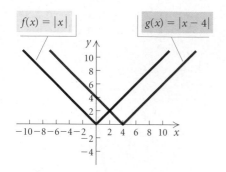

$f(x) = |x|$ $g(x) = |x - 4|$

FIGURE 2

x	$f(x) = |x|$	$(x, f(x))$	$g(x) = |x - 4|$	$(x, g(x))$
-4	4	$(-4, 4)$	8	$(-4, 8)$
-2	2	$(-2, 2)$	6	$(-2, 6)$
0	0	$(0, 0)$	4	$(0, 4)$
2	2	$(2, 2)$	2	$(2, 2)$
4	4	$(4, 4)$	0	$(4, 0)$

Observing points on f and g, we note that the x-coordinate of a point on the graph of g is 4 more than the x-coordinate of the corresponding point on f. Let's compare some ordered pairs.

Point on f Corresponding point on g
$(-4, 4) \longrightarrow (0, 4)$
$(-2, 2) \longrightarrow (2, 2)$
$(0, 0) \longrightarrow (4, 0)$

c) To graph $g(x) = \sqrt{x + 2}$, think of the graph of $f(x) = \sqrt{x}$. Since $g(x) = f(x + 2)$, the graph of $g(x) = \sqrt{x + 2}$ is the graph of $f(x) = \sqrt{x}$, shifted *left* 2 units (see Fig. 3).

d) To graph $h(x) = \sqrt{x + 2} - 3$, think of the graph of $f(x) = \sqrt{x}$. In part (c), we found that the graph of $g(x) = \sqrt{x + 2}$ is the graph of $f(x) = \sqrt{x}$ shifted left 2 units. Since $h(x) = g(x) - 3$, we shift the graph of $g(x) = \sqrt{x + 2}$ *down* 3 units. Together, the graph of $f(x) = \sqrt{x}$ is shifted *left* 2 units and *down* 3 units (see Fig. 4).

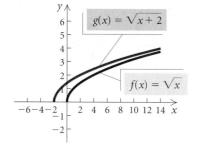

$g(x) = \sqrt{x + 2}$

$f(x) = \sqrt{x}$

FIGURE 3

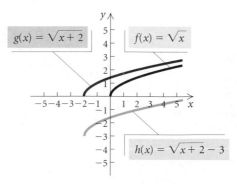

$g(x) = \sqrt{x + 2}$ $f(x) = \sqrt{x}$

$h(x) = \sqrt{x + 2} - 3$

FIGURE 4

Reflections

Suppose that we have a function given by $y = f(x)$. Let's explore the graphs of the new functions $y = -f(x)$ and $y = f(-x)$.

Compare the functions $y = f(x)$ and $y = -f(x)$ by observing the graphs of $y = \frac{1}{5}x^4$ and $y = -\frac{1}{5}x^4$ shown on the left below. What do you see? Test your observation with some other functions y_1 and y_2 where $y_2 = -y_1$.

Compare the functions $y = f(x)$ and $y = f(-x)$ by observing the graphs of $y = 2x^3 - x^4 + 5$ and $y = 2(-x)^3 - (-x)^4 + 5$ shown on the right below. What do you see? Test your observation with some other functions in which x is replaced with $-x$.

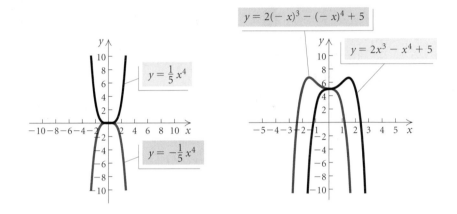

Given the graph of $y = f(x)$, we can reflect each point *across the x-axis* to obtain the graph of $y = -f(x)$. We can reflect each point of y *across the y-axis* to obtain the graph of $y = f(-x)$. The new graphs are called **reflections** of $y = f(x)$.

Reflections

The graph of $y = -f(x)$ is the **reflection** of the graph of $y = f(x)$ across the x-axis.

The graph of $y = f(-x)$ is the **reflection** of the graph of $y = f(x)$ across the y-axis.

If a point (x, y) is on the graph of $y = f(x)$, then $(x, -y)$ is on the graph of $y = -f(x)$, and $(-x, y)$ is on the graph of $y = f(-x)$.

EXAMPLE 5   Graph each of the following. Before doing so, describe how each graph can be obtained from the graph of $f(x) = x^3 - 4x^2$.

a) $g(x) = (-x)^3 - 4(-x)^2$ **b)** $h(x) = 4x^2 - x^3$

Solution

a) We first note that

$$f(-x) = (-x)^3 - 4(-x)^2 = g(x).$$

Thus the graph of g is a *reflection* of the graph of f across the y-axis (see Fig. 1). If (x, y) is on the graph of f, then $(-x, y)$ is on the graph of g. For example, $(2, -8)$ is on f and $(-2, -8)$ is on g.

b) We first note that

$$
\begin{aligned}
-f(x) &= -(x^3 - 4x^2) \\
&= -x^3 + 4x^2 \\
&= 4x^2 - x^3 \\
&= h(x).
\end{aligned}
$$

Thus the graph of h is a reflection of the graph of f across the x-axis (see Fig. 2). If (x, y) is on the graph of f, then $(x, -y)$ is on the graph of h. For example, $(2, -8)$ is on f and $(2, 8)$ is on h.

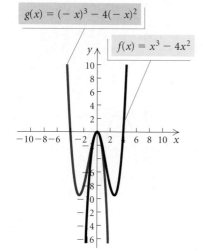

$g(x) = (-x)^3 - 4(-x)^2$

$f(x) = x^3 - 4x^2$

FIGURE 1

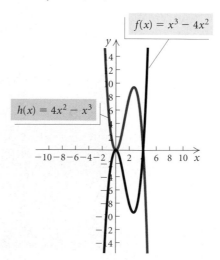

$f(x) = x^3 - 4x^2$

$h(x) = 4x^2 - x^3$

FIGURE 2

Vertical and Horizontal Stretchings and Shrinkings

Suppose that we have a function given by $y = f(x)$. Let's explore the graphs of the new functions $y = af(x)$ and $y = f(cx)$.

Consider the functions $y = x^3 - x$, $y = \frac{1}{10}(x^3 - x)$, $y = 2(x^3 - x)$, and $y = -2(x^3 - x)$ and compare their graphs. What pattern do you observe? Test it with some other graphs.

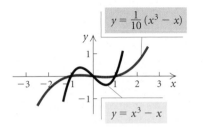

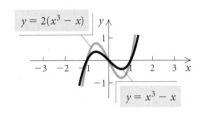

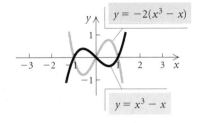

Consider any function f given by $y = f(x)$. Multiplying $f(x)$ by any constant a, where $|a| > 1$, to obtain $g(x) = af(x)$ will *stretch* the graph vertically away from the x-axis. If $0 < |a| < 1$, then the graph will be flattened or *shrunk* vertically toward the x-axis. If $a < 0$, the graph is also reflected across the x-axis.

> ### *Vertical Stretching and Shrinking*
>
> The graph of $y = af(x)$ can be obtained from the graph of $y = f(x)$ by
>
> > stretching vertically for $|a| > 1$, or
> > shrinking vertically for $0 < |a| < 1$.
>
> For $a < 0$, the graph is also reflected across the x-axis.
> (The y-coordinates of the graph of $y = af(x)$ can be obtained by multiplying the y-coordinates of $y = f(x)$ by a.)

Consider the functions $y = x^3 - x$, $y = (2x)^3 - (2x)$, $y = \left(\frac{1}{2}x\right)^3 - \left(\frac{1}{2}x\right)$, and $y = \left(-\frac{1}{2}x\right)^3 - \left(-\frac{1}{2}x\right)$ and compare their graphs. What pattern do you observe? Test it with some other graphs.

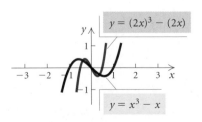

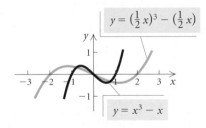

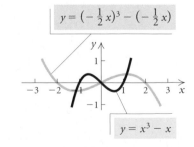

The constant c in the equation $g(x) = f(cx)$ will *stretch* the graph of $y = f(x)$ horizontally away from the y-axis if $0 < |c| < 1$. If $|c| > 1$, the graph will be *shrunk* horizontally toward the y-axis. If $c < 0$, the graph is also reflected across the y-axis.

Horizontal Stretching and Shrinking

The graph of $y = f(cx)$ can be obtained from the graph of $y = f(x)$ by

> shrinking horizontally for $|c| > 1$, or
>
> stretching horizontally for $0 < |c| < 1$.

For $c < 0$, the graph is also reflected across the y-axis.
(The x-coordinates of the graph of $y = f(cx)$ can be obtained by dividing the x-coordinates of the graph of $y = f(x)$ by c.)

It is instructive to use these concepts to create transformations of a given graph.

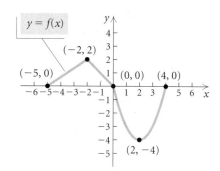

EXAMPLE 6 Shown at left is a graph of $y = f(x)$ for some function f. No formula for f is given. Graph each of the following.

a) $g(x) = 2f(x)$ **b)** $h(x) = \frac{1}{2}f(x)$

c) $r(x) = f(2x)$ **d)** $s(x) = f\left(\frac{1}{2}x\right)$

e) $t(x) = f\left(-\frac{1}{2}x\right)$

Solution

a) Since $|2| > 1$, the graph of $g(x) = 2f(x)$ is a vertical stretching of the graph of $y = f(x)$ by a factor of 2. We can consider the key points $(-5, 0)$, $(-2, 2)$, $(0, 0)$, $(2, -4)$, and $(4, 0)$ on the graph of $y = f(x)$. The transformation multiplies each y-coordinate by 2 to obtain the key points $(-5, 0)$, $(-2, 4)$, $(0, 0)$, $(2, -8)$, and $(4, 0)$ on the graph of $g(x) = 2f(x)$. The graph is shown below.

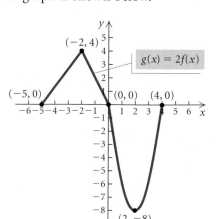

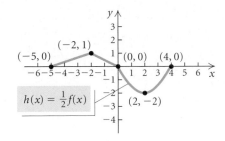

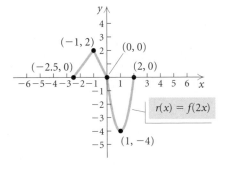

b) Since $\left|\frac{1}{2}\right| < 1$, the graph of $h(x) = \frac{1}{2} f(x)$ is a vertical shrinking of the graph of $y = f(x)$ by a factor of $\frac{1}{2}$. We again consider the key points $(-5, 0)$, $(-2, 2)$, $(0, 0)$, $(2, -4)$, and $(4, 0)$ on the graph of $y = f(x)$. The transformation multiplies each y-coordinate by $\frac{1}{2}$ to obtain the key points $(-5, 0)$, $(-2, 1)$, $(0, 0)$, $(2, -2)$, and $(4, 0)$ on the graph of $h(x) = \frac{1}{2} f(x)$. The graph is shown at left.

c) Since $|2| > 1$, the graph of $r(x) = f(2x)$ is a horizontal shrinking of the graph of $y = f(x)$. We consider the key points $(-5, 0)$, $(-2, 2)$, $(0, 0)$, $(2, -4)$, and $(4, 0)$ on the graph of $y = f(x)$. The transformation divides each x-coordinate by 2 to obtain the key points $(-2.5, 0)$, $(-1, 2)$, $(0, 0)$, $(1, -4)$, and $(2, 0)$ on the graph of $r(x) = f(2x)$. The graph is shown at left.

d) Since $\left|\frac{1}{2}\right| < 1$, the graph of $s(x) = f\left(\frac{1}{2} x\right)$ is a horizontal stretching of the graph of $y = f(x)$. We consider the key points $(-5, 0)$, $(-2, 2)$, $(0, 0)$, $(2, -4)$, and $(4, 0)$ on the graph of $y = f(x)$. The transformation divides each x-coordinate by $\frac{1}{2}$ (which is the same as multiplying by 2) to obtain the key points $(-10, 0)$, $(-4, 2)$, $(0, 0)$, $(4, -4)$, and $(8, 0)$ on the graph of $s(x) = f\left(\frac{1}{2} x\right)$. The graph is shown below.

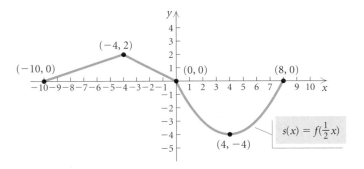

e) The graph of $t(x) = f\left(-\frac{1}{2} x\right)$ can be obtained by reflecting the graph in part (d) across the y-axis.

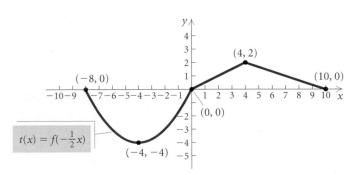

EXAMPLE 7 Use the graph of $y = f(x)$ shown at left to graph $y = -2f(x - 3) + 1$.

Solution

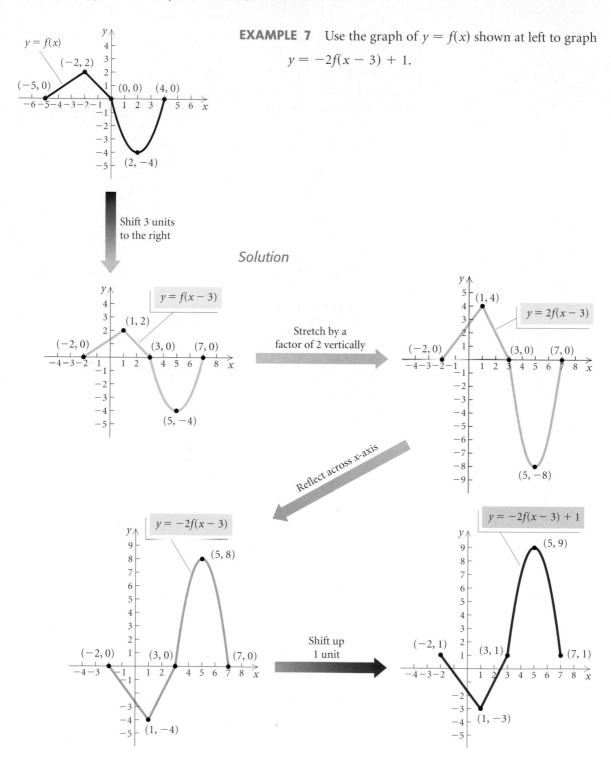

Summary of Transformations of $y = f(x)$

Vertical Translation: $y = f(x) \pm b$

For $b > 0$,

> the graph of $y = f(x) + b$ is the graph of $y = f(x)$ shifted *up* b units;

> the graph of $y = f(x) - b$ is the graph of $y = f(x)$ shifted *down* b units.

Horizontal Translation: $y = f(x \mp d)$

For $d > 0$,

> the graph of $y = f(x - d)$ is the graph of $y = f(x)$ shifted *right* d units;

> the graph of $y = f(x + d)$ is the graph of $y = f(x)$ shifted *left* d units.

Reflections

Across the x-axis: The graph of $y = -f(x)$ is the reflection of the graph of $y = f(x)$ across the x-axis.

Across the y-axis: The graph of $y = f(-x)$ is the reflection of the graph of $y = f(x)$ across the y-axis.

Vertical Stretching or Shrinking: $y = af(x)$

The graph of $y = af(x)$ can be obtained from the graph of $y = f(x)$ by

> stretching vertically for $|a| > 1$, or

> shrinking vertically for $0 < |a| < 1$.

For $a < 0$, the graph is also reflected across the x-axis.

Horizontal Stretching or Shrinking: $y = f(cx)$

The graph of $y = f(cx)$ can be obtained from the graph of $y = f(x)$ by

> shrinking horizontally for $|c| > 1$, or

> stretching horizontally for $0 < |c| < 1$.

For $c < 0$, the graph is also reflected across the y-axis.

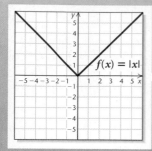

$f(x) = |x|$

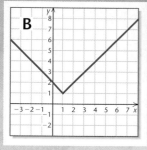

A

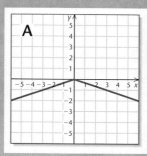

B

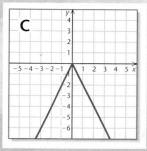

C

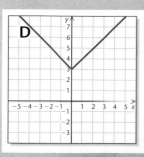

D

Visualizing the Graph

Match the function with its graph. Use transformation graphing techniques to obtain the graph of g from the basic function $f(x) = |x|$ shown at top left.

1. $g(x) = -2|x|$

2. $g(x) = |x - 1| + 1$

3. $g(x) = -\left|\dfrac{1}{3}x\right|$

4. $g(x) = |2x|$

5. $g(x) = |x + 2|$

6. $g(x) = |x| + 3$

7. $g(x) = -\dfrac{1}{2}|x - 4|$

8. $g(x) = \dfrac{1}{2}|x| - 3$

9. $g(x) = -|x| - 2$

Answers on page A-9

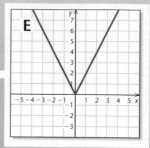

E

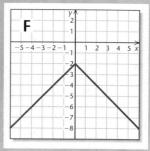

F

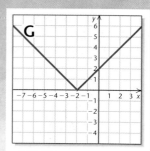

G

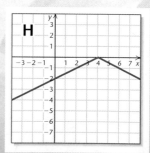

H

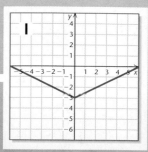

I

1.7

Exercise Set

Determine visually whether the graph is symmetric with respect to the x-axis, the y-axis, and/or the origin.

1.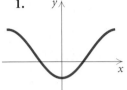

2.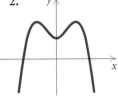

3.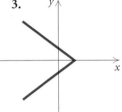

4.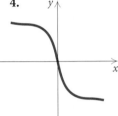

5.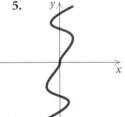

6.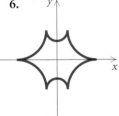

Determine whether the graph is symmetric with respect to the x-axis, the y-axis, and/or the origin.

7. $5x - 5y = 0$

8. $6x + 7y = 0$

9. $3x^2 - 2y^2 = 3$

10. $5y = 7x^2 - 2x$

11. $y = |2x|$

12. $y^3 = 2x^2$

13. $2x^4 + 3 = y^2$

14. $2y^2 = 5x^2 + 12$

15. $3y^3 = 4x^3 + 2$

16. $3x = |y|$

17. $xy = 12$

18. $xy - x^2 = 3$

Find the point that is symmetric to the given point with respect to the x-axis; the y-axis; the origin.

19. $(-5, 6)$

20. $\left(\frac{7}{2}, 0\right)$

21. $(-10, -7)$

22. $\left(1, \frac{3}{8}\right)$

23. $(0, -4)$

24. $(8, -3)$

Determine visually whether the function is even, odd, or neither even nor odd.

25.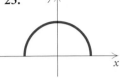

26.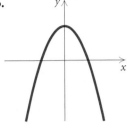

27.

28.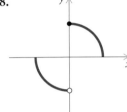

29.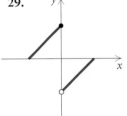

30.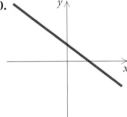

Determine whether the function is even, odd, or neither even nor odd.

31. $f(x) = -3x^3 + 2x$

32. $f(x) = 7x^3 + 4x - 2$

33. $f(x) = 5x^2 + 2x^4 - 1$

34. $f(x) = x + \frac{1}{x}$

35. $f(x) = x^{17}$

36. $f(x) = \sqrt[3]{x}$

37. $f(x) = \frac{1}{x^2}$

38. $f(x) = x - |x|$

39. $f(x) = 8$

40. $f(x) = \sqrt{x^2 + 1}$

Describe how the graph of the function can be obtained from one of the basic graphs on pages 141 and 142. Then graph the function.

41. $f(x) = (x - 3)^2$

42. $g(x) = x^2 + \frac{1}{2}$

43. $g(x) = x - 3$

44. $g(x) = -x - 2$

45. $h(x) = -\sqrt{x}$

46. $g(x) = \sqrt{x - 1}$

47. $h(x) = \frac{1}{x} + 4$

48. $g(x) = \frac{1}{x - 2}$

49. $h(x) = -3x + 3$

50. $f(x) = 2x + 1$

51. $h(x) = \frac{1}{2}|x| - 2$

52. $g(x) = -|x| + 2$

53. $g(x) = -(x - 2)^3$

54. $f(x) = (x + 1)^3$

55. $g(x) = (x + 1)^2 - 1$

56. $h(x) = -x^2 - 4$

57. $g(x) = \frac{1}{3}x^3 + 2$

58. $h(x) = (-x)^3$

59. $f(x) = \sqrt{x + 2}$

60. $f(x) = -\frac{1}{2}\sqrt{x - 1}$

61. $f(x) = \sqrt[3]{x} - 2$

62. $h(x) = \sqrt[3]{x + 1}$

Describe how the graph of the function can be obtained from one of the basic graphs on pages 141 and 142.

63. $g(x) = |3x|$

64. $f(x) = \frac{1}{2}\sqrt[3]{x}$

65. $h(x) = \frac{2}{x}$

66. $f(x) = |x - 3| - 4$

67. $f(x) = 3\sqrt{x} - 5$

68. $f(x) = 5 - \frac{1}{x}$

69. $g(x) = \left|\frac{1}{3}x\right| - 4$

70. $f(x) = \frac{2}{3}x^3 - 4$

71. $f(x) = -\frac{1}{4}(x - 5)^2$

72. $f(x) = (-x)^3 - 5$

73. $f(x) = \frac{1}{x + 3} + 2$

74. $g(x) = \sqrt{-x} + 5$

75. $h(x) = -(x - 3)^2 + 5$

76. $f(x) = 3(x + 4)^2 - 3$

The point $(-12, 4)$ is on the graph of $y = f(x)$. Find a point on the graph of $y = g(x)$.

77. $g(x) = \frac{1}{2}f(x)$

78. $g(x) = f(x - 2)$

79. $g(x) = f(-x)$

80. $g(x) = f(4x)$

81. $g(x) = f(x) - 2$

82. $g(x) = f\left(\frac{1}{2}x\right)$

83. $g(x) = 4f(x)$

84. $g(x) = -f(x)$

Write an equation for a function that has a graph with the given characteristics.

85. The shape of $y = x^2$, but upside-down and shifted right 8 units

86. The shape of $y = \sqrt{x}$, but shifted left 6 units and down 5 units

87. The shape of $y = |x|$, but shifted left 7 units and up 2 units

88. The shape of $y = x^3$, but upside-down and shifted right 5 units

89. The shape of $y = 1/x$, but shrunk horizontally by a factor of 2 and shifted down 3 units

90. The shape of $y = x^2$, but shifted right 6 units and up 2 units

91. The shape of $y = x^2$, but upside-down and shifted right 3 units and up 4 units

92. The shape of $y = |x|$, but stretched horizontally by a factor of 2 and shifted down 5 units

93. The shape of $y = \sqrt{x}$, but reflected across the y-axis and shifted left 2 units and down 1 unit

94. The shape of $y = 1/x$, but reflected across the x-axis and shifted up 1 unit

A graph of $y = f(x)$ follows. No formula for f is given. In Exercises 95–102, graph the given equation.

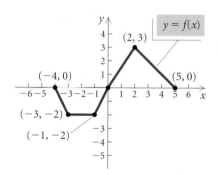

95. $g(x) = -2f(x)$

96. $g(x) = \frac{1}{2}f(x)$

97. $g(x) = f\left(-\frac{1}{2}x\right)$

98. $g(x) = f(2x)$

99. $g(x) = -\frac{1}{2}f(x - 1) + 3$

100. $g(x) = -3f(x + 1) - 4$

101. $g(x) = f(-x)$

102. $g(x) = -f(x)$

A graph of $y = g(x)$ follows. No formula for g is given. In Exercises 103–106, graph the given equation.

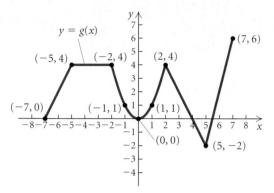

103. $h(x) = -g(x + 2) + 1$

104. $h(x) = \frac{1}{2}g(-x)$

105. $h(x) = g(2x)$

106. $h(x) = 2g(x - 1) - 3$

The graph of the function f is shown in figure (a). In Exercises 107–114, match the function g with one of the graphs (a)–(h), which follow. Some graphs may be used more than once.

a)

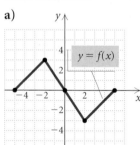

b)

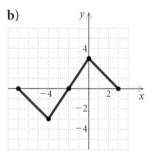

c)

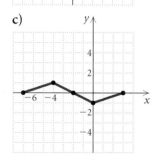

d)

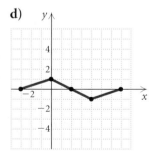

e)

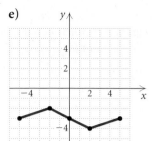

f)

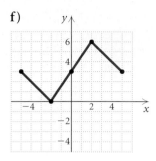

g)

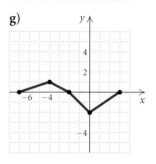

h)

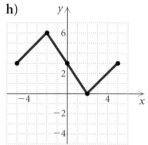

107. $g(x) = f(-x) + 3$

108. $g(x) = f(x) + 3$

109. $g(x) = -f(x) + 3$

110. $g(x) = -f(-x)$

111. $g(x) = \frac{1}{3}f(x - 2)$

112. $g(x) = \frac{1}{3}f(x) - 3$

113. $g(x) = \frac{1}{3}f(x + 2)$

114. $g(x) = -f(x + 2)$

For each pair of functions, determine if $g(x) = f(-x)$.

115. $f(x) = 2x^4 - 35x^3 + 3x - 5$,
$g(x) = 2x^4 + 35x^3 - 3x - 5$

116. $f(x) = \frac{1}{4}x^4 + \frac{1}{5}x^3 - 81x^2 - 17$,
$g(x) = \frac{1}{4}x^4 + \frac{1}{5}x^3 + 81x^2 - 17$

A graph of the function $f(x) = x^3 - 3x^2$ is shown below. Exercises 117–120 show graphs of functions transformed from this one. Find a formula for each function.

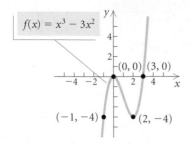

117.

118.

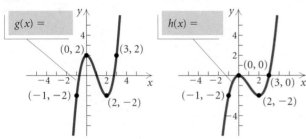

119.

120.

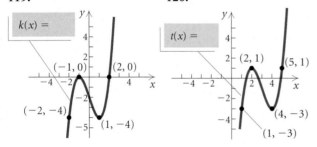

Collaborative Discussion and Writing

121. Consider the constant function $f(x) = 0$. Determine whether the graph of this function is symmetric with respect to the x-axis, the y-axis, and/or the origin. Determine whether this function is even or odd.

122. Describe conditions under which you would know whether a polynomial function $f(x) = a_n x^n + a_{n-1} x^{n-1} + \cdots + a_2 x^2 + a_1 x + a_0$ is even or odd without using an algebraic procedure. Explain.

123. Explain in your own words why the graph of $y = f(-x)$ is a reflection of the graph of $y = f(x)$ across the y-axis.

124. Without drawing the graph, describe what the graph of $f(x) = |x^2 - 9|$ looks like.

Skill Maintenance

125. Given $f(x) = 5x^2 - 7$, find each of the following.
 a) $f(-3)$
 b) $f(3)$
 c) $f(a)$
 d) $f(-a)$

126. Given $f(x) = 4x^3 - 5x$, find each of the following.
 a) $f(2)$
 b) $f(-2)$
 c) $f(a)$
 d) $f(-a)$

127. Write an equation of the line perpendicular to the graph of the line $8x - y = 10$ and containing the point $(-1, 1)$.

128. Find the slope and the y-intercept of the line with equation $2x - 9y + 1 = 0$.

Synthesis

Use the graph of the function f shown below in Exercises 129 and 130.

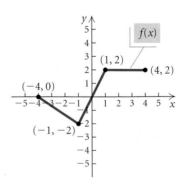

129. Graph: $y = |f(x)|$.

130. Graph: $y = f(|x|)$.

Use the graph of the function g shown below in Exercises 131 and 132.

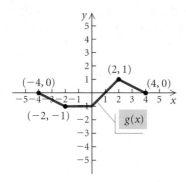

131. Graph: $y = |g(x)|$.

132. Graph: $y = g(|x|)$.

Determine whether the function is even, odd, or neither even nor odd.

133. $f(x) = x\sqrt{10 - x^2}$ **134.** $f(x) = \dfrac{x^2 + 1}{x^3 - 1}$

Determine whether the graph is symmetric with respect to the x-axis, the y-axis, and/or the origin.

135. $x^3 = y^2(2 - x)$

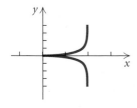

136. $(x^2 + y^2)^2 = 2xy$

137. $y^2 + 4xy^2 - y^4 = x^4 - 4x^3 + 3x^2 + 2x^2y^2$

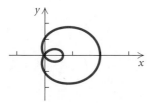

138. The graph of $f(x) = |x|$ passes through the points $(-3, 3)$, $(0, 0)$, and $(3, 3)$. Transform this function to one whose graph passes through the points $(5, 1)$, $(8, 4)$, and $(11, 1)$.

139. If $(-1, 5)$ is a point on the graph of $y = f(x)$, find b such that $(2, b)$ is on the graph of $y = f(x - 3)$.

State whether each of the following is true or false.

140. The product of two odd functions is odd.

141. The sum of two even functions is even.

142. The product of an even function and an odd function is odd.

143. Show that if f is *any* function, then the function E defined by

$$E(x) = \frac{f(x) + f(-x)}{2}$$

is even.

144. Show that if f is *any* function, then the function O defined by

$$O(x) = \frac{f(x) - f(-x)}{2}$$

is odd.

145. Consider the functions E and O of Exercises 143 and 144.
   a) Show that $f(x) = E(x) + O(x)$. This means that every function can be expressed as the sum of an even and an odd function.
   b) Let $f(x) = 4x^3 - 11x^2 + \sqrt{x} - 10$. Express f as a sum of an even function and an odd function.

Chapter 1 Summary and Review

Important Properties and Formulas

The Distance Formula

$$d = \sqrt{(x_2 - x_1)^2 + (y_2 - y_1)^2}$$

The Midpoint Formula

$$\left(\frac{x_1 + x_2}{2}, \frac{y_1 + y_2}{2} \right)$$

Equation of a Circle

$$(x - h)^2 + (y - k)^2 = r^2$$

Terminology about Lines

Slope: $m = \dfrac{y_2 - y_1}{x_2 - x_1}$

The Slope–Intercept Equation:
$$y = mx + b$$

The Point–Slope Equation:
$$y - y_1 = m(x - x_1)$$

Horizontal Lines: $y = b$

Vertical Lines: $x = a$

Parallel Lines: $m_1 = m_2,\ b_1 \neq b_2$

Perpendicular Lines:
$$m_1 m_2 = -1, \text{ or}$$
$$x = a,\ y = b$$

The Algebra of Functions

The Sum of Two Functions:
$(f + g)(x) = f(x) + g(x)$

The Difference of Two Functions:
$(f - g)(x) = f(x) - g(x)$

The Product of Two Functions:
$(fg)(x) = f(x) \cdot g(x)$

The Quotient of Two Functions:
$(f/g)(x) = f(x)/g(x),\ g(x) \neq 0$

The Composition of Two Functions:
$(f \circ g)(x) = f(g(x))$

Tests for Symmetry

x-axis: If replacing y with $-y$ produces an equivalent equation, then the graph is symmetric with respect to the x-axis.

y-axis: If replacing x with $-x$ produces an equivalent equation, then the graph is symmetric with respect to the y-axis.

Origin: If replacing x with $-x$ and y with $-y$ produces an equivalent equation, then the graph is symmetric with respect to the origin.

Even Function: $f(-x) = f(x)$

Odd Function: $f(-x) = -f(x)$

Transformations

Vertical Translation:	$y = f(x) \pm b$
Horizontal Translation:	$y = f(x \mp d)$
Reflection across the x-axis:	$y = -f(x)$
Reflection across the y-axis:	$y = f(-x)$

Vertical Stretching or Shrinking:
$$y = af(x)$$

Horizontal Stretching or Shrinking:
$$y = f(cx)$$

Review Exercises

Answers for all of the review exercises appear in the answer section. If you get an incorrect answer, return to the section of the textbook indicated in the answer.

Use substitution to determine whether the given ordered pairs are solutions of the given equation.

1. $\left(3, \frac{24}{9}\right), (0, -9);\ 2x - 9y = -18$

2. $(0, 7), (7, 1);\ y = 7$

Graph the equation.

3. $y = -\frac{2}{3}x + 1$

4. $2x - 4y = 8$

5. $y = 2 - x^2$

6. Find the distance between $(3, 7)$ and $(-2, 4)$.

7. Find the midpoint of the segment with endpoints $(3, 7)$ and $(-2, 4)$.

8. Find an equation of the circle with center $(-2, 6)$ and radius $\sqrt{13}$.

Find the center and the radius of the circle.

9. $(x + 1)^2 + (y - 3)^2 = 16$

10. $x^2 - 6x + y^2 + 10y + 33 = 0$

11. Find an equation of the circle having a diameter with endpoints $(-3, 5)$ and $(7, 3)$.

Determine whether the relation is a function. Identify the domain and the range.

12. $\{(3, 1), (5, 3), (7, 7), (3, 5)\}$

13. $\{(2, 7), (-2, -7), (7, -2), (0, 2), (1, -4)\}$

Determine whether the graph is that of a function.

14.

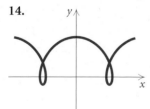

15.

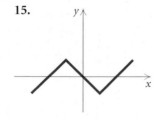

16.

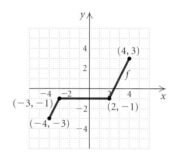

17.

18. A graph of a function is shown. Find $f(2)$, $f(-3)$, and $f(0)$.

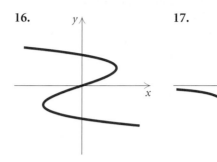

Find the domain of the function.

19. $f(x) = 4 - 5x + x^2$

20. $f(x) = \dfrac{3}{x} + 2$

21. $f(x) = \dfrac{1}{x^2 - 6x + 5}$

22. $f(x) = \dfrac{-5x}{|16 - x^2|}$

Graph the function. Then visually estimate the domain and the range.

23. $f(x) = \sqrt{16 - x^2}$

24. $g(x) = |x - 5|$

25. $f(x) = x^3 - 7$

26. $h(x) = x^4 + x^2$

27. Given that $f(x) = x^2 - x - 3$, find each of the following.

a) $f(0)$

b) $f(-3)$

c) $f(a - 1)$

d) $\dfrac{f(x + h) - f(x)}{h}$

In Exercises 28 and 29, the table of data contains input–output values for a function. Answer the following questions.

a) *Is the change in the inputs, x, the same?*
b) *Is the change in the outputs, y, the same?*
c) *Is the function linear?*

28.

x	y
−3	8
−2	11
−1	14
0	17
1	20
2	22
3	26

29.

x	y
20	11.8
30	24.2
40	36.6
50	49.0
60	61.4
70	73.8
80	86.2

Find the slope of the line containing the given points.

30. $(2, -11), (5, -6)$

31. $(5, 4), (-3, 4)$

32. $\left(\frac{1}{2}, 3\right), \left(\frac{1}{2}, 0\right)$

33. Find the slope and the *y*-intercept of the graph of
$$-2x - y = 7.$$

34. *Cost of a Formal Wedding.* In 1995, the average cost of a formal wedding was $17,634. This cost rose to $22,360 in 2003. (*Sources: Bride's Magazine; Washington Post*) Find the average rate of change in the cost of a formal wedding from 1995 to 2003.

35. Graph $y = -\frac{1}{4}x + 3$ using the slope and the *y*-intercept.

Write a slope–intercept equation for a line with the following characteristics.

36. $m = -\frac{2}{3}$, *y*-intercept $(0, -4)$

37. $m = 3$, passes through $(-2, -1)$

38. Passes through $(4, 1)$ and $(-2, -1)$

Determine whether the lines are parallel, perpendicular, or neither.

39. $3x - 2y = 8,$
$6x - 4y = 2$

40. $y - 2x = 4,$
$2y - 3x = -7$

41. $y = \frac{3}{2}x + 7,$
$y = -\frac{2}{3}x - 4$

Given the point $(1, -1)$ and the line $2x + 3y = 4$:

42. Find an equation of the line containing the given point and parallel to the given line.

43. Find an equation of the line containing the given point and perpendicular to the given line.

44. *Total Cost.* Clear County Cable Television charges a $25 installation fee and $20 per month for "basic" service. Write an equation that can be used to determine the total cost, $C(t)$, of t months of basic cable television service. Find the total cost of 6 months of service.

45. *Temperature and Depth of the Earth.* The function T given by $T(d) = 10d + 20$ can be used to determine the temperature T, in degrees Celsius, at a depth d, in kilometers, inside the earth.
a) Find $T(5)$, $T(20)$, and $T(1000)$.
b) The radius of the earth is about 5600 km. Use this fact to determine the domain of the function.

46. *Motor Vehicle Production.* The data in the following table show the increase in world motor vehicle production since 1950.

YEAR, x	NUMBER OF VEHICLES PRODUCED IN THE WORLD, y (IN MILLIONS)
1950, 0	10.6
1960, 10	16.5
1970, 20	29.4
1980, 30	38.6
1990, 40	48.6
1995, 45	50.0
2000, 50	57.5
2001, 51	56.3

Source: New York Times Almanac, 2003

Model the data with a linear function and, using this function, estimate the number of vehicles that will be produced in 2010. Answers will vary depending on the data points used.

Graph each of the following.

47. $f(x) = \begin{cases} x^3, & \text{for } x < -2, \\ |x|, & \text{for } -2 \le x \le 2, \\ \sqrt{x - 1}, & \text{for } x > 2 \end{cases}$

48. $f(x) = \begin{cases} \dfrac{x^2 - 1}{x + 1}, & \text{for } x \neq -1, \\ 3, & \text{for } x = -1 \end{cases}$

49. $f(x) = [\![x]\!]$

50. $f(x) = [\![x - 3]\!]$

51. For the function in Exercise 47, find $f(-1)$, $f(5)$, $f(-2)$, and $f(-3)$.

52. *Inscribed Rectangle.* A rectangle is inscribed in a semicircle of radius 2, as shown. The variable $x =$ half the length of the rectangle. Express the area of the rectangle as a function of x.

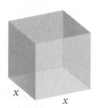

53. *Minimizing Surface Area.* A container firm is designing an open-top rectangular box, with a square base, that will hold 108 in³. Let $x =$ the length of a side of the base.

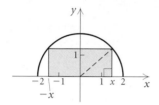

a) Express the surface area as a function of x.
b) Find the domain of the function.
c) Using the graph shown below, determine the dimensions that will minimize the surface area of the box.

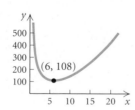

Given that $f(x) = \sqrt{x - 2}$ and $g(x) = x^2 - 1$, find each of the following if it exists.

54. $(f - g)(6)$

55. $(fg)(2)$

56. $(f + g)(-1)$

For each pair of functions in Exercises 57 and 58:
a) Find the domain of f, g, $f + g$, $f - g$, fg, and f/g.
b) Find $(f + g)(x)$, $(f - g)(x)$, $fg(x)$, and $(f/g)(x)$.

57. $f(x) = \dfrac{4}{x^2}$; $g(x) = 3 - 2x$

58. $f(x) = 3x^2 + 4x$; $g(x) = 2x - 1$

59. Given the total-revenue and total-cost functions $R(x) = 120x - 0.5x^2$ and $C(x) = 15x + 6$, find the total-profit function $P(x)$.

60. For the function
$$f(x) = 3 - x^2,$$
construct and simplify the difference quotient.

In Exercises 61 and 62, for the pair of functions:
a) Find the domain of $f \circ g$ and $g \circ f$.
b) Find $(f \circ g)(x)$ and $(g \circ f)(x)$.

61. $f(x) = \dfrac{4}{x^2}$; $g(x) = 3 - 2x$

62. $f(x) = 3x^2 + 4x$; $g(x) = 2x - 1$

Find $f(x)$ and $g(x)$ such that $h(x) = (f \circ g)x$.

63. $h(x) = \sqrt{5x + 2}$

64. $h(x) = 4(5x - 1)^2 + 9$

Graph the given equation and determine visually whether it is symmetric with respect to the x-axis, the y-axis, and/or the origin. Then verify your assertion algebraically.

65. $x^2 + y^2 = 4$

66. $y^2 = x^2 + 3$

67. $x + y = 3$

68. $y = x^2$

69. $y = x^3$

70. $y = x^4 - x^2$

Determine visually whether the function is even, odd, or neither even nor odd.

71.

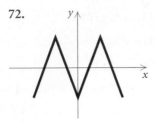

72.

73.

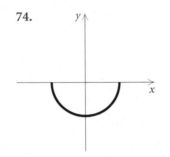

74.

Test whether the function is even, odd, or neither even nor odd.

75. $f(x) = 9 - x^2$

76. $f(x) = x^3 - 2x + 4$

77. $f(x) = x^7 - x^5$

78. $f(x) = |x|$

79. $f(x) = \sqrt{16 - x^2}$

80. $f(x) = \dfrac{10x}{x^2 + 1}$

Write an equation for a function that has a graph with the given characteristics.

81. The shape of $y = x^2$, but shifted left 3 units

82. The shape of $y = \sqrt{x}$, but upside-down and shifted right 3 units and up 4 units

83. The shape of $y = |x|$, but stretched vertically by a factor of 2 and shifted right 3 units

A graph of $y = f(x)$ is shown at right. No formula for f is given. Graph each of the following.

$y = f(x)$

84. $y = f(x - 1)$

85. $y = f(2x)$

86. $y = -2f(x)$

87. $y = 3 + f(x)$

Technology Connection

88. a) Fit a regression line to the data in Exercise 46 and use it to estimate the number of vehicles produced in 2010.

b) What is the correlation coefficient for the regression line? How close a fit is the regression line?

Collaborative Discussion and Writing

89. Given that $f(x) = 4x^3 - 2x + 7$, find each of the following. Then discuss how each expression differs from the other.

a) $f(x) + 2$ **b)** $f(x + 2)$ **c)** $f(x) + f(2)$

90. Given the graph of $y = f(x)$, explain and contrast the effect of the constant c on the graphs of $y = f(cx)$ and $y = cf(x)$.

91. a) Graph several functions of the type $y_1 = f(x)$ and $y_2 = |f(x)|$. Describe a procedure, involving transformations, for creating the graph of y_2 from y_1.

b) Describe a procedure, involving transformations, for creating the graph of $y_2 = f(|x|)$ from $y_1 = f(x)$.

Synthesis

Find the domain.

92. $f(x) = \dfrac{\sqrt{1 - x}}{x - |x|}$

93. $f(x) = (x - 9x^{-1})^{-1}$

94. Prove that the sum of two odd functions is odd.

95. Describe how the graph of $y = -f(-x)$ is obtained from the graph of $y = f(x)$.

Chapter 1 Test

1. Graph: $5x - 2y = -10$.

2. Find the distance between $(5, 8)$ and $(-1, 5)$.

3. Find the midpoint of the segment with endpoints $(-2, 6)$ and $(-4, 3)$.

4. Find an equation of the circle with center $(-1, 2)$ and radius $\sqrt{5}$.

5. Find the center and the radius of the circle
$$x^2 + 8x + y^2 - 10y + 5 = 0.$$

6. **a)** Determine whether the relation
$$\{(-4, 7), (3, 0), (1, 5), (0, 7)\}$$
is a function. Answer yes or no.
   b) Find the domain of the relation.
   c) Find the range of the relation.

7. Given that $f(x) = 2x^2 - x + 5$, find each of the following.
   a) $f(-1)$
   b) $f(a + 2)$

8. **a)** Graph: $f(x) = |x - 2| + 3$.
   b) Visually estimate the domain of $f(x)$.
   c) Visually estimate the range of $f(x)$.

Find the domain of the function.

9. $f(x) = \dfrac{1}{x - 4}$

10. $g(x) = x^3 + 2$

11. $h(x) = \sqrt{25 - x^2}$

12. Determine whether each graph is that of a function. Answer yes or no.
   a)

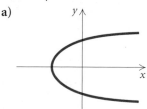

   b)

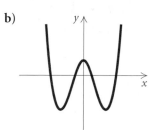

Find the slope of the line containing the given points.

13. $\left(-2, \frac{2}{3}\right), (-2, 5)$

14. $(4, -10), (-8, 12)$

15. $(-5, 6), \left(\frac{3}{4}, 6\right)$

16. *Debit-Card Transactions.* The total volume (in U.S. dollars) of debit-card transactions was \$8.2 billion in 1990. In 1999, the total was \$157.9 billion. Find the average rate of change in total debit-card transactions from 1990 to 1999.

17. Find the slope and the y-intercept of the graph of $-3x + 2y = 5$.

18. Write an equation for the line with $m = -\frac{5}{8}$ and y-intercept $(0, -5)$.

19. Write an equation for the line that passes through $(-5, 4)$ and $(3, -2)$.

20. Find an equation of the line containing the point $(-1, 3)$ and parallel to the line $x + 2y = -6$.

21. Determine whether the lines are parallel, perpendicular, or neither.
$$2x + 3y = -12,$$
$$2y - 3x = 8$$

22. The table below shows the per capita consumption of coffee in the United States in several years.

YEAR, x	PER CAPITA CONSUMPTION OF COFFEE (IN GALLONS)
1997, 0	23.3
1998, 1	23.9
1999, 2	25.1
2000, 3	26.3

Source: U.S. Department of Agriculture, Economic Research Service

Model the data listed in the table with a linear function and use that function to predict the per capita coffee consumption in 2005. Answers will vary depending on the data points used.

23. Graph $f(x)$:

$$f(x) = \begin{cases} x^2, & \text{for } x < -1, \\ |x|, & \text{for } -1 \le x \le 1, \\ \sqrt{x-1}, & \text{for } x > 1. \end{cases}$$

24. For the function in Exercise 23, find $f\left(-\frac{7}{8}\right)$, $f(5)$, and $f(-4)$.

25. Given that $f(x) = x^2 - 4x + 3$ and $g(x) = \sqrt{3-x}$, find $(f-g)(-1)$.

26. For $f(x) = x^2$ and $g(x) = \sqrt{x-3}$, find each of the following.

  a) The domain of f
  b) The domain of g
  c) $(f-g)(x)$
  d) $(fg)(x)$
  e) The domain of $(f/g)(x)$

27. Construct and simplify the difference quotient for

$$f(x) = x^2 + 4.$$

For $f(x) = \sqrt{x-5}$ and $g(x) = x^2 + 1$:

28. Find $(f \circ g)(x)$ and $(g \circ f)(x)$.

29. Find the domain of $(f \circ g)(x)$ and $(g \circ f)(x)$.

30. Determine whether the graph of $y = x^4 - 2x^2$ is symmetric with respect to the x-axis, the y-axis, and/or the origin.

31. Test whether the function

$$f(x) = \frac{2x}{x^2 + 1}$$

is even, odd, or neither even nor odd. Show your work.

32. Write an equation for a function that has the shape of $y = x^2$, but shifted right 2 units and down 1 unit.

33. Write an equation for a function that has the shape of $y = x^2$, but shifted left 2 units and down 3 units.

34. The graph of a function $y = f(x)$ is shown below. No formula for f is given. Graph $y = -\frac{1}{2}f(x)$.

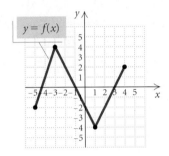

Synthesis

35. If $(-3, 1)$ is a point on the graph of $y = f(x)$, what point do you know is on the graph of $y = f(3x)$?

Functions, Equations, and Inequalities

*T*he number of U.S. households with broadband Internet access is projected to be 54.8 million in 2006. This is 25.6 million more households than in 2003. (*Source*: Forrester Research) How many U.S. households had broadband Internet access in 2003?

This problem appears as Exercise 29 in Section 2.1.

2.1

Linear Equations, Functions, and Models

- *Solve linear equations.*
- *Solve applied problems using linear models.*
- *Find zeros of linear functions.*
- *Solve a formula for a given variable.*

An **equation** is a statement that two expressions are equal. To **solve** an equation in one variable is to find all the values of the variable that make the equation true. Each of these numbers is a **solution** of the equation. The set of all solutions of an equation is its **solution set.** Some examples of **equations in one variable** are

$$2x + 3 = 5, \quad 3(x - 1) = 4x + 5, \quad x^2 - 3x + 2 = 0, \quad \text{and} \quad \frac{x - 3}{x + 4} = 1.$$

Linear Equations

The first two equations above are *linear equations* in one variable. We define such equations as follows.

> A **linear equation in one variable** is an equation that is equivalent to one of the form $mx + b = 0$, where m and b are real numbers and $m \neq 0$.

Equations that have the same solution set are **equivalent equations.** For example, $2x + 3 = 5$ and $x = 1$ are equivalent equations because 1 is the solution of each equation. On the other hand, $x^2 - 3x + 2 = 0$ and $x = 1$ are not equivalent equations because 1 and 2 are both solutions of $x^2 - 3x + 2 = 0$ but 2 is not a solution of $x = 1$.

To solve an equation, we find an equivalent equation in which the variable is isolated. The following principles allow us to solve linear equations.

Note to the student and the instructor: We assume that students come to a College Algebra course with some equation-solving skills from Intermediate Algebra. Thus a portion of the material presented in this section might be considered by some to be review in nature. We present this material here in order to use linear functions, with which students are familiar, to lay the groundwork for zeros of higher-order polynomial functions and their connection to solutions of equations and *x*-intercepts of graphs.

> ### Equation-Solving Principles
> For any real numbers a, b, and c:
> ***The Addition Principle:*** If $a = b$ is true, then $a + c = b + c$ is true.
> ***The Multiplication Principle:*** If $a = b$ is true, then $ac = bc$ is true.

EXAMPLE 1 Solve: $6x - 4 = 1$.

Algebraic Solution

We have

$$6x - 4 = 1$$

$$6x - 4 + 4 = 1 + 4 \qquad \textbf{Using the addition principle to add 4 on both sides}$$

$$6x = 5$$

$$\frac{6x}{6} = \frac{5}{6} \qquad \textbf{Using the multiplication principle to multiply by } \frac{1}{6}, \textbf{ or divide by 6, on both sides}$$

$$x = \frac{5}{6}. \qquad \textbf{Note that } 6x - 4 = 1 \textbf{ and } x = \frac{5}{6} \textbf{ are equivalent equations.}$$

CHECK:
$$\begin{array}{c|c} 6x - 4 = 1 \\ \hline 6 \cdot \frac{5}{6} - 4 \;?\; 1 & \textbf{Substituting } \frac{5}{6} \textbf{ for } x \\ 5 - 4 \\ 1 & 1 \quad \text{TRUE} \end{array}$$

The solution is $\frac{5}{6}$.

Visualizing the Solution

We graph $y = 6x - 4$ and $y = 1$. The first coordinate of the point of intersection of the graphs is the value of x for which $6x - 4 = 1$ and is thus the solution of the equation.

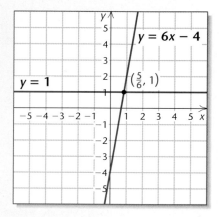

The solution is $\frac{5}{6}$.

Technology Connection

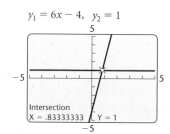

We can use the INTERSECT feature on a graphing calculator to solve equations. We call this the **Intersect method.** To use the Intersect method to solve the equation in Example 1, for instance, we graph $y_1 = 6x - 4$ and $y_2 = 1$. The value of x for which $y_1 = y_2$ is the solution of the equation. This value of x is the first coordinate of the point of intersection of the graphs of y_1 and y_2. Using the INTERSECT feature, we find that the first coordinate of this point is approximately 0.83333333.

If the solution is a rational number, we can find fraction notation for the exact solution by using the ▶FRAC feature. The solution is $\frac{5}{6}$.

EXAMPLE 2 Solve: $2(5 - 3x) = 8 - 3(x + 2)$.

Solution We have

$$2(5 - 3x) = 8 - 3(x + 2)$$

$10 - 6x = 8 - 3x - 6$	**Using the distributive property**
$10 - 6x = 2 - 3x$	**Combining like terms**
$10 - 6x + 6x = 2 - 3x + 6x$	**Using the addition principle to add $6x$ on both sides**
$10 = 2 + 3x$	
$10 - 2 = 2 + 3x - 2$	**Using the addition principle to add -2, or subtract 2, on both sides**
$8 = 3x$	
$\dfrac{8}{3} = \dfrac{3x}{3}$	**Using the multiplication principle to multiply by $\frac{1}{3}$, or divide by 3, on both sides**
$\dfrac{8}{3} = x.$	

Technology Connection

We can use the TABLE feature on a graphing calculator, set in ASK mode, to check the solutions of equations. In Example 2, for instance, let $y_1 = 2(5 - 3x)$ and $y_2 = 8 - 3(x + 2)$. When $\frac{8}{3}$ is entered for x, we see that y_1 and y_2 are both -6. (Note that the calculator converts $\frac{8}{3}$ to decimal notation.)

X	Y1	Y2
2.6667	−6	−6

X =

CHECK:

$$2(5 - 3x) = 8 - 3(x + 2)$$

$2\left(5 - 3 \cdot \frac{8}{3}\right)$	?	$8 - 3\left(\frac{8}{3} + 2\right)$
$2(5 - 8)$		$8 - 3\left(\frac{14}{3}\right)$
$2(-3)$		$8 - 14$
-6		-6

Substituting $\frac{8}{3}$ for x

TRUE

The solution is $\frac{8}{3}$.

Applications Using Linear Models

Mathematical techniques are used to answer questions arising from real-world situations. Linear equations and functions *model* many of these situations.

The following strategy is of great assistance in problem solving.

Five Steps for Problem Solving

1. **Familiarize** yourself with the problem situation. If the problem is presented in words, this means to read carefully. Some or all of the following can also be helpful.

   a) Make a drawing, if it makes sense to do so.

   b) Make a written list of the known facts and a list of what you wish to find out.

   c) Assign variables to represent unknown quantities.

   d) Organize the information in a chart or a table.

   e) Find further information. Look up a formula, consult a reference book or an expert in the field, or do research on the Internet.

   f) Guess or estimate the answer and check your guess or estimate.

2. **Translate** the problem situation to mathematical language or symbolism. For most of the problems you will encounter in algebra, this means to write one or more equations, but sometimes an inequality or some other mathematical symbolism may be appropriate.

3. **Carry out** some type of mathematical manipulation. Use your mathematical skills to find a possible solution. In algebra, this usually means to solve an equation, an inequality, or a system of equations.

4. **Check** to see whether your possible solution actually fits the problem situation and is thus really a solution of the problem. You might be able to solve an equation, but the solution(s) of the equation might not be solution(s) of the original problem.

5. **State** the answer clearly using a complete sentence.

EXAMPLE 3   *Time Devoted to Media.*   In 2003, each person in the United States devoted an average of 3587 hr to TV, music, books, movies, and the Internet. This was a 5% increase over the amount of time devoted to these media in 1998. (*Source*: Veronis, Suhler & Associates) How much time did each person devote to media, on average, in 1998?

Solution

1. **Familiarize.** Let's estimate that an average of 3400 hr was devoted to media in 1998. Then the number of hours devoted to media in 2003 would be

$$3400 + 5\% \cdot 3400 = 1(3400) + 0.05(3400)$$
$$= 1.05(3400)$$
$$= 3570.$$

Since 3587 hr per person were actually devoted to media in 2003, the estimate is close but a little low. Nevertheless, the calculations performed in checking the estimate indicate how we can translate this problem to an equation. We let m = the number of hours each person devoted to media, on average, in 1998. Then $m + 5\%m$, or $1 \cdot m + 0.05 \cdot m$, or $1.05m$, is the number of hours devoted to media in 2003.

2. **Translate.** We translate to an equation.

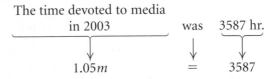

3. **Carry out.** We solve the equation, as follows:

$$1.05m = 3587$$

$$m = \frac{3587}{1.05} \qquad \textbf{Dividing by 1.05 on both sides}$$

$$m \approx 3416.$$

4. **Check.** We see that 5% of 3416 is about 171 and $3416 + 171 = 3587$, so the answer checks.

5. **State.** On average, each person in the United States devoted about 3416 hr to media in 1998. ▬

In some applications, we need to use a formula that describes the relationships among variables. When a situation involves distance, rate (also called speed or velocity), and time, for example, we use the following formula.

> ### The Motion Formula
> The distance d traveled by an object moving at rate r in time t is given by
> $$d = r \cdot t.$$

EXAMPLE 4 *Airplane Speed.* America West Airlines' fleet includes Boeing 737-200's, each with a cruising speed of 500 mph, and Bombardier deHavilland Dash 8-200's, each with a cruising speed of 302 mph (*Source:* America West Airlines). Suppose that a Dash 8-200 takes off and travels at its cruising speed. One hour later, a 737-200 takes off and follows the same route, traveling at its cruising speed. How long will it take the 737-200 to overtake the Dash 8-200?

Solution

1. **Familiarize.** We make a drawing showing both the known and the unknown information. We let t = the time, in hours, that the 737-200 travels before it overtakes the Dash 8-200. Since the Dash 8-200 takes

off 1 hr before the 737, it will travel for $t + 1$ hr before being over-taken. The planes will have traveled the same distance, d, when one overtakes the other.

737-200
500 mph t hr d mi

737-200
overtakes
Dash 8-200   Dash 8-200
302 mph $t + 1$ hr d mi   here.

We can also organize the information in a table, as follows.

$$d = r \cdot t$$

	DISTANCE	RATE	TIME	
737-200	d	500	t	$\longrightarrow d = 500t$
DASH 8-200	d	302	$t + 1$	$\longrightarrow d = 302(t + 1)$

2. **Translate.** Using the formula $d = rt$ in each row of the table, we get two expressions for d:

$$d = 500t \quad \text{and} \quad d = 302(t + 1).$$

Since the distances are the same, we have the following equation:

$$500t = 302(t + 1).$$

3. **Carry out.** We solve the equation, as follows:

$500t = 302(t + 1)$

$500t = 302t + 302$ **Using the distributive property**

$198t = 302$ **Subtracting 302t on both sides**

$t \approx 1.53.$ **Dividing by 198 on both sides and rounding to the nearest hundredth**

4. **Check.** If the 737-200 travels for about 1.53 hr, then the Dash 8-200 travels for about $1.53 + 1$, or 2.53 hr. In 2.53 hr, the Dash 8-200 travels 302(2.53), or 764.06 mi, and in 1.53 hr, the 737-200 travels 500(1.53), or 765 mi. Since 764.06 mi $\approx$ 765 mi, the answer checks. (Remember that we rounded the value of t.)

5. **State.** About 1.53 hr after the 737-200 has taken off, it will overtake the Dash 8-200.

For some applications, we need to use a formula to find the amount of interest earned by an investment or the amount of interest due on a loan.

> ### The Simple-Interest Formula
> The **simple interest** I on a principal of P dollars at interest rate r for t years is given by
> $$I = Prt.$$

EXAMPLE 5 *Student Loans.* Jared's two student loans total $12,000. One loan is at 5% simple interest and the other is at 8% simple interest. After 1 yr, Jared owes $750 in interest. What is the amount of each loan?

Solution

1. **Familiarize.** We let $x =$ the amount borrowed at 5% interest. Then the remainder of the $12,000, or $12,000 - x$, is borrowed at 8%. We organize the information in a table, keeping in mind the formula $I = Prt$.

	AMOUNT BORROWED	INTEREST RATE	TIME	AMOUNT OF INTEREST
5% LOAN	x	5%, or 0.05	1 yr	$x(0.05)(1)$, or $0.05x$
8% LOAN	$12,000 - x$	8%, or 0.08	1 yr	$(12,000 - x)(0.08)(1)$, or $0.08(12,000 - x)$
TOTAL	12,000			750

2. **Translate.** The total amount of interest on the two loans is $750. Thus we write the following equation.

$$
\underbrace{\text{Interest on 5\% loan}}_{0.05x} \;\underbrace{\text{plus}}_{+}\; \underbrace{\text{interest on 8\% loan}}_{0.08(12{,}000 - x)} \;\underbrace{\text{is}}_{=}\; \underbrace{\$750.}_{750}
$$

3. **Carry out.** We solve the equation, as follows:

$$0.05x + 0.08(12{,}000 - x) = 750$$
$$0.05x + 960 - 0.08x = 750 \qquad \text{Using the distributive property}$$
$$-0.03x + 960 = 750 \qquad \text{Collecting like terms}$$
$$-0.03x = -210 \qquad \text{Subtracting 960 on both sides}$$
$$x = 7000. \qquad \text{Dividing by } -0.03 \text{ on both sides}$$

If $x = 7000$, then $12{,}000 - x = 12{,}000 - 7000 = 5000$.

4. **Check.** The interest on $7000 at 5% for 1 yr is $7000(0.05)(1), or $350. The interest on $5000 at 8% for 1 yr is $5000(0.08)(1), or $400. Since $350 + $400 = $750, the answer checks.

5. **State.** Jared borrowed $7000 at 5% interest and $5000 at 8% interest.

Sometimes we use formulas from geometry in solving applied problems. In the following example, we use the formula for the perimeter P of a rectangle with length l and width w: $P = 2l + 2w$. There is a summary of geometric formulas at the back of the book.

EXAMPLE 6 *Soccer Fields.* The length of the largest regulation soccer field is 30 yd greater than the width and the perimeter is 460 yd. Find the length and the width.

Solution

1. **Familiarize.** We first make a drawing. Since the length of the field is described in terms of the width, we let $w =$ the width, in yards. Then $w + 30 =$ the length, in yards.

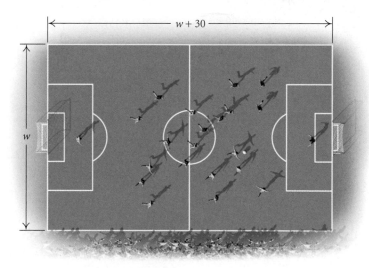

2. **Translate.** We use the formula for the perimeter of a rectangle:

$$P = 2l + 2w$$
$$460 = 2(w + 30) + 2w. \qquad \text{Substituting 460 for } P \text{ and } w + 30 \text{ for } l$$

3. **Carry out.** We solve the equation:

$$460 = 2(w + 30) + 2w$$
$$460 = 2w + 60 + 2w \qquad \text{Using the distributive property}$$
$$460 = 4w + 60 \qquad \text{Collecting like terms}$$
$$400 = 4w \qquad \text{Subtracting 60 on both sides}$$
$$100 = w. \qquad \text{Dividing by 4 on both sides}$$

If $w = 100$, then $w + 30 = 100 + 30 = 130$.

4. **Check.** The length, 130 yd, is 30 yd more than the width, 100 yd. Also,

$$2 \cdot 130 \text{ yd} + 2 \cdot 100 \text{ yd} = 260 \text{ yd} + 200 \text{ yd} = 460 \text{ yd}.$$

The answer checks.

5. **State.** The length of the largest regulation soccer field is 130 yd and the width is 100 yd.

EXAMPLE 7 *Cab Fare.* Metro Taxi charges a $1.25 pickup fee and $2 per mile traveled. Cecilia's cab fare from the airport to her hotel is $31.25. How many miles did she travel in the cab?

Solution

1. **Familiarize.** Let's guess that Cecilia traveled 12 mi in the cab. Then her fare would be

$$\$1.25 + \$2 \cdot 12 = \$1.25 + \$24 = \$25.25.$$

We see that our guess is low, but the calculation we did shows us how to translate the problem to an equation. We let $m =$ the number of miles that Cecilia traveled in the cab.

2. **Translate.** We translate to an equation.

Pickup fee	plus	cost per mile	times	number of miles traveled	is	total charge.
1.25	+	2	·	m	=	31.25

3. **Carry out.** We solve the equation:

$$1.25 + 2 \cdot m = 31.25$$
$$2m = 30 \qquad \text{Subtracting 1.25 on both sides}$$
$$m = 15. \qquad \text{Dividing by 2 on both sides}$$

4. **Check.** If Cecilia travels 15 mi in the cab, the mileage charge is $2 \cdot 15$, or $30. Then, with the $1.25 pickup fee included, her total charge is $1.25 + $30, or $31.25. The answer checks.

5. **State.** Cecilia traveled 15 mi in the cab.

Zeros of Linear Functions

An input for which a function's output is 0 is called a **zero** of the function. We will restrict our attention in this section to zeros of linear functions. This allows us to become familiar with the concept of a zero and it lays the groundwork for working with zeros of other types of functions later in this chapter and in succeeding chapters.

Zeros of Functions

An input c of a function f is called a **zero** of the function, if the output for c is 0. That is, c is a zero of f if $f(c) = 0$.

LINEAR FUNCTIONS
REVIEW SECTION 1.3.

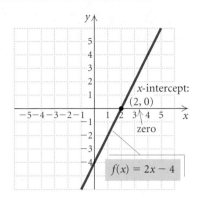

Recall that a linear function is given by $f(x) = mx + b$, where m and b are constants. For the linear function $f(x) = 2x - 4$, we have $f(2) = 2 \cdot 2 - 4 = 0$, so 2 is a **zero** of the function. In fact, 2 is the *only* zero of this function. In general, a **linear function $f(x) = mx + b$, with $m \neq 0$, has exactly one zero.**

Consider the graph of $f(x) = 2x - 4$, shown at left. We see from the graph that the zero, 2, is the first coordinate of the point at which the graph crosses the x-axis. This point, $(2, 0)$, is the **x-intercept** of the graph. Thus when we find the zero of a linear function, we are also finding the first coordinate of the x-intercept of the graph of the function.

For every linear function $f(x) = mx + b$, there is an associated linear equation $mx + b = 0$. When we find the zero of a function $f(x) = mx + b$, we are also finding the solution of the equation $mx + b = 0$.

EXAMPLE 8   Find the zero of $f(x) = 5x - 9$.

Algebraic Solution

We find the value of x for which $f(x) = 0$:

$5x - 9 = 0$ **Setting $f(x) = 0$**

$5x = 9$ **Adding 9 on both sides**

$x = \dfrac{9}{5}$, or 1.8.   **Dividing by 5 on both sides**

The zero is $\frac{9}{5}$, or 1.8. This means that $f\left(\frac{9}{5}\right) = 0$, or $f(1.8) = 0$. Note that the *zero* of the function $f(x) = 5x - 9$ is the *solution* of the equation $5x - 9 = 0$.

Visualizing the Solution

We graph $f(x) = 5x - 9$.

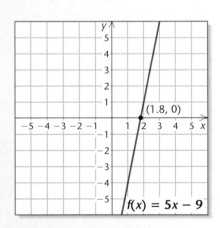

The x-intercept of the graph is $\left(\frac{9}{5}, 0\right)$, or $(1.8, 0)$. Thus, $\frac{9}{5}$, or 1.8, is the zero of the function.

Technology Connection

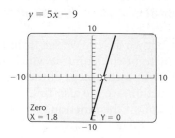

$y = 5x - 9$

We can use the ZERO feature on a graphing calculator to find the zeros of a function $f(x)$ and to solve the corresponding equation $f(x) = 0$. We call this the **Zero method.** To use the Zero method in Example 8, for instance, we graph $y = 5x - 9$ and use the ZERO feature to find the coordinates of the x-intercept of the graph. Note that the x-intercept must appear in the window when the ZERO feature is used. We see that the zero of the function is 1.8.

CONNECTING THE CONCEPTS

ZEROS, SOLUTIONS, AND INTERCEPTS

The zero of a linear function $f(x) = mx + b$, with $m \neq 0$, is the solution of the linear equation $mx + b = 0$ and is the first coordinate of the x-intercept of the graph of $f(x) = mx + b$. To find the zero of $f(x) = mx + b$, we solve $f(x) = 0$, or $mx + b = 0$.

FUNCTION	ZERO OF THE FUNCTION; SOLUTION OF THE EQUATION	X-INTERCEPT OF THE GRAPH

Linear Function

$f(x) = 2x - 4$, or

$\quad y = 2x - 4$

To find the **zero** of $f(x)$, we solve $f(x) = 0$:

$$2x - 4 = 0$$
$$2x = 4$$
$$x = 2.$$

The **solution** of $2x - 4 = 0$ is 2. This is the zero of the function $f(x) = 2x - 4$. That is, $f(2) = 0$.

The zero of $f(x)$ is the first coordinate of the **x-intercept** of the graph of $y = f(x)$.

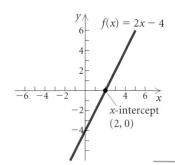

Formulas

A **formula** is an equation that can be used to *model* a situation. For example, the formula $P = 2l + 2w$ in Example 6 gives the perimeter of a rectangle with length l and width w. We also used the motion formula, $d = r \cdot t$, in Example 4 and the simple-interest formula, $I = Prt$, in Example 5.

The equation-solving principles presented earlier can be used to solve a formula for a given variable.

EXAMPLE 9   Solve $P = 2l + 2w$ for l.

Solution   We have

$$P = 2l + 2w \qquad \text{We want to isolate } l.$$

$$P - 2w = 2l \qquad \text{Subtracting } 2w \text{ on both sides}$$

$$\frac{P - 2w}{2} = l. \qquad \text{Dividing by 2 on both sides}$$

The formula $l = \dfrac{P - 2w}{2}$ can be used to determine a rectangle's length if we are given its perimeter and its width.

EXAMPLE 10   The formula $A = P + Prt$ gives the amount A to which a principal of P dollars will grow when invested at simple interest rate r for t years. Solve the formula for P.

Solution   We have

$$A = P + Prt \qquad \text{We want to isolate } P.$$

$$A = P(1 + rt) \qquad \text{Factoring}$$

$$\frac{A}{1 + rt} = \frac{P(1 + rt)}{1 + rt} \qquad \text{Dividing by } 1 + rt \text{ on both sides}$$

$$\frac{A}{1 + rt} = P.$$

The formula $P = \dfrac{A}{1 + rt}$ can be used to determine how much should be invested at simple interest rate r in order to have A dollars t years later.

Technology
   Connection

CONNECTING THE CONCEPTS

THE INTERSECT AND ZERO METHODS

An equation $f(x) = g(x)$ can be solved using the Intersect method by graphing $y_1 = f(x)$ and $y_2 = g(x)$ and using the INTERSECT feature to find the first coordinate of the point of intersection of the graphs. The equation can also be solved using the Zero method by writing it with 0 on one side of the equals sign and then using the ZERO feature.

Solve: $x - 1 = 2x - 6$.

The Intersect Method

Graph

$$f(x) = y_1 = x - 1$$

and

$$g(x) = y_2 = 2x - 6.$$

Point of intersection: $(5, 4)$
Solution: 5

The Zero Method

First add $-2x$ and 6 on both sides of the equation to get 0 on one side.

$$x - 1 = 2x - 6$$
$$x - 1 - 2x + 6 = 0$$

Graph

$$y = x - 1 - 2x + 6.$$

Zero: 5
Solution: 5

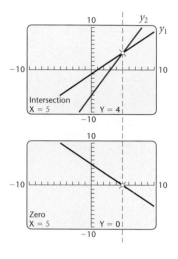

2.1

Exercise Set

Solve.

1. $4x + 5 = 21$

2. $2y - 1 = 3$

3. $4x + 3 = 0$

4. $3x - 16 = 0$

5. $3 - x = 12$

6. $4 - x = -5$

7. $8 = 5x - 3$

8. $9 = 4x - 8$

9. $y + 1 = 2y - 7$

10. $5 - 4x = x - 13$

11. $2x + 7 = x + 3$

12. $5x - 4 = 2x + 5$

13. $3x - 5 = 2x + 1$

14. $4x + 3 = 2x - 7$

15. $4x - 5 = 7x - 2$

16. $5x + 1 = 9x - 7$

17. $5x - 2 + 3x = 2x + 6 - 4x$

18. $5x - 17 - 2x = 6x - 1 - x$

19. $7(3x + 6) = 11 - (x + 2)$

20. $4(5y + 3) = 3(2y - 5)$

21. $3(x + 1) = 5 - 2(3x + 4)$

22. $4(3x + 2) - 7 = 3(x - 2)$

23. $2(x - 4) = 3 - 5(2x + 1)$

24. $3(2x - 5) + 4 = 2(4x + 3)$

25. *Rising Tuition Costs.* Tuition at public colleges and universities is rising more quickly each year. In 2001, the average increase was 7.7%, and in 2002, it was 9.6%, or $356 (*Source*: The College Board). What was the average tuition at a public college or university in 2001?

26. *Where the Textbook Dollar Goes.* Of each dollar spent on textbooks at college bookstores, 23.2 cents goes to the college store for profit, store operations, and personnel. On average, a college student spends $501 per year for textbooks. (*Source*: National Association of College Stores) How much of this expenditure goes to the college store?

27. *DVD Sales.* In 2002, $12.4 billion was spent on copies of films for home viewing with $8.1 billion of this amount being spent for DVDs (*Source*: Adams Media Research). What percent of the spending for films went for DVDs?

28. Training Day *Earnings.* The police thriller *Training Day* starring Denzel Washington earned $86.1 million on sales and rentals of DVDs through June, 2002. This is $9.8 million more than it earned in its initial theatrical release. (*Source*: Image Entertainment) How much did the film earn in its initial theatrical release?

29. *High-Speed Internet Access.* The number of U.S. households with broadband Internet access is projected to be 54.8 million in 2006. This is 25.6 million more households than in 2003. (*Source*: Forrester Research) How many U.S. households had broadband Internet access in 2003?

30. *Fast-Food Nutrition Information.* Together, a Big Mac and an order of Super-Size fries at McDonald's contain 1200 calories. The fries contain 20 calories more than the Big Mac. (*Source*: *American Journal of Public Health*, February 2002) How many calories are in each?

31. *Nutrition.* A slice of carrot cake from the popular restaurant The Cheesecake Factory contains 1560 calories. This is three-fourths of the average daily calorie requirement for many adults (*Source*: The Center for Science in the Public Interest). Find the average daily calorie requirement for these adults.

32. *Nielsen Ratings.* Nielsen Media Research surveys TV-watching habits and provides, among other things, a list of the 20 most-watched TV programs each week. Each rating point in the survey represents 1,067,000 households. One week "60 Minutes" had a rating of 9.4. How many households did this represent?

33. *Amount Borrowed.* Tamisha borrowed money from her father at 5% simple interest to help pay her tuition at Wellington Community College.

At the end of 1 yr, she owed a total of $1365 in principal and interest. How much did she borrow?

34. *Amount of an Investment.*   Khalid makes an investment at 4% simple interest. At the end of 1 yr, the total value of the investment is $1560. How much was originally invested?

35. *Sales Commission.*   Ryan, a consumer electronics salesperson, earns a base salary of $1500 per month and a commission of 8% on the amount of sales he makes. One month Ryan received a $2284paycheck. Find the amount of his sales for the month.

36. *Commission vs. Salary.*   Juliet has a choice between receiving an $1800 monthly salary from Pearson's Furniture or a base salary of $1600 and a 4% commission on the amount of furniture she sells during the month. For what amount of sales will the two choices be equal?

37. *Cab Fare.*   City Cabs charges a $1.75 pickup fee and $1.50 per mile traveled. Diego's fare for a cross-town cab ride is $19.75. How far did he travel in the cab?

38. *Hourly Wage.*   Soledad worked 48 hr one week and earned a $442 paycheck. She earns time and a half (1.5 times her regular hourly wage) for the hours she works in excess of 40. What is Soledad's regular hourly wage?

39. *Angle Measure.*   In triangle ABC, angle B is five times as large as angle A. The measure of angle C is 2° less than that of angle A. Find the measures of the angles. (*Hint*: The sum of the angle measures is 180°.)

40. *Angle Measure.*   In triangle ABC, angle B is twice as large as angle A. Angle C measures 20° more than angle A. Find the measures of the angles.

41. *Test Plot Dimensions.*   Morgan's Seeds has a rectangular test plot with a perimeter of 322 m. The length is 25 m more than the width. Find the dimensions of the plot.

42. *Garden Dimensions.*   The children at Tiny Tots Day Care plant a rectangular vegetable garden with a perimeter of 39 m. The length is twice the width. Find the dimensions of the garden.

43. *Soccer Field Dimensions.*   The width of the soccer field recommended for players under the age of 12 is 35 yd less than the length. The perimeter of the field is 330 yd. (*Source*: U.S. Youth Soccer) Find the dimensions of the field.

44. *Poster Dimensions.*   Marissa is designing a poster to promote the Talbot Street Art Fair. The width of the poster will be two-thirds of its height and its perimeter will be 100 in. Find the dimensions of the poster.

45. *Water Weight.*   Water accounts for 50% of a woman's weight (*Source*: National Institute for Fitness and Sport). Kimiko weighs 135 lb. How much of her body weight is water?

46. *Water Weight.*   Water accounts for 60% of a man's weight (*Source*: National Institute for Fitness and Sport). Emilio weighs 186 lb. How much of his body weight is water?

47. *Train Speeds.*   The speed of an Amtrak passenger train is 14 mph faster than the speed of a Central Railway freight train. The passenger train travels 400 mi in the same time it takes the freight train to travel 330 mi. Find the speed of each train.

48. *Distance Traveled.*   A private airplane leaves Midway Airport and flies due east at a speed of 180 km/h. Two hours later, a jet leaves Midway and flies due east at a speed of 900 km/h. How far from the airport will the jet overtake the private plane?

49. *Traveling Upstream.* A kayak moves at a rate of 12 mph in still water. If the river's current flows at a rate of 4 mph, how long does it take the boat to travel 36 mi upstream?

50. *Flying into a Headwind.* An airplane that travels 450 mph in still air encounters a 30-mph headwind. How long will it take the plane to travel 1050 mi into the wind?

51. *Traveling Downstream.* Angelo's kayak travels 14 km/h in still water. If the river's current flows at a rate of 2 km/h, how long will it take him to travel 20 km downstream?

52. *Flying with a Tailwind.* An airplane that can travel 375 mph in still air is flying with a 25-mph tailwind. How long will it take the plane to travel 700 mi with the wind?

53. *Investment Income.* Erica invested a total of $5000, part at 3% simple interest and part at 4% simple interest. At the end of 1 yr, the investments had earned $176 interest. How much was invested at each rate?

54. *Student Loans.* Dimitri's two student loans total $9000. One loan is at 5% simple interest and the other is at 6% simple interest. At the end of 1 yr, Dimitri owes $492 in interest. What is the amount of each loan?

55. *NCAA Violations.* Colleges and universities are responsible for self-reporting their secondary violations to the National Collegiate Athletic Association. (Most secondary violations are honest mistakes for which there is rarely a penalty.) Together, Division I and Division II schools reported 1989 secondary violations in 2001.

Division I schools reported about 6.5 times as many secondary violations as Division II schools. (*Source*: NCAA) How many secondary violations did each division report?

56. *Working Pharmacists.* It is estimated that there will be 224,500 working pharmacists in the United States in 2010. This is about 1.84 times the number of working pharmacists in 1975. (*Source*: U.S. Department of Health and Human Services) Find the number of working pharmacists in the United States in 1975.

57. *Instant Messenger Services.* AOL's Instant Messenger Service and Microsoft's MSN Messenger Service had a total of 81.9 million users in a recent month. AOL had 23.1 million more users than Microsoft. (*Source*: ComScore Media Matrix) How many users did each service have?

58. *Vanity Plates.* More vanity plates (automobile license plates personalized by the owner) are issued in Florida than in any other state. A total of 1,017,866 vanity plates were sold in Florida in a recent year. These plates accounted for 5.6% of all plates sold in Florida. (*Source: The Fredericksburg, VA Free Lance-Star*) How many license plates in all were sold in Florida?

59. *Erosion.* Because of erosion, Horseshoe Falls, one of the two falls that make up Niagara Falls, is migrating upstream at a rate of 2 ft per year (*Source: Indianapolis Star,* February 14, 1999). At this rate, how long will it take the falls to move one-fourth mile?

60. *Volcanic Activity.* A volcano that is currently about one-half mile below the surface of the Pacific Ocean near the Big Island of Hawaii will eventually become a new Hawaiian island, Loihi. The volcano will break the surface of the ocean in about 50,000 yr. (*Source:* U.S. Geological Survey) On average, how many inches does the volcano rise in a year?

Find the zero of the linear function.

61. $f(x) = x + 5$

62. $f(x) = 5x + 20$

63. $f(x) = -x + 18$

64. $f(x) = 8 + x$

65. $f(x) = 16 - x$

66. $f(x) = -2x + 7$

67. $f(x) = x + 12$

68. $f(x) = 8x + 2$

69. $f(x) = -x + 6$

70. $f(x) = 4 + x$

71. $f(x) = 20 - x$

72. $f(x) = -3x + 13$

73. $f(x) = x - 6$

74. $f(x) = 3x - 9$

75. $f(x) = -x + 15$

76. $f(x) = 4 - x$

In Exercises 77–82, use the given graph to find each of the following: (**a**) *the x-intercept and* (**b**) *the zero of the function.*

77.

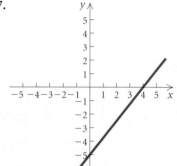

78.

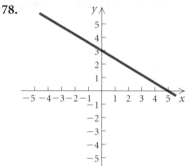

79.

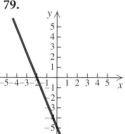

80.

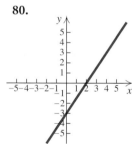

81.

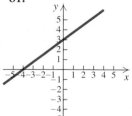

82.

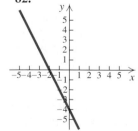

Solve.

83. $A = \frac{1}{2}bh$, for b
(Area of a triangle)

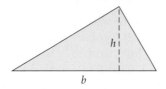

84. $A = \pi r^2$, for π
(Area of a circle)

85. $P = 2l + 2w$, for w
(Perimeter of a rectangle)

86. $A = P + Prt$, for r
(Simple interest)

87. $A = \frac{1}{2}h(b_1 + b_2)$, for h
(Area of a trapezoid)

88. $A = \frac{1}{2}h(b_1 + b_2)$, for b_2

89. $V = \frac{4}{3}\pi r^3$, for π
(Volume of a sphere)

90. $V = \frac{4}{3}\pi r^3$, for r^3

91. $F = \frac{9}{5}C + 32$, for C
(Temperature conversion)

92. $Ax + By = C$, for y
(Standard linear equation)

93. $Ax + By = C$, for A

94. $2w + 2h + l = p$, for w

95. $2w + 2h + l = p$, for h

96. $3x + 4y = 12$, for y

97. $2x - 3y = 6$, for y

98. $T = \frac{3}{10}(I - 12{,}000)$, for I

99. $a = b + bcd$, for b

100. $q = p - np$, for p

101. $z = xy - xy^2$, for x

102. $st = t - 4$, for t

Technology Connection

103. Use a graphing calculator to solve the equations in Exercises 1–8.

104. Use a graphing calculator to find the zeros of the functions in Exercises 61–68.

Collaborative Discussion and Writing

105. Explain in your own words why a linear function $f(x) = mx + b$, with $m \neq 0$, has exactly one zero.

106. The formula in Exercise 91, $F = \frac{9}{5}C + 32$, can be used to convert Celsius temperature to Fahrenheit temperature. Under what circumstances would it be useful to solve this formula for C?

Skill Maintenance

107. Write a slope–intercept equation for the line containing the point $(-1, 4)$ and parallel to the line $3x + 4y = 7$.

108. Write an equation of the line containing the points $(-5, 4)$ and $(3, -2)$.

Given that $f(x) = 2x - 1$ and $g(x) = 3x + 6$, find each of the following.

109. The domain of $f + g$

110. The domain of f/g

111. $(f - g)(x)$

112. $fg(-1)$

Synthesis

State whether each of the following is a linear function.

113. $f(x) = 7 - \dfrac{3}{2}x$

114. $f(x) = \dfrac{3}{2x} + 5$

115. $f(x) = x^2 + 1$

116. $f(x) = \dfrac{3}{4}x - (2.4)^2$

Solve.

117. $2x - \{x - [3x - (6x + 5)]\} = 4x - 1$

118. $14 - 2[3 + 5(x - 1)] = 3\{x - 4[1 + 6(2 - x)]\}$

119. *Packaging and Price.* Dannon recently replaced its 8-oz cup of yogurt with a 6-oz cup and reduced the suggested retail price from 89 cents to 71 cents (*Source*: IRI). Was the price per ounce reduced by the same percent as the size of the cup? If not, find the price difference per ounce in terms of a percent.

120. *Packaging and Price.* Wisk laundry detergent recently replaced its 100-oz container with an 80-oz container and reduced the suggested retail price from $6.99 to $5.75 (*Source*: IRI). Was the price per ounce reduced by the same percent as the size of the container? If not, find the price difference per ounce in terms of a percent.

121. *Running vs. Walking.* A 150-lb person who runs at 6 mph for 1 hr uses about 720 calories. The same person, walking at 4 mph for 90 min, uses about 480 calories. (*Source*: FitSmart, *USA Weekend*, July 19–21, 2002) Suppose a 150-lb person runs at 6 mph for 75 min. How far would the person have to walk at 4 mph in order to burn the same number of calories used running?

122. *Best-Sellers.* One week the popular diet book *Dr. Atkins' New Diet Revolution* sold 10 copies for every 6.3 copies of John Grisham's *King of Torts* sold (*Source*: *USA Today* Best-Selling Books). If a total of 4075 copies of the two books was sold, how many copies of each were sold?

2.2

The Complex Numbers

• *Perform computations involving complex numbers.*

Some functions have zeros that are not real numbers. In order to find the zeros of such functions, we must consider the **complex-number system.**

The Complex-Number System

We know that the square root of a negative number is not a real number. For example, $\sqrt{-1}$ is not a real number because there is no real number x such that $x^2 = -1$. This means that certain equations, like $x^2 = -1$, or $x^2 + 1 = 0$, do not have real-number solutions and certain functions, like $f(x) = x^2 + 1$, do not have real-number zeros. Consider the graph of $f(x) = x^2 + 1$.

RADICAL EXPRESSIONS
REVIEW SECTION R.6.

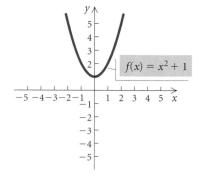

We see that the graph does not cross the x-axis and thus has no x-intercepts. This illustrates that the function $f(x) = x^2 + 1$ has no real-number zeros. Thus there are no real-number solutions of the corresponding equation $x^2 + 1 = 0$.

We can define a number that is a solution of the equation $x^2 + 1 = 0$.

The Number i
The number i is defined such that
$$i = \sqrt{-1} \quad \text{and} \quad i^2 = -1.$$

To express roots of negative numbers in terms of i, we can use the fact that
$$\sqrt{-p} = \sqrt{-1 \cdot p} = \sqrt{-1} \cdot \sqrt{p} = i\sqrt{p}$$
when p is a positive real number.

Study Tip

Don't be hesitant to ask questions in class at appropriate times. Most instructors welcome questions and encourage students to ask them. Other students in your class probably have the same questions you do.

EXAMPLE 1 Express each number in terms of i.

a) $\sqrt{-7}$ b) $\sqrt{-16}$ c) $-\sqrt{-13}$
d) $-\sqrt{-64}$ e) $\sqrt{-48}$

Solution

a) $\sqrt{-7} = \sqrt{-1 \cdot 7} = \sqrt{-1} \cdot \sqrt{7}$
$= i\sqrt{7}, \text{ or } \sqrt{7}i \leftarrow$ | i is *not* under the radical.

b) $\sqrt{-16} = \sqrt{-1 \cdot 16} = \sqrt{-1} \cdot \sqrt{16}$
$= i \cdot 4 = 4i$

c) $-\sqrt{-13} = -\sqrt{-1 \cdot 13} = -\sqrt{-1} \cdot \sqrt{13}$
$= -i\sqrt{13}, \text{ or } -\sqrt{13}i \leftarrow$

d) $-\sqrt{-64} = -\sqrt{-1 \cdot 64} = -\sqrt{-1} \cdot \sqrt{64}$
$= -i \cdot 8 = -8i$

e) $\sqrt{-48} = \sqrt{-1 \cdot 48} = \sqrt{-1} \cdot \sqrt{48}$
$= i\sqrt{16 \cdot 3} = i \cdot 4\sqrt{3}$
$= 4i\sqrt{3}, \text{ or } 4\sqrt{3}i \leftarrow$

The complex numbers are formed by adding real numbers and multiples of i.

Complex Numbers
A **complex number** is a number of the form $a + bi$, where a and b are real numbers. The number a is said to be the **real part** of $a + bi$ and the number b is said to be the **imaginary part** of $a + bi$.*

---

*Sometimes bi is considered to be the imaginary part.

Note that either a or b or both can be 0. When $b = 0$, $a + bi = a + 0i = a$, so every real number is a complex number. A complex number like $3 + 4i$ or $17i$, for which $b \neq 0$, is called an **imaginary number.** A complex number like $17i$ or $-4i$, in which $a = 0$ and $b \neq 0$, is sometimes called a **pure imaginary number.** The relationships among various types of complex numbers are shown in the figure below.

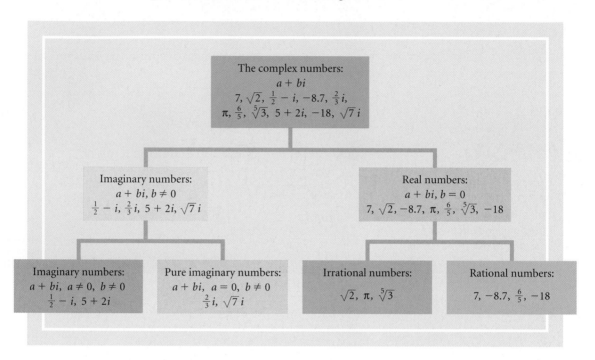

Addition and Subtraction

Technology Connection

When set in $a + bi$ mode, most graphing calculators perform operations on complex numbers. The operations in Example 2 are shown in the window below. Some calculators will express a complex number in the form (a, b) rather than $a + bi$.

```
(8+6i)+(3+2i)
                    11+8i
(4+5i)−(6−3i)
                    −2+8i
```

The complex numbers obey the commutative, associative, and distributive laws. Thus we can add and subtract them as we do binomials. We collect the real parts and the imaginary parts of complex numbers just as we collect like terms in binomials.

EXAMPLE 2 Add or subtract and simplify each of the following.

a) $(8 + 6i) + (3 + 2i)$ **b)** $(4 + 5i) - (6 - 3i)$

Solution

a) $(8 + 6i) + (3 + 2i) = (8 + 3) + (6i + 2i)$

Collecting the real parts and the imaginary parts

$$= 11 + (6 + 2)i = 11 + 8i$$

b) $(4 + 5i) - (6 - 3i) = (4 - 6) + [5i - (-3i)]$

Note that 6 and $-3i$ are both being subtracted.

$$= -2 + 8i$$

Multiplication

When $\sqrt{a}$ and $\sqrt{b}$ are real numbers, $\sqrt{a} \cdot \sqrt{b} = \sqrt{ab}$, but this is not true when $\sqrt{a}$ and $\sqrt{b}$ are not real numbers. Thus,

$$\sqrt{-2} \cdot \sqrt{-5} = \sqrt{-1} \cdot \sqrt{2} \cdot \sqrt{-1} \cdot \sqrt{5}$$
$$= i\sqrt{2} \cdot i\sqrt{5}$$
$$= i^2\sqrt{10} = -1\sqrt{10} = -\sqrt{10} \quad \text{is correct!}$$

But

$$\sqrt{-2} \cdot \sqrt{-5} = \sqrt{(-2)(-5)} = \sqrt{10} \quad \text{is wrong!}$$

Keeping this and the fact that $i^2 = -1$ in mind, we multiply in much the same way that we do with real numbers.

Technology Connection

We can multiply complex numbers on a graphing calculator set in $a + bi$ mode. The products found in Example 3 are shown below.

```
√(−16)√(−25)
                      −20
(1+2i)(1+3i)
                    −5+5i
(3−7i)²
                  −40−42i
```

EXAMPLE 3   Multiply and simplify each of the following.

a) $\sqrt{-16} \cdot \sqrt{-25}$ b) $(1 + 2i)(1 + 3i)$ c) $(3 - 7i)^2$

Solution

a) $\sqrt{-16} \cdot \sqrt{-25} = \sqrt{-1} \cdot \sqrt{16} \cdot \sqrt{-1} \cdot \sqrt{25}$
$$= i \cdot 4 \cdot i \cdot 5$$
$$= i^2 \cdot 20$$
$$= -1 \cdot 20 \qquad i^2 = -1$$
$$= -20$$

b) $(1 + 2i)(1 + 3i) = 1 + 3i + 2i + 6i^2$ **Multiplying each term of one number by every term of the other (FOIL)**
$$= 1 + 3i + 2i - 6 \qquad i^2 = -1$$
$$= -5 + 5i \qquad \text{Collecting like terms}$$

c) $(3 - 7i)^2 = 3^2 - 2 \cdot 3 \cdot 7i + (7i)^2$ **Recall that $(A - B)^2 = A^2 - 2AB + B^2$**
$$= 9 - 42i + 49i^2$$
$$= 9 - 42i - 49 \qquad i^2 = -1$$
$$= -40 - 42i$$

Recall that -1 raised to an *even* power is 1, and -1 raised to an *odd* power is -1. Simplifying powers of i can then be done by using the fact that $i^2 = -1$ and expressing the given power of i in terms of i^2. Consider the following:

$$i = \sqrt{-1},$$
$$i^2 = -1,$$
$$i^3 = i^2 \cdot i = (-1)i = -i,$$
$$i^4 = (i^2)^2 = (-1)^2 = 1,$$

$$i^5 = i^4 \cdot i = (i^2)^2 \cdot i = (-1)^2 \cdot i = 1 \cdot i = i,$$
$$i^6 = (i^2)^3 = (-1)^3 = -1,$$
$$i^7 = i^6 \cdot i = (i^2)^3 \cdot i = (-1)^3 \cdot i = -i,$$
$$i^8 = (i^2)^4 = (-1)^4 = 1.$$

Note that the powers of i cycle through the values i, -1, $-i$, and 1.

EXAMPLE 4   Simplify each of the following.

a) i^{37} b) i^{58}

c) i^{75} d) i^{80}

Solution

a) $i^{37} = i^{36} \cdot i = (i^2)^{18} \cdot i = (-1)^{18} \cdot i = 1 \cdot i = i$

b) $i^{58} = (i^2)^{29} = (-1)^{29} = -1$

c) $i^{75} = i^{74} \cdot i = (i^2)^{37} \cdot i = (-1)^{37} \cdot i = -1 \cdot i = -i$

d) $i^{80} = (i^2)^{40} = (-1)^{40} = 1$ ▬

These powers of i can also be simplified in terms of i^4 rather than i^2. Consider i^{37} in Example 4. When we divide 37 by 4, we get 9 with a remainder of 1. Then $37 = 4 \cdot 9 + 1$, so

$$i^{37} = (i^4)^9 \cdot i = 1^9 \cdot i = 1 \cdot i = i.$$

The other examples shown above can be done in a similar manner.

Conjugates and Division

Conjugates of complex numbers are defined as follows.

> ### Conjugate of a Complex Number
>
> The **conjugate** of a complex number $a + bi$ is $a - bi$. The numbers $a + bi$ and $a - bi$ are **complex conjugates.**

Each of the following pairs of numbers are complex conjugates:

$$-3 + 7i \text{ and } -3 - 7i; \quad 14 - 5i \text{ and } 14 + 5i; \quad \text{and} \quad 8i \text{ and } -8i.$$

The product of a complex number and its conjugate is a real number.

EXAMPLE 5   Multiply each of the following.

a) $(5 + 7i)(5 - 7i)$ b) $(8i)(-8i)$

Solution

a) $(5 + 7i)(5 - 7i) = 5^2 - (7i)^2$ **Using $(A + B)(A - B) = A^2 - B^2$**

$$= 25 - 49i^2$$
$$= 25 - 49(-1)$$
$$= 25 + 49$$
$$= 74$$

b) $(8i)(-8i) = -64i^2$

$$= -64(-1)$$
$$= 64$$ ▬

Conjugates are used when we divide complex numbers.

EXAMPLE 6 Divide $2 - 5i$ by $1 - 6i$.

Solution We write fraction notation and then multiply by 1, using the conjugate of the denominator to form the symbol for 1.

$$\frac{2 - 5i}{1 - 6i} = \frac{2 - 5i}{1 - 6i} \cdot \frac{1 + 6i}{1 + 6i}$$ Note that $1 + 6i$ is the conjugate of the divisor, $1 - 6i$.

$$= \frac{(2 - 5i)(1 + 6i)}{(1 - 6i)(1 + 6i)}$$

$$= \frac{2 + 7i - 30i^2}{1 - 36i^2}$$

$$= \frac{2 + 7i + 30}{1 + 36}$$ $i^2 = -1$

$$= \frac{32 + 7i}{37}$$

$$= \frac{32}{37} + \frac{7}{37}i.$$ Writing the quotient in the form $a + bi$

Technology Connection

With a graphing calculator set in $a + bi$ mode, we can divide complex numbers and express the real and imaginary parts in fraction form, just as we did in Example 6.

```
(2−5i)/(1−6i) ▶Frac
                32/37+7/37i
```

2.2

Exercise Set

Simplify. Write answers in the form $a + bi$, where a and b are real numbers.

1. $(-5 + 3i) + (7 + 8i)$

2. $(-6 - 5i) + (9 + 2i)$

3. $(4 - 9i) + (1 - 3i)$

4. $(7 - 2i) + (4 - 5i)$

5. $(12 + 3i) + (-8 + 5i)$

6. $(-11 + 4i) + (6 + 8i)$

7. $(-1 - i) + (-3 - i)$

8. $(-5 - i) + (6 + 2i)$

9. $\left(3 + \sqrt{-16}\right) + \left(2 + \sqrt{-25}\right)$

10. $\left(7 - \sqrt{-36}\right) + \left(2 + \sqrt{-9}\right)$

11. $(10 + 7i) - (5 + 3i)$

12. $(-3 - 4i) - (8 - i)$

13. $(13 + 9i) - (8 + 2i)$

14. $(-7 + 12i) - (3 - 6i)$

15. $(6 - 4i) - (-5 + i)$

16. $(8 - 3i) - (9 - i)$

17. $(-5 + 2i) - (-4 - 3i)$

18. $(-6 + 7i) - (-5 - 2i)$

19. $(4 - 9i) - (2 + 3i)$

20. $(10 - 4i) - (8 + 2i)$

21. $7i(2 - 5i)$ **22.** $3i(6 + 4i)$

23. $-2i(-8 + 3i)$ **24.** $-6i(-5 + i)$

25. $(1 + 3i)(1 - 4i)$

26. $(1 - 2i)(1 + 3i)$

27. $(2 + 3i)(2 + 5i)$

28. $(3 - 5i)(8 - 2i)$

29. $(-4 + i)(3 - 2i)$

30. $(5 - 2i)(-1 + i)$

31. $(8 - 3i)(-2 - 5i)$

32. $(7 - 4i)(-3 - 3i)$

33. $\left(3 + \sqrt{-16}\right)\left(2 + \sqrt{-25}\right)$

34. $\left(7 - \sqrt{-16}\right)\left(2 + \sqrt{-9}\right)$

35. $(5 - 4i)(5 + 4i)$

36. $(5 + 9i)(5 - 9i)$

37. $(3 + 2i)(3 - 2i)$

38. $(8 + i)(8 - i)$

39. $(7 - 5i)(7 + 5i)$

40. $(6 - 8i)(6 + 8i)$

41. $(4 + 2i)^2$

42. $(5 - 4i)^2$

43. $(-2 + 7i)^2$

44. $(-3 + 2i)^2$

45. $(1 - 3i)^2$

46. $(2 - 5i)^2$

47. $(-1 - i)^2$

48. $(-4 - 2i)^2$

49. $(3 + 4i)^2$

50. $(6 + 5i)^2$

51. $\dfrac{3}{5 - 11i}$

52. $\dfrac{i}{2 + i}$

53. $\dfrac{5}{2 + 3i}$

54. $\dfrac{-3}{4 - 5i}$

55. $\dfrac{4 + i}{-3 - 2i}$

56. $\dfrac{5 - i}{-7 + 2i}$

57. $\dfrac{5 - 3i}{4 + 3i}$

58. $\dfrac{6 + 5i}{3 - 4i}$

59. $\dfrac{2 + \sqrt{3}i}{5 - 4i}$

60. $\dfrac{\sqrt{5} + 3i}{1 - i}$

61. $\dfrac{1 + i}{(1 - i)^2}$

62. $\dfrac{1 - i}{(1 + i)^2}$

63. $\dfrac{4 - 2i}{1 + i} + \dfrac{2 - 5i}{1 + i}$

64. $\dfrac{3 + 2i}{1 - i} + \dfrac{6 + 2i}{1 - i}$

Simplify.

65. i^{11}

66. i^7

67. i^{35}

68. i^{24}

69. i^{64}

70. i^{42}

71. $(-i)^{71}$

72. $(-i)^6$

73. $(5i)^4$

74. $(2i)^5$

Technology Connection

75. Use a graphing calculator to perform the operations in Exercises 1, 9, 11, 21, 25, 41, and 55.

76. Use a graphing calculator to perform the operations in Exercises 2, 10, 12, 14, 22, 26, and 56.

Collaborative Discussion and Writing

77. Is the sum of two imaginary numbers always an imaginary number? Explain your answer.

78. Is the product of two imaginary numbers always an imaginary number? Explain your answer.

Skill Maintenance

79. Write a slope–intercept equation for the line containing the point $(3, -5)$ and perpendicular to the line $3x - 6y = 7$.

Given that $f(x) = x^2 + 4$ and $g(x) = 3x + 5$, find each of the following.

80. The domain of $f - g$

81. The domain of f/g

82. $(f - g)(x)$

83. $(f/g)(2)$

84. For the function $f(x) = x^2 - 3x + 4$, construct and simplify the difference quotient

$$\frac{f(x + h) - f(x)}{h}.$$

Synthesis

Determine whether each of the following is true or false.

85. The sum of two numbers that are conjugates of each other is always a real number.

86. The conjugate of a sum is the sum of the conjugates of the individual complex numbers.

87. The conjugate of a product is the product of the conjugates of the individual complex numbers.

Let $z = a + bi$ and $\bar{z} = a - bi$.

88. Find a general expression for $1/z$.

89. Find a general expression for $z\bar{z}$.

90. Solve $z + 6\bar{z} = 7$ for z.

2.3

Quadratic Equations, Functions, and Models

- *Find zeros of quadratic functions and solve quadratic equations by using the principle of zero products, by using the principle of square roots, by completing the square, and by using the quadratic formula.*
- *Solve equations that are reducible to quadratic.*
- *Solve applied problems using quadratic equations.*

Quadratic Equations and Quadratic Functions

In this section, we will explore the relationship between the solutions of quadratic equations and the zeros of quadratic functions. We define quadratic equations and functions as follows.

> ### Quadratic Equations
> A **quadratic equation** is an equation equivalent to
> $$ax^2 + bx + c = 0, \quad a \neq 0,$$
> where a, b, and c are real numbers.
>
> ### Quadratic Functions
> A **quadratic function** f is a function that can be written in the form
> $$f(x) = ax^2 + bx + c, \quad a \neq 0,$$
> where a, b, and c are real numbers.

A quadratic equation written in the form $ax^2 + bx + c = 0$ is said to be in **standard** form.

The *zeros* of a quadratic function $f(x) = ax^2 + bx + c$ are the *solutions* of the associated quadratic equation $ax^2 + bx + c = 0$. (These solutions are sometimes called *roots* of the equation.) Quadratic functions can have real-number or imaginary-number zeros and quadratic equations can have real-number or imaginary-number solutions. If the zeros or solutions are real numbers, they are also the first coordinates of the x-intercepts of the graph of the quadratic function.

ZEROS OF A FUNCTION

REVIEW SECTION 2.1.

The following principles allow us to solve many quadratic equations.

> ### Equation-Solving Principles
>
> **The Principle of Zero Products:** If $ab = 0$ is true, then $a = 0$ or $b = 0$, and if $a = 0$ or $b = 0$, then $ab = 0$.
>
> **The Principle of Square Roots:** If $x^2 = k$, then $x = \sqrt{k}$ or $x = -\sqrt{k}$.

FACTORING TRINOMIALS
REVIEW SECTION R.4.

EXAMPLE 1 Solve: $2x^2 - x = 3$.

Algebraic Solution

We have

$$2x^2 - x = 3$$
$$2x^2 - x - 3 = 0 \qquad \text{Subtracting 3 on both sides}$$
$$(x + 1)(2x - 3) = 0 \qquad \text{Factoring}$$
$$x + 1 = 0 \quad or \quad 2x - 3 = 0 \qquad \text{Using the principle of zero products}$$
$$x = -1 \quad or \qquad 2x = 3$$
$$x = -1 \quad or \qquad x = \tfrac{3}{2}.$$

CHECK:　For $x = -1$:

$$\frac{2x^2 - x = 3}{2(-1)^2 - (-1) \ ?\ 3}$$
$$2 \cdot 1 + 1$$
$$2 + 1$$
$$3 \ \big|\ 3 \quad \text{TRUE}$$

For $x = \tfrac{3}{2}$:

$$\frac{2x^2 - x = 3}{2\left(\tfrac{3}{2}\right)^2 - \tfrac{3}{2} \ ?\ 3}$$
$$2 \cdot \tfrac{9}{4} - \tfrac{3}{2}$$
$$\tfrac{9}{2} - \tfrac{3}{2}$$
$$\tfrac{6}{2}$$
$$3 \ \big|\ 3 \quad \text{TRUE}$$

The solutions are -1 and $\tfrac{3}{2}$.

Visualizing the Solution

The solutions of the equation $2x^2 - x = 3$, or the equivalent equation $2x^2 - x - 3 = 0$, are the zeros of the function $f(x) = 2x^2 - x - 3$. They are also the first coordinates of the x-intercepts of the graph of $f(x) = 2x^2 - x - 3$.

$$f(x) = 2x^2 - x - 3$$

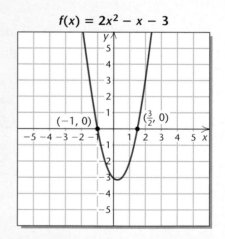

The solutions are -1 and $\tfrac{3}{2}$.

EXAMPLE 2 Solve: $2x^2 - 10 = 0$.

Algebraic Solution

We have

$$2x^2 - 10 = 0$$
$$2x^2 = 10 \qquad \text{Adding 10 on both sides}$$
$$x^2 = 5 \qquad \text{Dividing by 2 on both sides}$$
$$x = \sqrt{5} \quad \text{or} \quad x = -\sqrt{5}. \qquad \text{Using the principle of square roots}$$

CHECK:
$$\begin{array}{c}
2x^2 - 10 = 0 \\
\hline
2(\pm\sqrt{5})^2 - 10 \ ? \ 0 \qquad \text{We can check} \\
2 \cdot 5 - 10 \qquad \text{both solutions} \\
10 - 10 \qquad \text{at once.} \\
0 \ | \ 0 \quad \text{TRUE}
\end{array}$$

The solutions are $\sqrt{5}$ and $-\sqrt{5}$, or $\pm\sqrt{5}$.

Visualizing the Solution

The solutions of the equation $2x^2 - 10 = 0$ are the zeros of the function $f(x) = 2x^2 - 10$. Note that they are also the first coordinates of the x-intercepts of the graph of $f(x) = 2x^2 - 10$.

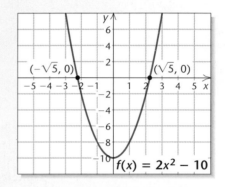

The solutions are $-\sqrt{5}$ and $\sqrt{5}$.

We have seen that some quadratic equations can be solved by factoring and using the principle of zero products. For example, consider the equation $x^2 - 3x - 4 = 0$:

$$x^2 - 3x - 4 = 0$$
$$(x + 1)(x - 4) = 0 \qquad \text{Factoring}$$
$$x + 1 = 0 \quad \text{or} \quad x - 4 = 0 \qquad \text{Using the principle of zero products}$$
$$x = -1 \quad \text{or} \quad x = 4.$$

The equation $x^2 - 3x - 4 = 0$ has *two real-number* solutions, -1 and 4. These are the zeros of the associated quadratic function $f(x) = x^2 - 3x - 4$ and the first coordinates of the x-intercepts of the graph of this function. (See Fig. 1.)

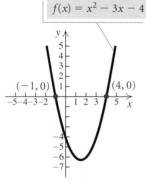

Two real-number zeros
Two x-intercepts

FIGURE 1

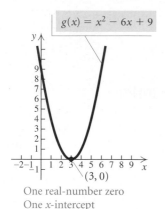

One real-number zero
One x-intercept

FIGURE 2

Next, consider the equation $x^2 - 6x + 9 = 0$. Again, we factor and use the principle of zero products:

$$x^2 - 6x + 9 = 0$$
$$(x - 3)(x - 3) = 0 \qquad \text{Factoring}$$
$$x - 3 = 0 \quad \text{or} \quad x - 3 = 0 \qquad \text{Using the principle of zero products}$$
$$x = 3 \quad \text{or} \qquad x = 3.$$

The equation $x^2 - 6x + 9 = 0$ has *one real-number* solution, 3. It is the zero of the quadratic function $g(x) = x^2 - 6x + 9$ and the first coordinate of the x-intercept of the graph of this function. (See Fig. 2.)

The principle of square roots can be used to solve quadratic equations like $x^2 + 13 = 0$:

$$x^2 + 13 = 0$$
$$x^2 = -13$$
$$x = \pm\sqrt{-13} \qquad \text{Using the principle of square roots}$$
$$x = \pm\sqrt{13}i.$$

The equation has *two imaginary-number* solutions, $-\sqrt{13}i$ and $\sqrt{13}i$. These are the zeros of the associated quadratic function $h(x) = x^2 + 13$. Since the zeros are not real numbers, the graph of the function has no x-intercepts. (See Fig. 3.)

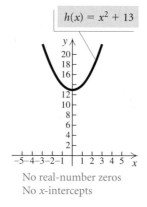

No real-number zeros
No x-intercepts

FIGURE 3

Completing the Square

Neither the principle of zero products nor the principle of square roots would yield the *exact* zeros of a function like $f(x) = x^2 - 6x - 10$ or the *exact* solutions of the associated equation $x^2 - 6x - 10 = 0$. If we wish to find

Technology Connection

Approximations for the zeros of the quadratic function $f(x) = x^2 - 6x - 10$ in Example 3 can be found using the Zero method.

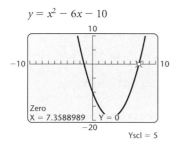

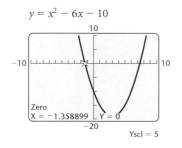

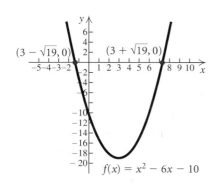

exact zeros or solutions, we can use a procedure called **completing the square** and then use the principle of square roots. (Recall that we completed the square in order to write the equation of a circle in standard form in Section 1.1. Here we complete the square in order to solve quadratic equations.)

EXAMPLE 3   Find the zeros of $f(x) = x^2 - 6x - 10$ by completing the square.

Solution   We find the values of x for which $f(x) = 0$. That is, we solve the associated equation $x^2 - 6x - 10 = 0$. Our goal is to find an equivalent equation of the form $x^2 + bx + c = d$ in which $x^2 + bx + c$ is a perfect square. Since

$$x^2 + bx + \left(\frac{b}{2}\right)^2 = \left(x + \frac{b}{2}\right)^2,$$

the number c is found by taking half the coefficient of the x-term and squaring it. Then for the equation $x^2 - 6x - 10 = 0$, we have

$$x^2 - 6x - 10 = 0$$
$$x^2 - 6x \quad\quad = 10 \quad\quad \textbf{Adding 10}$$
$$x^2 - 6x + 9 = 10 + 9 \quad\quad \textbf{Adding 9 to complete the square:}$$
$$\left(\frac{b}{2}\right)^2 = \left(\frac{-6}{2}\right)^2 = (-3)^2 = 9$$
$$x^2 - 6x + 9 = 19.$$

Because $x^2 - 6x + 9$ is a perfect square, we are able to write it as $(x - 3)^2$, the square of a binomial. We can then use the principle of square roots to finish the solution:

$$(x - 3)^2 = 19 \quad\quad \textbf{Factoring}$$
$$x - 3 = \pm\sqrt{19} \quad\quad \textbf{Using the principle of square roots}$$
$$x = 3 \pm \sqrt{19}. \quad\quad \textbf{Adding 3}$$

Therefore, the solutions of the equation are $3 + \sqrt{19}$ and $3 - \sqrt{19}$, or simply $3 \pm \sqrt{19}$. The zeros of $f(x) = x^2 - 6x - 10$ are also $3 + \sqrt{19}$ and $3 - \sqrt{19}$, or $3 \pm \sqrt{19}$.

Decimal approximations for $3 \pm \sqrt{19}$ can be found using a calculator:

$$3 + \sqrt{19} \approx 7.359 \quad \text{and} \quad 3 - \sqrt{19} \approx -1.359.$$

The zeros are approximately 7.359 and -1.359.   ▬

Before we can complete the square, the coefficient of the x^2-term must be 1. When it is not, we divide both sides of the equation by the x^2-coefficient.

EXAMPLE 4   Solve: $2x^2 - 1 = 3x$.

Solution   We have

$$2x^2 - 1 = 3x$$

$$2x^2 - 3x - 1 = 0 \qquad \text{Subtracting } 3x. \text{ We are unable to factor the result.}$$

$$2x^2 - 3x = 1 \qquad \text{Adding 1}$$

$$x^2 - \frac{3}{2}x = \frac{1}{2} \qquad \text{Dividing by 2 to make the } x^2\text{-coefficient 1}$$

$$x^2 - \frac{3}{2}x + \frac{9}{16} = \frac{1}{2} + \frac{9}{16} \qquad \text{Completing the square: } \frac{1}{2}\left(-\frac{3}{2}\right) = -\frac{3}{4} \text{ and } \left(-\frac{3}{4}\right)^2 = \frac{9}{16}; \text{ adding } \frac{9}{16}$$

$$\left(x - \frac{3}{4}\right)^2 = \frac{17}{16} \qquad \text{Factoring and simplifying}$$

$$x - \frac{3}{4} = \pm\frac{\sqrt{17}}{4} \qquad \text{Using the principle of square roots and the quotient rule for radicals}$$

$$x = \frac{3}{4} \pm \frac{\sqrt{17}}{4} \qquad \text{Adding } \frac{3}{4}$$

$$x = \frac{3 \pm \sqrt{17}}{4}.$$

The solutions are

$$\frac{3 + \sqrt{17}}{4} \quad \text{and} \quad \frac{3 - \sqrt{17}}{4}, \quad \text{or} \quad \frac{3 \pm \sqrt{17}}{4}.$$

Study Tip

The examples in the text are carefully chosen to prepare you for success with the exercise sets. Study the step-by-step solutions of the examples, noting that substitutions and explanations appear in red. The time you spend studying the examples will save you valuable time when you do your homework.

To solve a quadratic equation by completing the square:

1. Isolate the terms with variables on one side of the equation and arrange them in descending order.
2. Divide by the coefficient of the squared term if that coefficient is not 1.
3. Complete the square by taking half the coefficient of the first-degree term and adding its square on both sides of the equation.
4. Express one side of the equation as the square of a binomial.
5. Use the principle of square roots.
6. Solve for the variable.

Using the Quadratic Formula

Because completing the square works for *any* quadratic equation, it can be used to solve the general quadratic equation $ax^2 + bx + c = 0$ for x. The result will be a formula that can be used to solve any quadratic equation quickly.

Consider any quadratic equation in standard form:

$$ax^2 + bx + c = 0, \quad a \neq 0.$$

For now, we assume that $a > 0$ and solve by completing the square. As the steps are carried out, compare them with those of Example 4.

$$ax^2 + bx + c = 0 \qquad \text{Standard form}$$

$$ax^2 + bx = -c \qquad \text{Adding } -c$$

$$x^2 + \frac{b}{a}x = -\frac{c}{a} \qquad \text{Dividing by } a$$

Half of $\dfrac{b}{a}$ is $\dfrac{b}{2a}$ and $\left(\dfrac{b}{2a}\right)^2 = \dfrac{b^2}{4a^2}$. Thus we add $\dfrac{b^2}{4a^2}$:

$$x^2 + \frac{b}{a}x + \frac{b^2}{4a^2} = -\frac{c}{a} + \frac{b^2}{4a^2} \qquad \text{Adding } \frac{b^2}{4a^2} \text{ to complete the square}$$

$$\left(x + \frac{b}{2a}\right)^2 = -\frac{4ac}{4a^2} + \frac{b^2}{4a^2} \qquad \begin{array}{l}\text{Factoring and finding a common} \\ \text{denominator:} \\ -\dfrac{c}{a} = -\dfrac{4a}{4a} \cdot \dfrac{c}{a} = -\dfrac{4ac}{4a^2}\end{array}$$

$$\left(x + \frac{b}{2a}\right)^2 = \frac{b^2 - 4ac}{4a^2}$$

$$x + \frac{b}{2a} = \pm\frac{\sqrt{b^2 - 4ac}}{2a} \qquad \begin{array}{l}\textbf{Using the principle of square roots} \\ \textbf{and the quotient rule for radicals.} \\ \textbf{Since } a > 0, \sqrt{4a^2} = 2a.\end{array}$$

$$x = -\frac{b}{2a} \pm \frac{\sqrt{b^2 - 4ac}}{2a} \qquad \text{Adding } -\frac{b}{2a}$$

$$x = \frac{-b \pm \sqrt{b^2 - 4ac}}{2a}.$$

It can also be shown that this result holds if $a < 0$.

The Quadratic Formula

The solutions of $ax^2 + bx + c = 0$, $a \neq 0$, are given by

$$x = \frac{-b \pm \sqrt{b^2 - 4ac}}{2a}.$$

EXAMPLE 5   Solve $3x^2 + 2x = 7$. Find exact solutions and approximate solutions rounded to the nearest thousandth.

Solution   After finding standard form, we are unable to factor, so we identify a, b, and c in order to use the quadratic formula:

$$3x^2 + 2x - 7 = 0;$$

$$a = 3, \quad b = 2, \quad c = -7.$$

Technology Connection

We can solve the equation $3x^2 + 2x = 7$ in Example 5 using the Intersect method. We graph $y_1 = 3x^2 + 2x$ and $y_2 = 7$ and use the INTERSECT feature to find the coordinates of the points of intersection. The first coordinates of these points are the solutions of the equation $y_1 = y_2$, or $3x^2 + 2x = 7$.

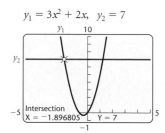

$y_1 = 3x^2 + 2x, \quad y_2 = 7$

Intersection
X = −1.896805   Y = 7

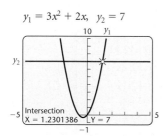

$y_1 = 3x^2 + 2x, \quad y_2 = 7$

Intersection
X = 1.2301386   Y = 7

The solutions are approximately -1.897 and 1.230. We could also write the equation in standard form, $3x^2 + 2x - 7 = 0$, and use the Zero method.

We then use the quadratic formula:

$$x = \frac{-b \pm \sqrt{b^2 - 4ac}}{2a}$$

$$= \frac{-2 \pm \sqrt{2^2 - 4(3)(-7)}}{2(3)} \qquad \text{Substituting}$$

$$= \frac{-2 \pm \sqrt{4 + 84}}{6} = \frac{-2 \pm \sqrt{88}}{6}$$

$$= \frac{-2 \pm \sqrt{4 \cdot 22}}{6} = \frac{-2 \pm 2\sqrt{22}}{6} = \frac{2(-1 \pm \sqrt{22})}{2 \cdot 3}$$

$$= \frac{2}{2} \cdot \frac{-1 \pm \sqrt{22}}{3} = \frac{-1 \pm \sqrt{22}}{3}.$$

The exact solutions are

$$\frac{-1 - \sqrt{22}}{3} \quad \text{and} \quad \frac{-1 + \sqrt{22}}{3}.$$

Using a calculator, we approximate the solutions to be -1.897 and 1.230.

EXAMPLE 6   Solve: $x^2 + 5x + 8 = 0$.

Algebraic Solution

To find the solutions, we use the quadratic formula. Here

$$a = 1, \quad b = 5, \quad c = 8;$$

$$x = \frac{-b \pm \sqrt{b^2 - 4ac}}{2a}$$

$$= \frac{-5 \pm \sqrt{5^2 - 4(1)(8)}}{2 \cdot 1} \qquad \text{Substituting}$$

$$= \frac{-5 \pm \sqrt{-7}}{2} \qquad \text{Simplifying}$$

$$= \frac{-5 \pm \sqrt{7}i}{2}.$$

The solutions are $-\dfrac{5}{2} - \dfrac{\sqrt{7}}{2}i$ and $-\dfrac{5}{2} + \dfrac{\sqrt{7}}{2}i$.

Visualizing the Solution

The graph of the function $f(x) = x^2 + 5x + 8$ has no x-intercepts.

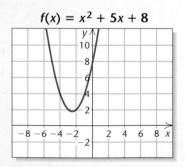

$$f(x) = x^2 + 5x + 8$$

Thus the function has no real-number zeros and there are no real-number solutions of the associated equation $x^2 + 5x + 8 = 0$.

The Discriminant

From the quadratic formula, we know that the solutions x_1 and x_2 of a quadratic equation are given by

$$x_1 = \frac{-b + \sqrt{b^2 - 4ac}}{2a} \quad \text{and} \quad x_2 = \frac{-b - \sqrt{b^2 - 4ac}}{2a}.$$

The expression $b^2 - 4ac$ shows the nature of the solutions. This expression is called the **discriminant.** If it is 0, then it makes no difference whether we choose the plus or the minus sign in the formula. That is, $x_1 = -\dfrac{b}{2a} = x_2$, so there is just one solution. In this case, we sometimes say that there is one repeated real solution. If the discriminant is positive, there will be two real solutions. If it is negative, we will be taking the square root of a negative number; hence there will be two imaginary-number solutions, and they will be complex conjugates.

> ### Discriminant
>
> For $ax^2 + bx + c = 0$:
>
> $b^2 - 4ac = 0 \longrightarrow$ One real-number solution;
>
> $b^2 - 4ac > 0 \longrightarrow$ Two different real-number solutions;
>
> $b^2 - 4ac < 0 \longrightarrow$ Two different imaginary-number solutions, complex conjugates.

In Example 5, the discriminant, 88, is positive, indicating that there are two different real-number solutions. If the discriminant is negative, as it is in Example 6, we know that there are two different imaginary-number solutions.

Equations Reducible to Quadratic

Some equations can be treated as quadratic, provided that we make a suitable substitution. For example, consider the following:

$$x^4 - 5x^2 + 4 = 0$$
$$(x^2)^2 - 5x^2 + 4 = 0 \qquad x^4 = (x^2)^2$$
$$u^2 - 5u + 4 = 0. \qquad \textbf{Substituting } u \textbf{ for } x^2$$

The equation $u^2 - 5u + 4 = 0$ can be solved for u by factoring or using the quadratic formula. Then we can reverse the substitution, replacing u with x^2, and solve for x. Equations like the one above are said to be **reducible to quadratic,** or **quadratic in form.**

EXAMPLE 7 Solve: $x^4 - 5x^2 + 4 = 0$.

Algebraic Solution

We let $u = x^2$ and substitute:

$u^2 - 5u + 4 = 0$ **Substituting u for x^2**

$(u - 1)(u - 4) = 0$ **Factoring**

$u - 1 = 0$ *or* $u - 4 = 0$ **Using the principle of zero products**

$u = 1$ *or* $u = 4$.

Don't stop here! We must solve for the original variable. We substitute x^2 for u and solve for x:

$x^2 = 1$ *or* $x^2 = 4$

$x = \pm 1$ *or* $x = \pm 2$.   **Using the principle of square roots**

The solutions are $-1, 1, -2,$ and 2.

Visualizing the Solution

The solutions of the given equation are the zeros of $f(x) = x^4 - 5x^2 + 4$. Note that the zeros occur at the x-values $-2, -1, 1,$ and 2.

$$f(x) = x^4 - 5x^2 + 4$$

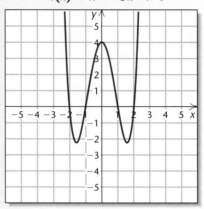

Technology Connection

We can use the Zero method to solve the equation in Example 7, $x^4 - 5x^2 + 4 = 0$. We graph the function $y = x^4 - 5x^2 + 4$ and use the ZERO feature to find the zeros.

$$y = x^4 - 5x^2 + 4$$

Zero
X = -2   Y = 0

The leftmost zero is -2. Using the ZERO feature three more times, we find that the other zeros are $-1, 1,$ and 2.

Applications

Some applied problems can be translated to quadratic equations.

EXAMPLE 8 *Time of a Free Fall.* The Petronas Towers in Kuala Lumpur, Malaysia, are 1482 ft tall. How long would it take an object dropped from the top to reach the ground?

Solution

1. **Familiarize.** The formula $s = 16t^2$ is used to approximate the distance s, in feet, that an object falls freely from rest in t seconds. In this case, the distance is 1482 ft.

2. **Translate.** We substitute 1482 for s in the formula:

   $$1482 = 16t^2.$$

3. **Carry out.** We use the principle of square roots:

   $$1482 = 16t^2$$
   $$\frac{1482}{16} = t^2 \qquad \text{Dividing by 16}$$
   $$\sqrt{\frac{1482}{16}} = t \qquad \begin{array}{l}\text{Taking the positive square root. Time} \\ \text{cannot be negative in this application.}\end{array}$$
   $$9.624 \approx t.$$

4. **Check.** In 9.624 sec, a dropped object would travel a distance of $16(9.624)^2$, or about 1482 ft. The answer checks.

5. **State.** It would take about 9.624 sec for an object dropped from the top of the Petronas Towers to reach the ground.

EXAMPLE 9 *Bicycling Speed.* Logan and Cassidy leave a campsite, Logan biking due north and Cassidy biking due east. Logan bikes 7 km/h slower than Cassidy. After 4 hr, they are 68 km apart. Find the speed of each bicyclist.

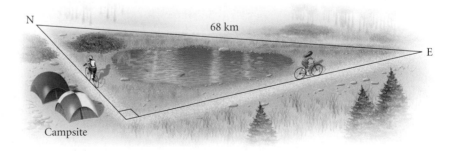

Solution

1. **Familiarize.** We let $r =$ Cassidy's speed, in kilometers per hour. Then $r - 7 =$ Logan's speed, in kilometers per hour. We will use the motion formula $d = rt$, where d is the distance, r is the rate (or speed), and t is the time. Then, after 4 hr, Cassidy has traveled $4r$ km and Logan has traveled $4(r - 7)$ km. We add these distances to the drawing, as shown below.

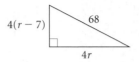

THE PYTHAGOREAN THEOREM
REVIEW SECTION R.6.

2. **Translate.** We use the Pythagorean theorem, $a^2 + b^2 = c^2$, where a and b are the lengths of the legs of a right triangle and c is the length of the hypotenuse:

$$(4r)^2 + [4(r - 7)]^2 = 68^2.$$

3. **Carry out.** We solve the equation:

$$(4r)^2 + [4(r - 7)]^2 = 68^2$$
$$16r^2 + 16(r^2 - 14r + 49) = 4624$$
$$16r^2 + 16r^2 - 224r + 784 = 4624$$

$$32r^2 - 224r - 3840 = 0 \qquad \text{Subtracting 4624}$$
$$r^2 - 7r - 120 = 0 \qquad \text{Dividing by 32}$$
$$(r + 8)(r - 15) = 0 \qquad \text{Factoring}$$

$$r + 8 = 0 \quad or \quad r - 15 = 0 \qquad \text{Principle of zero products}$$

$$r = -8 \quad or \qquad\quad r = 15.$$

4. **Check.** Since speed cannot be negative, we need to check only 15. If Cassidy's speed is 15 km/h, then Logan's speed is $15 - 7$, or 8 km/h. In 4 hr, Cassidy travels $4 \cdot 15$, or 60 km, and Logan travels $4 \cdot 8$, or 32 km. Then they are $\sqrt{60^2 + 32^2}$, or 68 km apart. The answer checks.

5. **State.** Cassidy's speed is 15 km/h, and Logan's speed is 8 km/h.   ▬

CONNECTING THE CONCEPTS

ZEROS, SOLUTIONS, AND INTERCEPTS

The zeros of a function $y = f(x)$ are also the solutions of the equation $f(x) = 0$, and the real-number zeros are the first coordinates of the x-intercepts of the graph of the function.

FUNCTION	ZEROS OF THE FUNCTION; SOLUTIONS OF THE EQUATION	X-INTERCEPTS OF THE GRAPH

Linear Function

$f(x) = 2x - 4$, or
$\quad y = 2x - 4$

To find the **zero** of $f(x)$, we solve $f(x) = 0$:

$$2x - 4 = 0$$
$$2x = 4$$
$$x = 2.$$

The **solution** of $2x - 4 = 0$ is 2. This is the zero of the function $f(x) = 2x - 4$. That is, $f(2) = 0$.

The zero of $f(x)$ is the first coordinate of the **x-intercept** of the graph of $y = f(x)$.

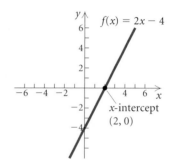

Quadratic Function

$g(x) = x^2 - 3x - 4$, or
$\quad y = x^2 - 3x - 4$

To find the **zeros** of $g(x)$, we solve $g(x) = 0$:

$$x^2 - 3x - 4 = 0$$
$$(x + 1)(x - 4) = 0$$
$$x + 1 = 0 \quad or \quad x - 4 = 0$$
$$x = -1 \quad or \quad \quad x = 4.$$

The **solutions** of $x^2 - 3x - 4 = 0$ are -1 and 4. They are the zeros of the function $g(x)$. That is, $g(-1) = 0$ and $g(4) = 0$.

The real-number zeros of $g(x)$ are the first coordinates of the **x-intercepts** of the graph of $y = g(x)$.

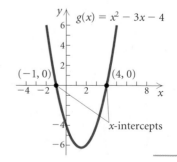

2.3

Exercise Set

Solve.

1. $(2x - 3)(3x - 2) = 0$

2. $(5x - 2)(2x + 3) = 0$

3. $x^2 - 8x - 20 = 0$

4. $x^2 + 6x + 8 = 0$

5. $3x^2 + x - 2 = 0$

6. $10x^2 - 16x + 6 = 0$

7. $4x^2 - 12 = 0$

8. $6x^2 = 36$

9. $3x^2 = 21$

10. $2x^2 - 10 = 0$

11. $5x^2 + 10 = 0$

12. $4x^2 + 12 = 0$

13. $2x^2 - 34 = 0$

14. $3x^2 = 33$

15. $2x^2 = 6x$

16. $18x + 9x^2 = 0$

17. $3y^3 - 5y^2 - 2y = 0$

18. $3t^3 + 2t = 5t^2$

19. $7x^3 + x^2 - 7x - 1 = 0$
    (*Hint*: Factor by grouping.)

20. $3x^3 + x^2 - 12x - 4 = 0$
    (*Hint*: Factor by grouping.)

In Exercises 21–26, use the given graph to find each of the following: (a) the x-intercepts and (b) the zeros of the function.

21.

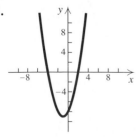

22.

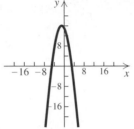

23.

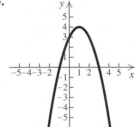

24.

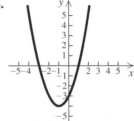

25.

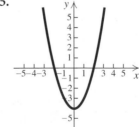

26.

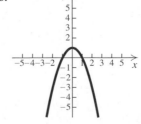

Solve by completing the square to obtain exact solutions.

27. $x^2 + 6x = 7$ **28.** $x^2 + 8x = -15$

29. $x^2 = 8x - 9$ **30.** $x^2 = 22 + 10x$

31. $x^2 + 8x + 25 = 0$ **32.** $x^2 + 6x + 13 = 0$

33. $3x^2 + 5x - 2 = 0$ **34.** $2x^2 - 5x - 3 = 0$

Use the quadratic formula to find exact solutions.

35. $x^2 - 2x = 15$ **36.** $x^2 + 4x = 5$

37. $5m^2 + 3m = 2$ **38.** $2y^2 - 3y - 2 = 0$

39. $3x^2 + 6 = 10x$ **40.** $3t^2 + 8t + 3 = 0$

41. $x^2 + x + 2 = 0$ **42.** $x^2 + 1 = x$

43. $5t^2 - 8t = 3$ **44.** $5x^2 + 2 = x$

45. $3x^2 + 4 = 5x$ **46.** $2t^2 - 5t = 1$

47. $x^2 - 8x + 5 = 0$ **48.** $x^2 - 6x + 3 = 0$

49. $3x^2 + x = 5$ **50.** $5x^2 + 3x = 1$

51. $2x^2 + 1 = 5x$ **52.** $4x^2 + 3 = x$

53. $5x^2 + 2x = -2$ **54.** $3x^2 + 3x = -4$

For each of the following, find the discriminant, $b^2 - 4ac$, and then determine whether one real-number solution, two different real-number solutions, or two different imaginary-number solutions exist.

55. $4x^2 = 8x + 5$ **56.** $4x^2 - 12x + 9 = 0$

57. $x^2 + 3x + 4 = 0$ **58.** $x^2 - 2x + 4 = 0$

59. $5t^2 - 7t = 0$ **60.** $5t^2 - 4t = 11$

Find the zeros of the function.

61. $f(x) = x^2 + 6x + 5$ **62.** $f(x) = x^2 - x - 2$

63. $f(x) = x^2 - 3x - 3$ **64.** $f(x) = 3x^2 + 8x + 2$

65. $f(x) = x^2 - 5x + 1$ **66.** $f(x) = x^2 - 3x - 7$

67. $f(x) = x^2 + 2x - 5$ **68.** $f(x) = x^2 - x - 4$

69. $f(x) = 2x^2 - x + 4$ **70.** $f(x) = 2x^2 + 3x + 2$

71. $f(x) = 3x^2 - x - 1$ **72.** $f(x) = 3x^2 + 5x + 1$

73. $f(x) = 5x^2 - 2x - 1$ **74.** $f(x) = 4x^2 - 4x - 5$

75. $f(x) = 4x^2 + 3x - 3$ **76.** $f(x) = x^2 + 6x - 3$

Solve.

77. $x^4 - 3x^2 + 2 = 0$

78. $x^4 + 3 = 4x^2$

79. $x^4 + 3x^2 = 10$

80. $x^4 - 8x^2 = 9$

81. $y^6 - 9y^3 + 8 = 0$

82. $y^6 - 26y^3 - 27 = 0$

83. $x - 3\sqrt{x} - 4 = 0$
(*Hint:* Let $u = \sqrt{x}$.)

84. $2x - 9\sqrt{x} + 4 = 0$

85. $m^{2/3} - 2m^{1/3} - 8 = 0$
(*Hint:* Let $u = m^{1/3}$.)

86. $t^{2/3} + t^{1/3} - 6 = 0$

87. $x^{1/2} - 3x^{1/4} + 2 = 0$

88. $x^{1/2} - 4x^{1/4} = -3$

89. $(2x - 3)^2 - 5(2x - 3) + 6 = 0$
(*Hint:* Let $u = 2x - 3$.)

90. $(3x + 2)^2 + 7(3x + 2) - 8 = 0$

91. $(2t^2 + t)^2 - 4(2t^2 + t) + 3 = 0$

92. $12 = (m^2 - 5m)^2 + (m^2 - 5m)$

Time of a Free Fall. The formula $s = 16t^2$ is used to approximate the distance s, in feet, that an object falls freely from rest in t seconds. Use this formula for Exercises 93 and 94.

93. The Warszawa Radio Mast in Poland, at 2120 ft, is the world's tallest structure (*Source: The Cambridge Fact Finder*). How long would it take an object falling freely from the top to reach the ground?

94. The tallest structure in the United States, at 2063 ft, is the KTHI-TV tower in North Dakota (*Source: The Cambridge Fact Finder*). How long would it take an object falling freely from the top to reach the ground?

Self-employed Workers. The function $w(x) = -0.01x^2 + 0.27x + 8.60$ can be used to estimate the number of self-employed workers in the United States, in millions, x years after 1980 (*Source: U.S. Bureau of Labor Statistics*). Use this function for Exercises 95 and 96.

95. For what years were there (or will there be) 9.7 million self-employed workers in the United States?

96. For what years were there (or will there be) 9.1 million self-employed workers in the United States?

97. The length of a rectangular poster is 1 ft more than the width and a diagonal of the poster is 5 ft. Find the length and the width.

98. One leg of a right triangle is 7 cm less than the length of the other leg. The length of the hypotenuse is 13 cm. Find the lengths of the legs.

99. One number is 5 greater than another. The product of the numbers is 36. Find the numbers.

100. One number is 6 less than another. The product of the numbers is 72. Find the numbers.

101. *Box Construction.* An open box is made from a 10-cm by 20-cm piece of tin by cutting a square from each corner and folding up the edges. The area of the resulting base is 96 cm². What is the length of the sides of the squares?

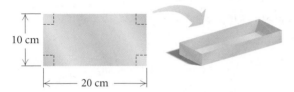

102. *Picture Frame Dimensions.* The frame of a picture is 28 cm by 32 cm outside and is of uniform width. What is the width of the frame if 192 cm² of the picture shows?

103. *Dimensions of a Rug.* Find the dimensions of a Persian rug whose perimeter is 28 ft and whose area is 48 ft².

104. *Petting Zoo Dimensions.* The director of the Glen Island Zoo wants to use 170 m of fencing to enclose a petting area of 1750 m². Find the dimensions of the petting area.

State whether the function is linear or quadratic.

105. $f(x) = 4 - 5x$

106. $f(x) = 4 - 5x^2$

107. $f(x) = 7x^2$

108. $f(x) = 23x + 6$

109. $f(x) = 1.2x - (3.6)^2$

110. $f(x) = 2 - x - x^2$

Technology Connection

Solve graphically. Round solutions to three decimal places, where appropriate.

111. $x^2 - 8x + 12 = 0$

112. $5x^2 + 42x + 16 = 0$

113. $7x^2 - 43x + 6 = 0$

114. $10x^2 - 23x + 12 = 0$

115. $6x + 1 = 4x^2$

116. $3x^2 + 5x = 3$

Use a graphing calculator to find the zeros of the function. Round to three decimal places.

117. $f(x) = 2x^2 - 5x - 4$

118. $f(x) = 4x^2 - 3x - 2$

119. $f(x) = 3x^2 + 2x - 4$

120. $f(x) = 9x^2 - 8x - 7$

121. $f(x) = 5.02x^2 - 4.19x - 2.057$

122. $f(x) = 1.21x^2 - 2.34x - 5.63$

Collaborative Discussion and Writing

123. Is it possible for a quadratic function to have one real zero and one imaginary zero? Why or why not?

124. The graph of a quadratic function can have 0, 1, or 2 *x*-intercepts. How can you predict the number of *x*-intercepts without drawing the graph or (completely) solving an equation?

Skill Maintenance

Associate's Degrees Conferred. The function $a(x) = 9096x + 387,725$ can be used to estimate the number of associate's degrees conferred x years after 1980 (Source: U.S. National Center for Education Statistics). Use this function for Exercises 125 and 126.

125. Estimate the number of associate's degrees conferred in 1998.

126. Predict the number of associate's degrees that will be conferred in 2010.

Determine whether the graph is symmetric with respect to the x-axis, the y-axis, and the origin.

127. $3x^2 + 4y^2 = 5$ **128.** $y^3 = 6x^2$

Determine whether the function is even, odd, or neither even nor odd.

129. $f(x) = 2x^3 - x$ **130.** $f(x) = 4x^2 + 2x - 3$

Synthesis

For each equation in Exercises 131–134, under the given condition: (a) Find k and (b) find a second solution.

131. $kx^2 - 17x + 33 = 0$; one solution is 3

132. $kx^2 - 2x + k = 0$; one solution is -3

133. $x^2 - kx + 2 = 0$; one solution is $1 + i$

134. $x^2 - (6 + 3i)x + k = 0$; one solution is 3

Solve.

135. $(x - 2)^3 = x^3 - 2$

136. $(x + 1)^3 = (x - 1)^3 + 26$

137. $(6x^3 + 7x^2 - 3x)(x^2 - 7) = 0$

138. $\left(x - \frac{1}{5}\right)\left(x^2 - \frac{1}{4}\right) + \left(x - \frac{1}{5}\right)\left(x^2 + \frac{1}{8}\right) = 0$

139. $x^2 + x - \sqrt{2} = 0$

140. $x^2 + \sqrt{5}x - \sqrt{3} = 0$

141. $2t^2 + (t - 4)^2 = 5t(t - 4) + 24$

142. $9t(t + 2) - 3t(t - 2) = 2(t + 4)(t + 6)$

143. $\sqrt{x - 3} - \sqrt[4]{x - 3} = 2$

144. $x^6 - 28x^3 + 27 = 0$

145. $\left(y + \frac{2}{y}\right)^2 + 3y + \frac{6}{y} = 4$

146. $x^2 + 3x + 1 - \sqrt{x^2 + 3x + 1} = 8$

147. Solve $\frac{1}{2}at^2 + v_0 t + x_0 = 0$ for t.

2.4

Analyzing Graphs of Quadratic Functions

- *Find the vertex, the axis of symmetry, and the maximum or minimum value of a quadratic function using the method of completing the square.*
- *Graph quadratic functions.*
- *Solve applied problems involving maximum and minimum function values.*

Graphing Quadratic Functions of the Type $f(x) = a(x - h)^2 + k$

The graph of a quadratic function is called a **parabola.** The graph of every parabola evolves from the graph of the squaring function $f(x) = x^2$ using transformations.

TRANSFORMATIONS

REVIEW SECTION 1.7.

We get the graph of $f(x) = a(x - h)^2 + k$ from the graph of $f(x) = x^2$ as follows:

$$f(x) = x^2$$
$$\downarrow$$
$$f(x) = ax^2$$ **Vertical stretching or shrinking with a reflection across the x-axis if $a < 0$**
$$\downarrow$$
$$f(x) = a(x - h)^2$$ **Horizontal translation**
$$\downarrow$$
$$f(x) = a(x - h)^2 + k.$$ **Vertical translation**

Consider the following graphs of the form $f(x) = a(x - h)^2 + k$. The point (h, k) at which the graph turns is called the **vertex.** The maximum or minimum value of $f(x)$ occurs at the vertex. Each graph has a line $x = h$ that is called the **axis of symmetry.**

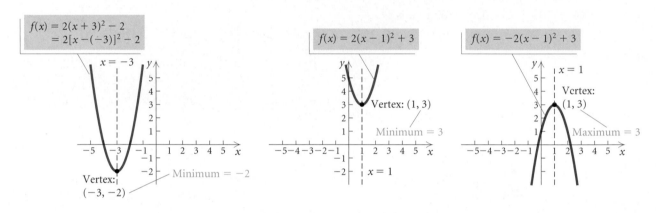

$f(x) = 2(x + 3)^2 - 2$
$\quad = 2[x - (-3)]^2 - 2$

$x = -3$

Vertex: $(-3, -2)$

Minimum $= -2$

$f(x) = 2(x - 1)^2 + 3$

Vertex: $(1, 3)$

Minimum $= 3$

$x = 1$

$f(x) = -2(x - 1)^2 + 3$

$x = 1$

Vertex: $(1, 3)$

Maximum $= 3$

CONNECTING THE CONCEPTS

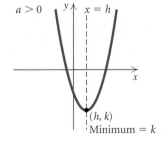

$a > 0$ $x = h$

(h, k)
Minimum $= k$

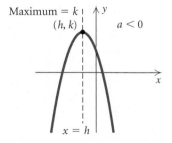

Maximum $= k$
(h, k) $a < 0$

$x = h$

GRAPHING QUADRATIC FUNCTIONS

The graph of $f(x) = a(x - h)^2 + k$ is a parabola that:

- opens up if $a > 0$ and down if $a < 0$;
- has (h, k) as the vertex;
- has $x = h$ as the axis of symmetry;
- has k as a minimum value (output) if $a > 0$
- has k as a maximum value if $a < 0$.

As we saw in Section 1.7, the constant a serves to stretch or shrink the graph vertically. As a parabola is stretched vertically, it becomes narrower, and as it is shrunk vertically, it becomes wider. That is, as $|a|$ increases, the graph becomes narrower, and as $|a|$ gets close to 0, the graph becomes wider.

If the equation is in the form $f(x) = a(x - h)^2 + k$, we can learn a great deal about the graph without graphing.

FUNCTION	$f(x) = 3\left(x - \frac{1}{4}\right)^2 - 2$ $= 3\left(x - \frac{1}{4}\right)^2 + (-2)$	$g(x) = -3(x + 5)^2 + 7$ $= -3[x - (-5)]^2 + 7$
VERTEX	$\left(\frac{1}{4}, -2\right)$	$(-5, 7)$
AXIS OF SYMMETRY	$x = \frac{1}{4}$	$x = -5$
MAXIMUM	No (3 > 0, graph opens up.)	Yes, 7 ($-3 < 0$, graph opens down.)
MINIMUM	Yes, -2 (3 > 0, graph opens up.)	No ($-3 < 0$, graph opens down.)

Note that the vertex (h, k) is used to find the maximum or the minimum value of the function. The maximum or minimum value is the number k, *not* the ordered pair (h, k).

Graphing Quadratic Functions of the Type $f(x) = ax^2 + bx + c, a \neq 0$

We now use a modification of the method of completing the square as an aid in graphing and analyzing quadratic functions of the form $f(x) = ax^2 + bx + c, a \neq 0$.

EXAMPLE 1   Find the vertex, the axis of symmetry, and the maximum or minimum value of $f(x) = x^2 + 10x + 23$. Then graph the function.

Solution   To express $f(x) = x^2 + 10x + 23$ in the form $f(x) = a(x - h)^2 + k$, we complete the square on the terms involving x. To do so, we take half the coefficient of x and square it, obtaining $(10/2)^2$, or 25. We now add and subtract that number on the right side:

$$f(x) = x^2 + 10x + 23 = x^2 + 10x + 25 - 25 + 23.$$

Since $25 - 25 = 0$, the new expression for the function is equivalent to the original expression. Note that this process differs from the one we used to complete the square in order to solve a quadratic equation, where we added the same number on both sides of the equation to obtain an equivalent equation. Instead, when we complete the square to write a function in the form $f(x) = a(x - h)^2 + k$, we add and subtract the same number on the right side. The entire process is shown below:

$$f(x) = x^2 + 10x + 23$$
Note that 25 completes the square for $x^2 + 10x$.

$$= x^2 + 10x + 25 - 25 + 23$$
Adding $25 - 25$, or 0, to the right side

$$= (x^2 + 10x + 25) - 25 + 23$$
Regrouping

$$= (x + 5)^2 - 2$$
Factoring and simplifying

$$= [x - (-5)]^2 + (-2).$$
Writing in the form $f(x) = a(x - h)^2 + k$

Keeping in mind that this function will have a minimum value since $a > 0$ ($a = 1$), from this form of the function we know the following:

Vertex: $(-5, -2)$;

Axis of symmetry: $x = -5$;

Minimum value of the function: -2.

To graph the function, we first plot the vertex and find several points on either side of it. Then we plot these points and connect them with a smooth curve.

x	$f(x)$	
-5	-2	← Vertex
-4	-1	
-2	7	
-7	2	
-8	7	

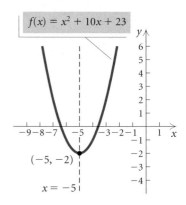

The graph of $f(x) = x^2 + 10x + 23$, or $[x - (-5)]^2 + (-2)$, shown above, is a shift of the graph of $y = x^2$ left 5 units and down 2 units. ▬

Keep in mind that the axis of symmetry is not part of the graph; it is a characteristic of the graph. If you fold the graph on its axis of symmetry, the two halves of the graph will coincide.

EXAMPLE 2 Find the vertex, the axis of symmetry, and the maximum or minimum value of $g(x) = x^2/2 - 4x + 8$. Then graph the function.

Solution We complete the square in order to write the function in the form $g(x) = a(x - h)^2 + k$. First, we factor $\frac{1}{2}$ out of the first two terms. This makes the coefficient of x^2 within the parentheses 1:

$$g(x) = \frac{x^2}{2} - 4x + 8$$

$$= \frac{1}{2}(x^2 - 8x) + 8.$$
Factoring $\frac{1}{2}$ out of the first two terms:
$\dfrac{x^2}{2} - 4x = \dfrac{1}{2} \cdot x^2 - \dfrac{1}{2} \cdot 8x$

Next, we complete the square inside the parentheses: Half of -8 is -4, and $(-4)^2 = 16$. We add and subtract 16 inside the parentheses:

$$g(x) = \tfrac{1}{2}(x^2 - 8x + 16 - 16) + 8$$

$$= \tfrac{1}{2}(x^2 - 8x + 16) - \tfrac{1}{2} \cdot 16 + 8 \qquad \text{Using the distributive law to remove } -16 \text{ from within the parentheses}$$

$$= \tfrac{1}{2}(x - 4)^2 + 0, \text{ or } \tfrac{1}{2}(x - 4)^2. \qquad \text{Factoring and simplifying}$$

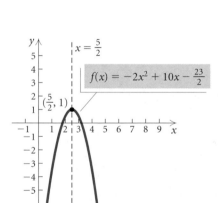

$g(x) = \dfrac{x^2}{2} - 4x + 8$

$x = 4$

$(4, 0)$

We know the following:

Vertex: $(4, 0)$;

Axis of symmetry: $x = 4$;

Minimum value of the function: 0.

Finally, we plot the vertex and several points on either side of it and draw the graph of the function. The graph of g is a vertical shrinking of the graph of $y = x^2$ along with a shift 4 units to the right.

EXAMPLE 3 Find the vertex, the axis of symmetry, and the maximum or minimum value of $f(x) = -2x^2 + 10x - \tfrac{23}{2}$. Then graph the function.

Solution We have

$$f(x) = -2x^2 + 10x - \tfrac{23}{2}$$

$$= -2(x^2 - 5x) - \tfrac{23}{2} \qquad \text{Factoring } -2 \text{ out of the first two terms}$$

$$= -2\left(x^2 - 5x + \tfrac{25}{4} - \tfrac{25}{4}\right) - \tfrac{23}{2} \qquad \text{Completing the square inside the parentheses}$$

$$= -2\left(x^2 - 5x + \tfrac{25}{4}\right) - 2\left(-\tfrac{25}{4}\right) - \tfrac{23}{2} \qquad \text{Using the distributive law to remove } -\tfrac{25}{4} \text{ from within the parentheses}$$

$$= -2\left(x - \tfrac{5}{2}\right)^2 + \tfrac{25}{2} - \tfrac{23}{2}$$

$$= -2\left(x - \tfrac{5}{2}\right)^2 + 1.$$

This form of the function yields the following:

Vertex: $\left(\tfrac{5}{2}, 1\right)$;

Axis of symmetry: $x = \tfrac{5}{2}$;

Maximum value of the function: 1.

$x = \dfrac{5}{2}$

$\left(\tfrac{5}{2}, 1\right)$

$f(x) = -2x^2 + 10x - \dfrac{23}{2}$

The graph is found by shifting the graph of $f(x) = x^2$ to the right $\tfrac{5}{2}$ units, reflecting it across the x-axis, stretching it vertically, and shifting it up 1 unit.

In many situations, we want to find the coordinates of the vertex directly from the equation $f(x) = ax^2 + bx + c$ using a formula. One way to develop such a formula is to observe that the x-coordinate of the vertex is centered between the x-intercepts, or zeros, of the function. By averaging the two solutions of $ax^2 + bx + c = 0$, we find a formula for the x-coordinate of the vertex:

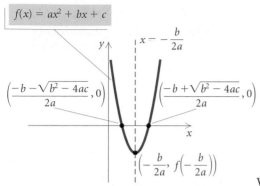

$f(x) = ax^2 + bx + c$

$$x\text{-coordinate of vertex} = \frac{\dfrac{-b - \sqrt{b^2 - 4ac}}{2a} + \dfrac{-b + \sqrt{b^2 - 4ac}}{2a}}{2}$$

$$= \frac{\dfrac{-2b}{2a}}{2} = \frac{-\dfrac{b}{a}}{2}$$

$$= -\frac{b}{a} \cdot \frac{1}{2} = -\frac{b}{2a}.$$

We use this value of x to find the y-coordinate of the vertex, $f\left(-\dfrac{b}{2a}\right)$.

The Vertex of a Parabola

The **vertex** of the graph of $f(x) = ax^2 + bx + c$ is

$$\left(-\frac{b}{2a},\, f\left(-\frac{b}{2a}\right)\right).$$

We calculate the x-coordinate.   We substitute to find the y-coordinate.

EXAMPLE 4   For the function $f(x) = -x^2 + 14x - 47$:

a) Find the vertex.

b) Determine whether there is a maximum or minimum value and find that value.

c) Find the range.

d) On what intervals is the function increasing? decreasing?

Solution   There is no need to graph the function.

a) The x-coordinate of the vertex is

$$-\frac{b}{2a} = -\frac{14}{2(-1)}, \text{ or } 7.$$

Since

$$f(7) = -7^2 + 14 \cdot 7 - 47 = 2,$$

the vertex is $(7, 2)$.

b) Since a is negative ($a = -1$), the graph opens down so the second coordinate of the vertex, 2, is the maximum value of the function.

c) The range is $(-\infty, 2]$.

d) Since the graph opens down, function values increase to the left of the vertex and decrease to the right of the vertex. Thus the function is increasing on the interval $(-\infty, 7)$ and decreasing on $(7, \infty)$.

Technology Connection

We can use a graphing calculator to do Example 4. Once we have graphed $y = -x^2 + 14x - 47$, we see that the graph opens down and thus has a maximum value. We can use the MAXIMUM feature to find the coordinates of the vertex. Using these coordinates, we can then find the maximum value and the range of the function along with the intervals on which the function is increasing or decreasing.

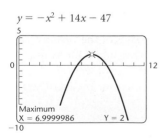

$y = -x^2 + 14x - 47$

Maximum
X = 6.9999986   Y = 2

Applications

Many real-world situations involve finding the maximum or minimum value of a quadratic function.

EXAMPLE 5 *Maximizing Area.*   A stone mason has enough stones to enclose a rectangular patio with 60 ft of stone wall. If the house forms one side of the rectangle, what is the maximum area that the mason can enclose? What should the dimensions of the patio be in order to yield this area?

PROBLEM-SOLVING STRATEGY

REVIEW SECTION 2.1.

Solution   We will use the five-step problem-solving strategy.

1. **Familiarize.** We make a drawing of the situation, using w to represent the width of the patio, in feet. Then $(60 - 2w)$ feet of stone is available for the length. Suppose the patio were 10 ft wide. It would then be $60 - 2 \cdot 10 = 40$ ft long. The area would be $(10 \text{ ft})(40 \text{ ft}) = 400 \text{ ft}^2$. If the patio were 12 ft wide, it would be $60 - 2 \cdot 12 = 36$ ft long. The area would be $(12 \text{ ft})(36 \text{ ft}) = 432 \text{ ft}^2$. If it were 16 ft wide, it would be $60 - 2 \cdot 16 = 28$ ft long and the area would be $(16 \text{ ft})(28 \text{ ft}) = 448 \text{ ft}^2$. There are more combinations of length and width than we could possibly try. Instead we will find a function that represents the area and then determine the maximum value of the function.

2. **Translate.** Since the area of a rectangle is given by length times width, we have

$$A(w) = (60 - 2w)w \qquad A = lw; l = 60 - 2w$$
$$= -2w^2 + 60w,$$

where $A(w)$ is the area of the patio, in square feet, as a function of the width, w.

3. **Carry out.** To solve this problem, we need to determine the maximum value of $A(w)$ and find the dimensions for which that maximum occurs. Since A is a quadratic function and w^2 has a negative coefficient, we know that the function has a maximum value that occurs at the vertex of the graph of the function. The first coordinate of the vertex, $(w, A(w))$, is

$$w = -\frac{b}{2a} = -\frac{60}{2(-2)} = 15 \text{ ft}.$$

Thus, if $w = 15$ ft, then the length $l = 60 - 2 \cdot 15 = 30$ ft; and the area is $15 \cdot 30$, or 450 ft^2.

4. **Check.** As a partial check, we note that $450 \text{ ft}^2 > 448 \text{ ft}^2$, which is the largest area we found in a guess in the *Familiarize* step.

5. **State.** The maximum possible area is 450 ft^2 when the patio is 15 ft wide and 30 ft long.

Technology Connection

As a more complete check in Example 5, assuming that the function $A(w)$ is correct, we could examine a table of values for

$$A(w) = (60 - 2w)w$$

and/or examine its graph.

X	Y₁
14.7	449.82
14.8	449.92
14.9	449.98
15	**450**
15.1	449.98
15.2	449.92
15.3	449.82

X = 15

$y = (60 - 2x)x$

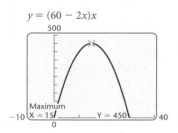

EXAMPLE 6 *Height of a Rocket.* A model rocket is launched with an initial velocity of 100 ft/sec from the top of a hill that is 20 ft high. Its height t seconds after it has been launched is given by the function $s(t) = -16t^2 + 100t + 20$. Determine the time at which the rocket reaches its maximum height and find the maximum height.

Solution

1., 2. Familiarize and **Translate.** We are given the function in the statment of the problem.

3. Carry out. We need to find the maximum value of the function and the value of t for which it occurs. Since $s(t)$ is a quadratic function and t^2 has a negative coefficient, we know that the maximum value of the function occurs at the vertex of the graph of the function. The first coordinate of the vertex gives the time t at which the rocket reaches its maximum height. It is

$$t = -\frac{b}{2a} = -\frac{100}{2(-16)} = 3.125.$$

The second coordinate of the vertex gives the maximum height of the rocket. We substitute in the function to find it:

$$s(3.125) = -16(3.125)^2 + 100(3.125) + 20 = 176.25.$$

4. Check. As a check, we can complete the square to write the function in the form $s(t) = a(t - h)^2 + k$ and determine the coordinates of the vertex from this form of the function. We get

$$s(t) = -16(t - 3.125)^2 + 176.25.$$

This confirms that the vertex is $(3.125, 176.25)$, so the answer checks.

5. State. The rocket reaches a maximum height of 176.25 ft 3.125 sec after it has been launched. ▬

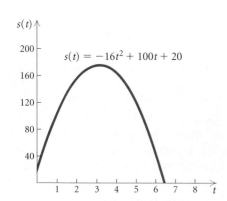

$s(t)$

200

160

120

80

40

$s(t) = -16t^2 + 100t + 20$

1 2 3 4 5 6 7 8 t

EXAMPLE 7 *Finding the Depth of a Well.* Two seconds after a chlorine tablet has been dropped into a well, a splash is heard. The speed of sound is 1100 ft/sec. How far is the top of the well from the water?

Solution

1. Familiarize. We first make a drawing and label it with known and unknown information. We let s = the depth of the well, in feet, t_1 = the time, in seconds, that it takes for the tablet to hit the water, and t_2 = the time, in seconds, that it takes for the sound to reach the top of the well. This gives us the equation

$$t_1 + t_2 = 2. \tag{1}$$

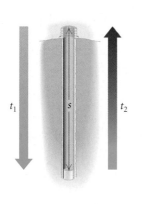

t_1 s t_2

2. **Translate.** Can we find any relationship between the two times and the distance s? Often in problem solving you may need to look up related formulas in a physics book, another mathematics book, or on the Internet. We find that the formula

$$s = 16t^2$$

gives the distance, in feet, that a dropped object falls in t seconds. The time t_1 that it takes the tablet to hit the water can be found as follows:

$$s = 16t_1^2, \quad \text{or} \quad \frac{s}{16} = t_1^2, \text{ so } t_1 = \frac{\sqrt{s}}{4}. \qquad \begin{array}{l}\textbf{Taking the positive}\\ \textbf{square root}\end{array} \qquad \textbf{(2)}$$

To find an expression for t_2, the time it takes the sound to travel to the top of the well, recall that *Distance = Rate · Time*. Thus,

$$s = 1100t_2, \quad \text{or} \quad t_2 = \frac{s}{1100}. \qquad\qquad \textbf{(3)}$$

We now have expressions for t_1 and t_2, both in terms of s. Substituting into equation (1), we obtain

$$t_1 + t_2 = 2, \quad \text{or} \quad \frac{\sqrt{s}}{4} + \frac{s}{1100} = 2. \qquad \textbf{(4)}$$

3. **Carry out.** We solve equation (4) for s. Multiplying by 1100, we get

$$275\sqrt{s} + s = 2200, \quad \text{or} \quad s + 275\sqrt{s} - 2200 = 0.$$

This equation is reducible to quadratic with $u = \sqrt{s}$. Substituting, we get

$$u^2 + 275u - 2200 = 0.$$

Using the quadratic formula, we can solve for u:

$$u = \frac{-b \pm \sqrt{b^2 - 4ac}}{2a}$$

$$= \frac{-275 + \sqrt{275^2 - 4 \cdot 1 \cdot (-2200)}}{2 \cdot 1} \qquad \begin{array}{l}\textbf{We want only the}\\ \textbf{positive solution.}\end{array}$$

$$= \frac{-275 + \sqrt{84{,}425}}{2}$$

$$\approx 7.78.$$

Since $u \approx 7.78$, we have

$$\sqrt{s} = 7.78$$

$$s \approx 60.5. \qquad \textbf{Squaring both sides}$$

4. **Check.** To check, we can substitute 60.5 for s in equation (4) and see that $t_1 + t_2 \approx 2$. We leave the mathematics for the student.

5. **State.** The top of the well is about 60.5 ft above the water.

Technology Connection

We can solve the equation in Example 7 graphically using the Intersect method. It will probably require some trial and error to determine an appropriate window.

$$y_1 = \frac{\sqrt{x}}{4} + \frac{x}{1100}, \quad y_2 = 2$$

Intersection
X = 60.526879   Y = 2

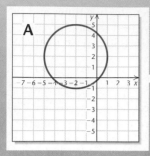

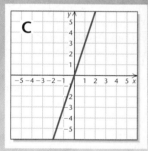

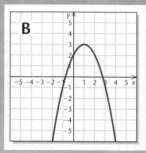

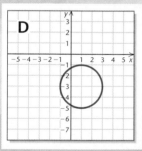

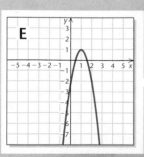

Visualizing the Graph

Match the equation with its graph.

1. $y = 3x$

2. $y = -(x - 1)^2 + 3$

3. $x^2 + 4x + y^2 - 4y - 1 = 0$

4. $y = 3$

5. $2x - 3y = 6$

6. $(x - 1)^2 + (y + 3)^2 = 4$

7. $y = -2x + 1$

8. $y = 2x^2 - x - 4$

9. $x = -2$

10. $y = -3x^2 + 6x - 2$

Answers on page A-15

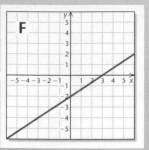

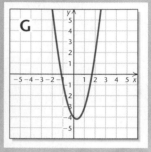

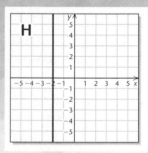

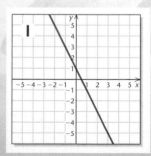

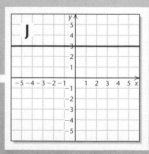

2.4

Exercise Set

In Exercises 1 and 2, use the given graph to find each of the following: **(a)** the vertex; **(b)** the axis of symmetry; and **(c)** the maximum or minimum value of the function.

1.

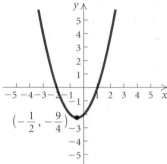

$\left(-\dfrac{1}{2}, -\dfrac{9}{4}\right)$

2.

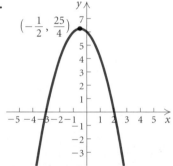

$\left(-\dfrac{1}{2}, \dfrac{25}{4}\right)$

In Exercises 3–14, **(a)** find the vertex; **(b)** find the axis of symmetry; **(c)** determine whether there is a maximum or minimum value and find that value; and **(d)** graph the function.

3. $f(x) = x^2 - 8x + 12$

4. $g(x) = x^2 + 7x - 8$

5. $f(x) = x^2 - 7x + 12$

6. $g(x) = x^2 - 5x + 6$

7. $f(x) = x^2 + 4x + 5$

8. $f(x) = x^2 + 2x + 6$

9. $f(x) = -x^2 - 6x + 3$

10. $f(x) = -x^2 - 8x + 5$

11. $g(x) = 2x^2 + 6x + 8$

12. $f(x) = 2x^2 - 10x + 14$

13. $g(x) = -2x^2 + 2x + 1$

14. $f(x) = -3x^2 - 3x + 1$

In Exercises 15–21, match the equation with one of the figures (a)–(h), which follow.

a)

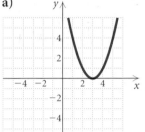

b)

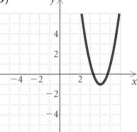

c)

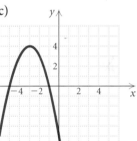

d)

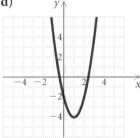

e)

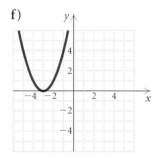

f)

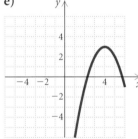

g)

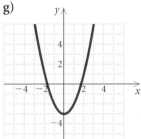

h)

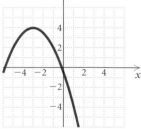

15. $y = (x + 3)^2$

16. $y = -(x - 4)^2 + 3$

17. $y = 2(x - 4)^2 - 1$

18. $y = x^2 - 3$

19. $y = -\frac{1}{2}(x + 3)^2 + 4$ **20.** $y = (x - 3)^2$

21. $y = -(x + 3)^2 + 4$ **22.** $y = 2(x - 1)^2 - 4$

In Exercises 23–32:

a) *Find the vertex.*
b) *Determine whether there is a maximum or minimum value and find that value.*
c) *Find the range.*
d) *Find the intervals on which the function is increasing and the intervals on which the function is decreasing.*

23. $f(x) = x^2 - 6x + 5$

24. $f(x) = x^2 + 4x - 5$

25. $f(x) = 2x^2 + 4x - 16$

26. $f(x) = \frac{1}{2}x^2 - 3x + \frac{5}{2}$

27. $f(x) = -\frac{1}{2}x^2 + 5x - 8$

28. $f(x) = -2x^2 - 24x - 64$

29. $f(x) = 3x^2 + 6x + 5$

30. $f(x) = -3x^2 + 24x - 49$

31. $g(x) = -4x^2 - 12x + 9$

32. $g(x) = 2x^2 - 6x + 5$

33. *Height of a Ball.* A ball is thrown directly upward from a height of 6 ft with an initial velocity of 20 ft/sec. The function $s(t) = -16t^2 + 20t + 6$ gives the height of the ball t seconds after it has been thrown. Determine the time at which the ball reaches its maximum height and find the maximum height.

34. *Height of a Projectile.* A stone is thrown directly upward from a height of 30 ft with an initial velocity of 60 ft/sec. The height of the stone t seconds after it has been thrown is given by the function $s(t) = -16t^2 + 60t + 30$. Determine the time at which the stone reaches its maximum height and find the maximum height.

35. *Height of a Rocket.* A model rocket is launched with an initial velocity of 120 ft/sec from a height of 80 ft. The height of the rocket t seconds after it has been launched is given by the function $s(t) = -16t^2 + 120t + 80$. Determine the time at which the rocket reaches its maximum height and find the maximum height.

36. *Height of a Rocket.* A model rocket is launched with an initial velocity of 150 ft/sec from a height of 40 ft. The function $s(t) = -16t^2 + 150t + 40$ gives the height of the rocket t seconds after it has been launched. Determine the time at which the rocket reaches its maximum height and find the maximum height.

37. *Maximizing Volume.* Mendoza Manufacturing plans to produce a one-compartment vertical file by bending the long side of a 10-in. by 18-in. sheet of plastic along two lines to form a U-shape. How tall should the file be in order to maximize the volume that the file can hold?

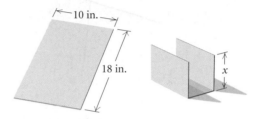

38. *Maximizing Area.* A fourth-grade class decides to enclose a rectangular garden, using the side of the school as one side of the rectangle. What is the maximum area that the class can enclose with 32 ft of fence? What should the dimensions of the garden be in order to yield this area?

39. *Maximizing Area.* The sum of the base and the height of a triangle is 20 cm. Find the dimensions for which the area is a maximum.

40. *Maximizing Area.* The sum of the base and the height of a parallelogram is 69 cm. Find the dimensions for which the area is a maximum.

41. *Minimizing Cost.* Aki's Bicycle Designs has determined that when x hundred bicycles are built, the average cost per bicycle is given by

$$C(x) = 0.1x^2 - 0.7x + 2.425,$$

where $C(x)$ is in hundreds of dollars. How many bicycles should be built in order to minimize the average cost per bicycle?

Maximizing Profit. *In business, profit is the difference between revenue and cost, that is,*

$$Total\ profit = Total\ revenue - Total\ cost,$$
$$P(x) = R(x) - C(x),$$

where x is the number of units sold. Find the maximum profit and the number of units that must be sold in order to yield the maximum profit for each of the following.

42. $R(x) = 5x, \ C(x) = 0.001x^2 + 1.2x + 60$

43. $R(x) = 50x - 0.5x^2, \ C(x) = 10x + 3$

44. $R(x) = 20x - 0.1x^2, \ C(x) = 4x + 2$

45. *Finding the Height of an Elevator Shaft.* Jenelle dropped a screwdriver from the top of an elevator shaft. Exactly 5 sec later, she hears the sound of the screwdriver hitting the bottom of the shaft. How tall is the elevator shaft? (*Hint*: See Example 7.)

46. *Finding the Height of a Cliff.* A water balloon is dropped from a cliff. Exactly 3 sec later, the sound of the balloon hitting the ground reaches the top of the cliff. How high is the cliff? (*Hint*: See Example 7.)

47. *Maximizing Area.* A rancher needs to enclose two adjacent rectangular corrals, one for cattle and one for sheep. If a river forms one side of the corrals and

240 yd of fencing is available, what is the largest total area that can be enclosed?

48. *Norman Window.* A Norman window is a rectangle with a semicircle on top. Sky Blue Windows is designing a Norman window that will require 24 ft of trim on the outer edges. What dimensions will allow the maximum amount of light to enter a house?

A Norman window

Technology Connection

Use a graphing calculator to check the answers for each of the following.

49. Exercises 3 and 13

50. Exercises 4 and 14

51. Exercises 23 and 27

52. Exercises 24 and 28

Collaborative Discussion and Writing

53. Write a problem for a classmate to solve. Design it so that it is a maximum or minimum problem using a quadratic function.

54. Discuss two ways in which we used completing the square in this chapter.

55. Suppose that the graph of $f(x) = ax^2 + bx + c$ has x-intercepts $(x_1, 0)$ and $(x_2, 0)$. What are the x-intercepts of $g(x) = -ax^2 - bx - c$? Explain.

Skill Maintenance

For each function f, construct and simplify the difference quotient

$$\frac{f(x + h) - f(x)}{h}.$$

56. $f(x) = 3x - 7$

57. $f(x) = 2x^2 - x + 4$

A graph of $y = f(x)$ follows. No formula is given for f. Make a hand-drawn graph of each of the following.

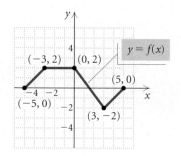

58. $g(x) = f(2x)$ **59.** $g(x) = -2f(x)$

Synthesis

60. Find b such that
$$f(x) = -4x^2 + bx + 3$$
has a maximum value of 50.

61. Find c such that
$$f(x) = -0.2x^2 - 3x + c$$
has a maximum value of -225.

62. Find a quadratic function with vertex $(4, -5)$ and containing the point $(-3, 1)$.

63. Graph: $f(x) = (|x| - 5)^2 - 3$.

64. *Minimizing Area.* A 24-in. piece of string is cut into two pieces. One piece is used to form a circle while the other is used to form a square. How should the string be cut so that the sum of the areas is a minimum?

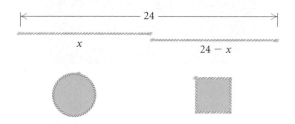

2.5

More Equation Solving

•   *Solve rational and radical equations and equations with absolute value.*

Rational Equations

Equations containing rational expressions are called **rational equations.** Solving such equations involves multiplying both sides by the least common denominator (LCD) to *clear the equation of fractions.*

EXAMPLE 1 Solve: $\dfrac{x-8}{3} + \dfrac{x-3}{2} = 0$.

Algebraic Solution

We have

$$\dfrac{x-8}{3} + \dfrac{x-3}{2} = 0 \qquad \text{The LCD is } 3 \cdot 2, \text{ or } 6.$$

$$6\left(\dfrac{x-8}{3} + \dfrac{x-3}{2}\right) = 6 \cdot 0 \qquad \begin{array}{l}\text{Multiplying by} \\ \text{the LCD on} \\ \text{both sides to} \\ \text{clear fractions}\end{array}$$

$$6 \cdot \dfrac{x-8}{3} + 6 \cdot \dfrac{x-3}{2} = 0$$

$$2(x-8) + 3(x-3) = 0$$

$$2x - 16 + 3x - 9 = 0$$

$$5x - 25 = 0$$

$$5x = 25$$

$$x = 5.$$

The possible solution is 5.

CHECK: $\dfrac{x-8}{3} + \dfrac{x-3}{2} = 0$

$$\dfrac{5-8}{3} + \dfrac{5-3}{2} \;?\; 0$$

$$\dfrac{-3}{3} + \dfrac{2}{2}$$

$$-1 + 1$$

$$0 \;\Big|\; 0 \quad \text{TRUE}$$

The solution is 5.

Visualizing the Solution

The solution of the given equation is the zero of the function

$$f(x) = \dfrac{x-8}{3} + \dfrac{x-3}{2}.$$

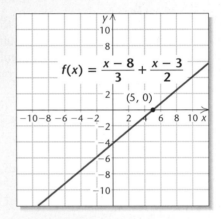

The zero of the function is 5. Thus the solution of the equation is 5.

CAUTION! Clearing fractions is a valid procedure when solving rational equations but not when adding, subtracting, multiplying, or dividing rational expressions. A rational expression may have operation signs but no equals sign. A rational equation always has an equals sign. For example, $\dfrac{x-8}{3} + \dfrac{x-3}{2}$ is a rational expression but $\dfrac{x-8}{3} + \dfrac{x-3}{2} = 0$ is a rational equation. To *simplify* the rational *expression* $\dfrac{x-8}{3} + \dfrac{x-3}{2}$, we first find the LCD and write each fraction with that denominator. The final result is usually a rational expression. To *solve* the rational *equation* $\dfrac{x-8}{3} + \dfrac{x-3}{2} = 0$, we first multiply both sides by the LCD to clear fractions. The final result is one or more numbers. As we will see in Example 2, these numbers must be checked in the original equation.

Technology Connection

We can use the Zero method to solve the equation in Example 1. We find the zero of the function

$$f(x) = \frac{x-8}{3} + \frac{x-3}{2}.$$

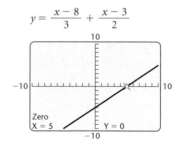

$$y = \frac{x-8}{3} + \frac{x-3}{2}$$

The zero of the function is 5. Thus the solution of the equation is 5.

When we use the multiplication principle to multiply (or divide) on both sides of an equation by an expression with a variable, we might not obtain an equivalent equation. We must check the possible solutions obtained in this manner by substituting them in the original equation. The next example illustrates this.

EXAMPLE 2 Solve: $\dfrac{x^2}{x-3} = \dfrac{9}{x-3}$.

Solution The LCD is $x - 3$.

$$(x - 3) \cdot \frac{x^2}{x-3} = (x-3) \cdot \frac{9}{x-3}$$
$$x^2 = 9$$
$$x = -3 \quad or \quad x = 3 \qquad \textbf{Using the principle of square roots}$$

The possible solutions are -3 and 3. We check.

CHECK: For -3:

$$\frac{x^2}{x-3} = \frac{9}{x-3}$$

$$\frac{(-3)^2}{-3-3} \; ? \; \frac{9}{-3-3}$$

$$\frac{9}{-6} \; \Big| \; \frac{9}{-6} \qquad \text{TRUE}$$

For 3:

$$\frac{x^2}{x-3} = \frac{9}{x-3}$$

$$\frac{3^2}{3-3} \; ? \; \frac{9}{3-3}$$

$$\frac{9}{0} \; \Big| \; \frac{9}{0} \qquad \text{NOT DEFINED}$$

The number -3 checks, so it is a solution. Since division by 0 is not defined, 3 is not a solution. Note that 3 is not in the domain of either $x^2/(x-3)$ or $9/(x-3)$. ▬

Radical Equations

A **radical equation** is an equation in which variables appear in one or more radicands. For example,

$$\sqrt{2x - 5} - \sqrt{x - 3} = 1$$

is a radical equation. The following principle is used to solve such equations.

> ### *The Principle of Powers*
> For any positive integer n:
>
> If $a = b$ is true, then $a^n = b^n$ is true.

EXAMPLE 3   Solve: $5 + \sqrt{x + 7} = x$.

Algebraic Solution

We first isolate the radical and then use the principle of powers.

$$5 + \sqrt{x + 7} = x$$

$$\sqrt{x + 7} = x - 5 \qquad \text{Subtracting 5 on both sides}$$

$$\left(\sqrt{x + 7}\right)^2 = (x - 5)^2 \qquad \text{Using the principle of powers; squaring both sides}$$

$$x + 7 = x^2 - 10x + 25$$

$$0 = x^2 - 11x + 18 \qquad \text{Subtracting } x \text{ and 7}$$

$$0 = (x - 9)(x - 2) \qquad \text{Factoring}$$

$$x - 9 = 0 \quad or \quad x - 2 = 0$$

$$x = 9 \quad or \quad x = 2$$

The possible solutions are 9 and 2.

CHECK: For 9:

$$5 + \sqrt{x + 7} = x$$

$$\overline{5 + \sqrt{9 + 7}} \; ? \; 9$$

$$5 + \sqrt{16}$$

$$5 + 4$$

$$9 \;\bigg|\; 9 \quad \text{TRUE}$$

For 2:

$$5 + \sqrt{x + 7} = x$$

$$\overline{5 + \sqrt{2 + 7}} \; ? \; 2$$

$$5 + \sqrt{9}$$

$$5 + 3$$

$$8 \;\bigg|\; 2 \quad \text{FALSE}$$

Since 9 checks but 2 does not, the only solution is 9.

Visualizing the Solution

When we graph $y = 5 + \sqrt{x + 7}$ and $y = x$, we find that the first coordinate of the point of intersection of the graphs is 9. Thus the solution of $5 + \sqrt{x + 7} = x$ is 9.

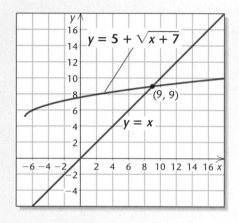

Note that the graphs show that the equation has only one solution.

Technology Connection

We can use a graphing calculator to solve the equation in Example 3. We can graph $y_1 = 5 + \sqrt{x + 7}$ and $y_2 = x$. Using the INTERSECT feature, we see in the window on the left below that the solution is 9.

We can also use the ZERO feature to get this result, as shown in the window on the right. To do so, we first write the equivalent equation $5 + \sqrt{x + 7} - x = 0$. The zero of the function $f(x) = 5 + \sqrt{x + 7} - x$ is 9, so the solution of the original equation is 9.

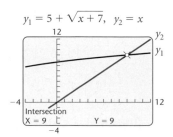

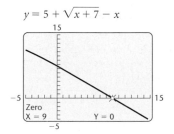

When we raise both sides of an equation to an even power, the resulting equation can have solutions that the original equation does not. This is because the converse of the principle of powers is not necessarily true. That is, if $a^n = b^n$ is true, we do not know that $a = b$ is true. For example, $(-2)^2 = 2^2$, but $-2 \neq 2$. Thus, as we see in Example 3, it is necessary to check the possible solutions in the original equation when the principle of powers is used to raise both sides of an equation to an even power.

When a radical equation has two radical terms on one side, we isolate one of them and then use the principle of powers. If, after doing so, a radical term remains, we repeat these steps.

Study Tip

Consider forming a study group with some of your fellow students. Exchange e-mail addresses, telephone numbers, and schedules so you can coordinate study time for homework and tests.

EXAMPLE 4   Solve: $\sqrt{x - 3} + \sqrt{x + 5} = 4$.

Solution   We have

$$\sqrt{x - 3} = 4 - \sqrt{x + 5} \qquad \text{Isolating one radical}$$

$$\left(\sqrt{x - 3}\right)^2 = \left(4 - \sqrt{x + 5}\right)^2 \qquad \text{Using the principle of powers; squaring both sides}$$

$$x - 3 = 16 - 8\sqrt{x + 5} + (x + 5)$$

$$x - 3 = 21 - 8\sqrt{x + 5} + x \qquad \text{Combining like terms}$$

$$-24 = -8\sqrt{x + 5} \qquad \text{Isolating the remaining radical; subtracting } x \text{ and 21 on both sides}$$

$$3 = \sqrt{x + 5} \qquad \text{Dividing by } -8 \text{ on both sides}$$

$$3^2 = \left(\sqrt{x + 5}\right)^2 \qquad \text{Using the principle of powers; squaring both sides}$$

$$9 = x + 5$$

$$4 = x.$$

The number 4 checks and is the solution.

ABSOLUTE VALUE
REVIEW SECTION R.1.

Equations with Absolute Value

Recall that the absolute value of a number is its distance from 0 on the number line. We use this concept to solve equations with absolute value.

> For $a > 0$ and an algebraic expression X:
>
> $|X| = a$ is equivalent to $X = -a$ or $X = a$.

EXAMPLE 5   Solve each of the following.

a) $|x| = 5$ b) $|x - 3| = 2$

a) Algebraic Solution

We have

$$|x| = 5$$
$$x = -5 \quad or \quad x = 5. \qquad \textbf{Writing an equivalent statement}$$

The solutions are -5 and 5.

Visualizing the Solution

The first coordinates of the points of intersection of the graphs of $y = |x|$ and $y = 5$ are -5 and 5. These are the solutions of the equation $|x| = 5$.

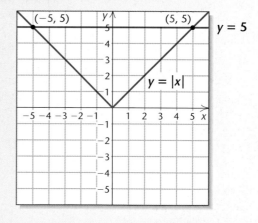

b)
$$|x - 3| = 2$$
$$x - 3 = -2 \quad or \quad x - 3 = 2 \qquad \textbf{Writing an equivalent statement}$$
$$x = 1 \quad or \quad x = 5 \qquad \textbf{Adding 3}$$

The solutions are 1 and 5.

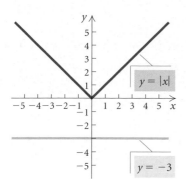

When $a = 0$, $|X| = a$ is equivalent to $X = 0$. Note that for $a < 0$, $|X| = a$ has no solution, because the absolute value of an expression is never negative. We can use a graph to illustrate the last statement for a specific value of a. For example, if we let $a = -3$ and graph $y = |x|$ and $y = -3$, we see that the graphs do not intersect, as shown at left. Thus the equation $|x| = -3$ has no solution. The solution set is the **empty set,** denoted $\varnothing$.

2.5

Exercise Set

Solve.

1. $\dfrac{1}{4} + \dfrac{1}{5} = \dfrac{1}{t}$

2. $\dfrac{1}{3} - \dfrac{5}{6} = \dfrac{1}{x}$

3. $\dfrac{x+2}{4} - \dfrac{x-1}{5} = 15$

4. $\dfrac{t+1}{3} - \dfrac{t-1}{2} = 1$

5. $\dfrac{1}{2} + \dfrac{2}{x} = \dfrac{1}{3} + \dfrac{3}{x}$

6. $\dfrac{1}{t} + \dfrac{1}{2t} + \dfrac{1}{3t} = 5$

7. $\dfrac{3x}{x+2} + \dfrac{6}{x} = \dfrac{12}{x^2 + 2x}$

8. $\dfrac{5x}{x-4} - \dfrac{20}{x} = \dfrac{80}{x^2 - 4x}$

9. $\dfrac{4}{x^2 - 1} - \dfrac{2}{x-1} = \dfrac{3}{x+1}$

10. $\dfrac{3y+5}{y^2 + 5y} + \dfrac{y+4}{y+5} = \dfrac{y+1}{y}$

11. $\dfrac{490}{x^2 - 49} = \dfrac{5x}{x-7} - \dfrac{35}{x+7}$

12. $\dfrac{3}{m+2} + \dfrac{2}{m} = \dfrac{4m-4}{m^2 - 4}$

13. $\dfrac{1}{x-6} - \dfrac{1}{x} = \dfrac{6}{x^2 - 6x}$

14. $\dfrac{8}{x^2 - 4} = \dfrac{x}{x-2} - \dfrac{2}{x+2}$

15. $\dfrac{x}{x-4} - \dfrac{4}{x+4} = \dfrac{32}{x^2 - 16}$

16. $\dfrac{x}{x-1} - \dfrac{1}{x+1} = \dfrac{2}{x^2 - 1}$

17. $\dfrac{x}{x+3} + \dfrac{3}{x-3} = \dfrac{18}{x^2 - 9}$

18. $\dfrac{x}{x-5} - \dfrac{5}{x+5} = \dfrac{50}{x^2 - 25}$

19. $\dfrac{1}{5x+20} - \dfrac{1}{x^2 - 16} = \dfrac{3}{x-4}$

20. $\dfrac{1}{4x+12} - \dfrac{1}{x^2 - 9} = \dfrac{5}{x-3}$

21. $\dfrac{2}{5x+5} - \dfrac{3}{x^2 - 1} = \dfrac{4}{x-1}$

22. $\dfrac{1}{3x+6} - \dfrac{1}{x^2 - 4} = \dfrac{3}{x-2}$

23. $\dfrac{8}{x^2 - 2x + 4} = \dfrac{x}{x + 2} + \dfrac{24}{x^3 + 8}$

24. $\dfrac{18}{x^2 - 3x + 9} - \dfrac{x}{x + 3} = \dfrac{81}{x^3 + 27}$

25. $\sqrt{3x - 4} = 1$

26. $\sqrt{4x + 1} = 3$

27. $\sqrt{2x - 5} = 2$

28. $\sqrt{3x + 2} = 6$

29. $\sqrt{7 - x} = 2$

30. $\sqrt{5 - x} = 1$

31. $\sqrt{1 - 2x} = 3$

32. $\sqrt{2 - 7x} = 2$

33. $\sqrt[3]{5x - 2} = -3$

34. $\sqrt[3]{2x + 1} = -5$

35. $\sqrt[4]{x^2 - 1} = 1$

36. $\sqrt[5]{3x + 4} = 2$

37. $\sqrt{y - 1} + 4 = 0$

38. $\sqrt{m + 1} - 5 = 8$

39. $\sqrt{b + 3} - 2 = 1$

40. $\sqrt{x - 4} + 1 = 5$

41. $\sqrt{z + 2} + 3 = 4$

42. $\sqrt{y - 5} - 2 = 3$

43. $\sqrt{2x + 1} - 3 = 3$

44. $\sqrt{3x - 1} + 2 = 7$

45. $\sqrt{2 - x} - 4 = 6$

46. $\sqrt{5 - x} + 2 = 8$

47. $\sqrt[3]{6x + 9} + 8 = 5$

48. $\sqrt[5]{2x - 3} - 1 = 1$

49. $\sqrt{x + 4} + 2 = x$

50. $\sqrt{x + 1} + 1 = x$

51. $\sqrt{x - 3} + 5 = x$

52. $\sqrt{x + 3} - 1 = x$

53. $\sqrt{x + 7} = x + 1$

54. $\sqrt{6x + 7} = x + 2$

55. $\sqrt{3x + 3} = x + 1$

56. $\sqrt{2x + 5} = x - 5$

57. $\sqrt{5x + 1} = x - 1$

58. $\sqrt{7x + 4} = x + 2$

59. $\sqrt{x - 3} + \sqrt{x + 2} = 5$

60. $\sqrt{x} - \sqrt{x - 5} = 1$

61. $\sqrt{3x - 5} + \sqrt{2x + 3} + 1 = 0$

62. $\sqrt{2m - 3} = \sqrt{m + 7} - 2$

63. $\sqrt{x} - \sqrt{3x - 3} = 1$

64. $\sqrt{2x + 1} - \sqrt{x} = 1$

65. $\sqrt{2y - 5} - \sqrt{y - 3} = 1$

66. $\sqrt{4p + 5} + \sqrt{p + 5} = 3$

67. $\sqrt{y + 4} - \sqrt{y - 1} = 1$

68. $\sqrt{y + 7} + \sqrt{y + 16} = 9$

69. $\sqrt{x + 5} + \sqrt{x + 2} = 3$

70. $\sqrt{6x + 6} = 5 + \sqrt{21 - 4x}$

71. $x^{1/3} = -2$ **72.** $t^{1/5} = 2$

73. $t^{1/4} = 3$ **74.** $m^{1/2} = -7$

75. $|x| = 7$ **76.** $|x| = 4.5$

77. $|x| = -10.7$ **78.** $|x| = -\frac{3}{5}$

79. $|x - 1| = 4$ **80.** $|x - 7| = 5$

81. $|3x| = 1$ **82.** $|5x| = 4$

83. $|x| = 0$ **84.** $|6x| = 0$

85. $|3x + 2| = 1$ **86.** $|7x - 4| = 8$

87. $\left|\frac{1}{2}x - 5\right| = 17$ **88.** $\left|\frac{1}{3}x - 4\right| = 13$

89. $|x - 1| + 3 = 6$ **90.** $|x + 2| - 5 = 9$

91. $|x + 3| - 2 = 8$ **92.** $|x - 4| + 3 = 9$

93. $|3x + 1| - 4 = -1$ **94.** $|2x - 1| - 5 = -3$

95. $|4x - 3| + 1 = 7$ **96.** $|5x + 4| + 2 = 5$

97. $12 - |x + 6| = 5$ **98.** $9 - |x - 2| = 7$

Solve.

99. $\dfrac{P_1 V_1}{T_1} = \dfrac{P_2 V_2}{T_2}$, for T_1

(A chemistry formula for gases)

100. $\dfrac{1}{F} = \dfrac{1}{m} + \dfrac{1}{p}$, for F
(A formula from optics)

101. $\dfrac{1}{R} = \dfrac{1}{R_1} + \dfrac{1}{R_2}$, for R_2
(Resistance)

102. $A = P(1 + i)^2$, for i
(Compound interest)

103. $\dfrac{1}{F} = \dfrac{1}{m} + \dfrac{1}{p}$, for p
(A formula from optics)

Technology Connection

104. Use a graphing calculator to do Exercises 4, 24, 54, 66, and 90.

105. Use a graphing calculator to do Exercises 5, 23, 63, 65, and 89.

Collaborative Discussion and Writing

106. Explain why it is necessary to check the possible solutions of a rational equation.

107. Explain in your own words why it is necessary to check the possible solutions when the principle of powers is used to solve an equation.

Skill Maintenance

Find the zero of the function.

108. $f(x) = -3x + 9$ **109.** $f(x) = 15 - 2x$

110. In 1990, the U.S. per-capita consumption of bananas was 24.4 lb. In 2000, this number rose to 28.4 lb. (*Source*: U.S. Department of Agriculture) Find the percent of increase.

111. In 2000, 501,000 adults earned high school equivalency diplomas by passing the General Educational Development (GED) test. This was 82,000 more than the number who passed the test in 1990. (*Source*: U.S. National Center for Education Statistics) How many adults passed the GED test in 1990?

Synthesis

Solve.

112. $\dfrac{x + 3}{x + 2} - \dfrac{x + 4}{x + 3} = \dfrac{x + 5}{x + 4} - \dfrac{x + 6}{x + 5}$

113. $(x - 3)^{2/3} = 2$

114. $\sqrt{15 + \sqrt{2x + 80}} = 5$

115. $\sqrt{x + 5} + 1 = \dfrac{6}{\sqrt{x + 5}}$

116. $x^{2/3} = x$

2.6

Solving Linear Inequalities

- *Solve linear inequalities.*
- *Solve compound inequalities.*
- *Solve inequalities with absolute value.*
- *Solve applied problems using inequalities.*

An **inequality** is a sentence with $<$, $>$, $\le$, or $\ge$ as its verb. An example is $3x - 5 < 6 - 2x$. To **solve** an inequality is to find all values of the variable that make the inequality true. Each of these numbers is a **solution** of the inequality, and the set of all such solutions is its **solution set.** Inequalities that have the same solution set are called **equivalent inequalities.**

Linear Inequalities

The principles for solving inequalities are similar to those for solving equations.

Principles for Solving Inequalities

For any real numbers a, b, and c:

The Addition Principle for Inequalities: If $a < b$ is true, then $a + c < b + c$ is true.

The Multiplication Principle for Inequalities: If $a < b$ and $c > 0$ are true, then $ac < bc$ is true. If $a < b$ and $c < 0$ are true, then $ac > bc$ is true.

Similar statements hold for $a \leq b$.

When both sides of an inequality are multiplied by a negative number, we must reverse the inequality sign.

First-degree inequalities with one variable, like those in Example 1 below, are **linear inequalities.**

EXAMPLE 1   Solve each of the following. Then graph the solution set.

a) $3x - 5 < 6 - 2x$ **b)** $13 - 7x \geq 10x - 4$

Solution

a) $3x - 5 < 6 - 2x$

$\qquad 5x - 5 < 6 \qquad$ Using the addition principle for inequalities; adding $2x$

$\qquad\qquad 5x < 11 \qquad$ Using the addition principle for inequalities; adding 5

$\qquad\qquad x < \frac{11}{5} \qquad$ Using the multiplication principle for inequalities; multiplying by $\frac{1}{5}$ or dividing by 5

Any number less than $\frac{11}{5}$ is a solution. The solution set is $\left\{ x \mid x < \frac{11}{5} \right\}$, or $\left(-\infty, \frac{11}{5} \right)$. The graph of the solution set is shown below.

b) $13 - 7x \geq 10x - 4$

$\qquad 13 - 17x \geq -4 \qquad$ **Subtracting $10x$**

$\qquad\qquad -17x \geq -17 \qquad$ **Subtracting 13**

$\qquad\qquad x \leq 1 \qquad$ **Dividing by -17 and reversing the inequality sign**

The solution set is $\{ x \mid x \leq 1 \}$, or $(-\infty, 1]$. The graph of the solution set is shown below.

Technology Connection

To check Example 1(a) graphically, we graph
$y_1 = 3x - 5$ and
$y_2 = 6 - 2x$. The graph shows that for $x < 2.2$, or $x < \frac{11}{5}$, the graph of y_1 lies below the graph of y_2, or $y_1 < y_2$.

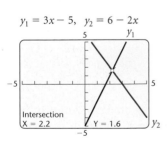

Compound Inequalities

When two inequalities are joined by the word *and* or the word *or*, a **compound inequality** is formed. A compound inequality like

$$-3 < 2x + 5 \quad and \quad 2x + 5 \le 7$$

is called a **conjunction,** because it uses the word *and.* The sentence $-3 < 2x + 5 \le 7$ is an abbreviation for the preceding conjunction.

Compound inequalities can be solved using the addition and multiplication principles for inequalities.

EXAMPLE 2 Solve $-3 < 2x + 5 \le 7$. Then graph the solution set.

Solution   We have

$$-3 < 2x + 5 \le 7$$
$$-8 < 2x \le 2 \qquad \text{Subtracting 5}$$
$$-4 < x \le 1. \qquad \text{Dividing by 2}$$

The solution set is $\{x \mid -4 < x \le 1\}$, or $(-4, 1]$. The graph of the solution set is shown below.

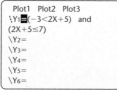

Technology
Connection

We can perform a partial check of the solution of Example 2 graphically using operations from the TEST menu of a graphing calculator. We graph $y_1 = (-3 < 2x + 5) \; and \; (2x + 5 \le 7)$ in DOT mode. The calculator graphs a segment 1 unit above the x-axis for the values of x for which this expression for y is true. Here the number 1 corresponds to "true."

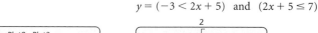

$$y = (-3 < 2x + 5) \quad and \quad (2x + 5 \le 7)$$

The segment extends from -4 to 1, confirming that all x-values from -4 to 1 are in the solution set. The algebraic solution indicates that the endpoint 1 is also in the solution set.

A compound inequality like $2x - 5 \le -7 \; or \; 2x - 5 > 1$ is called a **disjunction,** because it contains the word *or.* Unlike some conjunctions, it cannot be abbreviated; that is, it cannot be written without the word *or.*

To check Example 3 graphically, we graph $y_1 = 2x - 5$, $y_2 = -7$, and $y_3 = 1$. Note that for $\{x \mid x \le -1 \ or \ x > 3\}$, $y_1 \le y_2 \ or \ y_1 > y_3$.

$y_1 = 2x - 5$, $y_2 = -7$, $y_3 = 1$

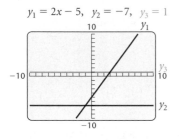

EXAMPLE 3   Solve $2x - 5 \le -7 \ or \ 2x - 5 > 1$. Then graph the solution set.

Solution   We have

$$2x - 5 \le -7 \quad or \quad 2x - 5 > 1$$
$$2x \le -2 \quad or \quad 2x > 6 \qquad \text{Adding 5}$$
$$x \le -1 \quad or \quad x > 3. \qquad \text{Dividing by 2}$$

The solution set is $\{x \mid x \le -1 \ or \ x > 3\}$. We can also write the solution using interval notation and the symbol $\cup$ for the **union** or inclusion of both sets: $(-\infty, -1] \cup (3, \infty)$. The graph of the solution set is shown below.

Inequalities with Absolute Value

Inequalities sometimes contain absolute-value notation. The following properties are used to solve them.

For $a > 0$ and an algebraic expression X:

$|X| < a$ is equivalent to $-a < X < a$.

$|X| > a$ is equivalent to $X < -a \ or \ X > a$.

Similar statements hold for $|X| \le a$ and $|X| \ge a$.

For example,

$|x| < 3$ is equivalent to $-3 < x < 3$;

$|y| \ge 1$ is equivalent to $y \le -1 \ or \ y \ge 1$; and

$|2x + 3| \le 4$ is equivalent to $-4 \le 2x + 3 \le 4$.

EXAMPLE 4   Solve each of the following. Then graph the solution set.

a) $|3x + 2| < 5$ **b)** $|5 - 2x| \ge 1$

Solution

a) $|3x + 2| < 5$

$$-5 < 3x + 2 < 5 \qquad \text{Writing an equivalent inequality}$$
$$-7 < 3x < 3 \qquad \text{Subtracting 2}$$
$$-\tfrac{7}{3} < x < 1 \qquad \text{Dividing by 3}$$

The solution set is $\left\{x \mid -\tfrac{7}{3} < x < 1\right\}$, or $\left(-\tfrac{7}{3}, 1\right)$. The graph of the solution set is shown below.

b) $|5 - 2x| \geq 1$

$$5 - 2x \leq -1 \quad or \quad 5 - 2x \geq 1 \qquad \text{Writing an equivalent inequality}$$
$$-2x \leq -6 \quad or \quad -2x \geq -4 \qquad \text{Subtracting 5}$$
$$x \geq 3 \quad or \quad x \leq 2 \qquad \text{Dividing by } -2 \text{ and reversing the inequality signs}$$

The solution set is $\{x \,|\, x \leq 2 \text{ or } x \geq 3\}$, or $(-\infty, 2] \cup [3, \infty)$. The graph of the solution set is shown below.

An Application

EXAMPLE 5 *Income Plans.* For his house-painting job, Eric can be paid in one of two ways:

Plan A: $250 plus $10 per hour;

Plan B: $20 per hour.

Suppose that a job takes n hours. For what values of n is plan B better for Eric?

Solution

1. **Familiarize.** Suppose that a job takes 20 hr. Then $n = 20$, and under plan A, Eric would earn $250 + $10 \cdot 20$, or $250 + 200, or $450. His earnings under plan B would be $20 \cdot 20$, or $400. This shows that plan A is better for Eric if a job takes 20 hr. Similarly, if a job takes 30 hr, then $n = 30$, and under plan A, Eric would earn $250 + $10 \cdot 30$, or $250 + 300, or $550. Under plan B, he would earn $20 \cdot 30$, or $600, so plan B is better in this case. To determine *all* values of n for which plan B is better for Eric, we solve an inequality. Our work in this step helps us write the inequality.

2. **Translate.** We translate to an equation.

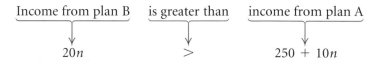

Income from plan B	is greater than	income from plan A
$20n$	$>$	$250 + 10n$

3. **Carry out.** We solve the inequality:

$$20n > 250 + 10n$$
$$10n > 250 \qquad \text{Subtracting } 10n \text{ on both sides}$$
$$n > 25. \qquad \text{Dividing by 10 on both sides}$$

4. **Check.** For $n = 25$, the income from plan A is $250 + $10 \cdot 25$, or $250 + 250, or $500, and the income from plan B is $20 \cdot 25$, or $500.

This shows that for a job that takes 25 hr to complete, the income is the same under either plan. In the *Familiarize* step, we saw that plan B pays more for a 30-hr job. Since $30 > 25$, this provides a partial check of the result. We cannot check all values of n.

5. **State.** For values of n greater than 25 hr, plan B is better for Eric. ▬

2.6

Exercise Set

Solve and graph the solution set.

1. $x + 6 < 5x - 6$

2. $3 - x < 4x + 7$

3. $3x - 3 + 2x \geq 1 - 7x - 9$

4. $5y - 5 + y \leq 2 - 6y - 8$

5. $14 - 5y \leq 8y - 8$

6. $8x - 7 < 6x + 3$

7. $-\frac{3}{4}x \geq -\frac{5}{8} + \frac{2}{3}x$

8. $-\frac{5}{6}x \leq \frac{3}{4} + \frac{8}{3}x$

9. $4x(x - 2) < 2(2x - 1)(x - 3)$

10. $(x + 1)(x + 2) > x(x + 1)$

Solve and write interval notation for the solution set. Then graph the solution set.

11. $-2 \leq x + 1 < 4$

12. $-3 < x + 2 \leq 5$

13. $5 \leq x - 3 \leq 7$

14. $-1 < x - 4 < 7$

15. $-3 \leq x + 4 \leq 3$

16. $-5 < x + 2 < 15$

17. $-2 < 2x + 1 < 5$

18. $-3 \leq 5x + 1 \leq 3$

19. $-4 \leq 6 - 2x < 4$

20. $-3 < 1 - 2x \leq 3$

21. $-5 < \frac{1}{2}(3x + 1) \leq 7$

22. $\frac{2}{3} \leq -\frac{4}{5}(x - 3) < 1$

23. $3x \leq -6 \text{ or } x - 1 > 0$

24. $2x < 8 \text{ or } x + 3 \geq 10$

25. $2x + 3 \leq -4 \text{ or } 2x + 3 \geq 4$

26. $3x - 1 < -5 \text{ or } 3x - 1 > 5$

27. $2x - 20 < -0.8 \text{ or } 2x - 20 > 0.8$

28. $5x + 11 \leq -4 \text{ or } 5x + 11 \geq 4$

29. $x + 14 \leq -\frac{1}{4} \text{ or } x + 14 \geq \frac{1}{4}$

30. $x - 9 < -\frac{1}{2} \text{ or } x - 9 > \frac{1}{2}$

31. $|x| < 7$

32. $|x| \leq 4.5$

33. $|x| \geq 4.5$

34. $|x| > 7$

35. $|x + 8| < 9$

36. $|x + 6| \leq 10$

37. $|x + 8| \geq 9$

38. $|x + 6| > 10$

39. $\left|x - \frac{1}{4}\right| < \frac{1}{2}$

40. $|x - 0.5| \leq 0.2$

41. $|3x| < 1$

42. $|5x| \leq 4$

43. $|2x + 3| \leq 9$

44. $|3x + 4| < 13$

45. $|x - 5| > 0.1$

46. $|x - 7| \geq 0.4$

47. $|6 - 4x| \leq 8$

48. $|5 - 2x| > 10$

49. $\left|x + \frac{2}{3}\right| \leq \frac{5}{3}$

50. $\left|x + \frac{3}{4}\right| < \frac{1}{4}$

51. $\left|\dfrac{2x + 1}{3}\right| > 5$

52. $\left|\dfrac{2x - 1}{3}\right| \geq \dfrac{5}{6}$

53. $|2x - 4| < -5$

54. $|3x + 5| < 0$

55. *Cost of Business on the Internet.* The equation $y = 12.7x + 15.2$ estimates the amount that businesses will spend, in billions of dollars, on Internet software to conduct transactions via the Web, where x is the number of years after 2002 (*Source*: IDC). For what years will the spending be more than $66 billion?

56. *Digital Hubs.* The equation $y = 5x + 5$ estimates the number of U.S. households, in millions, expected to install devices that receive and manage broadband

TV and Internet content to the home, where *x* is the number of years after 2002 (*Source*: Forrester Research). For what years will there be at least 20 million homes with these devices?

57. *Moving Costs.* Acme Movers charges $100 plus $30 per hour to move a household across town. Hank's Movers charges $55 per hour. For what lengths of time does it cost less to hire Hank's Movers?

58. *Investment Income.* Gina plans to invest $12,000, part at 4% simple interest and the rest at 6% simple interest. What is the most she can invest at 4% and still be guaranteed at least $650 in interest per year?

59. *Investment Income.* Kyle plans to invest $7500, part at 4% simple interest and the rest at 5% simple interest. What is the most that he can invest at 4% and still be guaranteed at least $325 in interest per year?

60. *Checking-account Plans.* The Addison Bank offers two checking-account plans. The Smart Checking plan charges 20¢ per check whereas the Consumer Checking plan costs $6 per month plus 5¢ per check. For what number of checks per month will the Smart Checking plan cost less?

61. *Checking-account Plans.* Parson's Bank offers two checking-account plans. The No Frills plan charges 35¢ per check whereas the Simple Checking plan costs $5 per month plus 10¢ per check. For what number of checks per month will the Simple Checking plan cost less?

62. *Income Plans.* Karen can be paid in one of two ways for selling insurance policies:

Plan A: A salary of $750 per month, plus a commission of 10% of sales;
Plan B: A salary of $1000 per month, plus a commission of 8% of sales in excess of $2000.

For what amount of monthly sales is plan A better than plan B if we can assume that sales are always more than $2000?

63. *Income Plans.* Curt can be paid in one of two ways for the furniture he sells:

Plan A: A salary of $900 per month, plus a commission of 10% of sales;
Plan B: A salary of $1200 per month, plus a commission of 15% of sales in excess of $8000.

For what amount of monthly sales is plan B better than plan A if we can assume that Curt's sales are always more than $8000?

64. *Income Plans.* Jeanette can be paid in one of two ways for painting a house:

Plan A: $200 plus $12 per hour;
Plan B: $20 per hour.

Suppose a job takes *n* hours to complete. For what values of *n* is plan A better for Jeanette?

Technology Connection

65. Use a graphing calculator to check your answers to Exercises 9, 21, and 41.

66. Use a graphing calculator to check your answers to Exercises 10, 22, and 42.

Collaborative Discussion and Writing

67. Explain why $|x| < p$ has no solution for $p \leq 0$.

68. Explain why all real numbers are solutions of $|x| > p$, for $p < 0$.

Skill Maintenance

In each of Exercises 69–76, fill in the blank with the correct term. Some of the given choices will not be used.

distance formula
midpoint formula
function
relation
x-intercept
y-intercept
perpendicular
parallel
horizontal lines
vertical lines
symmetric with respect to the x-axis
symmetric with respect to the y-axis
symmetric with respect to the origin
increasing
decreasing
constant

69. A(n) _____ is a point $(0, b)$.

70. The _____ is $d = \sqrt{(x_2 - x_1)^2 + (y_2 - y_1)^2}$.

71. A(n) _____ is a correspondence such that each member of the domain corresponds to at least one member of the range.

72. A(n) _____ is a correspondence such that each member of the domain corresponds to exactly one member of the range.

73. _____ are given by equations of the type $y = b$, or $f(x) = b$.

74. Nonvertical lines are _____ if and only if they have the same slope and different y-intercepts.

75. A function f is said to be _____ on an open interval I if, for all a and b in that interval, $a < b$ implies $f(a) > f(b)$.

76. For an equation $y = f(x)$, if replacing x with $-x$ produces an equivalent equation, then the graph is _____ .

Synthesis

Solve.

77. $2x \leq 5 - 7x < 7 + x$

78. $x \leq 3x - 2 \leq 2 - x$

79. $|3x - 1| > 5x - 2$

80. $|x + 2| \leq |x - 5|$

81. $|p - 4| + |p + 4| < 8$

82. $|x| + |x + 1| < 10$

83. $|x - 3| + |2x + 5| > 6$

Chapter 2 Summary and Review

Important Properties and Formulas

Equation-Solving Principles

The Addition Principle:

If $a = b$ is true, then $a + c = b + c$ is true.

The Multiplication Principle:

If $a = b$ is true, then $ac = bc$ is true.

The Principle of Zero Products:

If $ab = 0$ is true, then $a = 0$ or $b = 0$, and

if $a = 0$ or $b = 0$, then $ab = 0$.

The Principle of Square Roots:

If $x^2 = k$, then $x = \sqrt{k}$ or $x = -\sqrt{k}$.

The Principle of Powers:

For any positive integer n, if $a = b$ is true, then $a^n = b^n$ is true.

Five Steps for Problem Solving

1. Familiarize.
2. Translate.
3. Carry out.
4. Check.
5. State.

Zero of a Function:

An input c of a function f is a zero of f if $f(c) = 0$.

Complex Number: $a + bi$, a, b real, $i^2 = -1$

Imaginary Number: $a + bi$, $b \neq 0$

Complex Conjugates: $a + bi$, $a - bi$

Quadratic Equation:

$$ax^2 + bx + c = 0, \ a \neq 0, \ a, b, c \text{ real}$$

Quadratic Function:

$$f(x) = ax^2 + bx + c, \ a \neq 0, \ a, b, c \text{ real}$$

Quadratic Formula:

For $ax^2 + bx + c = 0, a \neq 0,$

$$x = \frac{-b \pm \sqrt{b^2 - 4ac}}{2a}.$$

Principles for Solving Inequalities

The Addition Principle for Inequalities:

If $a < b$ is true, then $a + c < b + c$ is true.

The Multiplication Principle for Inequalities:

If $a < b$ and $c > 0$ are true, then $ac < bc$ is true.

If $a < b$ and $c < 0$ are true, then $ac > bc$ is true.

Similar statements hold for $\leq$.

Equations and Inequalities with Absolute Value

For $a > 0$,

$$|X| = a \longrightarrow X = -a \ \text{ or } \ X = a$$
$$|X| < a \longrightarrow -a < X < a$$
$$|X| > a \longrightarrow X < -a \ \text{ or } \ X > a$$

Review Exercises

Solve.

1. $4y - 5 = 1$

2. $3x - 4 = 5x + 8$

3. $5(3x + 1) = 2(x - 4)$

4. $2(n - 3) = 3(n + 5)$

5. $(2y + 5)(3y - 1) = 0$

6. $x^2 + 4x - 5 = 0$

7. $3x^2 + 2x = 8$

8. $5x^2 = 15$

9. $x^2 - 10 = 0$

Find the zero(s) of the function.

10. $f(x) = 6x - 18$

11. $f(x) = x - 4$

12. $f(x) = 2 - 10x$

13. $f(x) = 8 - 2x$

14. $f(x) = x^2 - 2x + 1$

15. $f(x) = x^2 + 2x - 15$

16. $f(x) = 2x^2 - x - 5$

17. $f(x) = 3x^2 + 2x - 3$

Solve.

18. $\dfrac{5}{2x + 3} + \dfrac{1}{x - 6} = 0$

19. $\dfrac{3}{8x + 1} + \dfrac{8}{2x + 5} = 1$

20. $\sqrt{5x + 1} - 1 = \sqrt{3x}$

21. $\sqrt{x - 1} - \sqrt{x - 4} = 1$

22. $|x - 4| = 3$

23. $|2y + 7| = 9$

Solve and write interval notation for the solution set. Then graph the solution set.

24. $-3 \le 3x + 1 < 5$

25. $-2 < 5x - 4 \le 6$

26. $2x < -1 \text{ or } x + 3 > 0$

27. $3x + 7 \le 2 \text{ or } 2x + 3 \ge 5$

28. $|6x - 1| < 5$

29. $|x + 4| \ge 2$

30. Solve $V = lwh$ for h.

31. Solve $M = n + 0.3s$ for s.

32. Solve $v = \sqrt{2gh}$ for h.

33. Solve $\dfrac{1}{a} + \dfrac{1}{b} = \dfrac{1}{t}$ for t.

Express in terms of i.

34. $-\sqrt{-40}$

35. $\sqrt{-12} \cdot \sqrt{-20}$

36. $\dfrac{\sqrt{-49}}{-\sqrt{-64}}$

Simplify each of the following. Write the answer in the form $a + bi$, where a and b are real numbers.

37. $(6 + 2i)(-4 - 3i)$ **38.** $\dfrac{2 - 3i}{1 - 3i}$

39. $(3 - 5i) - (2 - i)$ **40.** $(6 + 2i) + (-4 - 3i)$

41. i^{23} **42.** $(-3i)^{28}$

Solve by completing the square to obtain exact solutions. Show your work.

43. $x^2 - 3x = 18$ **44.** $3x^2 - 12x - 6 = 0$

Solve. Give exact solutions.

45. $3x^2 + 10x = 8$ **46.** $r^2 - 2r + 10 = 0$

47. $x^2 = 18 + 3x$ **48.** $x = 2\sqrt{x} - 1$

49. $y^4 - 3y^2 + 1 = 0$

50. $(x^2 - 1)^2 \quad (x^2 - 1) - 2 = 0$

51. $(p - 3)(3p + 2)(p + 2) = 0$

52. $x^3 + 5x^2 - 4x - 20 = 0$

In Exercises 53 and 54, complete the square to:

a) *find the vertex;*
b) *find the axis of symmetry;*
c) *determine whether there is a maximum or minimum value and find that value;*
d) *find the range; and*
e) *graph the function.*

53. $f(x) = -4x^2 + 3x - 1$ **54.** $f(x) = 5x^2 - 10x + 3$

In Exercises 55–58, match the equation with one of the figures (a)–(d), which follow.

a)

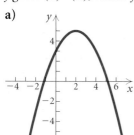

b)

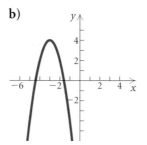

c)

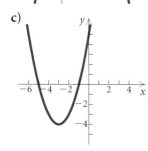

d)

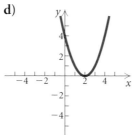

55. $y = (x - 2)^2$

56. $y = (x + 3)^2 - 4$

57. $y = -2(x + 3)^2 + 4$

58. $y = -\frac{1}{2}(x - 2)^2 + 5$

59. *Legs of a Right Triangle.*   The hypotenuse of a right triangle is 50 ft. One leg is 10 ft longer than the other. What are the lengths of the legs?

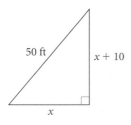

60. *Motion.*   A Riverboat Cruise Line boat travels 8 mi upstream and 8 mi downstream. The total time for both parts of the trip is 3 hr. The speed of the stream is 2 mph. What is the speed of the boat in still water?

61. *Motion.*   Two freight trains leave the same city at right angles. The first train travels at a speed of 60 km/h. In 1 hr, the trains are 100 km apart. How fast is the second train traveling?

62. *Sidewalk Width.*   A 60-ft by 80-ft parking lot is torn up to install a sidewalk of uniform width around its perimeter. The new area of the parking lot is two thirds of the old area. How wide is the sidewalk?

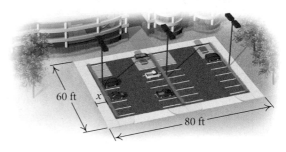

63. *Maximizing Volume.*   The Berniers have 24 ft of flexible fencing with which to build a rectangular "toy corral." If the fencing is 2 ft high, what dimensions should the corral have in order to maximize its volume?

64. *Dimensions of a Box.*   An open box is made from a 10-cm by 20-cm piece of aluminum by cutting a square from each corner and folding up the edges. The area of the resulting base is 90 cm^2. What is the length of the sides of the squares?

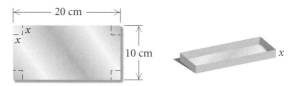

65. *Faculty at Two-Year Colleges.*   The equation $y = 6x + 121$ estimates the number of faculty members at two-year colleges, in thousands, where x is the number of years after 1970 (*Source:* U.S. National Center for Education Statistics). For what years will there be more than 325 thousand faculty members?

66. *Temperature Conversion.*   The formula $C = \frac{5}{9}(F - 32)$ can be used to convert Fahrenheit temperatures F to Celsius temperatures C. For what Fahrenheit temperatures is the Celsius temperature lower than 45°C?

Technology Connection

Use a graphing calculator to do each of the following.

67. Exercises 2, 6, 18, 20, and 22

68. Exercises 1, 7, 19, 21, and 23

69. Exercises 10 and 14

70. Exercises 13 and 15

Collaborative Discussion and Writing

71. As the first step in solving
$$3x - 1 = 8,$$
Stella multiplies by $\frac{1}{3}$ on both sides. What advice would you give her about the procedure for solving equations?

72. If the graphs of
$$f(x) = a_1(x - h_1)^2 + k_1$$
and
$$g(x) = a_2(x - h_2)^2 + k_2$$
have the same shape, what, if anything, can you conclude about the a's, the h's, and the k's? Explain your answer.

Synthesis

Solve.

73. $\sqrt{\sqrt{\sqrt{\sqrt{x}}}} = 2$

74. $(x - 1)^{2/3} = 4$

75. $(t - 4)^{4/5} = 3$

76. $\sqrt{x + 2} + \sqrt[4]{x + 2} - 2 = 0$

77. $(2y - 2)^2 + y - 1 = 5$

78. Find b such that $f(x) = -3x^2 + bx - 1$ has a maximum value of 2.

79. At the beginning of the year, $3500 was deposited in a savings account. One year later, $4000 was deposited in another account. The interest rate was the same for both accounts. At the end of the second year, there was a total of $8518.35 in the accounts. What was the annual interest rate?

Chapter 2 Test

Solve. Find exact solutions.

1. $6x + 7 = 1$

2. $3y - 4 = 5y + 6$

3. $2(4x + 1) = 8 - 3(x - 5)$

4. $(2x - 1)(x + 5) = 0$

5. $6x^2 - 36 = 0$

6. $x^2 + 4 = 0$

7. $x^2 - 2x - 3 = 0$

8. $x^2 - 5x + 3 = 0$

9. $2t^2 - 3t + 4 = 0$

10. $x + 5\sqrt{x} - 36 = 0$

11. $\dfrac{3}{3x + 4} + \dfrac{2}{x - 1} = 2$

12. $\sqrt{x + 4} - 2 = 1$

13. $\sqrt{x + 4} - \sqrt{x - 4} = 2$

14. $|4y - 3| = 5$

Solve and write interval notation for the solution set. Then graph the solution set.

15. $-7 < 2x + 3 < 9$

16. $2x - 1 \le 3 \text{ or } 5x + 6 \ge 26$

17. $|x + 3| \le 4$

18. $|x + 5| > 2$

19. Solve $V = \frac{2}{3}\pi r^2 h$ for h.

20. Solve $R = \sqrt{3np}$ for n.

21. Solve $x^2 + 4x = 1$ by completing the square. Find the exact solutions. Show your work.

22. *Parking Lot Dimensions.* The parking lot behind Kai's Kafé has a perimeter of 210 m. The width is three-fourths of the length. What are the dimensions of the parking lot?

23. *River Current.* Deke's boat travels 12 km/h in still water. Deke travels 45 km downstream and then returns 45 km upstream in a total time of 8 hr. Find the speed of the current.

24. *Pricing.* Jessie's Juice Bar prices its bottled juices by raising the wholesale price 50% and then adding 25¢. What is the wholesale price of a bottle of juice that sells for $2.95?

Express in terms of i.

25. $\sqrt{-43}$

26. $-\sqrt{-25}$

Simplify.

27. $(5 - 2i) - (2 + 3i)$

28. $(3 + 4i)(2 - i)$

29. $\dfrac{1 - i}{6 + 2i}$

30. i^{33}

Find the zero(s) of each function.

31. $f(x) = 3x + 9$

32. $f(x) = 4x^2 - 11x - 3$

33. $f(x) = 2x^2 - x - 7$

34. For the graph of the function $f(x) = -x^2 + 2x + 8$:
   a) Find the vertex.
   b) Find the axis of symmetry.
   c) State whether there is a maximum or minimum value and find that value.
   d) Find the range.
   e) Graph the function.

35. *Maximizing Area.* A homeowner wants to fence a rectangular play yard using 60 ft of fencing. The side of the house will be used as one side of the rectangle. Find the dimensions for which the area is a maximum.

36. *Moving Costs.* Morgan Movers charges $90 plus $25 per hour to move households across town. McKinley Movers charges $40 per hour for crosstown moves. For what lengths of time does it cost less to hire Morgan Movers?

Synthesis

37. Find a such that $f(x) = ax^2 - 4x + 3$ has a maximum value of 12.

Polynomial and Rational Functions

3

APPLICATION

During winter, bald eagles travel south to find open water in states that have milder winter temperatures. The number of bald eagles spotted in Indiana during the winters from 1996 to 2002 can be modeled by the quartic function

$$f(x) = -3.394x^4 + 35.838x^3 - 98.955x^2 + 41.930x + 174.974,$$

where x is the number of years since 1996 (*Source*: Indiana Department of Natural Resources). Find the number of bald eagles in Indiana in the winters of 1997, 1999, 2001, and 2002.

This problem appears as Exercise 79 in Section 3.1.

243

3.1

Polynomial Functions and Models

- *Determine the behavior of the graph of a polynomial function using the leading-term test.*
- *Factor polynomial functions and find the zeros and their multiplicities.*
- *Graph polynomial functions.*
- *Use the intermediate value theorem to determine whether a function has a real zero between two given real numbers.*
- *Solve applied problems using polynomial models.*

There are many different kinds of functions. The constant, linear, and quadratic functions that we studied in Chapters 1 and 2 are part of a larger group of functions called *polynomial functions.*

Polynomial Function

A **polynomial function** P is given by

$$P(x) = a_n x^n + a_{n-1} x^{n-1} + a_{n-2} x^{n-2} + \cdots + a_1 x + a_0,$$

where the coefficients $a_n, a_{n-1}, \ldots, a_1, a_0$ are real numbers and the exponents are whole numbers.

The first nonzero coefficient, a_n, is called the **leading coefficient.** The term $a_n x^n$ is called the **leading term.** The **degree** of the polynomial function is n. Some examples of polynomial functions are as follows.

POLYNOMIAL FUNCTION	DEGREE	EXAMPLE
Constant	0	$f(x) = 3$
Linear	1	$f(x) = \frac{2}{3}x + 1$
Quadratic	2	$f(x) = 2x^2 - x + 3$
Cubic	3	$f(x) = x^3 + 2x^2 + x - 5$
Quartic	4	$f(x) = -x^4 - 1.1x^3 + 0.3x^2 - 2.8x - 1.7$

The function $f(x) = 0$ can be described in many ways:

$$f(x) = 0 = 0x^2 = 0x^{15} = 0x^{48},$$

and so on. For this reason, we say that the constant function $f(x) = 0$ has no degree.

From our study of functions in Chapters 1 and 2, we know how to find or at least estimate many characteristics of a polynomial function. Let's consider two examples for review.

Quadratic Function

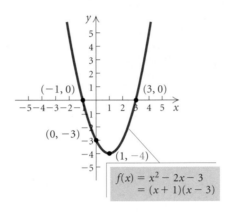

$f(x) = x^2 - 2x - 3$
$= (x + 1)(x - 3)$

Function: $f(x) = x^2 - 2x - 3$
$\qquad\qquad = (x + 1)(x - 3)$

Zeros: $-1, 3$

x-intercepts: $(-1, 0), (3, 0)$

y-intercept: $(0, -3)$

Minimum: -4 at $x = 1$

Maximum: None

Domain: All real numbers, $(-\infty, \infty)$

Range: $[-4, \infty)$

Cubic Function

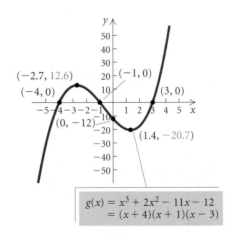

$g(x) = x^3 + 2x^2 - 11x - 12$
$= (x + 4)(x + 1)(x - 3)$

Function: $g(x) = x^3 + 2x^2 - 11x - 12$
$\qquad\qquad = (x + 4)(x + 1)(x - 3)$

Zeros: $-4, -1, 3$

x-intercepts: $(-4, 0), (-1, 0), (3, 0)$

y-intercept: $(0, -12)$

Relative minimum: -20.7 at $x = 1.4$

Relative maximum: 12.6 at $x = -2.7$

Domain: All real numbers, $(-\infty, \infty)$

Range: All real numbers, $(-\infty, \infty)$

All graphs of polynomial functions have some characteristics in common. Compare the following graphs. How do the graphs of polynomial functions differ from the graphs of nonpolynomial functions? Describe some characteristics of the graphs of polynomial functions that you observe.

Polynomial Functions

$f(x) = x^2 + 3x + 1$

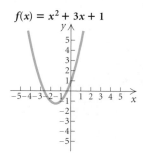

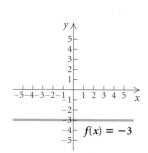

$f(x) = -3$

$f(x) = 2x^3 + x^2 + x - 1$

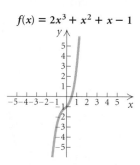

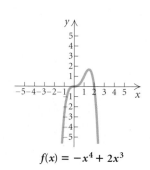

$f(x) = -x^4 + 2x^3$

Nonpolynomial Functions

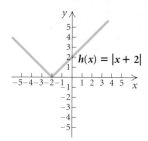

$h(x) = |x + 2|$

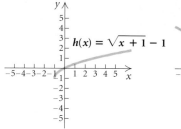

$h(x) = \sqrt{x + 1} - 1$

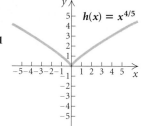

$h(x) = x^{4/5}$

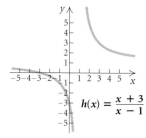

$h(x) = \dfrac{x + 3}{x - 1}$

You probably noted that the graph of a polynomial function is *continuous*, that is, it has no holes or breaks. It is also smooth; there are no sharp corners. Furthermore, the *domain* of a polynomial function is the set of all real numbers, $(-\infty, \infty)$.

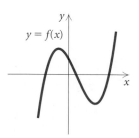

A continuous function

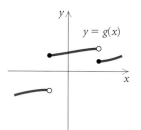

A discontinuous function

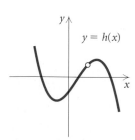

A discontinuous function

The Leading-Term Test

The behavior of the graph of a polynomial function as x becomes very large $(x \to \infty)$ or very small $(x \to -\infty)$ is referred to as the end behavior of the graph. The leading term of a polynomial function determines its end behavior.

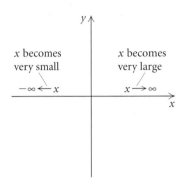

Using the graphs shown below, let's see if we can discover some general patterns by comparing the end behavior of even- and odd-degree functions. We also observe the effect of positive and negative leading coefficients.

Even Degree

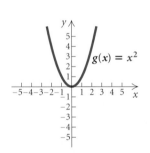

$g(x) = x^2$

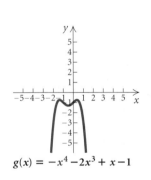

$g(x) = -x^4 - 2x^3 + x - 1$

$g(x) = \frac{1}{2}x^6 + 3$

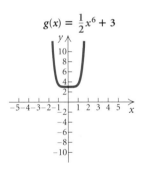

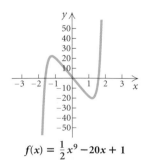

$g(x) = 1 - x - x^{10}$

Odd Degree

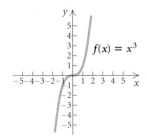

$f(x) = x^3$

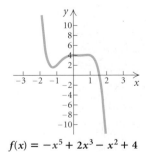

$f(x) = -x^5 + 2x^3 - x^2 + 4$

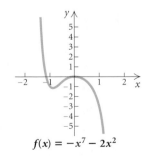

$f(x) = -x^7 - 2x^2$

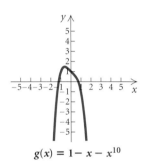

$f(x) = \frac{1}{2}x^9 - 20x + 1$

We can summarize our observations as follows.

The Leading-Term Test

If $a_n x^n$ is the leading term of a polynomial, then the behavior of the graph as $x \to \infty$ or as $x \to -\infty$ can be described in one of the four following ways.

If n is even, and $a_n > 0$:

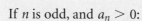

If n is even, and $a_n < 0$:

If n is odd, and $a_n > 0$:

If n is odd, and $a_n < 0$:

The ⌇⌇⌇ portion of the graph is not determined by this test.

EXAMPLE 1 Using the leading-term test, match each of the following functions with one of the graphs A–D, which follow.

a) $f(x) = 3x^4 - 2x^3 + 3$

b) $f(x) = -5x^3 - x^2 + 4x + 2$

c) $f(x) = x^5 + \frac{1}{4}x + 1$

d) $f(x) = -x^6 + x^5 - 4x^3$

A.

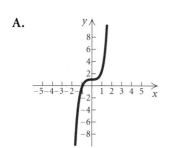

B.

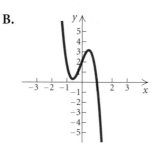

C.

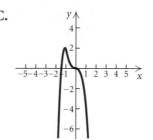

D.

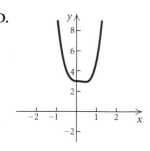

Solution

	LEADING TERM	DEGREE OF LEADING TERM	SIGN OF LEADING COEFFICIENT	GRAPH
a)	$3x^4$	Even	Positive	D
b)	$-5x^3$	Odd	Negative	B
c)	x^5	Odd	Positive	A
d)	$-x^6$	Even	Negative	C

Finding Zeros of Factored Polynomial Functions

Let's review the meaning of the real zeros of a function and their connection to the x-intercepts of the function's graph.

CONNECTING THE CONCEPTS

ZEROS, SOLUTIONS, AND INTERCEPTS

FUNCTION	REAL ZEROS OF THE FUNCTION; SOLUTIONS OF THE EQUATION	X-INTERCEPTS OF THE GRAPH

Quadratic Polynomial

$g(x) = x^2 - 2x - 8$
$\quad = (x + 2)(x - 4),$

or

$\quad y = (x + 2)(x - 4)$

To find the **zeros** of $g(x)$, we solve $g(x) = 0$:

$\quad x^2 - 2x - 8 = 0$

$(x + 2)(x - 4) = 0$

$\quad x + 2 = 0 \quad or \quad x - 4 = 0$

$\quad\quad x = -2 \quad or \quad\quad x = 4.$

The **solutions** of $x^2 - 2x - 8 = 0$ are -2 and 4. They are the zeros of the function $g(x)$. That is,

$g(-2) = 0 \quad$ and $\quad g(4) = 0.$

The zeros of $g(x)$ are the x-coordinates of the **x-intercepts** of the graph of $y = g(x)$.

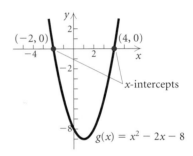

Cubic Polynomial

$h(x)$
$\quad = x^3 + 2x^2 - 5x - 6$
$\quad = (x + 3)(x + 1)(x - 2),$

or

$\quad y = (x + 3)(x + 1)(x - 2)$

To find the **zeros** of $h(x)$, we solve $h(x) = 0$:

$\quad\quad x^3 + 2x^2 - 5x - 6 = 0$

$\quad\quad (x + 3)(x + 1)(x - 2) = 0$

$x + 3 = 0 \quad or \quad x + 1 = 0 \quad or \quad x - 2 = 0$

$\quad x = -3 \quad or \quad\quad x = -1 \quad or \quad\quad x = 2.$

The **solutions** of $x^3 + 2x^2 - 5x - 6 = 0$ are -3, -1, and 2. They are the zeros of the function $h(x)$. That is,

$h(-3) = 0,$
$h(-1) = 0, \quad$ and
$\quad h(2) = 0.$

The zeros of $h(x)$ are the x-coordinates of the **x-intercepts** of the graph of $y = h(x)$.

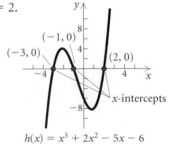

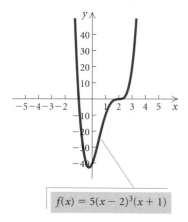

$P(x) = x^3 + x^2 - 17x + 15$

The connection between the zeros of a function and the x-intercepts of the graph of the function is easily seen in the examples above. If c is a real zero of a function (that is, $f(c) = 0$), then $(c, 0)$ is an x-intercept of the graph of the function.

EXAMPLE 2   Consider $P(x) = x^3 + x^2 - 17x + 15$. Determine whether each of the numbers 2 and -5 is a zero of $P(x)$.

Solution   We have

$$P(2) = (2)^3 + (2)^2 - 17(2) + 15 = -7. \qquad \textbf{Substituting 2 into the polynomial}$$

Since $P(2) \neq 0$, we know that 2 is *not* a zero of the polynomial.
    We also have

$$P(-5) = (-5)^3 + (-5)^2 - 17(-5) + 15 = 0. \qquad \textbf{Substituting } -5 \textbf{ into the polynomial}$$

Since $P(-5) = 0$, we know that -5 is a zero of $P(x)$. ▬

Let's take a closer look at the polynomial

$$h(x) = x^3 + 2x^2 - 5x - 6$$

(see Connecting the Concepts on page 249). Is there a connection between the factors of the polynomial and the zeros of the function? The factors of $h(x)$ are

$$x + 3, \qquad x + 1, \quad \text{and} \quad x - 2,$$

and the zeros are

$$-3, \qquad -1, \quad \text{and} \quad 2.$$

We note that when the polynomial is expressed as a product of linear factors, each factor determines a zero of the function. Thus if we know the linear factors of a polynomial, we can easily find the zeros of the equation $f(x) = 0$ using the principle of zero products.

PRINCIPLE OF ZERO PRODUCTS

REVIEW SECTION 2.3.

EXAMPLE 3   Find the zeros of

$$f(x) = 5(x - 2)(x - 2)(x - 2)(x + 1)$$
$$= 5(x - 2)^3(x + 1).$$

Solution   To solve the equation $f(x) = 0$, we use the principle of zero products, solving $x - 2 = 0$ and $x + 1 = 0$. The zeros of $f(x)$ are 2 and -1. (See Fig. 1.) ▬

$f(x) = 5(x - 2)^3(x + 1)$

FIGURE 1

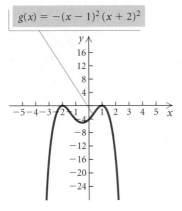

$g(x) = -(x-1)^2(x+2)^2$

FIGURE 2

EXAMPLE 4 Find the zeros of
$$g(x) = -(x-1)(x-1)(x+2)(x+2)$$
$$= -(x-1)^2(x+2)^2.$$

Solution To solve the equation $g(x) = 0$, we use the principle of zero products, solving $x - 1 = 0$ and $x + 2 = 0$. The zeros of $g(x)$ are -2 and 1. (See Fig. 2.)

Let's consider the occurrences of the zeros in the functions in Examples 3 and 4 and their relationship to the graphs of those functions. In Example 3, the factor $x - 2$ occurs three times. In a case like this, we say that the zero we obtain from this factor, 2, has a **multiplicity** of 3. The factor $x + 1$ occurs one time. The zero we obtain from this factor, -1, has a *multiplicity* of 1.

In Example 4, the factors $x - 1$ and $x + 2$ each occur two times. Thus both zeros, 1 and -2, have a *multiplicity* of 2.

Note, in Example 3, that the zeros have odd multiplicities and the graph crosses the x-axis at both -1 and 2. But in Example 4, the zeros have even multiplicities and the graph is tangent to (touches but does not cross) the x-axis at -2 and 1. This leads us to the following generalization.

Even and Odd Multiplicity

If $(x - c)^k$, $k \geq 1$, is a factor of a polynomial function $P(x)$ and:

- k is odd, then the graph crosses the x-axis at $(c, 0)$;
- k is even, then the graph is tangent to the x-axis at $(c, 0)$.

Technology Connection

Using the Zero method, we can determine the zeros of the function in Example 5.

The window below shows the calculator display when we find the leftmost zero. The other zeros, 2 and 3, can be found in the same manner.

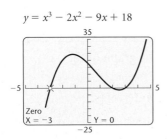

$y = x^3 - 2x^2 - 9x + 18$

Some polynomials can be factored by grouping.

EXAMPLE 5 Find the zeros of
$$f(x) = x^3 - 2x^2 - 9x + 18.$$

Solution We factor by grouping, as follows:

$$f(x) = x^3 - 2x^2 - 9x + 18$$
$$= x^2(x - 2) - 9(x - 2) \qquad \text{Grouping } x^3 \text{ with } -2x^2 \text{ and } -9x \text{ with } 18 \text{ and factoring each group}$$
$$= (x - 2)(x^2 - 9) \qquad \text{Factoring out } x - 2$$
$$= (x - 2)(x + 3)(x - 3). \qquad \text{Factoring } x^2 - 9$$

Then, by the principle of zero products, the solutions of the equation $f(x) = 0$ are 2, -3, and 3. These are the zeros of $f(x)$.

Other factoring techniques can also be used.

EXAMPLE 6   Find the zeros of

$$f(x) = x^4 + 4x^2 - 45.$$

Solution   We factor as follows:

$$f(x) = x^4 + 4x^2 - 45 = (x^2 - 5)(x^2 + 9).$$

We now solve the equation $f(x) = 0$ to determine the zeros. We use the principle of zero products:

$$(x^2 - 5)(x^2 + 9) = 0$$

$$x^2 - 5 = 0 \quad or \quad x^2 + 9 = 0$$

$$x^2 = 5 \quad or \quad x^2 = -9$$

$$x = \pm\sqrt{5} \quad or \quad x = \pm\sqrt{-9} = \pm3i.$$

The solutions are $\pm\sqrt{5}$ and $\pm3i$. These are the zeros of $f(x)$.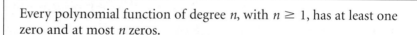

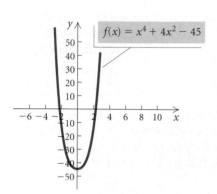

Only the real-number zeros of a function correspond to the *x*-intercepts of its graph. For instance, the real-number zeros of the function in Example 6, $-\sqrt{5}$ and $\sqrt{5}$, can be seen on the graph of the function at left, but the nonreal zeros, $-3i$ and $3i$, cannot.

Every polynomial function of degree n, with $n \geq 1$, has at least one zero and at most n zeros.

This is often stated as follows: "Every polynomial function of degree n, with $n \geq 1$, has *exactly* n zeros." This statement is compatible with the preceding statement, if one takes multiplicities into account.

Graphing Polynomial Functions

In addition to using the leading-term test and finding the zeros of the function, it is helpful to consider the following facts when graphing a polynomial function.

If $P(x)$ is a polynomial function of degree n, the graph of the function has:

- at most n real zeros, and thus at most n *x*-intercepts;
- at most $n - 1$ turning points.

(Turning points on a graph, also called relative maxima and minima, occur when the function changes from decreasing to increasing or from increasing to decreasing.)

EXAMPLE 7   Graph the polynomial function $h(x) = -2x^4 + 3x^3$.

Solution

1. First, we use the leading-term test to determine the end behavior of the graph. The leading term is $-2x^4$. The degree, 4, is even, and the coefficient, -2, is negative. Thus the end behavior of the graph as $x \to \infty$ and as $x \to -\infty$ can be sketched as follows.

2. The zeros of the function are the first coordinates of the x-intercepts of the graph. To find the zeros, we solve $h(x) = 0$ by factoring and using the principle of zero products.

    $$-2x^4 + 3x^3 = 0$$
    $$-x^3(2x - 3) = 0 \qquad \text{Factoring}$$
    $$-x^3 = 0 \quad or \quad 2x - 3 = 0 \qquad \text{Using the principle of zero products}$$
    $$x = 0 \quad or \qquad x = \frac{3}{2}.$$

    The zeros of the function are 0 and $\frac{3}{2}$. Note that the multiplicity of 0 is 3 and the multiplicity of $\frac{3}{2}$ is 1. The x-intercepts are $(0,0)$ and $\left(\frac{3}{2},0\right)$.

3. The zeros divide the x-axis into three intervals:

    $$(-\infty, 0), \qquad \left(0, \frac{3}{2}\right), \quad \text{and} \quad \left(\frac{3}{2}, \infty\right).$$

    The sign of $h(x)$ is the same for all values of x in a specific interval. That is, $h(x)$ is positive for all x-values in an interval or $h(x)$ is negative for all x-values in an interval. To determine which, we choose a test value for x from each interval and find $h(x)$.

INTERVAL	TEST VALUE, x	FUNCTION VALUE, $h(x)$	SIGN OF $h(x)$	LOCATION OF POINTS ON GRAPH
$(-\infty, 0)$	-1	-5	$-$	Below x-axis
$\left(0, \frac{3}{2}\right)$	1	1	$+$	Above x-axis
$\left(\frac{3}{2}, \infty\right)$	2	-8	$-$	Below x-axis

This test-point procedure also gives us three points to plot. In this case, we have $(-1, -5)$, $(1, 1)$, and $(2, -8)$.

4. To determine the y-intercept, we find $h(0)$:

$$h(x) = -2x^4 + 3x^3$$
$$h(0) = -2 \cdot 0^4 + 3 \cdot 0^3 = 0.$$

The y-intercept is $(0, 0)$.

5. A few additional points are helpful when completing the graph.

x	$h(x)$
-1.5	-20.25
-0.5	-0.5
0.5	0.25
2.5	-31.25

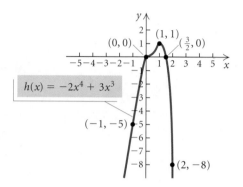

6. The degree of h is 4. The graph of h can have at most 4 x-intercepts and at most 3 turning points. In fact, it has 2 x-intercepts and 1 turning point. The zeros, 0 and $\frac{3}{2}$, each have odd multiplicities, 3 for 0 and 1 for $\frac{3}{2}$. Since the multiplicities are odd, the graph crosses the x-axis at 0 and $\frac{3}{2}$. The end behavior of the graph is what we described in step (1). As $x \to \infty$ and $x \to -\infty$, $h(x) \to -\infty$. The graph appears to be correct.

The following is a procedure that we can follow when graphing polynomial functions.

To graph a polynomial function:

1. Use the leading-term test to determine the end behavior.

2. Find the zeros of the function by solving $f(x) = 0$. Any real zeros are the first coordinates of the x-intercepts.

3. Use the x-intercepts (zeros) to divide the x-axis into intervals and choose a test point in each interval to determine the sign of all function values in that interval.

4. Find $f(0)$. This gives the y-intercept of the function.

5. If necessary, find additional function values to determine the general shape of the graph and then draw the graph.

6. As a partial check, use the facts that the graph has at most n x-intercepts and at most $n - 1$ turning points. Multiplicity of zeros can also be considered in order to check where the graph crosses or is tangent to the x-axis.

EXAMPLE 8   Graph the polynomial function

$$f(x) = 2x^3 + x^2 - 8x - 4.$$

Solution

1. The leading term is $2x^3$. The degree, 3, is odd, and the coefficient, 2, is positive. Thus the end behavior of the graph will appear as follows.

2. To find the zeros, we solve $f(x) = 0$. Here we can use factoring by grouping.

$$2x^3 + x^2 - 8x - 4 = 0$$
$$x^2(2x + 1) - 4(2x + 1) = 0 \qquad \text{Factoring by grouping}$$
$$(2x + 1)(x^2 - 4) = 0$$
$$(2x + 1)(x + 2)(x - 2) = 0 \qquad \text{Factoring a difference of squares}$$

The zeros are $-\frac{1}{2}$, -2, and 2. The x-intercepts are $(-2, 0)$, $\left(-\frac{1}{2}, 0\right)$, and $(2, 0)$.

3. The zeros divide the x-axis into four intervals:

$$(-\infty, -2), \qquad \left(-2, -\frac{1}{2}\right), \qquad \left(-\frac{1}{2}, 2\right), \quad \text{and} \quad (2, \infty).$$

We choose a test value for x from each interval and find $f(x)$.

INTERVAL	TEST VALUE, x	FUNCTION VALUE, $f(x)$	SIGN OF $f(x)$	LOCATION OF POINTS ON GRAPH
$(-\infty, -2)$	-3	-25	$-$	Below x-axis
$\left(-2, -\frac{1}{2}\right)$	-1	3	$+$	Above x-axis
$\left(-\frac{1}{2}, 2\right)$	1	-9	$-$	Below x-axis
$(2, \infty)$	3	35	$+$	Above x-axis

The test values and corresponding function values also give us four points on the graph: $(-3, -25)$, $(-1, 3)$, $(1, -9)$, and $(3, 35)$.

4. To determine the y-intercept, we find $f(0)$:

$$f(x) = 2x^3 + x^2 - 8x - 4$$
$$f(0) = 2 \cdot 0^3 + 0^2 - 8 \cdot 0 - 4 = -4.$$

The y-intercept is $(0, -4)$.

5. We find a few additional points and complete the graph.

x	$f(x)$
-2.5	-9
-1.5	3.5
0.5	-7.5
1.5	-7

$f(x) = 2x^3 + x^2 - 8x - 4$

6. The degree of f is 3. The graph of f can have at most 3 x-intercepts and at most 2 turning points. It has 3 x-intercepts and 2 turning points. Each zero has a multiplicity of 1; thus the graph crosses the x-axis at -2, $-\frac{1}{2}$, and 2. The graph has the end behavior described in step (1). As $x \to -\infty$, $h(x) \to -\infty$, and as $x \to \infty$, $h(x) \to \infty$. The graph appears to be correct.

Some polynomials are difficult to factor. In the next example, the polynomial is given in factored form. In Sections 3.2 and 3.3, we will learn methods that facilitate determining factors of such polynomials.

EXAMPLE 9   Graph the polynomial function

$$g(x) = x^4 - 7x^3 + 12x^2 + 4x - 16$$
$$= (x + 1)(x - 2)^2(x - 4).$$

Solution

1. The leading term is x^4. The degree, 4, is even, and the coefficient, 1, is positive. The sketch below shows the end behavior.

2. To find the zeros, we solve $g(x) = 0$:

$$(x + 1)(x - 2)^2(x - 4) = 0.$$

The zeros are -1, 2, and 4. The x-intercepts are $(-1, 0)$, $(2, 0)$, and $(4, 0)$.

3. The zeros divide the x-axis into four intervals:

$$(-\infty, -1), \quad (-1, 2), \quad (2, 4), \quad \text{and} \quad (4, \infty).$$

We choose a test value for x from each interval and find $g(x)$.

INTERVAL	TEST VALUE, x	FUNCTION VALUE, $g(x)$	SIGN OF $g(x)$	LOCATION OF POINTS ON GRAPH
$(-\infty, -1)$	-1.25	≈ 13.9	$+$	Above x-axis
$(-1, 2)$	1	-6	$-$	Below x-axis
$(2, 4)$	3	-4	$-$	Below x-axis
$(4, \infty)$	4.25	≈ 6.6	$+$	Above x-axis

The test values and corresponding function values also give us four points on the graph: $(-1.25, 13.9)$, $(1, -6)$, $(3, -4)$, and $(4.25, 6.6)$.

4. To determine the y-intercept, we find $g(0)$:

$$g(x) = x^4 - 7x^3 + 12x^2 + 4x - 16$$
$$g(0) = 0^4 - 7 \cdot 0^3 + 12 \cdot 0^2 + 4 \cdot 0 - 16 = -16.$$

The y-intercept is $(0, -16)$.

5. We find a few additional points and draw the graph.

x	$g(x)$
-0.5	-14.1
0.5	-11.8
1.5	-1.6
2.5	-1.3
3.5	-5.1

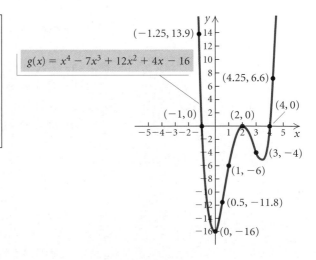

6. The degree of g is 4. The graph of g can have at most 4 x-intercepts and at most 3 turning points. It has 3 x-intercepts and 3 turning points. One of the zeros, 2, has a multiplicity of 2, so the graph is tangent to the x-axis at 2. The other zeros, -1 and 4, each have a multiplicity of 1 so the graph crosses the x-axis at -1 and 4. The graph has the end behavior described in step (1). As $x \to \infty$ and as $x \to -\infty$, $g(x) \to \infty$. The graph appears to be correct. ▬

The Intermediate Value Theorem

Polynomial functions P are continuous, hence their graphs are unbroken. The domain of a polynomial function, unless restricted by the statement of the function, is $(-\infty, \infty)$. Suppose two function values $P(a)$ and $P(b)$ have opposite signs. Since P is continuous, its graph must be a curve from $(a, P(a))$ to $(b, P(b))$ without a break. Then it follows that the curve must cross the x-axis at some point c between a and b—that is, the function has a zero at c between a and b.

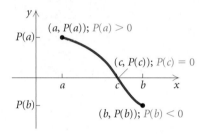

The Intermediate Value Theorem

For any polynomial function $P(x)$ with real coefficients, suppose that for $a \neq b$, $P(a)$ and $P(b)$ are of opposite signs. Then the function has a real zero between a and b.

EXAMPLE 10   Using the intermediate value theorem, determine, if possible, whether the function has a real zero between a and b.

a) $f(x) = x^3 + x^2 - 6x$; $a = -4, b = -2$

b) $f(x) = x^3 + x^2 - 6x$; $a = -1, b = 3$

c) $g(x) = \frac{1}{3}x^4 - x^3$; $a = -\frac{1}{2}, b = \frac{1}{2}$

d) $g(x) = \frac{1}{3}x^4 - x^3$; $a = 1, b = 2$

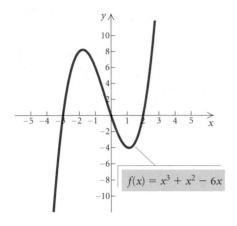

$f(x) = x^3 + x^2 - 6x$

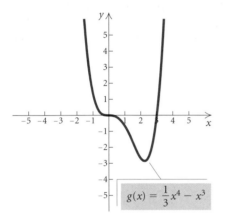

$g(x) = \frac{1}{3}x^4 - x^3$

Solution   We find $f(a)$ and $f(b)$ or $g(a)$ and $g(b)$ and determine whether they differ in sign. The graphs of $f(x)$ and $g(x)$ at left provide a visual check of the solutions.

a) $f(-4) = (-4)^3 + (-4)^2 - 6(-4) = -24$,

$f(-2) = (-2)^3 + (-2)^2 - 6(-2) = 8$

Note that $f(-4)$ is negative and $f(-2)$ is positive. By the intermediate value theorem, since $f(-4)$ and $f(-2)$ have opposite signs, then $f(x)$ has a zero between -4 and -2.

b) $f(-1) = (-1)^3 + (-1)^2 - 6(-1) = 6$,

$f(3) = 3^3 + 3^2 - 6(3) = 18$

Both $f(-1)$ and $f(3)$ are positive. Thus the intermediate value theorem does not allow us to determine whether there is a real zero between -1 and 3. Note that the graph of $f(x)$ shows that there are two zeros between -1 and 3.

c) $g\left(-\frac{1}{2}\right) = \frac{1}{3}\left(-\frac{1}{2}\right)^4 - \left(-\frac{1}{2}\right)^3 = \frac{7}{48}$,

$g\left(\frac{1}{2}\right) = \frac{1}{3}\left(\frac{1}{2}\right)^4 - \left(\frac{1}{2}\right)^3 = -\frac{5}{48}$

Since $g\left(-\frac{1}{2}\right)$ and $g\left(\frac{1}{2}\right)$ have opposite signs, $g(x)$ has a zero between $-\frac{1}{2}$ and $\frac{1}{2}$.

d) $g(1) = \frac{1}{3}(1)^4 - 1^3 = -\frac{2}{3}$,

$g(2) = \frac{1}{3}(2)^4 - 2^3 = -\frac{8}{3}$

Both $g(1)$ and $g(2)$ are negative. This does not necessarily mean that there is not a zero between 1 and 2. The graph of $g(x)$ does show that there are no zeros between 1 and 2, but the function values $-\frac{2}{3}$ and $-\frac{8}{3}$ do not allow us to use the intermediate value theorem to determine this.

Polynomial Models

Polynomial functions have many uses as models in science, engineering, and business. The simplest use of polynomial functions in applied problems occurs when we merely evaluate a polynomial function. In such cases, a model has already been developed.

EXAMPLE 11   *Ibuprofen in the Bloodstream.*   The polynomial function

$$M(t) = 0.5t^4 + 3.45t^3 - 96.65t^2 + 347.7t$$

can be used to estimate the number of milligrams of the pain relief medication ibuprofen in the bloodstream t hours after 400 mg of the medication has been taken. Find the number of milligrams in the bloodstream at $t = 0, 0.5, 1, 1.5$, and so on, up to 6 hr. Round the function values to the nearest tenth.

Solution   Using a calculator, we compute the function values.

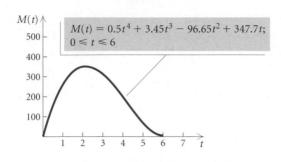

$M(0) = 0,$

$M(0.5) = 150.2,$

$M(1) = 255,$

$M(1.5) = 318.3,$

$M(2) = 344.4,$

$M(2.5) = 338.6,$

$M(3) = 306.9,$

$M(3.5) = 255.9,$

$M(4) = 193.2,$

$M(4.5) = 126.9,$

$M(5) = 66,$

$M(5.5) = 20.2,$

$M(6) = 0.$

Recall that the domain of a polynomial function, unless restricted by a statement of the function, is $(-\infty, \infty)$. The implications of the application in Example 11 restrict the domain of the function. If we assume that a patient had not taken any of the medication before, it seems reasonable that $M(0) = 0$; that is, at time 0, there is 0 mg of the medication in the bloodstream. After the medication has been taken, $M(t)$ will be positive for a period of time and eventually decrease back to 0 when $t = 6$ and not increase again (unless another dose is taken). Thus the restricted domain is $[0, 6]$.

Technology Connection

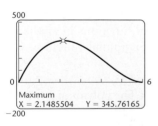

We can evaluate the function in Example 11 with the TABLE feature of a graphing calculator set in AUTO mode. We start at 0 and use a step-value of 0.5.

As discussed above, the domain of $M(t)$ is $[0, 6]$. To determine the range, we find the relative maximum value of the function using the MAXIMUM feature.

The maximum is about 345.76 mg. It occurs approximately 2.15 hr, or 2 hr 9 min, after the initial dose has been taken. The range is about $[0, 345.76]$.

Technology Connection

Previously, we used regression to model data with linear functions. We now expand that procedure to include quadratic, cubic, and quartic models.

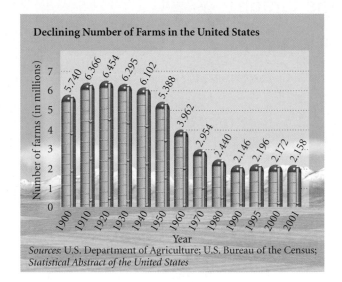

Declining Number of Farms in the United States

Sources: U.S. Department of Agriculture; U.S. Bureau of the Census; *Statistical Abstract of the United States*

Declining Number of Farms in the United States. Today U.S. farm acreage is about the same as it was in the early part of the twentieth century, but the number of farms has decreased.

Looking at the graph at left, we note that the data could be modeled with a cubic or a quartic function. Here we use the quartic model, where x is the number of years since 1900.

Using the REGRESSION feature with DIAGNOSTIC turned on, we get the following.

```
QuarticReg
 y = ax4 + bx3 + ... + e
 a = 1.4403278E−7
 b = −7.340139E−6
 c = −.0018779144
 d = .0822189806
↓e = 5.714950699
```

```
QuarticReg
 y = ax4 + bx3 + ... + e
↑b = −7.340139E−6
 c = −.0018779144
 d = .0822189806
 e = 5.714950699
 R2 = .9923821873
```

Note that a and b are given in scientific notation on the graphing calculator, but we convert to decimal notation when we write the function:

$$f(x) = 0.00000014403278x^4 - 0.000007340139x^3 - 0.0018779144x^2 + 0.0822189806x + 5.714950699.$$

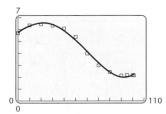

The scatterplot and graph are shown at left. With this function, we can estimate that there were about 5.533 million farms in 1945 and about 2.779 million in 1975. Looking at the bar graph shown above, we see that these estimates appear to be fairly accurate.

If we use the function to estimate the number of farms in 2005 and in 2010, we get about 2.654 million and 3.354 million, respectively. These estimates are probably not realistic. The quartic model has a high value for R^2, approximately 0.992, over the domain of the data, but this number does not reflect the degree of accuracy for extended values. It is always important when using regression to evaluate predictions with common sense and knowledge of current trends.

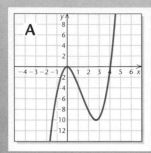

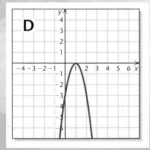

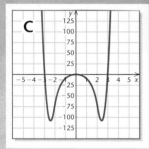

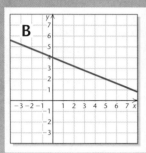

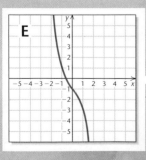

Visualizing the Graph

Match the function with its graph.

1. $f(x) = -x^4 - x + 5$

2. $f(x) = -3x^2 + 6x - 3$

3. $f(x) = x^4 - 4x^3 + 3x^2 + 4x - 4$

4. $f(x) = -\dfrac{2}{5}x + 4$

5. $f(x) = x^3 - 4x^2$

6. $f(x) = x^6 - 9x^4$

7. $f(x) = x^5 - 3x^3 + 2$

8. $f(x) = -x^3 - x - 1$

9. $f(x) = x^2 + 7x + 6$

10. $f(x) = \dfrac{7}{2}$

Answers on page A-19

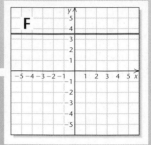

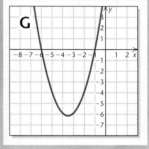

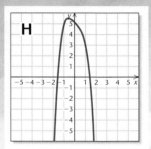

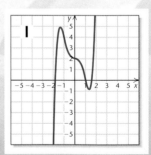

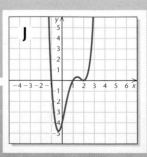

3.1

Exercise Set

Classify the polynomial as constant, linear, quadratic, cubic, or quartic and determine the leading term, the leading coefficient, and the degree of the polynomial.

1. $g(x) = \frac{1}{2}x^3 - 10x + 8$

2. $f(x) = 15x^2 - 10 + 0.11x^4 - 7x^3$

3. $h(x) = 0.9x - 0.13$

4. $f(x) = -6$

5. $g(x) = 305x^4 + 4021$

6. $h(x) = 2.4x^3 + 5x^2 - x + \frac{7}{8}$

7. $h(x) = -5x^2 + 7x^3 + x^4$

8. $f(x) = 2 - x^2$

9. $g(x) = 4x^3 - \frac{1}{2}x^2 + 8$

10. $f(x) = 12 + x$

In Exercises 11–18, select one of the following four sketches to describe the end behavior of the graph of the function.

a)   **b)**

c)   **d)**

11. $f(x) = -3x^3 - x + 4$

12. $f(x) = \frac{1}{4}x^4 + \frac{1}{2}x^3 - 6x^2 + x - 5$

13. $f(x) = -x^6 + \frac{3}{4}x^4$

14. $f(x) = \frac{2}{5}x^5 - 2x^4 + x^3 - \frac{1}{2}x + 3$

15. $f(x) = -3.5x^4 + x^6 + 0.1x^7$

16. $f(x) = -x^3 + x^5 - 0.5x^6$

17. $f(x) = 10 + \frac{1}{10}x^4 - \frac{2}{5}x^3$

18. $f(x) = 2x + x^3 - x^5$

In Exercises 19–24, use the leading-term test and your knowledge of y-intercepts to match the function with one of graphs (a)–(f), which follow.

a) 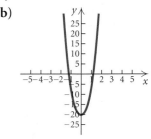   **b)**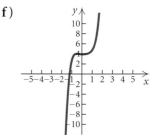

c) 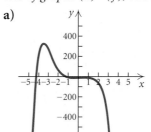   **d)**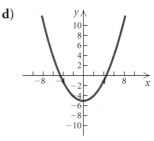

e) 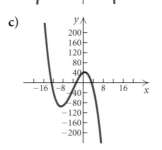   **f)**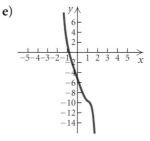

19. $f(x) = \frac{1}{4}x^2 - 5$

20. $f(x) = -0.5x^6 - x^5 + 4x^4 - 5x^3 - 7x^2 + x - 3$

21. $f(x) = x^5 - x^4 + x^2 + 4$

22. $f(x) = -\frac{1}{3}x^3 - 4x^2 + 6x + 42$

23. $f(x) = x^4 - 2x^3 + 12x^2 + x - 20$

24. $f(x) = -0.3x^7 + 0.11x^6 - 0.25x^5 + x^4 + x^3 - 6x - 5$

25. Use substitution to determine whether 4, 5, and -2 are zeros of

$$f(x) = x^3 - 9x^2 + 14x + 24.$$

26. Use substitution to determine whether 2, 3, and -1 are zeros of
$$f(x) = 2x^3 - 3x^2 + x + 6.$$

27. Use substitution to determine whether 2, 3, and -1 are zeros of
$$g(x) = x^4 - 6x^3 + 8x^2 + 6x - 9.$$

28. Use substitution to determine whether 1, -2, and 3 are zeros of
$$g(x) = x^4 - x^3 - 3x^2 + 5x - 2.$$

Find the zeros of the polynomial function and state the multiplicity of each.

29. $f(x) = (x + 3)^2(x - 1)$

30. $f(x) = (x + 5)^3(x - 4)(x + 1)^2$

31. $f(x) = -2(x - 4)(x - 4)(x - 4)(x + 6)$

32. $f(x) = \left(x + \frac{1}{2}\right)(x + 7)(x + 7)(x + 5)$

33. $f(x) = (x^2 - 9)^3$

34. $f(x) = (x^2 - 4)^2$

35. $f(x) = x^3(x - 1)^2(x + 4)$

36. $f(x) = x^2(x + 3)^2(x - 4)(x + 1)^4$

37. $f(x) = -8(x - 3)^2(x + 4)^3 x^4$

38. $f(x) = (x^2 - 5x + 6)^2$

39. $f(x) = x^4 - 4x^2 + 3$

40. $f(x) = x^4 - 10x^2 + 9$

41. $f(x) = x^3 + 3x^2 - x - 3$

42. $f(x) = x^3 - x^2 - 2x + 2$

43. $f(x) = 2x^3 - x^2 - 8x + 4$

44. $f(x) = 3x^3 + x^2 - 48x - 16$

Sketch the graph of each polynomial function. Follow the steps outlined in the procedure on page 254.

45. $f(x) = -x^3 - 2x^2$

46. $g(x) = x^4 - 4x^3 + 3x^2$

47. $h(x) = x^5 - 4x^3$

48. $g(x) = -x(x - 1)^2(x + 4)^2$

49. $h(x) = x(x - 4)(x + 1)(x - 2)$

50. $g(x) = -x^4 - 2x^3$

51. $f(x) = \frac{1}{2}x^3 + \frac{5}{2}x^2$

52. $h(x) = x^3 - 3x^2$

53. $g(x) = (x - 2)^3(x + 3)$

54. $f(x) = -\frac{1}{2}(x - 2)(x + 1)^2(x - 1)$

55. $f(x) = x^3 - x$

56. $h(x) = -x(x - 3)(x - 3)(x + 2)$

57. $f(x) = (x - 2)^2(x + 1)^4$

58. $g(x) = x^4 - 9x^2$

59. $g(x) = -(x - 1)^4$

60. $h(x) = (x + 2)^3$

61. $h(x) = x^3 + 3x^2 - x - 3$

62. $g(x) = -x^3 + 2x^2 + 4x - 8$

63. $f(x) = 6x^3 - 8x^2 - 54x + 72$

64. $h(x) = x^5 - 5x^3 + 4x$

Using the intermediate value theorem, determine if possible whether the function f has a real zero between a and b.

65. $f(x) = x^3 + 3x^2 - 9x - 13; a = -5, b = -4$

66. $f(x) = x^3 + 3x^2 - 9x - 13; a = 1, b = 2$

67. $f(x) = 3x^2 - 2x - 11; a = -3, b = -2$

68. $f(x) = 3x^2 - 2x - 11; a = 2, b = 3$

69. $f(x) = x^4 - 2x^2 - 6; a = 2, b = 3$

70. $f(x) = 2x^5 - 7x + 1; a = 1, b = 2$

71. $f(x) = x^3 - 5x^2 + 4; a = 4, b = 5$

72. $f(x) = x^4 - 3x^2 + x - 1; a = -3, b = -2$

73. *Tobacco Acreage.* Tobacco was America's first export, but tobacco fields are now disappearing due to lawsuits, a decrease in smoking, and competition from imports.

The quartic function

$$f(x) = 0.854x^4 - 16.115x^3 + 86.769x^2 \\ - 139.235x + 735.664,$$

where x is the number of years since 1993, can be used to estimate the acreage (in thousands) of tobacco harvested from 1993 to 2003 (*Source*: National Agriculture Statistics Source, USDA). Find the average harvested in 1995 and in 2000.

74. *Projectile Motion.* A stone thrown downward with an initial velocity of 34.3 m/sec will travel a distance of s meters, where

$$s(t) = 4.9t^2 + 34.3t$$

and t is in seconds. If a stone is thrown downward at 34.3 m/sec from a height of 294 m, how long will it take the stone to hit the ground?

75. *Vertical Leap.* A formula relating an athlete's vertical leap V, in inches, to hang time T, in seconds, is

$$V = 48T^2.$$

Anfernee Hardaway of the Phoenix Suns has a vertical leap of 36 in. (*Source*: National Basketball Association). What is his hang time?

76. *Investments in China.* Total foreign direct investment in China, in billions of dollars, over the years 1992–2002 is modeled by the cubic function

$$f(x) = 0.556x^3 - 7.575x^2 + 22.112x + 71.346,$$

where x is the number of years since 1992 (*Source*: Ministry of Foreign Trade and Economic Cooperation). Find the foreign investment in China in 1993, 1996, 2000, and 2002. Then use this model to estimate the investment in 2003.

77. *Games in a Sports League.* If there are x teams in a sports league and all the teams play each other twice, a total of $N(x)$ games are played, where

$$N(x) = x^2 - x.$$

A softball league has 9 teams, each of which plays the others twice. If the league pays $45 per game for the field and umpires, how much will it cost to play the entire schedule?

78. *Windmill Power.* Under certain conditions, the power P, in watts per hour, generated by a windmill with winds blowing v miles per hour is given by

$$P(v) = 0.015v^3.$$

a) Find the power generated by 15-mph winds.
b) How fast must the wind blow in order to generate 120 watts of power in 1 hr?

79. *Bald Eagles.* During winter, bald eagles travel south to find open water in states that have milder winter temperatures. The number of bald eagles spotted in Indiana during the winters from 1996 to 2002 can be modeled by the quartic function

$$f(x) = -3.394x^4 + 35.838x^3 - 98.955x^2 \\ + 41.930x + 174.974,$$

where x is the number of years since 1996 (*Source*: Indiana Department of Natural Resources). Find the number of bald eagles in Indiana in the winters of 1997, 1999, 2001, and 2002.

80. *French Language Registrations.* Higher education French language registrations in the United States from 1970 to 1998 can be modeled with the quartic function

$$f(x) = 0.004x^4 - 0.268x^3 + 5.836x^2 \\ - 46.121x + 360.046,$$

where f is in thousands and x is the number of years since 1970. Use this model to find the number of higher education French language registrations in 1970, 1980, 1990, and 1998. Then use this function to estimate the French language registrations in 2000. Round answers to the nearest thousand.

81. *Interest Compounded Annually.* When P dollars is invested at interest rate i, compounded annually, for t years, the investment grows to A dollars, where

$$A = P(1 + i)^t.$$

a) Find the interest rate i if $4000 grows to $4368.10 in 2 yr.

b) Find the interest rate i if $10,000 grows to $13,310 in 3 yr.

82. *Threshold Weight.* In a study performed by Alvin Shemesh, it was found that the **threshold weight W**, defined as the weight above which the risk of death rises dramatically, is given by

$$W(h) = \left(\frac{h}{12.3}\right)^3,$$

where W is in pounds and h is a person's height, in inches. Find the threshold weight of a person who is 5 ft, 7 in. tall.

Technology Connection

Using a graphing calculator, estimate the relative maxima and minima and the range of the polynomial function.

83. $g(x) = x^3 - 1.2x + 1$

84. $h(x) = -\frac{1}{2}x^4 + 3x^3 - 5x^2 + 3x + 6$

85. $f(x) = x^6 - 3.8$

86. $h(x) = 2x^3 - x^4 + 20$

87. $f(x) = x^2 + 10x - x^5$

88. $f(x) = 2x^4 - 5.6x^2 + 10$

89. *U.S. Farms.* As the number of farms has decreased in the United States, the average size of the remaining farms has grown larger, as shown below.

YEAR	AVERAGE ACREAGE PER FARM
1900	147
1910	139
1920	149
1930	157
1940	175
1950	216
1960	297
1970	373
1980	426
1990	460
2000	434

Source: U.S. Departments of Agriculture and of Commerce

a) Use a graphing calculator to fit linear, quadratic, cubic, and quartic functions to the data. Let x represent the number of years since 1900.

b) With each function found in part (a), estimate the average acreage in 2005 and in 2010 and determine which function gives the most realistic estimates.

90. *Funeral Costs.* Since the Federal Trade Commission began regulating funeral directors in 1984, the average cost of a funeral for an adult has greatly increased, as shown in the following table.

YEAR	AVERAGE COST OF FUNERAL
1980	$1809
1985	2737
1991	3742
1995	4624
1996	4782
1998	5020
2001	5180

Source: National Funeral Directors Association

a) Use a graphing calculator to fit linear, quadratic, cubic, and quartic functions to the data. Let x represent the number of years since 1980.

b) With each function found in part (a), estimate the cost of a funeral in 2005 and in 2010 and determine which function gives the most realistic estimates.

91. *Mortgage Debt.* Mortgage debt is mounting in the United States, as shown in the table below.

Year	Mortgage Debt (in billions)
1992	$4254
1993	4209
1994	4381
1995	4577
1996	4865
1997	5203
1998	5723
1999	6360
2000	6887
2001	7596

a) Use a graphing calculator to fit linear and quadratic functions to the data. Let x represent the number of years since 1992.

b) Use the functions found in part (a) to estimate the debt in 2005. Compare the estimates and determine which model gives the most realistic estimate.

92. *Dog Years.* A dog's life span is typically much shorter than that of a human. Age equivalents for dogs and humans are given in the following table.

Age of Dog, x (in years)	Human Age, $h(x)$ (in years)
0.25	5
0.5	10
1	15
2	24
4	32
6	40
8	48
10	56
14	72
18	91
21	106

Source: Country, December 1992, p. 60

a) Use a graphing calculator to fit linear and cubic functions to the data. Which function has the better fit?

b) Use the function found in part (a) to estimate the equivalent human age for dogs that are 5, 10, and 15 years old.

Collaborative Discussion and Writing

93. How is the range of a polynomial function related to the degree of the polynomial?

94. Polynomial functions are continuous. Discuss what "continuous" means in terms of the domains of the functions and the characteristics of their graphs.

Skill Maintenance

Find the distance between the pair of points.

95. $(3, -5)$ and $(0, -1)$

96. $(4, 2)$ and $(-2, -4)$

97. Find the center and the radius of the circle $(x - 3)^2 + (y + 5)^2 = 49$.

98. The diameter of a circle connects the points $(-6, 5)$ and $(-2, 1)$ on the circle. Find the coordinates of the center of the circle and the length of the radius.

Solve.

99. $2y - 3 \geq 1 - y + 5$

100. $(x - 2)(x + 5) > x(x - 3)$

101. $|x + 6| \geq 7$

102. $\left| x + \frac{1}{4} \right| \leq \frac{2}{3}$

Synthesis

103. In early 1995, $2000 was deposited at a certain interest rate compounded annually. One year later, $1200 was deposited in another account at the same rate. At the end of that year, there was a total of $3573.80 in both accounts. What is the annual interest rate?

3.2

Polynomial Division; The Remainder and Factor Theorems

- *Perform long division with polynomials and determine whether one polynomial is a factor of another.*
- *Use synthetic division to divide a polynomial by $x - c$.*
- *Use the remainder theorem to find a function value $f(c)$.*
- *Use the factor theorem to determine whether $x - c$ is a factor of $f(x)$.*

In general, finding exact zeros of many polynomial functions is neither easy nor straightforward. In this section and the one that follows, we develop concepts that help us find exact zeros of certain polynomial functions with degree 3 or greater.

Consider the polynomial

$$h(x) = x^3 + 2x^2 - 5x - 6 = (x + 3)(x + 1)(x - 2).$$

The factors are

$$x + 3, \qquad x + 1, \quad \text{and} \quad x - 2,$$

and the zeros are

$$-3, \qquad -1, \quad \text{and} \quad 2.$$

When a polynomial is expressed in factored form, each factor determines a zero of the function. Thus if we know the factors of a polynomial, we can easily find the zeros. We now show how this idea can be "reversed" so that if we know the zeros of a polynomial function, we can find the factors of the polynomial.

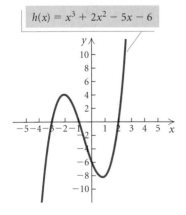

$h(x) = x^3 + 2x^2 - 5x - 6$

Division and Factors

When we divide one polynomial by another, we obtain a quotient and a remainder. If the remainder is 0, then the divisor is a factor of the dividend.

EXAMPLE 1 Divide to determine whether $x + 1$ and $x - 3$ are factors of

$$x^3 + 2x^2 - 5x - 6.$$

Solution We have

$$
\begin{array}{r}
\overbrace{x^2 + \ x \ - 6}^{\textbf{Quotient}} \\[2pt]
x + 1 \overline{)\ x^3 + 2x^2 - 5x - 6} \\
\underline{x^3 + \ x^2} \\
x^2 - 5x \\
\underline{x^2 + \ x} \\
-6x - 6 \\
\underline{-6x - 6} \\
0
\end{array}
$$

$x + 1 \overline{)\ x^3 + 2x^2 - 5x - 6} \longleftarrow$ **Dividend**

Divisor

$0 \longleftarrow$ **Remainder**

Since the remainder is 0, we know that $x + 1$ is a factor of $x^3 + 2x^2 - 5x - 6$. In fact, we know that

$$x^3 + 2x^2 - 5x - 6 = (x + 1)(x^2 + x - 6).$$

We also have

$$
\begin{array}{r}
x^2 + 5x + 10 \\
x - 3 \overline{)\ x^3 + 2x^2 - \ 5x - \ 6} \\
\underline{x^3 - 3x^2} \\
5x^2 - \ 5x \\
\underline{5x^2 - 15x} \\
10x - \ 6 \\
\underline{10x - 30} \\
24
\end{array}
$$

$24 \longleftarrow$ **Remainder**

Since the remainder is not 0, we know that $x - 3$ is *not* a factor of $x^3 + 2x^2 - 5x - 6$.

When we divide a polynomial $P(x)$ by a divisor $d(x)$, a polynomial $Q(x)$ is the quotient and a polynomial $R(x)$ is the remainder. The remainder must either be 0 or have degree less than that of $d(x)$.

As in arithmetic, to check a division, we multiply the quotient by the divisor and add the remainder, to see if we get the dividend. Thus these polynomials are related as follows:

$$P(x) = d(x) \cdot Q(x) + R(x)$$

Dividend   Divisor   Quotient   Remainder

For example, if $P(x) = x^3 + 2x^2 - 5x - 6$ and $d(x) = x + 1$, as in Example 1, then $Q(x) = x^2 + x - 6$ and $R(x) = 0$, and

$$
\underbrace{x^3 + 2x^2 - 5x - 6}_{P(x)} = \underbrace{(x + 1)}_{d(x)} \cdot \underbrace{(x^2 + x - 6)}_{Q(x)} + \underbrace{0}_{R(x)}.
$$

The Remainder Theorem and Synthetic Division

Consider the function

$$h(x) = x^3 + 2x^2 - 5x - 6.$$

When we divided $h(x)$ by $x + 1$ and $x - 3$ in Example 1, the remainders were 0 and 24, respectively. Let's now find the function values $h(-1)$ and $h(3)$:

$$h(-1) = (-1)^3 + 2(-1)^2 - 5(-1) - 6 = 0;$$
$$h(3) = (3)^3 + 2(3)^2 - 5(3) - 6 = 24.$$

Note that the function values are the same as the remainders. This suggests the following theorem.

The Remainder Theorem

If a number c is substituted for x in the polynomial $f(x)$, then the result $f(c)$ is the remainder that would be obtained by dividing $f(x)$ by $x - c$. That is, if $f(x) = (x - c) \cdot Q(x) + R$, then $f(c) = R$.

PROOF (OPTIONAL).   The equation $f(x) = d(x) \cdot Q(x) + R(x)$, where $d(x) = x - c$, is the basis of this proof. If we divide $f(x)$ by $x - c$, we obtain a quotient $Q(x)$ and a remainder $R(x)$ related as follows:

$$f(x) = (x - c) \cdot Q(x) + R(x).$$

The remainder $R(x)$ must either be 0 or have degree less than $x - c$. Thus, $R(x)$ must be a constant. Let's call this constant R. The equation above is true for any replacement of x, so we replace x with c. We get

$$\begin{aligned} f(c) &= (c - c) \cdot Q(c) + R \\ &= 0 \cdot Q(c) + R \\ &= R. \end{aligned}$$

Thus the function value $f(c)$ is the remainder obtained when we divide $f(x)$ by $x - c$.

The remainder theorem motivates us to find a rapid way of dividing by $x - c$ in order to find function values. To streamline division, we can arrange the work so that duplicate and unnecessary writing is avoided. Consider the following:

$$(4x^3 - 3x^2 + x + 7) \div (x - 2).$$

A.
$$\begin{array}{r} 4x^2 + 5x + 11 \\ x - 2 \overline{\smash{\big)}\ 4x^3 - 3x^2 + x + 7} \\ \underline{4x^3 - 8x^2} \\ 5x^2 + x \\ \underline{5x^2 - 10x} \\ 11x + 7 \\ \underline{11x - 22} \\ 29 \end{array}$$

B.
$$\begin{array}{r} 4 \quad\ 5 \quad\ 11 \\ 1 - 2 \overline{\smash{\big)}\ 4 - 3 + 1 + 7} \\ \underline{4 - 8} \\ 5 + 1 \\ \underline{5 - 10} \\ 11 + 7 \\ \underline{11 - 22} \\ 29 \end{array}$$

The division in (B) is the same as that in (A), but we wrote only the coefficients. The color numerals are duplicated, so we look for an arrangement in which they are not duplicated. In place of the divisor in the form $x - c$, we can simply use c and then add rather than subtract. When the procedure is "collapsed," we have the algorithm known as **synthetic division.**

C. *Synthetic Division*

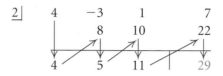

The divisor is $x - 2$; thus we use 2 in synthetic division.

We "bring down" the 4. Then we multiply it by the 2 to get 8 and add to get 5. We then multiply 5 by 2 to get 10, add, and so on. The last number, 29, is the remainder. The others, 4, 5, and 11, are the coefficients of the quotient, $4x^2 + 5x + 11$.

When using synthetic division, we write a 0 for a missing term in the dividend.

EXAMPLE 2 Use synthetic division to find the quotient and the remainder:

$$(2x^3 + 7x^2 - 5) \div (x + 3).$$

Solution First, we note that $x + 3 = x - (-3)$.

$$
\begin{array}{r|rrrr}
-3 & 2 & 7 & 0 & -5 \\
   &   & -6 & -3 & 9 \\
\hline
   & 2 & 1 & -3 & 4
\end{array}
$$

Note: **We must write a 0 for the missing term.**

The quotient is $2x^2 + x - 3$. The remainder is 4. (Note that the degree of the quotient is 1 less than the degree of the dividend when the degree of the divisor is 1.)

We can now use synthetic division to find polynomial function values.

EXAMPLE 3 Given that $f(x) = 2x^5 - 3x^4 + x^3 - 2x^2 + x - 8$, find $f(10)$.

Solution By the remainder theorem, $f(10)$ is the remainder when $f(x)$ is divided by $x - 10$. We use synthetic division to find that remainder.

$$
\begin{array}{r|rrrrrr}
10 & 2 & -3 & 1 & -2 & 1 & -8 \\
   &   & 20 & 170 & 1710 & 17{,}080 & 170{,}810 \\
\hline
   & 2 & 17 & 171 & 1708 & 17{,}081 & 170{,}802
\end{array}
$$

Thus, $f(10) = 170{,}802$.

Compare the computations in Example 3 with those in a direct substitution:

$$f(10) = 2(10)^5 - 3(10)^4 + (10)^3 - 2(10)^2 + 10 - 8$$
$$= 170{,}802.$$

EXAMPLE 4 Determine whether -4 is a zero of $f(x)$, where

$$f(x) = x^3 + 8x^2 + 8x - 32.$$

Solution We use synthetic division and the remainder theorem to find $f(-4)$.

$$
\begin{array}{r|rrrr}
-4 & 1 & 8 & 8 & -32 \\
   &   & -4 & -16 & 32 \\
\hline
   & 1 & 4 & -8 & \big|\ \ \ 0
\end{array}
$$

Since $f(-4) = 0$, the number -4 is a zero of $f(x)$. ▬

Finding Factors of Polynomials

We now consider a useful result that follows from the remainder theorem.

> **The Factor Theorem**
>
> For a polynomial $f(x)$, if $f(c) = 0$, then $x - c$ is a factor of $f(x)$.

PROOF (OPTIONAL). If we divide $f(x)$ by $x - c$, we obtain a quotient and a remainder, related as follows:

$$f(x) = (x - c) \cdot Q(x) + f(c).$$

Then if $f(c) = 0$, we have

$$f(x) = (x - c) \cdot Q(x),$$

so $x - c$ is a factor of $f(x)$.

The factor theorem is very useful in factoring polynomials, and hence in solving polynomial equations and finding zeros of polynomial functions.

EXAMPLE 5 Let $f(x) = x^3 + 2x^2 - 5x - 6$. Factor $f(x)$ and solve the equation $f(x) = 0$.

Solution We look for linear factors of the form $x - c$. Let's try $x - 1$. (In the next section, we will learn a method for choosing the numbers to try for c.) We use synthetic division to determine whether $f(1) = 0$.

$$
\begin{array}{r|rrrr}
1 & 1 & 2 & -5 & -6 \\
  &   & 1 & 3 & -2 \\
\hline
  & 1 & 3 & -2 & \big|\ -8
\end{array}
$$

Since $f(1) \neq 0$, we know that $x - 1$ is *not a factor* of $f(x)$. We try $x + 1$ or $x - (-1)$.

$$
\begin{array}{r|rrrr}
-1 & 1 & 2 & -5 & -6 \\
   &   & -1 & -1 & 6 \\
\hline
   & 1 & 1 & -6 & \big|\ \ \ 0
\end{array}
$$

Technology Connection

In Example 5, we can use a table set in ASK mode to check the solutions of the equation

$$x^3 + 2x^2 - 5x - 6 = 0.$$

We check the solutions, -1, -3, and 2, of $f(x) = 0$ by evaluating $f(-1)$, $f(-3)$, and $f(2)$.

X	Y₁
−1	0
−3	0
2	0

X = 2

Since $f(-1) = 0$, we know that $x + 1$ *is one factor* and the quotient, $x^2 + x - 6$, is another. Thus,

$$f(x) = (x + 1)(x^2 + x - 6).$$

The trinomial is easily factored in this case, so we have

$$f(x) = (x + 1)(x + 3)(x - 2).$$

Our goal is to solve the equation $f(x) = 0$. To do so, we use the principle of zero products:

$$(x + 1)(x + 3)(x - 2) = 0$$

$$x + 1 = 0 \quad or \quad x + 3 = 0 \quad or \quad x - 2 = 0$$

$$x = -1 \quad or \quad x = -3 \quad or \quad x = 2.$$

The solutions of the equation $x^3 + 2x^2 - 5x - 6 = 0$ are $-1, -3$, and 2. They are also the zeros of the function $f(x) = x^3 + 2x^2 - 5x - 6$. ▬

3.2

Exercise Set

1. For the function

$$f(x) = x^4 - 6x^3 + x^2 + 24x - 20,$$

use long division to determine which of the following are factors of $f(x)$.

a) $x + 1$ b) $x - 2$ c) $x + 5$

2. For the function

$$h(x) = x^3 - x^2 - 17x - 15,$$

use long division to determine which of the following are factors of $h(x)$.

a) $x + 5$ b) $x + 1$ c) $x + 3$

3. For the function

$$g(x) = x^3 - 2x^2 - 11x + 12,$$

use long division to determine which of the following are factors of $g(x)$.

a) $x - 4$ b) $x - 3$ c) $x - 1$

4. For the function

$$f(x) = x^4 + 8x^3 + 5x^2 - 38x + 24,$$

use long division to determine which of the following are factors of $f(x)$.

a) $x + 6$ b) $x + 1$ c) $x - 4$

In each of the following, a polynomial $P(x)$ and a divisor $d(x)$ are given. Use long division to find the quotient $Q(x)$ and the remainder $R(x)$ when $P(x)$ is divided by $d(x)$, and express $P(x)$ in the form $d(x) \cdot Q(x) + R(x)$.

5. $P(x) = x^3 - 8$,
    $d(x) = x + 2$

6. $P(x) = 2x^3 - 3x^2 + x - 1$,
    $d(x) = x - 3$

7. $P(x) = x^3 + 6x^2 - 25x + 18$,
    $d(x) = x + 9$

8. $P(x) = x^3 - 9x^2 + 15x + 25$,
    $d(x) = x - 5$

9. $P(x) = x^4 - 2x^2 + 3$,
    $d(x) = x + 2$

10. $P(x) = x^4 + 6x^3$,
    $d(x) = x - 1$

Use synthetic division to find the quotient and the remainder.

11. $(2x^4 + 7x^3 + x - 12) \div (x + 3)$

12. $(x^3 - 7x^2 + 13x + 3) \div (x - 2)$

13. $(x^3 - 2x^2 - 8) \div (x + 2)$

14. $(x^3 - 3x + 10) \div (x - 2)$

15. $(3x^3 - x^2 + 4x - 10) \div (x + 1)$

16. $(4x^4 - 2x + 5) \div (x + 3)$

17. $(x^5 + x^3 - x) \div (x - 3)$

18. $(x^7 - x^6 + x^5 - x^4 + 2) \div (x + 1)$

19. $(x^4 - 1) \div (x - 1)$

20. $(x^5 + 32) \div (x + 2)$

21. $(2x^4 + 3x^2 - 1) \div \left(x - \frac{1}{2}\right)$

22. $(3x^4 - 2x^2 + 2) \div \left(x - \frac{1}{4}\right)$

Use synthetic division to find the function values.

23. $f(x) = x^3 - 6x^2 + 11x - 6$; find $f(1)$, $f(-2)$, and $f(3)$.

24. $f(x) = x^3 + 7x^2 - 12x - 3$; find $f(-3)$, $f(-2)$, and $f(1)$.

25. $f(x) = x^4 - 3x^3 + 2x + 8$; find $f(-1)$, $f(4)$, and $f(-5)$.

26. $f(x) = 2x^4 + x^2 - 10x + 1$; find $f(-10)$, $f(2)$, and $f(3)$.

27. $f(x) = 2x^5 - 3x^4 + 2x^3 - x + 8$; find $f(20)$ and $f(-3)$.

28. $f(x) = x^5 - 10x^4 + 20x^3 - 5x - 100$; find $f(-10)$ and $f(5)$.

29. $f(x) = x^4 - 16$; find $f(2)$, $f(-2)$, $f(3)$, and $f\left(1 - \sqrt{2}\right)$.

30. $f(x) = x^5 + 32$; find $f(2)$, $f(-2)$, $f(3)$, and $f(2 + 3i)$.

Using synthetic division, determine whether the numbers are zeros of the polynomial function.

31. $-3, 2$; $f(x) = 3x^3 + 5x^2 - 6x + 18$

32. $-4, 2$; $f(x) = 3x^3 + 11x^2 - 2x + 8$

33. $-3, 1$; $h(x) = x^4 + 4x^3 + 2x^2 - 4x - 3$

34. $2, -1$; $g(x) = x^4 - 6x^3 + x^2 + 24x - 20$

35. $i, -2i$; $g(x) = x^3 - 4x^2 + 4x - 16$

36. $\frac{1}{3}, 2$; $h(x) = x^3 - x^2 - \frac{1}{9}x + \frac{1}{9}$

37. $-3, \frac{1}{2}$; $f(x) = x^3 - \frac{7}{2}x^2 + x - \frac{3}{2}$

38. $i, -i, -2$; $f(x) = x^3 + 2x^2 + x + 2$

Factor the polynomial $f(x)$. Then solve the equation $f(x) = 0$.

39. $f(x) = x^3 + 4x^2 + x - 6$

40. $f(x) = x^3 + 5x^2 - 2x - 24$

41. $f(x) = x^3 - 6x^2 + 3x + 10$

42. $f(x) = x^3 + 2x^2 - 13x + 10$

43. $f(x) = x^3 - x^2 - 14x + 24$

44. $f(x) = x^3 - 3x^2 - 10x + 24$

45. $f(x) = x^4 - 7x^3 + 9x^2 + 27x - 54$

46. $f(x) = x^4 - 4x^3 - 7x^2 + 34x - 24$

47. $f(x) = x^4 - x^3 - 19x^2 + 49x - 30$

48. $f(x) = x^4 + 11x^3 + 41x^2 + 61x + 30$

Sketch the graph of the polynomial function. Follow the procedure outlined on page 254. Use synthetic division and the remainder theorem to find the zeros.

49. $f(x) = x^4 - x^3 - 7x^2 + x + 6$

50. $f(x) = x^4 + x^3 - 3x^2 - 5x - 2$

51. $f(x) = x^3 - 7x + 6$

52. $f(x) = x^3 - 12x + 16$

53. $f(x) = -x^3 + 3x^2 + 6x - 8$

54. $f(x) = -x^4 + 2x^3 + 3x^2 - 4x - 4$

Technology Connection

55. Use a graphing calculator to find the function values in Exercises 23–30.

Collaborative Discussion and Writing

56. Can an nth-degree polynomial function have more than n zeros? Why or why not?

57. Let $P(x)$ be a polynomial function with $p(x)$ a factor of $P(x)$. If $p(2) = 0$, does it follow that $P(2) = 0$? Why or why not? If $P(2) = 0$, does it follow that $p(2) = 0$? Why or why not?

Skill Maintenance

Find exact solutions.

58. $2x - 7 = 5x + 8$

59. $2x^2 + 12 = 5x$

60. $7x^2 + 4x = 3$

Consider the function

$$g(x) = x^2 + 5x - 14$$

in Exercises 61–63.

61. What are the inputs if the output is -14?

62. What is the output if the input is 3?

63. Given an output of -20, find the corresponding inputs.

64. *Catalog Sales.* Catalog retailers have witnessed a continual linear gain in sales for over a decade. Catalog sales have risen from \$27.2 billion in 1989 to \$72 billion in 2001 (*Source*: The Direct Marketing Association). Using these two data points, find a linear function, $f(x) = mx + b$, that models the data. Let x represent the number of years since 1989. Then use this function to estimate catalog sales for 2000, 2005, and 2010.

65. The sum of the base and the height of a triangle is 30 in. Find the dimensions for which the area is a maximum.

Synthesis

In Exercises 66 and 67, a graph of a polynomial function is given. On the basis of the graph:

a) *Find as many factors of the polynomial as you can.*
b) *Construct a polynomial function with the zeros shown in the graph.*
c) *Can you find any other polynomial functions with the given zeros?*
d) *Can you find any other polynomial functions with the given zeros and the same graph?*

66.

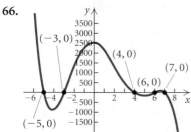

67.

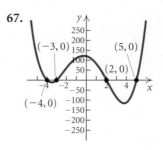

68. For what values of k will the remainder be the same when $x^2 + kx + 4$ is divided by $x - 1$ and $x + 1$?

69. Find k such that $x + 2$ is a factor of $x^3 - kx^2 + 3x + 7k$.

70. *Beam Deflection.* A beam rests at two points A and B and has a concentrated load applied to its center. Let $y =$ the deflection, in feet, of the beam at a distance of x feet from A. Under certain conditions, this deflection is given by

$$y = \frac{1}{13}x^3 - \frac{1}{14}x.$$

Find the zeros of the polynomial in the interval $[0, 2]$.

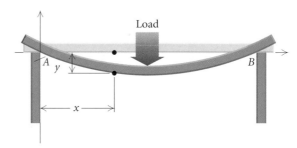

Solve.

71. $\dfrac{2x^2}{x^2 - 1} + \dfrac{4}{x + 3} = \dfrac{32}{x^3 + 3x^2 - x - 3}$

72. $\dfrac{6x^2}{x^2 + 11} + \dfrac{60}{x^3 - 7x^2 + 11x - 77} = \dfrac{1}{x - 7}$

73. Find a 15th-degree polynomial for which $x - 1$ is a factor. Answers may vary.

Use synthetic division to divide.

74. $(x^2 - 4x - 2) \div [x - (3 + 2i)]$

75. $(x^2 - 3x + 7) \div (x - i)$

3.3

Theorems about Zeros of Polynomial Functions

- *Find a polynomial with specified zeros.*
- *For a polynomial function with integer coefficients, find the rational zeros and the other zeros, if possible.*
- *Use Descartes' rule of signs to find information about the number of real zeros of a polynomial function with real coefficients.*

We will now allow the coefficients of a polynomial to be complex numbers. In certain cases, we will restrict the coefficients to be real numbers, rational numbers, or integers, as shown in the following examples.

POLYNOMIAL	TYPE OF COEFFICIENT
$5x^3 - 3x^2 + (2 + 4i)x + i$	Complex
$5x^3 - 3x^2 + \sqrt{2}x - \pi$	Real
$5x^3 - 3x^2 + \frac{2}{3}x - \frac{7}{4}$	Rational
$5x^3 - 3x^2 + 8x - 11$	Integer

The Fundamental Theorem of Algebra

A linear, or first-degree, polynomial function $f(x) = mx + b$ (where $m \neq 0$) has just one zero, $-b/m$. It can be shown that any quadratic polynomial function with complex numbers for coefficients has at least one, and at most two, complex zeros. The following theorem is a generalization. No proof is given in this text.

> ### The Fundamental Theorem of Algebra
>
> Every polynomial function of degree n, with $n \geq 1$, has at least one zero in the system of complex numbers.

Note that although the fundamental theorem of algebra guarantees that a zero exists, it does not tell how to find it. Recall that the zeros of a polynomial function $f(x)$ are the solutions of the polynomial equation $f(x) = 0$. We now develop some concepts that can help in finding zeros. First, we consider one of the results of the fundamental theorem of algebra.

> Every polynomial function f of degree n, with $n \geq 1$, can be factored into n linear factors (not necessarily unique); that is,
> $$f(x) = a_n(x - c_1)(x - c_2) \cdots (x - c_n).$$

Finding Polynomials with Given Zeros

Given several numbers, we can find a polynomial function with those numbers as its zeros.

EXAMPLE 1   Find a polynomial function of degree 3, having the zeros $1, 3i$, and $-3i$.

Solution   Such a polynomial has factors $x - 1$, $x - 3i$, and $x + 3i$, so we have

$$f(x) = a_n(x - 1)(x - 3i)(x + 3i).$$

The number a_n can be any nonzero number. The simplest polynomial will be obtained if we let it be 1. If we then multiply the factors, we obtain

$$f(x) = (x - 1)(x^2 + 9) \qquad \text{Multiplying } (x - 3i)(x + 3i)$$
$$= x^3 - x^2 + 9x - 9.$$

EXAMPLE 2   Find a polynomial function of degree 5 with -1 as a zero of multiplicity 3, 4 as a zero of multiplicity 1, and 0 as a zero of multiplicity 1.

Solution   Proceeding as in Example 1, letting $a_n = 1$, we obtain

$$f(x) = (x + 1)^3(x - 4)(x - 0)$$
$$= x^5 - x^4 - 9x^3 - 11x^2 - 4x.$$

Zeros of Polynomial Functions with Real Coefficients

Consider the quadratic equation $x^2 - 2x + 2 = 0$, with real coefficients. Its solutions are $1 + i$ and $1 - i$. Note that they are complex conjugates. This generalizes to any polynomial with real coefficients.

> If a complex number $a + bi$, $b \neq 0$, is a zero of a polynomial function $f(x)$ with *real* coefficients, then its conjugate, $a - bi$, is also a zero. (Nonreal zeros occur in conjugate pairs.)

In order for the preceding to be true, it is essential that the coefficients be real numbers.

Rational Coefficients

When a polynomial has rational numbers for coefficients, certain irrational zeros also occur in pairs, as described in the following theorem.

> If $a + c\sqrt{b}$, where a and c are rational and b is not a perfect square, is a zero of a polynomial function $f(x)$ with *rational* coefficients, then $a - c\sqrt{b}$ is also a zero.

EXAMPLE 3 Suppose that a polynomial function of degree 6 with rational coefficients has $-2 + 5i$, $-2i$, and $1 - \sqrt{3}$ as three of its zeros. Find the other zeros.

Solution The other zeros are the conjugates of the given zeros, $-2 - 5i$, $2i$, and $1 + \sqrt{3}$. There are no other zeros because a polynomial of degree 6 can have at most 6 zeros. ■

EXAMPLE 4 Find a polynomial function of lowest degree with rational coefficients that has $1 - \sqrt{2}$ and $1 + 2i$ as two of its zeros.

Solution The function must also have the zeros $1 + \sqrt{2}$ and $1 - 2i$. Because we want to find the polynomial function of lowest degree with the given zeros, we will not include additional zeros. That is, we will write a polynomial function of degree 4. Thus, if we let $a_n = 1$, the polynomial function is

$$f(x) = \left[x - \left(1 - \sqrt{2}\right)\right]\left[x - \left(1 + \sqrt{2}\right)\right]\left[x - \left(1 + 2i\right)\right]\left[x - \left(1 - 2i\right)\right]$$
$$= (x^2 - 2x - 1)(x^2 - 2x + 5)$$
$$= x^4 - 4x^3 + 8x^2 - 8x - 5.$$ ■

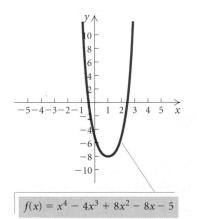

$f(x) = x^4 - 4x^3 + 8x^2 - 8x - 5$

Integer Coefficients and the Rational Zeros Theorem

It is not always easy to find the zeros of a polynomial function. However, if a polynomial function has integer coefficients, there is a procedure that will yield all the rational zeros.

> ### The Rational Zeros Theorem
> Let
> $$P(x) = a_n x^n + a_{n-1}x^{n-1} + \cdots + a_1 x + a_0,$$
> where all the coefficients are integers. Consider a rational number denoted by p/q, where p and q are relatively prime (having no common factor besides -1 and 1). If p/q is a zero of $P(x)$, then p is a factor of a_0 and q is a factor of a_n.

EXAMPLE 5 Given $f(x) = 3x^4 - 11x^3 + 10x - 4$:

a) Find the rational zeros and then the other zeros; that is, solve $f(x) = 0$.

b) Factor $f(x)$ into linear factors.

Solution

a) Because the degree of $f(x)$ is 4, there are at most 4 distinct zeros. The rational zeros theorem says that if p/q is a zero of $f(x)$, then p must be

We can narrow the list of possibilities for the rational zeros of a function by examining its graph.

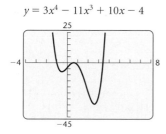

$y = 3x^4 - 11x^3 + 10x - 4$

From the graph of the function in Example 5, we see that of the possibilities for rational zeros, only the numbers

$$-1, \quad \tfrac{1}{3}, \quad \tfrac{2}{3}$$

might actually be rational zeros. We can check these with the TABLE feature. Note that the graphing calculator converts $\tfrac{1}{3}$ and $\tfrac{2}{3}$ to decimal notation.

X	Y₁
−1	0
.33333	−1.037
.66667	0
X =	

Since $f(-1) = 0$ and $f\left(\tfrac{2}{3}\right) = 0$, -1 and $\tfrac{2}{3}$ are zeros. Since $f\left(\tfrac{1}{3}\right) \neq 0$, $\tfrac{1}{3}$ is not a zero.

a factor of -4 and q must be a factor of 3. Thus the possibilities for p/q are

$$\frac{\text{Possibilities for } p}{\text{Possibilities for } q}: \quad \frac{\pm 1, \pm 2, \pm 4}{\pm 1, \pm 3};$$

Possibilities for p/q: $1, -1, 2, -2, 4, -4, \tfrac{1}{3}, -\tfrac{1}{3}, \tfrac{2}{3}, -\tfrac{2}{3}, \tfrac{4}{3}, -\tfrac{4}{3}.$

To find which are zeros, we could use substitution, but synthetic division is usually more efficient. It is easier to consider the integers first. Then we consider the fractions, if the integers do not produce all the zeros.

We try 1:

$$
\begin{array}{r|rrrrr}
1 & 3 & -11 & 0 & 10 & -4 \\
  &   & 3 & -8 & -8 & 2 \\
\hline
  & 3 & -8 & -8 & 2 & \big|\ -2.
\end{array}
$$

Since $f(1) = -2$, 1 is not a zero.
   We try -1:

$$
\begin{array}{r|rrrrr}
-1 & 3 & -11 & 0 & 10 & -4 \\
   &   & -3 & 14 & -14 & 4 \\
\hline
   & 3 & -14 & 14 & -4 & \big|\ 0.
\end{array}
$$

We have $f(-1) = 0$, so -1 is a zero. Thus, $x + 1$ is a factor. Using the results of the second division above, we can express $f(x)$ as

$$f(x) = (x + 1)(3x^3 - 14x^2 + 14x - 4).$$

We now consider the factor $3x^3 - 14x^2 + 14x - 4$ and check the other possible zeros. We use synthetic division again, to determine whether -1 is a zero of multiplicity 2 of $f(x)$:

$$
\begin{array}{r|rrrr}
-1 & 3 & -14 & 14 & -4 \\
   &   & -3 & 17 & -31 \\
\hline
   & 3 & -17 & 31 & \big|\ -35.
\end{array}
$$

We see that -1 is not a double zero. We leave it to the student to verify that $2, -2, 4,$ and -4 are not zeros. There are no other possible zeros that are integers, so we start checking the fractions.
   Let's now try $\tfrac{2}{3}$.

$$
\begin{array}{r|rrrr}
2/3 & 3 & -14 & 14 & -4 \\
    &   & 2 & -8 & 4 \\
\hline
    & 3 & -12 & 6 & \big|\ 0
\end{array}
$$

Since the remainder is 0, we know that $x - \tfrac{2}{3}$ is a factor of $3x^3 - 14x^2 + 14x - 4$ and is also a factor of $f(x)$. Thus, $\tfrac{2}{3}$ is a zero of $f(x)$.
   Using the results of the synthetic division, we can factor further:

$$f(x) = (x + 1)\left(x - \tfrac{2}{3}\right)(3x^2 - 12x + 6) \qquad \text{Using the results of the last synthetic division}$$

$$= (x + 1)\left(x - \tfrac{2}{3}\right) \cdot 3 \cdot (x^2 - 4x + 2). \qquad \text{Removing a factor of 3}$$

The quadratic formula can be used to find the values of x for which $x^2 - 4x + 2 = 0$. Those values are also zeros of $f(x)$:

$$x = \frac{-b \pm \sqrt{b^2 - 4ac}}{2a}$$

$$= \frac{-(-4) \pm \sqrt{(-4)^2 - 4 \cdot 1 \cdot 2}}{2 \cdot 1}$$

$$= \frac{4 \pm \sqrt{8}}{2} = \frac{4 \pm 2\sqrt{2}}{2} = \frac{2(2 \pm \sqrt{2})}{2}$$

$$= 2 \pm \sqrt{2}.$$

The rational zeros are -1 and $\frac{2}{3}$. The other zeros are $2 \pm \sqrt{2}$.

b) The complete factorization of $f(x)$ is

$$f(x) = 3(x + 1)\left(x - \tfrac{2}{3}\right)\left[x - \left(2 - \sqrt{2}\right)\right]\left[x - \left(2 + \sqrt{2}\right)\right].$$

EXAMPLE 6 Given $f(x) = 2x^5 - x^4 - 4x^3 + 2x^2 - 30x + 15$:

a) Find the rational zeros and then the other zeros; that is, solve $f(x) = 0$.

b) Factor $f(x)$ into linear factors.

Solution

a) Because the degree of $f(x)$ is 5, there are at most 5 distinct zeros. According to the rational zeros theorem, any rational zero of f must be of the form p/q, where p is a factor of 15 and q is a factor of 2. The possibilities are

$$\frac{\text{Possibilities for } p}{\text{Possibilities for } q} : \frac{\pm 1, \pm 3, \pm 5, \pm 15}{\pm 1, \pm 2};$$

Possibilities for p/q: $1, -1, 3, -3, 5, -5, 15, -15, \frac{1}{2}, -\frac{1}{2}, \frac{3}{2}, -\frac{3}{2},$
$\frac{5}{2}, -\frac{5}{2}, \frac{15}{2}, -\frac{15}{2}.$

We use synthetic division to check each of the possibilities. Let's try 1 and -1.

$$
\begin{array}{r|rrrrrr}
1 & 2 & -1 & -4 & 2 & -30 & 15 \\
  &   & 2 & 1 & -3 & -1 & -31 \\
\hline
  & 2 & 1 & -3 & -1 & -31 & -16 \\
\end{array}
$$

$$
\begin{array}{r|rrrrrr}
-1 & 2 & -1 & -4 & 2 & -30 & 15 \\
   &   & -2 & 3 & 1 & -3 & 33 \\
\hline
   & 2 & -3 & -1 & 3 & -33 & 48 \\
\end{array}
$$

Since $f(1) = -16$ and $f(-1) = 48$, neither 1 nor -1 is a zero of the function. We leave it to the student to verify that the other integer possibilities are not zeros. We now check $\frac{1}{2}$.

$$
\begin{array}{r|rrrrrr}
\frac{1}{2} & 2 & -1 & -4 & 2 & -30 & 15 \\
            &   & 1 & 0 & -2 & 0 & -15 \\
\hline
            & 2 & 0 & -4 & 0 & -30 & 0 \\
\end{array}
$$

Technology Connection

We can inspect the graph of the function in Example 6 for zeros that appear to be near any of the possible rational zeros.

$y = 2x^5 - x^4 - 4x^3 + 2x^2$
$\quad - 30x + 15$

We see that, of the possibilities in the list, only the numbers

$$-\tfrac{5}{2}, \quad \tfrac{1}{2}, \quad \text{and} \quad \tfrac{5}{2}$$

might be rational zeros. With the TABLE feature, we can determine that only $\frac{1}{2}$ is actually a rational zero.

This means that $x - \frac{1}{2}$ is a factor of $f(x)$. We write the factorization and try to factor further:

$$f(x) = \left(x - \tfrac{1}{2}\right)(2x^4 - 4x^2 - 30)$$
$$= \left(x - \tfrac{1}{2}\right) \cdot 2 \cdot (x^4 - 2x^2 - 15) \qquad \textbf{Factoring out the 2}$$
$$= \left(x - \tfrac{1}{2}\right) \cdot 2 \cdot (x^2 - 5)(x^2 + 3). \qquad \textbf{Factoring the trinomial}$$

We now solve the equation $f(x) = 0$ to determine the zeros. We use the principle of zero products:

$$\left(x - \tfrac{1}{2}\right) \cdot 2 \cdot (x^2 - 5)(x^2 + 3) = 0$$

$$x - \tfrac{1}{2} = 0 \quad or \quad x^2 - 5 = 0 \qquad or \quad x^2 + 3 = 0$$
$$x = \tfrac{1}{2} \quad or \qquad x^2 = 5 \qquad or \qquad x^2 = -3$$
$$x = \tfrac{1}{2} \quad or \qquad x = \pm\sqrt{5} \quad or \qquad x = \pm\sqrt{3}i.$$

There is only one rational zero, $\frac{1}{2}$. The other zeros are $\pm\sqrt{5}$ and $\pm\sqrt{3}i$.

b) The factorization into linear factors is

$$f(x) = 2\left(x - \tfrac{1}{2}\right)\left(x + \sqrt{5}\right)\left(x - \sqrt{5}\right)\left(x + \sqrt{3}i\right)\left(x - \sqrt{3}i\right). \quad \blacksquare$$

Descartes' Rule of Signs

The development of a rule that helps determine the number of positive real zeros and the number of negative real zeros of a polynomial function is credited to the French mathematician René Descartes. To use the rule, we must have the polynomial arranged in descending or ascending order, with no zero terms written in, the leading coefficient positive, and the *constant term* not 0. Then we determine the number of *variations of sign*, that is, the number of times, in reading through the polynomial, that successive coefficients are of different signs.

EXAMPLE 7 Determine the number of variations of sign in the polynomial function $P(x) = 2x^5 - 3x^2 + x + 4$.

Solution   We have

$$P(x) = 2x^5 - 3x^2 + x + 4$$

From positive to negative; a variation

From negative to positive; a variation

Both positive; no variation

The number of variations of sign is 2. $\blacksquare$

Note the following:

$$P(-x) = 2(-x)^5 - 3(-x)^2 + (-x) + 4$$
$$= -2x^5 - 3x^2 - x + 4.$$

We see that the number of variations of sign in $P(-x)$ is 1.

We now state Descartes' rule, without proof.

Descartes' Rule of Signs

Let $P(x)$ be a polynomial function with real coefficients and a nonzero constant term. The number of positive real zeros of $P(x)$ is either:

1. The same as the number of variations of sign in $P(x)$, or
2. Less than the number of variations of sign in $P(x)$ by a positive even integer.

The number of negative real zeros of $P(x)$ is either:

3. The same as the number of variations of sign in $P(-x)$, or
4. Less than the number of variations of sign in $P(-x)$ by a positive even integer.

A zero of multiplicity m must be counted m times.

In each of Examples 8–10, what does Descartes' rule of signs tell you about the number of positive real zeros and the number of negative real zeros?

EXAMPLE 8 $P(x) = 2x^5 - 5x^2 - 3x + 6$

Solution   The number of variations of sign in $P(x)$ is 2. Therefore, the number of positive real zeros is either 2 or less than 2 by 2, 4, 6, and so on. Thus the number of positive real zeros is either 2 or 0, since a negative number of zeros has no meaning.

$$P(-x) = -2x^5 - 5x^2 + 3x + 6$$

The number of variations of sign in $P(-x)$ is 1. Thus there is exactly 1 negative real zero. Since nonreal, complex conjugates occur in pairs, we also know the possible ways in which nonreal zeros might occur, as summarized in the table shown at left.   ▬

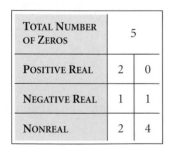

TOTAL NUMBER OF ZEROS	5	
POSITIVE REAL	2	0
NEGATIVE REAL	1	1
NONREAL	2	4

EXAMPLE 9 $P(x) = 5x^4 - 3x^3 + 7x^2 - 12x + 4$

Solution   There are 4 variations of sign. Thus the number of positive real zeros is either

$$4 \quad \text{or} \quad 4 - 2 \quad \text{or} \quad 4 - 4.$$

That is, the number of positive real zeros is 4, 2, or 0.

$$P(-x) = 5x^4 + 3x^3 + 7x^2 + 12x + 4$$

There are 0 changes in sign, so there are no negative real zeros.   ▬

EXAMPLE 10 $P(x) = 6x^6 - 2x^2 - 5x$

Solution   As written, the polynomial does not satisfy the conditions of Descartes' rule of signs because the constant term is 0. But because x is a factor of every term, we know that the polynomial has 0 as a zero. We can then factor as follows:

$$P(x) = x(6x^5 - 2x - 5).$$

Now we analyze $Q(x) = 6x^5 - 2x - 5$ and $Q(-x) = -6x^5 + 2x - 5$. The number of variations of sign in $Q(x)$ is 1. Therefore, there is exactly 1 positive real zero. The number of variations of sign in $Q(-x)$ is 2. Thus the number of negative real zeros is 2 or 0. The same results apply to $P(x)$.   ▬

3.3

Exercise Set

Find a polynomial function of degree 3 with the given numbers as zeros.

1. $-2, 3, 5$

2. $-1, 0, 4$

3. $-3, 2i, -2i$

4. $2, i, -i$

5. $\sqrt{2}, -\sqrt{2}, 3$

6. $-5, \sqrt{3}, -\sqrt{3}$

7. $1 - \sqrt{3}, 1 + \sqrt{3}, -2$

8. $-4, 1 - \sqrt{5}, 1 + \sqrt{5}$

9. $1 + 6i, 1 - 6i, -4$

10. $1 + 4i, 1 - 4i, -1$

11. $-\frac{1}{3}, 0, 2$

12. $-3, 0, \frac{1}{2}$

13. Find a polynomial function of degree 5 with -1 as a zero of multiplicity 3, 0 as a zero of multiplicity 1, and 1 as a zero of multiplicity 1.

14. Find a polynomial function of degree 4 with -2 as a zero of multiplicity 1, 3 as a zero of multiplicity 2, and -1 as a zero of multiplicity 1.

15. Find a polynomial of degree 4 with -1 as a zero of multiplicity 3 and 0 as a zero of multiplicity 1.

16. Find a polynomial of degree 5 with $-\frac{1}{2}$ as a zero of multiplicity 2, 0 as a zero of multiplicity 1, and 1 as a zero of multiplicity 2.

Suppose that a polynomial function of degree 4 with rational coefficients has the given numbers as zeros. Find the other zero(s).

17. $-1, \sqrt{3}, \frac{11}{3}$

18. $-\sqrt{2}, -1, \frac{4}{5}$

19. $-i, 2 - \sqrt{5}$

20. $i, -3 + \sqrt{3}$

21. $3i, 0, -5$

22. $3, 0, -2i$

23. $-4 - 3i, 2 - \sqrt{3}$

24. $6 - 5i, -1 + \sqrt{7}$

Suppose that a polynomial function of degree 5 with rational coefficients has the given numbers as zeros. Find the other zero(s).

25. $-\frac{1}{2}, \sqrt{5}, -4i$

26. $\frac{3}{4}, -\sqrt{3}, 2i$

27. $-5, 0, 2 - i, 4$

28. $-2, 3, 4, 1 - i$

29. $6, -3 + 4i, 4 - \sqrt{5}$

30. $-3 - 3i, 2 + \sqrt{13}, 6$

31. $-\frac{3}{4}, \frac{3}{4}, 0, 4 - i$

32. $-0.6, 0, 0.6, -3 + \sqrt{2}$

Find a polynomial function of lowest degree with rational coefficients that has the given numbers as some of its zeros.

33. $1 + i, 2$

34. $2 - i, -1$

35. $4i$

36. $-5i$

37. $-4i, 5$

38. $3, -i$

39. $1 - i, -\sqrt{5}$

40. $2 - \sqrt{3}, 1 + i$

41. $\sqrt{5}, -3i$

42. $-\sqrt{2}, 4i$

Given that the polynomial function has the given zero, find the other zeros.

43. $f(x) = x^3 + 5x^2 - 2x - 10; \ -5$

44. $f(x) = x^3 - x^2 + x - 1; \ 1$

45. $f(x) = x^4 - 5x^3 + 7x^2 - 5x + 6; \ -i$

46. $f(x) = x^4 - 16; \ 2i$

47. $f(x) = x^3 - 6x^2 + 13x - 20; \ 4$

48. $f(x) = x^3 - 8; \ 2$

List all possible rational zeros.

49. $f(x) = x^5 - 3x^2 + 1$

50. $f(x) = x^7 + 37x^5 - 6x^2 + 12$

51. $f(x) = 2x^4 - 3x^3 - x + 8$

52. $f(x) = 3x^3 - x^2 + 6x - 9$

53. $f(x) = 15x^6 + 47x^2 + 2$

54. $f(x) = 10x^{25} + 3x^{17} - 35x + 6$

For each polynomial function:

a) *Find the rational zeros and then the other zeros; that is, solve $f(x) = 0$.*
b) *Factor $f(x)$ into linear factors.*

55. $f(x) = x^3 + 3x^2 - 2x - 6$

56. $f(x) = x^3 - x^2 - 3x + 3$

57. $f(x) = x^3 - 3x + 2$

58. $f(x) = x^3 - 2x + 4$

59. $f(x) = x^3 - 5x^2 + 11x + 17$

60. $f(x) = 2x^3 + 7x^2 + 2x - 8$

61. $f(x) = 5x^4 - 4x^3 + 19x^2 - 16x - 4$

62. $f(x) = 3x^4 - 4x^3 + x^2 + 6x - 2$

63. $f(x) = x^4 - 3x^3 - 20x^2 - 24x - 8$

64. $f(x) = x^4 + 5x^3 - 27x^2 + 31x - 10$

65. $f(x) = x^3 - 4x^2 + 2x + 4$

66. $f(x) = x^3 - 8x^2 + 17x - 4$

67. $f(x) = x^3 + 8$

68. $f(x) = x^3 - 8$

69. $f(x) = \frac{1}{3}x^3 - \frac{1}{2}x^2 - \frac{1}{6}x + \frac{1}{6}$

70. $f(x) = \frac{2}{3}x^3 - \frac{1}{2}x^2 + \frac{2}{3}x - \frac{1}{2}$

Find only the rational zeros.

71. $f(x) = x^4 + 32$

72. $f(x) = x^6 + 8$

73. $f(x) = x^3 - x^2 - 4x + 3$

74. $f(x) = 2x^3 + 3x^2 + 2x + 3$

75. $f(x) = x^4 + 2x^3 + 2x^2 - 4x - 8$

76. $f(x) = x^4 + 6x^3 + 17x^2 + 36x + 66$

77. $f(x) = x^5 - 5x^4 + 5x^3 + 15x^2 - 36x + 20$

78. $f(x) = x^5 - 3x^4 - 3x^3 + 9x^2 - 4x + 12$

What does Descartes' rule of signs tell you about the number of positive real zeros and the number of negative real zeros of the function?

79. $f(x) = 3x^5 - 2x^2 + x - 1$

80. $g(x) = 5x^6 - 3x^3 + x^2 - x$

81. $h(x) = 6x^7 + 2x^2 + 5x + 4$

82. $P(x) = -3x^5 - 7x^3 - 4x - 5$

83. $F(p) = 3p^{18} + 2p^4 - 5p^2 + p + 3$

84. $H(t) = 5t^{12} - 7t^4 + 3t^2 + t + 1$

85. $C(x) = 7x^6 + 3x^4 - x - 10$

86. $g(z) = -z^{10} + 8z^7 + z^3 + 6z - 1$

87. $h(t) = -4t^5 - t^3 + 2t^2 + 1$

88. $P(x) = x^6 + 2x^4 - 9x^3 - 4$

89. $f(y) = y^4 + 13y^3 - y + 5$

90. $Q(x) = x^4 - 2x^2 + 12x - 8$

91. $r(x) = x^4 - 6x^2 + 20x - 24$

92. $f(x) = x^5 - 2x^3 - 8x$

93. $R(x) = 3x^5 - 5x^3 - 4x$

94. $f(x) = x^4 - 9x^2 - 6x + 4$

Sketch the graph of the polynomial function. Follow the procedure outlined on page 254. Use the rational zeros theorem when finding the zeros.

95. $f(x) = 4x^3 + x^2 - 8x - 2$

96. $f(x) = 3x^3 - 4x^2 - 5x + 2$

97. $f(x) = 2x^4 - 3x^3 - 2x^2 + 3x$

98. $f(x) = 4x^4 - 37x^2 + 9$

Technology Connection

99. Use a graphing calculator to narrow the list of possible rational zeros in Exercises 60–63. Then determine the rational zeros by evaluating the function using the TABLE feature.

Collaborative Discussion and Writing

100. If $Q(x) = -P(x)$, do $P(x)$ and $Q(x)$ have the same zeros? Why or why not?

101. Is it possible for a third-degree polynomial with rational coefficients to have no real zeros? Why or why not?

Skill Maintenance

For Exercises 102 and 103, complete the square to:

a) *find the vertex;*
b) *find the axis of symmetry; and*
c) *determine whether there is a maximum or minimum value and find that value.*

102. $f(x) = 3x^2 - 6x - 1$

103. $f(x) = x^2 - 8x + 10$

Find the zeros of the function.

104. $g(x) = x^2 - 8x - 33$

105. $f(x) = -\frac{4}{5}x + 8$

Classify the polynomial function as constant, linear, quadratic, cubic, or quartic and determine the leading term, the leading coefficient, and the degree of the polynomial. Then describe the end behavior of the function's graph.

106. $f(x) = -x^2 - 3x + 6$

107. $g(x) = -x^3 - 2x^2$

108. $h(x) = x - 2$

109. $f(x) = -\frac{4}{9}$

110. $h(x) = x^3 + \frac{1}{2}x^2 - 4x - 3$

111. $g(x) = x^4 - 2x^3 + x^2 - x + 2$

Synthesis

Use synthetic division to find the quotient and the remainder.

112. $(x^3 + 3ix^2 - 4ix - 2) \div (x + i)$

113. $(x^4 - y^4) \div (x - y)$

114. Use the rational zeros theorem and the equation $x^4 - 12 = 0$ to show that $\sqrt[4]{12}$ is irrational.

115. Consider $f(x) = 2x^3 - 5x^2 - 4x + 3$. Find the solutions of each equation.

    a) $f(x) = 0$ b) $f(x - 1) = 0$
    c) $f(x + 2) = 0$ d) $f(2x) = 0$

Find the rational zeros.

116. $P(x) = x^6 - 6x^5 - 72x^4 - 81x^2 + 486x + 5832$

117. $P(x) = 2x^5 - 33x^4 - 84x^3 + 2203x^2 - 3348x - 10,080$

3.4

Rational Functions

- *Graph a rational function, identifying all asymptotes.*
- *Solve applied problems involving rational functions.*

Now we turn our attention to functions that represent the quotient of two polynomials. Whereas the sum, difference, or product of two polynomials is a polynomial, in general the quotient of two polynomials is *not* itself a polynomial.

A *rational number* can be expressed as the quotient of two integers, p/q, where $q \neq 0$. A *rational function* is formed by the quotient of two polynomials, $p(x)/q(x)$, where $q(x) \neq 0$. Here are some examples of rational functions and their graphs.

$$f(x) = \frac{1}{x} \qquad\qquad f(x) = \frac{1}{x^2} \qquad\qquad f(x) = \frac{x-3}{x^2+x-2}$$

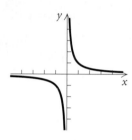
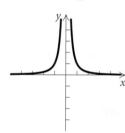
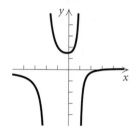

$$f(x) = \frac{2x+5}{2x-6} \qquad\qquad f(x) = \frac{x^2+2x-3}{x^2-x-2} \qquad\qquad f(x) = \frac{-x^2}{x+1}$$

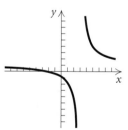
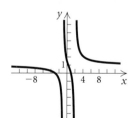
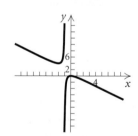

Rational Function

A **rational function** is a function f that is a quotient of two polynomials, that is,

$$f(x) = \frac{p(x)}{q(x)},$$

where $p(x)$ and $q(x)$ are polynomials and where $q(x)$ is not the zero polynomial. The domain of f consists of all inputs x for which $q(x) \neq 0$.

The Domain of a Rational Function

DOMAINS OF FUNCTIONS

REVIEW SECTION 1.2.

EXAMPLE 1 Consider

$$f(x) = \frac{1}{x - 3}.$$

Find the domain and graph f.

Solution When the denominator $x - 3$ is 0, we have $x = 3$, so the only input that results in a denominator of 0 is 3. Thus the domain is

$$\{x \mid x \neq 3\}, \text{ or } (-\infty, 3) \cup (3, \infty).$$

The graph of this function is the graph of $y = 1/x$ translated to the right 3 units.

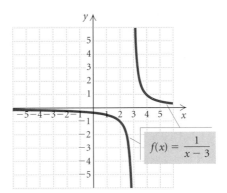

Technology Connection

$y = \dfrac{1}{x - 3}$

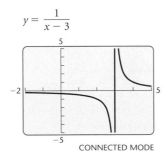

CONNECTED MODE

$y = \dfrac{1}{x - 3}$

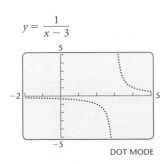

DOT MODE

The graphs at left show the function $f(x) = 1/(x - 3)$ graphed in CONNECTED mode and in DOT mode. Using CONNECTED mode can lead to an incorrect graph. In CONNECTED mode, a graphing calculator connects plotted points with line segments. In DOT mode, it simply plots unconnected points. In the first graph, the graphing calculator has connected the points plotted on either side of the x-value 3 with a line that appears to be the vertical line $x = 3$. (It is not actually vertical since it connects the last point to the left of $x = 3$ with the first point to the right of $x = 3$.) Since 3 is not in the domain of the function, the vertical line $x = 3$ cannot be part of the graph. We will see later in this section that vertical lines like $x = 3$, although not part of the graph, are important in the construction of graphs. If you have a choice when graphing rational functions, use DOT mode.

EXAMPLE 2   Determine the domain of each of the functions illustrated at the beginning of this section.

Solution   The domain of each rational function will be the set of all real numbers except those values that make the denominator 0. To determine those exceptions, we set the denominator equal to 0 and solve for x.

FUNCTION	DOMAIN
$f(x) = \dfrac{1}{x}$	$\{x \mid x \neq 0\}$, or $(-\infty, 0) \cup (0, \infty)$
$f(x) = \dfrac{1}{x^2}$	$\{x \mid x \neq 0\}$, or $(-\infty, 0) \cup (0, \infty)$
$f(x) = \dfrac{x - 3}{x^2 + x - 2} = \dfrac{x - 3}{(x + 2)(x - 1)}$	$\{x \mid x \neq -2 \text{ and } x \neq 1\}$, or $(-\infty, -2) \cup (-2, 1) \cup (1, \infty)$
$f(x) = \dfrac{2x + 5}{2x - 6} = \dfrac{2x + 5}{2(x - 3)}$	$\{x \mid x \neq 3\}$, or $(-\infty, 3) \cup (3, \infty)$
$f(x) = \dfrac{x^2 + 2x - 3}{x^2 - x - 2} = \dfrac{x^2 + 2x - 3}{(x + 1)(x - 2)}$	$\{x \mid x \neq -1 \text{ and } x \neq 2\}$, or $(-\infty, -1) \cup (-1, 2) \cup (2, \infty)$
$f(x) = \dfrac{-x^2}{x + 1}$	$\{x \mid x \neq -1\}$, or $(-\infty, -1) \cup (-1, \infty)$

As a partial check of the domains, we can observe the discontinuities (breaks) in the graphs of these functions (see page 286). ▬

Asymptotes

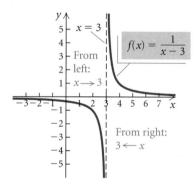

Vertical asymptote: $x = 3$

Look at the graph of $f(x) = 1/(x - 3)$, shown at left (also see Example 1). Let's explore what happens as x-values get closer and closer to 3 from the left. We then explore what happens as x-values get closer and closer to 3 from the right.

From left:

x	2	$2\frac{1}{2}$	$2\frac{99}{100}$	$2\frac{9999}{10,000}$	$2\frac{999,999}{1,000,000}$	$\longrightarrow 3$
$f(x)$	-1	-2	-100	$-10,000$	$-1,000,000$	$\longrightarrow -\infty$

From right:

x	4	$3\frac{1}{2}$	$3\frac{1}{100}$	$3\frac{1}{10,000}$	$3\frac{1}{1,000,000}$	$\longrightarrow 3$
$f(x)$	1	2	100	10,000	1,000,000	$\longrightarrow \infty$

We see that as x-values get closer and closer to 3 from the left, the function values (y-values) decrease without bound (that is, they approach negative infinity, $-\infty$). Similarly, as the x-values approach 3 from the right, the function values increase without bound (that is, they approach positive infinity, ∞). We write this as

$$f(x) \to -\infty \text{ as } x \to 3^- \quad \text{and} \quad f(x) \to \infty \text{ as } x \to 3^+.$$

We read "$f(x) \to -\infty$ as $x \to 3^-$" as "$f(x)$ decreases without bound as x approaches 3 from the left." We read "$f(x) \to \infty$ as $x \to 3^+$" as "$f(x)$ increases without bound as x approaches 3 from the right." The notation $x \to 3$ means that x gets as close to 3 as possible without being equal to 3. The vertical line $x = 3$ is said to be a *vertical asymptote* for this curve.

The line $x = a$ is a **vertical asymptote** for the graph of f if any of the following is true:

$$f(x) \to \infty \text{ as } x \to a^- \quad \text{or} \quad f(x) \to -\infty \text{ as } x \to a^-, \quad \text{or}$$
$$f(x) \to \infty \text{ as } x \to a^+ \quad \text{or} \quad f(x) \to -\infty \text{ as } x \to a^+.$$

The following figures show the four ways in which a vertical asymptote can occur.

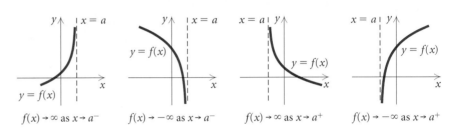

$f(x) \to \infty$ as $x \to a^-$ $\qquad$ $f(x) \to -\infty$ as $x \to a^-$ $\qquad$ $f(x) \to \infty$ as $x \to a^+$ $\qquad$ $f(x) \to -\infty$ as $x \to a^+$

The vertical asymptotes of a rational function $f(x) = p(x)/q(x)$ are found by determining the zeros of $q(x)$ that are not also zeros of $p(x)$. If $p(x)$ and $q(x)$ are polynomials with no common factors other than constants, we need determine only the zeros of the denominator $q(x)$.

Determining Vertical Asymptotes

For a rational function $f(x) = p(x)/q(x)$, where $p(x)$ and $q(x)$ are polynomials with no common factors other than constants, if a is a zero of the denominator, then the line $x = a$ is a vertical asymptote for the graph of the function.

EXAMPLE 3 Determine the vertical asymptotes of each of the following functions.

a) $f(x) = \dfrac{2x - 11}{x^2 + 2x - 8}$ $\qquad\qquad$ **b)** $g(x) = \dfrac{x - 2}{x^3 - 5x}$

Solution

a) We factor to find the zeros of the denominator:

$$x^2 + 2x - 8 = (x + 4)(x - 2).$$

The zeros of the denominator are -4 and 2. Thus the vertical asymptotes are the lines $x = -4$ and $x = 2$ (see Fig. 1).

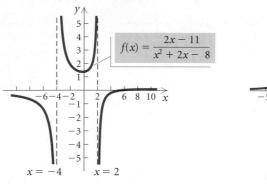

$$f(x) = \frac{2x - 11}{x^2 + 2x - 8}$$

$x = -4$ $x = 2$

FIGURE 1

$$g(x) = \frac{x - 2}{x^3 - 5x}$$

$x = -\sqrt{5}$ $x = 0$ $x = \sqrt{5}$

FIGURE 2

b) We factor to find the zeros of the denominator:

$$x^3 - 5x = x(x^2 - 5).$$

Setting $x(x^2 - 5) = 0$ and solving for x, we get

$$
\begin{aligned}
x = 0 &\quad or \quad x^2 - 5 = 0 \\
x = 0 &\quad or \quad x^2 = 5 \\
x = 0 &\quad or \quad x = \pm\sqrt{5}.
\end{aligned}
$$

The zeros of the denominator are 0, $\sqrt{5}$, and $-\sqrt{5}$. Thus the vertical asymptotes are the lines $x = 0$, $x = \sqrt{5}$, and $x = -\sqrt{5}$ (see Fig. 2).

$$f(x) = \frac{1}{x - 3}$$

$y = 0$

$x \longrightarrow \infty$

$-\infty \longleftarrow x$

Horizontal asymptote: $y = 0$

Looking again at the graph of $f(x) = 1/(x - 3)$, shown at left (also see Example 1), let's explore what happens to $f(x) = 1/(x - 3)$ as x increases without bound (approaches positive infinity, ∞) and as x decreases without bound (approaches negative infinity, $-\infty$).

x increases without bound:

x	100	5000	1,000,000	$\longrightarrow \infty$
$f(x)$	≈ 0.0103	≈ 0.0002	≈ 0.000001	$\longrightarrow 0$

x decreases without bound:

x	-300	-8000	$-1,000,000$	$\longrightarrow -\infty$
$f(x)$	≈ -0.0033	≈ -0.0001	≈ -0.000001	$\longrightarrow 0$

We see that

$$\frac{1}{x - 3} \to 0 \text{ as } x \to \infty \quad \text{and} \quad \frac{1}{x - 3} \to 0 \text{ as } x \to -\infty.$$

Since $y = 0$ is the equation of the x-axis, we say that the curve approaches the x-axis asymptotically and that the x-axis is a *horizontal asymptote* for the curve.

The line $y = b$ is a **horizontal asymptote** for the graph of f if either or both of the following are true:

$$f(x) \to b \text{ as } x \to \infty \quad \text{or} \quad f(x) \to b \text{ as } x \to -\infty.$$

The following figures illustrate four ways in which horizontal asymptotes can occur. In each case, the curve gets close to the line $y = b$ either as $x \to \infty$ or as $x \to -\infty$. Keep in mind that the symbols ∞ and $-\infty$ convey the idea of increasing without bound and decreasing without bound, respectively.

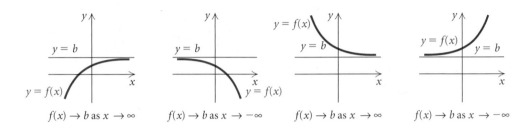

$f(x) \to b$ as $x \to \infty$ $f(x) \to b$ as $x \to -\infty$ $f(x) \to b$ as $x \to \infty$ $f(x) \to b$ as $x \to -\infty$

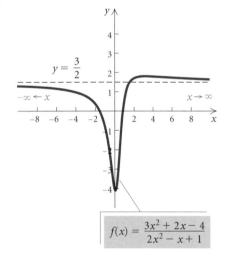

$$f(x) = \frac{3x^2 + 2x - 4}{2x^2 - x + 1}$$

How can we determine a horizontal asymptote? As x gets very large or very small, the value of the polynomial function $p(x)$ is dominated by the function's leading term. Because of this, if $p(x)$ and $q(x)$ have the *same* degree, the value of $p(x)/q(x)$ as $x \to \infty$ or as $x \to -\infty$ is dominated by the ratio of the numerator's leading coefficient to the denominator's leading coefficient.

For $f(x) = (3x^2 + 2x - 4)/(2x^2 - x + 1)$, we see that

$$\frac{3x^2 + 2x - 4}{2x^2 - x + 1} \to \frac{3}{2}, \text{ or } 1.5, \text{ as } x \to \infty, \quad \text{and}$$

$$\frac{3x^2 + 2x - 4}{2x^2 - x + 1} \to \frac{3}{2}, \text{ or } 1.5, \text{ as } x \to -\infty.$$

We say that the curve approaches the horizontal line $y = \frac{3}{2}$ asymptotically and that $y = \frac{3}{2}$ is a *horizontal asymptote* for the curve.

It follows that when the numerator and the denominator of a rational function have the same degree, the line $y = a/b$ is the horizontal asymptote, where a and b are the leading coefficients of the numerator and the denominator, respectively.

EXAMPLE 4 Find the horizontal asymptote: $f(x) = \dfrac{-7x^4 - 10x^2 + 1}{11x^4 + x - 2}$.

Solution The numerator and the denominator have the same degree. The ratio of the leading coefficients is $-\frac{7}{11}$, so the line $y = -\frac{7}{11}$ is the horizontal asymptote.

Technology Connection

We can check Example 4 by using a graphing calculator to evaluate the function for a very large and a very small value of x.

X	Y₁
100000	−.6364
−80000	−.6364

X =

For each number, the function value is close to $-0.\overline{63}$, or $-\frac{7}{11}$.

To check Example 4, we could evaluate the function for a very large and a very small value of x. Another check, one that is useful in calculus, is to multiply by 1, using $(1/x^4)/(1/x^4)$:

$$f(x) = \frac{-7x^4 - 10x^2 + 1}{11x^4 + x - 2} \cdot \frac{\dfrac{1}{x^4}}{\dfrac{1}{x^4}} = \frac{\dfrac{-7x^4}{x^4} - \dfrac{10x^2}{x^4} + \dfrac{1}{x^4}}{\dfrac{11x^4}{x^4} + \dfrac{x}{x^4} - \dfrac{2}{x^4}}$$

$$= \frac{-7 - \dfrac{10}{x^2} + \dfrac{1}{x^4}}{11 + \dfrac{1}{x^3} - \dfrac{2}{x^4}}.$$

As $|x|$ becomes very large, each expression with a power of x in the denominator tends toward 0. Specifically, as $x \to \infty$ or as $x \to -\infty$, we have

$$f(x) \to \frac{-7 - 0 + 0}{11 + 0 - 0}, \quad \text{or} \quad f(x) \to -\frac{7}{11}.$$

The horizontal asymptote is $y = -\frac{7}{11}$.

We now investigate the occurrence of a horizontal asymptote when the degree of the numerator is less than the degree of the denominator.

EXAMPLE 5 Find the horizontal asymptote: $f(x) = \dfrac{2x + 3}{x^3 - 2x^2 + 4}$.

Solution We let $p(x) = 2x + 3$, $q(x) = x^3 - 2x^2 + 4$, and $f(x) = p(x)/q(x)$. Note that as $x \to \infty$, the value of $q(x)$ grows much faster than the value of $p(x)$. Because of this, the ratio $p(x)/q(x)$ shrinks toward 0. As $x \to -\infty$, the ratio $p(x)/q(x)$ behaves in a similar manner. The horizontal asymptote is $y = 0$, the x-axis. This is the case for all rational functions for which the degree of the numerator is less than the degree of the denominator. Note in Example 1 that $y = 0$, the x-axis, is the horizontal asymptote of $f(x) = 1/(x - 3)$.

The following statements describe the two ways in which a horizontal asymptote occurs.

Determining a Horizontal Asymptote

- When the numerator and the denominator of a rational function have the same degree, the line $y = a/b$ is the horizontal asymptote, where a and b are the leading coefficients of the numerator and the denominator, respectively.

- When the degree of the numerator of a rational function is less than the degree of the denominator, the x-axis, or $y = 0$, is the horizontal asymptote.

(continued)

> • When the degree of the numerator of a rational function is greater than the degree of the denominator, there is no horizontal asymptote.

The following statements are also true.

> The graph of a rational function never crosses a vertical asymptote.

> The graph of a rational function might cross a horizontal asymptote but does not necessarily do so.

EXAMPLE 6   Graph

$$g(x) = \frac{2x^2 + 1}{x^2}.$$

Include and label all asymptotes.

Solution   Since 0 is the zero of the denominator, the y-axis, $x = 0$, is the vertical asymptote. Note also that the degree of the numerator is the same as the degree of the denominator. Thus, $y = 2/1$, or 2, is the horizontal asymptote.

  To draw the graph, we first draw the asymptotes with dashed lines. Then we compute and plot some ordered pairs and draw the two branches of the curve.

x	$g(x)$
-2	$2\frac{1}{4}$
$-1\frac{1}{2}$	$2\frac{4}{9}$
-1	3
$-\frac{1}{2}$	6
$\frac{1}{2}$	6
1	3
$1\frac{1}{2}$	$2\frac{4}{9}$
2	$2\frac{1}{4}$

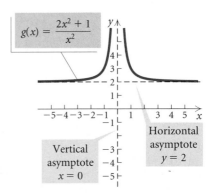

Sometimes a line that is neither horizontal nor vertical is an asymptote. Such a line is called an **oblique asymptote,** or a **slant asymptote.**

EXAMPLE 7   Find all the asymptotes of

$$f(x) = \frac{2x^2 - 3x - 1}{x - 2}.$$

Solution   The line $x = 2$ is the vertical asymptote because 2 is the zero of the denominator. There is no horizontal asymptote because the degree of the numerator is greater than the degree of the denominator. When the degree of the numerator is 1 greater than the degree of the denominator, we divide to find an equivalent expression:

$$\frac{2x^2 - 3x - 1}{x - 2} = (2x + 1) + \frac{1}{x - 2}.$$

$$
\begin{array}{r}
2x + 1 \\
x - 2 \overline{)\,2x^2 - 3x - 1} \\
\underline{2x^2 - 4x} \\
x - 1 \\
\underline{x - 2} \\
1
\end{array}
$$

Now we see that when $x \to \infty$ or $x \to -\infty$, $1/(x - 2) \to 0$ and the value of $f(x) \to 2x + 1$. This means that as $|x|$ becomes very large, the graph of $f(x)$ gets very close to the graph of $y = 2x + 1$. Thus the line $y = 2x + 1$ is the oblique asymptote.

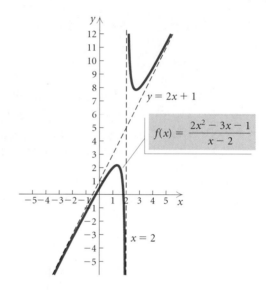

Occurrence of Lines as Asymptotes of Rational Functions

For a rational function $f(x) = p(x)/q(x)$, where $p(x)$ and $q(x)$ have no common factors other than constants:

Vertical asymptotes occur at any x-values that make the denominator 0.

The x-axis is the horizontal asymptote when the degree of the numerator is less than the degree of the denominator.

A horizontal asymptote other than the x-axis occurs when the numerator and the denominator have the same degree.

An oblique asymptote occurs when the degree of the numerator is 1 greater than the degree of the denominator.

There can be only one horizontal asymptote or one oblique asymptote and never both.

An asymptote is *not* part of the graph of the function.

The following is an outline of a procedure that we can follow to create accurate graphs of rational functions.

To graph a rational function $f(x) = p(x)/q(x)$, where $p(x)$ and $q(x)$ have no common factor other than constants:

1. Find the real zeros of the denominator. Determine the domain of the function and sketch any vertical asymptotes.

2. Find the horizontal or the oblique asymptote, if there is one, and sketch it.

3. Find the zeros of the function. The zeros are found by determining the zeros of the numerator. These are the first coordinates of the x-intercepts of the graph.

4. Find $f(0)$. This gives the y-intercept $(0, f(0))$, of the function.

5. Find other function values to determine the general shape. Then draw the graph.

EXAMPLE 8   Graph: $f(x) = \dfrac{2x + 3}{3x^2 + 7x - 6}$.

Solution

1. We find the zeros of the denominator by solving $3x^2 + 7x - 6 = 0$. Since

   $$3x^2 + 7x - 6 = (3x - 2)(x + 3),$$

   the zeros are $\frac{2}{3}$ and -3. Thus the domain excludes $\frac{2}{3}$ and -3 and is

   $$(-\infty, -3) \cup \left(-3, \tfrac{2}{3}\right) \cup \left(\tfrac{2}{3}, \infty\right).$$

   The graph has vertical asymptotes $x = -3$ and $x = \frac{2}{3}$. We sketch these as dashed lines.

2. Because the degree of the numerator is less than the degree of the denominator, the x-axis, $y = 0$, is the horizontal asymptote.

3. To find the zeros of the numerator, we solve $2x + 3 = 0$ and get $x = -\frac{3}{2}$. Thus, $-\frac{3}{2}$ is the zero of the function, and the pair $\left(-\frac{3}{2}, 0\right)$ is the x-intercept.

4. We find $f(0)$:

   $$f(0) = \frac{2 \cdot 0 + 3}{3 \cdot 0^2 + 7 \cdot 0 - 6}$$

   $$= \frac{3}{-6} = -\frac{1}{2}.$$

   Thus, $\left(0, -\frac{1}{2}\right)$ is the y-intercept.

5. We find other function values to determine the general shape and then draw the graph. Note that the graph of this function crosses its horizontal asymptote at $x = -\frac{3}{2}$.

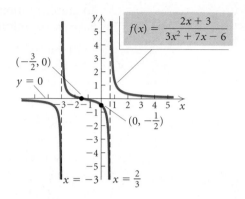

$$f(x) = \frac{2x + 3}{3x^2 + 7x - 6}$$

EXAMPLE 9 Graph: $g(x) = \dfrac{x^2 - 1}{x^2 + x - 6}$.

Solution

1. We find the zeros of the denominator by solving $x^2 + x - 6 = 0$. Since

$$x^2 + x - 6 = (x + 3)(x - 2),$$

the zeros are -3 and 2. Thus the domain excludes the x-values -3 and 2 and is

$$(-\infty, -3) \cup (-3, 2) \cup (2, \infty).$$

The graph has vertical asymptotes $x = -3$ and $x = 2$. We sketch these as dashed lines.

2. The numerator and the denominator have the same degree, so the horizontal asymptote is determined by the ratio of the leading coefficients: $1/1$, or 1. Thus $y = 1$ is the horizontal asymptote. We sketch it with a dashed line.

3. To find the zeros of the numerator, we solve $x^2 - 1 = 0$. The solutions are -1 and 1. Thus, -1 and 1 are the zeros of the function and the pairs $(-1, 0)$ and $(1, 0)$ are the x-intercepts.

4. We find $g(0)$:

$$g(0) = \frac{0^2 - 1}{0^2 + 0 - 6} = \frac{-1}{-6} = \frac{1}{6}.$$

Thus, $\left(0, \frac{1}{6}\right)$ is the y-intercept.

5. We find other function values to determine the general shape and then draw the graph.

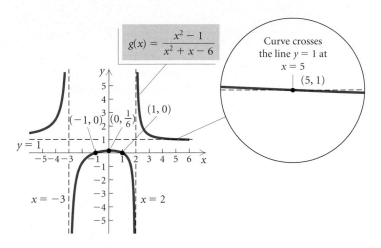

The magnified portion of the graph in Example 9 at left shows another situation in which a graph can cross its horizontal asymptote. The point where $g(x)$ crosses $y = 1$ can be found by setting $g(x) = 1$ and solving for x:

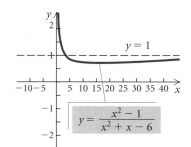

$$\frac{x^2 - 1}{x^2 + x - 6} = 1$$

$$x^2 - 1 = x^2 + x - 6$$

$$-1 = x - 6 \qquad \text{Subtracting } x^2$$

$$5 = x. \qquad \text{Adding 6}$$

The point of intersection is $(5, 1)$. Let's observe the behavior of the curve after it crosses the horizontal asymptote at $x = 5$. (See the graph at left.) It continues to decrease for a short interval and then begins to increase, getting closer and closer to $y = 1$ as $x \to \infty$.

Graphs of rational functions can also cross an oblique asymptote. The graph of

$$f(x) = \frac{2x^3}{x^2 + 1}$$

shown below crosses its oblique asymptote $y = 2x$. **Remember, graphs can cross horizontal or oblique asymptotes, but they cannot cross vertical asymptotes.**

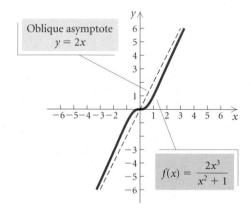

Let's now graph a rational function $f(x) = p(x)/q(x)$, where $p(x)$ and $q(x)$ have a common factor.

EXAMPLE 10   Graph: $g(x) = \dfrac{x - 2}{x^2 - x - 2}$.

Solution   We first express the denominator in factored form:

$$g(x) = \frac{x - 2}{x^2 - x - 2} = \frac{x - 2}{(x + 1)(x - 2)}.$$

The domain of the function is $\{x \mid x \neq -1 \ and \ x \neq 2\}$, or $(-\infty, -1) \cup (-1, 2) \cup (2, \infty)$. The zeros of the denominator are -1 and 2, and the zero of the numerator is 2. Since -1 is the only zero of the denominator that is *not* a zero of the numerator, the graph of the function has $x = -1$ as its only vertical asymptote. The degree of the numerator is less than the degree of the denominator, so $y = 0$ is the horizontal asymptote. There are no zeros of the function and thus no x-intercepts, because 2 is the only zero of the numerator and 2 is not in the domain of the function. Since $g(0) = 1$, $(0, 1)$ is the y-intercept. We draw the graph indicating the "hole" when $x = 2$ with an open circle.

The rational expression $(x - 2)/((x + 1)(x - 2))$ can be simplified. Thus,

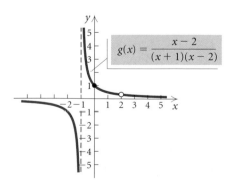

$$g(x) = \frac{x - 2}{(x + 1)(x - 2)} = \frac{1}{x + 1}, \quad \text{where } x \neq -1 \text{ and } x \neq 2.$$

The graph of $g(x)$ is the graph of $y = 1/(x + 1)$ with the point $\left(2, \frac{1}{3}\right)$ missing.

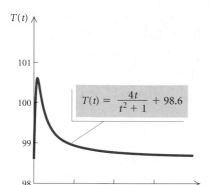

$T(t) = \dfrac{4t}{t^2 + 1} + 98.6$

Technology Connection

The maximum temperature T during the illness described in Example 11 can be found using the MAXIMUM feature in the CALC menu.

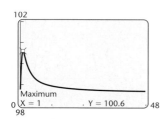

The maximum temperature was 100.6° at $t = 1$ hr.

Applications

EXAMPLE 11 *Temperature During an Illness.* The temperature T, in degrees Fahrenheit, of a person during an illness is given by the function

$$T(t) = \frac{4t}{t^2 + 1} + 98.6,$$

where time t is given in hours since the onset of the illness. The graph of this function is shown at left.

a) Find the temperature at $t = 0, 1, 2, 5, 12,$ and 24.

b) Find the horizontal asymptote of the graph of $T(t)$. Complete:

$$T(t) \rightarrow \boxed{} \text{ as } t \rightarrow \infty.$$

c) Give the meaning of the answer to part (b) in terms of the application.

Solution

a) We have

$$T(0) = 98.6, \qquad T(1) = 100.6, \qquad T(2) = 100.2,$$
$$T(5) \approx 99.369, \qquad T(12) \approx 98.931, \quad \text{and} \quad T(24) \approx 98.766.$$

b) Since

$$T(t) = \frac{4t}{t^2 + 1} + 98.6$$
$$= \frac{98.6t^2 + 4t + 98.6}{t^2 + 1},$$

the horizontal asymptote is $y = 98.6/1$, or 98.6. Then it follows that $T(t) \rightarrow 98.6$ as $t \rightarrow \infty$.

c) As time goes on, the temperature returns to "normal," which is 98.6°.

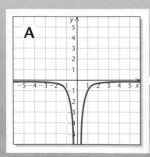

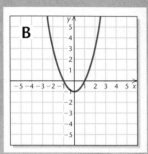

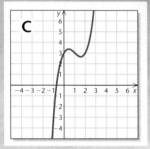

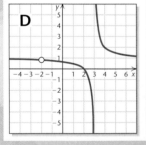

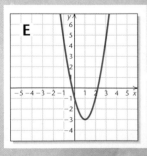

Visualizing the Graph

Match the function with its graph.

1. $f(x) = -\dfrac{1}{x^2}$

2. $f(x) = x^3 - 3x^2 + 2x + 3$

3. $f(x) = \dfrac{x^2 - 4}{x^2 - x - 6}$

4. $f(x) = -x^2 + 4x - 1$

5. $f(x) = \dfrac{x - 3}{x^2 + x - 6}$

6. $f(x) = \dfrac{3}{4}x + 2$

7. $f(x) = x^2 - 1$

8. $f(x) = x^4 - 2x^2 - 5$

9. $f(x) = \dfrac{8x - 4}{3x + 6}$

10. $f(x) = 2x^2 - 4x - 1$

Answers on page A-21

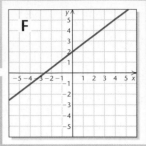

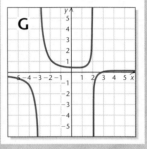

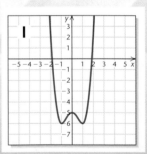

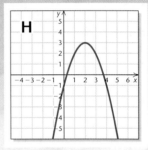

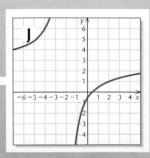

3.4

Exercise Set

In Exercises 1–6, use your knowledge of asymptotes and intercepts to match the equation with one of the graphs (a)–(f), which follow. List all asymptotes.

a)

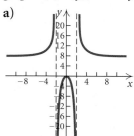

b)

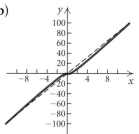

c)

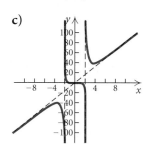

d)

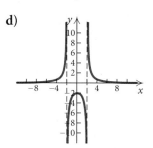

e)

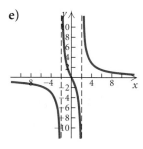

f)

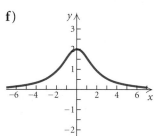

1. $f(x) = \dfrac{8}{x^2 - 4}$

2. $f(x) = \dfrac{8}{x^2 + 4}$

3. $f(x) = \dfrac{8x}{x^2 - 4}$

4. $f(x) = \dfrac{8x^2}{x^2 - 4}$

5. $f(x) = \dfrac{8x^3}{x^2 - 4}$

6. $f(x) = \dfrac{8x^3}{x^2 + 4}$

Determine the vertical asymptotes of the graph of each of the following functions.

7. $g(x) = \dfrac{1}{x^2}$

8. $f(x) = \dfrac{4}{x + 10}$

9. $h(x) = \dfrac{x + 7}{2 - x}$

10. $g(x) = \dfrac{x^4 + 2}{x}$

11. $f(x) = \dfrac{3 - x}{(x - 4)(x + 6)}$

12. $h(x) = \dfrac{x^2 + 4}{x(x + 5)(x - 2)}$

13. $g(x) = \dfrac{x^2}{2x^2 - x - 3}$

14. $f(x) = \dfrac{x + 5}{x^2 + 4x - 32}$

Determine the horizontal asymptote of the graph of each of the following functions.

15. $f(x) = \dfrac{3x^2 + 5}{4x^2 - 3}$

16. $g(x) = \dfrac{x + 6}{x^3 + 2x^2}$

17. $h(x) = \dfrac{x^2 - 4}{2x^4 + 3}$

18. $f(x) = \dfrac{x^5}{x^5 + x}$

19. $g(x) = \dfrac{x^3 - 2x^2 + x - 1}{x^2 - 16}$

20. $h(x) = \dfrac{8x^4 + x - 2}{2x^4 - 10}$

Determine the oblique asymptote of the graph of each of the following functions.

21. $g(x) = \dfrac{x^2 + 4x - 1}{x + 3}$

22. $f(x) = \dfrac{x^2 - 6x}{x - 5}$

23. $h(x) = \dfrac{x^4 - 2}{x^3 + 1}$

24. $g(x) = \dfrac{12x^3 - x}{6x^2 + 4}$

25. $f(x) = \dfrac{x^3 - x^2 + x - 4}{x^2 + 2x - 1}$

26. $h(x) = \dfrac{5x^3 - x^2 + x - 1}{x^2 - x + 2}$

Graph each of the following. Be sure to label all the asymptotes. List the domain and the x- and y-intercepts.

27. $f(x) = \dfrac{1}{x}$ **28.** $g(x) = \dfrac{1}{x^2}$

29. $h(x) = -\dfrac{4}{x^2}$ **30.** $f(x) = -\dfrac{6}{x}$

31. $g(x) = \dfrac{x^2 - 4x + 3}{x + 1}$ **32.** $h(x) = \dfrac{2x^2 - x - 3}{x - 1}$

33. $f(x) = \dfrac{1}{x + 3}$ **34.** $f(x) = \dfrac{1}{x - 5}$

35. $f(x) = \dfrac{-2}{x - 5}$ **36.** $f(x) = \dfrac{3}{3 - x}$

37. $f(x) = \dfrac{2x + 1}{x}$ **38.** $f(x) = \dfrac{3x - 1}{x}$

39. $f(x) = \dfrac{1}{(x - 2)^2}$ **40.** $f(x) = \dfrac{-2}{(x - 3)^2}$

41. $f(x) = -\dfrac{1}{x^2}$ **42.** $f(x) = \dfrac{1}{3x^2}$

43. $f(x) = \dfrac{1}{x^2 + 3}$ **44.** $f(x) = \dfrac{-1}{x^2 + 2}$

45. $f(x) = \dfrac{x^2 - 4}{x - 2}$ **46.** $f(x) = \dfrac{x^2 - 9}{x + 3}$

47. $f(x) = \dfrac{x - 1}{x + 2}$ **48.** $f(x) = \dfrac{x - 2}{x + 1}$

49. $f(x) = \dfrac{x + 3}{2x^2 - 5x - 3}$ **50.** $f(x) = \dfrac{3x}{x^2 + 5x + 4}$

51. $f(x) = \dfrac{x^2 - 9}{x + 1}$

52. $f(x) = \dfrac{x^2 - 4}{x - 1}$

53. $f(x) = \dfrac{x^2 + x - 2}{2x^2 + 1}$

54. $f(x) = \dfrac{x^2 - 2x - 3}{3x^2 + 2}$

55. $g(x) = \dfrac{3x^2 - x - 2}{x - 1}$

56. $f(x) = \dfrac{2x + 1}{2x^2 - 5x - 3}$

57. $f(x) = \dfrac{x - 1}{x^2 - 2x - 3}$

58. $f(x) = \dfrac{x + 2}{x^2 + 2x - 15}$

59. $f(x) = \dfrac{x - 3}{(x + 1)^3}$

60. $f(x) = \dfrac{x + 2}{(x - 1)^3}$

61. $f(x) = \dfrac{x^3 + 1}{x}$

62. $f(x) = \dfrac{x^3 - 1}{x}$

63. $f(x) = \dfrac{x^3 + 2x^2 - 15x}{x^2 - 5x - 14}$

64. $f(x) = \dfrac{x^3 + 2x^2 - 3x}{x^2 - 25}$

65. $f(x) = \dfrac{5x^4}{x^4 + 1}$

66. $f(x) = \dfrac{x + 1}{x^2 + x - 6}$

67. $f(x) = \dfrac{x^2}{x^2 - x - 2}$

68. $f(x) = \dfrac{x^2 - x - 2}{x + 2}$

Find a rational function that satisfies the given conditions for each of the following. Answers may vary, but try to give the simplest answer possible.

69. Vertical asymptotes $x = -4$, $x = 5$

70. Vertical asymptotes $x = -4$, $x = 5$; x-intercept $(-2, 0)$

71. Vertical asymptotes $x = -4$, $x = 5$; horizontal asymptote $y = \frac{3}{2}$; x-intercept $(-2, 0)$

72. Oblique asymptote $y = x - 1$

73. *Medical Dosage.* The function

$$N(t) = \frac{0.8t + 1000}{5t + 4}, \quad t \geq 15$$

gives the body concentration $N(t)$, in parts per million, of a certain dosage of medication after time t, in hours.

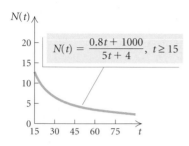

a) Find the horizontal asymptote of the graph and complete the following:

$$N(t) \rightarrow \boxed{} \text{ as } t \rightarrow \infty.$$

b) Explain the meaning of the answer to part (a) in terms of the application.

74. *Average Cost.* The average cost per tape, in dollars, for a company to produce x videotapes on exercising is given by the function

$$A(x) = \frac{2x + 100}{x}, \quad x > 0.$$

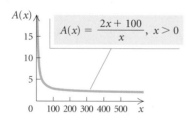

a) Find the horizontal asymptote of the graph and complete the following:

$$A(x) \rightarrow \boxed{} \text{ as } x \rightarrow \infty.$$

b) Explain the meaning of the answer to part (a) in terms of the application.

75. *Population Growth.* The population P, in thousands, of Lordsburg is given by

$$P(t) = \frac{500t}{2t^2 + 9},$$

where t is the time, in months.

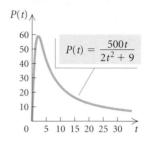

a) Find the population at $t = 0$, 1, 3, and 8 months.

b) Find the horizontal asymptote of the graph and complete the following:

$$P(t) \rightarrow \boxed{} \text{ as } t \rightarrow \infty.$$

c) Explain the meaning of the answer to part (b) in terms of the application.

Technology Connection

76. *Minimizing Surface Area.* The Hold-It Container Co. is designing an open-top rectangular box, with a square base, that will hold 108 cubic centimeters.

a) Express the surface area S as a function of the length x of a side of the base.

b) Graph the function on the interval $(0, \infty)$.

c) Estimate the minimum surface area and the value of x that will yield it.

77. For Exercise 75, find the maximum population and the value of t that will yield it.

78. Graph

$$y_1 = \frac{x^3 + 4}{x} \quad \text{and} \quad y_2 = x^2$$

using the same viewing window. Explain how the parabola $y_2 = x^2$ can be thought of as a nonlinear asymptote for y_1.

Collaborative Discussion and Writing

79. Under what circumstances will a rational function have a domain consisting of all real numbers?

80. Explain why the graph of a rational function cannot have both a horizontal and an oblique asymptote.

Skill Maintenance

In each of Exercises 81–88, fill in the blank with the correct term. Some of the given choices will not be used. Others will be used more than once.

> *x*-intercept
> *y*-intercept
> odd function
> even function
> domain
> range
> slope
> distance formula
> midpoint formula
> horizontal lines
> vertical lines
> point–slope equation
> slope–intercept equation
> difference quotient
> $f(x) = f(-x)$
> $f(-x) = -f(x)$

81. The _____ of a line containing (x_1, y_1) and (x_2, y_2) is given by $(y_2 - y_1)/(x_2 - x_1)$.

82. The _____ of the line with slope m and y-intercept $(0, b)$ is $y = mx + b$.

83. The _____ of the line with slope m passing through (x_1, y_1) is $y - y_1 = m(x - x_1)$.

84. A function is a correspondence between a first set, called the _____, and a second set, called the _____, such that each member of the _____ corresponds to exactly one member of the _____.

85. For each x in the domain of an odd function f, _____.

86. A(n) _____ is a point $(a, 0)$.

87. The _____ is $\left(\dfrac{x_1 + x_2}{2}, \dfrac{y_1 + y_2}{2} \right)$.

88. _____ are given by equations of the type $x = a$.

Synthesis

Find the nonlinear asymptotes of the function.

89. $f(x) = \dfrac{x^5 + 2x^3 + 4x^2}{x^2 + 2}$

90. $f(x) = \dfrac{x^4 + 3x^2}{x^2 + 1}$

Graph the function.

91. $f(x) = \dfrac{2x^3 + x^2 - 8x - 4}{x^3 + x^2 - 9x - 9}$

92. $f(x) = \dfrac{x^3 + 4x^2 + x - 6}{x^2 - x - 2}$

Find the domain.

93. $f(x) = \sqrt{\dfrac{72}{x^2 - 4x - 21}}$

94. $f(x) = \sqrt{x^2 - 4x - 21}$

3.5

Polynomial and Rational Inequalities

• *Solve polynomial and rational inequalities.*

We will use a combination of algebraic and graphical methods to solve polynomial and rational inequalities.

Polynomial Inequalities

Just as a quadratic equation can be written in the form $ax^2 + bx + c = 0$, a **quadratic inequality** can be written in the form $ax^2 + bx + c \ \square \ 0$, where $\square$ is $<$, $>$, $\leq$, or $\geq$. Here are some examples of quadratic inequalities:

$$3x^2 - 2x - 5 > 0, \qquad -\tfrac{1}{2}x^2 + 4x - 7 \leq 0.$$

Quadratic inequalities are one type of **polynomial inequality.** Other examples of polynomial inequalities are

$$-2x^4 + x^2 - 3 < 7, \qquad \tfrac{2}{3}x + 4 \geq 0, \quad \text{and} \quad 4x^3 - 2x^2 > 5x + 7.$$

When the inequality symbol in a polynomial inequality is replaced with an equals sign, a **related equation** is formed. Polynomial inequalities can be easily solved once the related equation has been solved.

EXAMPLE 1 Solve: $x^3 - x > 0$.

Solution We are asked to find all x-values for which $x^3 - x > 0$. To locate these values, we graph $f(x) = x^3 - x$. Then we note that whenever the function changes sign, its graph passes through an x-intercept. Thus to solve $x^3 - x > 0$, we first solve the related equation $x^3 - x = 0$ to find all zeros of the function:

$$x^3 - x = 0$$
$$x(x^2 - 1) = 0$$
$$x(x + 1)(x - 1) = 0.$$

The zeros are -1, 0, and 1. Thus the x-intercepts of the graph are $(-1, 0)$, $(0, 0)$, and $(1, 0)$, as shown in the figure at left. The zeros divide the x-axis into four intervals:

$$(-\infty, -1), \qquad (-1, 0), \qquad (0, 1), \quad \text{and} \quad (1, \infty).$$

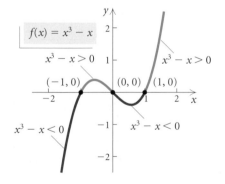

For *all* x-values within a given interval, the sign of $x^3 - x$ must be either *positive* or *negative*. To determine which, we choose a test value for x from each interval and find $f(x)$. We can also determine the sign of $f(x)$ in each interval by simply looking at the graph of the function.

INTERVAL	TEST VALUE	SIGN OF $f(x)$
$(-\infty, -1)$	$f(-2) = -6$	Negative
$(-1, 0)$	$f(-0.5) = 0.375$	Positive
$(0, 1)$	$f(0.5) = -0.375$	Negative
$(1, \infty)$	$f(2) = 6$	Positive

Since we are solving $x^3 - x > 0$, the solution set consists of only two of the four intervals, those in which the sign of $f(x)$ is *positive*. We see that the solution set is $(-1, 0) \cup (1, \infty)$, or $\{x \,|\, -1 < x < 0 \text{ or } x > 1\}$. ▬

To solve a polynomial inequality:

1. Find an equivalent inequality with 0 on one side.
2. Solve the related polynomial equation.
3. Use the solutions to divide the x-axis into intervals. Then select a test value from each interval and determine the polynomial's sign on the interval.
4. Determine the intervals for which the inequality is satisfied and write interval notation or set-builder notation for the solution set. Include the endpoints of the intervals in the solution set if the inequality symbol is $\leq$ or $\geq$.

EXAMPLE 2   Solve: $3x^4 + 10x \leq 11x^3 + 4$.

Solution   By subtracting $11x^3 + 4$, we form the equivalent inequality

$$3x^4 - 11x^3 + 10x - 4 \leq 0.$$

Algebraic Solution

To solve the related equation

$$3x^4 - 11x^3 + 10x - 4 = 0,$$

we need to use the theorems of Section 3.3. We solved this equation in Example 5 in Section 3.3. The solutions are

$$-1, \quad 2 - \sqrt{2}, \quad \tfrac{2}{3}, \quad \text{and} \quad 2 + \sqrt{2},$$

or approximately

$$-1, \quad 0.586, \quad 0.667, \quad \text{and} \quad 3.414.$$

These numbers divide the x-axis into five intervals: $(-\infty, -1)$, $\left(-1, 2 - \sqrt{2}\right)$, $\left(2 - \sqrt{2}, \tfrac{2}{3}\right)$, $\left(\tfrac{2}{3}, 2 + \sqrt{2}\right)$, and $\left(2 + \sqrt{2}, \infty\right)$.

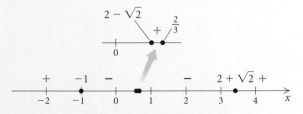

We then let $f(x) = 3x^4 - 11x^3 + 10x - 4$ and, using test values for $f(x)$, determine the sign of $f(x)$ in each interval:

INTERVAL	TEST VALUE	SIGN OF $f(x)$
$(-\infty, -1)$	$f(-2) = 112$	$+$
$\left(-1, 2 - \sqrt{2}\right)$	$f(0) = -4$	$-$
$\left(2 - \sqrt{2}, 2/3\right)$	$f(0.6) = 0.0128$	$+$
$\left(2/3, 2 + \sqrt{2}\right)$	$f(1) = -2$	$-$
$\left(2 + \sqrt{2}, \infty\right)$	$f(4) = 100$	$+$

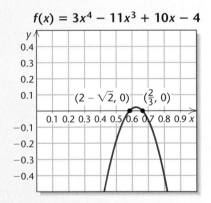

Function values are negative in the intervals $\left(-1, 2 - \sqrt{2}\right)$ and $\left(\tfrac{2}{3}, 2 + \sqrt{2}\right)$. Since the inequality sign is $\leq$, we include the endpoints of the intervals in the solution set. The solution set is

$$\left[-1, 2 - \sqrt{2}\right] \cup \left[\tfrac{2}{3}, 2 + \sqrt{2}\right], \quad \text{or}$$

$$\left\{x \mid -1 \leq x \leq 2 - \sqrt{2} \text{ or } \tfrac{2}{3} \leq x \leq 2 + \sqrt{2}\right\}.$$

Visualizing the Solution

Observing the graph of the function

$$f(x) = 3x^4 - 11x^3 + 10x - 4$$

and a closeup view of the graph on the interval $(0, 1)$, we see the intervals on which $f(x) \leq 0$. The values of $f(x)$ are less than or equal to 0 in two intervals.

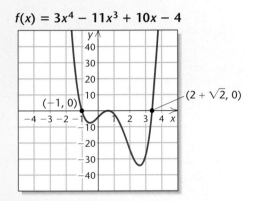

The solution set of the inequality

$$3x^4 - 11x^3 + 10x - 4 \leq 0$$

is

$$\left[-1, 2 - \sqrt{2}\right] \cup \left[\tfrac{2}{3}, 2 + \sqrt{2}\right].$$

Technology Connection

Polynomial inequalities can be solved quickly with a graphing calculator. Consider the inequality in Example 2:

$$3x^4 + 10x \le 11x^3 + 4, \quad \text{or}$$
$$3x^4 - 11x^3 + 10x - 4 \le 0.$$

We graph the related equation

$$y = 3x^4 - 11x^3 + 10x - 4$$

and use the ZERO feature. We see in the window on the left below that two of the zeros are -1 and approximately 3.414 $\left(2 + \sqrt{2} \approx 3.414\right)$. However, this window leaves us uncertain about the number of zeros of the function in the interval $[0, 1]$.

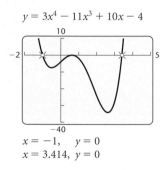

$y = 3x^4 - 11x^3 + 10x - 4$

$x = -1, \quad y = 0$
$x = 3.414, \quad y = 0$

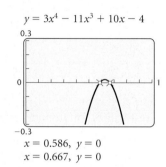

$y = 3x^4 - 11x^3 + 10x - 4$

$x = 0.586, \quad y = 0$
$x = 0.667, \quad y = 0$

The window on the right above shows another view of the zeros in the interval $[0, 1]$. Those zeros are about 0.586 and 0.667 $\left(2 - \sqrt{2} \approx 0.586; \frac{2}{3} \approx 0.667\right)$. The intervals to be considered are $(-\infty, -1)$, $(-1, 0.586)$, $(0.586, 0.667)$, $(0.667, 3.414)$, and $(3.414, \infty)$. We note on the graph where the function is negative. Then, including appropriate endpoints, we find that the solution set is approximately

$$[-1, 0.586] \cup [0.667, 3.414].$$

Rational Inequalities

Some inequalities involve rational expressions and functions. These are called **rational inequalities.** To solve rational inequalities, we need to make some adjustments to the preceding method.

EXAMPLE 3 Solve: $\dfrac{x - 3}{x + 4} \ge \dfrac{x + 2}{x - 5}$.

Solution We first subtract $(x + 2)/(x - 5)$ in order to find an equivalent inequality with 0 on one side:

$$\frac{x - 3}{x + 4} - \frac{x + 2}{x - 5} \ge 0.$$

Algebraic Solution

We look for all values of x for which the related function

$$f(x) = \frac{x-3}{x+4} - \frac{x+2}{x-5}$$

is not defined or is 0. These are called **critical values.**

A look at the denominators shows that $f(x)$ is not defined for $x = -4$ and $x = 5$. Next, we solve $f(x) = 0$:

$$\frac{x-3}{x+4} - \frac{x+2}{x-5} = 0$$

$$(x+4)(x-5)\left(\frac{x-3}{x+4} - \frac{x+2}{x-5}\right) = (x+4)(x-5)\cdot 0$$

$$(x-5)(x-3) - (x+4)(x+2) = 0$$

$$(x^2 - 8x + 15) - (x^2 + 6x + 8) = 0$$

$$-14x + 7 = 0$$

$$x = \tfrac{1}{2}.$$

The critical values are $-4, \tfrac{1}{2}$, and 5. These values divide the x-axis into four intervals:

$$(-\infty, -4), \quad \left(-4, \tfrac{1}{2}\right), \quad \left(\tfrac{1}{2}, 5\right), \quad \text{and} \quad (5, \infty).$$

We then use a test value to determine the sign of $f(x)$ in each interval.

INTERVAL	TEST VALUE	SIGN OF $f(x)$
$(-\infty, -4)$	$f(-5) = 7.7$	$+$
$\left(-4, \tfrac{1}{2}\right)$	$f(-2) = -2.5$	$-$
$\left(\tfrac{1}{2}, 5\right)$	$f(3) = 2.5$	$+$
$(5, \infty)$	$f(6) = -7.7$	$-$

Function values are positive in the intervals $(-\infty, -4)$ and $\left(\tfrac{1}{2}, 5\right)$. Since $f\left(\tfrac{1}{2}\right) = 0$ and the inequality symbol is $\geq$, we know that $\tfrac{1}{2}$ must be in the solution set. Note that since neither -4 nor 5 is in the domain of f, they cannot be part of the solution set.

The solution set is $(-\infty, -4) \cup \left[\tfrac{1}{2}, 5\right)$.

Visualizing the Solution

The graph of the related function

$$f(x) = \frac{x-3}{x+4} - \frac{x+2}{x-5}$$

confirms the three critical values found algebraically: -4 and 5 where $f(x)$ is not defined and $\tfrac{1}{2}$ where $f(x) = 0$.

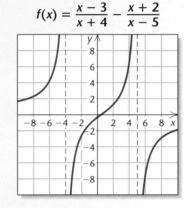

$$f(x) = \frac{x-3}{x+4} - \frac{x+2}{x-5}$$

The graph shows where $f(x)$ is positive and where it is negative. Note that -4 and 5 cannot be in the solution set since y is not defined for these values. We do include $\tfrac{1}{2}$, however, since the inequality symbol is $\geq$ and $f\left(\tfrac{1}{2}\right) = 0$. The solution set is

$$(-\infty, -4) \cup \left[\tfrac{1}{2}, 5\right).$$

The following is a method for solving rational inequalities.

> **To solve a rational inequality:**
>
> 1. Find an equivalent inequality with 0 on one side.
> 2. Change the inequality symbol to an equals sign and solve the related equation.
> 3. Find values of the variable for which the related rational function is not defined.
> 4. The numbers found in steps (2) and (3) are called critical values. Use the critical values to divide the x-axis into intervals. Then test an x-value from each interval to determine the function's sign in that interval.
> 5. Select the intervals for which the inequality is satisfied and write interval notation or set-builder notation for the solution set. If the inequality symbol is $\leq$ or $\geq$, then the solutions to step (2) should be included in the solution set. The x-values found in step (3) are never included in the solution set.

3.5

Exercise Set

For the function
$$f(x) = x^2 + 2x - 15,$$
solve each of the following.

1. $f(x) = 0$ **2.** $f(x) < 0$

3. $f(x) \leq 0$ **4.** $f(x) > 0$

5. $f(x) \geq 0$

For the function
$$g(x) = x^5 - 9x^3,$$
solve each of the following.

6. $g(x) = 0$ **7.** $g(x) < 0$

8. $g(x) \leq 0$ **9.** $g(x) > 0$

10. $g(x) \geq 0$

In Exercises 11–14, a related function is graphed. Solve the given inequality.

11. $x^3 + 6x^2 < x + 30$

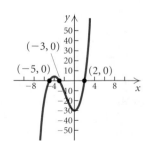

12. $x^4 - 27x^2 - 14x + 120 \geq 0$

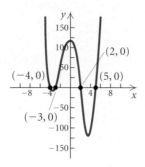

13. $\dfrac{8x}{x^2 - 4} \geq 0$

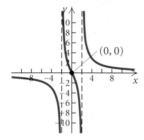

14. $\dfrac{8}{x^2 - 4} < 0$

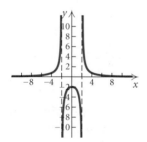

Solve.

15. $(x - 1)(x + 4) < 0$

16. $(x + 3)(x - 5) < 0$

17. $(x - 4)(x + 2) \geq 0$

18. $(x - 2)(x + 1) \geq 0$

19. $x^2 + x - 2 > 0$

20. $x^2 - x - 6 > 0$

21. $x^2 > 25$

22. $x^2 \leq 1$

23. $4 - x^2 \leq 0$

24. $11 - x^2 \geq 0$

25. $6x - 9 - x^2 < 0$

26. $x^2 + 2x + 1 \leq 0$

27. $x^2 + 12 < 4x$

28. $x^2 - 8 > 6x$

29. $4x^3 - 7x^2 \leq 15x$

30. $2x^3 - x^2 < 5x$

31. $x^3 + 3x^2 - x - 3 \geq 0$

32. $x^3 + x^2 - 4x - 4 \geq 0$

33. $x^3 - 2x^2 < 5x - 6$

34. $x^3 + x \leq 6 - 4x^2$

35. $x^5 + x^2 \geq 2x^3 + 2$

36. $x^5 + 24 > 3x^3 + 8x^2$

37. $2x^3 + 6 \leq 5x^2 + x$

38. $2x^3 + x^2 < 10 + 11x$

39. $x^3 + 5x^2 - 25x \leq 125$

40. $x^3 - 9x + 27 \geq 3x^2$

41. $\dfrac{1}{x + 4} > 0$

42. $\dfrac{1}{x - 3} \leq 0$

43. $\dfrac{-4}{2x + 5} < 0$

44. $\dfrac{-2}{5 - x} \geq 0$

45. $\dfrac{x - 4}{x + 3} - \dfrac{x + 2}{x - 1} \leq 0$

46. $\dfrac{x + 1}{x - 2} + \dfrac{x - 3}{x - 1} < 0$

47. $\dfrac{2x - 1}{x + 3} \geq \dfrac{x + 1}{3x + 1}$

48. $\dfrac{x + 5}{x - 4} > \dfrac{3x + 2}{2x + 1}$

49. $\dfrac{x + 1}{x - 2} \geq 3$

50. $\dfrac{x}{x - 5} < 2$

51. $x - 2 > \dfrac{1}{x}$

52. $4 \geq \dfrac{4}{x} + x$

53. $\dfrac{2}{x^2 - 4x + 3} \leq \dfrac{5}{x^2 - 9}$

54. $\dfrac{3}{x^2 - 4} \leq \dfrac{5}{x^2 + 7x + 10}$

55. $\dfrac{3}{x^2 + 1} \geq \dfrac{6}{5x^2 + 2}$

56. $\dfrac{4}{x^2 - 9} < \dfrac{3}{x^2 - 25}$

57. $\dfrac{5}{x^2 + 3x} < \dfrac{3}{2x + 1}$

58. $\dfrac{2}{x^2 + 3} > \dfrac{3}{5 + 4x^2}$

59. $\dfrac{5x}{7x - 2} > \dfrac{x}{x + 1}$

60. $\dfrac{x^2 - x - 2}{x^2 + 5x + 6} < 0$

61. $\dfrac{x}{x^2 + 4x - 5} + \dfrac{3}{x^2 - 25} \leq \dfrac{2x}{x^2 - 6x + 5}$

62. $\dfrac{2x}{x^2 - 9} + \dfrac{x}{x^2 + x - 12} \geq \dfrac{3x}{x^2 + 7x + 12}$

63. *Temperature During an Illness.* The temperature T, in degrees Fahrenheit, of a person during an illness is given by the function

$$T(t) = \dfrac{4t}{t^2 + 1} + 98.6,$$

where t is the time, in hours. Find the interval on which the temperature was over $100°$. (See Example 11 in Section 3.4.)

64. *Population Growth.* The population P, in thousands, of Lordsburg is given by

$$P(t) = \dfrac{500t}{2t^2 + 9},$$

where t is the time, in months. Find the interval on which the population was 40 thousand or greater. (See Exercise 75 in Exercise Set 3.4.)

65. *Total Profit.* Flexl, Inc., determines that its total profit is given by the function

$$P(x) = -3x^2 + 630x - 6000.$$

a) Flexl makes a profit for those nonnegative values of x for which $P(x) > 0$. Find the values of x for which Flexl makes a profit.

b) Flexl loses money for those nonnegative values of x for which $P(x) < 0$. Find the values of x for which Flexl loses money.

66. *Height of a Thrown Object.* The function

$$S(t) = -16t^2 + 32t + 1920$$

gives the height S, in feet, of an object thrown from a cliff that is 1920 ft high. Here t is the time, in seconds, that the object is in the air.

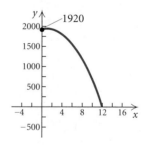

a) For what times is the height greater than 1920 ft?

b) For what times is the height less than 640 ft?

67. *Number of Diagonals.* A polygon with n sides has D diagonals, where D is given by the function

$$D(n) = \dfrac{n(n - 3)}{2}.$$

Find the number of sides n if

$$27 \leq D \leq 230.$$

68. *Number of Handshakes.*   If there are n people in a room, the number N of possible handshakes by all the people in the room is given by the function

$$N(n) = \frac{n(n-1)}{2}.$$

For what number n of people is

$$66 \le N \le 300?$$

Technology Connection

69. Use a graphing calculator and the ZERO feature to check your answers to Exercises 15, 21, 29, 37, and 49.

Collaborative Discussion and Writing

70. Why, when solving rational inequalities, do we need to find values for which the function is undefined as well as zeros of the function?

71. Under what circumstances would a quadratic inequality have a solution set that is a closed interval? Under what circumstances would a quadratic inequality have an empty solution set?

Skill Maintenance

Find an equation for a circle satisfying the given conditions.

72. Center: $(0, -3)$; diameter of length $\frac{7}{2}$

73. Center: $(-2, 4)$; radius of length 3

In Exercises 74 and 75:

a) *Find the vertex.*
b) *Determine whether there is a maximum or minimum value and find that value.*
c) *Find the range.*

74. $g(x) = x^2 - 10x + 2$

75. $h(x) = -2x^2 + 3x - 8$

Synthesis

Solve.

76. $x^4 - 6x^2 + 5 > 0$

77. $x^4 + 3x^2 > 4x - 15$

78. $\left| \dfrac{x+3}{x-4} \right| < 2$

79. $|x^2 - 5| = 5 - x^2$

80. $(7 - x)^{-2} < 0$

81. $2|x|^2 - |x| + 2 \le 5$

82. $\left| 2 - \dfrac{1}{x} \right| \le 2 + \left| \dfrac{1}{x} \right|$

83. $\left| 1 + \dfrac{1}{x} \right| < 3$

84. $|1 + 5x - x^2| \ge 5$

85. $|x^2 + 3x - 1| < 3$

86. Write a polynomial inequality for which the solution set is $[-4, 3] \cup [7, \infty)$.

87. Write a quadratic inequality for which the solution set is $(-4, 3)$.

3.6

Variation and Applications

• *Find equations of direct, inverse, and combined variation given values of the variables.*
• *Solve applied problems involving variation.*

We now extend our study of formulas and functions by considering applications involving variation.

Direct Variation

Suppose an executive chef earns $23 per hour. In 1 hr, $23 is earned; in 2 hr, $46 is earned; in 3 hr, $69 is earned; and so on. This gives rise to a set of ordered pairs of numbers:

$$(1, 23), \qquad (2, 46), \qquad (3, 69), \qquad (4, 92),$$

and so on. Note that the ratio of the second coordinate to the first is the same number for each pair:

$$\frac{23}{1} = 23, \qquad \frac{46}{2} = 23, \qquad \frac{69}{3} = 23, \qquad \frac{92}{4} = 23, \quad \text{and so on.}$$

Whenever a situation produces pairs of numbers in which the *ratio is constant*, we say that there is **direct variation.** Here the amount earned E varies directly as the time worked t:

$$\frac{E}{t} = 23 \text{ (a constant)}, \quad \text{or} \quad E = 23t,$$

or, using function notation, $E(t) = 23t$. This equation is an equation of **direct variation.** The coefficient, 23 in the situation above, is called the **variation constant.** In this case, it is the rate of change of earnings with respect to time.

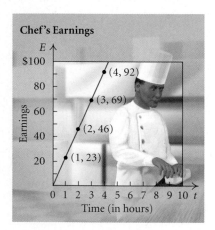

Chef's Earnings

Earnings vs. Time (in hours); points (1, 23), (2, 46), (3, 69), (4, 92).

The graph of $y = kx$, $k > 0$, always goes through the origin and rises from left to right. Note that as x increases, y increases; that is, the function is increasing on the interval $(0, \infty)$. The constant k is also the slope of the line.

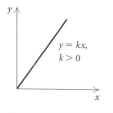

$y = kx$,
$k > 0$

Direct Variation

If a situation gives rise to a linear function $f(x) = kx$, or $y = kx$, where k is a positive constant, we say that we have **direct variation,** or that **y varies directly as x,** or that **y is directly proportional to x.** The number k is called the **variation constant,** or **constant of proportionality.**

EXAMPLE 1 Find the variation constant and an equation of variation in which y varies directly as x, and $y = 32$ when $x = 2$.

Solution We know that $(2, 32)$ is a solution of $y = kx$. Thus,

$$y = kx$$
$$32 = k \cdot 2 \qquad \textbf{Substituting}$$
$$\frac{32}{2} = k, \text{ or } k = 16. \qquad \textbf{Solving for } k$$

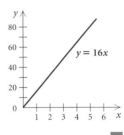

The variation constant, 16, is the rate of change of y with respect to x. The equation of variation is $y = 16x$.

EXAMPLE 2 *Water from Melting Snow.* The number of centimeters W of water produced from melting snow varies directly as S, the number of centimeters of snow. Meteorologists have found that 150 cm of snow will melt to 16.8 cm of water. To how many centimeters of water will 200 cm of snow melt?

Solution We can express the amount of water as a function of the amount of snow. Thus, $W(S) = kS$, where k is the variation constant. We first find k using the given data and then find an equation of variation:

$$W(S) = kS \qquad \textbf{\textit{W} varies directly as \textit{S}.}$$
$$W(150) = k \cdot 150 \qquad \textbf{Substituting 150 for \textit{S}}$$
$$16.8 = k \cdot 150 \qquad \textbf{Replacing \textit{W}(150) with 16.8}$$
$$\frac{16.8}{150} = k \qquad \textbf{Solving for \textit{k}}$$
$$0.112 = k. \qquad \textbf{This is the variation constant.}$$

The equation of variation is $W(S) = 0.112S$.

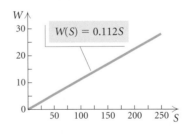

Next, we use the equation to find how many centimeters of water will result from melting 200 cm of snow:

$$W(S) = 0.112S$$
$$W(200) = 0.112(200) \qquad \textbf{Substituting}$$
$$= 22.4.$$

Thus, 200 cm of snow will melt to 22.4 cm of water.

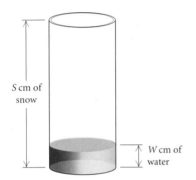

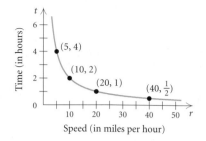

Time (in hours)

(5, 4)

(10, 2)

(20, 1)

$(40, \frac{1}{2})$

Speed (in miles per hour)

Inverse Variation

Suppose a bus is traveling a distance of 20 mi. At a speed of 5 mph, the trip will take 4 hr; at 10 mph, it will take 2 hr; at 20 mph, it will take 1 hr; at 40 mph, it will take $\frac{1}{2}$ hr; and so on. We plot this information on a graph, using speed as the first coordinate and time as the second coordinate to determine a set of ordered pairs:

$$(5, 4), \qquad (10, 2), \qquad (20, 1), \qquad \left(40, \tfrac{1}{2}\right), \quad \text{and so on.}$$

Note that the products of the coordinates are all the same number:

$$5 \cdot 4 = 20, \qquad 10 \cdot 2 = 20, \qquad 20 \cdot 1 = 20, \qquad 40 \cdot \tfrac{1}{2} = 20, \quad \text{and so on.}$$

Whenever a situation produces pairs of numbers in which the *product is constant*, we say that there is **inverse variation.** Here the time varies inversely as the speed:

$$rt = 20 \ (\text{a constant}), \quad \text{or} \quad t = \frac{20}{r},$$

or, using function notation, $t(r) = 20/r$. This equation is an equation of **inverse variation.** The coefficient, 20 in the situation above, is called the **variation constant.** Note that as the first number increases, the second number decreases.

It is helpful to look at the graph of $y = k/x, k > 0$. The graph is like the one shown below for positive values of x. Note that as x increases, y decreases; that is, the function is decreasing on the interval $(0, \infty)$.

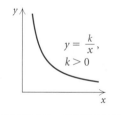

$y = \dfrac{k}{x},$
$k > 0$

Inverse Variation

If a situation gives rise to a function $f(x) = k/x$, or $y = k/x$, where k is a positive constant, we say that we have **inverse variation,** or that **y varies inversely as x,** or that **y is inversely proportional to x.** The number k is called the **variation constant,** or **constant of proportionality.**

EXAMPLE 3 Find the variation constant and an equation of variation in which y varies inversely as x, and $y = 16$ when $x = 0.3$.

Solution   We know that $(0.3, 16)$ is a solution of $y = k/x$. We substitute:

$$y = \frac{k}{x}$$

$$16 = \frac{k}{0.3} \qquad \text{Substituting}$$

$$(0.3)16 = k \qquad \text{Solving for } k$$

$$4.8 = k.$$

$y = \dfrac{4.8}{x}$

The variation constant is 4.8. The equation of variation is $y = 4.8/x$.

There are many problems that translate to an equation of inverse variation.

EXAMPLE 4 *Framing a House.* The time t required to do a job varies inversely as the number of people P who work on the job (assuming that all work at the same rate). If it takes 72 hr for 9 people to frame a house, how long will it take 12 people to complete the same job?

Solution We can express the amount of time required, in hours, as a function of the number of people working. Thus we have $t(P) = k/P$. We first find k using the given information and then find an equation of variation:

$$t(P) = \frac{k}{P} \qquad \text{t varies inversely as P.}$$

$$t(9) = \frac{k}{9} \qquad \text{Substituting 9 for P}$$

$$72 = \frac{k}{9} \qquad \text{Replacing $t(9)$ with 72}$$

$$9 \cdot 72 = k \qquad \text{Solving for k}$$

$$648 = k. \qquad \text{This is the variation constant.}$$

The equation of variation is $t(P) = 648/P$.

Next, we use the equation to find the time that it would take 12 people to do the job. We compute $t(12)$:

$$t(P) = \frac{648}{P}$$

$$t(12) = \frac{648}{12} \qquad \text{Substituting}$$

$$t = 54.$$

Thus it would take 54 hr for 12 people to complete the job.

Combined Variation

We now look at other kinds of variation.

> y varies **directly as the nth power of x** if there is some positive constant k such that
>
> $$y = kx^n.$$
>
> y varies **inversely as the nth power of x** if there is some positive constant k such that
>
> $$y = \frac{k}{x^n}.$$
>
> y varies **jointly as x and z** if there is some positive constant k such that
>
> $$y = kxz.$$

There are other types of combined variation as well. Consider the formula for the volume of a right circular cylinder, $V = \pi r^2 h$, in which V, r, and h are variables and π is a constant. We say that V varies jointly as h and the square of r.

EXAMPLE 5 Find an equation of variation in which y varies directly as the square of x, and $y = 12$ when $x = 2$.

Solution We write an equation of variation and find k:

$$y = kx^2$$
$$12 = k \cdot 2^2 \qquad \text{Substituting}$$
$$12 = k \cdot 4$$
$$3 = k.$$

Thus, $y = 3x^2$.

EXAMPLE 6 Find an equation of variation in which y varies jointly as x and z, and $y = 42$ when $x = 2$ and $z = 3$.

Solution We have

$$y = kxz$$
$$42 = k \cdot 2 \cdot 3 \qquad \text{Substituting}$$
$$42 = k \cdot 6$$
$$7 = k.$$

Thus, $y = 7xz$.

EXAMPLE 7 Find an equation of variation in which y varies jointly as x and z and inversely as the square of w, and $y = 105$ when $x = 3$, $z = 20$, and $w = 2$.

Solution We have

$$y = k \cdot \frac{xz}{w^2}$$
$$105 = k \cdot \frac{3 \cdot 20}{2^2} \qquad \text{Substituting}$$
$$105 = k \cdot 15$$
$$7 = k.$$

Thus, $y = 7\dfrac{xz}{w^2}$.

Many applied problems can be modeled using equations of combined variation.

EXAMPLE 8 *Volume of a Tree.* The volume of wood V in a tree varies jointly as the height h and the square of the girth g (girth is distance around). If the volume of a redwood tree is 216 m³ when the height is 30 m and the girth is 1.5 m, what is the height of a tree whose volume is 960 m^3 and girth is 2 m?

Solution   We first find k using the first set of data. Then we solve for h using the second set of data.

$$V = khg^2$$
$$216 = k \cdot 30 \cdot 1.5^2$$
$$216 = k \cdot 30 \cdot 2.25$$
$$216 = k \cdot 67.5$$
$$3.2 = k$$

Then the equation of variation is $V = 3.2hg^2$. We substitute the second set of data into the equation:

$$960 = 3.2 \cdot h \cdot 2^2$$
$$960 = 12.8 \cdot h$$
$$75 = h.$$

Thus the height of the tree is 75 m.

3.6

Exercise Set

Find the variation constant and an equation of variation for the given situation.

1. y varies directly as x, and $y = 54$ when $x = 12$

2. y varies directly as x, and $y = 0.1$ when $x = 0.2$

3. y varies inversely as x, and $y = 3$ when $x = 12$

4. y varies inversely as x, and $y = 12$ when $x = 5$

5. y varies directly as x, and $y = 1$ when $x = \frac{1}{4}$

6. y varies inversely as x, and $y = 0.1$ when $x = 0.5$

7. y varies inversely as x, and $y = 32$ when $x = \frac{1}{8}$

8. y varies directly as x, and $y = 3$ when $x = 33$

9. y varies directly as x, and $y = \frac{3}{4}$ when $x = 2$

10. y varies inversely as x, and $y = \frac{1}{5}$ when $x = 35$

11. y varies inversely as x, and $y = 1.8$ when $x = 0.3$

12. y varies directly as x, and $y = 0.9$ when $x = 0.4$

13. *Work Rate.*   The time T required to do a job varies inversely as the number of people P working. It takes 5 hr for 7 bricklayers to build a park wall (see the graph below). How long will it take 10 bricklayers to complete the job?

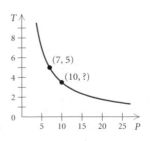

14. *Weekly Allowance.* According to Fidelity Investments *Investment Vision Magazine*, the average weekly allowance *A* of children varies directly as their grade level *G*. It is known that the average allowance of a 9th-grade student is $9.66 per week. What then is the average allowance of a 4th-grade student?

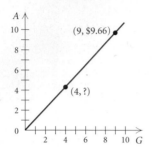

15. *Fat Intake.* The maximum number of grams of fat that should be in a diet varies directly as a person's weight. A person weighing 120 lb should have no more than 60 g of fat per day. What is the maximum daily fat intake for a person weighing 180 lb?

16. *Rate of Travel.* The time *t* required to drive a fixed distance varies inversely as the speed *r*. It takes 5 hr at a speed of 80 km/h to drive a fixed distance. How long will it take to drive the same distance at a speed of 70 km/h?

17. *Beam Weight.* The weight *W* that a horizontal beam can support varies inversely as the length *L* of the beam. Suppose an 8-m beam can support 1200 kg. How many kilograms can a 14-m beam support?

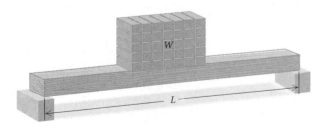

18. *House of Representatives.* The number of representatives *N* that each state has varies directly as the number of people *P* living in the state. If New York, with 19,011,000 residents, has 29 representatives, how many representatives does Colorado, with a population of 4,418,000, have?

19. *Weight on Mars.* The weight *M* of an object on Mars varies directly as its weight *E* on Earth. A person who weighs 95 lb on Earth weighs 38 lb on Mars. How much would a 100-lb person weigh on Mars?

20. *Pumping Rate.* The time *t* required to empty a tank varies inversely as the rate *r* of pumping. If a pump can empty a tank in 45 min at the rate of 600 kL/min, how long will it take the pump to empty the same tank at the rate of 1000 kL/min?

21. *Hooke's Law.* Hooke's law states that the distance *d* that a spring will stretch varies directly as the mass *m* of an object hanging from the spring. If a 3-kg mass stretches a spring 40 cm, how far will a 5-kg mass stretch the spring?

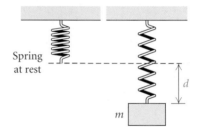

22. *Relative Aperture.* The relative aperture, or f-stop, of a 23.5-mm diameter lens is directly proportional to the focal length *F* of the lens. If a 150-mm focal length has an f-stop of 6.3, find the f-stop of a 23.5-mm diameter lens with a focal length of 80 mm.

23. *Musical Pitch.* The pitch *P* of a musical tone varies inversely as its wavelength *W*. One tone has a pitch of 330 vibrations per second and a wavelength of 3.2 ft. Find the wavelength of another tone that has a pitch of 550 vibrations per second.

24. *Recycling Rechargeable Batteries.* The Indiana Household Hazardous Waste Task Force was awarded the 2002 National Community Recycling Leadership Award, with special recognition given to Monroe County Solid Waste Management District. Monroe County, Indiana, with a population of 9880, collected 4445 lb of rechargeable batteries in a recent recycling effort. (*Source*: Rechargeable Battery Recycling Corporation) If the number of pounds collected varies directly as the population, how many pounds of rechargeable batteries could a county with a population of 74,650 collect for recycling?

Find an equation of variation for the given situation.

25. y varies inversely as the square of x, and $y = 0.15$ when $x = 0.1$.

26. y varies inversely as the square of x, and $y = 6$ when $x = 3$.

27. y varies directly as the square of x, and $y = 0.15$ when $x = 0.1$.

28. y varies directly as the square of x, and $y = 6$ when $x = 3$.

29. y varies jointly as x and z, and $y = 56$ when $x = 7$ and $z = 8$.

30. y varies directly as x and inversely as z, and $y = 4$ when $x = 12$ and $z = 15$.

31. y varies jointly as x and the square of z, and $y = 105$ when $x = 14$ and $z = 5$.

32. y varies jointly as x and z and inversely as w, and $y = \frac{3}{2}$ when $x = 2$, $z = 3$, and $w = 4$.

33. y varies jointly as x and z and inversely as the product of w and p, and $y = \frac{3}{28}$ when $x = 3$, $z = 10$, $w = 7$, and $p = 8$.

34. y varies jointly as x and z and inversely as the square of w, and $y = \frac{12}{5}$ when $x = 16$, $z = 3$, and $w = 5$.

35. *Intensity of Light.* The intensity I of light from a light bulb varies inversely as the square of the distance d from the bulb. Suppose that I is 90 W/m² (watts per square meter) when the distance is 5 m. How much *farther* would it be to a point where the intensity is 40 W/m²?

36. *Atmospheric Drag.* Wind resistance, or atmospheric drag, tends to slow down moving objects.

Atmospheric drag varies jointly as an object's surface area A and velocity v. If a car traveling at a speed of 40 mph with a surface area of 37.8 ft² experiences a drag of 222 N (Newtons), how fast must a car with 51 ft² of surface area travel in order to experience a drag force of 430 N?

37. *Stopping Distance of a Car.* The stopping distance d of a car after the brakes have been applied varies directly as the square of the speed r. If a car traveling 60 mph can stop in 200 ft, how fast can a car travel and still stop in 72 ft?

38. *Weight of an Astronaut.* The weight W of an object varies inversely as the square of the distance d from the center of the earth. At sea level (3978 mi from the center of the earth), an astronaut weighs 220 lb. Find his weight when he is 200 mi above the surface of the earth and the spacecraft is not in motion.

39. *Earned-Run Average.* A pitcher's earned-run average E varies directly as the number R of earned runs allowed and inversely as the number I of innings pitched. In a recent year, Kevin Brown of the Los Angeles Dodgers had an earned-run average of 2.24. He gave up 28 earned runs in $112\frac{1}{3}$ innings. How many earned runs would he have given up had he pitched 300 innings with the same average? Round to the nearest whole number.

40. *Boyle's Law.* The volume V of a given mass of a gas varies directly as the temperature T and inversely as the pressure P. If $V = 231$ cm³ when $T = 42°$ and $P = 20$ kg/cm², what is the volume when $T = 30°$ and $P = 15$ kg/cm²?

Collaborative Discussion and Writing

41. If y varies directly as x^2, explain why doubling x would not cause y to be doubled as well.

42. If y varies directly as x and x varies inversely as z, how does y vary with regard to z? Why?

Skill Maintenance

43. Graph: $f(x) = \begin{cases} x - 2, & \text{for } x \le -1, \\ 3, & \text{for } -1 < x \le 2, \\ x, & \text{for } x > 2. \end{cases}$

Determine algebraically whether the graph is symmetric with respect to the x-axis, the y-axis, and the origin.

44. $y = 3x^4 - 3$

45. $y^2 = x$

46. $2x - 5y = 0$

Synthesis

47. *Volume and Cost.* An 18-oz jar of peanut butter in the shape of a right circular cylinder is 5 in. high and 3 in. in diameter and sells for $1.80. For the same brand in the same store, a 28-oz jar that is $5\frac{1}{2}$ in. high and $3\frac{1}{4}$ in. in diameter sells for $3.20. If we assume that cost is directly proportional to volume, what should the price of the larger jar be?

If we assume that cost is directly proportional to weight, what should the price of the larger jar be?

48. In each of the following equations, state whether y varies directly as x, inversely as x, or neither directly nor inversely as x.

a) $7xy = 14$

b) $x - 2y = 12$

c) $-2x + 3y = 0$

d) $x = \dfrac{3}{4} y$

e) $\dfrac{x}{y} = 2$

49. *Area of a Circle.* The area of a circle varies directly as the square of the length of a diameter. What is the variation constant?

50. Describe in words the variation given by the equation

$$Q = \frac{kp^2}{q^3}.$$

Chapter 3 Summary and Review

Important Properties and Formulas

Polynomial Function:

$$P(x) = a_n x^n + a_{n-1}x^{n-1} + a_{n-2}x^{n-2} + \cdots \\ + a_1 x + a_0$$

The Leading-Term Test: If $a_n x^n$ is the leading term of a polynomial function, then the behavior of the graph as $x \to \infty$ and as $x \to -\infty$ can be described in one of the four following ways.

If n is even,
and $a_n > 0$:

If n is even,
and $a_n < 0$:

If n is odd,
and $a_n > 0$:

If n is odd,
and $a_n < 0$:

The Intermediate Value Theorem: For any polynomial function $P(x)$ with real coefficients, suppose that $a \ne b$ and that $P(a)$ and $P(b)$ are of opposite signs. Then the function has a real zero between a and b.

Polynomial Division:

$$P(x) = D(x) \cdot Q(x) + R(x)$$

Dividend   Divisor   Quotient   Remainder

The Remainder Theorem: The remainder found by dividing $P(x)$ by $x - c$ is $P(c)$.

The Factor Theorem: For a polynomial $f(x)$, if $f(c) = 0$, then $x - c$ is a factor of $f(x)$.

The Fundamental Theorem of Algebra: Every polynomial of degree n, $n \geq 1$, with complex coefficients has at least one complex-number zero.

The Rational Zeros Theorem: Consider the polynomial equation

$$a_n x^n + a_{n-1} x^{n-1} + a_{n-2} x^{n-2} + \cdots + a_1 x + a_0 = 0,$$

where all the coefficients are integers and $n \geq 1$. Also, consider a rational number p/q, where p and q have no common factor other than -1 and 1. If p/q is a solution of the polynomial equation, then p is a factor of a_0 and q is a factor of a_n.

Descartes' Rule of Signs

Let $P(x)$ be a polynomial function with real coefficients and a nonzero constant term. The number of positive real zeros of $P(x)$ is either:

1. The same as the number of variations of sign in $P(x)$, or
2. Less than the number of variations of sign in $P(x)$ by a positive even integer.

The number of negative real zeros of $P(x)$ is either:

3. The same as the number of variations of sign in $P(-x)$, or
4. Less than the number of variations of sign in $P(-x)$ by a positive even integer.

A zero of multiplicity m must be counted m times.

Rational Function:

$$f(x) = \frac{p(x)}{q(x)},$$

where $p(x)$ and $q(x)$ are polynomials and where $q(x)$ is not the zero polynomial. The domain of $f(x)$ consists of all x for which $q(x) \neq 0$.

Occurrence of Lines as Asymptotes

For a rational function $p(x)/q(x)$, where $p(x)$ and $q(x)$ have no common factors other than constants:

Vertical asymptotes occur at any x-values that make the denominator 0.

The x-axis is the horizontal asymptote when the degree of the numerator is less than the degree of the denominator.

A horizontal asymptote other than the x-axis occurs when the numerator and the denominator have the same degree.

An oblique asymptote occurs when the degree of the numerator is 1 greater than the degree of the denominator.

Variation

Direct: $y = kx$

Inverse: $y = \dfrac{k}{x}$

Joint: $y = kxz$

Review Exercises

Classify the polynomial as constant, linear, quadratic, cubic, or quartic and determine the leading term, the leading coefficient, and the degree of the polynomial.

1. $f(x) = 7x^2 - 5 + 0.45x^4 - 3x^3$

2. $h(x) = -25$

3. $g(x) = 6 - 0.5x$

4. $f(x) = \frac{1}{3}x^3 - 2x + 3$

Use the leading-term test to describe the end behavior of the graph of the function.

5. $f(x) = -\frac{1}{2}x^4 + 3x^2 + x - 6$

6. $f(x) = x^5 + 2x^3 - x^2 + 5x + 4$

Find the zeros of the polynomial function and state the multiplicity of each.

7. $g(x) = \left(x - \frac{2}{3}\right)(x + 2)^3(x - 5)^2$

8. $f(x) = x^4 - 26x^2 + 25$

9. $h(x) = x^3 + 4x^2 - 9x - 36$

Sketch the graph of each polynomial function.

10. $f(x) = -x^4 + 2x^3$

11. $g(x) = (x - 1)^3(x + 2)^2$

12. $h(x) = x^3 + 3x^2 - x - 3$

13. $f(x) = x^4 - 5x^3 + 6x^2 + 4x - 8$

14. $g(x) = 2x^3 + 7x^2 - 14x + 5$

Using the intermediate value theorem, determine, if possible, whether the function f has a zero between a and b.

15. $f(x) = 4x^2 - 5x - 3;\ a = 1, b = 2$

16. $f(x) = x^3 - 4x^2 + \frac{1}{2}x + 2;\ a = 0, b = 1$

17. *Interest Compounded Annually.* When P dollars is invested at interest rate i, compounded annually, for t years, the investment grows to A dollars, where
$$A = P(1 + i)^t.$$
a) Find the interest rate i if $6250 grows to $6760 in 2 yr.

b) Find the interest rate i if $1,000,000 grows to $1,215,506.25 in 4 yr.

In each of the following, a polynomial P(x) and a divisor d(x) are given. Use long division to find the quotient Q(x) and the remainder R(x) when P(x) is divided by d(x), and express P(x) in the form $d(x) \cdot Q(x) + R(x)$.

18. $P(x) = 6x^3 - 2x^2 + 4x - 1,$
$d(x) = x - 3$

19. $P(x) = x^4 - 2x^3 + x + 5,$
$d(x) = x + 1$

Use synthetic division to find the quotient and the remainder.

20. $(x^3 + 2x^2 - 13x + 10) \div (x - 5)$

21. $(x^4 + 3x^3 + 3x^2 + 3x + 2) \div (x + 2)$

22. $(x^5 - 2x) \div (x + 1)$

Use synthetic division to find the indicated function value.

23. $f(x) = x^3 + 2x^2 - 13x + 10;\ f(5)$

24. $f(x) = x^4 - 16;\ f(-2)$

25. $f(x) = x^5 - 4x^4 + x^3 - x^2 + 2x - 100;\ f(-10)$

Using synthetic division, determine whether the numbers are zeros of the polynomial function.

26. $-i, -5;\ f(x) = x^3 - 5x^2 + x - 5$

27. $-1, -2;\ f(x) = x^4 - 4x^3 - 3x^2 + 14x - 8$

28. $\frac{1}{3}, 1;\ f(x) = x^3 - \frac{4}{3}x^2 - \frac{5}{3}x + \frac{2}{3}$

29. $2, -\sqrt{3};\ f(x) = x^4 - 5x^2 + 6$

Factor the polynomial f(x). Then solve the equation $f(x) = 0$.

30. $f(x) = x^3 + 2x^2 - 7x + 4$

31. $f(x) = x^3 + 4x^2 - 3x - 18$

32. $f(x) = x^4 - 4x^3 - 21x^2 + 100x - 100$

33. $f(x) = x^4 - 3x^2 + 2$

Find a polynomial function of degree 3 with the given numbers as zeros.

34. $-4, -1, 2$

35. $-3, 1 - i, 1 + i$

36. $\frac{1}{2}, 1 - \sqrt{2}, 1 + \sqrt{2}$

37. Find a polynomial function of degree 4 with -5 as a zero of multiplicity 3 and $\frac{1}{2}$ as a zero of multiplicity of 1.

38. Find a polynomial function of degree 5 with -3 as a zero of multiplicity 2, 2 as a zero of multiplicity 1, and 0 as a zero of multiplicity 2.

Suppose that a polynomial function of degree 5 with rational coefficients has the given zeros. Find the other zero(s).

39. $-\frac{2}{3}, \sqrt{5}, i$

40. $0, 1 + \sqrt{3}, -\sqrt{3}$

41. $-\sqrt{2}, \frac{1}{2}, 1, 2$

Find a polynomial function of lowest degree with rational coefficients and the following as some of its zeros.

42. $\sqrt{11}$

43. $-i, 6$

44. $-1, 4, 1 + i$

45. $\sqrt{5}, -2i$

46. $\frac{1}{3}, 0, -3$

List all possible rational zeros.

47. $h(x) = 4x^5 - 2x^3 + 6x - 12$

48. $g(x) = 3x^4 - x^3 + 5x^2 - x + 1$

49. $f(x) = x^3 - 2x^2 + x - 24$

For each polynomial function:

a) *Find the rational zeros and then the other zeros; that is, solve $f(x) = 0$.*

b) *Factor $f(x)$ into linear factors.*

50. $f(x) = 3x^5 + 2x^4 - 25x^3 - 28x^2 + 12x$

51. $f(x) = x^3 - 2x^2 - 3x + 6$

52. $f(x) = x^4 - 6x^3 + 9x^2 + 6x - 10$

53. $f(x) = x^3 + 3x^2 - 11x - 5$

54. $f(x) = 3x^3 - 8x^2 + 7x - 2$

55. $f(x) = x^5 - 8x^4 + 20x^3 - 8x^2 - 32x + 32$

56. $f(x) = x^6 + x^5 - 28x^4 - 16x^3 + 192x^2$

57. $f(x) = 2x^5 - 13x^4 + 32x^3 - 38x^2 + 22x - 5$

What does Descartes' rule of signs tell you about the number of positive real zeros and the number of negative real zeros of each of the following polynomial functions?

58. $f(x) = 2x^6 - 7x^3 + x^2 - x$

59. $h(x) = -x^8 + 6x^5 - x^3 + 2x - 2$

60. $g(x) = 5x^5 - 4x^2 + x - 1$

Graph each of the following. Be sure to label all the asymptotes. List the domain and the x- and y-intercepts.

61. $f(x) = \dfrac{x^2 - 5}{x + 2}$

62. $f(x) = \dfrac{5}{(x - 2)^2}$

63. $f(x) = \dfrac{x^2 + x - 6}{x^2 - x - 20}$

64. $f(x) = \dfrac{x - 2}{x^2 - 2x - 15}$

In Exercises 65 and 66, find a rational function that satisfies the given conditions. Answers may vary, but try to give the simplest answer possible.

65. Vertical asymptotes $x = -2, x = 3$

66. Vertical asymptotes $x = -2, x = 3$; horizontal asymptote $y = 4$; x-intercept $(-3, 0)$

67. *Medical Dosage.* The function

$$N(t) = \frac{0.7t + 2000}{8t + 9}, \quad t \geq 5$$

gives the body concentration $N(t)$, in parts per million, of a certain dosage of medication after time t, in hours.

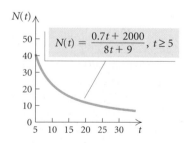

a) Find the horizontal asymptote of the graph and complete the following:

$$N(t) \rightarrow \boxed{} \text{ as } t \rightarrow \infty.$$

b) Explain the meaning of the answer to part (a) in terms of the application.

Solve.

68. $x^2 - 9 < 0$

69. $2x^2 > 3x + 2$

70. $(1 - x)(x + 4)(x - 2) \leq 0$

71. $\dfrac{x - 2}{x + 3} < 4$

72. *Height of a Rocket.* The function

$$S(t) = -16t^2 + 80t + 224$$

gives the height S, in feet, of a model rocket launched with a velocity of 80 ft/sec from a hill that is 224 ft high, where t is the time, in seconds.

a) Determine when the rocket reaches the ground.
b) On what interval is the height greater than 320 ft?

73. *Population Growth.* The population P, in thousands, of Novi is given by

$$P(t) = \frac{8000t}{4t^2 + 10},$$

where t is the time, in months. Find the interval on which the population was 400,000 or greater.

74. Find an equation of variation in which y varies directly as x, and $y = 100$ when $x = 25$.

75. Find an equation of variation in which y varies inversely as x, and $y = 100$ when $x = 25$.

76. *Pumping Time.* The time t required to empty a tank varies inversely as the rate r of pumping. If a pump can empty a tank in 35 min at the rate of 800 kL per minute, how long will it take the pump to empty the same tank at the rate of 1400 kL per minute?

77. *Power of Electric Current.* The power P expended by heat in an electric circuit of fixed resistance varies directly as the square of the current C in the circuit. A circuit expends 180 watts when a current of 6 amperes is flowing. What is the amount of heat expended when the current is 10 amperes?

78. *Test Score.* The score N on a test varies directly as the number of correct responses a. Ellen answers

28 questions correctly and earns a score of 87. What would Ellen's score have been if she had answered 25 questions correctly?

Find an equation of variation for the given situation.

79. y varies inversely as the square of x, and $y = 12$ when $x = 2$.

80. y varies jointly as x and the square of z and inversely as w, and $y = 2$ when $x = 16$, $w = 0.2$, and $z = \frac{1}{2}$.

Technology Connection

Use a graphing calculator to graph the polynomial function. Then estimate the function's **(a)** *zeros,* **(b)** *relative maxima,* **(c)** *relative minima, and* **(d)** *domain and range.*

81. $f(x) = -2x^2 - 3x + 6$

82. $f(x) = x^3 + 3x^2 - 2x - 6$

83. $f(x) = x^4 - 3x^3 + 2x^2$

84. *Cholesterol Level and the Risk of Heart Attack.* The data in the following table show the relationship of cholesterol level in men to the risk of a heart attack.

Cholesterol Level	Number of Men per 100,000, Who Suffer a Heart Attack
100	30
200	65
250	100
275	130

Source: Nutrition Action Healthletter

a) Use regression on a graphing calculator to fit linear, quadratic, and cubic functions to the data.
b) It is also known that 180 of 10,000 men with a cholesterol level of 300 have a heart attack. Which function in part (a) would best make this prediction?
c) Use the answer to part (b) to predict the heart attack rate for men with cholesterol levels of 350 and of 400.

Collaborative Discussion and Writing

85. Explain the difference between a polynomial function and a rational function.

86. Explain and contrast the three types of asymptotes considered for rational functions.

Synthesis

87. *Interest Rate.* In early 2000, $3500 was deposited at a certain interest rate. One year later, $4000 was deposited in another account at the same rate. At the end of that year, there was a total of $8518.35 in both accounts. What is the annual interest rate?

Solve.

88. $x^2 \geq 5 - 2x$

89. $\left| 1 - \dfrac{1}{x^2} \right| < 3$

90. $x^4 - 2x^3 + 3x^2 - 2x + 2 = 0$

91. $(x - 2)^{-3} < 0$

92. Express $x^3 - 1$ as a product of linear factors.

93. Find k such that $x + 3$ is a factor of $x^3 + kx^2 + kx - 15$.

94. When $x^2 - 4x + 3k$ is divided by $x + 5$, the remainder is 33. Find the value of k.

Find the domain of the function.

95. $f(x) = \sqrt{x^2 + 3x - 10}$

96. $f(x) = \sqrt{x^2 - 3.1x + 2.2} + 1.75$

97. $f(x) = \dfrac{1}{\sqrt{5 - |7x + 2|}}$

Chapter 3 Test

Classify the polynomial as constant, linear, quadratic, cubic, or quartic and determine the leading term, the leading coefficient, and the degree of the polynomial.

1. $f(x) = 2x^3 + 6x^2 - x^4 + 11$

2. $h(x) = -4.7x + 29$

3. Find the zeros of the polynomial function and state the multiplicity of each:
$$f(x) = x(3x - 5)(x - 3)^2(x + 1)^3.$$

Sketch the graph of each polynomial function.

4. $f(x) = x^3 - 5x^2 + 2x + 8$

5. $f(x) = -2x^4 + 3x^3 + 3x^2 - 6x + 3$

Using the intermediate value theorem, determine, if possible, whether the function has a zero between a and b.

6. $f(x) = -5x^2 + 3$; $a = 0, b = 2$

7. $g(x) = 2x^3 + 6x^2 - 3$; $a = -2, b = -1$

8. From 1980 to 2000, nearly 90,000 young people under age 20 were killed by gunfire in the United States. The quartic function
$$f(x) = 0.152x^4 - 9.786x^3 + 173.317x^2 - 831.220x + 4053.134$$
can be used to estimate the number of people $f(x)$ under age 20 who were killed x years after 1980. Find the number who were killed in 1986, in 1993, and in 1999.

9. Use long division to find the quotient $Q(x)$ and the remainder $R(x)$ when $P(x)$ is divided by $d(x)$, and express $P(x)$ in the form $d(x) \cdot Q(x) + R(x)$. Show your work.
$$P(x) = x^4 + 3x^3 + 2x - 5,$$
$$d(x) = x - 1$$

10. Use synthetic division to find the quotient and remainder. Show your work.
$$(3x^3 - 12x + 7) \div (x - 5)$$

11. Use synthetic division to find $P(-3)$ for $P(x) = 2x^3 - 6x^2 + x - 4$. Show your work.

12. Use synthetic division to determine whether -2 is a zero of $f(x) = x^3 + 4x^2 + x - 6$. Answer yes or no. Show your work.

13. Find a polynomial of degree 4 with -3 as a zero of multiplicity 2 and 0 and 6 as zeros of multiplicity 1.

14. Suppose that a polynomial function of degree 5 with rational coefficients has 1, $\sqrt{3}$, and $2 - i$ as zeros. Find the other zeros.

Find a polynomial function of lowest degree with rational coefficients and the following as some of its zeros.

15. $-10, 3i$

16. $0, -\sqrt{3}, 1 - i$

List all possible rational zeros.

17. $f(x) = 2x^3 + x^2 - 2x + 12$

18. $h(x) = 10x^4 - x^3 + 2x - 5$

For each polynomial function:

a) *Find the rational zeros and then the other zeros; that is, solve $f(x) = 0$.*
b) *Factor $f(x)$ into linear factors.*

19. $f(x) = x^3 + x^2 - 5x - 5$

20. $g(x) = 2x^4 - 11x^3 + 16x^2 - x - 6$

21. $h(x) = x^3 + 4x^2 + 4x + 16$

22. $f(x) = 3x^4 - 11x^3 + 15x^2 - 9x + 2$

23. What does Descartes' rule of signs tell you about the number of positive real zeros and the number of negative real zeros of the following function?
$$g(x) = -x^8 + 2x^6 - 4x^3 - 1$$

Graph each of the following. Be sure to label all the asymptotes. List the domain and the x- and y-intercepts.

24. $f(x) = \dfrac{2}{(x - 3)^2}$

25. $f(x) = \dfrac{x + 3}{x^2 - 3x - 4}$

26. Find a rational function that has vertical asymptotes $x = -1$ and $x = 2$ and x-intercept $(-4, 0)$.

Solve.

27. $2x^2 > 5x + 3$

28. $\dfrac{x + 1}{x - 4} \leq 3$

29. The function $S(t) = -16t^2 + 64t + 192$ gives the height S, in feet, of a model rocket launched with a velocity of 64 ft/sec from a hill that is 192 ft high.
   a) Determine how long it will take the rocket to reach the ground.
   b) Find the interval on which the height of the rocket is greater than 240 ft.

30. Find an equation of variation in which y varies inversely as x, and $y = 5$ when $x = 6$.

31. The stopping distance d of a car after the brakes have been applied varies directly as the square of the speed r. If a car traveling 60 mph can stop in 200 ft, how long will it take a car traveling 30 mph to stop?

32. Find an equation of variation where y varies jointly as x and the square of z and inversely as w, and $y = 100$ when $x = 0.1$, $z = 10$, and $w = 5$.

Synthesis

33. Find the domain of $f(x) = \sqrt{x^2 + x - 12}$.

Exponential and Logarithmic Functions

APPLICATION

Vietnam has recently taken some major steps in opening its economy. Both U.S. exports to Vietnam and U.S. imports from Vietnam have increased exponentially since 1997. The following functions model the U.S. trade with Vietnam:

U.S. exports
to Vietnam: $E(t) = 256.43(1.22)^t,$

U.S. imports
from Vietnam: $I(t) = 464.38(1.42)^t,$

where t is the number of years since 1997 and $E(t)$ and $I(t)$ are in millions of dollars (*Source*: U.S. Bureau of the Census). Find the exports and the imports for 2003, 2005, and 2008.

This problem appears as Exercise 59 in Exercise Set 4.2.

4.1

Inverse Functions

- *Determine whether a function is one-to-one, and if it is, find a formula for its inverse.*
- *Simplify expressions of the type $(f \circ f^{-1})(x)$ and $(f^{-1} \circ f)(x)$.*

RELATIONS

REVIEW SECTION 1.2.

Inverses

When we go from an output of a function back to its input or inputs, we get an inverse relation. When that relation is a function, we have an inverse function.

Consider the relation h given as follows:

$$h = \{(-8, 5), (4, -2), (-7, 1), (3.8, 6.2)\}.$$

Suppose we *interchange* the first and second coordinates. The relation we obtain is called the **inverse** of the relation h and is given as follows:

$$\text{Inverse of } h = \{(5, -8), (-2, 4), (1, -7), (6.2, 3.8)\}.$$

> ### Inverse Relation
>
> Interchanging the first and second coordinates of each ordered pair in a relation produces the **inverse relation.**

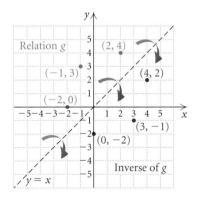

EXAMPLE 1 Consider the relation g given by

$$g = \{(2, 4), (-1, 3), (-2, 0)\}.$$

Graph the relation in blue. Find the inverse and graph it in red.

Solution   The relation g is shown in blue in the figure at left. The inverse of the relation is

$$\{(4, 2), (3, -1), (0, -2)\}$$

and is shown in red. The pairs in the inverse are reflections across the line $y = x$.   ▬

> ### Inverse Relation
>
> If a relation is defined by an equation, interchanging the variables produces an equation of the **inverse relation.**

EXAMPLE 2   Find an equation for the inverse of the relation:

$$y = x^2 - 5x.$$

Solution   We interchange x and y and obtain an equation of the inverse:

$$x = y^2 - 5y.$$

━

If a relation is given by an equation, then the solutions of the inverse can be found from those of the original equation by interchanging the first and second coordinates of each ordered pair. Thus the graphs of a relation and its inverse are always reflections of each other across the line $y = x$. This is illustrated with the equations of Example 2 in the tables and graph below. We will explore inverses and their graphs later in this section.

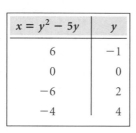

$x = y^2 - 5y$	y
6	-1
0	0
-6	2
-4	4

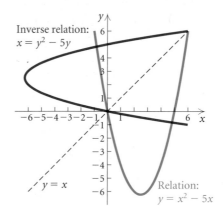

x	$y = x^2 - 5x$
-1	6
0	0
2	-6
4	-4

Inverses and One-to-One Functions

Let's consider the following two functions.

YEAR (DOMAIN)	FIRST-CLASS POSTAGE COST, IN CENTS (RANGE)
1978 ⟶	15
1983 ⟶	20
1984 ⟶	
1989 ⟶	25
1991 ⟶	29
1995 ⟶	32
1999 ⟶	33
2001 ⟶	34
2003 ⟶	37

NUMBER (DOMAIN)	CUBE (RANGE)
-3 ⟶	-27
-2 ⟶	-8
-1 ⟶	-1
0 ⟶	0
1 ⟶	1
2 ⟶	8
3 ⟶	27

Source: U.S. Postal Service

Suppose we reverse the arrows. Are these inverse relations functions?

Year (Range)	First-Class Postage Cost, in Cents (Domain)		Number (Range)	Cube (Domain)
1978 ←	15		−3 ←	−27
1983 ←	20		−2 ←	−8
1984 ←			−1 ←	−1
1989 ←	25		0 ←	0
1991 ←	29		1 ←	1
1995 ←	32		2 ←	8
1999 ←	33		3 ←	27
2001 ←	34			
2003 ←	37			

Source: U.S. Postal Service

We see that the inverse of the postage function is not a function. Like all functions, each input in the postage function has exactly one output. However, the output for both 1983 and 1984 is 20. Thus in the inverse of the postage function, the input 20 has *two* outputs, 1983 and 1984. When the same output of a function comes from two or more different inputs, the inverse cannot be a function. In the cubing function, each output corresponds to exactly one input, so its inverse is also a function. The cubing function is an example of a **one-to-one function.**

If the inverse of a function f is also a function, it is named f^{-1} (read "f-inverse").

The -1 in f^{-1} is *not* an exponent!

Do *not* misinterpret the -1 in f^{-1} as a negative exponent: f^{-1} does *not* mean the reciprocal of f and $f^{-1}(x)$ is *not* equal to $\dfrac{1}{f(x)}$.

One-to-One Functions

A function f is **one-to-one** if different inputs have different outputs—that is,

$$\text{if} \quad a \neq b, \quad \text{then} \quad f(a) \neq f(b).$$

Or a function f is **one-to-one** if when the outputs are the same, the inputs are the same—that is,

$$\text{if} \quad f(a) = f(b), \quad \text{then} \quad a = b.$$

Properties of One-to-One Functions and Inverses

- If a function is one-to-one, then its inverse is a function.
- The domain of a one-to-one function f is the range of the inverse f^{-1}.
- The range of a one-to-one function f is the domain of the inverse f^{-1}.
- A function that is increasing over its domain or is decreasing over its domain is a one-to-one function.

EXAMPLE 3 Given the function f described by $f(x) = 2x - 3$, prove that f is one-to-one (that is, it has an inverse that is a function).

Solution To show that f is one-to-one, we show that if $f(a) = f(b)$, then $a = b$. Assume that $f(a) = f(b)$ for a and b in the domain of f. Since $f(a) = 2a - 3$ and $f(b) = 2b - 3$, we have

$$2a - 3 = 2b - 3$$
$$2a = 2b \qquad \text{Adding 3}$$
$$a = b. \qquad \text{Dividing by 2}$$

Thus, if $f(a) = f(b)$, then $a = b$. This shows that f is one-to-one. —

EXAMPLE 4 Given the function g described by $g(x) = x^2$, prove that g is not one-to-one.

Solution We can prove that g is not one-to-one by finding two numbers a and b for which $a \neq b$ and $g(a) = g(b)$. Two such numbers are -3 and 3, because $-3 \neq 3$ and $g(-3) = g(3) = 9$. Thus g is not one-to-one. —

The following graph shows a function, in blue, and its inverse, in red. To determine whether the inverse is a function, we can apply the vertical-line test to its graph. By reflecting each such vertical line back across the line $y = x$, we obtain an equivalent **horizontal-line test** for the original function.

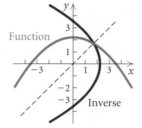

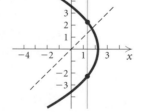

The vertical-line test shows that the inverse is not a function.

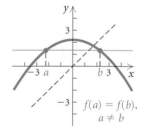

The horizontal-line test shows that the function is not one-to-one.

> ### *Horizontal-Line Test*
>
> If it is possible for a horizontal line to intersect the graph of a function more than once, then the function is *not* one-to-one and its inverse is *not* a function.

EXAMPLE 5 From the graph shown, determine whether each function is one-to-one and thus has an inverse that is a function.

a)

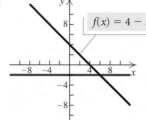

$f(x) = 4 - x$

b)

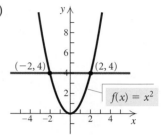

$(-2, 4)$ $(2, 4)$ $f(x) = x^2$

c)

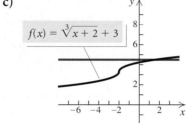

$f(x) = \sqrt[3]{x + 2} + 3$

d)

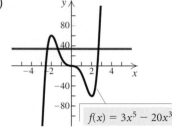

$f(x) = 3x^5 - 20x^3$

Solution For each function, we apply the horizontal-line test.

a) No horizontal line intersects the graph more than once, so the function is one-to-one. It has an inverse that is a function.

b) There are many horizontal lines that intersect the graph more than once. Note that where the line $y = 4$ intersects the graph, the first co-ordinates are -2 and 2. Although these are different inputs, they have the same output, 4. Thus the function is not one-to-one. The inverse is not a function.

c) No horizontal line intersects the graph more than once, so the function is one-to-one. It has an inverse that is a function.

d) There are many horizontal lines that intersect the graph more than once. Thus the function is not one-to-one. The inverse is not a function.

Finding Formulas for Inverses

Suppose that a function is described by a formula. If it has an inverse that is a function, we proceed as follows to find a formula for f^{-1}.

> ### *Obtaining a Formula for an Inverse*
> If a function f is one-to-one, a formula for its inverse can generally be found as follows:
>
> 1. Replace $f(x)$ with y.
> 2. Interchange x and y.
> 3. Solve for y.
> 4. Replace y with $f^{-1}(x)$.

EXAMPLE 6   Determine whether the function $f(x) = 2x - 3$ is one-to-one, and if it is, find a formula for $f^{-1}(x)$.

Solution   The graph of f is shown at left. It passes the horizontal-line test. Thus it is one-to-one and its inverse is a function. We also proved that f is one-to-one in Example 3. We find a formula for $f^{-1}(x)$.

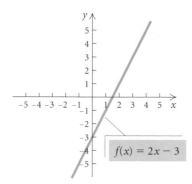

$f(x) = 2x - 3$

1. Replace $f(x)$ with y: $\qquad\qquad y = 2x - 3$
2. Interchange x and y: $\qquad\qquad x = 2y - 3$
3. Solve for y: $\qquad\qquad\qquad x + 3 = 2y$

$$\frac{x + 3}{2} = y$$

4. Replace y with $f^{-1}(x)$: $f^{-1}(x) = \dfrac{x + 3}{2}.$

Consider

$$f(x) = 2x - 3 \quad \text{and} \quad f^{-1}(x) = \frac{x + 3}{2}$$

from Example 6. For the input 5, we have

$$f(5) = 2 \cdot 5 - 3 = 10 - 3 = 7.$$

The output is 7. Now we use 7 for the input in the inverse:

$$f^{-1}(7) = \frac{7 + 3}{2} = \frac{10}{2} = 5.$$

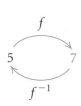

The function f takes the number 5 to 7. The inverse function f^{-1} takes the number 7 back to 5.

EXAMPLE 7   Graph

$$f(x) = 2x - 3 \quad \text{and} \quad f^{-1}(x) = \frac{x + 3}{2}$$

using the same set of axes. Then compare the two graphs.

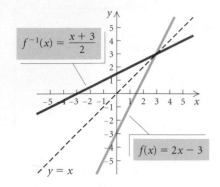

$$f^{-1}(x) = \frac{x + 3}{2}$$

$$f(x) = 2x - 3$$

$$y = x$$

Technology Connection

On some graphing calculators, we can graph the inverse of a function after graphing the function itself by accessing a drawing feature. Consult your user's manual or the *Graphing Calculator Manual* that accompanies this text for the procedure.

Solution   The graphs of f and f^{-1} are shown at left. The solutions of the inverse function can be found from those of the original function by interchanging the first and second coordinates of each ordered pair.

x	$f(x) = 2x - 3$
-1	-5
0	-3 ← y-intercept
2	1
3	3

x	$f^{-1}(x) = \dfrac{x + 3}{2}$
-5	-1
-3	0 ← x-intercept
1	2
3	3

When we interchange x and y in finding a formula for the inverse of $f(x) = 2x - 3$, we are in effect reflecting the graph of that function across the line $y = x$. For example, when the coordinates of the y-intercept, $(0, -3)$, of the graph of f are reversed, we get the x-intercept, $(-3, 0)$, of the graph of f^{-1}. If we were to graph $f(x) = 2x - 3$ in wet ink and fold along the line $y = x$, the graph of $f^{-1}(x) = (x + 3)/2$ would be formed by the ink transferred from f.

> The graph of f^{-1} is a reflection of the graph of f across the line $y = x$.

EXAMPLE 8   Consider $g(x) = x^3 + 2$.

a) Determine whether the function is one-to-one.

b) If it is one-to-one, find a formula for its inverse.

c) Graph the function and its inverse.

Solution

a) The graph of $g(x) = x^3 + 2$ is shown at left. It passes the horizontal-line test and thus has an inverse that is a function. We also know that $g(x)$ is one-to-one because it is an increasing function over its entire domain.

b) We follow the procedure for finding an inverse.

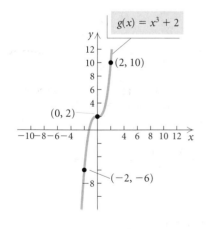

$$g(x) = x^3 + 2$$

$(2, 10)$

$(0, 2)$

$(-2, -6)$

1. Replace $g(x)$ with y: $\qquad y = x^3 + 2$

2. Interchange x and y: $\qquad x = y^3 + 2$

3. Solve for y: $\qquad\qquad x - 2 = y^3$

$$\sqrt[3]{x - 2} = y$$

4. Replace y with $g^{-1}(x)$: $\quad g^{-1}(x) = \sqrt[3]{x - 2}.$

c) To find the graph, we reflect the graph of $g(x) = x^3 + 2$ across the line $y = x$. This can be done by plotting points.

x	$g(x)$
-2	-6
-1	1
0	2
1	3
2	10

x	$g^{-1}(x)$
-6	-2
1	-1
2	0
3	1
10	2

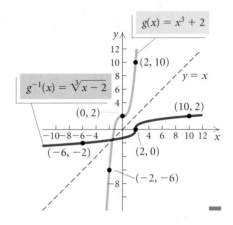

Inverse Functions and Composition

Suppose that we were to use some input a for a one-to-one function f and find its output, $f(a)$. The function f^{-1} would then take that output back to a. Similarly, if we began with an input b for the function f^{-1} and found its output, $f^{-1}(b)$, the original function f would then take that output back to b. This is summarized as follows.

> If a function f is one-to-one, then f^{-1} is the unique function such that each of the following holds:
>
> $$(f^{-1} \circ f)(x) = f^{-1}(f(x)) = x, \quad \text{for each } x \text{ in the domain of } f, \text{ and}$$
>
> $$(f \circ f^{-1})(x) = f(f^{-1}(x)) = x, \quad \text{for each } x \text{ in the domain of } f^{-1}.$$

EXAMPLE 9   Given that $f(x) = 5x + 8$, use composition of functions to show that $f^{-1}(x) = (x - 8)/5$.

Solution   We find $(f^{-1} \circ f)(x)$ and $(f \circ f^{-1})(x)$ and check to see that each is x:

$$(f^{-1} \circ f)(x) = f^{-1}(f(x))$$
$$= f^{-1}(5x + 8) = \frac{(5x + 8) - 8}{5} = \frac{5x}{5} = x;$$

$$(f \circ f^{-1})(x) = f(f^{-1}(x))$$
$$= f\left(\frac{x - 8}{5}\right) = 5\left(\frac{x - 8}{5}\right) + 8 = x - 8 + 8 = x.$$

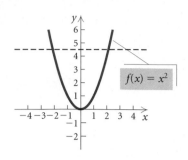

Restricting a Domain

In the case in which the inverse of a function is not a function, the domain of the function can be restricted to allow the inverse to be a function. We saw in Examples 4 and 5(b) that $f(x) = x^2$ is not one-to-one. The graph is shown at left.

Suppose that we had tried to find a formula for the inverse as follows:

$$y = x^2 \qquad \text{Replacing } f(x) \text{ with } y$$
$$x = y^2 \qquad \text{Interchanging } x \text{ and } y$$
$$\pm\sqrt{x} = y. \qquad \text{Solving for } y$$

This is not the equation of a function. An input of, say, 4 would yield two outputs, -2 and 2. In such cases, it is convenient to consider "part" of the function by restricting the domain of $f(x)$. For example, if we restrict the domain of $f(x) = x^2$ to nonnegative numbers, then its inverse is a function, as shown with the graphs of $f(x) = x^2$, $x \geq 0$, and $f^{-1}(x) = \sqrt{x}$ below.

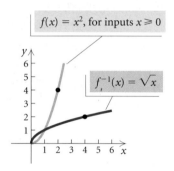

4.1

Exercise Set

Find the inverse of the relation.

1. $\{(7, 8), (-2, 8), (3, -4), (8, -8)\}$

2. $\{(0, 1), (5, 6), (-2, -4)\}$

3. $\{(-1, -1), (-3, 4)\}$

4. $\{(-1, 3), (2, 5), (-3, 5), (2, 0)\}$

Find an equation of the inverse relation.

5. $y = 4x - 5$

6. $2x^2 + 5y^2 = 4$

7. $x^3 y = -5$

8. $y = 3x^2 - 5x + 9$

9. $x = y^2 - 2y$

10. $x = \frac{1}{2}y + 4$

Graph the equation by substituting and plotting points. Then reflect the graph across the line $y = x$ to obtain the graph of its inverse.

11. $x = y^2 - 3$

12. $y = x^2 + 1$

13. $y = 3x - 2$

14. $x = -y + 4$

15. $y = |x|$

16. $x + 2 = |y|$

Given the function f, prove that f is one-to-one using the definition of a one-to-one function on page 332.

17. $f(x) = \frac{1}{3}x - 6$

18. $f(x) = 4 - 2x$

19. $f(x) = x^3 + \frac{1}{2}$

20. $f(x) = \sqrt[3]{x}$

Given the function g, prove that g is not one-to-one using the definition of a one-to-one function on page 332.

21. $g(x) = 1 - x^2$

22. $g(x) = 3x^2 + 1$

23. $g(x) = x^4 - x^2$

24. $g(x) = \dfrac{1}{x^6}$

Using the horizontal-line test, determine whether the function is one-to-one.

25. $f(x) = 2.7^x$

26. $f(x) = 2^{-x}$

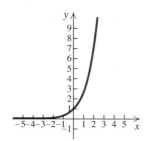

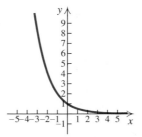

27. $f(x) = 4 - x^2$

28. $f(x) = x^3 - 3x + 1$

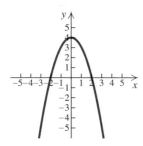

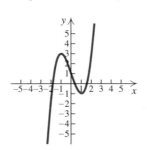

29. $f(x) = \dfrac{8}{x^2 - 4}$

30. $f(x) = \sqrt{\dfrac{10}{4 + x}}$

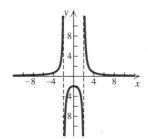

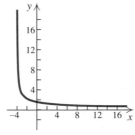

31. $f(x) = \sqrt[3]{x + 2} - 2$

32. $f(x) = \dfrac{8}{x}$

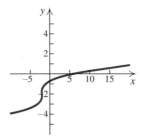

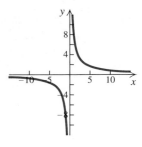

Graph the function and determine whether the function is one-to-one using the horizontal-line test.

33. $f(x) = 5x - 8$

34. $f(x) = 3 + 4x$

35. $f(x) = 1 - x^2$

36. $f(x) = |x| - 2$

37. $f(x) = |x + 2|$

38. $f(x) = -0.8$

39. $f(x) = -\dfrac{4}{x}$

40. $f(x) = \dfrac{2}{x + 3}$

41. $f(x) = \frac{2}{3}$

42. $f(x) = \frac{1}{2}x^2 + 3$

43. $f(x) = \sqrt{25 - x^2}$

44. $f(x) = -x^3 + 2$

In Exercises 45–60, for each function:

a) *Sketch the graph and use the graph to determine whether the function is one-to-one.*

b) *If the function is one-to-one, find a formula for the inverse.*

45. $f(x) = x + 4$

46. $f(x) = 7 - x$

47. $f(x) = 2x - 1$

48. $f(x) = 5x + 8$

49. $f(x) = \dfrac{4}{x + 7}$

50. $f(x) = -\dfrac{3}{x}$

51. $f(x) = \dfrac{x + 4}{x - 3}$

52. $f(x) = \dfrac{5x - 3}{2x + 1}$

53. $f(x) = x^3 - 1$

54. $f(x) = (x + 5)^3$

55. $f(x) = x\sqrt{4 - x^2}$

56. $f(x) = 2x^2 - x - 1$

57. $f(x) = 5x^2 - 2, \ x \geq 0$

58. $f(x) = 4x^2 + 3, \ x \geq 0$

59. $f(x) = \sqrt{x + 1}$

60. $f(x) = \sqrt[3]{x - 8}$

Find the inverse by thinking about the operations of the function and then reversing, or undoing, them. Check your work algebraically.

FUNCTION	INVERSE
61. $f(x) = 3x$	$f^{-1}(x) = $
62. $f(x) = \frac{1}{4}x + 7$	$f^{-1}(x) = $
63. $f(x) = -x$	$f^{-1}(x) = $
64. $f(x) = \sqrt[3]{x} - 5$	$f^{-1}(x) = $
65. $f(x) = \sqrt[3]{x - 5}$	$f^{-1}(x) = $
66. $f(x) = x^{-1}$	$f^{-1}(x) = $

Each graph in Exercises 67–72 is the graph of a one-to-one function f. Sketch the graph of the inverse function f^{-1}.

67.
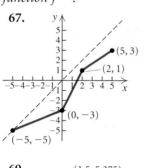

68.
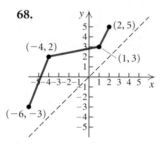

69.
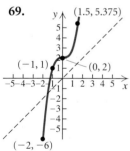

70.
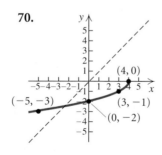

71.
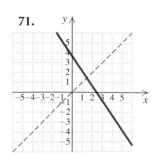

72.
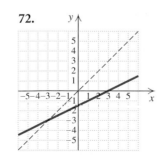

For the function f, use composition of functions to show that f^{-1} is as given.

73. $f(x) = \frac{7}{8}x, \ f^{-1}(x) = \frac{8}{7}x$

74. $f(x) = \dfrac{x + 5}{4}, \ f^{-1}(x) = 4x - 5$

75. $f(x) = \dfrac{1 - x}{x}, \ f^{-1}(x) = \dfrac{1}{x + 1}$

76. $f(x) = \sqrt[3]{x + 4}, \ f^{-1}(x) = x^3 - 4$

77. $f(x) = \dfrac{2}{5}x + 1, \ f^{-1}(x) = \dfrac{5x - 5}{2}$

78. $f(x) = \dfrac{x + 6}{3x - 4}, \ f^{-1}(x) = \dfrac{4x + 6}{3x - 1}$

Find the inverse of the given one-to-one function f. Give the domain and the range of f and f^{-1}, and then graph both f and f^{-1} on the same axes. (Hint: The domain of a one-to-one function is the range of its inverse. The range of a one-to-one function is the domain of its inverse.)

79. $f(x) = 5x - 3$ **80.** $f(x) = 2 - x$

81. $f(x) = \dfrac{2}{x}$ **82.** $f(x) = -\dfrac{3}{x + 1}$

83. $f(x) = \dfrac{1}{3}x^3 - 2$ **84.** $f(x) = \sqrt[3]{x} - 1$

85. $f(x) = \dfrac{x + 1}{x - 3}$ **86.** $f(x) = \dfrac{x - 1}{x + 2}$

87. Find $f(f^{-1}(5))$ and $f^{-1}(f(a))$:
$$f(x) = x^3 - 4.$$

88. Find $f^{-1}(f(p))$ and $f(f^{-1}(1253))$:
$$f(x) = \sqrt[5]{\frac{2x - 7}{3x + 4}}.$$

89. *Dress Sizes in the United States and Italy.* A function that will convert dress sizes in the United States to those in Italy is
$$g(x) = 2(x + 12).$$

a) Find the dress sizes in Italy that correspond to sizes 6, 8, 10, 14, and 18 in the United States.
b) Find a formula for the inverse of the function.
c) Use the inverse function to find the dress sizes in the United States that correspond to 36, 40, 44, 52, and 60 in Italy.

90. *Bus Chartering.* An organization determines that the cost per person of chartering a bus is given by the formula

$$C(x) = \frac{100 + 5x}{x},$$

where x is the number of people in the group and $C(x)$ is in dollars. Determine $C^{-1}(x)$ and explain what it represents.

Technology Connection

Graph the function and its inverse using a graphing calculator. Use an inverse drawing feature, if available. Find the domain and the range of f. Find the domain and the range of the inverse f^{-1}.

91. $f(x) = 0.8x + 1.7$

92. $f(x) = 2.7 - 1.08x$

93. $f(x) = \frac{1}{2}x - 4$

94. $f(x) = x^3 - 1$

95. $f(x) = \sqrt{x - 3}$

96. $f(x) = -\dfrac{2}{x}$

97. $f(x) = x^2 - 4, \; x \geq 0$

98. $f(x) = 3 - x^2, \; x \geq 0$

99. $f(x) = (3x - 9)^3$

100. $f(x) = \sqrt[3]{\dfrac{x - 3.2}{1.4}}$

101. *Reaction Distance.* You are driving a car when a deer suddenly darts across the road in front of you.

Your brain registers the emergency and sends a signal to your foot to hit the brake. The car travels a distance D, in feet, during this time, where D is a function of the speed r, in miles per hour, that the car is traveling when you see the deer. That reaction distance D is a linear function given by

$$D(r) = \frac{11r + 5}{10}.$$

a) Find $D(0)$, $D(10)$, $D(20)$, $D(50)$, and $D(65)$.
b) Find $D^{-1}(r)$ and explain what it represents.
c) Graph the function and its inverse.

102. *Bread Consumption.* The number N of 1-lb loaves of bread consumed in the United States per person per year t years after 1995 is given by the function

$$N(t) = 0.6514t + 53.1599$$

(*Source*: U.S. Department of Agriculture).

a) Find the bread consumption per person in 2005 and in 2010.
b) Graph the function and its inverse.
c) Explain what the inverse represents.

Collaborative Discussion and Writing

103. Explain why an even function f does not have an inverse f^{-1}.

104. The following formulas for the conversion between Fahrenheit and Celsius temperatures have been considered several times in this text:

$$C = \tfrac{5}{9}(F - 32)$$

and

$$F = \tfrac{9}{5}C + 32.$$

Discuss these formulas from the standpoint of inverses.

Skill Maintenance

Consider the following quadratic functions. Without graphing them, answer the questions below.

a) $f(x) = 2x^2$
b) $f(x) = -x^2$
c) $f(x) = \frac{1}{4}x^2$
d) $f(x) = -5x^2 + 3$
e) $f(x) = \frac{2}{3}(x - 1)^2 - 3$
f) $f(x) = -2(x + 3)^2 + 1$
g) $f(x) = (x - 3)^2 + 1$
h) $f(x) = -4(x + 1)^2 - 3$

105. Which functions have a maximum value?

106. Which graphs open up?

107. Consider (a) and (c). Which graph is narrower?

108. Consider (d) and (e). Which graph is narrower?

109. Which graph has vertex $(-3, 1)$?

110. For which is the line of symmetry $x = 0$?

Synthesis

111. The function $f(x) = x^2 - 3$ is not one-to-one. Restrict the domain of f so that its inverse is a function. Find the inverse and state the restriction on the domain of the inverse.

112. Consider the function f given by

$$f(x) = \begin{cases} x^3 + 2, & x \le -1, \\ x^2, & -1 < x < 1, \\ x + 1, & x \ge 1. \end{cases}$$

Does f have an inverse that is a function? Why or why not?

113. Find three examples of functions that are their own inverses, that is, $f = f^{-1}$.

114. Given the function $f(x) = ax + b$, $a \ne 0$, find the values of a and b for which $f^{-1}(x) = f(x)$.

115. Given the graph of a function $f(x)$, how can you determine graphically if $f^{-1}(x) = f(x)$?

4.2

- *Graph exponential equations and functions.*
- *Solve applied problems involving exponential functions and their graphs.*

Exponential Functions and Graphs

We now turn our attention to the study of a set of functions that are very rich in application. Consider the following graphs. Each one illustrates an *exponential function*. In this section, we consider such functions, their inverses, and some important applications of exponential functions.

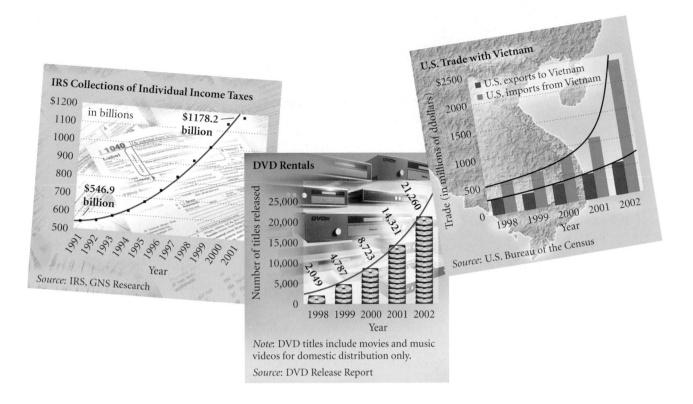

IRS Collections of Individual Income Taxes
Source: IRS, GNS Research

DVD Rentals

Note: DVD titles include movies and music videos for domestic distribution only.
Source: DVD Release Report

U.S. Trade with Vietnam
Source: U.S. Bureau of the Census

Graphing Exponential Functions

We now define exponential functions. We assume that a^x has meaning for any real number x and any positive real number a and that the laws of exponents still hold, though we will not prove them here.

EXPONENTS

REVIEW SECTIONS R.2 AND R.6.

Exponential Function

The function $f(x) = a^x$, where x is a real number, $a > 0$ and $a \neq 1$, is called the **exponential function,** base a.

We require the **base** to be positive in order to avoid the complex numbers that would occur by taking even roots of negative numbers—an example is $(-1)^{1/2}$, the square root of -1, which is not a real number. The restriction $a \neq 1$ is made to exclude the constant function $f(x) = 1^x = 1$, which does not have an inverse because it is not one-to-one.

The following are examples of exponential functions:

$$f(x) = 2^x, \qquad f(x) = \left(\frac{1}{2}\right)^x, \qquad f(x) = (3.57)^x.$$

Note that, in contrast to functions like $f(x) = x^5$ and $f(x) = x^{1/2}$ in which the variable is the base of an exponential expression, the variable in an exponential function is *in the exponent.*

Let's now consider graphs of exponential functions.

EXAMPLE 1   Graph the exponential function

$$y = f(x) = 2^x.$$

Solution   We compute some function values and list the results in a table (at left).

$$f(0) = 2^0 = 1;$$
$$f(1) = 2^1 = 2;$$
$$f(2) = 2^2 = 4;$$
$$f(3) = 2^3 = 8;$$
$$f(-1) = 2^{-1} = \frac{1}{2^1} = \frac{1}{2};$$
$$f(-2) = 2^{-2} = \frac{1}{2^2} = \frac{1}{4};$$
$$f(-3) = 2^{-3} = \frac{1}{2^3} = \frac{1}{8}.$$

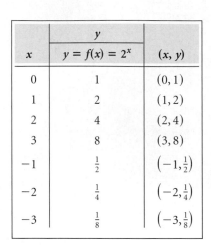

x	$y = f(x) = 2^x$	(x, y)
0	1	$(0, 1)$
1	2	$(1, 2)$
2	4	$(2, 4)$
3	8	$(3, 8)$
−1	$\frac{1}{2}$	$\left(-1, \frac{1}{2}\right)$
−2	$\frac{1}{4}$	$\left(-2, \frac{1}{4}\right)$
−3	$\frac{1}{8}$	$\left(-3, \frac{1}{8}\right)$

Next, we plot these points and connect them with a smooth curve. Be sure to plot enough points to determine how steeply the curve rises.

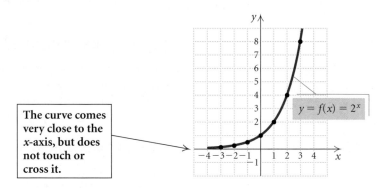

The curve comes very close to the x-axis, but does not touch or cross it.

$$y = f(x) = 2^x$$

Note that as x increases, the function values increase without bound. As x decreases, the function values decrease, getting close to 0. That is, as $x \to -\infty$, $y \to 0$. Thus the x-axis, or the line $y = 0$, is a horizontal asymptote. As the x-inputs decrease, the curve gets closer and closer to this line, but does not cross it.   ▬

HORIZONTAL ASYMPTOTES

REVIEW SECTION 3.4.

EXAMPLE 2   Graph the exponential function $y = f(x) = \left(\frac{1}{2}\right)^x$.

Solution   We compute some function values and list the results in a table, as shown on the following page. Before we plot these points and draw the curve, note that

$$y = f(x) = \left(\frac{1}{2}\right)^x = (2^{-1})^x = 2^{-x}.$$

	y	
x	$y = f(x) = 2^{-x}$	(x, y)
0	1	$(0, 1)$
1	$\frac{1}{2}$	$\left(1, \frac{1}{2}\right)$
2	$\frac{1}{4}$	$\left(2, \frac{1}{4}\right)$
3	$\frac{1}{8}$	$\left(3, \frac{1}{8}\right)$
-1	2	$(-1, 2)$
-2	4	$(-2, 4)$
-3	8	$(-3, 8)$

This tells us, before we begin graphing, that this graph is a reflection of the graph of $y = 2^x$ across the y-axis.

$$f(0) = 2^{-0} = 1; \qquad\qquad f(-1) = 2^{-(-1)} = 2^1 = 2;$$

$$f(1) = 2^{-1} = \frac{1}{2^1} = \frac{1}{2}; \qquad f(-2) = 2^{-(-2)} = 2^2 = 4;$$

$$f(2) = 2^{-2} = \frac{1}{2^2} = \frac{1}{4}; \qquad f(-3) = 2^{-(-3)} = 2^3 = 8.$$

$$f(3) = 2^{-3} = \frac{1}{2^3} = \frac{1}{8};$$

Next, we plot these points and connect them with a smooth curve.

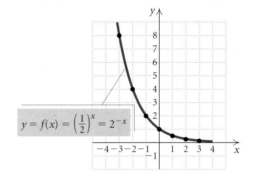

$$y = f(x) = \left(\frac{1}{2}\right)^x = 2^{-x}$$

Note that as x increases, the function values decrease, getting close to 0. The x-axis, $y = 0$, is the horizontal asymptote. As x decreases, the function values increase without bound.

Observe the following graphs of exponential functions and look for patterns in the graphs.

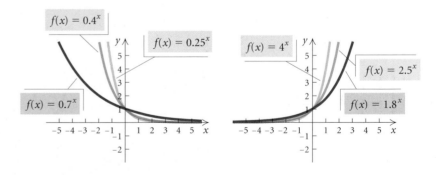

$f(x) = 0.4^x$

$f(x) = 0.25^x$

$f(x) = 0.7^x$

$f(x) = 4^x$

$f(x) = 2.5^x$

$f(x) = 1.8^x$

What relationship do you see between the base a and the shape of the resulting graph of $f(x) = a^x$? What do all the graphs have in common? How do they differ?

CONNECTING THE CONCEPTS

PROPERTIES OF EXPONENTIAL FUNCTIONS

Let's list and compare some characteristics of exponential functions, keeping in mind that the definition of an exponential function, $f(x) = a^x$, requires that a be positive and different from 1.

$f(x) = a^x$, $a > 0$, $a \neq 1$

Continuous

One-to-one

Domain: $(-\infty, \infty)$

Range: $(0, \infty)$

Increasing if $a > 1$

Decreasing if $0 < a < 1$

Horizontal asymptote is x-axis

y-intercept: $(0, 1)$

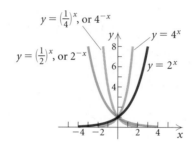

TRANSFORMATIONS OF FUNCTIONS

REVIEW SECTION 1.6.

To graph other types of exponential functions, keep in mind the ideas of translation, stretching, and reflection. All these concepts allow us to visualize the graph before drawing it.

EXAMPLE 3 Graph each of the following. Before doing so, describe how each graph can be obtained from the graph of $f(x) = 2^x$.

a) $f(x) = 2^{x-2}$

b) $f(x) = 2^x - 4$

c) $f(x) = 5 - 2^{-x}$

Solution

a) The graph of this function is the graph of $y = 2^x$ shifted *right* 2 units.

x	$f(x)$
-2	0.0625
-1	0.125
0	0.25
1	0.5
2	1
3	2
4	4
5	8

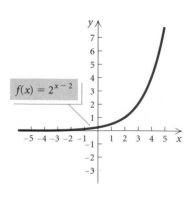

b) The graph is the graph of $y = 2^x$ shifted *down* 4 units.

x	$f(x)$
-2	-3.75
-1	-3.5
0	-3
1	-2
2	0
3	4

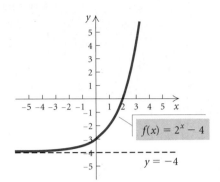

$f(x) = 2^x - 4$

$y = -4$

c) The graph is a reflection of the graph of $y = 2^x$ across the y-axis, followed by a reflection across the x-axis and then a shift *up* 5 units.

x	$f(x)$
-3	-3
-2	1
-1	3
0	4
1	4.5
2	4.75

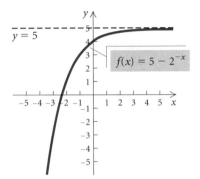

$y = 5$

$f(x) = 5 - 2^{-x}$

Applications

One of the most frequent applications of exponential functions occurs with compound interest.

EXAMPLE 4 *Compound Interest.* The amount of money A that a principal P will grow to after t years at interest rate r (in decimal form), compounded n times per year, is given by the formula

$$A = P\left(1 + \frac{r}{n}\right)^{nt}.$$

Suppose that $100,000 is invested at 6.5% interest, compounded semiannually.

a) Find a function for the amount of money after t years.

b) Find the amount of money in the account at $t = 0, 4, 8$, and 10 yr.

c) Graph the function.

Solution

a) Since $P = \$100{,}000$, $r = 6.5\% = 0.065$, and $n = 2$, we can substitute these values and write the following function:

$$A(t) = 100{,}000\left(1 + \frac{0.065}{2}\right)^{2 \cdot t} = \$100{,}000(1.0325)^{2t}.$$

b) We can compute function values with a calculator:

$$A(0) = 100{,}000(1.0325)^{2 \cdot 0} = \$100{,}000;$$
$$A(4) = 100{,}000(1.0325)^{2 \cdot 4} \approx \$129{,}157.75;$$
$$A(8) = 100{,}000(1.0325)^{2 \cdot 8} \approx \$166{,}817.25;$$
$$A(10) = 100{,}000(1.0325)^{2 \cdot 10} \approx \$189{,}583.79.$$

c) We use the function values computed in part (b) and others if we wish, and draw the graph as follows. Note that the axes are scaled differently because of the large numbers and that t is restricted to nonnegative values, because negative time values have no meaning here.

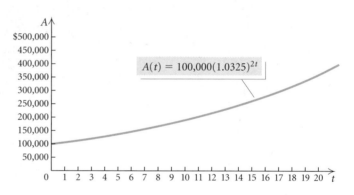

Technology Connection

We can find the function values in Example 4(b) using the VALUE feature from the CALC menu.

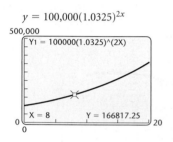

$$y = 100{,}000(1.0325)^{2x}$$

We could also use the TABLE feature.

The Number *e*

We now consider a very special number in mathematics. In 1741, Leonhard Euler named this number *e*. Though you may not have encountered it before, you will see here and in future mathematics that it has many important applications. To explain this number, we use the compound interest formula $A = P(1 + r/n)^{nt}$ in Example 4. Suppose that \$1 is an initial investment at 100% interest for 1 yr. (No bank would pay this.) The formula above becomes a function A defined in terms of the number of compounding periods n. Since $P = 1$, $r = 100\% = 1$, and $t = 1$,

$$A = P\left(1 + \frac{r}{n}\right)^{nt} = 1\left(1 + \frac{1}{n}\right)^{n \cdot 1} = \left(1 + \frac{1}{n}\right)^{n}.$$

Let's visualize this function with its graph shown below and explore the values of $A(n)$ as $n \to \infty$. Consider the graph for larger and larger values of n. Does this function have a horizontal asymptote?

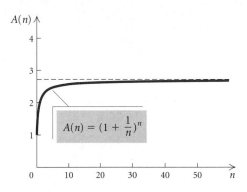

Let's find some function values using a calculator.

n, Number of Compounding Periods	$A(n) = \left(1 + \dfrac{1}{n}\right)^n$
1 (compounded annually)	$2.00
2 (compounded semiannually)	2.25
3	2.3704
4 (compounded quarterly)	2.4414
5	2.4883
100	2.7048
365 (compounded daily)	2.7146
8760 (compounded hourly)	2.7181

It appears from these values that the graph does have a horizontal asymptote, $y \approx 2.7$. As the values of n get larger and larger, the function values get closer and closer to the number Euler named e. Its decimal representation does not terminate or repeat; it is irrational.

$$e = 2.7182818284\ldots$$

EXAMPLE 5 Find each value of e^x, to four decimal places, using the $\boxed{e^x}$ key on a calculator.

a) e^3 **b)** $e^{-0.23}$

c) e^0 **d)** e^1

Solution

FUNCTION VALUE	READOUT	ROUNDED
a) e^3	e^(3) 20.08553692	20.0855
b) $e^{-0.23}$	e^(−.23) .7945336025	0.7945
c) e^0	e^(0) 1	1
d) e^1	e^(1) 2.718281828	2.7183

Graphs of Exponential Functions, Base *e*

We demonstrate ways in which to graph exponential functions.

EXAMPLE 6 Graph $f(x) = e^x$ and $g(x) = e^{-x}$.

Solution We can compute points for each equation using the $\boxed{e^x}$ key on a calculator. (See the tables at left.) Then we plot these points and draw the graphs of the functions.

x	$f(x) = e^x$	$g(x) = e^{-x}$
−2	0.135	7.389
−1	0.368	2.718
0	1	1
1	2.718	0.368
2	7.389	0.135

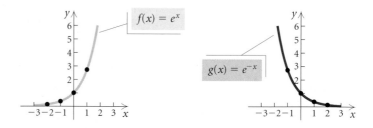

Note that the graph of *g* is a reflection of the graph of *f* across the *y*-axis.

EXAMPLE 7 Graph each of the following. Before doing so, describe how each graph can be obtained from the graph of $y = e^x$.

a) $f(x) = e^{-0.5x}$

b) $f(x) = 1 - e^{-2x}$

c) $f(x) = e^{x+3}$

Solution

a) We note that the graph of this function is a horizontal stretching of the graph of $y = e^x$ followed by a reflection across the y-axis.

x	$f(x)$
-2	2.718
-1	1.649
0	1
1	0.607
2	0.368

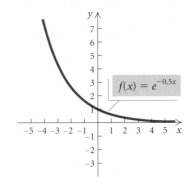

$f(x) = e^{-0.5x}$

b) The graph is a horizontal shrinking of the graph of $y = e^x$, followed by a reflection across the y-axis, then across the x-axis, followed by a translation up 1 unit.

x	$f(x)$
-1	-6.389
0	0
1	0.865
2	0.982
3	0.998

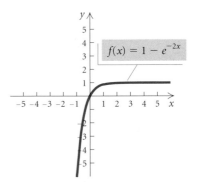

$f(x) = 1 - e^{-2x}$

c) The graph is a translation of the graph of $y = e^x$ left 3 units.

x	$f(x)$
-7	0.018
-5	0.135
-3	1
-1	7.389
0	20.086

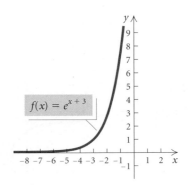

$f(x) = e^{x+3}$

4.2

Exercise Set

Find each of the following, to four decimal places, using a calculator.

1. e^4

2. e^{10}

3. $e^{-2.458}$

4. $\left(\dfrac{1}{e^3}\right)^2$

In Exercises 5–10, match the function with one of the graphs (a)–(f), which follow.

5. $f(x) = -2^x - 1$

6. $f(x) = -\left(\frac{1}{2}\right)^x$

7. $f(x) = e^x + 3$

8. $f(x) = e^{x+1}$

9. $f(x) = 3^{-x} - 2$

10. $f(x) = 1 - e^x$

a)
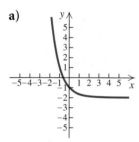

b)
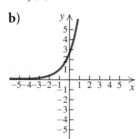

c)
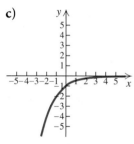

d)
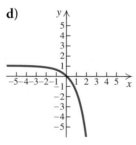

e)
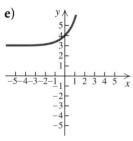

f)
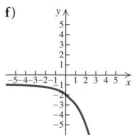

Graph the function.

11. $f(x) = 3^x$

12. $f(x) = 5^x$

13. $f(x) = 6^x$

14. $f(x) = 3^{-x}$

15. $f(x) = \left(\frac{1}{4}\right)^x$

16. $f(x) = \left(\frac{2}{3}\right)^x$

17. $y = -2^x$

18. $y = 3 - 3^x$

19. $f(x) = -0.25^x + 4$

20. $f(x) = 0.6^x - 3$

21. $f(x) = 1 + e^{-x}$

22. $f(x) = 2 - e^{-x}$

23. $y = \frac{1}{4}e^x$

24. $y = 2e^{-x}$

25. $f(x) = 1 - e^{-x}$

26. $f(x) = e^x - 2$

Sketch the graph of the function. Describe how each graph can be obtained from the graph of a basic exponential function.

27. $f(x) = 2^{x+1}$

28. $f(x) = 2^{x-1}$

29. $f(x) = 2^x - 3$

30. $f(x) = 2^x + 1$

31. $f(x) = 4 - 3^{-x}$

32. $f(x) = 2^{x-1} - 3$

33. $f(x) = \left(\frac{3}{2}\right)^{x-1}$

34. $f(x) = 3^{4-x}$

35. $f(x) = 2^{x+3} - 5$

36. $f(x) = -3^{x-2}$

37. $f(x) = e^{2x}$

38. $f(x) = e^{-0.2x}$

39. $y = e^{-x+1}$

40. $y = e^{2x} + 1$

41. $f(x) = 2(1 - e^{-x})$

42. $f(x) = 1 - e^{-0.01x}$

43. *Compound Interest.* Suppose that $82,000 is invested at $4\frac{1}{2}\%$ interest, compounded quarterly.
   a) Find the function for the amount of money after t years.
   b) Find the amount of money in the account at $t = 0, 2, 5$, and 10 yr.

44. *Compound Interest.* Suppose that $750 is invested at 7% interest, compounded semiannually.

   a) Find the function for the amount of money after t years.

   b) Find the amount of money in the account at $t = 1, 6, 10, 15,$ and 25 yr.

45. *Interest on a CD.* On Jacob's sixth birthday, his grandparents present him with a $3000 CD that earns 5% interest, compounded quarterly. If the CD matures on his sixteenth birthday, what amount will be available then?

46. *Interest in a College Trust Fund.* Following the birth of a child, Juan deposits $10,000 in a college trust fund where interest is 6.4%, compounded semiannually.

   a) Find a function for the amount in the account after t years.

   b) Find the amount of money in the account at $t = 0, 4, 8, 10,$ and 18 yr.

In Exercises 47–54, use the compound-interest formula to find the account balance with the given conditions.

   $P =$ principal,
   $r =$ interest rate,
   $n =$ number of compounding periods per year,
   $t =$ time, in years,
   $A =$ account balance.

	P	r	Compounded	n	t	A
47.	$3,000	4%	Semiannually		2	
48.	$12,500	3%	Quarterly		3	
49.	$120,000	2.5%	Annually		10	
50.	$120,000	2.5%	Quarterly		10	
51.	$53,500	$5\frac{1}{2}$%	Quarterly		$6\frac{1}{2}$	
52.	$6,250	$6\frac{3}{4}$%	Semiannually		$4\frac{1}{2}$	
53.	$17,400	8.1%	Daily		5	
54.	$900	7.3%	Daily		$7\frac{1}{4}$	

55. *Recycling Aluminum Cans.* It is estimated that two thirds of all aluminum cans distributed will be recycled each year (*Source*: Alcoa Corporation). A beverage company distributes 350,000 cans. The number still in use after time t, in years, is given by the exponential function

$$N(t) = 350,000\left(\tfrac{2}{3}\right)^t.$$

   a) How many cans are still in use after 0 yr? 1 yr? 4 yr? 10 yr?

   b) Graph the function.

56. *Growth of Bacteria* Escherichia coli. The bacteria *Escherichia coli* are commonly found in the human bladder. Suppose that 3000 of the bacteria are present at time $t = 0$. Then under certain conditions, t minutes later, the number of bacteria present is

$$N(t) = 3000(2)^{t/20}.$$

   a) How many bacteria will be present after 10 min? 20 min? 30 min? 40 min? 60 min?

   b) Graph the function.

57. *Cash Signing Bonuses.* The percentage of executives receiving cash signing bonuses is increasing. The percentage is given by the function

$$B(t) = 5.402(1.340)^t,$$

where t is the number of years since 1998 (*Source*: William M. Mercer analysis of the most recent proxy statements of 350 large U.S. firms). Find the percentage of executives receiving signing bonuses in 2003, in 2004, and in 2005.

58. *Botox® Injections.* There were nearly 6.9 million cosmetic surgical and nonsurgical procedures performed in the United States in 2002. Botulinum toxin injection tops the 2002 nonsurgical list at 1,658,667 (up 2446% since 1997). The number of Botox® procedures is given by the function

$$I(t) = 85,163.85(1.95)^t,$$

where t is the number of years since 1997 (*Source*: American Society for Aesthetic Plastic Surgery). Predict the number of procedures in 2005 and in 2008.

59. *U.S. Trade with Vietnam.* Vietnam has recently taken some major steps in opening its economy. Both U.S exports to Vietnam and U.S. imports from Vietnam have increased exponentially since 1997.

The following functions model the U.S. trade with Vietnam:

U.S. exports
to Vietnam: $E(t) = 256.43(1.22)^t$,
U.S. imports
from Vietnam: $I(t) = 464.38(1.42)^t$,

where t is the number of years since 1997 and $E(t)$ and $I(t)$ are in millions of dollars (*Source*: U.S. Bureau of the Census). Find the exports and the imports, in millions of dollars, for 2003, 2005, and 2008.

60. *Percentage of Children in the United States.* In 1999, there were 70.2 million children (26% of the total population) under age 18 in the United States. This percentage was down from 36% in 1964 and is expected to continue to decrease. The following function can be used to project the percentage of children in the United States:

$$K(t) = 35.37(0.99)^t,$$

where t is the number of years since 1964 (*Source*: www.childstats.gov). Project the percentage of the U.S. population under age 18 for 2005, 2010, and 2015.

61. *Salvage Value.* A top-quality fax–copying machine is purchased for $1800. Its value each year is about 80% of the value of the preceding year. After t years, its value, in dollars, is given by the exponential function

$$V(t) = 1800(0.8)^t.$$

Find the value of the machine after 0 yr, 1 yr, 2 yr, 5 yr, and 10 yr.

62. *Cellular Phone Boom.* The number of people using cellular phones has grown exponentially in recent years. The total number of cellular phone subscribers C, in millions, is given by

$$C(t) = 0.362(1.468)^t,$$

where t is the number of years since 1985.

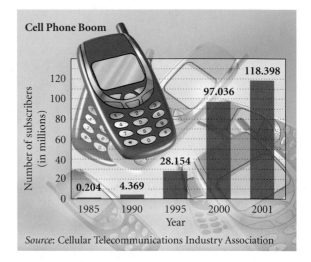

Source: Cellular Telecommunications Industry Association

Find the total number of subscribers in 1992, in 1999, and in 2004.

63. *Timber Demand.* World demand for timber is increasing exponentially. The demand N, in billions of cubic feet purchased, is given by

$$N(t) = 46.6(1.018)^t,$$

where t is the number of years since 1981 (*Sources*: U.N. Food and Agricultural Organization; American Forest and Paper Association). Find the demand for timber in 2005 and in 2010.

64. *Typing Speed.* Sarah is taking keyboarding at a community college. After she practices for t hours, her speed, in words per minute, is given by the function

$$S(t) = 200[1 - (0.86)^t].$$

What is Sarah's speed after practicing for 10 hr? 20 hr? 40 hr? 100 hr?

65. *Advertising.* A company begins a radio advertising campaign in New York City to market a new DVD video game. The percentage of the target market that buys a game is generally a function of the length of the advertising campaign. The estimated percentage is given by

$$f(t) = 100(1 - e^{-0.04t}),$$

where t is the number of days of the campaign. Find $f(25)$, the percentage of the target market that has bought the product after a 25-day advertising campaign.

66. *Growth of a Stock.* The value of a stock is given by the function

$$V(t) = 58(1 - e^{-1.1t}) + 20,$$

where V is the value of the stock after time t, in months. Find $V(1)$, $V(2)$, $V(4)$, $V(6)$, and $V(12)$.

Technology Connection

In Exercises 67–80, use a graphing calculator to match the equation with one of the figures (a)–(n), which follow.

a)

b)
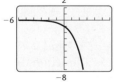

c)
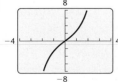

d)
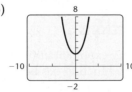

e)
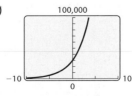

f)
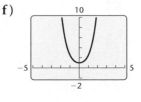

g)
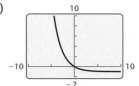

h)
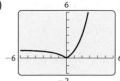

i)
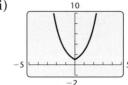

j)
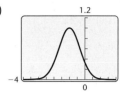

k)
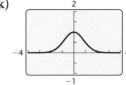

l)
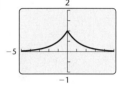

m)
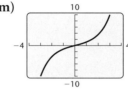

n)
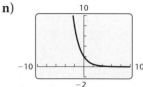

67. $y = 3^x - 3^{-x}$

68. $y = 3^{-(x+1)^2}$

69. $f(x) = -2.3^x$

70. $f(x) = 30{,}000(1.4)^x$

71. $y = 2^{-|x|}$

72. $y = 2^{-(x-1)}$

73. $f(x) = (0.58)^x - 1$

74. $y = 2^x + 2^{-x}$

75. $g(x) = e^{|x|}$

76. $f(x) = |2^x - 1|$

77. $y = 2^{-x^2}$

78. $y = |2^{x^2} - 8|$

79. $g(x) = \dfrac{e^x - e^{-x}}{2}$

80. $f(x) = \dfrac{e^x + e^{-x}}{2}$

Collaborative Discussion and Writing

81. Describe the differences between the graphs of $f(x) = x^3$ and $g(x) = 3^x$.

82. Suppose that $10,000 is invested for 8 yr at 6.4% interest, compounded annually. In what year will the most interest be earned? Why?

Graph the pair of equations using the same set of axes. Then compare the results.

83. $y = 3^x$, $x = 3^y$

84. $y = 1^x$, $x = 1^y$

Skill Maintenance

Simplify.

85. $(1 - 4i)(7 + 6i)$ **86.** $\dfrac{2 - i}{3 + i}$

Find the x-intercepts and the zeros of the function.

87. $f(x) = 2x^2 - 13x - 7$

88. $h(x) = x^3 - 3x^2 + 3x - 1$

89. $h(x) = x^4 - x^2$

90. $g(x) = x^3 + x^2 - 12x$

Solve.

91. $x^3 + 6x^2 - 16x = 0$ **92.** $3x^2 - 6 = 5x$

Synthesis

93. Which is larger, 7^{π} or π^7? 70^{80} or 80^{70}?

4.3

Logarithmic Functions and Graphs

- *Graph logarithmic functions.*
- *Convert between exponential and logarithmic equations.*
- *Find common and natural logarithms using a calculator.*

We now consider *logarithmic*, or *logarithm*, *functions*. These functions are inverses of exponential functions and have many applications.

Logarithmic Functions

We have noted that every exponential function (with $a > 0$ and $a \neq 1$) is one-to-one. Thus such a function has an inverse that is a function. In this section, we will name these inverse functions logarithmic functions and use them in applications. For now, we draw their graphs by interchanging x and y.

EXAMPLE 1 Graph: $x = 2^y$.

Solution Note that x is alone on one side of the equation. We can find ordered pairs that are solutions by choosing values for y and then computing the corresponding x-values.

For $y = 0$, $x = 2^0 = 1$.
For $y = 1$, $x = 2^1 = 2$.
For $y = 2$, $x = 2^2 = 4$.
For $y = 3$, $x = 2^3 = 8$.

For $y = -1$, $x = 2^{-1} = \dfrac{1}{2^1} = \dfrac{1}{2}$.

For $y = -2$, $x = 2^{-2} = \dfrac{1}{2^2} = \dfrac{1}{4}$.

For $y = -3$, $x = 2^{-3} = \dfrac{1}{2^3} = \dfrac{1}{8}$.

x $x = 2^y$	y	(x, y)
1	0	$(1, 0)$
2	1	$(2, 1)$
4	2	$(4, 2)$
8	3	$(8, 3)$
$\dfrac{1}{2}$	-1	$\left(\dfrac{1}{2}, -1\right)$
$\dfrac{1}{4}$	-2	$\left(\dfrac{1}{4}, -2\right)$
$\dfrac{1}{8}$	-3	$\left(\dfrac{1}{8}, -3\right)$

(1) Choose values for y.
(2) Compute values for x.

We plot the points and connect them with a smooth curve. Note that the curve does not touch or cross the y-axis. The y-axis is a vertical asymptote.

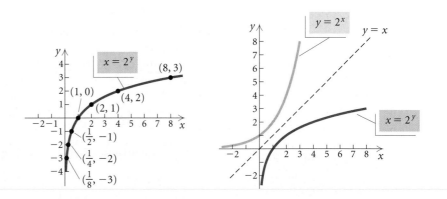

Note too that this curve looks just like the graph of $y = 2^x$, except that it is reflected across the line $y = x$, as we would expect for an inverse. The inverse of $y = 2^x$ is $x = 2^y$.

To find a formula for f^{-1} when $f(x) = 2^x$, we try to use the method of Section 4.1:

1. Replace $f(x)$ with y:
$$f(x) = 2^x$$
$$y = 2^x$$

2. Interchange x and y:
$$x = 2^y$$

3. Solve for y:
$$y = \text{the power to which we raise 2 to get } x$$

4. Replace y with $f^{-1}(x)$:
$$f^{-1}(x) = \text{the power to which we raise 2 to get } x.$$

Mathematicians have defined a new symbol to replace the words "the power to which we raise 2 to get x." That symbol is "$\log_2 x$," read "the logarithm, base 2, of x."

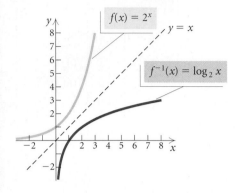

Logarithmic Function, Base 2

"$\log_2 x$," read "the logarithm, base 2, of x," means "the power to which we raise 2 to get x."

Thus if $f(x) = 2^x$, then $f^{-1}(x) = \log_2 x$. For example,

$$f^{-1}(8) = \log_2 8 = 3,$$

because

3 is the power to which we raise 2 to get 8.

Similarly, $\log_2 13$ is the power to which we raise 2 to get 13. As yet, we have no simpler way to say this other than

"$\log_2 13$ is the power to which we raise 2 to get 13."

Later, however, we will learn how to approximate this expression using a calculator.

For any exponential function $f(x) = a^x$, its inverse is called a **logarithmic function, base a**. The graph of the inverse can be obtained by reflecting the graph of $y = a^x$ across the line $y = x$, to obtain $x = a^y$. Then $x = a^y$ is equivalent to $y = \log_a x$. We read $\log_a x$ as "the logarithm, base a, of x."

The inverse of $f(x) = a^x$ is given by $f^{-1}(x) = \log_a x$.

Logarithmic Function, Base a

We define $y = \log_a x$ as that number y such that $x = a^y$, where $x > 0$ and a is a positive constant other than 1.

Let's look at the graphs of $f(x) = a^x$ and $f^{-1}(x) = \log_a x$ for $a > 1$ and $0 < a < 1$.

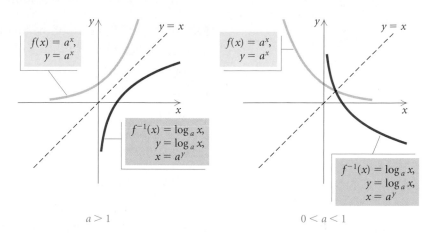

$a > 1$ $0 < a < 1$

Note that the graphs of $f(x)$ and $f^{-1}(x)$ are reflections of each other across the line $y = x$.

CONNECTING THE CONCEPTS

COMPARING EXPONENTIAL AND LOGARITHMIC FUNCTIONS

Generally, we use a number that is greater than 1 for a logarithmic base. In the following table, we compare exponential and logarithmic functions with bases a greater than 1. Similar statements could be made for a, where $0 < a < 1$. It is helpful to visualize the differences by carefully observing the graphs.

EXPONENTIAL FUNCTION

$y = a^x$
$f(x) = a^x$
$a > 1$
Continuous
One-to-one
Domain: All real
   numbers, $(-\infty, \infty)$
Range: All positive
   real numbers, $(0, \infty)$
Increasing
Horizontal asymptote is x-axis:
   $(a^x \to 0$ as $x \to -\infty)$
y-intercept: $(0, 1)$
There is no x-intercept.

LOGARITHMIC FUNCTION

$x = a^y$
$f^{-1}(x) = \log_a x$
$a > 1$
Continuous
One-to-one
Domain: All positive
   real numbers, $(0, \infty)$
Range: All real
   numbers, $(-\infty, \infty)$
Increasing
Vertical asymptote is y-axis:
   $(\log_a x \to -\infty$ as $x \to 0^+)$
x-intercept: $(1, 0)$
There is no y-intercept.

Finding Certain Logarithms

Let's use the definition of logarithms to find some logarithmic values.

EXAMPLE 2 Find each of the following logarithms.

a) $\log_{10} 10{,}000$ **b)** $\log_{10} 0.01$

c) $\log_2 8$ **d)** $\log_9 3$

e) $\log_6 1$ **f)** $\log_8 8$

Solution

a) The exponent to which we raise 10 to obtain 10,000 is 4; thus $\log_{10} 10{,}000 = 4$.

b) We have $0.01 = \dfrac{1}{100} = \dfrac{1}{10^2} = 10^{-2}$. The exponent to which we raise 10 to get 0.01 is -2, so $\log_{10} 0.01 = -2$.

c) $8 = 2^3$. The exponent to which we raise 2 to get 8 is 3, so $\log_2 8 = 3$.

d) $3 = \sqrt{9} = 9^{1/2}$. The exponent to which we raise 9 to get 3 is $\frac{1}{2}$, so $\log_9 3 = \frac{1}{2}$.

e) $1 = 6^0$. The exponent to which we raise 6 to get 1 is 0, so $\log_6 1 = 0$.

f) $8 = 8^1$. The exponent to which we raise 8 to get 8 is 1, so $\log_8 8 = 1$.

     Examples 2(e) and 2(f) illustrate two important properties of logarithms. The property $\log_a 1 = 0$ follows from the fact that $a^0 = 1$. Thus, $\log_5 1 = 0, \log_{10} 1 = 0$, and so on. The property $\log_a a = 1$ follows from the fact that $a^1 = a$. Thus, $\log_5 5 = 1, \log_{10} 10 = 1$, and so on.

> $\log_a 1 = 0$ and $\log_a a = 1$,   for any logarithmic base a.

Converting Between Exponential and Logarithmic Equations

It is helpful in dealing with logarithmic functions to remember that a logarithm of a number is an *exponent*. It is the exponent y in $x = a^y$. You might think to yourself, "the logarithm, base a, of a number x is the power to which a must be raised to get x."

     We are led to the following. (The symbol $\longleftrightarrow$ means that the two statements are equivalent; that is, when one is true, the other is true. The words "if and only if" can be used in place of $\longleftrightarrow$.)

> $\log_a x = y \longleftrightarrow x = a^y$ A logarithm is an exponent!

EXAMPLE 3 Convert each of the following to a logarithmic equation.

a) $16 = 2^x$ **b)** $10^{-3} = 0.001$ **c)** $e^t = 70$

Solution

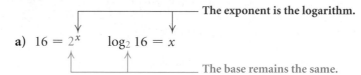

The exponent is the logarithm.

a) $16 = 2^x$ $\quad \log_2 16 = x$

The base remains the same.

b) $10^{-3} = 0.001 \rightarrow \log_{10} 0.001 = -3$

c) $e^t = 70 \rightarrow \log_e 70 = t$

EXAMPLE 4 Convert each of the following to an exponential equation.

a) $\log_2 32 = 5$ **b)** $\log_a Q = 8$ **c)** $x = \log_t M$

Solution

The logarithm is the exponent.

a) $\log_2 32 = 5$ $\quad 2^5 = 32$

The base remains the same.

b) $\log_a Q = 8 \rightarrow a^8 = Q$

c) $x = \log_t M \rightarrow t^x = M$

Finding Logarithms on a Calculator

Before calculators became so widely available, base-10 logarithms, or **common logarithms,** were used extensively to simplify complicated calculations. In fact, that is why logarithms were invented. The abbreviation **log,** with no base written, is used to represent common logarithms, or base-10 logarithms. Thus,

$\quad \log x \quad$ means $\quad \log_{10} x.$

For example, $\log 29$ means $\log_{10} 29$. Let's compare $\log 29$ with $\log 10$ and $\log 100$:

$$\left.\begin{array}{r} \log 10 = \log_{10} 10 = 1 \\ \log 29 = ? \\ \log 100 = \log_{10} 100 = 2 \end{array}\right\}$$ Since 29 is between 10 and 100, it seems reasonable that $\log 29$ is between 1 and 2.

On a calculator, the key for common logarithms is generally marked $\boxed{\text{LOG}}$. Using that key, we find that

$$\log 29 \approx 1.462397998 \approx 1.4624$$

rounded to four decimal places. Since $1 < 1.4624 < 2$, our answer seems reasonable. This also tells us that $10^{1.4624} \approx 29$.

EXAMPLE 5   Find each of the following common logarithms on a calculator. If you are using a graphing calculator, set the calculator in REAL mode. Round to four decimal places.

a) log 645,778   **b)** log 0.0000239   **c)** log (−3)

Solution

FUNCTION VALUE	READOUT	ROUNDED
a) log 645,778	log(645778) 5.810083246	5.8101
b) log 0.0000239	log(0.0000239) −4.621602099	−4.6216
c) log (−3)	ERR:NONREAL ANS *	Does not exist

Since 5.810083246 is the power to which we raise 10 to get 645,778, we can check part (a) by finding $10^{5.810083246}$. We can check part (b) in a similar manner. In part (c), log (−3) does not exist as a real number because there is no real-number power to which we can raise 10 to get −3. The number 10 raised to any real-number power is positive. The common logarithm of a negative number does not exist as a real number. Recall that logarithmic functions are inverses of exponential functions, and since the range of an exponential function is $(0, \infty)$, the domain of $f(x) = \log_a x$ is $(0, \infty)$.   ▬

Natural Logarithms

Logarithms, base e, are called **natural logarithms.** The abbreviation "ln" is generally used for natural logarithms. Thus,

$$\ln x \quad \text{means} \quad \log_e x.$$

For example, ln 53 means $\log_e 53$. On a calculator, the key for natural logarithms is generally marked $\boxed{\text{LN}}$. Using that key, we find that

$$\ln 53 \approx 3.970291914$$
$$\approx 3.9703$$

rounded to four decimal places. This also tells us that $e^{3.9703} \approx 53$.

---

*If the graphing calculator is set in $a + bi$ mode, the readout is .4771212547 + 1.364376354i.

EXAMPLE 6 Find each of the following natural logarithms on a calculator. If you are using a graphing calculator, set the calculator in REAL mode. Round to four decimal places.

a) ln 645,778   b) ln 0.0000239   c) ln (−5)
d) ln e e) ln 1

Solution

FUNCTION VALUE	READOUT	ROUNDED
a) ln 645,778	ln(645778) 13.37821107	13.3782
b) ln 0.0000239	ln(0.0000239) −10.6416321	−10.6416
c) ln (−5)	ERR:NONREAL ANS *	Does not exist
d) ln e	ln(e) 1	1
e) ln 1	ln(1) 0	0

Since 13.37821107 is the power to which we raise e to get 645,778, we can check part (a) by finding $e^{13.37821107}$. We can check parts (b), (d), and (e) in a similar manner. In parts (d) and (e), note that ln e = $\log_e e$ = 1 and ln 1 = $\log_e 1$ = 0.

ln 1 = 0 and ln e = 1, for the logarithmic base e.

Changing Logarithmic Bases

Most calculators give the values of both common logarithms and natural logarithms. To find a logarithm with a base other than 10 or e, we can use the following conversion formula.

---

*If the graphing calculator is set in $a + bi$ mode, the readout is 1.609437912 + 3.141592654i.

> **The Change-of-Base Formula**
>
> For any logarithmic bases a and b, and any positive number M,
>
> $$\log_b M = \frac{\log_a M}{\log_a b}.$$

We will prove this result in the next section.

EXAMPLE 7 Find $\log_5 8$ using common logarithms.

Solution First, we let $a = 10$, $b = 5$, and $M = 8$. Then we substitute into the change-of-base formula:

$$\log_5 8 = \frac{\log_{10} 8}{\log_{10} 5} \qquad \textbf{Substituting}$$

$$\approx 1.2920. \qquad \textbf{Using a calculator}$$

Since $\log_5 8$ is the power to which we raise 5 to get 8, we would expect this power to be greater than 1 ($5^1 = 5$) and less than 2 ($5^2 = 25$), so the result is reasonable. ▬

We can also use base e for a conversion.

EXAMPLE 8 Find $\log_5 8$ using natural logarithms.

Solution Substituting e for a, 5 for b, and 8 for M, we have

$$\log_5 8 = \frac{\log_e 8}{\log_e 5}$$

$$= \frac{\ln 8}{\ln 5} \approx 1.2920.$$ ▬

Graphs of Logarithmic Functions

Let's now consider graphs of logarithmic functions.

EXAMPLE 9 Graph: $y = f(x) = \log_5 x$.

Solution

The equation $y = \log_5 x$ is equivalent to $x = 5^y$. We can find ordered pairs that are solutions by choosing values for y and computing the x-values. We then plot points, remembering that x is still the first coordinate.

Technology Connection

To graph $y = \log_5 x$ in Example 9 with a graphing calculator, we must first change the base. Here we change from base 5 to base e:

$$y = \log_5 x = \frac{\ln x}{\ln 5}.$$

The graph is shown below.

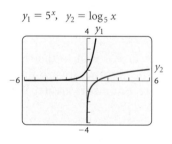

$$y = \log_5 x = \frac{\ln x}{\ln 5}$$

Some graphing calculators can graph inverses without the need to first find an equation of the inverse. If we begin with $y_1 = 5^x$, the graphs of both y_1 and its inverse, $y_2 = \log_5 x$, will be drawn as shown below.

$$y_1 = 5^x, \quad y_2 = \log_5 x$$

For $y = 0$, $x = 5^0 = 1$.
For $y = 1$, $x = 5^1 = 5$.
For $y = 2$, $x = 5^2 = 25$.
For $y = 3$, $x = 5^3 = 125$.

For $y = -1$, $x = 5^{-1} = \dfrac{1}{5}$.

For $y = -2$, $x = 5^{-2} = \dfrac{1}{25}$.

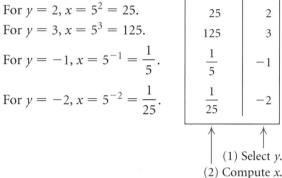

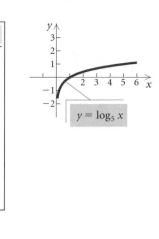

x, or 5^y	y
1	0
5	1
25	2
125	3
$\dfrac{1}{5}$	-1
$\dfrac{1}{25}$	-2

(1) Select y.
(2) Compute x.

EXAMPLE 10 Graph: $g(x) = \ln x$.

Solution To graph $y = g(x) = \ln x$, we select values for x and use a calculator to find the corresponding values of $\ln x$. We then plot points and draw the curve.

x	$g(x)$ $g(x) = \ln x$
0.5	-0.7
1	0
2	0.7
3	1.1
4	1.4
5	1.6

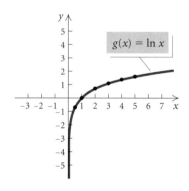

We could also write $g(x) = \ln x$, or $y = \ln x$, as $x = e^y$, select values for y, and use a calculator to find the corresponding values of x.

Recall that the graph of $f(x) = \log_a x$, for any base a, has the x-intercept $(1, 0)$. The domain is the set of positive real numbers, and the range is the set of all real numbers. The y-axis is the vertical asymptote.

EXAMPLE 11 Graph each of the following. Describe how each graph can be obtained from the graph of $y = \ln x$. Give the domain and the vertical asymptote of each function.

a) $f(x) = \ln(x + 3)$
b) $f(x) = 3 - \frac{1}{2}\ln x$
c) $f(x) = |\ln(x - 1)|$

Solution

a) The graph is a shift of the graph of $y = \ln x$ left 3 units. The domain is the set of all real numbers greater than -3, $(-3, \infty)$. The line $x = -3$ is the vertical asymptote.

x	$f(x)$
-2.9	-2.303
-2	0
0	1.099
2	1.609
4	1.946

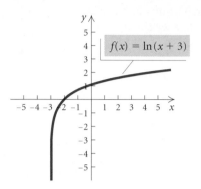

b) The graph is a vertical shrinking of the graph of $y = \ln x$, followed by a reflection across the x-axis, and then a translation up 3 units. The domain is the set of all positive real numbers, $(0, \infty)$. The y-axis is the vertical asymptote.

x	$f(x)$
0.1	4.151
1	3
3	2.451
6	2.104
9	1.901

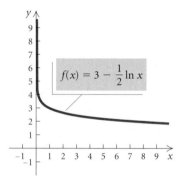

c) The graph is a translation of the graph of $y = \ln x$, 1 unit to the right. Then the absolute value has the effect of reflecting negative outputs across the x-axis. The domain is the set of all real numbers greater than 1, $(1, \infty)$. The line $x = 1$ is the vertical asymptote.

x	$f(x)$
1.1	2.303
2	0
4	1.099
6	1.609
8	1.946

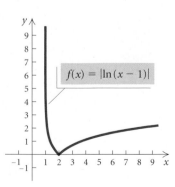

Applications

EXAMPLE 12 *Walking Speed.* In a study by psychologists Bornstein and Bornstein, it was found that the average walking speed w, in feet per second, of a person living in a city of population P, in thousands, is given by the function

$$w(P) = 0.37 \ln P + 0.05$$

(*Source*: *International Journal of Psychology*).

a) The population of Philadelphia, Pennsylvania, is 1,517,600. Find the average walking speed of people living in Philadelphia.

b) The population of Salem, Oregon, is 137,000. Find the average walking speed of people living in Salem.

Solution

a) Since 1,517,600 = 1517.6 thousand, we substitute 1517.6 for P, since P is in thousands:

$$w(1517.6) = 0.37 \ln 1517.6 + 0.05 \qquad \textbf{Substituting}$$
$$\approx 2.8 \text{ ft/sec.} \qquad \textbf{Finding the natural logarithm and simplifying}$$

The average walking speed of people living in Philadelphia is about 2.8 ft/sec.

b) We substitute 137 for P:

$$w(137) = 0.37 \ln 137 + 0.05 \qquad \textbf{Substituting}$$
$$\approx 1.9 \text{ ft/sec.}$$

The average walking speed of people living in Salem is about 1.9 ft/sec.

EXAMPLE 13 *Earthquake Magnitude.* The magnitude R, measured on the Richter scale, of an earthquake of intensity I is defined as

$$R = \log \frac{I}{I_0},$$

where I_0 is a minimum intensity used for comparison. We can think of I_0 as a threshold intensity that is the weakest earthquake that can be recorded on a seismograph. If one earthquake is 10 times as intense as another, its magnitude on the Richter scale is 1 greater than that of the other. If one earthquake is 100 times as intense as another, its magnitude on the Richter scale is 2 higher, and so on. Thus an earthquake whose magnitude is 7 on the Richter scale is 10 times as intense as an earthquake whose magnitude is 6. Earthquake intensities can be interpreted as multiples of the minimum intensity I_0.

The earthquake in Ahmedabad, India, on January 26, 2001, had an intensity of $10^{7.9} \cdot I_0$. What was its magnitude on the Richter scale?

Solution We substitute into the formula:

$$R = \log \frac{I}{I_0} = \log \frac{10^{7.9} I_0}{I_0} = \log 10^{7.9} = 7.9.$$

The magnitude of the earthquake was 7.9 on the Richter scale.

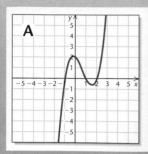

A

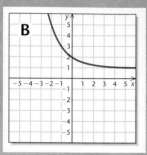

B

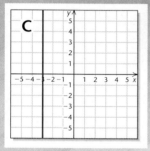

C

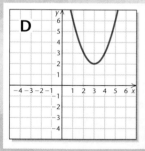

D

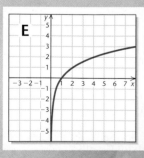

E

Visualizing the Graph

Match the equation or function with its graph.

1. $f(x) = 4^x$

2. $f(x) = \ln x - 3$

3. $(x + 3)^2 + y^2 = 9$

4. $f(x) = 2^{-x} + 1$

5. $f(x) = \log_2 x$

6. $f(x) = x^3 - 2x^2 - x + 2$

7. $x = -3$

8. $f(x) = e^x - 4$

9. $f(x) = (x - 3)^2 + 2$

10. $3x = 6 + y$

Answers on page A-31

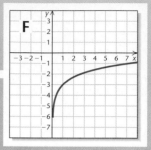

F

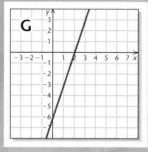

G

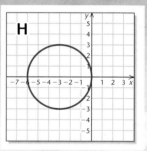

H

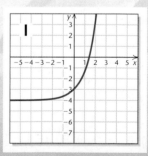

I

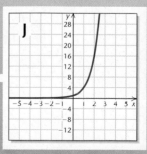

J

4.3

Exercise Set

Graph each of the following.

1. $x = 3^y$

2. $x = 4^y$

3. $x = \left(\frac{1}{2}\right)^y$

4. $x = \left(\frac{4}{3}\right)^y$

5. $y = \log_3 x$

6. $y = \log_4 x$

7. $f(x) = \log x$

8. $f(x) = \ln x$

Find each of the following. Do not use a calculator.

9. $\log_2 16$

10. $\log_3 9$

11. $\log_5 125$

12. $\log_2 64$

13. $\log 0.001$

14. $\log 100$

15. $\log_2 \frac{1}{4}$

16. $\log_8 2$

17. $\ln 1$

18. $\ln e$

19. $\log 10$

20. $\log 1$

21. $\log_5 5^4$

22. $\log \sqrt{10}$

23. $\log_3 \sqrt[4]{3}$

24. $\log 10^{8/5}$

25. $\log 10^{-7}$

26. $\log_5 1$

27. $\log_{49} 7$

28. $\log_3 3^{-2}$

29. $\ln e^{3/4}$

30. $\log_2 \sqrt{2}$

31. $\log_4 1$

32. $\ln e^{-5}$

33. $\ln \sqrt{e}$

34. $\log_{64} 4$

Convert to a logarithmic equation.

35. $10^3 = 1000$

36. $5^{-3} = \frac{1}{125}$

37. $8^{1/3} = 2$

38. $10^{0.3010} = 2$

39. $e^3 = t$

40. $Q^t = x$

41. $e^2 = 7.3891$

42. $e^{-1} = 0.3679$

43. $p^k = 3$

44. $e^{-t} = 4000$

Convert to an exponential equation.

45. $\log_5 5 = 1$

46. $t = \log_4 7$

47. $\log 0.01 = -2$

48. $\log 7 = 0.845$

49. $\ln 30 = 3.4012$

50. $\ln 0.38 = -0.9676$

51. $\log_a M = -x$

52. $\log_t Q = k$

53. $\log_a T^3 = x$

54. $\ln W^5 = t$

Find each of the following using a calculator. Round to four decimal places.

55. $\log 3$

56. $\log 8$

57. $\log 532$

58. $\log 93,100$

59. $\log 0.57$

60. $\log 0.082$

61. $\log(-2)$

62. $\ln 50$

63. $\ln 2$

64. $\ln(-4)$

65. $\ln 809.3$

66. $\ln 0.00037$

67. $\ln(-1.32)$

68. $\ln 0$

Find the logarithm using common logarithms and the change-of-base formula.

69. $\log_4 100$

70. $\log_3 20$

71. $\log_{100} 0.3$

72. $\log_\pi 100$

73. $\log_{200} 50$

74. $\log_{5.3} 1700$

Find the logarithm using natural logarithms and the change-of-base formula.

75. $\log_3 12$

76. $\log_4 25$

77. $\log_{100} 15$

78. $\log_9 100$

Graph the function and its inverse using the same set of axes.

79. $f(x) = 3^x, \ f^{-1}(x) = \log_3 x$

80. $f(x) = \log_4 x, \ f^{-1}(x) = 4^x$

81. $f(x) = \log x, \ f^{-1}(x) = 10^x$

82. $f(x) = e^x, \ f^{-1}(x) = \ln x$

For each of the following functions, briefly describe how the graph can be obtained from a basic logarithmic function. Then graph the function. Give the domain and the vertical asymptote of each function.

83. $f(x) = \log_2 (x + 3)$

84. $f(x) = \log_3 (x - 2)$

85. $y = \log_3 x - 1$

86. $y = 3 + \log_2 x$

87. $f(x) = 4 \ln x$

88. $f(x) = \frac{1}{2} \ln x$

89. $y = 2 - \ln x$

90. $y = \ln (x + 1)$

91. *Walking Speed.* Refer to Example 12. Various cities and their populations are given below. Find the average walking speed in each city.

   a) Columbus, Ohio: 711,500
   b) Las Vegas, Nevada: 478,400
   c) Corpus Christi, Texas: 277,000
   d) Independence, Missouri: 113,000
   e) Albuquerque, New Mexico: 448,600
   f) Anaheim, California: 328,000
   g) New York, New York: 8,008,300
   h) Chicago, Illinois: 2,896,000

92. *Earthquake Magnitude.* Refer to Example 13. Various locations of earthquakes and their intensities are given below. What was the magnitude on the Richter scale?

   a) Mexico City, 1978: $10^{7.85} \cdot I_0$
   b) San Francisco, 1906: $10^{8.25} \cdot I_0$
   c) Chile, 1960: $10^{9.6} \cdot I_0$
   d) Italy, 1980: $10^{7.85} \cdot I_0$
   e) San Francisco, 1989: $10^{6.9} \cdot I_0$

93. *Forgetting.* Students in an accounting class took a final exam and then took equivalent forms of the exam at monthly intervals thereafter. The average score $S(t)$, as a percent, after t months was found to be given by the function

$$S(t) = 78 - 15 \log (t + 1), \quad t \geq 0.$$

   a) What was the average score when the students initially took the test, $t = 0$?
   b) What was the average score after 4 months? 24 months?

94. *pH of Substances in Chemistry.* In chemistry, the pH of a substance is defined as

$$pH = -\log [H^+],$$

where H^+ is the hydrogen ion concentration, in moles per liter. Find the pH of each substance.

Litmus paper is used to test pH.

Substance	Hydrogen Ion Concentration
a) Pineapple juice	1.6×10^{-4}
b) Hair rinse	0.0013
c) Mouthwash	6.3×10^{-7}
d) Eggs	1.6×10^{-8}
e) Tomatoes	6.3×10^{-5}

95. Find the hydrogen ion concentration of each substance, given the pH (see Exercise 94). Express the answer in scientific notation.

Substance	pH
a) Tap water	7
b) Rainwater	5.4
c) Orange juice	3.2
d) Wine	4.8

96. *Advertising.* A model for advertising response is given by the function

$$N(a) = 1000 + 200 \ln a, \quad a \geq 1,$$

where $N(a)$ is the number of units sold when a is the amount spent on advertising, in thousands of dollars.

a) How many units were sold after spending $1000 ($a = 1$) on advertising?

b) How many units were sold after spending $5000?

97. *Loudness of Sound.* The **loudness L,** in bels (after Alexander Graham Bell), of a sound of intensity I is defined to be

$$L = \log \frac{I}{I_0},$$

where I_0 is the minimum intensity detectable by the human ear (such as the tick of a watch at 20 ft under quiet conditions). If a sound is 10 times as intense as another, its loudness is 1 bel greater than that of the other. If a sound is 100 times as intense as another, its loudness is 2 bels greater, and so on. The bel is a large unit, so a subunit, the **decibel,** is generally used. For L, in decibels, the formula is

$$L = 10 \log \frac{I}{I_0}.$$

Find the loudness, in decibels, of each sound with the given intensity.

Sound	Intensity
a) Library	$2510 \cdot I_0$
b) Dishwasher	$2,500,000 \cdot I_0$
c) Conversational speech	$10^6 \cdot I_0$
d) Heavy truck	$10^9 \cdot I_0$

Technology Connection

98. Use a graphing calculator that can graph inverses without the need to first find an equation of the inverse to graph the functions in Exercises 79–82.

Collaborative Discussion and Writing

99. If $\log b < 0$, what can you say about b?

100. Explain how the graph of $f(x) = \ln x$ can be used to obtain the graph of $g(x) = e^{x-2}$.

Skill Maintenance

Find the slope and the y-intercept of the line.

101. $y = 6$

102. $3x - 10y = 14$

103. $y = 2x - \frac{3}{13}$

104. $x = -4$

Use synthetic division to find the function values.

105. $g(x) = x^3 - 6x^2 + 3x + 10$; find $f(-5)$

106. $f(x) = x^4 - 2x^3 + x - 6$; find $f(-1)$

Find a polynomial function of degree 3 with the given numbers as zeros. Answers may vary.

107. $\sqrt{7}, -\sqrt{7}, 0$

108. $4i, -4i, 1$

Synthesis

Simplify.

109. $\dfrac{\log_5 8}{\log_5 2}$

110. $\dfrac{\log_3 64}{\log_3 16}$

Find the domain of the function.

111. $f(x) = \log_5 x^3$

112. $f(x) = \log_4 x^2$

113. $f(x) = \ln |x|$

114. $f(x) = \log (3x - 4)$

Solve.

115. $\log_2 (2x + 5) < 0$

116. $\log_2 (x - 3) \geq 4$

In Exercises 117–120, match the equation with one of figures (a)–(d), which follow.

a)

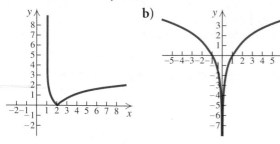

b)

c)

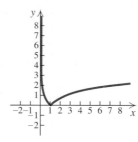

d)

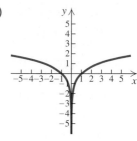

117. $f(x) = \ln|x|$

118. $f(x) = |\ln x|$

119. $f(x) = \ln x^2$

120. $g(x) = |\ln(x - 1)|$

4.4

Properties of Logarithmic Functions

• *Convert from logarithms of products, powers, and quotients to expressions in terms of individual logarithms, and conversely.*
• *Simplify expressions of the type $\log_a a^x$ and $a^{\log_a x}$.*

We now establish some properties of logarithmic functions. These properties are based on the corresponding rules for exponents.

Logarithms of Products

The first property of logarithms corresponds to the product rule for exponents: $a^m \cdot a^n = a^{m+n}$.

> ### The Product Rule
>
> For any positive numbers M and N and any logarithmic base a,
>
> $$\log_a MN = \log_a M + \log_a N.$$
>
> (The logarithm of a product is the sum of the logarithms of the factors.)

EXAMPLE 1 Express as a sum of logarithms: $\log_3(9 \cdot 27)$.

Solution   We have

$$\log_3(9 \cdot 27) = \log_3 9 + \log_3 27. \qquad \textbf{Using the product rule}$$

As a check, note that

$$\log_3(9 \cdot 27) = \log_3 243 = 5 \qquad \textbf{$3^5 = 243$}$$

and $\log_3 9 + \log_3 27 = 2 + 3 = 5.$ **$3^2 = 9; 3^3 = 27$**

EXAMPLE 2   Express as a single logarithm: $\log_2 p^3 + \log_2 q$.

Solution   We have

$$\log_2 p^3 + \log_2 q = \log_2(p^3 q).$$ —

A PROOF OF THE PRODUCT RULE: Let $\log_a M = x$ and $\log_a N = y$. Converting to exponential equations, we have $a^x = M$ and $a^y = N$. Then

$$MN = a^x \cdot a^y = a^{x+y}.$$

Converting back to a logarithmic equation, we get

$$\log_a MN = x + y.$$

Remembering what x and y represent, we know it follows that

$$\log_a MN = \log_a M + \log_a N.$$

Logarithms of Powers

The second property of logarithms corresponds to the power rule for exponents: $(a^m)^n = a^{mn}$.

> ### The Power Rule
>
> For any positive number M, any logarithmic base a, and any real number p,
>
> $$\log_a M^p = p \log_a M.$$
>
> (The logarithm of a power of M is the exponent times the logarithm of M.)

EXAMPLE 3   Express each of the following as a product.

a) $\log_a 11^{-3}$ **b)** $\log_a \sqrt[4]{7}$ **c)** $\ln x^6$

Solution

a) $\log_a 11^{-3} = -3 \log_a 11$ **Using the power rule**

b) $\log_a \sqrt[4]{7} = \log_a 7^{1/4}$ **Writing exponential notation**

$\qquad\qquad = \frac{1}{4} \log_a 7$ **Using the power rule**

c) $\ln x^6 = 6 \ln x$ **Using the power rule**   —

| RATIONAL EXPONENTS |
| REVIEW SECTION R.6. |

A PROOF OF THE POWER RULE: Let $x = \log_a M$. The equivalent exponential equation is $a^x = M$. Raising both sides to the power p, we obtain

$$(a^x)^p = M^p, \quad \text{or} \quad a^{xp} = M^p.$$

Converting back to a logarithmic equation, we get

$$\log_a M^p = xp.$$

But $x = \log_a M$, so substituting gives us

$$\log_a M^p = (\log_a M)p = p \log_a M.$$

Logarithms of Quotients

The third property of logarithms corresponds to the quotient rule for exponents: $a^m/a^n = a^{m-n}$.

The Quotient Rule

For any positive numbers M and N, and any logarithmic base a,

$$\log_a \frac{M}{N} = \log_a M - \log_a N.$$

(The logarithm of a quotient is the logarithm of the numerator minus the logarithm of the denominator.

EXAMPLE 4   Express as a difference of logarithms: $\log_t \dfrac{8}{w}$.

Solution

$$\log_t \frac{8}{w} = \log_t 8 - \log_t w \qquad \textbf{Using the quotient rule}$$

EXAMPLE 5   Express as a single logarithm: $\log_b 64 - \log_b 16$.

Solution

$$\log_b 64 - \log_b 16 = \log_b \frac{64}{16} = \log_b 4$$

A PROOF OF THE QUOTIENT RULE: The proof follows from both the product and the power rules:

$$\log_a \frac{M}{N} = \log_a MN^{-1}$$

$$= \log_a M + \log_a N^{-1} \qquad \textbf{Using the product rule}$$

$$= \log_a M + (-1) \log_a N \qquad \textbf{Using the power rule}$$

$$= \log_a M - \log_a N.$$

Common Errors

$\log_a MN \neq (\log_a M)(\log_a N)$ The logarithm of a product is *not* the product of the logarithms.

$\log_a(M + N) \neq \log_a M + \log_a N$ The logarithm of a sum is *not* the sum of the logarithms.

$\log_a \dfrac{M}{N} \neq \dfrac{\log_a M}{\log_a N}$ The logarithm of a quotient is *not* the quotient of the logarithms.

$(\log_a M)^P \neq P \log_a M$ The power of a logarithm is *not* the exponent times the logarithm.

Applying the Properties

EXAMPLE 6   Express each of the following in terms of sums and differences of logarithms.

a) $\log_a \dfrac{x^2 y^5}{z^4}$

b) $\log_a \sqrt[3]{\dfrac{a^2 b}{c^5}}$

c) $\log_b \dfrac{ay^5}{m^3 n^4}$

Solution

a) $\log_a \dfrac{x^2 y^5}{z^4} = \log_a(x^2 y^5) - \log_a z^4$ Using the quotient rule

$= \log_a x^2 + \log_a y^5 - \log_a z^4$ Using the product rule

$= 2 \log_a x + 5 \log_a y - 4 \log_a z$ Using the power rule

b) $\log_a \sqrt[3]{\dfrac{a^2 b}{c^5}} = \log_a \left(\dfrac{a^2 b}{c^5}\right)^{1/3}$ **Writing exponential notation**

$= \dfrac{1}{3} \log_a \dfrac{a^2 b}{c^5}$ **Using the power rule**

$= \dfrac{1}{3} (\log_a a^2 b - \log_a c^5)$ **Using the quotient rule. The parentheses are important.**

$= \dfrac{1}{3} (2 \log_a a + \log_a b - 5 \log_a c)$ **Using the product and power rules**

$= \dfrac{1}{3} (2 + \log_a b - 5 \log_a c)$ $\log_a a = 1$

$= \dfrac{2}{3} + \dfrac{1}{3} \log_a b - \dfrac{5}{3} \log_a c$ **Multiplying to remove parentheses**

c) $\log_b \dfrac{ay^5}{m^3 n^4} = \log_b ay^5 - \log_b m^3 n^4$ **Using the quotient rule**

$= (\log_b a + \log_b y^5) - (\log_b m^3 + \log_b n^4)$ **Using the product rule**

$= \log_b a + \log_b y^5 - \log_b m^3 - \log_b n^4$ **Removing parentheses**

$= \log_b a + 5 \log_b y - 3 \log_b m - 4 \log_b n$ **Using the power rule**

EXAMPLE 7 Express as a single logarithm:

$$5 \log_b x - \log_b y + \frac{1}{4} \log_b z.$$

Solution

$$5 \log_b x - \log_b y + \frac{1}{4} \log_b z = \log_b x^5 - \log_b y + \log_b z^{1/4}$$

Using the power rule

$$= \log_b \frac{x^5}{y} + \log_b z^{1/4}$$

Using the quotient rule

$$= \log_b \frac{x^5 z^{1/4}}{y}, \text{ or } \log_b \frac{x^5 \sqrt[4]{z}}{y}$$

Using the product rule

EXAMPLE 8 Given that $\log_a 2 \approx 0.301$ and $\log_a 3 \approx 0.477$, find each of the following.

a) $\log_a 6$ b) $\log_a \dfrac{2}{3}$ c) $\log_a 81$

d) $\log_a \dfrac{1}{4}$ e) $\log_a 5$ f) $\dfrac{\log_a 3}{\log_a 2}$

Solution

a) $\log_a 6 = \log_a(2 \cdot 3) = \log_a 2 + \log_a 3$ **Using the product rule**
$$\approx 0.301 + 0.477$$
$$\approx 0.778$$

b) $\log_a \frac{2}{3} = \log_a 2 - \log_a 3$ **Using the quotient rule**
$$\approx 0.301 - 0.477 \approx -0.176$$

c) $\log_a 81 = \log_a 3^4 = 4 \log_a 3$ **Using the power rule**
$$\approx 4(0.477) \approx 1.908$$

d) $\log_a \frac{1}{4} = \log_a 1 - \log_a 4$ **Using the quotient rule**
$$= 0 - \log_a 2^2$$ $\log_a 1 = 0$
$$= -2 \log_a 2$$ **Using the power rule**
$$\approx -2(0.301) \approx -0.602$$

e) $\log_a 5$ *cannot* be found using these properties and the given information.

$$(\log_a 5 \neq \log_a 2 + \log_a 3)$$ $\log_a 2 + \log_a 3 = \log_a 2 \cdot 3 = \log_a 6$

f) $\dfrac{\log_a 3}{\log_a 2} \approx \dfrac{0.477}{0.301} \approx 1.585$ **We simply divide, not using any of the properties.**

Simplifying Expressions of the Type $\log_a a^x$ and $a^{\log_a x}$

We have two final properties to consider. The first follows from the product rule: Since $\log_a a^x = x \log_a a = x \cdot 1 = x$, we have $\log_a a^x = x$. This property also follows from the definition of a logarithm: x is the power to which we raise a in order to get a^x.

> ### The Logarithm of a Base to a Power
>
> For any base a and any real number x,
>
> $$\log_a a^x = x.$$
>
> (The logarithm, base a, of a to a power is the power.)

EXAMPLE 9 Simplify each of the following.

a) $\log_a a^8$ **b)** $\ln e^{-t}$ **c)** $\log 10^{3k}$

Solution

a) $\log_a a^8 = 8$ 8 is the power to which we raise a in order to get a^8.

b) $\ln e^{-t} = \log_e e^{-t} = -t$ $\ln e^x = x$

c) $\log 10^{3k} = \log_{10} 10^{3k} = 3k$ ▬

    Let $M = \log_a x$. Then $a^M = x$. Substituting $\log_a x$ for M, we obtain $a^{\log_a x} = x$. This also follows from the definition of a logarithm: $\log_a x$ is the power to which a is raised in order to get x.

> ### A Base to a Logarithmic Power
>
> For any base a and any positive real number x,
>
> $$a^{\log_a x} = x.$$
>
> (The number a raised to the power $\log_a x$ is x.)

Study Tip

Immediately after each quiz or chapter test, write out a step-by-step solution to the questions you missed. Visit your instructor during office hours for help with problems that are still giving you trouble. When the week of the final examination arrives, you will be glad to have the excellent study guide these corrected tests provide.

EXAMPLE 10 Simplify each of the following.

a) $4^{\log_4 k}$ **b)** $e^{\ln 5}$ **c)** $10^{\log 7t}$

Solution

a) $4^{\log_4 k} = k$

b) $e^{\ln 5} = e^{\log_e 5} = 5$

c) $10^{\log 7t} = 10^{\log_{10} 7t} = 7t$ ▬

A PROOF OF THE CHANGE-OF-BASE FORMULA: We close this section by proving the change-of-base formula and summarizing the properties of logarithms considered thus far in this chapter. In Section 4.3, we used the change-of-base formula,

$$\log_b M = \frac{\log_a M}{\log_a b},$$

to make base conversions in order to find logarithmic values using a calculator. Let $x = \log_b M$. Then

$b^x = M$	Definition of logarithm
$\log_a b^x = \log_a M$	Taking the logarithm on both sides
$x \log_a b = \log_a M$	Using the power rule
$x = \dfrac{\log_a M}{\log_a b},$	Dividing by $\log_a b$

so

$$x = \log_b M = \frac{\log_a M}{\log_a b}.$$

Following is a summary of the properties of logarithms.

CHANGE-OF-BASE

REVIEW SECTION 4.3.

Summary of the Properties of Logarithms

The Product Rule:	$\log_a MN = \log_a M + \log_a N$
The Power Rule:	$\log_a M^p = p \log_a M$
The Quotient Rule:	$\log_a \dfrac{M}{N} = \log_a M - \log_a N$
The Change-of-Base Formula:	$\log_b M = \dfrac{\log_a M}{\log_a b}$
Other Properties:	$\log_a a = 1, \qquad \log_a 1 = 0,$
	$\log_a a^x = x, \qquad a^{\log_a x} = x$

4.4

Exercise Set

Express as a sum of logarithms.

1. $\log_3 (81 \cdot 27)$ **2.** $\log_2 (8 \cdot 64)$

3. $\log_5 (5 \cdot 125)$ **4.** $\log_4 (64 \cdot 32)$

5. $\log_t 8Y$ **6.** $\log 0.2x$

7. $\ln xy$ **8.** $\ln ab$

Express as a product.

9. $\log_b t^3$ **10.** $\log_a x^4$

11. $\log y^8$ **12.** $\ln y^5$

13. $\log_c K^{-6}$ **14.** $\log_b Q^{-8}$

15. $\ln \sqrt[3]{4}$ **16.** $\ln \sqrt{a}$

Express as a difference of logarithms.

17. $\log_t \dfrac{M}{8}$

18. $\log_a \dfrac{76}{13}$

19. $\log \dfrac{x}{y}$

20. $\ln \dfrac{a}{b}$

21. $\ln \dfrac{r}{s}$

22. $\log_b \dfrac{3}{w}$

Express in terms of sums and differences of logarithms.

23. $\log_a 6xy^5z^4$

24. $\log_a x^3y^2z$

25. $\log_b \dfrac{p^2q^5}{m^4b^9}$

26. $\log_b \dfrac{x^2y}{b^3}$

27. $\ln \dfrac{2}{3x^3y}$

28. $\log \dfrac{5a}{4b^2}$

29. $\log \sqrt{r^3t}$

30. $\ln \sqrt[3]{5x^5}$

31. $\log_a \sqrt{\dfrac{x^6}{p^5q^8}}$

32. $\log_c \sqrt[3]{\dfrac{y^3z^2}{x^4}}$

33. $\log_a \sqrt[4]{\dfrac{m^8n^{12}}{a^3b^5}}$

34. $\log_a \sqrt{\dfrac{a^6b^8}{a^2b^5}}$

Express as a single logarithm and, if possible, simplify.

35. $\log_a 75 + \log_a 2$

36. $\log 0.01 + \log 1000$

37. $\log 10{,}000 - \log 100$

38. $\ln 54 - \ln 6$

39. $\frac{1}{2} \log n + 3 \log m$

40. $\frac{1}{2} \log a - \log 2$

41. $\frac{1}{2} \log_a x + 4 \log_a y - 3 \log_a x$

42. $\frac{2}{5} \log_a x - \frac{1}{3} \log_a y$

43. $\ln x^2 - 2 \ln \sqrt{x}$

44. $\ln 2x + 3(\ln x - \ln y)$

45. $\ln(x^2 - 4) - \ln(x + 2)$

46. $\log(x^3 - 8) - \log(x - 2)$

47. $\log(x^2 - 5x - 14) - \log(x^2 - 4)$

48. $\log_a \dfrac{a}{\sqrt{x}} - \log_a \sqrt{ax}$

49. $\ln x - 3[\ln(x - 5) + \ln(x + 5)]$

50. $\frac{2}{3}[\ln(x^2 - 9) - \ln(x + 3)] + \ln(x + y)$

51. $\frac{3}{2} \ln 4x^6 - \frac{4}{5} \ln 2y^{10}$

52. $120\left(\ln \sqrt[5]{x^3} + \ln \sqrt[3]{y^2} - \ln \sqrt[4]{16z^5}\right)$

Given that $\log_a 2 \approx 0.301$, $\log_a 7 \approx 0.845$, and $\log_a 11 \approx 1.041$, find each of the following. Round the answer to the nearest thousandth.

53. $\log_a \frac{2}{11}$

54. $\log_a 14$

55. $\log_a 98$

56. $\log_a \frac{1}{7}$

57. $\dfrac{\log_a 2}{\log_a 7}$

58. $\log_a 9$

Given that $\log_b 2 \approx 0.693$, $\log_b 3 \approx 1.099$, and $\log_b 5 \approx 1.609$, find each of the following. Round the answer to the nearest thousandth.

59. $\log_b 125$

60. $\log_b \frac{5}{3}$

61. $\log_b \frac{1}{6}$

62. $\log_b 30$

63. $\log_b \dfrac{3}{b}$

64. $\log_b 15b$

Simplify.

65. $\log_p p^3$

66. $\log_t t^{2713}$

67. $\log_e e^{|x-4|}$

68. $\log_q q^{\sqrt{3}}$

69. $3^{\log_3 4x}$

70. $5^{\log_5 (4x-3)}$

71. $10^{\log w}$

72. $e^{\ln x^3}$

73. $\ln e^{8t}$

74. $\log 10^{-k}$

75. $\log_b \sqrt{b}$

76. $\log_b \sqrt{b^3}$

Collaborative Discussion and Writing

77. Given that $f(x) = a^x$ and $g(x) = \log_a x$, find $(f \circ g)(x)$ and $(g \circ f)(x)$. These results are alternative proofs of what properties of logarithms already proven in this section? Explain.

78. Explain the errors, if any, in the following:
$$\log_a ab^3 = (\log_a a)(\log_a b^3) = 3 \log_a b.$$

Skill Maintenance

In each of Exercises 79–88, classify the function as linear, quadratic, cubic, quartic, rational, exponential, or logarithmic.

79. $f(x) = 5 - x^2 + x^4$

80. $f(x) = 2^x$

81. $f(x) = -\frac{3}{4}$

82. $f(x) = 4^x - 8$

83. $f(x) = -\dfrac{3}{x}$

84. $f(x) = \log x + 6$

85. $f(x) = -\frac{1}{3}x^3 - 4x^2 + 6x + 42$

86. $f(x) = \dfrac{x^2 - 1}{x^2 + x - 6}$

87. $f(x) = \frac{1}{2}x + 3$

88. $f(x) = 2x^2 - 6x + 3$

Synthesis

Solve for x.

89. $5^{\log_5 8} = 2x$ **90.** $\ln e^{3x-5} = -8$

Express as a single logarithm and, if possible, simplify.

91. $\log_a(x^2 + xy + y^2) + \log_a(x - y)$

92. $\log_a(a^{10} - b^{10}) - \log_a(a + b)$

Express as a sum or a difference of logarithms.

93. $\log_a \dfrac{x - y}{\sqrt{x^2 - y^2}}$ **94.** $\log_a \sqrt{9 - x^2}$

95. Given that $\log_a x = 2$, $\log_a y = 3$, and $\log_a z = 4$, find

$$\log_a \frac{\sqrt[4]{y^2 z^5}}{\sqrt[4]{x^3 z^{-2}}}.$$

Determine whether each of the following is true. Assume that a, x, M, and N are positive.

96. $\log_a M + \log_a N = \log_a(M + N)$

97. $\log_a M - \log_a N = \log_a \dfrac{M}{N}$

98. $\dfrac{\log_a M}{\log_a N} = \log_a M - \log_a N$

99. $\dfrac{\log_a M}{x} = \log_a M^{1/x}$

100. $\log_a x^3 = 3 \log_a x$

101. $\log_a 8x = \log_a x + \log_a 8$

102. $\log_N(MN)^x = x \log_N M + x$

Suppose that $\log_a x = 2$. Find each of the following.

103. $\log_a\left(\dfrac{1}{x}\right)$ **104.** $\log_{1/a} x$

105. Simplify:

$\log_{10} 11 \cdot \log_{11} 12 \cdot \log_{12} 13 \cdots \log_{998} 999 \cdot \log_{999} 1000.$

Write each of the following without using logarithms.

106. $\log_a x + \log_a y - mz = 0$

107. $\ln a - \ln b + xy = 0$

Prove each of the following for any base a and any positive number x.

108. $\log_a\left(\dfrac{1}{x}\right) = -\log_a x = \log_{1/a} x$

109. $\log_a\left(\dfrac{x + \sqrt{x^2 - 5}}{5}\right) = -\log_a\left(x - \sqrt{x^2 - 5}\right)$

4.5

Solving Exponential and Logarithmic Equations

• *Solve exponential and logarithmic equations.*

Solving Exponential Equations

Equations with variables in the exponents, such as

$$3^x = 20 \quad \text{and} \quad 2^{5x} = 64,$$

are called **exponential equations.** We now consider solving exponential equations.

Sometimes, as is the case with the equation $2^{5x} = 64$, we can write each side as a power of the same number:

$$2^{5x} = 2^6.$$

We can then set the exponents equal and solve:

$$5x = 6$$
$$x = \tfrac{6}{5}, \text{ or } 1.2.$$

We use the following property.

Base–Exponent Property

For any $a > 0$, $a \neq 1$,

$$a^x = a^y \longleftrightarrow x = y.$$

ONE-TO-ONE FUNCTIONS

REVIEW SECTION 4.1.

This property follows from the fact that for any $a > 0$, $a \neq 1$, $f(x) = a^x$ is a one-to-one function. If $a^x = a^y$, then $f(x) = f(y)$. Then since f is one-to-one, it follows that $x = y$. Conversely, if $x = y$, it follows that $a^x = a^y$, since we are raising a to the same power.

EXAMPLE 1   Solve: $2^{3x-7} = 32$.

Algebraic Solution

Note that $32 = 2^5$. Thus we can write each side as a power of the same number:

$$2^{3x-7} = 2^5.$$

Since the bases are the same number, 2, we can use the base–exponent property and set the exponents equal:

$$3x - 7 = 5$$
$$3x = 12$$
$$x = 4.$$

CHECK:
$$\frac{2^{3x-7} = 32}{}$$
$$2^{3(4)-7} \;?\; 32$$
$$2^{12-7}$$
$$2^5$$
$$32 \;\bigg|\; 32 \quad \text{TRUE}$$

The solution is 4.

Visualizing the Solution

When we graph $y = 2^{3x-7}$ and $y = 32$, we find that the first coordinate of the point of intersection of the graphs is 4.

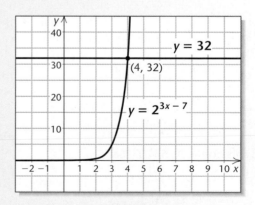

The solution of $2^{3x-7} = 32$ is 4.

Another property that is necessary when solving some exponential and logarithmic equations is as follows.

Property of Logarithmic Equality

For any $M > 0$, $N > 0$, $a > 0$, and $a \neq 1$,

$$\log_a M = \log_a N \longleftrightarrow M = N.$$

This property follows from the fact that for any $a > 0$, $a \neq 1$, $f(x) = \log_a x$ is a one-to-one function. If $\log_a x = \log_a y$, then $f(x) = f(y)$. Then since f is one-to-one, it follows that $x = y$. Conversely, if $x = y$, it follows that $\log_a x = \log_a y$, since we are taking the logarithm of the same number.

When it does not seem possible to write each side as a power of the same base, we can use the property of logarithmic equality and take the logarithm with any base on each side and then use the power rule for logarithms.

EXAMPLE 2 Solve: $3^x = 20$.

Algebraic Solution

We have

$$3^x = 20$$
$$\log 3^x = \log 20 \qquad \text{Taking the common logarithm on both sides}$$
$$x \log 3 = \log 20 \qquad \text{Using the power rule}$$
$$x = \frac{\log 20}{\log 3}. \qquad \text{Dividing by log 3}$$

This is an exact answer. We cannot simplify further, but we can approximate using a calculator:

$$x = \frac{\log 20}{\log 3} \approx 2.7268.$$

We can check this by finding $3^{2.7268}$:

$$3^{2.7268} \approx 20.$$

The solution is about 2.7268.

Visualizing the Solution

We graph $y = 3^x$ and $y = 20$. The first coordinate of the point of intersection of the graphs is the value of x for which $3^x = 20$ and is thus the solution of the equation.

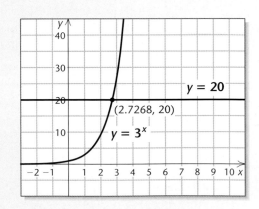

The solution is approximately 2.7268.

In Example 2, we took the common logarithm on both sides of the equation. Any base will give the same result. Let's try base 3. We have

$$3^x = 20$$
$$\log_3 3^x = \log_3 20$$
$$x = \log_3 20 \qquad \log_a a^x = x$$
$$x = \frac{\log 20}{\log 3} \approx 2.7268.$$

Note that we have to change the base to do the final calculation.

It will make our work easier if we take the natural logarithm when working with equations that have e as a base.

EXAMPLE 3 Solve: $e^{0.08t} = 2500$.

Algebraic Solution

We have

$$e^{0.08t} = 2500$$

$\ln e^{0.08t} = \ln 2500$ **Taking the natural logarithm on both sides**

$0.08t = \ln 2500$ **Finding the logarithm of a base to a power: $\log_a a^x = x$**

$t = \dfrac{\ln 2500}{0.08}$ **Dividing by 0.08**

$\approx 97.8.$

The solution is about 97.8.

Visualizing the Solution

The first coordinate of the point of intersection of the graphs of $y = e^{0.08t}$ and $y = 2500$ is about 97.8. This is the solution of the equation.

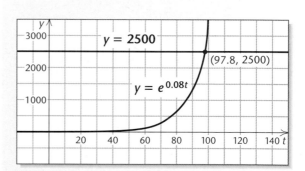

Technology Connection

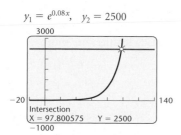

$y_1 = e^{0.08x}, \quad y_2 = 2500$

We can solve the equations in Examples 1–3 using the Intersect method. In Example 3, for instance, we graph $y = e^{0.08x}$ and $y = 2500$ and use the INTERSECT feature to find the coordinates of the point of intersection.

The first coordinate of the point of intersection is the solution of the equation $e^{0.08x} = 2500$. The solution is about 97.8. We could also write the equation in the form $e^{0.08x} - 2500 = 0$ and use the Zero method.

EXAMPLE 4   Solve: $e^x + e^{-x} - 6 = 0$.

Algebraic Solution

In this case, we have more than one term with x in the exponent:

$$e^x + e^{-x} - 6 = 0$$

$$e^x + \frac{1}{e^x} - 6 = 0 \qquad \begin{array}{l}\text{Rewriting } e^{-x} \text{ with}\\ \text{a positive exponent}\end{array}$$

$$e^{2x} + 1 - 6e^x = 0. \qquad \text{Multiplying both sides by } e^x$$

This equation is reducible to quadratic with $u = e^x$:

$$u^2 - 6u + 1 = 0.$$

Using the quadratic formula, we have

$$u = \frac{-(-6) \pm \sqrt{(-6)^2 - 4 \cdot 1 \cdot 1}}{2 \cdot 1}$$

$$u = \frac{6 \pm \sqrt{32}}{2} = \frac{6 \pm 4\sqrt{2}}{2}$$

$$u = 3 \pm 2\sqrt{2}$$

$$e^x = 3 \pm 2\sqrt{2}. \qquad \text{Replacing } u \text{ with } e^x$$

We now take the natural logarithm on both sides:

$$\ln e^x = \ln\left(3 \pm 2\sqrt{2}\right)$$

$$x = \ln\left(3 \pm 2\sqrt{2}\right). \qquad \text{Using } \ln e^x = x$$

Approximating each of the solutions, we obtain 1.76 and −1.76.

Visualizing the Solution

The solutions of the equation

$$e^x + e^{-x} - 6 = 0$$

are the zeros of the function

$$f(x) = e^x + e^{-x} - 6.$$

Note that the solutions are also the first coordinates of the x-intercepts of the graph of the function.

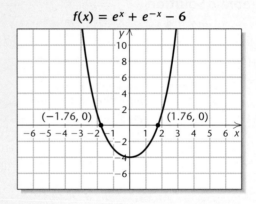

$$f(x) = e^x + e^{-x} - 6$$

The leftmost zero is about −1.76. The zero on the right is about 1.76. The solutions of the equation are approximately −1.76 and 1.76.

Technology Connection

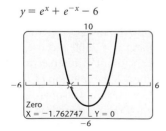

$$y = e^x + e^{-x} - 6$$

We can use the Zero method in Example 4 to solve the equation $e^x + e^{-x} - 6 = 0$. We graph the function $y = e^x + e^{-x} - 6$ and use the ZERO feature to find the zeros.

The leftmost zero is about −1.76. Using the ZERO feature one more time, we find that the other zero is about 1.76.

Solving Logarithmic Equations

Equations containing variables in logarithmic expressions, such as $\log_2 x = 4$ and $\log x + \log(x + 3) = 1$, are called **logarithmic equations.** To solve logarithmic equations algebraically, we first try to obtain a single logarithmic expression on one side and then write an equivalent exponential equation.

EXAMPLE 5 Solve: $\log_3 x = -2$.

Algebraic Solution

We have

$$\log_3 x = -2$$
$$3^{-2} = x \qquad \text{Converting to an exponential equation}$$
$$\frac{1}{3^2} = x$$
$$\frac{1}{9} = x.$$

CHECK:

$$\log_3 x = -2$$

$$\log_3 \frac{1}{9} \; ? \; -2$$
$$\log_3 3^{-2}$$
$$-2 \;\big|\; -2 \quad \text{TRUE}$$

The solution is $\frac{1}{9}$.

Visualizing the Solution

When we graph $y = \log_3 x$ and $y = -2$, we find that the first coordinate of the point of intersection of the graphs is $\frac{1}{9}$.

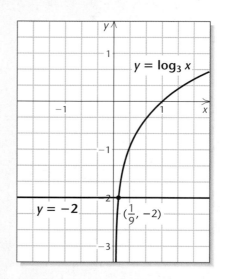

The solution of $\log_3 x = -2$ is $\frac{1}{9}$.

EXAMPLE 6   Solve: $\log x + \log (x + 3) = 1$.

Algebraic Solution

In this case, we have common logarithms. Writing the base of 10 will help us understand the problem:

$$\log_{10} x + \log_{10} (x + 3) = 1$$

$$\log_{10} [x(x + 3)] = 1 \qquad \text{Using the product rule to obtain a single logarithm}$$

$$x(x + 3) = 10^1 \qquad \text{Writing an equivalent exponential equation}$$

$$x^2 + 3x = 10$$

$$x^2 + 3x - 10 = 0$$

$$(x - 2)(x + 5) = 0 \qquad \text{Factoring}$$

$$x - 2 = 0 \quad or \quad x + 5 = 0$$

$$x = 2 \quad or \quad x = -5.$$

CHECK: For 2:

$$\dfrac{\log x + \log (x + 3) = 1}{}$$

$$\log 2 + \log (2 + 3) \ ? \ 1$$
$$\log 2 + \log 5$$
$$\log (2 \cdot 5)$$
$$\log 10$$
$$1 \ \Big| \ 1 \quad \text{TRUE}$$

For -5:

$$\dfrac{\log x + \log (x + 3) = 1}{}$$

$$\log (-5) + \log (-5 + 3) \ ? \ 1 \quad \text{FALSE}$$

The number -5 is not a solution because negative numbers do not have real-number logarithms. The solution is 2.

Visualizing the Solution

The solution of the equation

$$\log x + \log (x + 3) = 1$$

is the zero of the function

$$f(x) = \log x + \log (x + 3) - 1.$$

The solution is also the first coordinate of the x-intercept of the graph of the function.

$$f(x) = \log x + \log (x + 3) - 1$$

The solution of the equation is 2. From the graph, we can easily see that there is only one solution.

Technology Connection

In Example 6, we can graph the equations

$$y_1 = \log x + \log(x + 3)$$

and

$$y_2 = 1$$

and use the Intersect method. The first coordinate of the point of intersection is the solution of the equation.

We could also graph the function

$$y = \log x + \log(x + 3) - 1$$

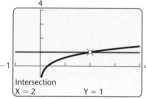
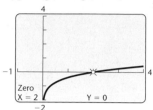

and use the Zero method. The zero of the function is the solution of the equation.

With either method, we see that the solution is 2. Note that the graphical solution gives only the one *true* solution.

EXAMPLE 7 Solve: $\log_3(2x - 1) - \log_3(x - 4) = 2$.

Algebraic Solution

$$\log_3(2x - 1) - \log_3(x - 4) = 2$$

$$\log_3 \frac{2x - 1}{x - 4} = 2 \qquad \textbf{Using the quotient rule}$$

$$\frac{2x - 1}{x - 4} = 3^2 \qquad \begin{array}{l}\textbf{Writing an equivalent}\\\textbf{exponential equation}\end{array}$$

$$\frac{2x - 1}{x - 4} = 9$$

$$(x - 4) \cdot \frac{2x - 1}{x - 4} = 9(x - 4) \qquad \begin{array}{l}\textbf{Multiplying by the}\\\textbf{LCD, } x - 4\end{array}$$

$$2x - 1 = 9x - 36$$

$$35 = 7x$$

$$5 = x.$$

CHECK: $\log_3(2x - 1) - \log_3(x - 4) = 2$

$$\begin{array}{c|c} \log_3(2 \cdot 5 - 1) - \log_3(5 - 4) \; ? \; 2 & \\ \log_3 9 - \log_3 1 & \\ 2 - 0 & \\ 2 & 2 \quad \text{TRUE} \end{array}$$

The solution is 5.

Visualizing the Solution

We see that the first coordinate of the point of intersection of the graphs of

$$y = \log_3(2x - 1) - \log_3(x - 4)$$

and

$$y = 2$$

is 5.

$$y = \log_3(2x - 1) - \log_3(x - 4)$$

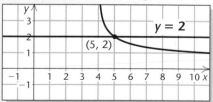

The solution is 5.

EXAMPLE 8 Solve: $\ln(4x + 6) - \ln(x + 5) = \ln x$.

Algebraic Solution

We have

$$\ln(4x + 6) - \ln(x + 5) = \ln x$$

$$\ln\frac{4x + 6}{x + 5} = \ln x \qquad \text{Using the quotient rule}$$

$$\frac{4x + 6}{x + 5} = x \qquad \begin{array}{l}\text{Using the property}\\ \text{of logarithmic}\\ \text{equality}\end{array}$$

$$(x + 5) \cdot \frac{4x + 6}{x + 5} = x(x + 5) \qquad \text{Multiplying by } x + 5$$

$$4x + 6 = x^2 + 5x$$

$$0 = x^2 + x - 6$$

$$0 = (x + 3)(x - 2) \qquad \text{Factoring}$$

$$x + 3 = 0 \quad \text{or} \quad x - 2 = 0$$

$$x = -3 \quad \text{or} \qquad x = 2.$$

The number -3 is not a solution because negative numbers do not have real-number logarithms. The value 2 checks and is the solution.

Visualizing the Solution

The solution of the equation

$$\ln(4x + 6) - \ln(x + 5) = \ln x$$

is the zero of the function

$$f(x) = \ln(4x + 6) - \ln(x + 5)$$
$$- \ln x.$$

The solution is also the first coordinate of the x-intercept of the graph of the function.

$f(x) = \ln(4x + 6) - \ln(x + 5) - \ln x$

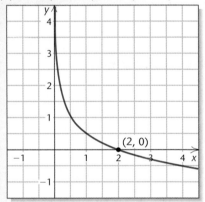

The solution of the equation is 2. From the graph, we can easily see that there is only one solution.

4.5

Exercise Set

Solve the exponential equation.

1. $3^x = 81$

2. $2^x = 32$

3. $2^{2x} = 8$

4. $3^{7x} = 27$

5. $2^x = 33$

6. $2^x = 40$

7. $5^{4x-7} = 125$

8. $4^{3x-5} = 16$

9. $27 = 3^{5x} \cdot 9^{x^2}$

10. $3^{x^2+4x} = \frac{1}{27}$

11. $84^x = 70$

12. $28^x = 10^{-3x}$

13. $e^{-c} = 5^{2c}$

14. $15^x = 30$

15. $e^t = 1000$

16. $e^{-t} = 0.04$

17. $e^{-0.03t} = 0.08$

18. $1000e^{0.09t} = 5000$

19. $3^x = 2^{x-1}$

20. $5^{x+2} = 4^{1-x}$

21. $(3.9)^x = 48$

22. $250 - (1.87)^x = 0$

23. $e^x + e^{-x} = 5$

24. $e^x - 6e^{-x} = 1$

25. $\dfrac{e^x + e^{-x}}{e^x - e^{-x}} = 3$

26. $\dfrac{5^x - 5^{-x}}{5^x + 5^{-x}} = 8$

Solve the logarithmic equation.

27. $\log_5 x = 4$

28. $\log_2 x = -3$

29. $\log x = -4$

30. $\log x = 1$

31. $\ln x = 1$

32. $\ln x = -2$

33. $\log_2 (10 + 3x) = 5$

34. $\log_5 (8 - 7x) = 3$

35. $\log x + \log (x - 9) = 1$

36. $\log_2 (x + 1) + \log_2 (x - 1) = 3$

37. $\log_2 (x + 20) - \log_2 (x + 2) = \log_2 x$

38. $\log (x + 5) - \log (x - 3) = \log 2$

39. $\log_8 (x + 1) - \log_8 x = 2$

40. $\log x - \log (x + 3) = -1$

41. $\log x + \log (x + 4) = \log 12$

42. $\ln x - \ln (x - 4) = \ln 3$

43. $\log_4 (x + 3) + \log_4 (x - 3) = 2$

44. $\ln (x + 1) - \ln x = \ln 4$

45. $\log (2x + 1) - \log (x - 2) = 1$

46. $\log_5 (x + 4) + \log_5 (x - 4) = 2$

47. $\ln (x + 8) + \ln (x - 1) = 2 \ln x$

48. $\log_3 x + \log_3 (x + 1) = \log_3 2 + \log_3 (x + 3)$

Technology Connection

Find approximate solution(s) of the equation.

49. $e^{7.2x} = 14.009$

50. $0.082e^{0.05x} = 0.034$

51. $xe^{3x} - 1 = 3$

52. $5e^{5x} + 10 = 3x + 40$

53. $4 \ln (x + 3.4) = 2.5$

54. $\ln x^2 = -x^2$

55. $\log_8 x + \log_8 (x + 2) = 2$

56. $\log_3 x + 7 = 4 - \log_5 x$

57. $\log_5 (x + 7) - \log_5 (2x - 3) = 1$

Approximate the point(s) of intersection of the pair of equations.

58. $y = \ln 3x, \quad y = 3x - 8$

59. $2.3x + 3.8y = 12.4, \quad y = 1.1 \ln (x - 2.05)$

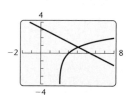

60. $y = 2.3 \ln (x + 10.7), \quad y = 10e^{-0.07x^2}$

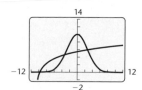

61. $y = 2.3 \ln (x + 10.7), \quad y = 10e^{-0.007x^2}$

Collaborative Discussion and Writing

62. In Example 3, we took the natural logarithm on both sides of the equation. What would have happened had we taken the common logarithm? Explain which approach seems better to you and why.

63. Explain how Exercises 31 and 32 could be solved using the graph of $f(x) = \ln x$.

Skill Maintenance

In Exercises 64–67:

a) *Find the vertex.*
b) *Find the axis of symmetry.*
c) *Determine whether there is a maximum or minimum value and find that value.*

64. $f(x) = -x^2 + 6x - 8$

65. $g(x) = x^2 - 6$

66. $H(x) = 3x^2 - 12x + 16$

67. $G(x) = -2x^2 - 4x - 7$

Synthesis

Solve using any method.

68. $\ln (\ln x) = 2$

69. $\ln (\log x) = 0$

70. $\ln \sqrt[4]{x} = \sqrt{\ln x}$

71. $\sqrt{\ln x} = \ln \sqrt{x}$

72. $\log_3 (\log_4 x) = 0$

73. $(\log_3 x)^2 - \log_3 x^2 = 3$

74. $(\log x)^2 - \log x^2 = 3$

75. $\ln x^2 = (\ln x)^2$

76. $e^{2x} - 9 \cdot e^x + 14 = 0$

77. $5^{2x} - 3 \cdot 5^x + 2 = 0$

78. $x \left(\ln \frac{1}{6} \right) = \ln 6$

79. $\log_3 |x| = 2$

80. $x^{\log x} = \dfrac{x^3}{100}$

81. $\ln x^{\ln x} = 4$

82. $\dfrac{(e^{3x+1})^2}{e^4} = e^{10x}$

83. $\dfrac{\sqrt{(e^{2x} \cdot e^{-5x})^{-4}}}{e^x \div e^{-x}} = e^7$

84. $e^x < \dfrac{4}{5}$

85. $|\log_5 x| + 3 \log_5 |x| = 4$

86. $|2^{x^2} - 8| = 3$

87. Given that $a = \log_8 225$ and $b = \log_2 15$, express a as a function of b.

88. Given that $a = (\log_{125} 5)^{\log_5 125}$, find the value of $\log_3 a$.

89. Given that
$$\log_2 [\log_3 (\log_4 x)] = \log_3 [\log_2 (\log_4 y)]$$
$$= \log_4 [\log_3 (\log_2 z)]$$
$$= 0,$$
find $x + y + z$.

90. Given that $f(x) = e^x - e^{-x}$, find $f^{-1}(x)$ if it exists.

4.6

Applications and Models: Growth and Decay

• *Solve applied problems involving exponential growth and decay.*

Exponential and logarithmic functions with base e are rich in applications to many fields such as business, science, psychology, and sociology.

Population Growth

The function

$$P(t) = P_0 e^{kt}, \quad k > 0$$

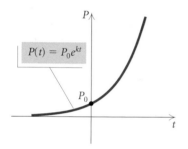

$$P(t) = P_0 e^{kt}$$

is a model of many kinds of population growth, whether it be a population of people, bacteria, cellular phones, or money. In this function, P_0 is the population at time 0, P is the population after time t, and k is called the **exponential growth rate.** The graph of such an equation is shown at left.

EXAMPLE 1 *Population Growth of India.* In 2001, the population of India was about 1030 million and the exponential growth rate was 1.4% per year (*Source: Statistical Abstract of the United States*).

a) Find the exponential growth function.

b) Estimate the population in 2005.

c) After how long will the population be double what it was in 2001?

Solution

a) At $t = 0$ (2001), the population was 1030 million. We substitute 1030 for P_0 and 1.4%, or 0.014, for k to obtain the exponential growth function

$$P(t) = 1030 e^{0.014t},$$

where t is the number of years after 2001 and $P(t)$ is in millions.

b) In 2005, $t = 4$; that is, 4 yr have passed since 2001. To find the population in 2005, we substitute 4 for t:

$$P(4) = 1030 e^{0.014(4)} = 1030 e^{0.056} \approx 1089.$$

The population will be about 1089 million, or 1,089,000,000, in 2005.

c) We are looking for the time T for which $P(T) = 2 \cdot 1030$, or 2060. The number T is called the **doubling time.** To find T, we solve the equation

$$2060 = 1030 e^{0.014T}.$$

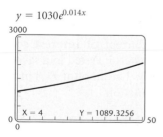

Algebraic Solution

We have

$$2060 = 1030e^{0.014T} \qquad \text{Substituting 2060 for } P(T)$$

$$2 = e^{0.014T} \qquad \text{Dividing by 1030}$$

$$\ln 2 = \ln e^{0.014T} \qquad \text{Taking the natural logarithm on both sides}$$

$$\ln 2 = 0.014T \qquad \ln e^x = x$$

$$\frac{\ln 2}{0.014} = T \qquad \text{Dividing by 0.014}$$

$$50 \approx T.$$

The population of India will be double what it was in 2001 about 50 yr after 2001.

Visualizing the Solution

From the graph of $y = 2060$ and $y = 1030e^{0.014T}$, we see that the first coordinate of the point of intersection of the graphs is about 50.

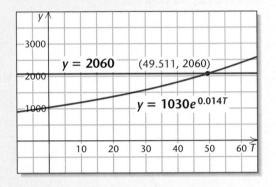

The solution of the equation is approximately 50.

Technology Connection

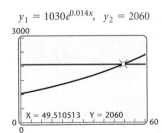

$y_1 = 1030e^{0.014x}, \quad y_2 = 2060$

Using the Intersect method in Example 1(c), we graph the equations

$$y_1 = 1030e^{0.014x} \quad \text{and} \quad y_2 = 2060$$

and find the first coordinate of their point of intersection. It is about 50, so the population of India will be double that of 2001 about 50 yr after 2001.

Interest Compounded Continuously

Here we explore the mathematics behind the concept of **interest compounded continuously.** Suppose that an amount P_0 is invested in a savings account at interest rate k *compounded continuously.* The amount $P(t)$ in the account after t years is given by the exponential function

$$P(t) = P_0 e^{kt}.$$

EXAMPLE 2 *Interest Compounded Continuously.* Suppose that $2000 is invested at interest rate k, compounded continuously, and grows to $2983.65 in 5 yr.

a) What is the interest rate?

b) Find the exponential growth function.

c) What will the balance be after 10 yr?

d) After how long will the $2000 have doubled?

Solution

a) At $t = 0$, $P(0) = P_0 = \$2000$. Thus the exponential growth function is of the form

$$P(t) = 2000e^{kt}.$$

We know that $P(5) = \$2983.65$. We substitute and solve for k:

$$2983.65 = 2000e^{k(5)} \qquad \textbf{Substituting 2983.65 for } P(t) \textbf{ and 5 for } t$$

$$2983.65 = 2000e^{5k}$$

$$\frac{2983.65}{2000} = e^{5k} \qquad \textbf{Dividing by 2000}$$

$$\ln \frac{2983.65}{2000} = \ln e^{5k} \qquad \textbf{Taking the natural logarithm}$$

$$\ln \frac{2983.65}{2000} = 5k \qquad \textbf{Using } \ln e^x = x$$

$$\frac{\ln \dfrac{2983.65}{2000}}{5} = k \qquad \textbf{Dividing by 5}$$

$$0.08 \approx k.$$

The interest rate is about 0.08, or 8%.

b) Substituting 0.08 for k in the function $P(t) = 2000e^{kt}$, we see that the exponential growth function is

$$P(t) = 2000e^{0.08t}.$$

c) The balance after 10 yr is

$$P(10) = 2000e^{0.08(10)}$$

$$= 2000e^{0.8}$$

$$\approx \$4451.08.$$

d) To find the doubling time T, we set $P(T) = 2 \cdot P_0 = 2 \cdot \$2000 = \$4000$ and solve for T.

Algebraic Solution

We have

$$4000 = 2000e^{0.08T}$$

$$2 = e^{0.08T} \qquad \text{Dividing by 2000}$$

$$\ln 2 = \ln e^{0.08T} \qquad \text{Taking the natural logarithm}$$

$$\ln 2 = 0.08T \qquad \ln e^x = x$$

$$\frac{\ln 2}{0.08} = T \qquad \text{Dividing by 0.08}$$

$$8.7 \approx T.$$

Thus the original investment of \$2000 will double in about 8.7 yr.

Visualizing the Solution

The solution of the equation

$$4000 = 2000e^{0.08T},$$

or

$$2000e^{0.08T} - 4000 = 0,$$

is the zero of the function

$$y = 2000e^{0.08T} - 4000.$$

Let's observe the zero from the graph shown here.

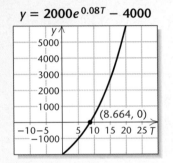

The zero is about 8.7. Thus the solution of the equation is approximately 8.7.

Technology Connection

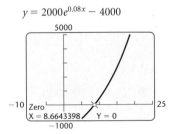

The money in Example 2 will have doubled when $P(t) = 2 \cdot P_0 = 4000$, or when $2000e^{0.08t} = 4000$. We use the Zero method. We graph the equation

$$y = 2000e^{0.08x} - 4000$$

and find the zero of the function. The zero of the function is the solution of the equation. The zero is about 8.7, so the original investment of \$2000 will double in about 8.7 yr.

We can find a general expression relating the growth rate k and the doubling time T by solving the following equation:

$$2P_0 = P_0 e^{kT} \qquad \text{Substituting } 2P_0 \text{ for } P \text{ and } T \text{ for } t$$

$$2 = e^{kT} \qquad \text{Dividing by } P_0$$

$$\ln 2 = \ln e^{kT} \qquad \text{Taking the natural logarithm}$$

$$\ln 2 = kT \qquad \text{Using } \ln e^x = x$$

$$\frac{\ln 2}{k} = T.$$

Growth Rate and Doubling Time

The **growth rate** k and the **doubling time** T are related by

$$kT = \ln 2, \quad \text{or} \quad k = \frac{\ln 2}{T}, \quad \text{or} \quad T = \frac{\ln 2}{k}.$$

Note that the relationship between k and T does not depend on P_0.

EXAMPLE 3 *World Population Growth.* The population of the world is now doubling every 57.8 yr. What is the exponential growth rate?

Solution We have

$$k = \frac{\ln 2}{T} = \frac{\ln 2}{57.8} \approx 1.2\%.$$

The growth rate of the world population is about 1.2% per year.

Models of Limited Growth

The model $P(t) = P_0 e^{kt}$, $k > 0$, has many applications involving unlimited population growth. However, in some populations, there can be factors that prevent a population from exceeding some limiting value—perhaps a limitation on food, living space, or other natural resources. One model of such growth is

$$P(t) = \frac{a}{1 + be^{-kt}}.$$

This is called a **logistic function.** This function increases toward a *limiting value a* as $t \to \infty$. Thus, $y = a$ is the horizontal asymptote of the graph of $P(t)$.

EXAMPLE 4 *Limited Population Growth.* A ship carrying 1000 passengers has the misfortune to be shipwrecked on a small island from which the passengers are never rescued. The natural resources of the island limit the population to 5780. The population gets closer and closer to this limiting

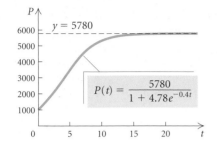

value, but never reaches it. The population of the island after time t, in years, is given by the logistic function

$$P(t) = \frac{5780}{1 + 4.78e^{-0.4t}}.$$

The graph of $P(t)$ is the curve shown at left. Note that this function increases toward a limiting value of 5780. The graph has $y = 5780$ as a horizontal asymptote. Find the population after 0, 1, 2, 5, 10, and 20 yr.

Solution   Using a calculator, we compute the function values. We find that

$$P(0) = 1000, \qquad P(5) \approx 3509.6,$$
$$P(1) \approx 1374.8, \qquad P(10) \approx 5314.7,$$
$$P(2) \approx 1836.2, \qquad P(20) \approx 5770.7.$$

Thus the population will be about 1000 after 0 yr, 1375 after 1 yr, 1836 after 2 yr, 3510 after 5 yr, 5315 after 10 yr, and 5771 after 20 yr.   ■

Another model of limited growth is provided by the function

$$P(t) = L(1 - e^{-kt}), \quad k > 0,$$

which is shown graphed at right. This function also increases toward a limiting value L, as $x \rightarrow \infty$, so $y = L$ is the horizontal asymptote of the graph of $P(t)$.

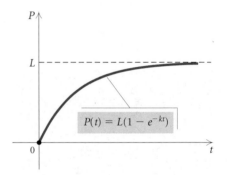

Exponential Decay

The function

$$P(t) = P_0 e^{-kt}, \quad k > 0$$

is an effective model of the decline, or decay, of a population. An example is the decay of a radioactive substance. In this case, P_0 is the amount of the substance at time $t = 0$, and $P(t)$ is the amount of the substance left after time t, where k is a positive constant that depends on the situation. The constant k is called the **decay rate.**

How can scientists determine that an animal bone has lost 30% of its carbon-14? The assumption is that the percentage of carbon-14 in the atmosphere and in living plants and animals is the same. When a plant or an animal dies, the amount of carbon-14 decays exponentially. The scientist burns the animal bone and uses a Geiger counter to determine the percentage of the smoke that is carbon-14. It is the amount that this varies from the percentage in the atmosphere that tells how much carbon-14 has been lost.

The process of carbon-14 dating was developed by the American chemist Willard E. Libby in 1952. It is known that the radioactivity in a living plant is 16 disintegrations per gram per minute. Since the half-life of carbon-14 is 5750 years, an object with an activity of 8 disintegrations per gram per minute is 5750 years old, one with an activity of 4 disintegrations per gram per minute is 11,500 years old, and so on. Carbon-14 dating can be used to measure the age of objects up to 40,000 years old. Beyond such an age, it is too difficult to measure the radioactivity and some other method would have to be used.

Carbon-14 was used to find the age of the Dead Sea Scrolls. It was also used to refute the authenticity of the Shroud of Turin, presumed to have covered the body of Christ.

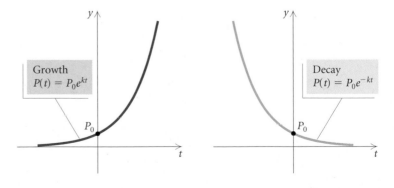

Study Tip

Newspapers and magazines are full of mathematical applications. The easiest ones to find are exponential. Find an application and share it with your class. As you obtain higher math skills, you will more readily observe the world from a mathematical perspective. Math courses become more interesting when we connect the concepts to the real world.

The **half-life** of bismuth is 5 days. This means that half of an amount of bismuth will cease to be radioactive in 5 days. The effect of half-life T is shown in the graph below for nonnegative inputs. The exponential function gets close to 0, but never reaches 0, as t gets very large. Thus, according to an exponential decay model, a radioactive substance never completely decays.

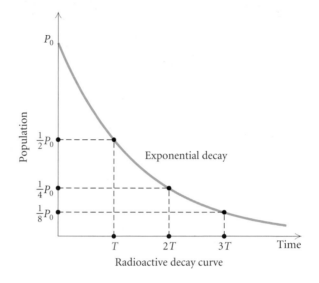

Radioactive decay curve

In 1947, a Bedouin youth looking for a stray goat climbed into a cave at Kirbet Qumran on the shores of the Dead Sea near Jericho and came upon earthenware jars containing an incalculable treasure of ancient manuscripts. Shown here are fragments of those so-called Dead Sea Scrolls, a portion of some 600 or so texts found so far and which concern the Jewish books of the Bible. Officials date them before 70 A.D., making them the oldest Biblical manuscripts by 1000 years.

EXAMPLE 5 *Carbon Dating.* The radioactive element carbon-14 has a half-life of 5750 yr. The percentage of carbon-14 present in the remains of organic matter can be used to determine the age of that organic matter. Archaeologists discovered that the linen wrapping from one of the Dead Sea Scrolls had lost 22.3% of its carbon-14 at the time it was found. How old was the linen wrapping?

Solution We first find k. When $t = 5750$ (the half-life), $P(t)$ will be half of P_0. We substitute $\frac{1}{2}P_0$ for $P(t)$ and 5750 for t and solve for k. Then

$$P(t) = P_0 e^{-kt}$$
$$\tfrac{1}{2}P_0 = P_0 e^{-k(5750)}$$
$$\tfrac{1}{2} = e^{-5750k} \qquad \textbf{Dividing by } P_0$$
$$\ln \tfrac{1}{2} = \ln e^{-5750k} \qquad \textbf{Taking the natural logarithm on both sides}$$
$$\ln 0.5 = -5750k.$$

Then

$$k = \frac{\ln 0.5}{-5750} \approx 0.00012.$$

Now we have the function

$$P(t) = P_0 e^{-0.00012t}.$$

(This function can be used for any subsequent carbon-dating problem.) If the linen wrapping has lost 22.3% of its carbon-14 from an initial amount

P_0, then $77.7\%P_0$ is the amount present. To find the age t of the wrapping, we solve the following equation for t:

$$77.7\%P_0 = P_0e^{-0.00012t} \qquad \textbf{Substituting } 77.7\%P_0 \textbf{ for } P$$

$$0.777 = e^{-0.00012t} \qquad \textbf{Dividing by } P_0 \textbf{ and writing } 77.7\% \textbf{ as } 0.777$$

$$\ln 0.777 = \ln e^{-0.00012t} \qquad \textbf{Taking the natural logarithm on both sides}$$

$$\ln 0.777 = -0.00012t \qquad \ln e^x = x$$

$$\frac{\ln 0.777}{-0.00012} = t \qquad \textbf{Dividing by } -0.00012$$

$$2103 \approx t.$$

Thus the linen wrapping on the Dead Sea Scrolls was about 2103 yr old when it was found.

Technology Connection

YEAR, x	DEBIT CARD VOLUME, y (IN BILLIONS)
1990, 0	$ 8.2
1991, 1	9.3
1992, 2	11.0
1993, 3	16.0
1994, 4	20.0
1995, 5	23.0
1996, 6	43.1
1997, 7	66.7
1998, 8	115.0
1999, 9	157.9

Sources: Credit Card Management, *ATM & Debit News*

```
ExpReg
y=a*b^x
a=6.073029319
b=1.407069275
r²=.9633789965
r = .9815187194
```

We now expand the regression procedure to include modeling data with an exponential function.

Debit Card Volume. The total debit card volume has increased dramatically in recent years, as shown in the table at left.

a) Use a graphing calculator to fit an exponential function to the data.

b) Predict the total debit card volume in 2005.

Solution

a) We will fit an equation of the type $y = a \cdot b^x$ to the data, where x is the number of years since 1990. Entering the data into the calculator and carrying out the regression procedure, we find that the equation is

$$y = 6.073029319(1.407069275)^x.$$

The correlation coefficient is very close to 1. This gives us an indication that the exponential function fits the data well.

b) We evaluate the function found in part (a) for $x = 15$ ($2005 - 1990 = 15$) and estimate that the total debit card volume in 2005 will be about $1019 billion, or $1,019,000,000,000.

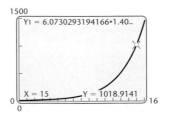

4.6

Exercise Set

1. *World Population Growth.* In 2001, the world population was 6.2 billion. The exponential growth rate was 1.2% per year.

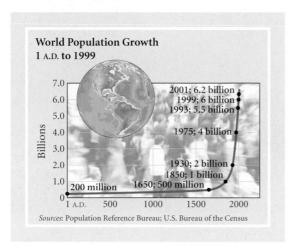

World Population Growth
1 A.D. to 1999

Billions

2001; 6.2 billion
1999; 6 billion
1993; 5.5 billion
1975; 4 billion
1930; 2 billion
1850; 1 billion
1650; 500 million
200 million

Sources: Population Reference Bureau; U.S. Bureau of the Census

a) Find the exponential growth function.
b) Estimate the population of the world in 2005 and in 2010.
c) When will the world population be 8 billion?
d) Find the doubling time.

2. *Population Growth of Rabbits.* Under ideal conditions, a population of rabbits has an exponential growth rate of 11.7% per day. Consider an initial population of 100 rabbits.

a) Find the exponential growth function.
b) What will the population be after 7 days? after 2 weeks?
c) Find the doubling time.

3. *Population Growth.* Complete the following table.

POPULATION	GROWTH RATE, k	DOUBLING TIME, T
a) Liberia	2.5% per year	
b) Japan		693 yr
c) Saudi Arabia	3.3% per year	
d) Germany	0.2% per year	

(continued)

POPULATION	GROWTH RATE, k	DOUBLING TIME, T
e) Egypt		46.2 yr
f) Philippines		36.5 yr
g) United States	0.9% per year	
h) China		99.0 yr
i) Turkey	1.1% per year	
j) Cuba	0.3% per year	

4. *Number of Physicians.* In 1950, the total number of physicians in the United States was 219,997. In 2000, it was 813,869.

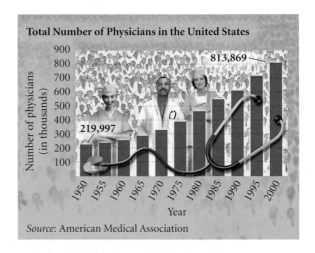

Total Number of Physicians in the United States

Number of physicians (in thousands)

813,869
219,997

Year

Source: American Medical Association

Assuming the exponential model applies:
a) Find the value of k, and write the function.
b) Estimate the number of physicians in the United States in 2003, in 2006, and in 2012.

5. *Population Growth of Israel.* The population of Israel has a growth rate of 1.3% per year. In 2001, the population was 5,938,000. The land area of Israel is 24,313,062,400 square yards. (*Source: Statistical Abstract of the United States*) Assuming this growth rate continues and is exponential, after how long will there be one person for every square yard of land?

6. *Value of Manhattan Island.* In 1626, Peter Minuit of the Dutch West India Company purchased Manhattan Island from Native Americans for $24. Assuming an exponential rate of inflation of 8% per year, how much will Manhattan be worth in 2005?

7. *Interest Compounded Continuously.* Suppose that $10,000 is invested at an interest rate of 5.4% per year, compounded continuously.

a) Find the exponential function that describes the amount in the account after time t, in years.

b) What is the balance after 1 yr? 2 yr? 5 yr? 10 yr?

c) What is the doubling time?

8. *Interest Compounded Continuously.* Complete the following table.

Initial Investment at $t = 0$, P_0	Interest Rate, k	Doubling Time, T	Amount After 5 Yr
a) $35,000	6.2%		
b) $5000			$ 7,130.90
c)	8.4%		$11,414.71
d)		11 yr	$17,539.32

9. *Carbon Dating.* A mummy discovered in the pyramid Khufu in Egypt has lost 46% of its carbon-14. Determine its age.

10. *Carbon Dating.* The statue of Zeus at Olympia in Greece is one of the Seven Wonders of the World. It is made of gold and ivory. The ivory was found to have lost 35% of its carbon-14. Determine the age of the statue.

11. *Radioactive Decay.* Complete the following table.

Radioactive Substance	Decay Rate, k	Half-life, T
a) Polonium		3 min
b) Lead		22 yr
c) Iodine-131	9.6% per day	
d) Krypton-85	6.3% per year	
e) Strontium-90		25 yr
f) Uranium-238		4560 yr
g) Plutonium		23,105 yr

12. *Number of Farms.* The number N of farms in the United States has declined continually since 1950. (See the Technology Connection, p. 261, in Section 3.1.) In 1950, there were 5,388,000 farms, and in 2001 that number had decreased to 2,158,000 (*Sources:* U.S. Department of Agriculture; U.S.

Bureau of the Census). Assuming the number of farms decreased according to the exponential model:

a) Find the value of k, and write an exponential function that describes the number of farms after time t, in years, where t is the number of years since 1950.

b) Estimate the number of farms in 2005 and in 2010.

c) At this decay rate, in what year will only 100,000 farms remain?

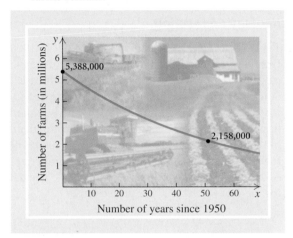

Number of years since 1950

13. *Declining Cellphone Prices.* Average cellphone prices have fallen sharply since their introduction to the market in 1983.

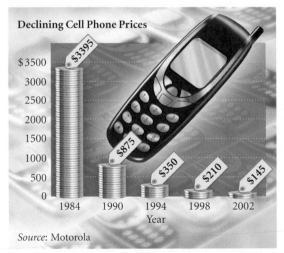

Source: Motorola

In 1984, the average price was $3395, and in 2002, it was only $145. (*Source*: Motorola) Assuming the average price of a cellphone decreased according to the exponential model:

a) Find the value of k, and write an exponential function that describes the average price of a cellphone after time t, in years, where t is the number of years since 1984.

b) Estimate the price of a cellphone in 2004 and in 2008.

c) At this decay rate, in what year will the price be $39?

14. *The Value of Mark McGwire's Baseball Card.* Collecting baseball cards has become a popular hobby. The card shown here is a photograph of Mark McGwire when he was a member of the Olympic USA Baseball Team in 1984.

In 1987, the value of the card was $8, and in 2003, it was $50 (*Source*: *Tuff Stuff*, A Krauss Publication, June 2003, p. 131). Assuming the value V_0 of the card has grown exponentially:

a) Find the value k, and determine the exponential growth function, assuming $V_0 = 8$.

b) Using the function found in part (a), estimate the value of the card in 2008.

c) What is the doubling time for the value of the card?

d) After how long will the value of the card be $2000, assuming there is no change in the growth rate?

15. *Spread of an Epidemic.* In a town whose population is 3500, a disease creates an epidemic. The number of people N infected t days after the disease has begun is given by the function

$$N(t) = \frac{3500}{1 + 19.9e^{-0.6t}}.$$

a) How many are initially infected with the disease ($t = 0$)?

b) Find the number infected after 2 days, 5 days, 8 days, 12 days, and 16 days.

c) Using this model, can you say whether all 3500 people will ever be infected? Explain.

16. *Limited Population Growth in a Lake.* A lake is stocked with 400 fish of a new variety. The size of the lake, the availability of food, and the number of other fish restrict the growth of that type of fish in the lake to a *limiting value* of 2500. The population of fish in the lake after time t, in months, is given by the function

$$P(t) = \frac{2500}{1 + 5.25e^{-0.32t}}.$$

Find the population after 0, 1, 5, 10, 15, and 20 months.

Newton's Law of Cooling. *Suppose that a body with temperature T_1 is placed in surroundings with temperature T_0 different from that of T_1. The body will either cool or warm to temperature $T(t)$ after time t, in minutes, where*

$$T(t) = T_0 + (T_1 - T_0)e^{-kt}.$$

Use this law in Exercises 17–20.

17. A cup of coffee with temperature 105°F is placed in a freezer with temperature 0°F. After 5 min, the temperature of the coffee is 70°F. What will its temperature be after 10 min?

18. A dish of lasagna baked at 375°F is taken out of the oven at 11:15 A.M. into a kitchen that is 72°F. What will the temperature of the lasagna be at 11:30 A.M.?

19. A chilled jello salad that has a temperature of 43°F is taken from the refrigerator and placed on the dining room table in a room that is 68°F. After 12 min, the temperature of the salad is 55°F. What will the temperature of the salad be after 20 min?

20. *When Was the Murder Committed?* The police discover the body of a murder victim. Critical to solving the crime is determining when the murder was committed. The coroner arrives at the murder scene at 12:00 P.M. She immediately takes the temperature of the body and finds it to be 94.6°. She then takes the temperature 1 hr later and finds it to be 93.4°. The temperature of the room is 70°. When was the murder committed?

Technology Connection

21. *Projected Number of Alzheimer's Patients.* In 1906, German psychiatrist Alois Alzheimer first described a disease that was later called Alzheimer's disease. Since life expectancy has significantly increased in the last century, the number of Alzheimer's patients has increased dramatically. The number of patients in the United States reached 4 million in 2000. The following table lists data regarding the number of Alzheimer's patients in years beyond 2000. The data for the years after 2000 are projections.

YEAR, x	NUMBER OF ALZHEIMER'S PATIENTS IN THE UNITED STATES, $A(x)$ (IN MILLIONS)
2000, 0	4.0
2010, 10	5.8
2020, 20	6.8
2030, 30	8.7
2040, 40	11.8
2050, 50	14.3

Source: "Alzheimer's, Unlocking the Mystery," by Geoffrey Cowley, *Newsweek*, January 31, 2000

a) Use a graphing calculator to fit the data with an exponential function and determine whether the function is a good fit.

b) Graph the function found in part (a) with a scatterplot of the data.

c) Estimate the number of Alzheimer's patients in 2005, in 2025, and in 2100.

22. *Forgetting.* In an art class, students were tested at the end of the course on a final exam. Then they were retested with an equivalent test at subsequent time intervals. Their scores after time *x*, in months, are given in the following table.

TIME, *x* (IN MONTHS)	SCORE, *y*
1	84.9%
2	84.6
3	84.4
4	84.2
5	84.1
6	83.9

a) Use a graphing calculator to fit a logarithmic function $y = a + b \ln x$ to the data.
b) Use the function to predict test scores after 8, 10, 24, and 36 months.
c) After how long will the test scores fall below 82%?

23. *Online Travel Revenue.* With the explosion of increased Internet use, more and more travelers are booking their travel reservations online. The following table lists the total online travel revenue for recent years. Most of the revenue is from airline tickets.

YEAR, *x*	ONLINE TRAVEL REVENUE, $T(x)$ (IN MILLIONS)
1996, 0	$ 276
1997, 1	827
1998, 2	1900
1999, 3	3200
2000, 4	4700
2001, 5	6500
2002, 6	8900

Source: Travel Industry Association of America

a) Create a scatterplot of the data. Let *x* = the number of years since 1996.
b) Use a graphing calculator to fit the data with linear, quadratic, and exponential functions. Determine which function has the best fit.
c) Graph all three functions found in part (b) with the scatterplot in part (a).
d) Use the functions found in part (b) to estimate the online travel revenue in 2010. Which function provides the most realistic prediction?

24. *Cost of Political Conventions.* The total cost of the Democratic and the Republican national conventions has increased 596% over the 20-yr period between 1980 and 2000. The following table lists the total cost, in millions of dollars, for selected years.

YEAR, *x*	TOTAL CONVENTIONS' COST, $C(x)$ (IN MILLIONS)
1980, 0	$ 23.1
1984, 4	31.8
1988, 8	44.4
1992, 12	58.8
1996, 16	90.6
2000, 20	160.8
2004*, 24	170.5

*estimate
Source: Campaign Finance Institute

a) Use a graphing calculator to fit the data with an exponential function.
b) Use the function to estimate the cost in 2007 and in 2010.
c) In what year will total cost exceed $400 million?

25. *Paper Recycling.* The percent of paper and paperboard that is recycled from municipal solid waste has grown exponentially. The following table lists the percentages for selected years.

YEAR, *x*	PERCENT OF PAPER RECYCLED, $R(x)$
1980, 0	21.6%
1990, 10	27.8
1994, 14	36.5
1995, 15	40.0
1996, 16	40.9
1997, 17	40.3
1998, 18	40.9
1999, 19	40.9
2000, 20	45.4

Source: Franklin Associates, Ltd., Prairie Village, KS, *Characterization of Municipal Solid Waste in the United States: 2000.* Prepared for the U.S. Environmental Protection Agency.

a) Use a graphing calculator to fit the data with an exponential function.

b) Use the function to estimate the percentage of paper that will be recycled in 2005 and in 2010.
c) In what year will the percentage exceed 70%?

26. *Effect of Advertising.* A company introduces a new software product on a trial run in a city. They advertised the product on television and found the following data relating the percent P of people who bought the product after x ads were run.

NUMBER OF ADS, x	PERCENTAGE WHO BOUGHT, P
0	0.2%
10	0.7
20	2.7
30	9.2
40	27.0
50	57.6
60	83.3
70	94.8
80	98.5
90	99.6

a) Use a graphing calculator to fit a logistic function

$$P(x) = \frac{a}{1 + be^{-kx}}$$

to the data.
b) What percent of people will buy the product when 55 ads are run? 100 ads?
c) Find the horizontal asymptote for the graph. Interpret the asymptote in terms of the advertising situation.

Collaborative Discussion and Writing

27. Browse through some newspapers or magazines to find some data that appear to fit an exponential model. Make a case for why such a fit is appropriate. Then fit an exponential function to the data and make some predictions.

28. *Atmospheric Pressure.* Atmospheric pressure P at an altitude a is given by

$$P = P_0 e^{0.00005a},$$

where P_0 is the pressure at sea level $\approx 14.7\ \text{lb/in}^2$ (pounds per square inch). Explain how a barometer,

or some device for measuring atmospheric pressure, can be used to find the height of a skyscraper.

Skill Maintenance

Fill in the blank with the correct name of the principle or rule.

> Principle of zero products
> Multiplication principle for equations
> Product rule
> Addition principle for inequalities
> Power rule
> Multiplication principle for inequalities
> Principle of square roots
> Quotient rule

29. For any real numbers a, b, and c: If $a < b$ and $c > 0$ are true, then $ac < bc$ is true. If $a < b$ and $c < 0$ are true, then $ac > bc$ is true. _____

30. For any positive numbers M and N and any logarithmic base a, $\log_a MN = \log_a M + \log_a N$. _____

31. If $ab = 0$ is true, then $a = 0$ or $b = 0$, and if $a = 0$ or $b = 0$, then $ab = 0$. _____

32. If $x^2 = k$, then $x = \sqrt{k}$ or $x = -\sqrt{k}$. _____

33. For any positive number M, any logarithmic base a, and any real number p, $\log_a M^p = p\log_a M$. _____

34. For any real numbers a, b, and c: If $a = b$ is true, then $ac = bc$ is true. _____

Synthesis

35. *Present Value.* Following the birth of a child, a parent wants to make an initial investment P_0 that will grow to $50,000 for the child's education at age 18. Interest is compounded continuously at 7%. What should the initial investment be? Such an amount is called the **present value** of $50,000 due 18 yr from now.

36. *Present Value.* Referring to Exercise 35:
   a) Solve $P = P_0 e^{kt}$ for P_0.
   b) Find the present value of $50,000 due 18 yr from now at interest rate 6.4%.

37. *Supply and Demand.* The supply and demand for the sale of a certain type of VCR are given by
$$S(p) = 480e^{-0.003p} \quad \text{and} \quad D(p) = 150e^{0.004p},$$
where $S(p)$ is the number of VCRs that the company is willing to sell at price p and $D(p)$ is the quantity that the public is willing to buy at price p. Find p, called the **equilibrium price,** such that $D(p) = S(p)$.

38. *Carbon Dating.* Recently, while digging in Chaco Canyon, New Mexico, archaeologists found corn pollen that was 4000 yr old (*Source: American Anthropologist*). This was evidence that Native Americans had been cultivating crops in the Southwest centuries earlier than scientists had thought. What percent of the carbon-14 had been lost from the pollen?

39. *Electricity.* The formula
$$i = \frac{V}{R}\left[1 - e^{-(R/L)t}\right]$$
occurs in the theory of electricity. Solve for t.

40. *The Beer–Lambert Law.* A beam of light enters a medium such as water or smog with initial intensity I_0. Its intensity decreases depending on the thickness (or concentration) of the medium. The intensity I at a depth (or concentration) of x units is given by
$$I = I_0 e^{-\mu x}.$$
The constant μ (the Greek letter "mu") is called the **coefficient of absorption,** and it varies with the medium. For sea water, $\mu = 1.4$.
   a) What percentage of light intensity I_0 remains at a depth of sea water that is 1 m? 3 m? 5 m? 50 m?
   b) Plant life cannot exist below 10 m. What percentage of I_0 remains at 10 m?

41. Given that $y = ae^x$, take the natural logarithm on both sides. Let $Y = \ln y$. Consider Y as a function of x. What kind of function is Y?

42. Given that $y = ax^b$, take the natural logarithm on both sides. Let $Y = \ln y$ and $X = \ln x$. Consider Y as a function of X. What kind of function is Y?

Chapter 4 Summary and Review

Important Properties and Formulas

One-to-One Function:	$f(a) = f(b) \rightarrow a = b$
Exponential Function:	$f(x) = a^x$
The Number e:	$e = 2.7182818284\ldots$
Logarithmic Function:	$f(x) = \log_a x$
A Logarithm Is an Exponent:	$\log_a x = y \longleftrightarrow x = a^y$
The Change-of-Base Formula:	$\log_b M = \dfrac{\log_a M}{\log_a b}$
The Product Rule:	$\log_a MN = \log_a M + \log_a N$
The Power Rule:	$\log_a M^p = p \log_a M$
The Quotient Rule:	$\log_a \dfrac{M}{N} = \log_a M - \log_a N$
Other Properties:	$\log_a a = 1, \qquad \log_a 1 = 0,$
	$\log_a a^x = x, \qquad a^{\log_a x} = x$
Base–Exponent Property:	$a^x = a^y \longleftrightarrow x = y,$ for $a > 0, a \neq 1$
Property of Logarithmic Equality:	$\log_a M = \log_a N \longleftrightarrow M = N,$ for $a > 0$ and $a \neq 1$

Exponential Growth Model:	$P(t) = P_0 e^{kt}, k > 0$
Exponential Decay Model:	$P(t) = P_0 e^{-kt}, k > 0$
Interest Compounded Continuously:	$P(t) = P_0 e^{kt}, k > 0$
Limited Growth:	$P(t) = \dfrac{a}{1 + be^{-kt}}, k > 0$
	$P(t) = L(1 - e^{-kt}), k > 0$

Review Exercises

1. Find the inverse of the relation
$$\{(1.3, -2.7), (8, -3), (-5, 3), (6, -3), (7, -5)\}.$$

2. Find an equation of the inverse relation.
   a) $y = -2x + 3$
   b) $y = 3x^2 + 2x - 1$
   c) $0.8x^3 - 5.4y^2 = 3x$

Graph the function and determine whether the function is one-to-one using the horizontal-line test.

3. $f(x) = -|x| + 3$

4. $f(x) = x^2 + 1$

5. $f(x) = 2x - \dfrac{3}{4}$

6. $f(x) = -\dfrac{6}{x + 1}$

In Exercises 7–12, given the function:

a) *Sketch the graph and determine whether the function is one-to-one.*

b) *If it is one-to-one, find a formula for the inverse.*

7. $f(x) = 2 - 3x$

8. $f(x) = \dfrac{x + 2}{x - 1}$

9. $f(x) = \sqrt{x - 6}$

10. $f(x) = x^3 - 8$

11. $f(x) = 3x^2 + 2x - 1$

12. $f(x) = e^x$

For the function f, use composition of functions to show that f^{-1} is as given.

13. $f(x) = 6x - 5, \ f^{-1}(x) = \dfrac{x + 5}{6}$

14. $f(x) = \dfrac{x + 1}{x}, \ f^{-1}(x) = \dfrac{1}{x - 1}$

Find the inverse of the given one-to-one function f. Give the domain and the range of f and of f^{-1} and then graph both f and f^{-1} on the same set of axes.

15. $f(x) = 2 - 5x$

16. $f(x) = \dfrac{x - 3}{x + 2}$

17. Find $f(f^{-1}(657))$: $f(x) = \dfrac{4x^5 - 16x^{37}}{119x}, x > 1.$

18. Find $f(f^{-1}(a))$: $f(x) = \sqrt[3]{3x - 4}.$

Graph each function.

19. $f(x) = \left(\dfrac{1}{3}\right)^x$

20. $f(x) = 1 + e^x$

21. $f(x) = -e^{-x}$

22. $f(x) = \log_2 x$

23. $f(x) = \dfrac{1}{2} \ln x$

24. $f(x) = \log x - 2$

In Exercises 25–30, match the equation with one of figures (a)–(f), which follow

a)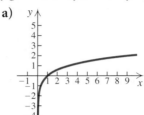

b)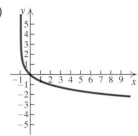

c)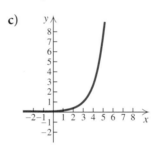

d)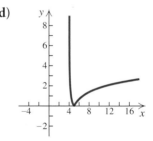

e)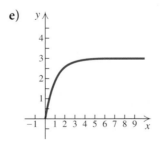

f)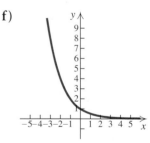

25. $f(x) = e^{x-3}$

26. $f(x) = \log_3 x$

27. $y = -\log_3 (x + 1)$

28. $y = \left(\dfrac{1}{2}\right)^x$

29. $f(x) = 3(1 - e^{-x}), \ x \geq 0$

30. $f(x) = |\ln (x - 4)|$

Find each of the following. Do not use a calculator.

31. $\log_5 125$ 32. $\log 100{,}000$

33. $\ln e$ 34. $\ln 1$

35. $\log 10^{1/4}$ 36. $\log_3 \sqrt{3}$

37. $\log 1$ 38. $\log 10$

39. $\log_2 \sqrt[3]{2}$ 40. $\log 0.01$

Convert to an exponential equation.

41. $\log_4 x = 2$ 42. $\log_a Q = k$

Convert to a logarithmic equation.

43. $4^{-3} = \frac{1}{64}$ 44. $e^x = 80$

Find each of the following using a calculator. Round to four decimal places.

45. $\log 11$ 46. $\log 0.234$

47. $\ln 3$ 48. $\ln 0.027$

49. $\log(-3)$ 50. $\ln 0$

Find the logarithm using the change-of-base formula.

51. $\log_5 24$ 52. $\log_8 3$

Express as a single logarithm and, if possible, simplify.

53. $3 \log_b x - 4 \log_b y + \frac{1}{2} \log_b z$

54. $\ln(x^3 - 8) - \ln(x^2 + 2x + 4) + \ln(x + 2)$

Express in terms of sums and differences of logarithms.

55. $\ln \sqrt[4]{wr^2}$ 56. $\log \sqrt[3]{\dfrac{M^2}{N}}$

Given that $\log_a 2 = 0.301$, $\log_a 5 = 0.699$, *and* $\log_a 6 = 0.778$, *find each of the following.*

57. $\log_a 3$ 58. $\log_a 50$

59. $\log_a \frac{1}{5}$ 60. $\log_a \sqrt[3]{5}$

Simplify.

61. $\ln e^{-5k}$ 62. $\log_5 5^{-6t}$

Solve.

63. $\log_4 x = 2$ 64. $3^{1-x} = 9^{2x}$

65. $e^x = 80$ 66. $4^{2x-1} - 3 = 61$

67. $\log_{16} 4 = x$ 68. $\log_x 125 = 3$

69. $\log_2 x + \log_2(x - 2) = 3$

70. $\log(x^2 - 1) - \log(x - 1) = 1$

71. $\log x^2 = \log x$

72. $e^{-x} = 0.02$

73. *Saving for College.* Following the birth of twins, the grandparents deposit \$16,000 in a college trust fund that earns 4.2% interest, compounded quarterly.
   a) Find a function for the amount in the account after t years.
   b) Find the amount of money in the account at $t = 0, 6, 12,$ and 18 yr.

74. *Giving Gift Certificates or Cash.* The percentage of consumers giving gift certificates or cash as presents during the Christmas holiday is soaring exponentially. The percentage is given by the function
$$G(t) = 28.75(1.20)^t,$$
where t is the number of years since 1999 (*Source:* America's Research Group). Find the percentage of consumers giving gift certificates or cash in 2003 and in 2004.

75. How long will it take an investment to double itself if it is invested at 8.6%, compounded continuously?

76. The population of a city doubled in 30 yr. What was the exponential growth rate?

77. How old is a skeleton that has lost 27% of its carbon-14?

78. The hydrogen ion concentration of milk is 2.3×10^{-6}. What is the pH? (See Exercise 94 in Exercise Set 4.3.)

79. The earthquake in Kashgar, China, on February 25, 2003, had an intensity of $10^{6.3} \cdot I_0$ (*Source:* U.S. Geological Survey). What is the magnitude on the Richter scale?

80. What is the loudness, in decibels, of a sound whose intensity is $1000 I_0$? (See Exercise 97 in Exercise Set 4.3.)

81. *Walking Speed.* The average walking speed w, in feet per second, of a person living in a city of population P, in thousands, is given by the function
$$w(P) = 0.37 \ln P + 0.05.$$
   a) The population of Dayton, Ohio, is 166,000. Find the average walking speed.   ft/sec
   b) A city's population has an average walking speed of 3.4 ft/sec. Find the population.

82. *Toll-free Numbers.* The use of toll-free numbers has grown exponentially. In 1967, there were 7 million calls to toll-free numbers, and in 1991, there were 10.2 billion such calls (*Source*: Federal Communication Commission).

a) Find the exponential growth rate k.
b) Find the exponential growth function.
c) Estimate the number of toll-free calls placed in 2005 and in 2010.
d) In what year will 1 trillion such calls be placed?

83. *The Population of Brazil.* The population of Brazil was 174.5 million in 2001, and the exponential growth rate was 0.8% per year (*Source*: U.S. Bureau of the Census, World Population Profile).

a) Find the exponential growth function.
b) What will the population be in 2004? in 2012?
c) When will the population be 200 million?
d) What is the doubling time?

Technology Connection

84. *Cholesterol Level and the Risk of Heart Attack.* The data in the following table show the relationship of cholesterol level in men to the risk of a heart attack.

Cholesterol Level, x	Men, per 10,000, Who Suffer a Heart Attack, y
100	30
200	65
250	100
275	130
300	180

Source: Nutrition Action Healthletter

a) Use a graphing calculator to fit an exponential function to the data.
b) Graph the function with a scatterplot of the data.
c) Predict the heart attack rate for men with cholesterol levels of 150, 350, and 400.

85. Using only a graphing calculator, determine whether the following functions are inverses of each other:

$$f(x) = \frac{4 + 3x}{x - 2}, \qquad g(x) = \frac{x + 4}{x - 3}.$$

86. a) Use a graphing calculator to graph $f(x) = 5e^{-x} \ln x$ in the viewing window $[-1, 10, -5, 5]$.
b) Estimate the relative maximum and minimum values of the function.

Collaborative Discussion and Writing

87. Suppose that you were trying to convince a fellow student that

$$\log_2 (x + 5) \neq \log_2 x + \log_2 5.$$

Give as many explanations as you can.

88. Describe the difference between $f^{-1}(x)$ and $[f(x)]^{-1}$.

Synthesis

Solve.

89. $|\log_4 x| = 3$ **90.** $\log x = \ln x$

91. $5^{\sqrt{x}} = 625$

92. Find the domain: $f(x) = \log_3 (\ln x)$.

Chapter 4 Test

1. Find the inverse of the relation
$$\{(-2,5),(4,3),(0,-1),(-6,-3)\}.$$

Determine whether the function is one-to-one. Answer yes or no.

2.

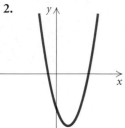

3.

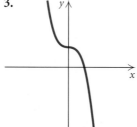

In Exercises 4–8, given the function:

a) *Sketch the graph and determine whether the function is one-to-one.*

b) *If it is one-to-one, find a formula for the inverse.*

4. $f(x) = x^3 + 1$

5. $f(x) = 1 - x$

6. $f(x) = \dfrac{x}{2 - x}$

7. $f(x) = x^2 + x - 3$

8. Use composition of functions to show that f^{-1} is as given:
$$f(x) = -4x + 3, \quad f^{-1}(x) = \frac{3 - x}{4}.$$

9. Find the inverse of the one-to-one function
$$f(x) = \frac{1}{x - 4}.$$

Give the domain and the range of f and of f^{-1} and then graph both f and f^{-1} on the same set of axes.

Graph each of the following functions.

10. $f(x) = 4^{-x}$

11. $f(x) = \log x$

12. $f(x) = e^x - 3$

13. $f(x) = \ln(x + 2)$

Find each of the following. Do not use a calculator.

14. $\log 0.00001$

15. $\ln e$

16. $\ln 1$

17. $\log_4 \sqrt[5]{4}$

18. Convert to an exponential equation: $\ln x = 4$.

19. Convert to a logarithmic equation: $3^x = 5.4$.

Find each of the following using a calculator. Round to four decimal places.

20. $\ln 16$

21. $\log 0.293$

22. Find $\log_6 10$ using the change-of-base formula.

23. Express as a single logarithm:
$$2 \log_a x - \log_a y + \tfrac{1}{2} \log_a z.$$

24. Express $\ln \sqrt[5]{x^2 y}$ in terms of sums and differences of logarithms.

25. Given that $\log_a 2 = 0.328$ and $\log_a 8 = 0.984$, find $\log_a 4$.

26. Simplify: $\ln e^{-4t}$.

Solve.

27. $\log_{25} 5 = x$

28. $\log_3 x + \log_3 (x + 8) = 2$

29. $3^{4-x} = 27^x$

30. $e^x = 65$

31. *Growth Rate.* A country's population doubled in 45 yr. What was the exponential growth rate?

32. *Compound Interest.* Suppose $1000 is invested at interest rate k, compounded continuously, and grows to $1144.54 in 3 yr.
a) Find the interest rate.
b) Find the exponential growth function.
c) Find the balance after 8 yr.
d) Find the doubling time.

Synthesis

33. Solve: $4^{\sqrt[3]{x}} = 8$.

Systems of Equations and Matrices

5

APPLICATION

A dmission to the Indianapolis Children's Museum costs $5.50 more for an adult than for a child (*Source*: Indianapolis Children's Museum). Admission to the museum for Barbara and Jeff Johnson and their five children costs $39. Find the cost of each adult's admission and each child's admission.

This problem appears as Exercise 43 in Section 5.1.

5.1

Systems of Equations in Two Variables

- *Solve a system of two linear equations in two variables by graphing.*
- *Solve a system of two linear equations in two variables using the substitution and the elimination methods.*
- *Use systems of two linear equations to solve applied problems.*

A **system of equations** is composed of two or more equations considered simultaneously. For example,

$$x - y = 5,$$
$$2x + y = 1$$

is a **system of two linear equations in two variables.** The solution set of this system consists of all ordered pairs that make *both* equations true. The ordered pair $(2, -3)$ is a solution of the system of equations above. We can verify this by substituting 2 for x and -3 for y in *each* equation.

$x - y = 5$		$2x + y = 1$	
$2 - (-3)$? 5		$2 \cdot 2 + (-3)$? 1	
$2 + 3$		$4 - 3$	
5 │ 5	TRUE	1 │ 1	TRUE

Solving Systems of Equations Graphically

Recall that the graph of a linear equation is a line that contains all the ordered pairs in the solution set of the equation. When we graph a system of linear equations, each point at which the equations intersect is a solution of *both* equations and therefore a **solution of the system of equations.**

EXAMPLE 1 Solve the following system of equations graphically.

$$x - y = 5,$$
$$2x + y = 1$$

Solution We graph the equations on the same set of axes, as shown below.

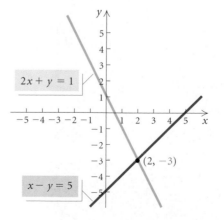

Technology Connection

To use a graphing calculator to solve the system of equations in Example 1, it might be necessary to write each equation in "$Y = \cdots$" form. If so, we would graph $y_1 = x - 5$ and $y_2 = -2x + 1$ and then use the INTERSECT feature.

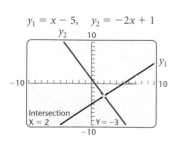

We see that the graphs intersect at a single point, $(2, -3)$, so $(2, -3)$ is the solution of the system of equations. To check this solution, we substitute 2 for x and -3 for y in both equations, as we did above.   —

The graphs of most of the systems of equations that we use to model applications intersect at a single point, like the system above. However, it is possible that the graphs will have no points in common or infinitely many points in common. Each of these possibilities is illustrated below.

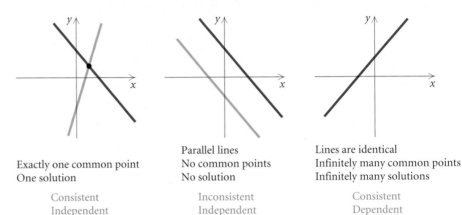

Exactly one common point
One solution

Consistent
Independent

Parallel lines
No common points
No solution

Inconsistent
Independent

Lines are identical
Infinitely many common points
Infinitely many solutions

Consistent
Dependent

If a system of equations has at least one solution, it is **consistent.** If the system has no solutions, it is **inconsistent.** In addition, if a system of two linear equations in two variables has an infinite number of solutions, the equations are **dependent.** Otherwise, they are **independent.**

The Substitution Method

Solving a system of equations graphically is not always accurate when the solutions are not integers. A solution like $\left(\frac{43}{27}, -\frac{19}{27}\right)$, for instance, will be difficult to determine from a hand-drawn graph.

Algebraic methods for solving systems of equations, when used correctly, always give accurate results. One such technique is the **substitution method.** It is used most often when a variable is alone on one side of an equation or when it is easy to solve for a variable. To apply the substitution method, we begin by using one of the equations to express one variable in terms of the other; then we substitute that expression in the other equation of the system.

EXAMPLE 2 Use the substitution method to solve the system

$$x - y = 5, \quad (1)$$
$$2x + y = 1. \quad (2)$$

Solution First, we solve equation (1) for x. (We could have solved for y instead.) We have

$$x - y = 5, \quad (1)$$
$$x = y + 5. \qquad \text{Solving for } x$$

Then we substitute $y + 5$ for x in equation (2). This gives an equation in one variable, which we know how to solve:

$$2x + y = 1 \qquad (2)$$
$$2(y + 5) + y = 1 \qquad \text{The parentheses are necessary.}$$
$$2y + 10 + y = 1 \qquad \text{Removing parentheses}$$
$$3y + 10 = 1 \qquad \text{Collecting like terms on the left}$$
$$3y = -9 \qquad \text{Subtracting 10 on both sides}$$
$$y = -3. \qquad \text{Dividing by 3 on both sides}$$

Now we substitute -3 for y in either of the original equations (this is called **back-substitution**) and solve for x. We choose equation (1):

$$x - y = 5, \quad (1)$$
$$x - (-3) = 5 \qquad \text{Substituting } -3 \text{ for } y$$
$$x + 3 = 5$$
$$x = 2. \qquad \text{Subtracting 3 on both sides}$$

We have previously checked the pair $(2, -3)$ in both equations. The solution of the system of equations is $(2, -3)$.

The Elimination Method

Another algebraic technique for solving systems of equations is the **elimination method.** With this method, we eliminate a variable by adding two equations. If the coefficients of a particular variable are opposites, we can eliminate that variable simply by adding the original equations. For example, if the x-coefficient is -3 in one equation and is 3 in the other equation, then the sum of the x-terms will be 0 and thus the variable x will be eliminated when we add the equations.

EXAMPLE 3 Use the elimination method to solve the system

$$2x + y = 2, \quad (1)$$
$$x - y = 7. \quad (2)$$

Algebraic Solution

Since the y-coefficients, 1 and -1, are opposites, we can eliminate y by adding the equations:

$$
\begin{array}{ll}
2x + y = 2 & (1) \\
\underline{x - y = 7} & (2) \\
3x \quad\;\; = 9 & \textbf{Adding} \\
x = 3. &
\end{array}
$$

We then back-substitute 3 for x in either equation and solve for y. We choose equation (1):

$$
\begin{array}{ll}
2x + y = 2 & (1) \\
2 \cdot 3 + y = 2 & \textbf{Substituting 3 for } x \\
6 + y = 2 & \\
y = -4. &
\end{array}
$$

We check the solution by substituting the pair $(3, -4)$ in both equations.

$$
\begin{array}{c|c}
2x + y = 2 & x - y = 7 \\
\hline
2 \cdot 3 + (-4) \;?\; 2 & 3 - (-4) \;?\; 7 \\
6 - 4 \;\Big|\; & 3 + 4 \;\Big|\; \\
2 \;\Big|\; 2 \quad \text{TRUE} & 7 \;\Big|\; 7 \quad \text{TRUE}
\end{array}
$$

The solution is $(3, -4)$.

Visualizing the Solution

We graph $2x + y = 2$ and $x - y = 7$.

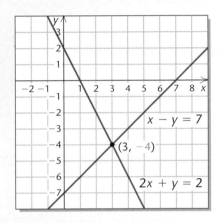

The graphs intersect at the point $(3, -4)$, so the solution of the system of equations is $(3, -4)$.

Before we add, it might be necessary to multiply one or both equations by suitable constants in order to find two equations in which coefficients are opposites.

EXAMPLE 4 Use the elimination method to solve the system

$$4x + 3y = 11, \quad (1)$$
$$-5x + 2y = 15. \quad (2)$$

Algebraic Solution

We can obtain x-coefficients that are opposites by multiplying the first equation by 5 and the second equation by 4:

$$\begin{array}{ll} 20x + 15y = 55 & \textbf{Multiplying equation (1) by 5} \\ \underline{-20x + 8y = 60} & \textbf{Multiplying equation (2) by 4} \\ 23y = 115 & \textbf{Adding} \\ y = 5. \end{array}$$

We then back-substitute 5 for y in either equation (1) or (2) and solve for x. We choose equation (1):

$$\begin{array}{ll} 4x + 3y = 11 & (1) \\ 4x + 3 \cdot 5 = 11 & \textbf{Substituting 5 for } y \\ 4x + 15 = 11 & \\ 4x = -4 & \\ x = -1. & \end{array}$$

We can check the pair $(-1, 5)$ by substituting in both equations. The solution is $(-1, 5)$.

Visualizing the Solution

We graph $4x + 3y = 11$ and $-5x + 2y = 15$.

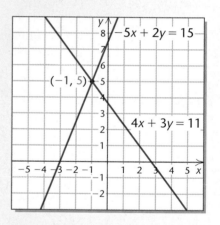

The graphs intersect at the point $(-1, 5)$, so the solution of the system of equations is $(-1, 5)$.

In Example 4, the two systems

$$\begin{array}{ll} 4x + 3y = 11, & \\ -5x + 2y = 15 & \end{array} \quad \text{and} \quad \begin{array}{ll} 20x + 15y = 55, & \\ -20x + 8y = 60 & \end{array}$$

are **equivalent** because they have exactly the same solutions. When we use the elimination method, we often multiply one or both equations by constants to find equivalent equations that allow us to eliminate a variable by adding.

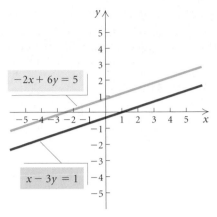

$-2x + 6y = 5$

$x - 3y = 1$

FIGURE 1

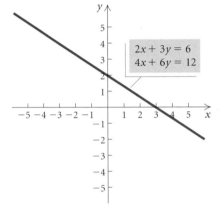

$2x + 3y = 6$
$4x + 6y = 12$

FIGURE 2

EXAMPLE 5 Solve each of the following systems using the elimination method.

a) $x - 3y = 1,$ (1)   **b)** $2x + 3y = 6,$ (1)

    $-2x + 6y = 5$ (2) $4x + 6y = 12$ (2)

Solution

a) We multiply equation (1) by 2 and add:

$$\begin{aligned} 2x - 6y &= 2 \qquad \textbf{Multiplying equation (1) by 2} \\ \underline{-2x + 6y} &= 5 \qquad \textbf{(2)} \\ 0 &= 7. \qquad \textbf{Adding} \end{aligned}$$

There are no values of x and y for which $0 = 7$ is true, so the system has *no solution*. The solution set is $\varnothing$. The system of equations is inconsistent. The graphs of the equations are parallel lines, as shown in Fig. 1.

b) We multiply equation (1) by -2 and add:

$$\begin{aligned} -4x - 6y &= -12 \qquad \textbf{Multiplying equation (1) by } \mathbf{-2} \\ \underline{4x + 6y} &= 12 \qquad \textbf{(2)} \\ 0 &= 0. \qquad \textbf{Adding} \end{aligned}$$

We obtain the equation $0 = 0$, which is true for all values of x and y. This tells us that the equations are dependent, so there are *infinitely many solutions*. That is, any solution of one equation of the system is also a solution of the other. The graphs of the equations are identical, as shown in Fig. 2.

Solving either equation for y, we have $y = -\frac{2}{3}x + 2$, so we can write the solutions of the system as ordered pairs (x, y), where y is expressed as $-\frac{2}{3}x + 2$. Thus the solutions can be written in the form $\left(x, -\frac{2}{3}x + 2\right)$. Any real value that we choose for x then gives us a value for y and thus an ordered pair in the solution set. For example,

if $x = -3$,   then $-\frac{2}{3}x + 2 = -\frac{2}{3}(-3) + 2 = 4,$

if $x = 0$,   then $-\frac{2}{3}x + 2 = -\frac{2}{3} \cdot 0 + 2 = 2,$ and

if $x = 6$,   then $-\frac{2}{3}x + 2 = -\frac{2}{3} \cdot 6 + 2 = -2.$

Thus some of the solutions are $(-3, 4)$, $(0, 2)$, and $(6, -2)$. Similarly, solving either equation for x, we have $x = -\frac{3}{2}y + 3$, so the solutions (x, y) can also be written, expressing x as $-\frac{3}{2}y + 3$, in the form $\left(-\frac{3}{2}y + 3, y\right)$.

Applications

Frequently the most challenging and time-consuming step in the problem-solving process is translating a situation to mathematical language. However, in many cases, this task is made easier if we translate to more than one equation in more than one variable.

EXAMPLE 6 *Snack Mixtures.* At SmartSnax.com, caramel corn worth $2.50 per pound is mixed with honey roasted mixed nuts worth $7.50 per pound in order to get 20 lb of a mixture worth $4.50 per pound. How much of each snack is used?

Solution We use the five-step problem-solving process.

1. **Familiarize.** Let's begin by making a guess. Suppose 16 lb of caramel corn and 4 lb of nuts are used. Then the total weight of the mixture would be 16 lb + 4 lb, or 20 lb, the desired weight. The total values of these amounts of ingredients are found by multiplying the price per pound by the number of pounds used:

   Caramel corn: $2.50(16) = $40

   Nuts: $7.50(4) = $30

   Total value: $70.

   The desired value of the mixture is $4.50 per pound, so the value of 20 lb would be $4.50(20), or $90. Thus we see that our guess, which led to a total of $70, is incorrect. Nevertheless, these calculations will help us to translate.

2. **Translate.** We organize the information in a table. We let x = the number of pounds of caramel corn in the mixture and y = the number of pounds of nuts.

	CARAMEL CORN	NUTS	MIXTURE	
PRICE PER POUND	$2.50	$7.50	$4.50	
NUMBER OF POUNDS	x	y	20	$\longrightarrow x + y = 20$
VALUE OF MIXTURE	$2.50x$	$7.50y$	4.50(20), or 90	$\longrightarrow 2.50x + 7.50y = 90$

From the second row of the table, we get one equation:

$$x + y = 20.$$

The last row of the table yields a second equation:

$$2.50x + 7.50y = 90, \quad \text{or} \quad 2.5x + 7.5y = 90.$$

We can multiply by 10 on both sides of the second equation to clear the decimals. This gives us the following system of equations:

$$x + y = 20, \qquad (1)$$
$$25x + 75y = 900. \qquad (2)$$

3. Carry out. We carry out the solution as follows.

Algebraic Solution

Using the elimination method, we multiply equation (1) by -25 and add it to equation (2):

$$-25x - 25y = -500$$
$$\underline{25x + 75y = 900}$$
$$50y = 400$$
$$y = 8.$$

Then we back-substitute to find x:

$$x + y = 20 \qquad (1)$$
$$x + 8 = 20 \qquad \textbf{Substituting 8 for } y$$
$$x = 12.$$

Visualizing the Solution

The solution of the system of equations is the point of intersection of the graphs of the equations.

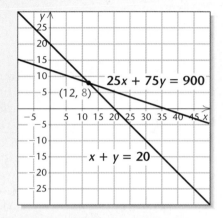

The graphs intersect at the point $(12, 8)$, so the solution of the system of equations is $(12, 8)$.

4. Check. If 12 lb of caramel corn and 8 lb of nuts are used, the mixture weighs $12 + 8$, or 20 lb. The value of the mixture is $\$2.50(12) + \$7.50(8)$, or $\$30 + \$60$, or $\$90$. Since the possible solution yields the desired weight and value of the mixture, our result checks.

5. State. The mixture should consist of 12 lb of caramel corn and 8 lb of honey roasted nuts.

EXAMPLE 7 *Boating.*   Kerry's motorboat takes 3 hr to make a downstream trip with a 3-mph current. The return trip against the same current takes 5 hr. Find the speed of the boat in still water.

Solution

1. Familiarize. We first make a drawing, letting $r =$ the speed of the boat in still water, in miles per hour, and $d =$ the distance traveled, in miles. When the boat is traveling downstream, the current adds to its speed, so the downstream speed is $r + 3$. On the other hand, the current slows the boat down when it travels upstream, so the upstream speed is $r - 3$.

Downstream:
Speed: $r + 3$
Time: 3 hr
Distance: d

Upstream:
Speed: $r - 3$
Time: 5 hr
Distance: d

2. **Translate.** We organize the information in a table. Using the formula *Distance = Rate* (or *Speed*) · *Time*, we find that each row of the table yields an equation.

	DISTANCE	RATE	TIME	
DOWNSTREAM	d	$r + 3$	3	$\longrightarrow d = (r + 3)3$
UPSTREAM	d	$r - 3$	5	$\longrightarrow d = (r - 3)5$

We have a system of equations:

$$d = (r + 3)3, \qquad (1)$$
$$d = (r - 3)5. \qquad (2)$$

3. **Carry out.** We solve the system of equations using substitution:

$(r + 3)3 = (r - 3)5$ **Substituting $(r + 3)3$ for d in equation (2)**

$3r + 9 = 5r - 15$ **Multiplying**

$9 = 2r - 15$ **Subtracting $3r$ on both sides**

$24 = 2r$ **Adding 15 on both sides**

$12 = r.$ **Dividing by 2 on both sides**

4. **Check.** If $r = 12$, then the boat's speed downstream is $r + 3 = 12 + 3$, or 15 mph, and the speed upstream is $r - 3 = 12 - 3$, or 9 mph. At 15 mph, in 3 hr the boat travels $15 \cdot 3 = 45$ mi; at 9 mph, in 5 hr the boat travels $9 \cdot 5 = 45$ mi. Since the distances are the same, the answer checks.

5. **State.** The speed of the boat in still water is 12 mph. ▬

EXAMPLE 8 *Supply and Demand.* Suppose that the price and the supply of the Star Station satellite radio are related by the equation

$$y = 90 + 30x,$$

where y is the price, in dollars, at which the seller is willing to supply x thousand units. Also suppose that the price and the demand for the same model of satellite radio are related by the equation

$$y = 200 - 25x,$$

where y is the price, in dollars, at which the consumer is willing to buy x thousand units.

The **equilibrium point** for this product is the pair (x, y) that is a solution of both equations. The **equilibrium price** is the price at which the amount of the product that the seller is willing to supply is the same as the amount demanded by the consumer. Find the equilibrium point for this product.

Solution

1., 2. Familiarize and **Translate.** We are given a system of equations in the statement of the problem, so no further translation is necessary.

$$y = 90 + 30x, \qquad (1)$$
$$y = 200 - 25x. \qquad (2)$$

We substitute some values for x in each equation to get an idea of the corresponding prices. When $x = 1$,

$$y = 90 + 30 \cdot 1 = 120, \qquad \text{Substituting in equation (1)}$$
$$y = 200 - 25 \cdot 1 = 175. \qquad \text{Substituting in equation (2)}$$

This indicates that the price when 1 thousand units are supplied is lower than the price when 1 thousand units are demanded.
When $x = 4$,

$$y = 90 + 30 \cdot 4 = 210, \qquad \text{Substituting in equation (1)}$$
$$y = 200 - 25 \cdot 4 = 100. \qquad \text{Substituting in equation (2)}$$

In this case, the price related to supply is higher than the price related to demand. It would appear that the x-value we are looking for is between 1 and 4.

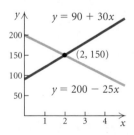

3. Carry out. We use the substitution method:

$$200 - 25x = 90 + 30x \qquad \begin{array}{l}\text{Substituting } 200 - 25x \text{ for } y \\ \text{in equation (1)}\end{array}$$
$$110 = 55x \qquad \text{Adding } 25x \text{ and subtracting } 90 \text{ on both sides}$$
$$2 = x. \qquad \text{Dividing by 55 on both sides}$$

We now back-substitute 2 for x in either equation and find y:

$$y = 200 - 25x \qquad (2)$$
$$y = 200 - 25 \cdot 2 \qquad \text{Substituting 2 for } x$$
$$y = 200 - 50$$
$$y = 150.$$

4. Check. We can check by substituting 2 for x and 150 for y in both equations. Also note that 2 is between 1 and 4 as expected from the *Familiarize* and *Translate* steps.

5. State. The equilibrium point is $(2, \$150)$. That is, the equilibrium supply is 2 thousand units and the equilibrium price is $\$150$.

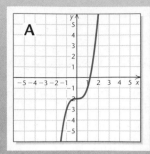

A

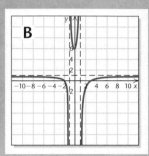

B

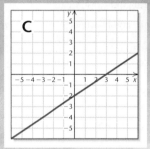

C

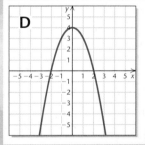

D

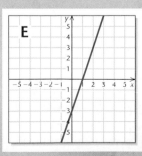

E

Visualizing the Graph

Match the equation or system of equations with its graph.

1. $2x - 3y = 6$

2. $f(x) = x^2 - 2x - 3$

3. $f(x) = -x^2 + 4$

4. $x^2 - 4x + y^2 + 6y + 4 = 0$

5. $f(x) = x^3 - 2$

6. $f(x) = -(x - 1)^2(x + 1)^2$

7. $f(x) = \dfrac{x - 1}{x^2 - 4}$

8. $f(x) = \dfrac{x^2 - x - 6}{x^2 - 1}$

9. $\begin{aligned} x - y &= -1, \\ 2x - y &= 2 \end{aligned}$

10. $\begin{aligned} 3x - y &= 3, \\ 2y &= 6x - 6 \end{aligned}$

Answers on page A-35

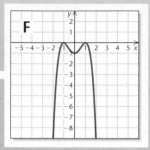

F

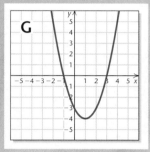

G

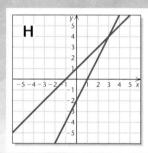

H

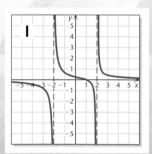

I

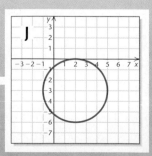

J

5.1

Exercise Set

In Exercises 1–6, match the system of equations with one of the graphs (a)–(f), which follow.

a)

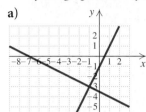

b)

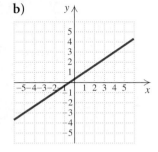

c)

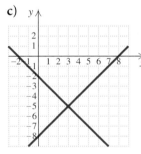

d)

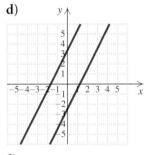

e)

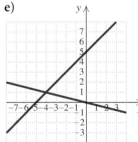

f)

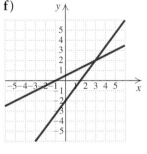

1. $x + y = -2,$
$y = x - 8$

2. $x - y = -5,$
$x = -4y$

3. $x - 2y = -1,$
$4x - 3y = 6$

4. $2x - y = 1,$
$x + 2y = -7$

5. $2x - 3y = -1,$
$-4x + 6y = 2$

6. $4x - 2y = 5,$
$6x - 3y = -10$

Solve graphically.

7. $x + y = 2,$
$3x + y = 0$

8. $x + y = 1,$
$3x + y = 7$

9. $x + 2y = 1,$
$x + 4y = 3$

10. $3x + 4y = 5,$
$x - 2y = 5$

11. $y + 1 = 2x,$
$y - 1 = 2x$

12. $2x - y = 1,$
$3y = 6x - 3$

13. $x - y = -6,$
$y = -2x$

14. $2x + y = 5,$
$x = -3y$

15. $2y = x - 1,$
$3x = 6y + 3$

16. $y = 3x + 2,$
$3x - y = -3$

Solve using the substitution method.

17. $x + y = 9,$
$2x - 3y = -2$

18. $3x - y = 5,$
$x + y = \frac{1}{2}$

19. $x - 2y = 7,$
$x = y + 4$

20. $x + 4y = 6,$
$x = -3y + 3$

21. $y = 2x - 6,$
$5x - 3y = 16$

22. $3x + 5y = 2,$
$2x - y = -3$

23. $x - 5y = 4,$
$y = 7 - 2x$

24. $5x + 3y = -1,$
$x + y = 1$

25. $2x - 3y = 5,$
$5x + 4y = 1$

26. $3x + 4y = 6,$
$2x + 3y = 5$

27. $x + 2y = 2,$
$4x + 4y = 5$

28. $2x - y = 2,$
$4x + y = 3$

Solve using the elimination method. Also determine whether each system is consistent or inconsistent and whether the equations are dependent or independent.

29. $x + 2y = 7,$
$x - 2y = -5$

30. $3x + 4y = -2,$
$-3x - 5y = 1$

31. $x - 3y = 2,$
$6x + 5y = -34$

32. $x + 3y = 0,$
$20x - 15y = 75$

33. $3x - 12y = 6,$
$2x - 8y = 4$

34. $2x + 6y = 7,$
$3x + 9y = 10$

35. $2x = 5 - 3y,$
$4x = 11 - 7y$

36. $7(x - y) = 14,$
$2x = y + 5$

37. $0.3x - 0.2y = -0.9,$
$0.2x - 0.3y = -0.6$
(*Hint*: Since each coefficient has one decimal place, first multiply each equation by 10 to clear the decimals.)

38. $0.2x - 0.3y = 0.3,$
$0.4x + 0.6y = -0.2$
(*Hint*: Since each coefficient has one decimal place, first multiply each equation by 10 to clear the decimals.)

39. $\frac{1}{5}x + \frac{1}{2}y = 6,$
$\frac{3}{5}x - \frac{1}{2}y = 2$
(*Hint*: First multiply by the least common denominator to clear fractions.)

40. $\frac{2}{3}x + \frac{3}{5}y = -17,$
$\frac{1}{2}x - \frac{1}{3}y = -1$
(*Hint*: First multiply by the least common denominator to clear fractions.)

41. *Daytona 500 Concessions.*   The amount of ice used in drinks consumed at the 2003 Daytona 500 NASCAR race was 50 times the amount of hot dogs consumed, by weight. A total of 204 tons of ice and hot dogs were consumed. (*Source*: Americrown Service Corporation) Find the number of tons of ice and of hot dogs consumed.

42. *Annual College Student Spending.*   College students spend $183 more each year on textbooks and course materials than on computer equipment. They spend a total of $819 on textbooks and course materials and computer equipment each year. (*Source*: National Association of College Stores) How much is spent each year on textbooks and course materials and how much is spent on computer equipment?

43. *Museum Admission Prices.*   Admission to the Indianapolis Children's Museum costs $5.50 more for an adult than for a child (*Source*: Indianapolis Children's Museum). Admission to the museum for

Barbara and Jeff Johnson and their five children costs $39. Find the cost of each adult's admission and each child's admission.

44. *Mail-Order Business.*   A mail-order lacrosse equipment business shipped 120 packages one day. Customers are charged $3.50 for each standard-delivery package and $7.50 for each express-delivery package. Total shipping charges for the day were $596. How many of each kind of package were shipped?

45. *Sales Promotions.*   During a one-month promotional campaign, Video Village gave either a free video rental or a 12-serving box of microwave popcorn to new members. It cost the store $1 for each free rental and $2 for each box of popcorn. In all, 48 new members were signed up and the store's cost for the incentives was $86. How many of each incentive were given away?

46. *Concert Ticket Prices.*   One evening 1500 concert tickets were sold for the Fairmont Summer Jazz Festival. Tickets cost $25 for covered pavilion seats and $15 for lawn seats. Total receipts were $28,500. How many of each type of ticket were sold?

47. *Supply and Demand.*   The supply and demand for a particular model of personal digital assistant are related to price by the equations
$$y = 70 + 2x,$$
$$y = 175 - 5x,$$
respectively, where y is the price, in dollars, and x is the number of units, in thousands. Find the equilibrium point for this product.

48. *Supply and Demand.*   The supply and demand for a particular model of treadmill are related to price by

the equations

$$y = 240 + 40x,$$
$$y = 500 - 25x,$$

respectively, where y is the price, in dollars, and x is the number of units, in thousands. Find the equilibrium point for this product.

The point at which a company's costs equal its revenues is the break-even point. In Exercises 49–52, C represents the production cost, in dollars, of x units of a product and R represents the revenue, in dollars, from the sale of x units. Find the number of units that must be produced and sold in order to break even. That is, find the value of x for which C = R.

49. $C = 14x + 350,$
$R = 16.5x$

50. $C = 8.5x + 75,$
$R = 10x$

51. $C = 15x + 12,000,$
$R = 18x - 6000$

52. $C = 3x + 400,$
$R = 7x - 600$

53. *Nutrition.* A one-cup serving of spaghetti with meatballs contains 260 Cal (calories) and 32 g of carbohydrates. A one-cup serving of chopped iceberg lettuce contains 5 Cal and 1 g of carbohydrates. (*Source: Home and Garden Bulletin No. 72*, U.S. Government Printing Office, Washington, D.C. 20402) How many servings of each would be required to obtain 400 Cal and 50 g of carbohydrates?

54. *Nutrition.* One serving of tomato soup contains 100 Cal and 18 g of carbohydrates. One slice of

whole wheat bread contains 70 Cal and 13 g of carbohydrates. (*Source: Home and Garden Bulletin No. 72*, U.S. Government Printing Office, Washington, D.C. 20402) How many servings of each would be required to obtain 230 Cal and 42 g of carbohydrates?

55. *Motion.* A Leisure Time Cruises riverboat travels 46 km downstream in 2 hr. It travels 51 km upstream in 3 hr. Find the speed of the boat and the speed of the stream.

56. *Motion.* A DC10 travels 3000 km with a tail wind in 3 hr. It travels 3000 km with a head wind in 4 hr. Find the speed of the plane and the speed of the wind.

57. *Investment.* Bernadette inherited $15,000 and invested it in two municipal bonds, which pay 7% and 9% simple interest. The annual interest is $1230. Find the amount invested at each rate.

58. *Tee Shirt Sales.* Mack's Tee Shirt Shack sold 36 shirts one day. All short-sleeved tee shirts cost $12 and all long-sleeved tee shirts cost $18. Total receipts for the day were $522. How many of each kind of shirt were sold?

59. *Coffee Mixtures.* The owner of The Daily Grind coffee shop mixes French roast coffee worth $9.00 per pound with Kenyan coffee worth $7.50 per pound in order to get 10 lb of a mixture worth $8.40 per pound. How much of each type of coffee was used?

60. *Motion.* A Boeing 747 flies the 3000-mi distance from Los Angeles to New York, with a tail wind, in 5 hr. The return trip, against the wind, takes 6 hr. Find the speed of the plane and the speed of the wind.

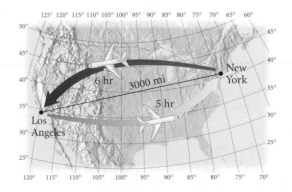

61. *Commissions.* Jackson Manufacturing offers its sales representatives a choice between being paid a commission of 8% of sales or being paid a monthly salary of $1500 plus a commission of 1% of sales. For what monthly sales do the two plans pay the same amount?

62. *Motion.* Two private airplanes travel toward each other from cities that are 780 km apart at speeds of 190 km/h and 200 km/h. They left at the same time. In how many hours will they meet?

Technology Connection

63. Use a graphing calculator to solve the systems of equations in Exercises 7, 17, and 29.

64. Use a graphing calculator to solve the systems of equations in Exercises 8, 18, and 30.

65. *Beef and Chicken Consumption.* The amount of beef consumed in the United States has remained fairly constant in recent years while the amount of chicken consumed has increased, as shown by the data in the following table.

Year	Per Capita Beef Consumption (in pounds)	Per Capita Chicken Consumption (in pounds)
1995	63.6	48.2
1998	63.6	49.8
1999	64.4	52.9
2000	64.4	52.9

Source: U.S. Department of Agriculture, Economic Research Service, *Food Consumption, Prices, and Expenditures, 1970–2000*

a) Find linear regression functions $b(x)$ and $c(x)$ that represent the per capita beef consumption and the per capita chicken consumption, respectively, in number of pounds x years after 1995.
b) Use the functions found in part (a) to estimate when chicken consumption will equal beef consumption.

66. *U.S. Labor Force.* The percentages of men and women in the civilian labor force in recent years are shown in the following table.

Year	Percentage of Men in the Labor Force	Percentage of Women in the Labor Force
1995	75.0	58.9
1999	74.7	60.0
2000	74.7	60.2
2001	74.4	60.1

Source: U.S. Bureau of Labor Statistics

a) Find linear regression functions $m(x)$ and $w(x)$ that represent the percentages of men and women in the civilian labor force, respectively, x years after 1995.
b) Use the functions found in part (a) to predict when the percentages of men and women in the labor force will be equal.

Collaborative Discussion and Writing

67. Explain in your own words when the elimination method for solving a system of equations is preferable to the substitution method.

68. Cassidy solves the equation $2x + 5 = 3x - 7$ by finding the point of intersection of the graphs of $y_1 = 2x + 5$ and $y_2 = 3x - 7$. She finds the same point when she solves the system of equations

$$y = 2x + 5,$$
$$y = 3x - 7.$$

Explain the difference between the solutions.

Skill Maintenance

69. In the first quarter of 2003, 232 million DVDs were shipped to retailers. This was a 93% increase over the number shipped in the same period in 2002. (*Source*: Consumer Electronics Association) Find the number of DVDs shipped to retailers in the first quarter of 2002.

70. The movie *Spider-Man* led box-office receipts in 2002 with gross ticket sales of $404 million. This was about one-third more than the second-place film *Star Wars: Episode II—Attack of the Clones.* (*Source*: *New York Times*, December 30, 2002) Find the gross ticket sales for *Star Wars* in 2002.

Consider the function

$$f(x) = x^2 - 4x + 3$$

in Exercises 71–74.

71. What are the inputs if the output is 15?

72. What is the output if the input is -2?

73. Given an output of 8, find the corresponding inputs.

74. Find the zeros of the function.

Synthesis

75. *Motion.* Nancy jogs and walks to campus each day. She averages 4 km/h walking and 8 km/h jogging. The distance from home to the campus is 6 km and she makes the trip in 1 hr. How far does she jog on each trip?

76. *e-Commerce.* shirts.com advertises a limited-time sale, offering 1 turtleneck for $15 and 2 turtlenecks for $25. A total of 1250 turtlenecks are sold and $16,750 is taken in. How many customers ordered 2 turtlenecks?

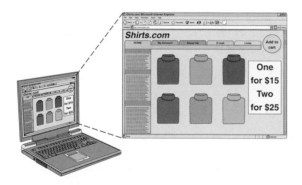

77. *Motion.* A train leaves Union Station for Central Station, 216 km away, at 9 A.M. One hour later, a train leaves Central Station for Union Station. They meet at noon. If the second train had started at 9 A.M. and the first train at 10:30 A.M., they would still have met at noon. Find the speed of each train.

78. *Antifreeze Mixtures.* An automobile radiator contains 16 L of antifreeze and water. This mixture is 30% antifreeze. How much of this mixture should be drained and replaced with pure antifreeze so that the final mixture will be 50% antifreeze?

79. Two solutions of the equation $Ax + By = 1$ are $(3, -1)$ and $(-4, -2)$. Find A and B.

80. *Ticket Line.* You are in line at a ticket window. There are 2 more people ahead of you in line than there are behind you. In the entire line, there are three times as many people as there are behind you. How many people are ahead of you?

81. *Gas Mileage.* The Honda Civic Hybrid vehicle with manual transmission, powered by gasoline–electric technology, gets 46 miles per gallon (mpg) in city driving and 51 mpg in highway driving (*Source*: Honda.com). The car is driven 621 mi on 13 gal of gasoline. How many miles were driven in the city and how many were driven on the highway?

82. *Motion.* Heather is hiking and is standing on a railroad bridge, as shown in the figure below. A train is approaching from the direction shown by the arrow. If Heather runs at a speed of 10 mph toward the train, she will reach point P on the bridge at the same moment that the train does. If she runs to point Q at the other end of the bridge at a speed of 10 mph, she will reach point Q also at the same moment that the train does. How fast, in miles per hour, is the train traveling?

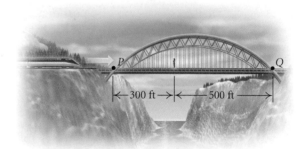

5.2

Systems of Equations in Three Variables

- *Solve systems of linear equations in three variables.*
- *Use systems of three equations to solve applied problems.*
- *Model a situation using a quadratic function.*

A **linear equation in three variables** is an equation equivalent to one of the form $Ax + By + Cz = D$, where A, B, C, and D are real numbers and A, B, and C are not all 0. A **solution of a system of three equations in three variables** is an ordered triple that makes all three equations true. For example, the triple $(2, -1, 0)$ is a solution of the system of equations

$$4x + 2y + 5z = 6,$$
$$2x - y + z = 5,$$
$$3x + 2y - z = 4.$$

We can verify this by substituting 2 for x, -1 for y, and 0 for z in each equation.

Solving Systems of Equations in Three Variables

We will solve systems of equations in three variables using an algebraic method called **Gaussian elimination,** named for the German mathematician Karl Friedrich Gauss (1777–1855). Our goal is to transform the original system to an equivalent one of the form

$$Ax + By + Cz = D,$$
$$Ey + Fz = G,$$
$$Hz = K.$$

Then we solve the third equation for z and back-substitute to find y and then x.

Each of the following operations can be used to transform the original system to an equivalent system in the desired form.

1. Interchange any two equations.
2. Multiply both sides of one of the equations by a nonzero constant.
3. Add a nonzero multiple of one equation to another equation.

EXAMPLE 1 Solve the following system:

$$x - 2y + 3z = 11, \quad (1)$$
$$4x + 2y - 3z = 4, \quad (2)$$
$$3x + 3y - z = 4. \quad (3)$$

Solution   First, we choose one of the variables to eliminate using two different pairs of equations. Let's eliminate x from equations (2) and (3). We multiply equation (1) by -4 and add it to equation (2). We also multiply equation (1) by -3 and add it to equation (3).

$$
\begin{array}{ll}
-4x + 8y - 12z = -44 & \textbf{Multiplying (1) by } -4 \\
\underline{4x + 2y - 3z = 4} & \text{(2)} \\
 10y - 15z = -40; & \text{(4)}
\end{array}
$$

$$
\begin{array}{ll}
-3x + 6y - 9z = -33 & \textbf{Multiplying (1) by } -3 \\
\underline{3x + 3y - z = 4} & \text{(3)} \\
 9y - 10z = -29. & \text{(5)}
\end{array}
$$

Now we have

$$
\begin{array}{ll}
x - 2y + 3z = 11, & \text{(1)} \\
10y - 15z = -40, & \text{(4)} \\
9y - 10z = -29. & \text{(5)}
\end{array}
$$

Next, we multiply equation (5) by 10 to make the y-coefficient a multiple of the y-coefficient in the equation above it:

$$
\begin{array}{ll}
x - 2y + 3z = 11, & \text{(1)} \\
10y - 15z = -40, & \text{(4)} \\
90y - 100z = -290. & \text{(6)}
\end{array}
$$

Next, we multiply equation (4) by -9 and add it to equation (6):

$$
\begin{array}{ll}
-90y + 135z = 360 & \textbf{Multiplying (4) by } -9 \\
\underline{90y - 100z = -290} & \text{(6)} \\
 35z = 70. & \text{(7)}
\end{array}
$$

We now have the system of equations

$$
\begin{array}{ll}
x - 2y + 3z = 11, & \text{(1)} \\
10y - 15z = -40, & \text{(4)} \\
35z = 70. & \text{(7)}
\end{array}
$$

Now we solve equation (7) for z:

$$
35z = 70
$$
$$
z = 2.
$$

Then we back-substitute 2 for z in equation (4) and solve for y:

$$
10y - 15 \cdot 2 = -40
$$
$$
10y - 30 = -40
$$
$$
10y = -10
$$
$$
y = -1.
$$

Finally, we back-substitute -1 for y and 2 for z in equation (1) and solve for x:

$$x - 2(-1) + 3 \cdot 2 = 11$$
$$x + 2 + 6 = 11$$
$$x = 3.$$

We can check the triple $(3, -1, 2)$ in each of the three original equations. Since it makes all three equations true, the solution is $(3, -1, 2)$.   —

EXAMPLE 2   Solve the following system:

$$x + y + z = 7, \qquad (1)$$
$$3x - 2y + z = 3, \qquad (2)$$
$$x + 6y + 3z = 25. \qquad (3)$$

Solution   We multiply equation (1) by -3 and add it to equation (2). We also multiply equation (1) by -1 and add it to equation (3).

$$x + y + z = 7, \qquad (1)$$
$$-5y - 2z = -18, \qquad (4)$$
$$5y + 2z = 18 \qquad (5)$$

Next, we add equation (4) to equation (5):

$$x + y + z = 7, \qquad (1)$$
$$-5y - 2z = -18, \qquad (4)$$
$$0 = 0.$$

The equation $0 = 0$ tells us that equation (3) of the original system is dependent on the first two equations. Thus the original system is equivalent to

$$x + y + z = 7, \qquad (1)$$
$$3x - 2y + z = 3. \qquad (2)$$

In this particular case, the original system has infinitely many solutions. (In some cases, a system containing dependent equations could be inconsistent.) To find an expression for these solutions, we first solve equation (4) for either y or z. We choose to solve for y:

$$-5y - 2z = -18 \qquad (4)$$
$$-5y = 2z - 18$$
$$y = -\tfrac{2}{5}z + \tfrac{18}{5}.$$

Then we back-substitute in equation (1) to find an expression for x in terms of z:

$$x - \tfrac{2}{5}z + \tfrac{18}{5} + z = 7 \qquad \text{Substituting } -\tfrac{2}{5}z + \tfrac{18}{5} \text{ for } y$$
$$x + \tfrac{3}{5}z + \tfrac{18}{5} = 7$$
$$x + \tfrac{3}{5}z = \tfrac{17}{5}$$
$$x = -\tfrac{3}{5}z + \tfrac{17}{5}.$$

The solutions of the system of equations are ordered triples of the form $\left(-\frac{3}{5}z + \frac{17}{5}, -\frac{2}{5}z + \frac{18}{5}, z\right)$, where z can be any real number. Any real number that we use for z then gives us values for x and y and thus an ordered triple in the solution set. For example, if we choose $z = 0$, we have the solution $\left(\frac{17}{5}, \frac{18}{5}, 0\right)$. If we choose $z = -1$, we have $(4, 4, -1)$. ▬

If we get a false equation, such as $0 = -5$, at some stage of the elimination process, we conclude that the original system is *inconsistent*. That is, it has no solutions.

Although systems of three linear equations in three variables do not lend themselves well to graphical solutions, it is of interest to picture some possible solutions. The graph of a linear equation in three variables is a plane. Thus the solution set of such a system is the intersection of three planes. Some possibilities are shown below.

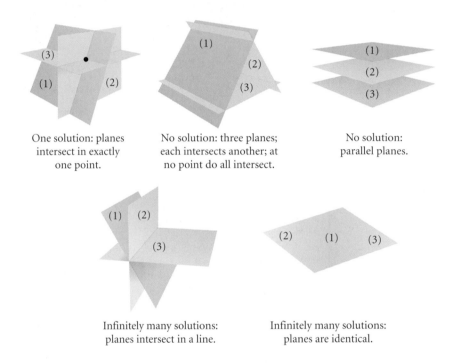

One solution: planes intersect in exactly one point.

No solution: three planes; each intersects another; at no point do all intersect.

No solution: parallel planes.

Infinitely many solutions: planes intersect in a line.

Infinitely many solutions: planes are identical.

Applications

Systems of equations in three or more variables allow us to solve many problems in fields such as business, the social and natural sciences, and engineering.

EXAMPLE 3 *Investment.* Moira inherited $15,000 and invested part of it in a money market account, part in municipal bonds, and part in a mutual fund. After 1 yr, she received a total of $730 in simple interest from the three investments. The money market account paid 4% annually, the bonds paid 5% annually, and the mutual fund paid 6% annually. There was $2000 more invested in the mutual fund than in bonds. Find the amount that Moira invested in each category.

Solution

1. **Familiarize.** We let x, y, and z represent the amounts invested in the money market account, the bonds, and the mutual fund, respectively. Then the amounts of income produced annually by each investment are given by 4%x, 5%y, and 6%z, or $0.04x$, $0.05y$, and $0.06z$.

2. **Translate.** The fact that a total of $15,000 is invested gives us one equation:

   $x + y + z = 15{,}000.$

   Since the total interest is $730, we have a second equation:

   $0.04x + 0.05y + 0.06z = 730.$

   Another statement in the problem gives us a third equation.

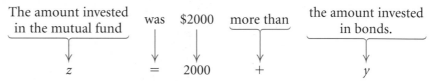

The amount invested in the mutual fund   was $2000   more than   the amount invested in bonds.

$$z \quad = \quad 2000 \quad + \quad y$$

We now have a system of three equations:

$$\begin{aligned} x + y + z &= 15{,}000, \\ 0.04x + 0.05y + 0.06z &= 730, \quad \text{or} \\ z &= 2000 + y; \end{aligned} \qquad \begin{aligned} x + y + z &= 15{,}000, \\ 4x + 5y + 6z &= 73{,}000, \\ -y + z &= 2000. \end{aligned}$$

3. **Carry out.** Solving the system of equations, we get

   $(7000, 3000, 5000).$

4. **Check.** The sum of the numbers is 15,000. The income produced is

   $0.04(7000) + 0.05(3000) + 0.06(5000) = 280 + 150 + 300, \quad \text{or} \quad \$730.$

   Also the amount invested in the mutual fund, $5000, is $2000 more than the amount invested in bonds, $3000. Our solution checks in the original problem.

5. **State.** Moira invested $7000 in a money market account, $3000 in municipal bonds, and $5000 in a mutual fund.

Mathematical Models and Applications

In a situation in which a quadratic function will serve as a mathematical model, we may wish to find an equation, or formula, for the function. For a linear model, we can find an equation if we know two data points. For a quadratic function, we need three data points.

EXAMPLE 4 *The Cost of Operating an Automobile at Various Speeds.* Under certain conditions, it is found that the cost of operating an automobile as a function of speed is approximated by a quadratic function. Use the data shown below to find an equation of the function. Then use the equation to determine the cost of operating the automobile at 60 mph and at 80 mph.

SPEED (IN MILES PER HOUR)	OPERATING COST PER MILE (IN CENTS)
10	22
20	20
50	20

Solution Letting $x =$ the speed and $f(x) =$ the cost per mile, in cents, we use the three data points $(10, 22)$, $(20, 20)$, and $(50, 20)$ to find a, b, and c in the equation $f(x) = ax^2 + bx + c$. First we substitute:

$$\begin{aligned} f(x) &= ax^2 + bx + c \\ \text{For } (10, 22):\quad 22 &= a \cdot 10^2 + b \cdot 10 + c; \\ \text{For } (20, 20):\quad 20 &= a \cdot 20^2 + b \cdot 20 + c; \\ \text{For } (50, 20):\quad 20 &= a \cdot 50^2 + b \cdot 50 + c. \end{aligned}$$

We now have a system of equations in the variables a, b, and c:

$$\begin{aligned} 100a + 10b + c &= 22, \\ 400a + 20b + c &= 20, \\ 2500a + 50b + c &= 20. \end{aligned}$$

Solving this system of equations, we obtain $(0.005, -0.35, 25)$. Thus,

$$f(x) = 0.005x^2 - 0.35x + 25.$$

To determine the cost of operating an automobile at 60 mph, we find $f(60)$:

$$\begin{aligned} f(60) &= 0.005(60)^2 - 0.35(60) + 25 \\ &= 22¢ \text{ per mile.} \end{aligned}$$

To find the cost of operating an automobile at 80 mph, we find $f(80)$:

$$\begin{aligned} f(80) &= 0.005(80)^2 - 0.35(80) + 25 \\ &= 29¢ \text{ per mile.} \end{aligned}$$

Technology Connection

The function in Example 4 can also be found using the QUADRATIC REGRESSION feature on a graphing calculator. Note that the method of Example 4 works when we have exactly three data points, whereas the QUADRATIC REGRESSION feature on a graphing calculator can be used for three or more points.

5.2

Exercise Set

Solve each of the following systems.

1. $x + y + z = 2,$
$6x - 4y + 5z = 31,$
$5x + 2y + 2z = 13$

2. $x + 6y + 3z = 4,$
$2x + y + 2z = 3,$
$3x - 2y + z = 0$

3. $x - y + 2z = -3,$
$x + 2y + 3z = 4,$
$2x + y + z = -3$

4. $x + y + z = 6,$
$2x - y - z = -3,$
$x - 2y + 3z = 6$

5. $x + 2y - z = 5,$
$2x - 4y + z = 0,$
$3x + 2y + 2z = 3$

6. $2x + 3y - z = 1,$
$x + 2y + 5z = 4,$
$3x - y - 8z = -7$

7. $x + 2y - z = -8,$
$2x - y + z = 4,$
$8x + y + z = 2$

8. $x + 2y - z = 4,$
$4x - 3y + z = 8,$
$5x - y = 12$

9. $2x + y - 3z = 1,$
$x - 4y + z = 6,$
$4x - 7y - z = 13$

10. $x + 3y + 4z = 1,$
$3x + 4y + 5z = 3,$
$x + 8y + 11z = 2$

11. $4a + 9b = 8,$
$8a + 6c = -1,$
$6b + 6c = -1$

12. $3p + 2r = 11,$
$q - 7r = 4,$
$p - 6q = 1$

13. $2x + z = 1,$
$3y - 2z = 6,$
$x - 2y = -9$

14. $3x + 4z = -11,$
$x - 2y = 5,$
$4y - z = -10$

15. $w + x + y + z = 2,$
$w + 2x + 2y + 4z = 1,$
$-w + x - y - z = -6,$
$-w + 3x + y - z = -2$

16. $w + x - y + z = 0,$
$-w + 2x + 2y + z = 5,$
$-w + 3x + y - z = -4,$
$-2w + x + y - 3z = -7$

17. *e-Commerce.* computerwarehouse.com charges $3 for shipping orders up to 10 lb, $5 for orders from 10 lb up to 15 lb, and $7.50 for orders of 15 lb or more. One day shipping charges for 150 orders totaled $680. The number of orders under 10 lb was

three times the number of orders weighing 15 lb or more. Find the number of packages shipped at each rate.

18. *Mail-Order Business.* Natural Fibers Clothing charges $4 for shipping orders of $25 or less, $8 for orders from $25.01 to $75, and $10 for orders over $75. One week shipping charges for 600 orders totaled $4280. Eighty more orders for $25 or less were shipped than orders for more than $75. Find the number of orders shipped at each rate.

19. *Nutrition.* A hospital dietician must plan a lunch menu that provides 485 Cal, 41.5 g of carbohydrates, and 35 mg of calcium. A 3-oz serving of broiled ground beef contains 245 Cal, 0 g of carbohydrates, and 9 mg of calcium. One baked potato contains 145 Cal, 34 g of carbohydrates, and 8 mg of calcium. A one-cup serving of strawberries contains 45 Cal, 10 g of carbohydrates, and 21 mg of calcium. (*Source: Home and Garden Bulletin No. 72*, U.S. Government Printing Office, Washington, D.C. 20402) How many servings of each are required to provide the desired nutritional values?

20. *Nutrition.* A diabetic patient wishes to prepare a meal consisting of roasted chicken breast, mashed potatoes, and peas. A 3-oz serving of roasted skinless chicken breast contains 140 Cal, 27 g of protein, and 64 mg of sodium. A one-cup serving of mashed potatoes contains 160 Cal, 4 g of protein, and 636 mg of sodium, and a one-cup serving of peas contains 125 Cal, 8 g of protein, and 139 mg of sodium. (*Source*: *Home and Garden Bulletin No 72*, U.S. Government Printing Office, Washington, D.C. 20402) How many servings of each should be used if the meal is to contain 415 Cal, 50.5 g of protein, and 553 mg of sodium?

21. *Investment.* Jamal earns a year-end bonus of $5000 and puts it in 3 one-year investments that pay $243 in simple interest. Part is invested at 3%, part at 4%, and part at 6%. There is $1500 more invested at 6% than at 3%. Find the amount invested at each rate.

22. *Investment.* Casey receives $126 per year in simple interest from three investments. Part is invested at 2%, part at 3%, and part at 4%. There is $500 more invested at 3% than at 2%. The amount invested at 4% is three times the amount invested at 3%. Find the amount invested at each rate.

23. *Price Increases.* Orange juice, a raisin bagel, and a cup of coffee from Kelly's Koffee Kart cost a total of $3. Kelly posts a notice announcing that, effective the following week, the price of orange juice will increase 50% and the price of bagels will increase 20%. After the increase, the same purchase will cost a total of $3.75, and orange juice will cost twice as much as coffee. Find the price of each item before the increase.

24. *Cost of Snack Food.* Martin and Eva pool their loose change to buy snacks on their coffee break. One day, they spent $5 on 1 carton of milk, 2 donuts, and 1 cup of coffee. The next day, they spent $5.50 on 3 donuts and 2 cups of coffee. The third day, they bought 1 carton of milk, 1 donut, and 2 cups of coffee and spent $5.25. On the fourth day, they have a total of $5.45 left. Is this enough to buy 2 cartons of milk and 2 donuts?

25. *Passenger Transportation.* The total volume of passenger traffic by private automobile, by bus (excluding school buses and urban transit buses), and by railroads in the United States in a recent year was 1899 billion passenger-miles. (One passenger-mile is the transportation of one passenger the distance of one mile.) The volume of bus traffic was 21 billion passenger-miles more than the volume of railroad traffic. The total volume of bus traffic was 1815 billion passenger-miles less than the volume of traffic by private automobile. (*Source*: *Transportation in America*, Eno Transportation Foundation, Inc., Washington, DC) What was the volume of each type of passenger traffic?

26. *Cheese Consumption.* The total per capita consumption (the average amount consumed per person) of cheddar, mozzarella, and Swiss cheese in the United States in a recent year was 20.4 lb. The total consumption of mozzarella and Swiss cheese was 0.2 lb more than that of cheddar cheese. The consumption of mozzarella cheese was 8.1 lb more than that of Swiss cheese. (*Source*: U.S. Department of Agriculture, Economic Research Service, *Food Consumption, Prices, and Expenditures*) What was the per capita consumption of each type of cheese?

27. *Golf.* On an 18-hole golf course, there are par-3 holes, par-4 holes, and par-5 holes. A golfer who shoots par on every hole has a score of 72. The sum of the number of par-3 holes and the number of par-5 holes is 8. How many of each type of hole are there on the golf course?

28. *Golf.* On an 18-hole golf course, there are par-3 holes, par-4 holes, and par-5 holes. A golfer who shoots par on every hole has a score of 70. There are twice as many par-4 holes as there are par-5 holes. How many of each type of hole are there on the golf course?

29. *Number of Marriages.* The following table shows the number of marriages, in thousands, in California, represented as years since 1990.

YEAR, x	NUMBER OF MARRIAGES (IN THOUSANDS)
1990, 0	237
1995, 5	200
2001, 11	224

Source: U.S. National Center for Health Statistics, *Vital Statistics of the United States*, annual, Monthly Vital Statistics Reports

a) Fit a quadratic function $f(x) = ax^2 + bx + c$ to the data.

b) Use the function to estimate the number of marriages in California in 2008.

30. *Red-Meat Consumption.* The following table shows the per capita consumption of red meat, in pounds, in the United States, represented as years since 1990.

YEAR, x	RED-MEAT CONSUMPTION (IN POUNDS)
1990, 0	112
1999, 9	115
2000, 10	114

Source: U.S. Department of Agriculture, Economic Research Service, *Food Consumption, Prices, and Expenditures*

a) Fit a quadratic function $f(x) = ax^2 + bx + c$ to the data.

b) Use the function to estimate the per capita consumption of red meat in 2010.

31. *Dairy Product Consumption.* The following table shows per capita consumption of dairy products, in pounds, in the United States, represented as years since 1985.

YEAR, x	MILK CONSUMPTION (IN POUNDS)
1985, 0	594
1995, 10	577
2000, 15	593

Source: U.S. Department of Agriculture, Economic Research Service, *Food Consumption, Prices, and Expenditures*

a) Fit a quadratic function $f(x) = ax^2 + bx + c$ to the data.

b) Use the function to estimate the per capita consumption of dairy products in 2007.

32. *Crude Steel Production.* The following table shows the world production of crude steel, in millions of metric tons, represented as years since 1990.

YEAR, x	CRUDE STEEL PRODUCTION (IN MILLIONS OF METRIC TONS)
1990, 0	773
1995, 5	686
2000, 10	739

Source: Statistical Division of the United Nations

a) Fit a quadratic function $f(x) = ax^2 + bx + c$ to the data.

b) Use the function to estimate the world production of crude steel in 2009.

Technology Connection

33. *Morning Newspapers.* The number of morning newspapers in the United States in various years is shown in the following table.

YEAR	NUMBER OF MORNING NEWSPAPERS
1920	437
1940	380
1960	312
1980	387
1990	559
2001	776

Source: *Editor & Publisher*

a) Use a graphing calculator to fit a quadratic function $f(x)$ to the data, where x is the number of years after 1920.

b) Use the function found in part (a) to estimate the number of morning newspapers in 2006 and in 2009.

34. *Nursery School Enrollment.* The number of children 3 to 5 years old enrolled in nursery school, in millions, in various years is shown in the following table.

YEAR	NUMBER OF CHILDREN ENROLLED IN NURSERY SCHOOL (IN MILLIONS)
1990	3.378
1995	4.331
1998	4.512
1999	4.506
2000	4.326

Source: U.S. Bureau of the Census, *Current Population Reports*

a) Use a graphing calculator to fit a quadratic function $n(x)$ to the data, where x is the number of years after 1990.

b) Use the function found in part (a) to estimate the number of children enrolled in nursery school in 2007 and in 2010.

Collaborative Discussion and Writing

35. Given two linear equations in three variables, $Ax + By + Cz = D$ and $Ex + Fy + Gz = H$, explain how you could find a third equation such that the system contains dependent equations.

36. Write a problem for a classmate to solve that can be translated to a system of three equations in three variables.

Skill Maintenance

In each of Exercises 37–44, fill in the blank with the correct term. Some of the given choices will not be used.

Descartes' rule of signs
the leading-term test
the intermediate value theorem
the fundamental theorem of algebra
a polynomial function
a rational function
a one-to-one function
a constant function
a horizontal asymptote
a vertical asymptote
an oblique asymptote
direct variation
inverse variation
a horizontal line
a vertical line
parallel
perpendicular

37. Two lines with slopes m_1 and m_2 are _____ if and only if the product of their slopes is -1.

38. We can use _____ to determine the behavior of the graph of a polynomial function as $x \rightarrow \infty$ or as $x \rightarrow -\infty$.

39. If it is possible for _____ to cross a graph more than once, then the graph is not the graph of a function.

40. A function is _____ if different inputs have different outputs.

41. _____ is a function that is a quotient of two polynomials.

42. If a situation gives rise to a function $f(x) = k/x$, or $y = k/x$, where k is a positive constant, we say that we have _____.

43. _____ of a rational function $p(x)/q(x)$, where $p(x)$ and $q(x)$ have no common factors other than constants, occurs at an x-value that makes the denominator 0.

44. When the numerator and the denominator of a rational function have the same degree, the graph of the function has _____.

Synthesis

In Exercises 45 and 46, let u represent $1/x$, *v represent* $1/y$, *and w represent* $1/z$. *Solve first for u, v, and w. Then solve the system.*

45. $\dfrac{2}{x} - \dfrac{1}{y} - \dfrac{3}{z} = -1,$

$\dfrac{2}{x} - \dfrac{1}{y} + \dfrac{1}{z} = -9,$

$\dfrac{1}{x} + \dfrac{2}{y} - \dfrac{4}{z} = 17$

46. $\dfrac{2}{x} + \dfrac{2}{y} - \dfrac{3}{z} = 3,$

$\dfrac{1}{x} - \dfrac{2}{y} - \dfrac{3}{z} = 9,$

$\dfrac{7}{x} - \dfrac{2}{y} + \dfrac{9}{z} = -39$

47. Find the sum of the angle measures at the tips of the star.

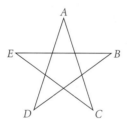

48. *Transcontinental Railroad.* Use the following facts to find the year in which the first U.S. transcontinental railroad was completed. The sum of the digits in the year is 24. The units digit is 1 more than the hundreds digit. Both the tens and the units digits are multiples of three.

In Exercises 49 and 50, three solutions of an equation are given. Use a system of three equations in three variables to find the constants and write the equation.

49. $Ax + By + Cz = 12;$
$\left(1, \frac{3}{4}, 3\right), \left(\frac{4}{3}, 1, 2\right),$ and $(2, 1, 1)$

50. $y = B - Mx - Nz;$
$(1, 1, 2), (3, 2, -6),$ and $\left(\frac{3}{2}, 1, 1\right)$

In Exercises 51 and 52, four solutions of the equation $y = ax^3 + bx^2 + cx + d$ *are given. Use a system of four equations in four variables to find the constants a, b, c, and d and write the equation.*

51. $(-2, 59), (-1, 13), (1, -1),$ and $(2, -17)$

52. $(-2, -39), (-1, -12), (1, -6),$ and $(3, 16)$

53. *Theater Attendance.* A performance at the Bingham Performing Arts Center was attended by 100 people. The audience consisted of adults, students, and children. The ticket prices were $10 for adults, $3 for students, and 50 cents for children. The total amount of money taken in was $100. How many adults, students, and children were in attendance? Does there seem to be some information missing? Do some careful reasoning.

5.3

Matrices and Systems of Equations

• *Solve systems of equations using matrices.*

Matrices and Row-Equivalent Operations

In this section, we consider additional techniques for solving systems of equations. You have probably observed that when we solve a system of equations, we perform computations with the coefficients and the constants and continually rewrite the variables. We can streamline the solution process by omitting the variables until a solution is found. For example, the system

$$2x - 3y = 7,$$
$$x + 4y = -2$$

can be written more simply as

$$\begin{bmatrix} 2 & -3 & | & 7 \\ 1 & 4 & | & -2 \end{bmatrix}.$$

The vertical line replaces the equals signs.

A rectangular array of numbers like the one above is called a **matrix** (pl., **matrices**). The matrix above is called an **augmented matrix** for the given system of equations, because it contains not only the coefficients but also the constant terms. The matrix

$$\begin{bmatrix} 2 & -3 \\ 1 & 4 \end{bmatrix}$$

is called the **coefficient matrix** of the system.

The **rows** of a matrix are horizontal, and the **columns** are vertical. The augmented matrix above has 2 rows and 3 columns, and the coefficient matrix has 2 rows and 2 columns. A matrix with m rows and n columns is said to be of **order** $m \times n$. Thus the order of the augmented matrix above is 2×3, and the order of the coefficient matrix is 2×2. When $m = n$, a matrix is said to be **square.** The coefficient matrix above is a square matrix. The numbers 2 and 4 lie on the **main diagonal** of the coefficient matrix. The numbers in a matrix are called **entries.**

Gaussian Elimination with Matrices

In Section 5.2, we described a series of operations that can be used to transform a system of equations to an equivalent system. Each of these operations corresponds to one that can be used to produce *row-equivalent matrices*.

Row-Equivalent Operations

1. Interchange any two rows.
2. Multiply each entry in a row by the same nonzero constant.
3. Add a nonzero multiple of one row to another row.

We can use these operations on the augmented matrix of a system of equations to solve the system.

EXAMPLE 1 Solve the following system:

$$2x - y + 4z = -3,$$
$$x - 2y - 10z = -6,$$
$$3x \quad\quad + 4z = 7.$$

Solution First, we write the augmented matrix, writing 0 for the missing *y*-term in the last equation:

$$\left[\begin{array}{ccc|c} 2 & -1 & 4 & -3 \\ 1 & -2 & -10 & -6 \\ 3 & 0 & 4 & 7 \end{array}\right].$$

Our goal is to find a row-equivalent matrix of the form

$$\left[\begin{array}{ccc|c} 1 & a & b & c \\ 0 & 1 & d & e \\ 0 & 0 & 1 & f \end{array}\right].$$

The variables can then be reinserted to form equations from which we can complete the solution. This is done by working from the bottom equation to the top and using back-substitution.

The first step is to multiply and/or interchange rows so that each number in the first column below the first number is a multiple of that number. In this case, we interchange the first and second rows to obtain a 1 in the upper left-hand corner.

$$\left[\begin{array}{ccc|c} 1 & -2 & -10 & -6 \\ 2 & -1 & 4 & -3 \\ 3 & 0 & 4 & 7 \end{array}\right]$$
   New row 1 = row 2
   New row 2 = row 1

Next, we multiply the first row by -2 and add it to the second row. We also multiply the first row by -3 and add it to the third row.

$$\left[\begin{array}{ccc|c} 1 & -2 & -10 & -6 \\ 0 & 3 & 24 & 9 \\ 0 & 6 & 34 & 25 \end{array}\right]$$
   Row 1 is unchanged.
   New row 2 = −2(row 1) + row 2
   New row 3 = −3(row 1) + row 3

Now we multiply the second row by $\frac{1}{3}$ to get a 1 in the second row, second column.

$$\left[\begin{array}{ccc|c} 1 & -2 & -10 & -6 \\ 0 & 1 & 8 & 3 \\ 0 & 6 & 34 & 25 \end{array}\right]$$
   New row 2 = $\frac{1}{3}$(row 2)

Then we multiply the second row by -6 and add it to the third row.

$$\left[\begin{array}{ccc|c} 1 & -2 & -10 & -6 \\ 0 & 1 & 8 & 3 \\ 0 & 0 & -14 & 7 \end{array}\right]$$
   New row 3 = −6(row 2) + row 3

Finally, we multiply the third row by $-\frac{1}{14}$ to get a 1 in the third row, third column.

$$\left[\begin{array}{ccc|c} 1 & -2 & -10 & -6 \\ 0 & 1 & 8 & 3 \\ 0 & 0 & 1 & -\frac{1}{2} \end{array}\right]$$
   New row 3 = $-\frac{1}{14}$(row 3)

Technology Connection

Row-equivalent operations can be performed on a graphing calculator. For example, to interchange the first and second rows of the augmented matrix, as we did in the first step in Example 1, we enter the matrix as matrix **A** and select "rowSwap" from the MATRIX MATH menu. Some graphing calculators will not automatically store the matrix produced using a row-equivalent operation, so when several operations are to be performed in succession, it is helpful to store the result of each operation as it is produced. In the window below, we see both the matrix produced by the rowSwap operation and the indication that this matrix is stored as matrix **B**.

```
rowSwap([A],1,2)→[B]
[[1  −2 −10 −6]
 [2  −1  4   −3]
 [3   0  4    7 ]]
```

Study Tip

The *Graphing Calculator Manual* that accompanies this text contains the keystrokes for performing all the row-equivalent operations in Example 1.

Now we can write the system of equations that corresponds to the last matrix above:

$$x - 2y - 10z = -6, \quad (1)$$
$$y + 8z = 3, \quad (2)$$
$$z = -\tfrac{1}{2}. \quad (3)$$

We back-substitute $-\frac{1}{2}$ for z in equation (2) and solve for y:

$$y + 8\left(-\tfrac{1}{2}\right) = 3$$
$$y - 4 = 3$$
$$y = 7.$$

Next, we back-substitute 7 for y and $-\frac{1}{2}$ for z in equation (1) and solve for x:

$$x - 2 \cdot 7 - 10\left(-\tfrac{1}{2}\right) = -6$$
$$x - 14 + 5 = -6$$
$$x - 9 = -6$$
$$x = 3.$$

The triple $\left(3, 7, -\frac{1}{2}\right)$ checks in the original system of equations, so it is the solution.

The procedure followed in Example 1 is called **Gaussian elimination with matrices.** The matrix at the bottom of the preceding page is in **row-echelon form.** To be in this form, a matrix must have the following properties.

Row-Echelon Form

1. If a row does not consist entirely of 0's, then the first nonzero element in the row is a 1 (called a **leading 1**).

2. For any two successive nonzero rows, the leading 1 in the lower row is farther to the right than the leading 1 in the higher row.

3. All the rows consisting entirely of 0's are at the bottom of the matrix.

If a fourth property is also satisfied, a matrix is said to be in **reduced row-echelon form:**

4. Each column that contains a leading 1 has 0's everywhere else.

EXAMPLE 2 Which of the following matrices are in row-echelon form? Which, if any, are in reduced row-echelon form?

a) $\begin{bmatrix} 1 & -3 & 5 & | & -2 \\ 0 & 1 & -4 & | & 3 \\ 0 & 0 & 1 & | & 10 \end{bmatrix}$ b) $\begin{bmatrix} 0 & -1 & | & 2 \\ 0 & 1 & | & 5 \end{bmatrix}$ c) $\begin{bmatrix} 1 & -2 & -6 & 4 & | & 7 \\ 0 & 3 & 5 & -8 & | & -1 \\ 0 & 0 & 1 & 9 & | & 2 \end{bmatrix}$

d) $\begin{bmatrix} 1 & 0 & 0 & | & -2.4 \\ 0 & 1 & 0 & | & 0.8 \\ 0 & 0 & 1 & | & 5.6 \end{bmatrix}$ e) $\begin{bmatrix} 1 & 0 & 0 & 0 & | & \frac{2}{3} \\ 0 & 1 & 0 & 0 & | & -\frac{1}{4} \\ 0 & 0 & 1 & 0 & | & \frac{6}{7} \\ 0 & 0 & 0 & 0 & | & 0 \end{bmatrix}$ f) $\begin{bmatrix} 1 & -4 & 2 & | & 5 \\ 0 & 0 & 0 & | & 0 \\ 0 & 1 & -3 & | & -8 \end{bmatrix}$

Solution   The matrices in (a), (d), and (e) satisfy the row-echelon criteria and, thus, are in row-echelon form. In (b) and (c), the first nonzero elements of the first and second rows, respectively, are not 1. In (f), the row consisting entirely of 0's is not at the bottom of the matrix. Thus the matrices in (b), (c), and (f) are not in row-echelon form. In (d) and (e), not only are the row-echelon criteria met but each column that contains a leading 1 also has 0's elsewhere, so these matrices are in reduced row-echelon form.   ▬

Gauss–Jordan Elimination

We have seen that with Gaussian elimination we perform row-equivalent operations on a matrix to obtain a row-equivalent matrix in row-echelon form. When we continue to apply these operations until we have a matrix in *reduced* row-echelon form, we are using **Gauss–Jordan elimination.** This method is named for Karl Friedrich Gauss and Wilhelm Jordan (1842–1899).

EXAMPLE 3 Use Gauss–Jordan elimination to solve the system of equations in Example 1.

Solution   Using Gaussian elimination in Example 1, we obtained the matrix

$$\begin{bmatrix} 1 & -2 & -10 & | & -6 \\ 0 & 1 & 8 & | & 3 \\ 0 & 0 & 1 & | & -\frac{1}{2} \end{bmatrix}.$$

We continue to perform row-equivalent operations until we have a matrix in reduced row-echelon form. We multiply the third row by 10 and add it to the first row. We also multiply the third row by -8 and add it to the second row.

$$\begin{bmatrix} 1 & -2 & 0 & | & -11 \\ 0 & 1 & 0 & | & 7 \\ 0 & 0 & 1 & | & -\frac{1}{2} \end{bmatrix}$$ New row 1 = 10(row 3) + row 1
New row 2 = −8(row 3) + row 2

Next, we multiply the second row by 2 and add it to the first row.

$$\left[\begin{array}{ccc|c} 1 & 0 & 0 & 3 \\ 0 & 1 & 0 & 7 \\ 0 & 0 & 1 & -\frac{1}{2} \end{array}\right] \qquad \text{New row 1} = 2(\text{row 2}) + \text{row 1}$$

Writing the system of equations that corresponds to this matrix, we have

$$
\begin{aligned}
x \qquad\qquad &= 3, \\
y \qquad &= 7, \\
z &= -\tfrac{1}{2}.
\end{aligned}
$$

We can actually read the solution, $\left(3, 7, -\frac{1}{2}\right)$, directly from the last column of the reduced row-echelon matrix.

EXAMPLE 4   Solve the following system:

$$
\begin{aligned}
3x - 4y - z &= 6, \\
2x - y + z &= -1, \\
4x - 7y - 3z &= 13.
\end{aligned}
$$

Solution   We write the augmented matrix and use Gauss–Jordan elimination.

$$\left[\begin{array}{ccc|c} 3 & -4 & -1 & 6 \\ 2 & -1 & 1 & -1 \\ 4 & -7 & -3 & 13 \end{array}\right]$$

We begin by multiplying the second and third rows by 3 so that each number in the first column below the first number, 3, is a multiple of that number.

$$\left[\begin{array}{ccc|c} 3 & -4 & -1 & 6 \\ 6 & -3 & 3 & -3 \\ 12 & -21 & -9 & 39 \end{array}\right] \qquad \begin{array}{l} \text{New row 2} = 3(\text{row 2}) \\ \text{New row 3} = 3(\text{row 3}) \end{array}$$

Next, we multiply the first row by -2 and add it to the second row. We also multiply the first row by -4 and add it to the third row.

$$\left[\begin{array}{ccc|c} 3 & -4 & -1 & 6 \\ 0 & 5 & 5 & -15 \\ 0 & -5 & -5 & 15 \end{array}\right] \qquad \begin{array}{l} \text{New row 2} = -2(\text{row 1}) + \text{row 2} \\ \text{New row 3} = -4(\text{row 1}) + \text{row 3} \end{array}$$

Now we add the second row to the third row.

$$\left[\begin{array}{ccc|c} 3 & -4 & -1 & 6 \\ 0 & 5 & 5 & -15 \\ 0 & 0 & 0 & 0 \end{array}\right] \qquad \text{New row 3} = \text{row 2} + \text{row 3}$$

We can stop at this stage because we have a row consisting entirely of 0's. The last row of the matrix corresponds to the equation $0 = 0$, which is true for all values of x, y, and z. Consequently, the equations are dependent and the system is equivalent to

$$3x - 4y - z = 6,$$
$$5y + 5z = -15.$$

This particular system has infinitely many solutions. (A system containing dependent equations could be inconsistent.)

Solving the second equation for y gives us

$$y = -z - 3.$$

Substituting $-z - 3$ for y in the first equation, we get

$$3x - 4(-z - 3) - z = 6$$
$$x = -z - 2.$$

Then the solutions of this system are of the form

$$(-z - 2, -z - 3, z),$$

where z can be any real number. ▬

Similarly, if we obtain a row whose only nonzero entry occurs in the last column, we have an inconsistent system of equations. For example, in the matrix

$$\begin{bmatrix} 1 & 0 & 3 & | & -2 \\ 0 & 1 & 5 & | & 4 \\ 0 & 0 & 0 & | & 6 \end{bmatrix},$$

the last row corresponds to the false equation $0 = 6$, so we know the original system of equations has no solution.

5.3

Exercise Set

Determine the order of the matrix.

1. $\begin{bmatrix} 1 & -6 \\ -3 & 2 \\ 0 & 5 \end{bmatrix}$

2. $\begin{bmatrix} 7 \\ -5 \\ -1 \\ 3 \end{bmatrix}$

3. $\begin{bmatrix} 2 & -4 & 0 & 9 \end{bmatrix}$

4. $\begin{bmatrix} -8 \end{bmatrix}$

5. $\begin{bmatrix} 1 & -5 & -8 \\ 6 & 4 & -2 \\ -3 & 0 & 7 \end{bmatrix}$

6. $\begin{bmatrix} 13 & 2 & -6 & 4 \\ -1 & 18 & 5 & -12 \end{bmatrix}$

Write the augmented matrix for the system of equations.

7. $2x - y = 7,$
$\quad x + 4y = -5$

8. $3x + 2y = 8,$
$\quad 2x - 3y = 15$

9. $x - 2y + 3z = 12,$
$2x \qquad - 4z = 8,$
$\qquad 3y + z = 7$

10. $x + y - z = 7,$
$\qquad 3y + 2z = 1,$
$-2x - 5y \qquad = 6$

Write the system of equations that corresponds to the augmented matrix.

11. $\begin{bmatrix} 3 & -5 & | & 1 \\ 1 & 4 & | & -2 \end{bmatrix}$

12. $\begin{bmatrix} 1 & 2 & | & -6 \\ 4 & 1 & | & -3 \end{bmatrix}$

13. $\begin{bmatrix} 2 & 1 & -4 & | & 12 \\ 3 & 0 & 5 & | & -1 \\ 1 & -1 & 1 & | & 2 \end{bmatrix}$

14. $\begin{bmatrix} -1 & -2 & 3 & | & 6 \\ 0 & 4 & 1 & | & 2 \\ 2 & -1 & 0 & | & 9 \end{bmatrix}$

Solve the system of equations using Gaussian elimination or Gauss–Jordan elimination.

15. $4x + 2y = 11,$
$3x - y = 2$

16. $2x + y = 1,$
$3x + 2y = -2$

17. $5x - 2y = -3,$
$2x + 5y = -24$

18. $2x + y = 1,$
$3x - 6y = 4$

19. $3x + 4y = 7,$
$-5x + 2y = 10$

20. $5x - 3y = -2,$
$4x + 2y = 5$

21. $3x + 2y = 6,$
$2x - 3y = -9$

22. $x - 4y = 9,$
$2x + 5y = 5$

23. $x - 3y = 8,$
$-2x + 6y = 3$

24. $4x - 8y = 12,$
$-x + 2y = -3$

25. $-2x + 6y = 4,$
$3x - 9y = -6$

26. $6x + 2y = -10,$
$-3x - y = 6$

27. $x + 2y - 3z = 9,$
$2x - y + 2z = -8,$
$3x - y - 4z = 3$

28. $x - y + 2z = 0,$
$x - 2y + 3z = -1,$
$2x - 2y + z = -3$

29. $4x - y - 3z = 1,$
$8x + y - z = 5,$
$2x + y + 2z = 5$

30. $3x + 2y + 2z = 3,$
$x + 2y - z = 5,$
$2x - 4y + z = 0$

31. $x - 2y + 3z = -4,$
$3x + y - z = 0,$
$2x + 3y - 5z = 1$

32. $2x - 3y + 2z = 2,$
$x + 4y - z = 9,$
$-3x + y - 5z = 5$

33. $2x - 4y - 3z = 3,$
$x + 3y + z = -1,$
$5x + y - 2z = 2$

34. $x + y - 3z = 4,$
$4x + 5y + z = 1,$
$2x + 3y + 7z = -7$

35. $p + q + r = 1,$
$p + 2q + 3r = 4,$
$4p + 5q + 6r = 7$

36. $m + n + t = 9,$
$m - n - t = -15,$
$3m + n + t = 2$

37. $a + b - c = 7,$
$a - b + c = 5,$
$3a + b - c = -1$

38. $a - b + c = 3,$
$2a + b - 3c = 5,$
$4a + b - c = 11$

39. $-2w + 2x + 2y - 2z = -10,$
$w + x + y + z = -5,$
$3w + x - y + 4z = -2,$
$w + 3x - 2y + 2z = -6$

40. $-w + 2x - 3y + z = -8,$
$-w + x + y - z = -4,$
$w + x + y + z = 22,$
$-w + x - y - z = -14$

Use Gaussian elimination or Gauss–Jordan elimination in Exercises 41–44.

41. *Time of Return.* The Houlihans pay their babysitter $5 per hour before 11 P.M. and $7.50 per hour after 11 P.M. One evening they went out for 5 hr and paid the sitter $30. What time did they come home?

42. *Advertising Expense.* eAuction.com spent a total of $11 million on advertising in fiscal years 2001, 2002, and 2003. The amount spent in 2003 was three times the amount spent in 2001. The amount spent in 2002 was $3 million less than the amount spent in 2003. How much was spent on advertising each year?

43. *Borrowing.* Gonzalez Manufacturing borrowed $30,000 to buy a new piece of equipment. Part of the money was borrowed at 8%, part at 10%, and part at 12%. The annual interest was $3040, and the total amount borrowed at 8% and 10% was twice the amount borrowed at 12%. How much was borrowed at each rate?

44. *Stamp Purchase.* Ricardo spent $20.10 on 37¢ and 23¢ stamps. He bought a total of 60 stamps. How many of each type did he buy?

Collaborative Discussion and Writing

45. Solve the following system of equations using matrices and Gaussian elimination. Then solve it again using Gauss–Jordan elimination. Do you prefer one method over the other? Why or why not?

$$3x + 4y + 2z = 0,$$
$$x - y - z = 10,$$
$$2x + 3y + 3z = -10$$

46. Explain in your own words why the augmented matrix below represents a system of dependent equations.

$$\begin{bmatrix} 1 & -3 & 2 & | & -5 \\ 0 & 1 & -4 & | & 8 \\ 0 & 0 & 0 & | & 0 \end{bmatrix}$$

Skill Maintenance

Classify the function as linear, quadratic, cubic, quartic, rational, exponential, or logarithmic.

47. $f(x) = 3^{x-1}$

48. $f(x) = 3x - 1$

49. $f(x) = \dfrac{3x - 1}{x^2 + 4}$

50. $f(x) = -\frac{3}{4}x^4 + \frac{9}{2}x^3 + 2x^2 - 4$

51. $f(x) = \ln(3x - 1)$

52. $f(x) = \frac{3}{4}x^3 - x$

53. $f(x) = 3$

54. $f(x) = 2 - x - x^2$

Synthesis

In Exercises 55 and 56, three solutions of the equation $y = ax^2 + bx + c$ are given. Use a system of three equations in three variables and Gaussian elimination or Gauss–Jordan elimination to find the constants a, b, and c and write the equation.

55. $(-3, 12), (-1, -7),$ and $(1, -2)$

56. $(-1, 0), (1, -3),$ and $(3, -22)$

57. Find two different row-echelon forms of

$$\begin{bmatrix} 1 & 5 \\ 3 & 2 \end{bmatrix}.$$

58. Consider the system of equations

$$x - y + 3z = -8,$$
$$2x + 3y - z = 5,$$
$$3x + 2y + 2kz = -3k.$$

For what value(s) of k, if any, will the system have

a) no solution?
b) exactly one solution?
c) infinitely many solutions?

Solve using matrices.

59. $y = x + z,$
$3y + 5z = 4,$
$x + 4 = y + 3z$

60. $x + y = 2z,$
$2x - 5z = 4,$
$x - z = y + 8$

61. $x - 4y + 2z = 7,$
$3x + y + 3z = -5$

62. $x - y - 3z = 3,$
$-x + 3y + z = -7$

63. $4x + 5y = 3,$
$-2x + y = 9,$
$3x - 2y = -15$

64. $2x - 3y = -1,$
$-x + 2y = -2,$
$3x - 5y = 1$

5.4

Matrix Operations

- *Add, subtract, and multiply matrices when possible.*
- *Write a matrix equation equivalent to a system of equations.*

In Section 5.3, we used matrices to solve systems of equations. Matrices are useful in many other types of applications as well. In this section, we study matrices and some of their properties.

A capital letter is generally used to name a matrix, and lower-case letters with double subscripts generally denote its entries. For example, a_{47}, read "a sub four seven," indicates the entry in the fourth row and the seventh column. A general term is represented by a_{ij}. The notation a_{ij} indicates the entry in row i and column j. In general, we can write a matrix as

$$\mathbf{A} = [a_{ij}] = \begin{bmatrix} a_{11} & a_{12} & a_{13} & \cdots & a_{1n} \\ a_{21} & a_{22} & a_{23} & \cdots & a_{2n} \\ a_{31} & a_{32} & a_{33} & \cdots & a_{3n} \\ \vdots & \vdots & \vdots & & \vdots \\ a_{m1} & a_{m2} & a_{m3} & \cdots & a_{mn} \end{bmatrix}.$$

The matrix above has m rows and n columns. That is, its order is $m \times n$.

Two matrices are **equal** if they have the same order and corresponding entries are equal.

Matrix Addition and Subtraction

To add or subtract matrices, we add or subtract their corresponding entries. The matrices must have the same order for this to be possible.

Addition and Subtraction of Matrices

Given two $m \times n$ matrices $\mathbf{A} = [a_{ij}]$ and $\mathbf{B} = [b_{ij}]$, their sum is

$$\mathbf{A} + \mathbf{B} = [a_{ij} + b_{ij}]$$

and their difference is

$$\mathbf{A} - \mathbf{B} = [a_{ij} - b_{ij}].$$

Addition of matrices is both commutative and associative.

EXAMPLE 1 Find $\mathbf{A} + \mathbf{B}$ for each of the following.

a) $\mathbf{A} = \begin{bmatrix} -5 & 0 \\ 4 & \frac{1}{2} \end{bmatrix}$, $\mathbf{B} = \begin{bmatrix} 6 & -3 \\ 2 & 3 \end{bmatrix}$

b) $\mathbf{A} = \begin{bmatrix} 1 & 3 \\ -1 & 5 \\ 6 & 0 \end{bmatrix}, \quad \mathbf{B} = \begin{bmatrix} -1 & -2 \\ 1 & -2 \\ -3 & 1 \end{bmatrix}$

Solution   We have a pair of 2×2 matrices in part (a) and a pair of 3×2 matrices in part (b). Since each pair of matrices has the same order, we can add the corresponding entries.

a) $\mathbf{A} + \mathbf{B} = \begin{bmatrix} -5 & 0 \\ 4 & \frac{1}{2} \end{bmatrix} + \begin{bmatrix} 6 & -3 \\ 2 & 3 \end{bmatrix}$

$= \begin{bmatrix} -5 + 6 & 0 + (-3) \\ 4 + 2 & \frac{1}{2} + 3 \end{bmatrix} = \begin{bmatrix} 1 & -3 \\ 6 & 3\frac{1}{2} \end{bmatrix}$

b) $\mathbf{A} + \mathbf{B} = \begin{bmatrix} 1 & 3 \\ -1 & 5 \\ 6 & 0 \end{bmatrix} + \begin{bmatrix} -1 & -2 \\ 1 & -2 \\ -3 & 1 \end{bmatrix}$

$= \begin{bmatrix} 1 + (-1) & 3 + (-2) \\ -1 + 1 & 5 + (-2) \\ 6 + (-3) & 0 + 1 \end{bmatrix} = \begin{bmatrix} 0 & 1 \\ 0 & 3 \\ 3 & 1 \end{bmatrix}$

EXAMPLE 2   Find $\mathbf{C} - \mathbf{D}$ for each of the following.

a) $\mathbf{C} = \begin{bmatrix} 1 & 2 \\ -2 & 0 \\ -3 & -1 \end{bmatrix}, \quad \mathbf{D} = \begin{bmatrix} 1 & -1 \\ 1 & 3 \\ 2 & 3 \end{bmatrix}$

b) $\mathbf{C} = \begin{bmatrix} 5 & -6 \\ -3 & 4 \end{bmatrix}, \quad \mathbf{D} = \begin{bmatrix} -4 \\ 1 \end{bmatrix}$

Solution

a) Since the order of each matrix is 3×2, we can subtract corresponding entries:

$\mathbf{C} - \mathbf{D} = \begin{bmatrix} 1 & 2 \\ -2 & 0 \\ -3 & -1 \end{bmatrix} - \begin{bmatrix} 1 & -1 \\ 1 & 3 \\ 2 & 3 \end{bmatrix}$

$= \begin{bmatrix} 1 - 1 & 2 - (-1) \\ -2 - 1 & 0 - 3 \\ -3 - 2 & -1 - 3 \end{bmatrix} = \begin{bmatrix} 0 & 3 \\ -3 & -3 \\ -5 & -4 \end{bmatrix}.$

b) $\mathbf{C}$ is a 2×2 matrix and $\mathbf{D}$ is a 2×1 matrix. Since the matrices do not have the same order, we cannot subtract.

The **opposite,** or **additive inverse,** of a matrix is obtained by replacing each entry with its opposite.

EXAMPLE 3 Find $-\mathbf{A}$ and $\mathbf{A} + (-\mathbf{A})$ for

$$\mathbf{A} = \begin{bmatrix} 1 & 0 & 2 \\ 3 & -1 & 5 \end{bmatrix}.$$

Solution To find $-\mathbf{A}$, we replace each entry of $\mathbf{A}$ with its opposite.

$$-\mathbf{A} = \begin{bmatrix} -1 & 0 & -2 \\ -3 & 1 & -5 \end{bmatrix},$$

$$\mathbf{A} + (-\mathbf{A}) = \begin{bmatrix} 1 & 0 & 2 \\ 3 & -1 & 5 \end{bmatrix} + \begin{bmatrix} -1 & 0 & -2 \\ -3 & 1 & -5 \end{bmatrix}$$

$$= \begin{bmatrix} 0 & 0 & 0 \\ 0 & 0 & 0 \end{bmatrix}.$$

A matrix having 0's for all its entries is called a **zero matrix.** When a zero matrix is added to a second matrix of the same order, the second matrix is unchanged. Thus a zero matrix is an **additive identity.** For example,

$$\begin{bmatrix} 0 & 0 & 0 \\ 0 & 0 & 0 \end{bmatrix}$$

is the additive identity for any 2×3 matrix.

Scalar Multiplication

When we find the product of a number and a matrix, we obtain a **scalar product.**

Scalar Product

The **scalar product** of a number k and a matrix $\mathbf{A}$ is the matrix denoted $k\mathbf{A}$, obtained by multiplying each entry of $\mathbf{A}$ by the number k. The number k is called a **scalar.**

Technology Connection

Scalar products, like those in Example 4, can be found using a graphing calculator.

```
3[A]
            [[-9   0]
             [ 12  15]]
(−1)[A]
            [[3    0 ]
             [−4  −5]]
```

EXAMPLE 4 Find $3\mathbf{A}$ and $(-1)\mathbf{A}$ for

$$\mathbf{A} = \begin{bmatrix} -3 & 0 \\ 4 & 5 \end{bmatrix}.$$

Solution We have

$$3\mathbf{A} = 3\begin{bmatrix} -3 & 0 \\ 4 & 5 \end{bmatrix} = \begin{bmatrix} 3(-3) & 3 \cdot 0 \\ 3 \cdot 4 & 3 \cdot 5 \end{bmatrix} = \begin{bmatrix} -9 & 0 \\ 12 & 15 \end{bmatrix},$$

$$(-1)\mathbf{A} = -1\begin{bmatrix} -3 & 0 \\ 4 & 5 \end{bmatrix} = \begin{bmatrix} -1(-3) & -1 \cdot 0 \\ -1 \cdot 4 & -1 \cdot 5 \end{bmatrix} = \begin{bmatrix} 3 & 0 \\ -4 & -5 \end{bmatrix}.$$

EXAMPLE 5 *Production.* Mitchell Fabricators, Inc., manufactures three styles of bicycle frames in its two plants. The following table shows the number of each style produced at each plant in April.

	MOUNTAIN BIKE	RACING BIKE	TOURING BIKE
NORTH PLANT	150	120	100
SOUTH PLANT	180	90	130

a) Write a 2 × 3 matrix **A** that represents the information in the table.

b) The manufacturer increased production by 20% in May. Find a matrix **M** that represents the increased production figures.

c) Find the matrix **A** + **M** and tell what it represents.

Solution

a) Write the entries in the table in a 2 × 3 matrix **A**.

$$\mathbf{A} = \begin{bmatrix} 150 & 120 & 100 \\ 180 & 90 & 130 \end{bmatrix}$$

b) The production in May will be represented by **A** + 20%**A**, or **A** + 0.2**A**, or 1.2**A**. Thus,

$$\mathbf{M} = (1.2) \begin{bmatrix} 150 & 120 & 100 \\ 180 & 90 & 130 \end{bmatrix} = \begin{bmatrix} 180 & 144 & 120 \\ 216 & 108 & 156 \end{bmatrix}.$$

c) $\mathbf{A} + \mathbf{M} = \begin{bmatrix} 150 & 120 & 100 \\ 180 & 90 & 130 \end{bmatrix} + \begin{bmatrix} 180 & 144 & 120 \\ 216 & 108 & 156 \end{bmatrix}$

$$= \begin{bmatrix} 330 & 264 & 220 \\ 396 & 198 & 286 \end{bmatrix}$$

The matrix **A** + **M** represents the total production of each of the three types of frames at each plant in April and May.   ▬

The properties of matrix addition and scalar multiplication are similar to the properties of addition and multiplication of real numbers.

Properties of Matrix Addition and Scalar Multiplication

For any $m \times n$ matrices **A**, **B**, and **C** and any scalars k and l:

$\mathbf{A} + \mathbf{B} = \mathbf{B} + \mathbf{A}.$ *Commutative Property of Addition*

$\mathbf{A} + (\mathbf{B} + \mathbf{C}) = (\mathbf{A} + \mathbf{B}) + \mathbf{C}.$ *Associative Property of Addition*

$(kl)\mathbf{A} = k(l\mathbf{A}).$ *Associative Property of Scalar Multiplication*

$k(\mathbf{A} + \mathbf{B}) = k\mathbf{A} + k\mathbf{B}.$ *Distributive Property*

$(k + l)\mathbf{A} = k\mathbf{A} + l\mathbf{A}.$ *Distributive Property*

There exists a unique matrix **0** such that:

$\mathbf{A} + \mathbf{0} = \mathbf{0} + \mathbf{A} = \mathbf{A}.$ *Additive Identity Property*

There exists a unique matrix $-\mathbf{A}$ such that:

$\mathbf{A} + (-\mathbf{A}) = -\mathbf{A} + \mathbf{A} = \mathbf{0}.$ *Additive Inverse Property*

Products of Matrices

Matrix multiplication is defined in such a way that it can be used in solving systems of equations and in many applications.

Matrix Multiplication

For an $m \times n$ matrix $\mathbf{A} = [a_{ij}]$ and an $n \times p$ matrix $\mathbf{B} = [b_{ij}]$, the **product** $\mathbf{AB} = [c_{ij}]$ is an $m \times p$ matrix, where

$$c_{ij} = a_{i1} \cdot b_{1j} + a_{i2} \cdot b_{2j} + a_{i3} \cdot b_{3j} + \cdots + a_{in} \cdot b_{nj}.$$

In other words, the entry c_{ij} in **AB** is obtained by multiplying the entries in row i of **A** by the corresponding entries in column j of **B** and adding the results.

Note that we can multiply two matrices only when the number of columns in the first matrix is equal to the number of rows in the second matrix.

EXAMPLE 6 For

$$A = \begin{bmatrix} 3 & 1 & -1 \\ 2 & 0 & 3 \end{bmatrix}, \quad B = \begin{bmatrix} 1 & 6 \\ 3 & -5 \\ -2 & 4 \end{bmatrix}, \quad \text{and} \quad C = \begin{bmatrix} 4 & -6 \\ 1 & 2 \end{bmatrix},$$

find each of the following.

a) **AB**   b) **BA**

c) **BC**   d) **AC**

Solution

a) **A** is a 2×3 matrix and **B** is a 3×2 matrix, so **AB** will be a 2×2 matrix.

$$AB = \begin{bmatrix} 3 & 1 & -1 \\ 2 & 0 & 3 \end{bmatrix} \begin{bmatrix} 1 & 6 \\ 3 & -5 \\ -2 & 4 \end{bmatrix}$$

$$= \begin{bmatrix} 3 \cdot 1 + 1 \cdot 3 + (-1)(-2) & 3 \cdot 6 + 1(-5) + (-1)(4) \\ 2 \cdot 1 + 0 \cdot 3 + 3(-2) & 2 \cdot 6 + 0(-5) + 3 \cdot 4 \end{bmatrix} = \begin{bmatrix} 8 & 9 \\ -4 & 24 \end{bmatrix}$$

b) **B** is a 3×2 matrix and **A** is a 2×3 matrix, so **BA** will be a 3×3 matrix.

$$BA = \begin{bmatrix} 1 & 6 \\ 3 & -5 \\ -2 & 4 \end{bmatrix} \begin{bmatrix} 3 & 1 & -1 \\ 2 & 0 & 3 \end{bmatrix}$$

$$= \begin{bmatrix} 1 \cdot 3 + 6 \cdot 2 & 1 \cdot 1 + 6 \cdot 0 & 1(-1) + 6 \cdot 3 \\ 3 \cdot 3 + (-5)(2) & 3 \cdot 1 + (-5)(0) & 3(-1) + (-5)(3) \\ -2 \cdot 3 + 4 \cdot 2 & -2 \cdot 1 + 4 \cdot 0 & -2(-1) + 4 \cdot 3 \end{bmatrix} = \begin{bmatrix} 15 & 1 & 17 \\ -1 & 3 & -18 \\ 2 & -2 & 14 \end{bmatrix}$$

Note in parts (a) and (b) that $AB \neq BA$. Multiplication of matrices is generally not commutative.

c) **B** is a 3×2 matrix and **C** is a 2×2 matrix, so **BC** will be a 3×2 matrix.

$$BC = \begin{bmatrix} 1 & 6 \\ 3 & -5 \\ -2 & 4 \end{bmatrix} \begin{bmatrix} 4 & -6 \\ 1 & 2 \end{bmatrix}$$

$$= \begin{bmatrix} 1 \cdot 4 + 6 \cdot 1 & 1(-6) + 6 \cdot 2 \\ 3 \cdot 4 + (-5)(1) & 3(-6) + (-5)(2) \\ 2 \cdot 4 + 4 \cdot 1 & -2(-6) + 4 \cdot 2 \end{bmatrix}$$

$$= \begin{bmatrix} 10 & 6 \\ 7 & -28 \\ -4 & 20 \end{bmatrix}$$

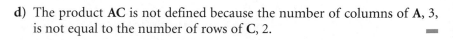

d) The product **AC** is not defined because the number of columns of **A**, 3, is not equal to the number of rows of **C**, 2. ▬

EXAMPLE 7 *Dairy Profit.* Dalton's Dairy produces no-fat ice cream and frozen yogurt. The following table shows the number of gallons of each product that are sold at the dairy's three retail outlets one week.

	STORE 1	STORE 2	STORE 3
NO-FAT ICE CREAM (IN GALLONS)	100	80	120
FROZEN YOGURT (IN GALLONS)	160	120	100

On each gallon of no-fat ice cream, the dairy's profit is $4, and on each gallon of frozen yogurt, it is $3. Use matrices to find the total profit on these items at each store for the given week.

Solution We can write the table showing the distribution of the products as a 2×3 matrix:

$$\mathbf{D} = \begin{bmatrix} 100 & 80 & 120 \\ 160 & 120 & 100 \end{bmatrix}.$$

The profit per gallon for each product can also be written as a matrix:

$$\mathbf{P} = \begin{bmatrix} 4 & 3 \end{bmatrix}.$$

Then the total profit at each store is given by the matrix product **PD**:

$$\mathbf{PD} = \begin{bmatrix} 4 & 3 \end{bmatrix} \begin{bmatrix} 100 & 80 & 120 \\ 160 & 120 & 100 \end{bmatrix}$$
$$= \begin{bmatrix} 4 \cdot 100 + 3 \cdot 160 & 4 \cdot 80 + 3 \cdot 120 & 4 \cdot 120 + 3 \cdot 100 \end{bmatrix}$$
$$= \begin{bmatrix} 880 & 680 & 780 \end{bmatrix}.$$

The total profit on no-fat ice cream and frozen yogurt for the given week was $880 at store 1, $680 at store 2, and $780 at store 3. ▬

A matrix that consists of a single row, like **P** in Example 7, is called a **row matrix.** Similarly, a matrix that consists of a single column, like

$$\begin{bmatrix} 8 \\ -3 \\ 5 \end{bmatrix},$$

is called a **column matrix.**

We have already seen that matrix multiplication is generally not commutative. Nevertheless, matrix multiplication does have some properties that are similar to those for multiplication of real numbers.

Properties of Matrix Multiplication

For matrices **A**, **B**, and **C**, assuming that the indicated operations are possible:

$$\mathbf{A(BC)} = \mathbf{(AB)C}. \qquad \textit{Associative Property of Multiplication}$$
$$\mathbf{A(B + C)} = \mathbf{AB} + \mathbf{AC}. \qquad \textit{Distributive Property}$$
$$\mathbf{(B + C)A} = \mathbf{BA} + \mathbf{CA}. \qquad \textit{Distributive Property}$$

Study Tip

The AW Math Tutor Center provides *free* tutoring to students using this text. Assisted by qualified mathematics instructors via telephone, fax, or e-mail, you can receive live tutoring on examples and exercises. This valuable resource provides you with immediate assistance when additional instruction is needed.

Matrix Equations

We can write a matrix equation equivalent to a system of equations.

EXAMPLE 8 Write a matrix equation equivalent to the following system of equations:

$$
\begin{aligned}
4x + 2y - z &= 3, \\
9x + z &= 5, \\
4x + 5y - 2z &= 1.
\end{aligned}
$$

Solution We write the coefficients on the left in a matrix. We then write the product of that matrix and the column matrix containing the variables, and set the result equal to the column matrix containing the constants on the right:

$$
\begin{bmatrix} 4 & 2 & -1 \\ 9 & 0 & 1 \\ 4 & 5 & -2 \end{bmatrix}
\begin{bmatrix} x \\ y \\ z \end{bmatrix}
=
\begin{bmatrix} 3 \\ 5 \\ 1 \end{bmatrix}.
$$

If we let

$$
\mathbf{A} = \begin{bmatrix} 4 & 2 & -1 \\ 9 & 0 & 1 \\ 4 & 5 & -2 \end{bmatrix}, \qquad
\mathbf{X} = \begin{bmatrix} x \\ y \\ z \end{bmatrix}, \quad \text{and} \quad
\mathbf{B} = \begin{bmatrix} 3 \\ 5 \\ 1 \end{bmatrix},
$$

we can write this matrix equation as $\mathbf{AX} = \mathbf{B}$.

In the next section, we will solve systems of equations using a matrix equation like the one in Example 8.

5.4

Exercise Set

Find x and y.

1. $[5 \quad x] = [y \quad -3]$

2. $\begin{bmatrix} 6x \\ 25 \end{bmatrix} = \begin{bmatrix} -9 \\ 5y \end{bmatrix}$

3. $\begin{bmatrix} 3 & 2x \\ y & -8 \end{bmatrix} = \begin{bmatrix} 3 & -2 \\ 1 & -8 \end{bmatrix}$

4. $\begin{bmatrix} x-1 & 4 \\ y+3 & -7 \end{bmatrix} = \begin{bmatrix} 0 & 4 \\ -2 & -7 \end{bmatrix}$

For Exercises 5–20, let

$$A = \begin{bmatrix} 1 & 2 \\ 4 & 3 \end{bmatrix}, \qquad B = \begin{bmatrix} -3 & 5 \\ 2 & -1 \end{bmatrix},$$

$$C = \begin{bmatrix} 1 & -1 \\ -1 & 1 \end{bmatrix}, \qquad D = \begin{bmatrix} 1 & 1 \\ 1 & 1 \end{bmatrix},$$

$$E = \begin{bmatrix} 1 & 3 \\ 2 & 6 \end{bmatrix}, \qquad F = \begin{bmatrix} 3 & 3 \\ -1 & -1 \end{bmatrix},$$

$$O = \begin{bmatrix} 0 & 0 \\ 0 & 0 \end{bmatrix}, \qquad I = \begin{bmatrix} 1 & 0 \\ 0 & 1 \end{bmatrix}.$$

Find each of the following.

5. A + B **6.** B + A **7.** E + O **8.** 2A

9. 3F **10.** (−1)D **11.** 3F + 2A **12.** A − B

13. B − A **14.** AB **15.** BA **16.** OF

17. CD **18.** EF **19.** AI **20.** IA

Find each product, if possible.

21. $\begin{bmatrix} -1 & 0 & 7 \\ 3 & -5 & 2 \end{bmatrix} \begin{bmatrix} 6 \\ -4 \\ 1 \end{bmatrix}$

22. $[6 \quad -1 \quad 2] \begin{bmatrix} 1 & 4 \\ -2 & 0 \\ 5 & -3 \end{bmatrix}$

23. $\begin{bmatrix} -2 & 4 \\ 5 & 1 \\ -1 & -3 \end{bmatrix} \begin{bmatrix} 3 & -6 \\ -1 & 4 \end{bmatrix}$

24. $\begin{bmatrix} 2 & -1 & 0 \\ 0 & 5 & 4 \end{bmatrix} \begin{bmatrix} -3 & 1 & 0 \\ 0 & 2 & -1 \\ 5 & 0 & 4 \end{bmatrix}$

25. $\begin{bmatrix} 1 \\ -5 \\ 3 \end{bmatrix} \begin{bmatrix} -6 & 5 & 8 \\ 0 & 4 & -1 \end{bmatrix}$

26. $\begin{bmatrix} 2 & 0 & 0 \\ 0 & -1 & 0 \\ 0 & 0 & 3 \end{bmatrix} \begin{bmatrix} 0 & -4 & 3 \\ 2 & 1 & 0 \\ -1 & 0 & 6 \end{bmatrix}$

27. $\begin{bmatrix} 1 & -4 & 3 \\ 0 & 8 & 0 \\ -2 & -1 & 5 \end{bmatrix} \begin{bmatrix} 3 & 0 & 0 \\ 0 & -4 & 0 \\ 0 & 0 & 1 \end{bmatrix}$

28. $\begin{bmatrix} 4 \\ -5 \end{bmatrix} \begin{bmatrix} 2 & 0 \\ 6 & -7 \\ 0 & -3 \end{bmatrix}$

29. *Budget.* For the month of June, Nelia budgets $150 for food, $80 for clothes, and $40 for entertainment.

a) Write a 1 × 3 matrix **B** that represents the amounts budgeted for these items.

b) After receiving a raise, Nelia increases the amount budgeted for each item in July by 5%. Find a matrix **R** that represents the new amounts.

c) Find **B** + **R** and tell what the entries represent.

30. *Produce.* The produce manager at Dugan's Market orders 40 lb of tomatoes, 20 lb of zucchini, and 30 lb of onions from a local farmer one week.

a) Write a 1 × 3 matrix **A** that represents the amount of each item ordered.

b) The following week the produce manager increases her order by 10%. Find a matrix **B** that represents this order.

c) Find **A** + **B** and tell what the entries represent.

31. *Nutrition.* A 3-oz serving of roasted, skinless chicken breast contains 140 Cal, 27 g of protein, 3 g of fat, 13 mg of calcium, and 64 mg of sodium. One-half cup of potato salad contains 180 Cal, 4 g of protein, 11 g of fat, 24 mg of calcium, and

662 mg of sodium. One broccoli spear contains 50 Cal, 5 g of protein, 1 g of fat, 82 mg of calcium, and 20 mg of sodium. (*Source: Home and Garden Bulletin No. 72*, U.S. Government Printing Office, Washington, D.C. 20402)

a) Write 1 × 5 matrices **C**, **P**, and **B** that represent the nutritional values of each food.

b) Find **C** + 2**P** + 3**B** and tell what the entries represent.

32. *Nutrition.* One slice of cheese pizza contains 290 Cal, 15 g of protein, 9 g of fat, and 39 g of carbohydrates. One-half cup of gelatin dessert contains 70 Cal, 2 g of protein, 0 g of fat, and 17 g of carbohydrates. One cup of whole milk contains 150 Cal, 8 g of protein, 8 g of fat, and 11 g of carbohydrates. (*Source: Home and Garden Bulletin No. 72*, U.S. Government Printing Office, Washington, D.C. 20402)

a) Write 1 × 4 matrices **P**, **G**, and **M** that represent the nutritional values of each food.

b) Find 3**P** + 2**G** + 2**M** and tell what the entries represent.

33. *Food Service Management.* The food service manager at a large hospital is concerned about maintaining reasonable food costs. The table below shows the cost per serving, in cents, for items on four menus.

MENU	MEAT	POTATO	VEGETABLE	SALAD	DESSERT
1	45.29	6.63	10.94	7.42	8.01
2	53.78	4.95	9.83	6.16	12.56
3	47.13	8.47	12.66	8.29	9.43
4	51.64	7.12	11.57	9.35	10.72

On a particular day, a dietician orders 65 meals from menu 1, 48 from menu 2, 93 from menu 3, and 57 from menu 4.

a) Write the information in the table as a 4 × 5 matrix **M**.

b) Write a row matrix **N** that represents the number of each menu ordered.

c) Find the product **NM**.

d) State what the entries of **NM** represent.

34. *Food Service Management.* A college food service manager uses a table like the one below to show the number of units of ingredients, by weight, required for various menu items.

	WHITE CAKE	BREAD	COFFEE CAKE	SUGAR COOKIES
FLOUR	1	2.5	0.75	0.5
MILK	0	0.5	0.25	0
EGGS	0.75	0.25	0.5	0.5
BUTTER	0.5	0	0.5	1

The cost per unit of each ingredient is 15 cents for flour, 28 cents for milk, 54 cents for eggs, and 83 cents for butter.

a) Write the information in the table as a 4 × 4 matrix **M**.

b) Write a row matrix **C** that represents the cost per unit of each ingredient.

c) Find the product **CM**.

d) State what the entries of **CM** represent.

35. *Production Cost.* Karin supplies two small campus coffee shops with homemade chocolate chip cookies, oatmeal cookies, and peanut butter cookies. The table below shows the number of each type of cookie, in dozens, that Karin sold in one week.

	MUGSY'S COFFEE SHOP	THE COFFEE CLUB
CHOCOLATE CHIP	8	15
OATMEAL	6	10
PEANUT BUTTER	4	3

Karin spends $3 for the ingredients for one dozen chocolate chip cookies, $1.50 for the ingredients for one dozen oatmeal cookies, and $2 for the ingredients for one dozen peanut butter cookies.

a) Write the information in the table as a 3 × 2 matrix **S**.

b) Write a row matrix **C** that represents the cost, per dozen, of the ingredients for each type of cookie.

c) Find the product **CS**.

d) State what the entries of **CS** represent.

36. *Profit.* A manufacturer produces exterior plywood, interior plywood, and fiberboard, which are shipped to two distributors. The table below shows the number of units of each type of product that are shipped to each warehouse.

	DISTRIBUTOR 1	DISTRIBUTOR 2
EXTERIOR PLYWOOD	900	500
INTERIOR PLYWOOD	450	1000
FIBERBOARD	600	700

The profits from each unit of exterior plywood, interior plywood, and fiberboard are $5, $8, and $4, respectively.

a) Write the information in the table as a 3×2 matrix **M**.

b) Write a row matrix **P** that represents the profit, per unit, of each type of product.

c) Find the product **PM**.

d) State what the entries of **PM** represent.

37. *Profit.* In Exercise 35, suppose that Karin's profits on one dozen chocolate chip, oatmeal, and peanut butter cookies are $6, $4.50, and $5.20, respectively.

a) Write a row matrix **P** that represents this information.

b) Use the matrices **S** and **P** to find Karin's total profit from each coffee shop.

38. *Production Cost.* In Exercise 36, suppose that the manufacturer's production costs for each unit of exterior plywood, interior plywood, and fiberboard are $20, $25, and $15, respectively.

a) Write a row matrix **C** that represents this information.

b) Use the matrices **M** and **C** to find the total production cost for the products shipped to each distributor.

Write a matrix equation equivalent to the system of equations.

39. $2x - 3y = 7,$
$\quad x + 5y = -6$

40. $-x + y = 3,$
$\quad 5x - 4y = 16$

41. $x + y - 2z = 6,$
$\quad 3x - y + z = 7,$
$\quad 2x + 5y - 3z = 8$

42. $3x - y + z = 1,$
$\quad x + 2y - z = 3,$
$\quad 4x + 3y - 2z = 11$

43. $3x - 2y + 4z = 17,$
$\quad 2x + y - 5z = 13$

44. $3x + 2y + 5z = 9,$
$\quad 4x - 3y + 2z = 10$

45. $-4w + x - y + 2z = 12,$
$\quad w + 2x - y - z = 0,$
$\quad -w + x + 4y - 3z = 1,$
$\quad 2w + 3x + 5y - 7z = 9$

46. $12w + 2x + 4y - 5z = 2,$
$\quad -w + 4x - y + 12z = 5,$
$\quad 2w - x + 4y = 13,$
$\quad 2x + 10y + z = 5$

Technology Connection

47. Use a graphing calculator to perform the operations in Exercises 5, 9, 13, and 15.

48. Use a graphing calculator to perform the operations in Exercises 6, 8, 12, and 18.

Collaborative Discussion and Writing

49. Is it true that if **AB** = **0**, for matrices **A** and **B**, then **A** = **0** or **B** = **0**? Why or why not?

50. Explain how Karin could use the matrix products found in Exercises 35 and 37 in making business decisions.

Skill Maintenance

In Exercises 51–54:

a) *Find the vertex.*
b) *Find the axis of symmetry.*
c) *Determine whether there is a maximum or minimum value and find that value.*
d) *Graph the function.*

51. $f(x) = x^2 - x - 6$

52. $f(x) = 2x^2 - 5x - 3$

53. $f(x) = -x^2 - 3x + 2$

54. $f(x) = -3x^2 + 4x + 4$

Synthesis

For Exercises 55–58, let

$$A = \begin{bmatrix} -1 & 0 \\ 2 & 1 \end{bmatrix} \quad and \quad B = \begin{bmatrix} 1 & -1 \\ 0 & 2 \end{bmatrix}.$$

55. Show that

$$(A + B)(A - B) \neq A^2 - B^2,$$

where

$$A^2 = AA \quad and \quad B^2 = BB.$$

56. Show that

$$(A + B)(A + B) \neq A^2 + 2AB + B^2.$$

57. Show that

$$(A + B)(A - B) = A^2 + BA - AB - B^2.$$

58. Show that

$$(A + B)(A + B) = A^2 + BA + AB + B^2.$$

In Exercises 59–63, let

$$A = \begin{bmatrix} a_{11} & a_{12} & a_{13} & \cdots & a_{1n} \\ a_{21} & a_{22} & a_{23} & \cdots & a_{2n} \\ a_{31} & a_{32} & a_{33} & \cdots & a_{3n} \\ \vdots & \vdots & \vdots & & \vdots \\ a_{m1} & a_{m2} & a_{m3} & \cdots & a_{mn} \end{bmatrix},$$

$$B = \begin{bmatrix} b_{11} & b_{12} & b_{13} & \cdots & b_{1n} \\ b_{21} & b_{22} & b_{23} & \cdots & b_{2n} \\ b_{31} & b_{32} & b_{33} & \cdots & b_{3n} \\ \vdots & \vdots & \vdots & & \vdots \\ b_{m1} & b_{m2} & b_{m3} & \cdots & b_{mn} \end{bmatrix},$$

and

$$C = \begin{bmatrix} c_{11} & c_{12} & c_{13} & \cdots & c_{1n} \\ c_{21} & c_{22} & c_{23} & \cdots & c_{2n} \\ c_{31} & c_{32} & c_{33} & \cdots & c_{3n} \\ \vdots & \vdots & \vdots & & \vdots \\ c_{m1} & c_{m2} & c_{m3} & \cdots & c_{mn} \end{bmatrix},$$

and let k and l be any scalars.

59. Prove that $A + B = B + A$.

60. Prove that $A + (B + C) = (A + B) + C$.

61. Prove that $(kl)A = k(lA)$.

62. Prove that $k(A + B) = kA + kB$.

63. Prove that $(k + l)A = kA + lA$.

5.5

Inverses of Matrices

- *Find the inverse of a square matrix, if it exists.*
- *Use inverses of matrices to solve systems of equations.*

In this section, we continue our study of matrix algebra, finding the **multiplicative inverse,** or simply **inverse,** of a square matrix, if it exists. Then we use such inverses to solve systems of equations.

The Identity Matrix

Recall that, for real numbers, $a \cdot 1 = 1 \cdot a = a$; 1 is the multiplicative identity. A multiplicative identity matrix is very similar to the number 1.

Identity Matrix

For any positive integer n, the $n \times n$ **identity matrix** is an $n \times n$ matrix with 1's on the main diagonal and 0's elsewhere and is denoted by

$$\mathbf{I} = \begin{bmatrix} 1 & 0 & 0 & \cdots & 0 \\ 0 & 1 & 0 & \cdots & 0 \\ 0 & 0 & 1 & \cdots & 0 \\ \vdots & \vdots & \vdots & & \vdots \\ 0 & 0 & 0 & \cdots & 1 \end{bmatrix}.$$

Then $\mathbf{AI} = \mathbf{IA} = \mathbf{A}$, for any $n \times n$ matrix $\mathbf{A}$.

EXAMPLE 1 For

$$\mathbf{A} = \begin{bmatrix} 4 & -7 \\ -3 & 2 \end{bmatrix} \quad \text{and} \quad \mathbf{I} = \begin{bmatrix} 1 & 0 \\ 0 & 1 \end{bmatrix},$$

find each of the following.

a) **AI**   b) **IA**

Solution

a) $\mathbf{AI} = \begin{bmatrix} 4 & -7 \\ -3 & 2 \end{bmatrix} \begin{bmatrix} 1 & 0 \\ 0 & 1 \end{bmatrix}$

$= \begin{bmatrix} 4 \cdot 1 - 7 \cdot 0 & 4 \cdot 0 - 7 \cdot 1 \\ -3 \cdot 1 + 2 \cdot 0 & -3 \cdot 0 + 2 \cdot 1 \end{bmatrix} = \begin{bmatrix} 4 & -7 \\ -3 & 2 \end{bmatrix} = \mathbf{A}$

b) $\mathbf{IA} = \begin{bmatrix} 1 & 0 \\ 0 & 1 \end{bmatrix} \begin{bmatrix} 4 & -7 \\ -3 & 2 \end{bmatrix}$

$$= \begin{bmatrix} 1 \cdot 4 + 0(-3) & 1(-7) + 0 \cdot 2 \\ 0 \cdot 4 + 1(-3) & 0(-7) + 1 \cdot 2 \end{bmatrix} = \begin{bmatrix} 4 & -7 \\ -3 & 2 \end{bmatrix} = \mathbf{A}$$

The Inverse of a Matrix

Recall that for every nonzero real number a, there is a multiplicative inverse $1/a$, or a^{-1}, such that $a \cdot a^{-1} = a^{-1} \cdot a = 1$. The multiplicative inverse of a matrix behaves in a similar manner.

> ### Inverse of a Matrix
> For an $n \times n$ matrix $\mathbf{A}$, if there is a matrix $\mathbf{A}^{-1}$ for which $\mathbf{A}^{-1} \cdot \mathbf{A} = \mathbf{I} = \mathbf{A} \cdot \mathbf{A}^{-1}$, then $\mathbf{A}^{-1}$ is the **inverse** of $\mathbf{A}$.

We read $\mathbf{A}^{-1}$ as "$\mathbf{A}$ inverse." Note that not every matrix has an inverse.

EXAMPLE 2 Verify that

$$\mathbf{B} = \begin{bmatrix} 4 & -3 \\ 3 & -2 \end{bmatrix} \quad \text{is the inverse of} \quad \mathbf{A} = \begin{bmatrix} -2 & 3 \\ -3 & 4 \end{bmatrix}.$$

Solution We show that $\mathbf{BA} = \mathbf{I} = \mathbf{AB}$.

$$\mathbf{BA} = \begin{bmatrix} 4 & -3 \\ 3 & -2 \end{bmatrix} \begin{bmatrix} -2 & 3 \\ -3 & 4 \end{bmatrix} = \begin{bmatrix} 1 & 0 \\ 0 & 1 \end{bmatrix}$$

$$\mathbf{AB} = \begin{bmatrix} -2 & 3 \\ -3 & 4 \end{bmatrix} \begin{bmatrix} 4 & -3 \\ 3 & -2 \end{bmatrix} = \begin{bmatrix} 1 & 0 \\ 0 & 1 \end{bmatrix}$$

We can find the inverse of a square matrix, if it exists, by using row-equivalent operations as in the Gauss–Jordan elimination method. For example, consider the matrix

$$\mathbf{A} = \begin{bmatrix} -2 & 3 \\ -3 & 4 \end{bmatrix}.$$

To find its inverse, we first form an **augmented matrix** consisting of **A** on the left side and the 2×2 identity matrix on the right side:

$$\begin{bmatrix} -2 & 3 & | & 1 & 0 \\ -3 & 4 & | & 0 & 1 \end{bmatrix}.$$

The 2×2 matrix **A**   The 2×2 identity matrix

Then we attempt to transform the augmented matrix to one of the form

$$\begin{bmatrix} 1 & 0 & | & a & b \\ 0 & 1 & | & c & d \end{bmatrix}.$$

The 2×2 identity matrix   The matrix $\mathbf{A}^{-1}$

If we can do this, the matrix on the right, $\begin{bmatrix} a & b \\ c & d \end{bmatrix}$, is $\mathbf{A}^{-1}$.

EXAMPLE 3 Find $\mathbf{A}^{-1}$, where

$$\mathbf{A} = \begin{bmatrix} -2 & 3 \\ -3 & 4 \end{bmatrix}.$$

Solution First, we write the augmented matrix. Then we transform it to the desired form.

$$\begin{bmatrix} -2 & 3 & | & 1 & 0 \\ -3 & 4 & | & 0 & 1 \end{bmatrix}$$

$$\begin{bmatrix} 1 & -\frac{3}{2} & | & -\frac{1}{2} & 0 \\ -3 & 4 & | & 0 & 1 \end{bmatrix} \qquad \text{New row 1} = -\tfrac{1}{2}(\text{row 1})$$

$$\begin{bmatrix} 1 & -\frac{3}{2} & | & -\frac{1}{2} & 0 \\ 0 & -\frac{1}{2} & | & -\frac{3}{2} & 1 \end{bmatrix} \qquad \text{New row 2} = 3(\text{row 1}) + \text{row 2}$$

$$\begin{bmatrix} 1 & -\frac{3}{2} & | & -\frac{1}{2} & 0 \\ 0 & 1 & | & 3 & -2 \end{bmatrix} \qquad \text{New row 2} = -2(\text{row 2})$$

$$\begin{bmatrix} 1 & 0 & | & 4 & -3 \\ 0 & 1 & | & 3 & -2 \end{bmatrix} \qquad \text{New row 1} = \tfrac{3}{2}(\text{row 2}) + \text{row 1}$$

Thus,

$$\mathbf{A}^{-1} = \begin{bmatrix} 4 & -3 \\ 3 & -2 \end{bmatrix},$$

which we verified in Example 2.

Technology Connection

The $\boxed{x^{-1}}$ key on a graphing calculator can be used to find the inverse of a matrix like the one in Example 3.

```
[A]⁻¹
           [[4  -3]
            [3  -2]]
```

EXAMPLE 4 Find A^{-1}, where

$$A = \begin{bmatrix} 1 & 2 & -1 \\ 3 & 5 & 3 \\ 2 & 4 & 3 \end{bmatrix}.$$

Solution First, we write the augmented matrix. Then we transform it to the desired form.

$$\left[\begin{array}{ccc|ccc} 1 & 2 & -1 & 1 & 0 & 0 \\ 3 & 5 & 3 & 0 & 1 & 0 \\ 2 & 4 & 3 & 0 & 0 & 1 \end{array}\right]$$

$$\left[\begin{array}{ccc|ccc} 1 & 2 & -1 & 1 & 0 & 0 \\ 0 & -1 & 6 & -3 & 1 & 0 \\ 0 & 0 & 5 & -2 & 0 & 1 \end{array}\right]$$
New row 2 = −3(row 1) + row 2
New row 3 = −2(row 1) + row 3

$$\left[\begin{array}{ccc|ccc} 1 & 2 & -1 & 1 & 0 & 0 \\ 0 & -1 & 6 & -3 & 1 & 0 \\ 0 & 0 & 1 & -\frac{2}{5} & 0 & \frac{1}{5} \end{array}\right]$$
New row 3 = $\frac{1}{5}$(row 3)

$$\left[\begin{array}{ccc|ccc} 1 & 2 & 0 & \frac{3}{5} & 0 & \frac{1}{5} \\ 0 & -1 & 0 & -\frac{3}{5} & 1 & -\frac{6}{5} \\ 0 & 0 & 1 & -\frac{2}{5} & 0 & \frac{1}{5} \end{array}\right]$$
New row 1 = row 3 + row 1
New row 2 = −6(row 3) + row 2

$$\left[\begin{array}{ccc|ccc} 1 & 0 & 0 & -\frac{3}{5} & 2 & -\frac{11}{5} \\ 0 & -1 & 0 & -\frac{3}{5} & 1 & -\frac{6}{5} \\ 0 & 0 & 1 & -\frac{2}{5} & 0 & \frac{1}{5} \end{array}\right]$$
New row 1 = 2(row 2) + row 1

$$\left[\begin{array}{ccc|ccc} 1 & 0 & 0 & -\frac{3}{5} & 2 & -\frac{11}{5} \\ 0 & 1 & 0 & \frac{3}{5} & -1 & \frac{6}{5} \\ 0 & 0 & 1 & -\frac{2}{5} & 0 & \frac{1}{5} \end{array}\right]$$
New row 2 = −1(row 2)

Thus,

$$A^{-1} = \begin{bmatrix} -\frac{3}{5} & 2 & -\frac{11}{5} \\ \frac{3}{5} & -1 & \frac{6}{5} \\ -\frac{2}{5} & 0 & \frac{1}{5} \end{bmatrix}.$$

Technology Connection

When we try to find the inverse of a noninvertible, or singular, matrix using a graphing calculator, the calculator returns an error message similar to ERR: SINGULAR MATRIX.

If a matrix has an inverse, we say that it is **invertible,** or **nonsingular.** When we cannot obtain the identity matrix on the left using the Gauss–Jordan method, then no inverse exists. This occurs when we obtain a row consisting entirely of 0's in either of the two matrices in the augmented matrix. In this case, we say that **A** is a **singular matrix.**

Solving Systems of Equations

MATRIX EQUATIONS

REVIEW SECTION 5.4.

We can write a system of n linear equations in n variables as a matrix equation $\mathbf{AX} = \mathbf{B}$. If $\mathbf{A}$ has an inverse, then the system of equations has a unique solution that can be found by solving for $\mathbf{X}$, as follows:

$$\mathbf{AX} = \mathbf{B}$$

$\mathbf{A}^{-1}(\mathbf{AX}) = \mathbf{A}^{-1}\mathbf{B}$ Multiplying by $\mathbf{A}^{-1}$ on the left on both sides

$(\mathbf{A}^{-1}\mathbf{A})\mathbf{X} = \mathbf{A}^{-1}\mathbf{B}$ Using the associative property of matrix multiplication

$\mathbf{IX} = \mathbf{A}^{-1}\mathbf{B}$ $\mathbf{A}^{-1}\mathbf{A} = \mathbf{I}$

$\mathbf{X} = \mathbf{A}^{-1}\mathbf{B}.$ $\mathbf{IX} = \mathbf{X}$

Matrix Solutions of Systems of Equations

For a system of n linear equations in n variables, $\mathbf{AX} = \mathbf{B}$, if $\mathbf{A}$ is an invertible matrix, then the unique solution of the system is given by

$$\mathbf{X} = \mathbf{A}^{-1}\mathbf{B}.$$

Since matrix multiplication is not commutative in general, care must be taken to multiply *on the left* by $\mathbf{A}^{-1}$.

EXAMPLE 5   Use an inverse matrix to solve the following system of equations:

$$-2x + 3y = 4,$$
$$-3x + 4y = 5.$$

Solution   We write an equivalent matrix equation, $\mathbf{AX} = \mathbf{B}$:

$$\begin{bmatrix} -2 & 3 \\ -3 & 4 \end{bmatrix} \cdot \begin{bmatrix} x \\ y \end{bmatrix} = \begin{bmatrix} 4 \\ 5 \end{bmatrix}$$

$$\quad \mathbf{A} \qquad \cdot \quad \mathbf{X} \ = \ \mathbf{B}$$

In Example 3, we found that

$$\mathbf{A}^{-1} = \begin{bmatrix} 4 & -3 \\ 3 & -2 \end{bmatrix}.$$

We also verified this in Example 2. Now we have

$$\mathbf{X} = \mathbf{A}^{-1}\mathbf{B}$$

$$\begin{bmatrix} x \\ y \end{bmatrix} = \begin{bmatrix} 4 & -3 \\ 3 & -2 \end{bmatrix} \begin{bmatrix} 4 \\ 5 \end{bmatrix} = \begin{bmatrix} 1 \\ 2 \end{bmatrix}.$$

The solution of the system of equations is $(1, 2)$.

Technology
Connection

To use a graphing calculator to solve the system of equations in Example 5, we enter $\mathbf{A}$ and $\mathbf{B}$ and then enter the notation $\mathbf{A}^{-1}\mathbf{B}$ on the home screen.

```
[A]⁻¹[B]
               [[1]
                [2]]
```

5.5

Exercise Set

Determine whether **B** *is the inverse of* **A.**

1. $A = \begin{bmatrix} 1 & -3 \\ -2 & 7 \end{bmatrix}$, $B = \begin{bmatrix} 7 & 3 \\ 2 & 1 \end{bmatrix}$

2. $A = \begin{bmatrix} 3 & 2 \\ 4 & 3 \end{bmatrix}$, $B = \begin{bmatrix} 3 & -2 \\ -4 & 3 \end{bmatrix}$

3. $A = \begin{bmatrix} -1 & -1 & 6 \\ 1 & 0 & -2 \\ 1 & 0 & -3 \end{bmatrix}$, $B = \begin{bmatrix} 2 & 3 & 2 \\ 3 & 3 & 4 \\ 1 & 1 & 1 \end{bmatrix}$

4. $A = \begin{bmatrix} -2 & 0 & -3 \\ 5 & 1 & 7 \\ -3 & 0 & 4 \end{bmatrix}$, $B = \begin{bmatrix} 4 & 0 & -3 \\ 1 & 1 & 1 \\ -3 & 0 & 2 \end{bmatrix}$

Use the Gauss–Jordan method to find A^{-1}, *if it exists. Check your answers by finding* $A^{-1}A$ *and* AA^{-1}.

5. $A = \begin{bmatrix} 3 & 2 \\ 5 & 3 \end{bmatrix}$

6. $A = \begin{bmatrix} 3 & 5 \\ 1 & 2 \end{bmatrix}$

7. $A = \begin{bmatrix} 6 & 9 \\ 4 & 6 \end{bmatrix}$

8. $A = \begin{bmatrix} -4 & -6 \\ 2 & 3 \end{bmatrix}$

9. $A = \begin{bmatrix} 4 & -3 \\ 1 & -2 \end{bmatrix}$

10. $A = \begin{bmatrix} 0 & -1 \\ 1 & 0 \end{bmatrix}$

11. $A = \begin{bmatrix} 3 & 1 & 0 \\ 1 & 1 & 1 \\ 1 & -1 & 2 \end{bmatrix}$

12. $A = \begin{bmatrix} 1 & 0 & 1 \\ 2 & 1 & 0 \\ 1 & -1 & 1 \end{bmatrix}$

13. $A = \begin{bmatrix} 1 & -4 & 8 \\ 1 & -3 & 2 \\ 2 & -7 & 10 \end{bmatrix}$

14. $A = \begin{bmatrix} -2 & 5 & 3 \\ 4 & -1 & 3 \\ 7 & -2 & 5 \end{bmatrix}$

15. $A = \begin{bmatrix} 2 & 3 & 2 \\ 3 & 3 & 4 \\ -1 & -1 & -1 \end{bmatrix}$

16. $A = \begin{bmatrix} 1 & 2 & 3 \\ 2 & -1 & -2 \\ -1 & 3 & 3 \end{bmatrix}$

17. $A = \begin{bmatrix} 1 & 2 & -1 \\ -2 & 0 & 1 \\ 1 & -1 & 0 \end{bmatrix}$

18. $A = \begin{bmatrix} 7 & -1 & -9 \\ 2 & 0 & -4 \\ -4 & 0 & 6 \end{bmatrix}$

19. $A = \begin{bmatrix} 1 & 3 & -1 \\ 0 & 2 & -1 \\ 1 & 1 & 0 \end{bmatrix}$

20. $A = \begin{bmatrix} -1 & 0 & -1 \\ -1 & 1 & 0 \\ 0 & 1 & 1 \end{bmatrix}$

21. $A = \begin{bmatrix} 1 & 2 & 3 & 4 \\ 0 & 1 & 3 & -5 \\ 0 & 0 & 1 & -2 \\ 0 & 0 & 0 & -1 \end{bmatrix}$

22. $A = \begin{bmatrix} -2 & -3 & 4 & 1 \\ 0 & 1 & 1 & 0 \\ 0 & 4 & -6 & 1 \\ -2 & -2 & 5 & 1 \end{bmatrix}$

23. $A = \begin{bmatrix} 1 & -14 & 7 & 38 \\ -1 & 2 & 1 & -2 \\ 1 & 2 & -1 & -6 \\ 1 & -2 & 3 & 6 \end{bmatrix}$

24. $A = \begin{bmatrix} 10 & 20 & -30 & 15 \\ 3 & -7 & 14 & -8 \\ -7 & -2 & -1 & 2 \\ 4 & 4 & -3 & 1 \end{bmatrix}$

In Exercises 25–28, a system of equations is given, together with the inverse of the coefficient matrix. Use the inverse of the coefficient matrix to solve the system of equations.

25. $\begin{aligned} 11x + 3y &= -4, \\ 7x + 2y &= 5; \end{aligned}$ $\quad A^{-1} = \begin{bmatrix} 2 & -3 \\ -7 & 11 \end{bmatrix}$

26. $\begin{aligned} 8x + 5y &= -6, \\ 5x + 3y &= 2; \end{aligned}$ $\quad A^{-1} = \begin{bmatrix} -3 & 5 \\ 5 & -8 \end{bmatrix}$

27. $\begin{aligned} 3x + y &= 2, \\ 2x - y + 2z &= -5, \\ x + y + z &= 5; \end{aligned}$ $\quad A^{-1} = \dfrac{1}{9}\begin{bmatrix} 3 & 1 & -2 \\ 0 & -3 & 6 \\ -3 & 2 & 5 \end{bmatrix}$

28. $\begin{aligned} y - z &= -4, \\ 4x + y &= -3, \\ 3x - y + 3z &= 1; \end{aligned}$ $\quad A^{-1} = \dfrac{1}{5}\begin{bmatrix} -3 & 2 & -1 \\ 12 & -3 & 4 \\ 7 & -3 & 4 \end{bmatrix}$

Solve the system of equations using the inverse of the coefficient matrix of the equivalent matrix equation.

29. $\begin{aligned} 4x + 3y &= 2, \\ x - 2y &= 6 \end{aligned}$

30. $\begin{aligned} 2x - 3y &= 7, \\ 4x + y &= -7 \end{aligned}$

31. $5x + y = 2,$
$3x - 2y = -4$

32. $x - 6y = 5,$
$-x + 4y = -5$

33. $x \quad\quad + z = 1,$
$2x + y \quad\quad = 3,$
$x - y + z = 4$

34. $x + 2y + 3z = -1,$
$2x - 3y + 4z = 2,$
$-3x + 5y - 6z = 4$

35. $2x + 3y + 4z = 2,$
$x - 4y + 3z = 2,$
$5x + y + z = -4$

36. $x + y \quad\quad = 2,$
$3x \quad\quad + 2z = 5,$
$2x + 3y - 3z = 9$

37. $2w - 3x + 4y - 5z = 0,$
$3w - 2x + 7y - 3z = 2,$
$w + x - y + z = 1,$
$-w - 3x - 6y + 4z = 6$

38. $5w - 4x + 3y - 2z = -6,$
$w + 4x - 2y + 3z = -5,$
$2w - 3x + 6y - 9z = 14,$
$3w - 5x + 2y - 4z = -3$

39. *Sales.* Stefan sold a total of 145 Italian sausages and hot dogs from his curbside pushcart and collected $242.05. He sold 45 more hot dogs than sausages. How many of each did he sell?

40. *Price of School Supplies.* Miranda bought 4 lab record books and 3 highlighters for $13.93. Victor bought 3 lab record books and 2 highlighters for $10.25. Find the price of each item.

41. *Cost.* Evergreen Landscaping bought 4 tons of topsoil, 3 tons of mulch, and 6 tons of pea gravel for $2825. The next week the firm bought 5 tons of topsoil, 2 tons of mulch, and 5 tons of pea gravel for $2663. Pea gravel costs $17 less per ton than topsoil. Find the price per ton for each item.

42. *Investment.* Selena receives $230 per year in simple interest from three investments totaling $8500. Part is invested at 2.2%, part at 2.65%, and the rest at 3.05%. There is $1500 more invested at 3.05% than at 2.2%. Find the amount invested at each rate.

Technology Connection

Use a graphing calculator to do the following.

43. Exercises 5, 13, and 21

44. Exercises 6, 14, and 24

45. Exercises 29, 35, and 37

46. Exercises 30, 34, and 38

Collaborative Discussion and Writing

47. For square matrices $\mathbf{A}$ and $\mathbf{B}$, is it true, in general, that $(\mathbf{A} + \mathbf{B})^{-1} = \mathbf{A}^{-1} + \mathbf{B}^{-1}$? Explain.

48. For square matrices $\mathbf{A}$ and $\mathbf{B}$, is it true, in general, that $(\mathbf{AB})^{-1} = \mathbf{A}^{-1}\mathbf{B}^{-1}$? Explain.

Skill Maintenance

Use synthetic division to find the function values.

49. $f(x) = x^3 - 6x^2 + 4x - 8;$ find $f(-2)$

50. $f(x) = 2x^4 - x^3 + 5x^2 + 6x - 4;$ find $f(3)$

Solve.

51. $2x^2 + x = 7$

52. $\dfrac{1}{x + 1} - \dfrac{6}{x - 1} = 1$

53. $\sqrt{2x + 1} - 1 = \sqrt{2x - 4}$

54. $x - \sqrt{x} - 6 = 0$

Factor the polynomial $f(x)$.

55. $f(x) = x^3 - 3x^2 - 6x + 8$

56. $f(x) = x^4 + 2x^3 - 16x^2 - 2x + 15$

Synthesis

State the conditions under which $\mathbf{A}^{-1}$ exists. Then find a formula for $\mathbf{A}^{-1}$.

57. $\mathbf{A} = [x]$

58. $\mathbf{A} = \begin{bmatrix} x & 0 \\ 0 & y \end{bmatrix}$

59. $\mathbf{A} = \begin{bmatrix} 0 & 0 & x \\ 0 & y & 0 \\ z & 0 & 0 \end{bmatrix}$

60. $\mathbf{A} = \begin{bmatrix} x & 1 & 1 & 1 \\ 0 & y & 0 & 0 \\ 0 & 0 & z & 0 \\ 0 & 0 & 0 & w \end{bmatrix}$

5.6

Determinants and Cramer's Rule

- *Evaluate determinants of square matrices.*
- *Use Cramer's rule to solve systems of equations.*

Determinants of Square Matrices

With every square matrix, we associate a number called its *determinant*.

Determinant of a 2 × 2 Matrix

The **determinant** of the matrix $\begin{bmatrix} a & c \\ b & d \end{bmatrix}$ is denoted $\begin{vmatrix} a & c \\ b & d \end{vmatrix}$ and is defined as

$$\begin{vmatrix} a & c \\ b & d \end{vmatrix} = ad - bc.$$

EXAMPLE 1 Evaluate: $\begin{vmatrix} \sqrt{2} & -3 \\ -4 & -\sqrt{2} \end{vmatrix}$.

Solution

$$\begin{vmatrix} \sqrt{2} & -3 \\ -4 & -\sqrt{2} \end{vmatrix} \qquad \text{The arrows indicate the products involved.}$$

$$= \sqrt{2}\left(-\sqrt{2}\right) - (-4)(-3)$$
$$= -2 - 12$$
$$= -14$$

We now consider a way to evaluate determinants of square matrices of order 3 × 3 or higher.

Evaluating Determinants Using Cofactors

Often we first find minors and cofactors of matrices in order to evaluate determinants.

Minor

For a square matrix $\mathbf{A} = [a_{ij}]$, the **minor** M_{ij} of an element a_{ij} is the determinant of the matrix formed by deleting the ith row and the jth column of $\mathbf{A}$.

EXAMPLE 2 For the matrix

$$\mathbf{A} = [a_{ij}] = \begin{bmatrix} -8 & 0 & 6 \\ 4 & -6 & 7 \\ -1 & -3 & 5 \end{bmatrix},$$

find each of the following.

a) M_{11} **b)** M_{23}

Solution

a) For M_{11}, we delete the first row and the first column and find the determinant of the 2×2 matrix formed by the remaining elements.

$$\begin{bmatrix} -8 & 0 & 6 \\ 4 & -6 & 7 \\ -1 & -3 & 5 \end{bmatrix} \qquad M_{11} = \begin{vmatrix} -6 & 7 \\ -3 & 5 \end{vmatrix}$$
$$= (-6) \cdot 5 - (-3) \cdot 7$$
$$= -30 - (-21)$$
$$= -30 + 21$$
$$= -9$$

b) For M_{23}, we delete the second row and the third column and find the determinant of the 2×2 matrix formed by the remaining elements.

$$\begin{bmatrix} -8 & 0 & 6 \\ 4 & -6 & 7 \\ -1 & -3 & 5 \end{bmatrix} \qquad M_{23} = \begin{vmatrix} -8 & 0 \\ -1 & -3 \end{vmatrix}$$
$$= -8(-3) - (-1)0$$
$$= 24$$

Cofactor

For a square matrix $\mathbf{A} = [a_{ij}]$, the **cofactor** A_{ij} of an element a_{ij} is given by

$$A_{ij} = (-1)^{i+j}M_{ij},$$

where M_{ij} is the minor of a_{ij}.

EXAMPLE 3 For the matrix given in Example 2, find each of the following.

a) A_{11} **b)** A_{23}

Solution

a) In Example 2, we found that $M_{11} = -9$. Then
$$A_{11} = (-1)^{1+1}(-9) = (1)(-9) = -9.$$

b) In Example 2, we found that $M_{23} = 24$. Then
$$A_{23} = (-1)^{2+3}(24) = (-1)(24) = -24.$$

Consider the matrix **A** given by

$$\mathbf{A} = \begin{bmatrix} a_{11} & a_{12} & a_{13} \\ a_{21} & a_{22} & a_{23} \\ a_{31} & a_{32} & a_{33} \end{bmatrix}.$$

The determinant of the matrix, denoted $|\mathbf{A}|$, can be found by multiplying each element of the first column by its cofactor and adding:

$$|\mathbf{A}| = a_{11}A_{11} + a_{21}A_{21} + a_{31}A_{31}.$$

Because

$$A_{11} = (-1)^{1+1}M_{11} = M_{11},$$
$$A_{21} = (-1)^{2+1}M_{21} = -M_{21},$$

and

$$A_{31} = (-1)^{3+1}M_{31} = M_{31},$$

we can write

$$|\mathbf{A}| = a_{11} \cdot \begin{vmatrix} a_{22} & a_{23} \\ a_{32} & a_{33} \end{vmatrix} - a_{21} \cdot \begin{vmatrix} a_{12} & a_{13} \\ a_{32} & a_{33} \end{vmatrix} + a_{31} \cdot \begin{vmatrix} a_{12} & a_{13} \\ a_{22} & a_{23} \end{vmatrix}.$$

It can be shown that we can determine $|\mathbf{A}|$ by choosing *any* row or column, multiplying each element in that row or column by its cofactor, and adding. This is called *expanding* across a row or down a column. We just expanded down the first column. We now define the determinant of a square matrix of any order.

Determinant of Any Square Matrix

For any square matrix **A** of order $n \times n$ ($n > 1$), we define the **determinant** of **A**, denoted $|\mathbf{A}|$, as follows. Choose any row or column. Multiply each element in that row or column by its cofactor and add the results. The determinant of a 1×1 matrix is simply the element of the matrix. The value of a determinant will be the same no matter which row or column is chosen.

EXAMPLE 4 Evaluate $|\mathbf{A}|$ by expanding across the third row.

$$\mathbf{A} = \begin{bmatrix} -8 & 0 & 6 \\ 4 & -6 & 7 \\ -1 & -3 & 5 \end{bmatrix}$$

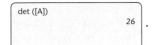
Solution We have

$$|\mathbf{A}| = (-1)A_{31} + (-3)A_{32} + 5A_{33}$$

$$= (-1)(-1)^{3+1} \cdot \begin{vmatrix} 0 & 6 \\ -6 & 7 \end{vmatrix} + (-3)(-1)^{3+2} \cdot \begin{vmatrix} -8 & 6 \\ 4 & 7 \end{vmatrix}$$

$$+ 5(-1)^{3+3} \cdot \begin{vmatrix} -8 & 0 \\ 4 & -6 \end{vmatrix}$$

$$= (-1) \cdot 1 \cdot [0 \cdot 7 - (-6)6] + (-3)(-1)[-8 \cdot 7 - 4 \cdot 6]$$

$$+ 5 \cdot 1 \cdot [-8(-6) - 4 \cdot 0]$$

$$= -[36] + 3[-80] + 5[48]$$

$$= -36 - 240 + 240 = -36.$$

The value of this determinant is -36 no matter which row or column we expand upon.

Cramer's Rule

Determinants can be used to solve systems of linear equations. Consider a system of two linear equations:

$$a_1x + b_1y = c_1,$$
$$a_2x + b_2y = c_2.$$

Solving this system using the elimination method, we obtain

$$x = \frac{c_1b_2 - c_2b_1}{a_1b_2 - a_2b_1}$$

and

$$y = \frac{a_1c_2 - a_2c_1}{a_1b_2 - a_2b_1}.$$

The numerators and denominators of these expressions can be written as determinants:

$$x = \frac{\begin{vmatrix} c_1 & b_1 \\ c_2 & b_2 \end{vmatrix}}{\begin{vmatrix} a_1 & b_1 \\ a_2 & b_2 \end{vmatrix}} \quad \text{and} \quad y = \frac{\begin{vmatrix} a_1 & c_1 \\ a_2 & c_2 \end{vmatrix}}{\begin{vmatrix} a_1 & b_1 \\ a_2 & b_2 \end{vmatrix}}.$$

If we let

$$D = \begin{vmatrix} a_1 & b_1 \\ a_2 & b_2 \end{vmatrix}, \quad D_x = \begin{vmatrix} c_1 & b_1 \\ c_2 & b_2 \end{vmatrix}, \quad \text{and} \quad D_y = \begin{vmatrix} a_1 & c_1 \\ a_2 & c_2 \end{vmatrix},$$

we have

$$x = \frac{D_x}{D} \quad \text{and} \quad y = \frac{D_y}{D}.$$

This procedure for solving systems of equations is known as *Cramer's rule.*

> ### Cramer's Rule for 2 × 2 Systems
>
> The solution of the system of equations
>
> $$a_1x + b_1y = c_1,$$
> $$a_2x + b_2y = c_2$$
>
> is given by
>
> $$x = \frac{D_x}{D}, \qquad y = \frac{D_y}{D},$$
>
> where
>
> $$D = \begin{vmatrix} a_1 & b_1 \\ a_2 & b_2 \end{vmatrix}, \qquad D_x = \begin{vmatrix} c_1 & b_1 \\ c_2 & b_2 \end{vmatrix},$$
>
> $$D_y = \begin{vmatrix} a_1 & c_1 \\ a_2 & c_2 \end{vmatrix}, \quad \text{and} \quad D \neq 0.$$

Note that the denominator D contains the coefficients of x and y, in the same position as in the original equations. For x, the numerator is obtained by replacing the x-coefficients in D (the a's) with the c's. For y, the numerator is obtained by replacing the y-coefficients in D (the b's) with the c's.

EXAMPLE 5   Solve using Cramer's rule:

$$2x + 5y = 7,$$
$$5x - 2y = -3.$$

Solution   We have

$$x = \frac{\begin{vmatrix} 7 & 5 \\ -3 & -2 \end{vmatrix}}{\begin{vmatrix} 2 & 5 \\ 5 & -2 \end{vmatrix}} = \frac{7(-2) - (-3)5}{2(-2) - 5 \cdot 5} = \frac{1}{-29} = -\frac{1}{29},$$

$$y = \frac{\begin{vmatrix} 2 & 7 \\ 5 & -3 \end{vmatrix}}{\begin{vmatrix} 2 & 5 \\ 5 & -2 \end{vmatrix}} = \frac{2(-3) - 5 \cdot 7}{-29} = \frac{-41}{-29} = \frac{41}{29}.$$

The solution is $\left(-\frac{1}{29}, \frac{41}{29}\right)$.

Technology Connection

To use Cramer's rule to solve the system of equations in Example 5 on a graphing calculator, we first enter the matrices corresponding to D, D_x, and D_y. We enter

$$\mathbf{A} = \begin{bmatrix} 2 & 5 \\ 5 & -2 \end{bmatrix}, \qquad \mathbf{B} = \begin{bmatrix} 7 & 5 \\ -3 & -2 \end{bmatrix}, \quad \text{and} \quad \mathbf{C} = \begin{bmatrix} 2 & 7 \\ 5 & -3 \end{bmatrix}.$$

Then

$$x = \frac{\det(\mathbf{B})}{\det(\mathbf{A})} \quad \text{and} \quad y = \frac{\det(\mathbf{C})}{\det(\mathbf{A})}.$$

```
det ([B])/det([A]
)▶Frac
                         -1/29
det ([C])/det([A]
)▶Frac
                         41/29
```

Cramer's rule works only when a system of equations has a unique solution. This occurs when $D \neq 0$. If $D = 0$ and D_x and D_y are also 0, then the equations are dependent. If $D = 0$ and D_x and/or D_y is not 0, then the system is inconsistent.

Cramer's rule can be extended to a system of n linear equations in n variables. We consider a 3×3 system.

Cramer's Rule for 3 × 3 Systems

The solution of the system of equations

$$a_1 x + b_1 y + c_1 z = d_1,$$
$$a_2 x + b_2 y + c_2 z = d_2,$$
$$a_3 x + b_3 y + c_3 z = d_3$$

is given by

$$x = \frac{D_x}{D}, \qquad y = \frac{D_y}{D}, \qquad z = \frac{D_z}{D},$$

where

$$D = \begin{vmatrix} a_1 & b_1 & c_1 \\ a_2 & b_2 & c_2 \\ a_3 & b_3 & c_3 \end{vmatrix}, \qquad D_x = \begin{vmatrix} d_1 & b_1 & c_1 \\ d_2 & b_2 & c_2 \\ d_3 & b_3 & c_3 \end{vmatrix},$$

$$D_y = \begin{vmatrix} a_1 & d_1 & c_1 \\ a_2 & d_2 & c_2 \\ a_3 & d_3 & c_3 \end{vmatrix}, \qquad D_z = \begin{vmatrix} a_1 & b_1 & d_1 \\ a_2 & b_2 & d_2 \\ a_3 & b_3 & d_3 \end{vmatrix}, \quad \text{and } D \neq 0.$$

Note that the determinant D_x is obtained from D by replacing the x-coefficients with d_1, d_2, and d_3. D_y and D_z are obtained in a similar manner. As with a system of two equations, Cramer's rule cannot be used if $D = 0$. If $D = 0$ and D_x, D_y, and D_z are 0, the equations are dependent. If $D = 0$ and one of D_x, D_y, or D_z is not 0, then the system is inconsistent.

EXAMPLE 6 Solve using Cramer's rule:

$$x - 3y + 7z = 13,$$
$$x + y + z = 1,$$
$$x - 2y + 3z = 4.$$

Solution We have

$$D = \begin{vmatrix} 1 & -3 & 7 \\ 1 & 1 & 1 \\ 1 & -2 & 3 \end{vmatrix} = -10, \qquad D_x = \begin{vmatrix} 13 & -3 & 7 \\ 1 & 1 & 1 \\ 4 & -2 & 3 \end{vmatrix} = 20,$$

$$D_y = \begin{vmatrix} 1 & 13 & 7 \\ 1 & 1 & 1 \\ 1 & 4 & 3 \end{vmatrix} = -6, \qquad D_z = \begin{vmatrix} 1 & -3 & 13 \\ 1 & 1 & 1 \\ 1 & -2 & 4 \end{vmatrix} = -24.$$

Then

$$x = \frac{D_x}{D} = \frac{20}{-10} = -2,$$

$$y = \frac{D_y}{D} = \frac{-6}{-10} = \frac{3}{5},$$

$$z = \frac{D_z}{D} = \frac{-24}{-10} = \frac{12}{5}.$$

The solution is $\left(-2, \frac{3}{5}, \frac{12}{5}\right)$. In practice, it is not necessary to evaluate D_z. When we have found values for x and y, we can substitute them into one of the equations to find z.

5.6

Exercise Set

Evaluate the determinant.

1. $\begin{vmatrix} -2 & -\sqrt{5} \\ -\sqrt{5} & 3 \end{vmatrix}$

2. $\begin{vmatrix} \sqrt{5} & -3 \\ 4 & 2 \end{vmatrix}$

3. $\begin{vmatrix} x & 4 \\ x & x^2 \end{vmatrix}$

4. $\begin{vmatrix} y^2 & -2 \\ y & 3 \end{vmatrix}$

5. $\begin{vmatrix} 3 & 1 & 2 \\ -2 & 3 & 1 \\ 3 & 4 & -6 \end{vmatrix}$

6. $\begin{vmatrix} 3 & -2 & 1 \\ 2 & 4 & 3 \\ -1 & 5 & 1 \end{vmatrix}$

7. $\begin{vmatrix} x & 0 & -1 \\ 2 & x & x^2 \\ -3 & x & 1 \end{vmatrix}$

8. $\begin{vmatrix} x & 1 & -1 \\ x^2 & x & x \\ 0 & x & 1 \end{vmatrix}$

Use the following matrix for Exercises 9–16:

$$\mathbf{A} = \begin{bmatrix} 7 & -4 & -6 \\ 2 & 0 & -3 \\ 1 & 2 & -5 \end{bmatrix}.$$

9. Find M_{11}, M_{32}, and M_{22}.

10. Find M_{13}, M_{31}, and M_{23}.

11. Find A_{11}, A_{32}, and A_{22}.

12. Find A_{13}, A_{31}, and A_{23}.

13. Evaluate $|\mathbf{A}|$ by expanding across the second row.

14. Evaluate $|\mathbf{A}|$ by expanding down the second column.

15. Evaluate $|\mathbf{A}|$ by expanding down the third column.

16. Evaluate $|\mathbf{A}|$ by expanding across the first row.

Use the following matrix for Exercises 17–22:

$$\mathbf{A} = \begin{bmatrix} 1 & 0 & 0 & -2 \\ 4 & 1 & 0 & 0 \\ 5 & 6 & 7 & 8 \\ -2 & -3 & -1 & 0 \end{bmatrix}$$

17. Find M_{41} and M_{33}.

18. Find M_{12} and M_{44}.

19. Find A_{24} and A_{43}.

20. Find A_{22} and A_{34}.

21. Evaluate $|\mathbf{A}|$ by expanding across the first row.

22. Evaluate $|\mathbf{A}|$ by expanding down the third column.

Solve using Cramer's rule.

23. $-2x + 4y = 3,$
$\quad 3x - 7y = 1$

24. $5x - 4y = -3,$
$\quad 7x + 2y = 6$

25. $2x - y = 5,$
$\quad x - 2y = 1$

26. $3x + 4y = -2,$
$\quad 5x - 7y = 1$

27. $2x + 9y = -2,$
$\quad 4x - 3y = 3$

28. $2x + 3y = -1,$
$\quad 3x + 6y = -0.5$

29. $2x + 5y = 7,$
$\quad 3x - 2y = 1$

30. $3x + 2y = 7,$
$\quad 2x + 3y = -2$

31. $3x + 2y - z = 4,$
$\quad 3x - 2y + z = 5,$
$\quad 4x - 5y - z = -1$

32. $\quad 3x - y + 2z = 1,$
$\quad\quad x - y + 2z = 3,$
$\quad -2x + 3y + z = 1$

33. $3x + 5y - z = -2,$
$\quad x - 4y + 2z = 13,$
$\quad 2x + 4y + 3z = 1$

34. $3x + 2y + 2z = 1,$
$\quad 5x - y - 6z = 3,$
$\quad 2x + 3y + 3z = 4$

35. $\quad x - 3y - 7z = 6,$
$\quad 2x + 3y + z = 9,$
$\quad 4x + y = 7$

36. $\quad x - 2y - 3z = 4,$
$\quad 3x - 2z = 8,$
$\quad 2x + y + 4z = 13$

37. $\quad\quad 6y + 6z = -1,$
$\quad 8x + 6z = -1,$
$\quad 4x + 9y = 8$

38. $3x + 5y = 2,$
$\quad 2x - 3z = 7,$
$\quad\quad 4y + 2z = -1$

Technology Connection

Use a graphing calculator to do the following.

39. Exercises 1 and 5   **40.** Exercises 2 and 6

For each matrix $\mathbf{A}$, evaluate $|\mathbf{A}|$ on a graphing calculator.

41. $\mathbf{A} = \begin{bmatrix} 1 & 0 & 0 & -2 \\ 4 & 1 & 0 & 0 \\ 5 & 6 & 7 & 8 \\ -2 & -3 & -1 & 0 \end{bmatrix}$

42. $\mathbf{A} = \begin{bmatrix} 5 & -4 & 2 & -2 \\ 3 & -3 & -4 & 7 \\ -2 & 3 & 2 & 4 \\ -8 & 9 & 5 & -5 \end{bmatrix}$

Use a graphing calculator to do the following.

43. Exercises 25 and 33   **44.** Exercises 26 and 34

Collaborative Discussion and Writing

45. Explain why the system of equations

$$a_1 x + b_1 y = c_1,$$
$$a_2 x + b_2 y = c_2$$

is either dependent or inconsistent when

$$\begin{vmatrix} a_1 & b_1 \\ a_2 & b_2 \end{vmatrix} = 0.$$

46. If the lines $a_1 x + b_1 y = c_1$ and $a_2 x + b_2 y = c_2$ are parallel, what can you say about the values of

$$\begin{vmatrix} a_1 & b_1 \\ a_2 & b_2 \end{vmatrix}, \qquad \begin{vmatrix} c_1 & b_1 \\ c_2 & b_2 \end{vmatrix}, \quad \text{and} \quad \begin{vmatrix} a_1 & c_1 \\ a_2 & c_2 \end{vmatrix}?$$

Skill Maintenance

Determine whether the function is one-to-one, and if it is, find a formula for $f^{-1}(x)$.

47. $f(x) = 3x + 2$

48. $f(x) = x^2 - 4$

49. $f(x) = |x| + 3$

50. $f(x) = \sqrt[3]{x} + 1$

Simplify. Write answers in the form $a + bi$, where a and b are real numbers.

51. $(3 - 4i) - (-2 - i)$

52. $(5 + 2i) + (1 - 4i)$

53. $(1 - 2i)(6 + 2i)$

54. $\dfrac{3 + i}{4 - 3i}$

Synthesis

Solve.

55. $\begin{vmatrix} x & 5 \\ -4 & x \end{vmatrix} = 24$

56. $\begin{vmatrix} y & 2 \\ 3 & y \end{vmatrix} = y$

57. $\begin{vmatrix} x & -3 \\ -1 & x \end{vmatrix} \geq 0$

58. $\begin{vmatrix} y & -5 \\ -2 & y \end{vmatrix} < 0$

59. $\begin{vmatrix} x + 3 & 4 \\ x - 3 & 5 \end{vmatrix} = -7$

60. $\begin{vmatrix} m + 2 & -3 \\ m + 5 & -4 \end{vmatrix} = 3m - 5$

61. $\begin{vmatrix} 2 & x & 1 \\ 1 & 2 & -1 \\ 3 & 4 & -2 \end{vmatrix} = -6$

62. $\begin{vmatrix} x & 2 & x \\ 3 & -1 & 1 \\ 1 & -2 & 2 \end{vmatrix} = -10$

Rewrite the expression using a determinant. Answers may vary.

63. $2L + 2W$

64. $\pi r + \pi h$

65. $a^2 + b^2$

66. $\frac{1}{2} h(a + b)$

67. $2\pi r^2 + 2\pi r h$

68. $x^2 y^2 - Q^2$

5.7

Systems of Inequalities and Linear Programming

- *Graph linear inequalities.*
- *Graph systems of linear inequalities.*
- *Solve linear programming problems.*

A graph of an inequality is a drawing that represents its solutions. We have already seen that an inequality in one variable can be graphed on a number line. An inequality in two variables can be graphed on a coordinate plane.

Graphs of Linear Inequalities

A statement like $5x - 4y < 20$ is a linear inequality in two variables.

Linear Inequality in Two Variables

A **linear inequality in two variables** is an inequality that can be written in the form

$$Ax + By < C,$$

where A, B, and C are real numbers and A and B are not both zero. The symbol $<$ may be replaced with $\leq$, $>$, or $\geq$.

A solution of a linear inequality in two variables is an ordered pair (x, y) for which the inequality is true. For example, $(1, 3)$ is a solution of $5x - 4y < 20$ because $5 \cdot 1 - 4 \cdot 3 < 20$, or $-7 < 20$, is true. On the other hand, $(2, -6)$ is not a solution of $5x - 4y < 20$ because $5 \cdot 2 - 4 \cdot (-6) \not< 20$, or $34 \not< 20$.

The **solution set** of an inequality is the set of all ordered pairs that make it true. The **graph of an inequality** represents its solution set.

EXAMPLE 1 Graph: $y < x + 3$.

Solution We begin by graphing the **related equation** $y = x + 3$. We use a dashed line because the inequality symbol is $<$. This indicates that the line itself is not in the solution set of the inequality.

Note that the line divides the coordinate plane into two regions called **half-planes.** One of these half-planes satisfies the inequality. Either *all* points in a half-plane are in the solution set of the inequality or *none* is.

To determine which half-plane satisfies the inequality, we try a test point in either region. The point $(0, 0)$ is usually a convenient choice so long as it does not lie on the line.

$$y < x + 3$$

0 ? $0 + 3$	
0 $\mid$ 3	TRUE

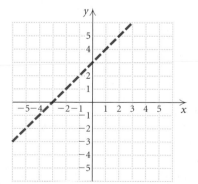

Since $(0, 0)$ satisfies the inequality, so do all points in the half-plane that contains $(0, 0)$. We shade this region to show the solution set of the inequality.

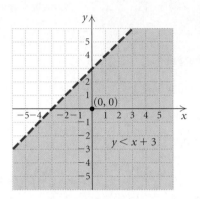

In general, we use the following procedure to graph linear inequalities in two variables.

To graph a linear inequality in two variables:

1. Replace the inequality symbol with an equals sign and graph this related equation. If the inequality symbol is $<$ or $>$, draw the line dashed. If the inequality symbol is $\leq$ or $\geq$, draw the line solid.

2. The graph consists of a half-plane on one side of the line and, if the line is solid, the line as well. To determine which half-plane to shade, test a point not on the line in the original inequality. If that point is a solution, shade the half-plane containing that point. If not, shade the opposite half-plane.

EXAMPLE 2   Graph: $3x + 4y \geq 12$.

Solution

1. First, we graph the related equation $3x + 4y = 12$. We use a solid line because the inequality symbol is $\geq$. This indicates that the line is included in the solution set.

2. To determine which half-plane to shade, we test a point in either region. We choose $(0, 0)$.

$$\begin{array}{c} 3x + 4y \geq 12 \\ \hline 3 \cdot 0 + 4 \cdot 0 \ ? \ 12 \\ 0 \ \mid \ 12 \quad \text{FALSE} \end{array}$$

Because $(0, 0)$ is *not* a solution, all the points in the half-plane that does *not* contain $(0, 0)$ are solutions. We shade that region, as shown in the figure at left.

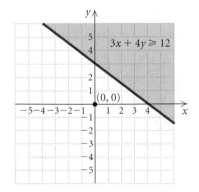

To graph the inequality in Example 2 on a graphing calculator, we first enter the related equation in the form $y = \dfrac{-3x + 12}{4}$. Then we select the "shade above" graph style.

$$y = \frac{-3x + 12}{4}$$

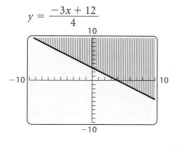

EXAMPLE 3 Graph $x > -3$ on a plane.

Solution

1. First, we graph the related equation $x = -3$. We use a dashed line because the inequality symbol is $>$. This indicates that the line is not included in the solution set.

2. The inequality tells us that all points (x, y) for which $x > -3$ are solutions. These are the points to the right of the line. We can also use a test point to determine the solutions. We choose $(5, 1)$.

$$\frac{x > -3}{5 \ ? \ -3} \quad \text{TRUE}$$

Because $(5, 1)$ is a solution, we shade the region containing that point—that is, the region to the right of the dashed line.

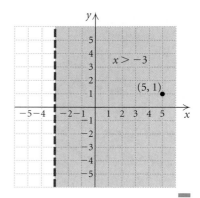

We can graph the inequality $y \le 4$ by first graphing $y = 4$ and then using the "shade below" graph style.

$$y = 4$$

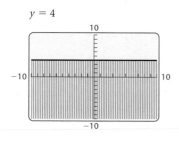

EXAMPLE 4 Graph $y \le 4$ on a plane.

Solution

1. First, we graph the related equation $y = 4$. We use a solid line because the inequality symbol is $\le$.

2. The inequality tells us that all points (x, y) for which $y \le 4$ are solutions of the inequality. These are the points on or below the line. We can also use a test point to determine the solutions. We choose $(-2, 5)$.

$$\frac{y \le 4}{5 \ ? \ 4} \quad \text{FALSE}$$

Because $(-2, 5)$ is not a solution, we shade the half-plane that does not contain that point.

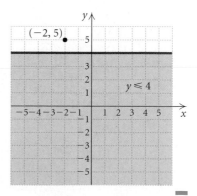

Systems of Linear Inequalities

A system of inequalities in two variables consists of two or more inequalities in two variables considered simultaneously. For example,

$$x + y \leq 4,$$
$$x - y \geq 2$$

is a system of two *linear* inequalities in two variables.

A solution of a system of inequalities is an ordered pair that is a solution of each inequality in the system. To graph a system of linear inequalities, we graph each inequality and determine the region that is common to *all* the solution sets.

EXAMPLE 5   Graph the solution set of the system

$$x + y \leq 4,$$
$$x - y \geq 2.$$

Solution   We graph $x + y \leq 4$ by first graphing the equation $x + y = 4$ using a solid line. Next, we choose $(0,0)$ as a test point and find that it is a solution of $x + y \leq 4$, so we shade the half-plane containing $(0,0)$ using red. Next, we graph $x - y = 2$ using a solid line. We find that $(0,0)$ is not a solution of $x - y \geq 2$, so we shade the half-plane that does not contain $(0,0)$ using green. The arrows at the ends of each line help to indicate the half-plane that contains each solution set.

**Technology
Connection**

We can use different shading patterns on a graphing calculator to graph the system of inequalities in Example 5. The solution set is the region shaded using both patterns.

$$y_1 = 4 - x, \quad y_2 = x - 2$$

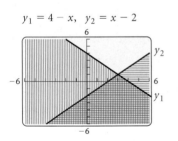

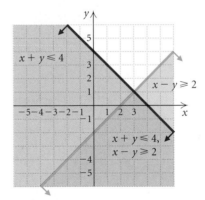

The solution set of the system of equations is the region shaded both red and green, or brown, including parts of the lines $x + y = 4$ and $x - y = 2$.

A system of inequalities may have a graph that consists of a polygon and its interior. As we will see later in this section, it is important in many applications to be able to find the vertices of such a polygon.

EXAMPLE 6   Graph the following system of inequalities and find the co-ordinates of any vertices formed:

$$3x - y \leq 6, \qquad (1)$$
$$y - 3 \leq 0, \qquad (2)$$
$$x + y \geq 0. \qquad (3)$$

Solution   We graph the related equations $3x - y = 6$, $y - 3 = 0$, and $x + y = 0$ using solid lines. The half-plane containing the solution set for each inequality is indicated by the arrows at the ends of each line. We shade the region common to all three solution sets.

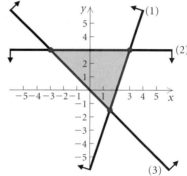

To find the vertices, we solve three systems of equations. The system of equations from inequalities (1) and (2) is

$$3x - y = 6,$$
$$y - 3 = 0.$$

Solving, we obtain the vertex $(3, 3)$.

The system of equations from inequalities (1) and (3) is

$$3x - y = 6,$$
$$x + y = 0.$$

Solving, we obtain the vertex $\left(\frac{3}{2}, -\frac{3}{2}\right)$.

The system of equations from inequalities (2) and (3) is

$$y - 3 = 0,$$
$$x + y = 0.$$

Solving, we obtain the vertex $(-3, 3)$.

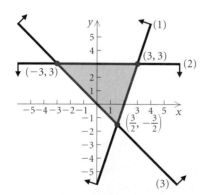

Applications: Linear Programming

In many applications, we want to find a maximum or minimum value. In business, for example, we might want to maximize profit and minimize cost. **Linear programming** can tell us how to do this.

In our study of linear programming, we will consider linear functions of two variables that are to be maximized or minimized subject to several conditions, or **constraints.** These constraints are expressed as inequalities. The solution set of the system of inequalities made up of the constraints contains all the **feasible solutions** of a linear programming problem. The function that we want to maximize or minimize is called the **objective function.**

It can be shown that the maximum and minimum values of the objective function occur at a vertex of the region of feasible solutions. Thus we have the following procedure.

Linear Programming Procedure

To find the maximum or minimum value of a linear objective function subject to a set of constraints:

1. Graph the region of feasible solutions.
2. Determine the coordinates of the vertices of the region.
3. Evaluate the objective function at each vertex. The largest and smallest of those values are the maximum and minimum values of the function, respectively.

EXAMPLE 7 *Maximizing Profit.* Dovetail Carpentry Shop makes bookcases and desks. Each bookcase requires 5 hr of woodworking and 4 hr of finishing. Each desk requires 10 hr of woodworking and 3 hr of finishing. Each month the shop has 600 hr of labor available for woodworking and 240 hr for finishing. The profit on each bookcase is $40 and on each desk is $75. How many of each product should be made each month in order to maximize profit?

Solution We let $x =$ the number of bookcases to be produced and $y =$ the number of desks. Then the profit P is given by the function

$$P = 40x + 75y.$$ **To emphasize that *P* is a function of two variables, we sometimes write $P(x, y) = 40x + 75y$.**

We know that x bookcases require $5x$ hr of woodworking and y desks require $10y$ hr of woodworking. Since there is no more than 600 hr of labor available for woodworking, we have one constraint:

$$5x + 10y \leq 600.$$

Similarly, the bookcases and desks require $4x$ hr and $3y$ hr of finishing, respectively. There is no more than 240 hr of labor available for finishing, so we have a second constraint:

$$4x + 3y \leq 240.$$

We also know that $x \geq 0$ and $y \geq 0$ because the carpentry shop cannot make a negative number of either product.

Thus we want to maximize the objective function

$$P = 40x + 75y$$

subject to the constraints

$$5x + 10y \leq 600,$$
$$4x + 3y \leq 240,$$
$$x \geq 0,$$
$$y \geq 0.$$

We graph the system of inequalities and determine the vertices.

Next, we evaluate the objective function P at each vertex.

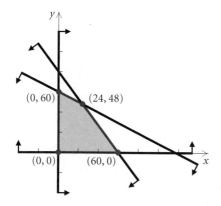

VERTICES (x, y)	PROFIT $P = 40x + 75y$	
$(0, 0)$	$P = 40 \cdot 0 + 75 \cdot 0 = 0$	
$(60, 0)$	$P = 40 \cdot 60 + 75 \cdot 0 = 2400$	
$(24, 48)$	$P = 40 \cdot 24 + 75 \cdot 48 = 4560$	⟵ Maximum
$(0, 60)$	$P = 40 \cdot 0 + 75 \cdot 60 = 4500$	

The carpentry shop will make a maximum profit of \$4560 when 24 bookcases and 48 desks are produced.

5.7

Exercise Set

In Exercises 1–8, match the inequality with one of the graphs (a)–(h), which follow.

a)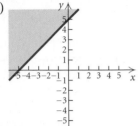

b)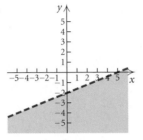

c)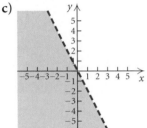

d)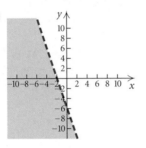

e)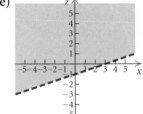

f)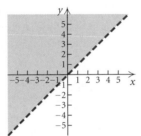

g)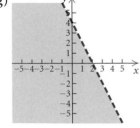

h)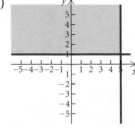

1. $y > x$

2. $y < -2x$

3. $y \leq x - 3$

4. $y \geq x + 5$

5. $2x + y < 4$

6. $3x + y < -6$

7. $2x - 5y > 10$

8. $3x - 9y < 9$

Graph.

9. $y > 2x$

10. $2y < x$

11. $y + x \geq 0$

12. $y - x < 0$

13. $y > x - 3$

14. $y \leq x + 4$

15. $x + y < 4$

16. $x - y \geq 5$

17. $3x - 2y \leq 6$

18. $2x - 5y < 10$

19. $3y + 2x \geq 6$

20. $2y + x \leq 4$

21. $3x - 2 \leq 5x + y$

22. $2x - 6y \geq 8 + 2y$

23. $x < -4$

24. $y \geq 5$

25. $y > -3$

26. $x \leq 5$

27. $-4 < y < -1$
  (*Hint*: Think of this as $-4 < y$ and $y < -1$.)

28. $-3 \leq x \leq 3$
  (*Hint*: Think of this as $-3 \leq x$ and $x \leq 3$.)

29. $y \geq |x|$

30. $y \leq |x + 2|$

In Exercises 31–36, match the system of inequalities with one of the graphs (a)–(f), which follow.

a)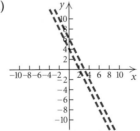

b)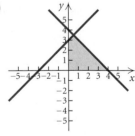

c)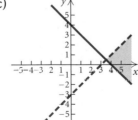

d)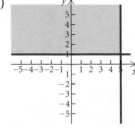

e)

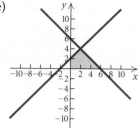

f)

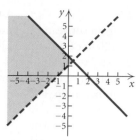

31. $y > x + 1,$
$\quad y \leq 2 - x$

32. $y < x - 3,$
$\quad y \geq 4 - x$

33. $2x + y < 4,$
$\quad 4x + 2y > 12$

34. $x \leq 5,$
$\quad y \geq 1$

35. $x + y \leq 4,$
$\quad x - y \geq -3,$
$\quad x \geq 0,$
$\quad y \geq 0$

36. $x - y \geq -2,$
$\quad x + y \leq 6,$
$\quad x \geq 0,$
$\quad y \geq 0$

Find a system of inequalities with the given graph. Answers may vary.

37.

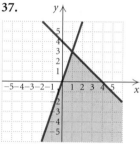

38.

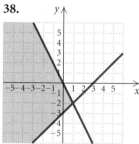

39.

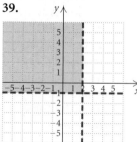

40.

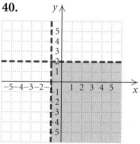

41.

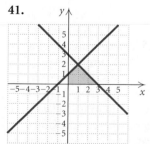

42.

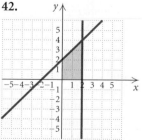

Graph the system of inequalities. Then find the coordinates of the vertices.

43. $y \leq x,$
$\quad y \geq 3 - x$

44. $y \geq x,$
$\quad y \leq x - 5$

45. $y \geq x,$
$\quad y \leq x - 4$

46. $y \geq x,$
$\quad y \leq 2 - x$

47. $y \geq -3,$
$\quad x \geq 1$

48. $y \leq -2,$
$\quad x \geq 2$

49. $x \leq 3,$
$\quad y \geq 2 - 3x$

50. $x \geq -2,$
$\quad y \leq 3 - 2x$

51. $x + y \leq 1,$
$\quad x - y \leq 2$

52. $y + 3x \geq 0,$
$\quad y + 3x \leq 2$

53. $2y - x \leq 2,$
$\quad y + 3x \geq -1$

54. $\quad y \leq 2x + 1,$
$\quad y \geq -2x + 1,$
$\quad x - 2 \leq 0$

55. $\quad x - y \leq 2,$
$\quad x + 2y \geq 8,$
$\quad y - 4 \leq 0$

56. $x + 2y \leq 12,$
$\quad 2x + y \leq 12,$
$\quad x \geq 0,$
$\quad y \geq 0$

57. $4y - 3x \geq -12,$
$\quad 4y + 3x \geq -36,$
$\quad y \leq 0,$
$\quad x \leq 0$

58. $8x + 5y \leq 40,$
$\quad x + 2y \leq 8,$
$\quad x \geq 0,$
$\quad y \geq 0$

59. $3x + 4y \geq 12,$
$\quad 5x + 6y \leq 30,$
$\quad 1 \leq x \leq 3$

60. $y - x \geq 1,$
$\quad y - x \leq 3,$
$\quad 2 \leq x \leq 5$

Find the maximum and the minimum values of the function and the values of x and y for which they occur.

61. $P = 17x - 3y + 60,$ subject to
$$6x + 8y \leq 48,$$
$$0 \leq y \leq 4,$$
$$0 \leq x \leq 7.$$

62. $Q = 28x - 4y + 72,$ subject to
$$5x + 4y \geq 20,$$
$$0 \leq y \leq 4,$$
$$0 \leq x \leq 3.$$

63. $F = 5x + 36y,$ subject to
$$5x + 3y \leq 34,$$
$$3x + 5y \leq 30,$$
$$x \geq 0,$$
$$y \geq 0.$$

64. $G = 16x + 14y$, subject to

$$3x + 2y \le 12,$$
$$7x + 5y \le 29,$$
$$x \ge 0,$$
$$y \ge 0.$$

65. *Maximizing Income.* Golden Harvest Foods makes jumbo biscuits and regular biscuits. The oven can cook at most 200 biscuits per day. Each jumbo biscuit requires 2 oz of flour, each regular biscuit requires 1 oz of flour, and there is 300 oz of flour available. The income from each jumbo biscuit is $0.10 and from each regular biscuit is $0.08. How many of each size biscuit should be made in order to maximize income? What is the maximum income?

66. *Maximizing Mileage.* Omar owns a car and a moped. He can afford 12 gal of gasoline to be split between the car and the moped. Omar's car gets 20 mpg and, with the fuel currently in the tank, can hold at most an additional 10 gal of gas. His moped gets 100 mpg and can hold at most 3 gal of gas. How many gallons of gasoline should each vehicle use if Omar wants to travel as far as possible? What is the maximum number of miles that he can travel?

20 mpg

100 mpg

67. *Maximizing Profit.* Norris Mill can convert logs into lumber and plywood. In a given week, the mill can turn out 400 units of production, of which 100 units of lumber and 150 units of plywood are required by regular customers. The profit is $20 per unit of lumber and $30 per unit of plywood. How many units of each should the mill produce in order to maximize the profit?

68. *Maximizing Profit.* Sunnydale Farm includes 240 acres of cropland. The farm owner wishes to plant this acreage in corn and oats. The profit per acre in corn production is $40 and in oats is $30. A total of 320 hr of labor is available. Each acre of corn requires 2 hr of labor, whereas each acre of oats requires 1 hr of labor. How should the land be divided between corn and oats in order to yield the maximum profit? What is the maximum profit?

69. *Minimizing Cost.* An animal feed to be mixed from soybean meal and oats must contain at least 120 lb of protein, 24 lb of fat, and 10 lb of mineral ash. Each 100-lb sack of soybean meal costs $15 and contains 50 lb of protein, 8 lb of fat, and 5 lb of mineral ash. Each 100-lb sack of oats costs $5 and contains 15 lb of protein, 5 lb of fat, and 1 lb of mineral ash. How many sacks of each should be used to satisfy the minimum requirements at minimum cost?

70. *Minimizing Cost.* Suppose that in the preceding problem the oats were replaced by alfalfa, which costs $8 per 100 lb and contains 20 lb of protein, 6 lb of fat, and 8 lb of mineral ash. How much of each is now required in order to minimize the cost?

71. *Maximizing Income.* Clayton is planning to invest up to $40,000 in corporate and municipal bonds. The least he is allowed to invest in corporate bonds is $6000, and he does not want to invest more than $22,000 in corporate bonds. He also does not want to invest more than $30,000 in municipal bonds. The interest is 8% on corporate bonds and $7\frac{1}{2}\%$ on municipal bonds. This is simple interest for one year. How much should he invest in each type of bond in order to maximize his income? What is the maximum income?

72. *Maximizing Income.* Margaret is planning to invest up to $22,000 in certificates of deposit at City Bank and People's Bank. She wants to invest at least $2000 but no more than $14,000 at City Bank. People's Bank does not insure more than a $15,000 investment, so she will invest no more than that in People's Bank. The interest is 6% at City Bank and $6\frac{1}{2}\%$ at People's Bank. This is simple interest for one year. How much should she invest in each bank in order to maximize her income? What is the maximum income?

73. *Minimizing Transportation Cost.* An airline with two types of airplanes, P_1 and P_2, has contracted with a tour group to provide transportation for a minimum of 2000 first-class, 1500 tourist-class, and 2400 economy-class passengers. For a certain trip, airplane P_1 costs $12 thousand to operate and can accommodate 40 first-class, 40 tourist-class, and 120 economy-class passengers, whereas airplane P_2 costs $10 thousand to operate and can accommodate 80 first-class, 30 tourist-class, and 40 economy-class passengers. How many of each type of airplane should be used in order to minimize the operating cost?

74. *Minimizing Transportation Cost.* Suppose that in the preceding problem a new airplane P_3 becomes available, having an operating cost for the same trip of $15 thousand and accommodating 40 first-class, 40 tourist-class, and 80 economy-class passengers. If airplane P_1 were replaced by airplane P_3, how many of P_2 and P_3 should be used in order to minimize the operating cost?

75. *Maximizing Profit.* It takes Just Sew 2 hr of cutting and 4 hr of sewing to make a knit suit. It takes 4 hr of cutting and 2 hr of sewing to make a worsted suit. At most 20 hr per day are available for cutting and at most 16 hr per day are available for sewing. The profit is $34 on a knit suit and $31 on a worsted suit. How many of each kind of suit should be made each day in order to maximize profit? What is the maximum profit?

76. *Maximizing Profit.* Cambridge Metal Works manufactures two sizes of gears. The smaller gear requires 4 hr of machining and 1 hr of polishing and yields a profit of $25. The larger gear requires 1 hr of machining and 1 hr of polishing and yields a profit of $10. The firm has available at most 24 hr per day for machining and 9 hr per day for polishing. How many of each type of gear should be produced each day in order to maximize profit? What is the maximum profit?

77. *Minimizing Nutrition Cost.* Suppose that it takes 12 units of carbohydrates and 6 units of protein to satisfy Jacob's minimum weekly requirements. A particular type of meat contains 2 units of carbohydrates and 2 units of protein per pound. A

particular cheese contains 3 units of carbohydrates and 1 unit of protein per pound. The meat costs $3.50 per pound and the cheese costs $4.60 per pound. How many pounds of each are needed in order to minimize the cost and still meet the minimum requirements?

78. *Minimizing Salary Cost.* The Spring Hill school board is analyzing education costs for Hill Top School. It wants to hire teachers and teacher's aides to make up a faculty that satisfies its needs at minimum cost. The average annual salary for a teacher is $35,000 and for a teacher's aide is $18,000. The school building can accommodate a faculty of no more than 50 but needs at least 20 faculty members to function properly. The school must have at least 12 aides, but the number of teachers must be at least twice the number of aides in order to accommodate the expectations of the community. How many teachers and teacher's aides should be hired in order to minimize salary costs?

79. *Maximizing Animal Support in a Forest.* A certain area of forest is populated by two species of animals, which scientists refer to as A and B for simplicity. The forest supplies two kinds of food, referred to as F_1 and F_2. For one year, each member of species A requires 1 unit of F_1 and 0.5 unit of F_2. Each member of species B requires 0.2 unit of F_1 and 1 unit of F_2. The forest can normally supply at most 600 units of F_1 and 525 units of F_2 per year. What is the maximum total number of these animals that the forest can support?

80. *Maximizing Animal Support in a Forest.* Refer to Exercise 79. If there is a wet spring, then supplies of food increase to 1080 units of F_1 and 810 units of F_2. In this case, what is the maximum total number of these animals that the forest can support?

Technology Connection

Use a graphing calculator to do the following.

81. Exercises 9, 13, and 25

82. Exercises 10, 14, and 24

Collaborative Discussion and Writing

83. Write an applied linear programming problem for a classmate to solve. Devise your problem so that the answer is "The bakery will make a maximum profit when 5 dozen pies and 12 dozen cookies are baked."

84. Describe how the graph of a linear inequality differs from the graph of a linear equation.

Skill Maintenance

Solve.

85. $-5 \le x + 2 < 4$

86. $|x - 3| \ge 2$

87. $x^2 - 2x \le 3$

88. $\dfrac{x - 1}{x + 2} > 4$

Synthesis

Graph the system of inequalities.

89. $y \ge x^2 - 2,$
$\quad y \le 2 - x^2$

90. $y < x + 1,$
$\quad y \ge x^2$

Graph the inequality.

91. $|x + y| \le 1$

92. $|x| + |y| \le 1$

93. $|x| > |y|$

94. $|x - y| > 0$

95. *Allocation of Resources.* Significant Sounds manufactures two types of speaker assemblies. The less expensive assembly, which sells for $350, consists of one midrange speaker and one tweeter. The more expensive speaker assembly, which sells for $600, consists of one woofer, one midrange speaker, and two tweeters. The manufacturer has in stock 44 woofers, 60 midrange speakers, and 90 tweeters. How many of each type of speaker assembly should be made in order to maximize income? What is the maximum income?

96. *Allocation of Resources.* Sitting Pretty Furniture produces chairs and sofas. The chairs require 20 ft of wood, 1 lb of foam rubber, and 2 yd² of fabric. The sofas require 100 ft of wood, 50 lb of foam rubber, and 20 yd² of fabric. The manufacturer has in stock 1900 ft of wood, 500 lb of foam rubber, and 240 yd² of fabric. The chairs can be sold for $80 each and the sofas for $300 each. How many of each should be produced in order to maximize income? What is the maximum income?

5.8

Partial Fractions

• *Decompose rational expressions into partial fractions.*

There are situations in calculus in which it is useful to write a rational expression as a sum of two or more simpler rational expressions. For example, in the equation

$$\frac{4x - 13}{2x^2 + x - 6} = \frac{3}{x + 2} + \frac{-2}{2x - 3},$$

each fraction on the right side is called a **partial fraction.** The expression on the right side is the **partial fraction decomposition** of the rational expression on the left side. In this section, we learn how such decompositions are created.

Partial Fraction Decompositions

The procedure for finding the partial fraction decomposition of a rational expression involves factoring its denominator into linear and quadratic factors.

Procedure for Decomposing a Rational Expression into Partial Fractions

Consider any rational expression $P(x)/Q(x)$ such that $P(x)$ and $Q(x)$ have no common factor other than 1 or -1.

1. If the degree of $P(x)$ is greater than or equal to the degree of $Q(x)$, divide to express $P(x)/Q(x)$ as a quotient + remainder/$Q(x)$ and follow steps (2)–(5) to decompose the resulting rational expression.

2. If the degree of $P(x)$ is less than the degree of $Q(x)$, factor $Q(x)$ into linear factors of the form $(px + q)^n$ and/or quadratic factors of the form $(ax^2 + bx + c)^m$. Any quadratic factor $ax^2 + bx + c$ must be *irreducible*, meaning that it cannot be factored into linear factors with rational coefficients.

3. Assign to each linear factor $(px + q)^n$ the sum of n partial fractions:

$$\frac{A_1}{px + q} + \frac{A_2}{(px + q)^2} + \cdots + \frac{A_n}{(px + q)^n}.$$

4. Assign to each quadratic factor $(ax^2 + bx + c)^m$ the sum of m partial fractions:

$$\frac{B_1 x + C_1}{ax^2 + bx + c} + \frac{B_2 x + C_2}{(ax^2 + bx + c)^2} + \cdots + \frac{B_m x + C_m}{(ax^2 + bx + c)^m}.$$

5. Apply algebraic methods, as illustrated in the following examples, to find the constants in the numerators of the partial fractions.

EXAMPLE 1 Decompose into partial fractions:

$$\frac{4x - 13}{2x^2 + x - 6}.$$

Solution The degree of the numerator is less than the degree of the denominator. We begin by factoring the denominator: $(x + 2)(2x - 3)$. We know that there are constants A and B such that

$$\frac{4x - 13}{(x + 2)(2x - 3)} = \frac{A}{x + 2} + \frac{B}{2x - 3}.$$

To determine A and B, we add the expressions on the right:

$$\frac{4x - 13}{(x + 2)(2x - 3)} = \frac{A(2x - 3) + B(x + 2)}{(x + 2)(2x - 3)}.$$

Next, we equate the numerators:

$$4x - 13 = A(2x - 3) + B(x + 2).$$

Since the last equation containing A and B is true for all x, we can substitute any value of x and still have a true equation. If we choose $x = \frac{3}{2}$, then $2x - 3 = 0$ and A will be eliminated when we make the substitution. This gives us

$$4\left(\tfrac{3}{2}\right) - 13 = A\left(2 \cdot \tfrac{3}{2} - 3\right) + B\left(\tfrac{3}{2} + 2\right)$$
$$-7 = 0 + \tfrac{7}{2}B.$$

Solving, we obtain $B = -2$.

If we choose $x = -2$, then $x + 2 = 0$ and B will be eliminated when we make the substitution. This gives us

$$4(-2) - 13 = A[2(-2) - 3] + B(-2 + 2)$$
$$-21 = -7A + 0.$$

Solving, we obtain $A = 3$.

The decomposition is as follows:

$$\frac{4x - 13}{2x^2 + x - 6} = \frac{3}{x + 2} + \frac{-2}{2x - 3}, \quad \text{or} \quad \frac{3}{x + 2} - \frac{2}{2x - 3}.$$

To check, we can add to see if we get the expression on the left.

EXAMPLE 2 Decompose into partial fractions:

$$\frac{7x^2 - 29x + 24}{(2x - 1)(x - 2)^2}.$$

Solution The degree of the numerator is 2 and the degree of the denominator is 3, so the degree of the numerator is less than the degree of the denominator. The denominator is given in factored form. The decomposition has the following form:

$$\frac{7x^2 - 29x + 24}{(2x - 1)(x - 2)^2} = \frac{A}{2x - 1} + \frac{B}{x - 2} + \frac{C}{(x - 2)^2}.$$

As in Example 1, we add the expressions on the right:

$$\frac{7x^2 - 29x + 24}{(2x - 1)(x - 2)^2} = \frac{A(x - 2)^2 + B(2x - 1)(x - 2) + C(2x - 1)}{(2x - 1)(x - 2)^2}.$$

Then we equate the numerators. This gives us

$$7x^2 - 29x + 24 = A(x - 2)^2 + B(2x - 1)(x - 2) + C(2x - 1).$$

Since the equation containing A, B, and C is true for all x, we can substitute any value of x and still have a true equation. In order to have $2x - 1 = 0$, we let $x = \frac{1}{2}$. This gives us

$$7\left(\tfrac{1}{2}\right)^2 - 29 \cdot \tfrac{1}{2} + 24 = A\left(\tfrac{1}{2} - 2\right)^2 + 0$$
$$\tfrac{45}{4} = \tfrac{9}{4}A.$$

Solving, we obtain $A = 5$.

   In order to have $x - 2 = 0$, we let $x = 2$. Substituting gives us

$$7(2)^2 - 29(2) + 24 = 0 + C(2 \cdot 2 - 1)$$
$$-6 = 3C.$$

Solving, we obtain $C = -2$.

   To find B, we choose any value for x except $\frac{1}{2}$ or 2 and replace A with 5 and C with -2. We let $x = 1$:

$$7 \cdot 1^2 - 29 \cdot 1 + 24 = 5(1 - 2)^2 + B(2 \cdot 1 - 1)(1 - 2)$$
$$+ (-2)(2 \cdot 1 - 1)$$
$$2 = 5 - B - 2$$
$$B = 1.$$

The decomposition is as follows:

$$\frac{7x^2 - 29x + 24}{(2x - 1)(x - 2)^2} = \frac{5}{2x - 1} + \frac{1}{x - 2} - \frac{2}{(x - 2)^2}.$$

EXAMPLE 3   Decompose into partial fractions:

$$\frac{6x^3 + 5x^2 - 7}{3x^2 - 2x - 1}.$$

Solution   The degree of the numerator is greater than that of the denominator. Thus we divide and find an equivalent expression:

$$
\begin{array}{r}
2x + 3 \\
3x^2 - 2x - 1 \overline{)6x^3 + 5x^2 - 7} \\
\underline{6x^3 - 4x^2 - 2x } \\
9x^2 + 2x - 7 \\
\underline{9x^2 - 6x - 3} \\
8x - 4
\end{array}
$$

The original expression is thus equivalent to

$$2x + 3 + \frac{8x - 4}{3x^2 - 2x - 1}.$$

We decompose the fraction to get

$$\frac{8x - 4}{(3x + 1)(x - 1)} = \frac{5}{3x + 1} + \frac{1}{x - 1}.$$

The final result is

$$2x + 3 + \frac{5}{3x + 1} + \frac{1}{x - 1}.$$

Systems of equations can be used to decompose rational expressions. Let's reconsider Example 2.

EXAMPLE 4 Decompose into partial fractions:

$$\frac{7x^2 - 29x + 24}{(2x - 1)(x - 2)^2}.$$

Solution The decomposition has the following form:

$$\frac{A}{2x - 1} + \frac{B}{x - 2} + \frac{C}{(x - 2)^2}.$$

We first add as in Example 2:

$$\frac{7x^2 - 29x + 24}{(2x - 1)(x - 2)^2} = \frac{A}{2x - 1} + \frac{B}{x - 2} + \frac{C}{(x - 2)^2}$$

$$= \frac{A(x - 2)^2 + B(2x - 1)(x - 2) + C(2x - 1)}{(2x - 1)(x - 2)^2}.$$

Then we equate numerators:

$$7x^2 - 29x + 24$$
$$= A(x - 2)^2 + B(2x - 1)(x - 2) + C(2x - 1)$$
$$= A(x^2 - 4x + 4) + B(2x^2 - 5x + 2) + C(2x - 1)$$
$$= Ax^2 - 4Ax + 4A + 2Bx^2 - 5Bx + 2B + 2Cx - C,$$

or

$$7x^2 - 29x + 24$$
$$= (A + 2B)x^2 + (-4A - 5B + 2C)x + (4A + 2B - C).$$

Next, we equate corresponding coefficients:

$7 = A + 2B,$	The coefficients of the x^2-terms must be the same.
$-29 = -4A - 5B + 2C,$	The coefficients of the x-terms must be the same.
$24 = 4A + 2B - C.$	The constant terms must be the same.

We now have a system of three equations. You should confirm that the solution of the system is

$$A = 5, \qquad B = 1, \quad \text{and} \quad C = -2.$$

The decomposition is as follows:

$$\frac{7x^2 - 29x + 24}{(2x - 1)(x - 2)^2} = \frac{5}{2x - 1} + \frac{1}{x - 2} - \frac{2}{(x - 2)^2}.$$

—

EXAMPLE 5 Decompose into partial fractions:

$$\frac{11x^2 - 8x - 7}{(2x^2 - 1)(x - 3)}.$$

Study Tip

The review icons in the text margins provide references to earlier sections in which you can find content related to the concept at hand. Reviewing this earlier content will add to your understanding of the current concept.

SYSTEMS OF EQUATIONS
IN THREE VARIABLES

REVIEW SECTION 5.2 OR 5.5.

Solution The decomposition has the following form:

$$\frac{11x^2 - 8x - 7}{(2x^2 - 1)(x - 3)} = \frac{Ax + B}{2x^2 - 1} + \frac{C}{x - 3}.$$

Adding and equating numerators, we get

$$11x^2 - 8x - 7 = (Ax + B)(x - 3) + C(2x^2 - 1)$$
$$= Ax^2 - 3Ax + Bx - 3B + 2Cx^2 - C,$$

or $\quad 11x^2 - 8x - 7 = (A + 2C)x^2 + (-3A + B)x + (-3B - C).$

We then equate corresponding coefficients:

$$11 = A + 2C, \qquad \text{The coefficients of the } x^2\text{-terms}$$
$$-8 = -3A + B, \qquad \text{The coefficients of the } x\text{-terms}$$
$$-7 = -3B - C. \qquad \text{The constant terms}$$

We solve this system of three equations and obtain

$$A = 3, \qquad B = 1, \quad \text{and} \quad C = 4.$$

The decomposition is as follows:

$$\frac{11x^2 - 8x - 7}{(2x^2 - 1)(x - 3)} = \frac{3x + 1}{2x^2 - 1} + \frac{4}{x - 3}.$$

5.8

Exercise Set

Decompose into partial fractions.

1. $\dfrac{x + 7}{(x - 3)(x + 2)}$

2. $\dfrac{2x}{(x + 1)(x - 1)}$

3. $\dfrac{7x - 1}{6x^2 - 5x + 1}$

4. $\dfrac{13x + 46}{12x^2 - 11x - 15}$

5. $\dfrac{3x^2 - 11x - 26}{(x^2 - 4)(x + 1)}$

6. $\dfrac{5x^2 + 9x - 56}{(x - 4)(x - 2)(x + 1)}$

7. $\dfrac{9}{(x + 2)^2(x - 1)}$

8. $\dfrac{x^2 - x - 4}{(x - 2)^3}$

9. $\dfrac{2x^2 + 3x + 1}{(x^2 - 1)(2x - 1)}$

10. $\dfrac{x^2 - 10x + 13}{(x^2 - 5x + 6)(x - 1)}$

11. $\dfrac{x^4 - 3x^3 - 3x^2 + 10}{(x + 1)^2(x - 3)}$

12. $\dfrac{10x^3 - 15x^2 - 35x}{x^2 - x - 6}$

13. $\dfrac{-x^2 + 2x - 13}{(x^2 + 2)(x - 1)}$

14. $\dfrac{26x^2 + 208x}{(x^2 + 1)(x + 5)}$

15. $\dfrac{6 + 26x - x^2}{(2x - 1)(x + 2)^2}$

16. $\dfrac{5x^3 + 6x^2 + 5x}{(x^2 - 1)(x + 1)^3}$

17. $\dfrac{6x^3 + 5x^2 + 6x - 2}{2x^2 + x - 1}$

18. $\dfrac{2x^3 + 3x^2 - 11x - 10}{x^2 + 2x - 3}$

19. $\dfrac{2x^2 - 11x + 5}{(x - 3)(x^2 + 2x - 5)}$

20. $\dfrac{3x^2 - 3x - 8}{(x - 5)(x^2 + x - 4)}$

21. $\dfrac{-4x^2 - 2x + 10}{(3x + 5)(x + 1)^2}$

22. $\dfrac{26x^2 - 36x + 22}{(x - 4)(2x - 1)^2}$

23. $\dfrac{36x + 1}{12x^2 - 7x - 10}$

24. $\dfrac{-17x + 61}{6x^2 + 39x - 21}$

25. $\dfrac{-4x^2 - 9x + 8}{(3x^2 + 1)(x - 2)}$

26. $\dfrac{11x^2 - 39x + 16}{(x^2 + 4)(x - 8)}$

Collaborative Discussion and Writing

27. Describe the two methods used to find the constants in a partial fraction decomposition.

28. What would you say to a classmate who tells you that the partial fraction decomposition of

$$\frac{3x^2 - 8x + 9}{(x + 3)(x^2 - 5x + 6)}$$

is

$$\frac{2}{x + 3} + \frac{x - 1}{x^2 - 5x + 6}?$$

Explain.

29. Explain the error in the following process.

$$\frac{x^2 + 4}{(x + 2)(x + 1)} = \frac{A}{x + 2} + \frac{B}{x + 1}$$

$$= \frac{A(x + 1) + B(x + 2)}{(x + 2)(x + 1)}$$

Then

$$x^2 + 4 = A(x + 1) + B(x + 2).$$

When $x = -1$:

$$(-1)^2 + 4 = A(-1 + 1) + B(-1 + 2)$$
$$5 = B.$$

When $x = -2$:

$$(-2)^2 + 4 = A(-2 + 1) + B(-2 + 2)$$
$$8 = -A$$
$$-8 = A.$$

Thus,

$$\frac{x^2 + 4}{(x + 2)(x + 1)} = \frac{-8}{x + 2} + \frac{5}{x + 1}.$$

Skill Maintenance

Find the zeros of the polynomial function.

30. $f(x) = x^3 - 3x^2 + x - 3$

31. $f(x) = x^3 + x^2 - 3x - 2$

32. $f(x) = x^4 - x^3 - 5x^2 - x - 6$

33. $f(x) = x^3 + 5x^2 + 5x - 3$

Synthesis

Decompose into partial fractions.

34. $\dfrac{9x^3 - 24x^2 + 48x}{(x - 2)^4(x + 1)}$

[*Hint*: Let the expression equal

$$\frac{A}{x + 1} + \frac{P(x)}{(x - 2)^4}$$

and find $P(x)$].

35. $\dfrac{x}{x^4 - a^4}$

36. $\dfrac{1}{e^{-x} + 3 + 2e^x}$

37. $\dfrac{1 + \ln x^2}{(\ln x + 2)(\ln x - 3)^2}$

Chapter 5 Summary and Review

Important Properties and Formulas

Row-Equivalent Operations

1. Interchange any two rows.
2. Multiply each entry in a row by the same nonzero constant.
3. Add a nonzero multiple of one row to another row.

Row-Echelon Form

1. If a row does not consist entirely of 0's, then the first nonzero element in the row is a 1 (called a leading 1).
2. For any two successive nonzero rows, the leading 1 in the lower row is farther to the right than the leading 1 in the higher row.

(continued)

3. All the rows consisting entirely of 0's are at the bottom of the matrix.

If a fourth property is also satisfied, a matrix is said to be in reduced row-echelon form:

4. Each column that contains a leading 1 has 0's everywhere else.

Properties of Matrix Addition and Scalar Multiplication

For matrices **A**, **B**, and **C** and any scalars k and l, assuming that the indicated operation is possible:

Commutative Property of Addition:

$$\mathbf{A} + \mathbf{B} = \mathbf{B} + \mathbf{A}.$$

Associative Property of Addition:

$$\mathbf{A} + (\mathbf{B} + \mathbf{C}) = (\mathbf{A} + \mathbf{B}) + \mathbf{C}.$$

Associative Property of Scalar Multiplication:

$$(kl)\mathbf{A} = k(l\mathbf{A}).$$

Additive Identity Property:
There exists a unique matrix **0** such that
$\mathbf{A} + \mathbf{0} = \mathbf{0} + \mathbf{A} = \mathbf{A}.$

Additive Inverse Property:
There exists a unique matrix $-\mathbf{A}$ such that
$\mathbf{A} + (-\mathbf{A}) = -\mathbf{A} + \mathbf{A} = \mathbf{0}.$

Distributive Properties:

$$k(\mathbf{A} + \mathbf{B}) = k\mathbf{A} + k\mathbf{B},$$
$$(k + l)\mathbf{A} = k\mathbf{A} + l\mathbf{A}.$$

Properties of Matrix Multiplication

For matrices **A**, **B**, and **C**, assuming that the indicated operation is possible:

Associative Property of Multiplication:

$$\mathbf{A}(\mathbf{BC}) = (\mathbf{AB})\mathbf{C}.$$

Distributive Properties:

$$\mathbf{A}(\mathbf{B} + \mathbf{C}) = \mathbf{AB} + \mathbf{AC},$$
$$(\mathbf{B} + \mathbf{C})\mathbf{A} = \mathbf{BA} + \mathbf{CA}.$$

Matrix Solutions of Systems of Equations

For a system of n linear equations in n variables, $\mathbf{AX} = \mathbf{B}$, if **A** is an invertible matrix, then the unique solution of the system is given by $\mathbf{X} = \mathbf{A}^{-1}\mathbf{B}.$

Determinant of a 2 × 2 Matrix

The **determinant** of the matrix $\begin{bmatrix} a & c \\ b & d \end{bmatrix}$ is

denoted $\begin{vmatrix} a & c \\ b & d \end{vmatrix}$ and is defined as

$$\begin{vmatrix} a & c \\ b & d \end{vmatrix} = ad - bc.$$

Determinant of Any Square Matrix

For any square matrix **A** of order $n \times n$ $(n > 1)$, we define the **determinant** of **A**, denoted $|\mathbf{A}|$, as follows. Choose any row or column. Multiply each element in that row or column by its cofactor and add the results. The determinant of a 1×1 matrix is simply the element of the matrix. The value of a determinant will be the same, no matter which row or column is chosen.

Cramer's Rule for 2 × 2 Systems

The solution of the system of equations

$$a_1 x + b_1 y = c_1,$$
$$a_2 x + b_2 y = c_2$$

is given by

$$x = \frac{D_x}{D}, \qquad y = \frac{D_y}{D},$$

where

$$D = \begin{vmatrix} a_1 & b_1 \\ a_2 & b_2 \end{vmatrix}, \qquad D_x = \begin{vmatrix} c_1 & b_1 \\ c_2 & b_2 \end{vmatrix},$$

$$D_y = \begin{vmatrix} a_1 & c_1 \\ a_2 & c_2 \end{vmatrix}, \quad \text{and} \quad D \neq 0.$$

Cramer's Rule for 3×3 Systems

The solution of the system of equations

$$a_1 x + b_1 y + c_1 z = d_1,$$
$$a_2 x + b_2 y + c_2 z = d_2,$$
$$a_3 x + b_3 y + c_3 z = d_3$$

is given by

$$x = \frac{D_x}{D}, \qquad y = \frac{D_y}{D}, \qquad z = \frac{D_z}{D},$$

where

$$D = \begin{vmatrix} a_1 & b_1 & c_1 \\ a_2 & b_2 & c_2 \\ a_3 & b_3 & c_3 \end{vmatrix}, \qquad D_x = \begin{vmatrix} d_1 & b_1 & c_1 \\ d_2 & b_2 & c_2 \\ d_3 & b_3 & c_3 \end{vmatrix},$$

$$D_y = \begin{vmatrix} a_1 & d_1 & c_1 \\ a_2 & d_2 & c_2 \\ a_3 & d_3 & c_3 \end{vmatrix}, \qquad D_z = \begin{vmatrix} a_1 & b_1 & d_1 \\ a_2 & b_2 & d_2 \\ a_3 & b_3 & d_3 \end{vmatrix},$$

and $D \neq 0$.

To Graph a Linear Inequality in Two Variables:

1. Replace the inequality symbol with an equals sign and graph this related equation. If the inequality symbol is $<$ or $>$, draw the line dashed. If the inequality symbol is $\leq$ or $\geq$, draw the line solid.
2. The graph consists of a half-plane on one side of the line and, if the line is solid, the line as well. To determine which half-plane to shade, test a point not on the line in the original inequality. If that point is a solution, shade the half-plane containing that point. If not, shade the opposite half-plane.

Linear Programming Procedure

To find the maximum or minimum value of a linear objective function subject to a set of constraints:

1. Graph the region of feasible solutions.
2. Determine the coordinates of the vertices of the region.

3. Evaluate the objective function at each vertex. The largest and smallest of those values are the maximum and minimum values of the function, respectively.

Procedure for Decomposing a Rational Expression into Partial Fractions

Consider any rational expression $P(x)/Q(x)$ such that $P(x)$ and $Q(x)$ have no common factor other than 1 or -1.

1. If the degree of $P(x)$ is greater than or equal to the degree of $Q(x)$, divide to express $P(x)/Q(x)$ as a quotient $+$ remainder$/Q(x)$ and follow steps (2)–(5) to decompose the resulting rational expression.
2. If the degree of $P(x)$ is less than the degree of $Q(x)$, factor $Q(x)$ into linear factors of the form $(px + q)^n$ and/or quadratic factors of the form $(ax^2 + bx + c)^m$. Any quadratic factor $ax^2 + bx + c$ must be irreducible, meaning that it cannot be factored into linear factors with rational coefficients.
3. Assign to each linear factor $(px + q)^n$ the sum of n partial fractions:

$$\frac{A_1}{px + q} + \frac{A_2}{(px + q)^2} + \cdots + \frac{A_n}{(px + q)^n}.$$

4. Assign to each quadratic factor $(ax^2 + bx + c)^m$ the sum of m partial fractions:

$$\frac{B_1 x + C_1}{ax^2 + bx + c} + \frac{B_2 x + C_2}{(ax^2 + bx + c)^2} + \cdots +$$
$$\frac{B_m x + C_m}{(ax^2 + bx + c)^m}.$$

5. Apply algebraic methods to find the constants in the numerators of the partial fractions.

Review Exercises

In Exercises 1–8, match the equations or inequalities with one of the graphs (a)–(h), which follow.

a)

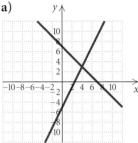

b)

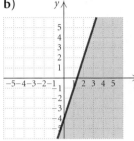

c)

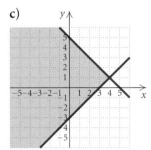

d)

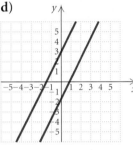

e)

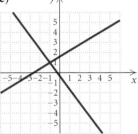

f)

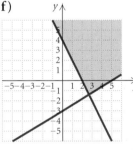

g)

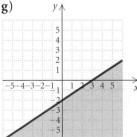

h)

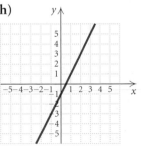

1. $x + y = 7,$
$2x - y = 5$

2. $3x - 5y = -8,$
$4x + 3y = -1$

3. $y = 2x - 1,$
$4x - 2y = 2$

4. $6x - 3y = 5,$
$y = 2x + 3$

5. $y \leq 3x - 4$

6. $2x - 3y \geq 6$

7. $x - y \leq 3,$
$x + y \leq 5$

8. $2x + y \geq 4,$
$3x - 5y \leq 15$

Solve.

9. $5x - 3y = -4,$
$3x - y = -4$

10. $2x + 3y = 2,$
$5x - y = -29$

11. $x + 5y = 12,$
$5x + 25y = 12$

12. $x + y = -2,$
$-3x - 3y = 6$

13. $2x - 4y + 3z = -3,$
$-5x + 2y - z = 7,$
$3x + 2y - 2z = 4$

14. $x + 5y + 3z = 0,$
$3x - 2y + 4z = 0,$
$2x + 3y - z = 0$

15. $x - y = 5,$
$y - z = 6,$
$z - w = 7,$
$x + w = 8$

16. Classify each of the systems in Exercises 9–15 as consistent or inconsistent.

17. Classify each of the systems in Exercises 9–15 as dependent or independent.

Solve the system of equations using Gaussian elimination or Gauss–Jordan elimination.

18. $x + 2y = 5,$
$2x - 5y = -8$

19. $3x + 4y + 2z = 3,$
$5x - 2y - 13z = 3,$
$4x + 3y - 3z = 6$

20. $3x + 5y + z = 0,$
$2x - 4y - 3z = 0,$
$x + 3y + z = 0$

21. $w + x + y + z = -2,$
$-3w - 2x + 3y + 2z = 10,$
$2w + 3x + 2y - z = -12,$
$2w + 4x - y + z = 1$

22. *Coins.*   The value of 75 coins, consisting of nickels and dimes, is $5.95. How many of each kind are there?

23. *Investment.*   The Mendez family invested $5000, part at 3% and the remainder at 3.5%. The annual income from both investments is $167. What is the amount invested at each rate?

24. *Nutrition.*   A dietician must plan a breakfast menu that provides 460 Cal, 9 g of fat, and 55 mg of calcium. One plain bagel contains 200 Cal, 2 g of fat, and 29 mg of calcium. A one-tablespoon serving of cream cheese contains 100 Cal, 10 g of fat, and 24 mg of calcium. One banana contains 105 Cal, 1 g of fat, and 7 g of calcium. (*Source: Home and Garden Bulletin No. 72*, U.S. Government Printing Office, Washington D.C. 20402) How many servings of each are required to provide the desired nutritional values?

25. *Test Scores.*   A student has a total of 225 on three tests. The sum of the scores on the first and second tests exceeds the score on the third test by 61. The first score exceeds the second by 6. Find the three scores.

26. *Lowfat Ice Cream Consumption.*   The table below shows the per capita consumption of lowfat ice cream, in pounds, in the United States, represented as years since 1995.

Year, x	Lowfat Ice Cream Consumption (in pounds)
1995, 0	7.4
1998, 3	8.1
2000, 5	7.3

Source: U.S. Department of Agriculture, Economic Research Service, *Food Consumption, Prices, and Expenditures*

a) Use a system of equations to fit a quadratic function $f(x) = ax^2 + bx + c$ to the data.
b) Use the function to estimate the per capita consumption of lowfat ice cream in 2003.

For Exercises 27–34, let

$$A = \begin{bmatrix} 1 & -1 & 0 \\ 2 & 3 & -2 \\ -2 & 0 & 1 \end{bmatrix},$$

$$B = \begin{bmatrix} -1 & 0 & 6 \\ 1 & -2 & 0 \\ 0 & 1 & -3 \end{bmatrix},$$

and

$$C = \begin{bmatrix} -2 & 0 \\ 1 & 3 \end{bmatrix}.$$

Find each of the following, if possible.

27. $A + B$ **28.** $-3A$

29. $-A$ **30.** AB

31. $B + C$ **32.** $A - B$

33. BA **34.** $A + 3B$

35. *Food Service Management.*   The table below shows the cost per serving, in cents, for items on four menus that are served at an elder-care facility.

Menu	Meat	Potato	Vegetable	Salad	Dessert
1	46.1	5.9	10.1	8.5	11.4
2	54.6	4.6	9.6	7.6	10.6
3	48.9	5.5	12.7	9.4	9.3
4	51.3	4.8	11.3	6.9	12.7

On a particular day, a dietician orders 32 meals from menu 1, 19 from menu 2, 43 from menu 3, and 38 from menu 4.

a) Write the information in the table as a 4×5 matrix **M**.
b) Write a row matrix **N** that represents the number of each menu ordered.
c) Find the product **NM**.
d) State what the entries of **NM** represent.

Find $\mathbf{A}^{-1}$, *if it exists.*

36. $\mathbf{A} = \begin{bmatrix} -2 & 0 \\ 1 & 3 \end{bmatrix}$

37. $\mathbf{A} = \begin{bmatrix} 0 & 0 & 3 \\ 0 & -2 & 0 \\ 4 & 0 & 0 \end{bmatrix}$

38. $\mathbf{A} = \begin{bmatrix} 1 & 0 & 0 & 0 \\ 0 & 4 & -5 & 0 \\ 0 & 2 & 2 & 0 \\ 0 & 0 & 0 & 1 \end{bmatrix}$

39. Write a matrix equation equivalent to this system of equations:

$$3x - 2y + 4z = 13,$$
$$x + 5y - 3z = 7,$$
$$2x - 3y + 7z = -8.$$

Solve the system of equations using the inverse of the coefficient matrix of the equivalent matrix equation.

40. $2x + 3y = 5,$
$3x + 5y = 11$

41. $5x - y + 2z = 17,$
$3x + 2y - 3z = -16,$
$4x - 3y - z = 5$

42. $w - x - y + z = -1,$
$2w + 3x - 2y - z = 2,$
$-w + 5x + 4y - 2z = 3,$
$3w - 2x + 5y + 3z = 4$

Evaluate the determinant.

43. $\begin{vmatrix} 1 & -2 \\ 3 & 4 \end{vmatrix}$

44. $\begin{vmatrix} \sqrt{3} & -5 \\ -3 & -\sqrt{3} \end{vmatrix}$

45. $\begin{vmatrix} 2 & -1 & 1 \\ 1 & 2 & -1 \\ 3 & 4 & -3 \end{vmatrix}$

46. $\begin{vmatrix} 1 & -1 & 2 \\ -1 & 2 & 0 \\ -1 & 3 & 1 \end{vmatrix}$

Solve using Cramer's rule.

47. $5x - 2y = 19,$
$7x + 3y = 15$

48. $x + y = 4,$
$4x + 3y = 11$

49. $3x - 2y + z = 5,$
$4x - 5y - z = -1,$
$3x + 2y - z = 4$

50. $2x - y - z = 2,$
$3x + 2y + 2z = 10,$
$x - 5y - 3z = -2$

Graph.

51. $y \leq 3x + 6$

52. $4x - 3y \geq 12$

53. Graph this system of inequalities and find the coordinates of any vertices formed.

$$2x + y \geq 9,$$
$$4x + 3y \geq 23,$$
$$x + 3y \geq 8,$$
$$x \geq 0,$$
$$y \geq 0$$

54. Find the maximum and minimum values of $T = 6x + 10y$ subject to

$$x + y \leq 10,$$
$$5x + 10y \geq 50,$$
$$x \geq 2,$$
$$y \geq 0.$$

55. *Maximizing a Test Score.* Marita is taking a test that contains questions in group A worth 7 points each and questions in group B worth 12 points each. The total number of questions answered must be at least 8. If Marita knows that group A questions take 8 min each and group B questions take 10 min each and the maximum time for the test is 80 min, how many questions from each group must she answer correctly in order to maximize her score? What is the maximum score?

Decompose into partial fractions.

56. $\dfrac{5}{(x + 2)^2(x + 1)}$

57. $\dfrac{-8x + 23}{2x^2 + 5x - 12}$

Collaborative Discussion and Writing

58. Write a problem for a classmate to solve that can be translated to a system of equations. Devise the problem so that the solution is "The caterer sold 20 cheese trays and 35 seafood trays."

59. For square matrices $\mathbf{A}$ and $\mathbf{B}$, is it true, in general, that $(\mathbf{AB})^2 = \mathbf{A}^2\mathbf{B}^2$? Explain.

Synthesis

60. One year, Don invested a total of $40,000, part at 4%, part at 5%, and the rest at $5\frac{1}{2}$%. The total amount of interest received on the investments was $1990. The interest received on the $5\frac{1}{2}$% investment was $590 more than the interest received on the 4% investment. How much was invested at each rate?

Solve.

61. $\dfrac{2}{3x} + \dfrac{4}{5y} = 8,$

$\dfrac{5}{4x} - \dfrac{3}{2y} = -6$

62. $\dfrac{3}{x} - \dfrac{4}{y} + \dfrac{1}{z} = -2,$

$\dfrac{5}{x} + \dfrac{1}{y} - \dfrac{2}{z} = 1,$

$\dfrac{7}{x} + \dfrac{3}{y} + \dfrac{2}{z} = 19$

Graph.

63. $|x| - |y| \leq 1$

64. $|xy| > 1$

Chapter 5 Test

Solve. Use any method. Also determine whether each system is consistent or inconsistent and whether the equations are dependent or independent.

1. $3x + 2y = 1,$
$2x - y = -11$

2. $2x - y = 3,$
$2y = 4x - 6$

3. $x - y = 4,$
$3y = 3x - 8$

4. $2x - 3y = 8,$
$5x - 2y = 9$

Solve.

5. $4x + 2y + z = 4,$
$3x - y + 5z = 4,$
$5x + 3y - 3z = -2$

6. *Ticket Sales.* One evening 750 tickets were sold for Shortridge Community College's spring musical. Tickets cost $3 for students and $5 for nonstudents. Total receipts were $3066. How many of each type of ticket were sold?

7. Tricia, Maria, and Antonio can process 352 telephone orders per day. Tricia and Maria together can process 224 orders per day while Tricia and Antonio together can process 248 orders per day. How many orders can each of them process alone?

For Exercises 8–13, let

$$A = \begin{bmatrix} 1 & -1 & 3 \\ -2 & 5 & 2 \end{bmatrix}, \quad B = \begin{bmatrix} -5 & 1 \\ -2 & 4 \end{bmatrix}, \quad \text{and}$$

$$C = \begin{bmatrix} 3 & -4 \\ -1 & 0 \end{bmatrix}.$$

Find each of the following, if possible.

8. $B + C$

9. $A - C$

10. CB

11. AB

12. $2A$

13. C^{-1}

14. *Food Service Management.* The table below shows the cost per serving, in cents, for items on three lunch menus served at a senior citizens' center.

MENU	MAIN DISH	SIDE DISH	DESSERT
1	49	10	13
2	43	12	11
3	51	8	12

On a particular day, 26 Menu 1 meals, 18 Menu 2 meals, and 23 Menu 3 meals are served.

a) Write the information in the table as a 3 × 3 matrix **M**.

b) Write a row matrix **N** that represents the number of each menu served.

c) Find the product **NM**.

d) State what the entries of **NM** represent.

15. Write a matrix equation equivalent to the system of equations

$$3x - 4y + 2z = -8,$$
$$2x + 3y + z = 7,$$
$$x - 5y - 3z = 3.$$

16. Solve the system of equations using the inverse of the coefficient matrix of the equivalent matrix equation.

$$3x + 2y + 6z = 2,$$
$$x + y + 2z = 1,$$
$$2x + 2y + 5z = 3$$

Evaluate the determinant.

17. $\begin{vmatrix} 3 & -5 \\ 8 & 7 \end{vmatrix}$

18. $\begin{vmatrix} 2 & -1 & 4 \\ -3 & 1 & -2 \\ 5 & 3 & -1 \end{vmatrix}$

19. Solve using Cramer's rule. Show your work.

$$5x + 2y = -1,$$
$$7x + 6y = 1$$

20. Graph: $3x + 4y \le -12$.

21. Find the maximum and minimum values of $Q = 2x + 3y$ subject to

$$x + y \le 6,$$
$$2x - 3y \ge -3,$$
$$x \ge 1,$$
$$y \ge 0.$$

22. *Maximizing Profit.* Casey's Cakes prepares pound cakes and carrot cakes. In a given week, at most 100 cakes can be prepared, of which 25 pound cakes and 15 carrot cakes are required by regular customers. The profit from each pound cake is $3 and the profit from each carrot cake is $4. How many of each type of cake should be prepared in order to maximize the profit?

23. Decompose into partial fractions:

$$\frac{3x - 11}{x^2 + 2x - 3}.$$

Synthesis

24. Three solutions of the equation $Ax - By = Cz - 8$ are $(2, -2, 2)$, $(-3, -1, 1)$, and $(4, 2, 9)$. Find A, B, and C.

Conic Sections

6

Marcia Graham, owner of Graham's Graphics, is designing an advertising brochure for the Art League's spring show. Each page of the brochure is rectangular with an area of 20 in^2 and a perimeter of 18 in. Find the dimensions of the brochure.

This problem appears as Exercise 57 in Section 6.4.

501

6.1

The Parabola

• *Given an equation of a parabola, complete the square, if necessary, and then find the vertex, the focus, and the directrix and graph the parabola.*

A **conic section** is formed when a right circular cone with two parts, called *nappes*, is intersected by a plane. One of four types of curves can be formed: a parabola, a circle, an ellipse, or a hyperbola.

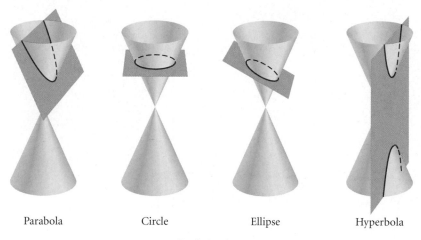

| Parabola | Circle | Ellipse | Hyperbola |

Conic Sections

Conic sections can be defined algebraically using second-degree equations of the form $Ax^2 + Bxy + Cy^2 + Dx + Ey + F = 0$. In addition, they can be defined geometrically as a set of points that satisfy certain conditions.

Parabolas

In Section 2.4, we saw that the graph of the quadratic function $f(x) = ax^2 + bx + c$, $a \neq 0$, is a parabola. A parabola can be defined geometrically.

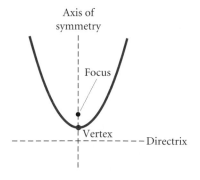

Parabola

A **parabola** is the set of all points in a plane equidistant from a fixed line (the **directrix**) and a fixed point not on the line (the **focus**).

The line that is perpendicular to the directrix and contains the focus is the **axis of symmetry.** The **vertex** is the midpoint of the segment between the focus and the directrix. (See the figure at left.)

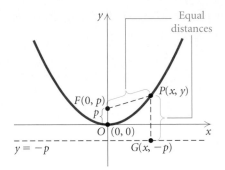

FIGURE 1

Let's derive the standard equation of a parabola with vertex $(0,0)$ and directrix $y = -p$, where $p > 0$. We place the coordinate axes as shown in Fig. 1. The y-axis is the axis of symmetry and contains the focus F. The distance from the focus to the vertex is the same as the distance from the vertex to the directrix. Thus the coordinates of F are $(0, p)$.

Let $P(x, y)$ be any point on the parabola and consider $\overline{PG}$ perpendicular to the line $y = -p$. The coordinates of G are $(x, -p)$. By the definition of a parabola,

$$PF = PG. \qquad \text{The distance from } P \text{ to the focus is the same as the distance from } P \text{ to the directrix.}$$

Then using the distance formula, we have

$$\sqrt{(x - 0)^2 + (y - p)^2} = \sqrt{(x - x)^2 + [y - (-p)]^2}$$
$$x^2 + y^2 - 2py + p^2 = y^2 + 2py + p^2 \qquad \text{Squaring both sides and squaring the binomials}$$
$$x^2 = 4py.$$

We have shown that if $P(x, y)$ is on the parabola shown in Fig. 1, then its coordinates satisfy this equation. The converse is also true, but we will not prove it here.

Note that if $p > 0$, as above, the graph opens up. If $p < 0$, the graph opens down.

The equation of a parabola with vertex $(0,0)$ and directrix $x = -p$ is derived similarly. Such a parabola opens either to the right ($p > 0$), as shown in Fig. 2, or to the left ($p < 0$).

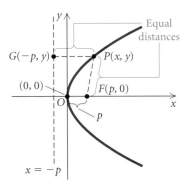

FIGURE 2

Standard Equation of a Parabola with Vertex at the Origin

The standard equation of a parabola with vertex $(0,0)$ and directrix $y = -p$ is

$$x^2 = 4py.$$

The focus is $(0, p)$ and the y-axis is the axis of symmetry.

The standard equation of a parabola with vertex $(0,0)$ and directrix $x = -p$ is

$$y^2 = 4px.$$

The focus is $(p, 0)$ and the x-axis is the axis of symmetry.

EXAMPLE 1 Find the focus and the directrix of the parabola $y = -\frac{1}{12}x^2$. Then graph the parabola.

Solution We write $y = -\frac{1}{12}x^2$ in the form $x^2 = 4py$:

$$-\frac{1}{12}x^2 = y \qquad \text{Given equation}$$
$$x^2 = -12y \qquad \text{Multiplying by } -12 \text{ on both sides}$$
$$x^2 = 4(-3)y. \qquad \text{Standard form}$$

Thus, $p = -3$, so the focus is $(0, p)$, or $(0, -3)$. The directrix is $y = -p = -(-3) = 3$.

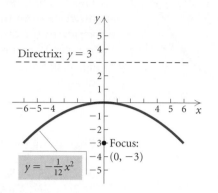

Directrix: $y = 3$

$y = -\frac{1}{12}x^2$

Focus: $(0, -3)$

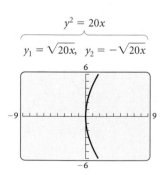
EXAMPLE 2 Find an equation of the parabola with vertex $(0, 0)$ and focus $(5, 0)$. Then graph the parabola.

Solution The focus is on the x-axis so the line of symmetry is the x-axis. Thus the equation is of the type

$$y^2 = 4px.$$

Since the focus $(5, 0)$ is 5 units to the right of the vertex, $p = 5$ and the equation is

$$y^2 = 4(5)x, \quad \text{or} \quad y^2 = 20x.$$

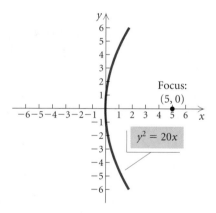

Focus: $(5, 0)$

$y^2 = 20x$

Finding Standard Form by Completing the Square

If a parabola with vertex at the origin is translated horizontally $|h|$ units and vertically $|k|$ units, it has an equation as follows.

Standard Equation of a Parabola with Vertex (h, k) and Vertical Axis of Symmetry

The standard equation of a parabola with vertex (h, k) and vertical axis of symmetry is

$$(x - h)^2 = 4p(y - k),$$

where the vertex is (h, k), the focus is ($h, k + p$), and the directrix is $y = k - p$.

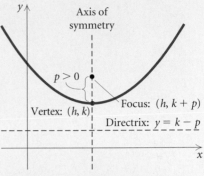

(When $p < 0$, the parabola opens down.)

Standard Equation of a Parabola with Vertex (h, k) and Horizontal Axis of Symmetry

The standard equation of a parabola with vertex (h, k) and horizontal axis of symmetry is

$$(y - k)^2 = 4p(x - h),$$

where the vertex is (h, k), the focus is ($h + p, k$), and the directrix is $x = h - p$.

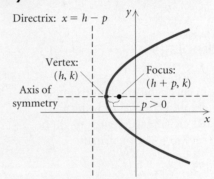

(When $p < 0$, the parabola opens to the left.)

COMPLETING THE SQUARE
REVIEW SECTION 2.3.

We can complete the square on equations of the form

$$y = ax^2 + bx + c \quad \text{or} \quad x = ay^2 + by + c$$

in order to write them in standard form.

EXAMPLE 3 For the parabola

$$x^2 + 6x + 4y + 5 = 0,$$

find the vertex, the focus, and the directrix. Then draw the graph.

Solution We first complete the square:

$$x^2 + 6x + 4y + 5 = 0$$

$$x^2 + 6x \quad\quad = -4y - 5 \qquad\text{Subtracting } 4y \text{ and } 5 \text{ on both sides}$$

$$x^2 + 6x + 9 = -4y - 5 + 9 \qquad\text{Adding 9 on both sides to complete the square on the left side}$$

$$x^2 + 6x + 9 = -4y + 4$$

$$(x + 3)^2 = -4(y - 1) \qquad\text{Factoring}$$

$$[x - (-3)]^2 = 4(-1)(y - 1). \qquad\text{Writing standard form: } (x - h)^2 = 4p(y - k)$$

We see that $h = -3$, $k = 1$, and $p = -1$, so we have the following:

Vertex (h, k): $\quad\quad\quad (-3, 1)$;

Focus $(h, k + p)$: $\quad\quad (-3, 1 + (-1))$, or $(-3, 0)$;

Directrix $y = k - p$: $\quad y = 1 - (-1)$, or $y = 2$.

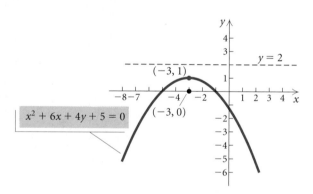

$x^2 + 6x + 4y + 5 = 0$

Technology Connection

We can check the graph in Example 3 on a graphing calculator using a squared viewing window. It might be necessary to solve for y first:

$$x^2 + 6x + 4y + 5 = 0$$

$$4y = -x^2 - 6x - 5$$

$$y = \tfrac{1}{4}(-x^2 - 6x - 5).$$

$y = \tfrac{1}{4}(-x^2 - 6x - 5)$

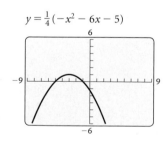

EXAMPLE 4 For the parabola

$$y^2 - 2y - 8x - 31 = 0,$$

find the vertex, the focus, and the directrix. Then draw the graph.

Solution We first complete the square:

$$y^2 - 2y - 8x - 31 = 0$$

$$y^2 - 2y \qquad = 8x + 31 \qquad \text{Adding } 8x \text{ and } 31 \text{ on both sides}$$

$$y^2 - 2y + 1 = 8x + 31 + 1 \qquad \text{Adding 1 on both sides to complete the square on the left side}$$

$$y^2 - 2y + 1 = 8x + 32$$

$$(y - 1)^2 = 8(x + 4) \qquad \text{Factoring}$$

$$(y - 1)^2 = 4(2)[x - (-4)]. \qquad \text{Writing standard form:} \atop (y - k)^2 = 4p(x - h)$$

We see that $h = -4$, $k = 1$, and $p = 2$, so we have the following:

Vertex (h, k): $\qquad (-4, 1)$;

Focus $(h + p, k)$: $\qquad (-4 + 2, 1)$, or $(-2, 1)$;

Directrix $x = h - p$: $\quad x = -4 - 2$, or $x = -6$.

$y^2 - 2y - 8x - 31 = 0$

Technology Connection

$$y^2 - 2y - 8x - 31 = 0$$

$$y_1 = \frac{2 + \sqrt{32x + 128}}{2},$$

$$y_2 = \frac{2 - \sqrt{32x + 128}}{2}$$

We can check the graph in Example 4 on a graphing calculator using a squared viewing window. We solve the original equation for y using the quadratic formula:

$$y^2 - 2y - 8x - 31 = 0$$

$$y^2 - 2y + (-8x - 31) = 0$$

$$a = 1, \quad b = -2, \quad c = -8x - 31$$

$$y = \frac{-(-2) \pm \sqrt{(-2)^2 - 4 \cdot 1(-8x - 31)}}{2 \cdot 1}$$

$$y = \frac{2 \pm \sqrt{32x + 128}}{2}.$$

We now graph

$$y_1 = \frac{2 + \sqrt{32x + 128}}{2} \quad \text{and} \quad y_2 = \frac{2 - \sqrt{32x + 128}}{2}.$$

Applications

Parabolas have many applications. For example, cross sections of car headlights, flashlights, and searchlights are parabolas. The bulb is located at the focus and light from that point is reflected outward parallel to the axis of symmetry. Satellite dishes and field microphones used at sporting events often have parabolic cross sections. Incoming radio waves or sound waves parallel to the axis are reflected into the focus. Cables hung between structures in suspension bridges, such as the Golden Gate Bridge, form parabolas. When a cable supports only its own weight, however, it forms a curve called a *catenary* rather than a parabola.

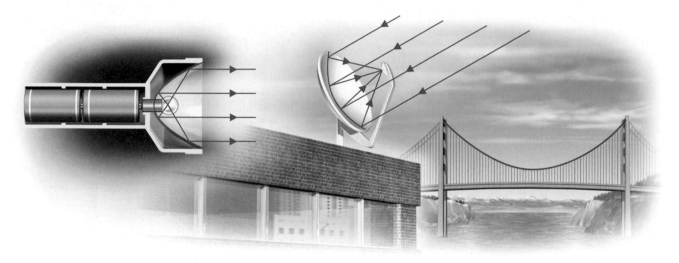

6.1

Exercise Set

In Exercises 1–6, match the equation with one of the graphs (a)–(f), which follow.

a)
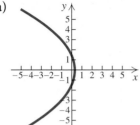

b)
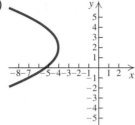

c)
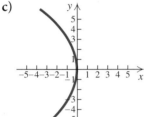

d)
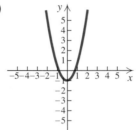

e)
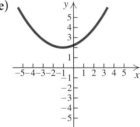

f)
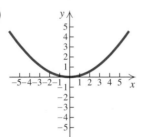

1. $x^2 = 8y$

2. $y^2 = -10x$

3. $(y - 2)^2 = -3(x + 4)$

4. $(x + 1)^2 = 5(y - 2)$

5. $13x^2 - 8y - 9 = 0$

6. $41x + 6y^2 = 12$

Find the vertex, the focus, and the directrix. Then draw the graph.

7. $x^2 = 20y$ **8.** $x^2 = 16y$

9. $y^2 = -6x$ **10.** $y^2 = -2x$

11. $x^2 - 4y = 0$ **12.** $y^2 + 4x = 0$

13. $x = 2y^2$ **14.** $y = \frac{1}{2}x^2$

Find an equation of a parabola satisfying the given conditions.

15. Focus $(4, 0)$, directrix $x = -4$

16. Focus $\left(0, \frac{1}{4}\right)$, directrix $y = -\frac{1}{4}$

17. Focus $(0, -\pi)$, directrix $y = \pi$

18. Focus $\left(-\sqrt{2}, 0\right)$, directrix $x = \sqrt{2}$

19. Focus $(3, 2)$, directrix $x = -4$

20. Focus $(-2, 3)$, directrix $y = -3$

Find the vertex, the focus, and the directrix. Then draw the graph.

21. $(x + 2)^2 = -6(y - 1)$

22. $(y - 3)^2 = -20(x + 2)$

23. $x^2 + 2x + 2y + 7 = 0$

24. $y^2 + 6y - x + 16 = 0$

25. $x^2 - y - 2 = 0$

26. $x^2 - 4x - 2y = 0$

27. $y = x^2 + 4x + 3$

28. $y = x^2 + 6x + 10$

29. $y^2 - y - x + 6 = 0$

30. $y^2 + y - x - 4 = 0$

31. *Satellite Dish.*   An engineer designs a satellite dish with a parabolic cross section. The dish is 15 ft wide at the opening and the focus is placed 4 ft from the vertex.

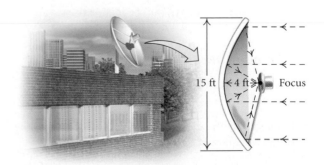

a) Position a coordinate system with the origin at the vertex and the *x*-axis on the parabola's axis of symmetry and find an equation of the parabola.

b) Find the depth of the satellite dish at the vertex.

32. *Headlight Mirror.* A car headlight mirror has a parabolic cross section with diameter 6 in. and depth 1 in.

6 in.   Focus   1 in.

a) Position a coordinate system with the origin at the vertex and the *x*-axis on the parabola's axis of symmetry and find an equation of the parabola.

b) How far from the vertex should the bulb be positioned if it is to be placed at the focus?

33. *Spotlight.* A spotlight has a parabolic cross section that is 4 ft wide at the opening and 1.5 ft deep at the vertex. How far from the vertex is the focus?

34. *Field Microphone.* A field microphone used at a football game has a parabolic cross section and is 18 in. deep. The focus is 4 in. from the vertex. Find the width of the microphone at the opening.

Technology Connection

Use a graphing calculator to find the vertex, the focus, and the directrix of each of the following.

35. $4.5x^2 - 7.8x + 9.7y = 0$

36. $134.1y^2 + 43.4x - 316.6y - 122.4 = 0$

Collaborative Discussion and Writing

37. Is a parabola always the graph of a function? Why or why not?

38. Explain how the distance formula is used to find the standard equation of a parabola.

Skill Maintenance

Consider the following linear equations. Without graphing them, answer the questions below.

a) $y = 2x$
b) $y = \frac{1}{3}x + 5$
c) $y = -3x - 2$
d) $y = -0.9x + 7$
e) $y = -5x + 3$
f) $y = x + 4$
g) $8x - 4y = 7$
h) $3x + 6y = 2$

39. Which has/have *x*-intercept $\left(\frac{2}{3}, 0\right)$?

40. Which has/have *y*-intercept $(0, 7)$?

41. Which slant up from left to right?

42. Which has the least steep slant?

43. Which has/have slope $\frac{1}{3}$?

44. Which, if any, contain the point $(3, 7)$?

45. Which, if any, are parallel?

46. Which, if any, are perpendicular?

Synthesis

47. Find an equation of the parabola with a vertical axis of symmetry and vertex $(-1, 2)$ and containing the point $(-3, 1)$.

48. Find an equation of a parabola with a horizontal axis of symmetry and vertex $(-2, 1)$ and containing the point $(-3, 5)$.

49. *Suspension Bridge.* The cables of a suspension bridge are 50 ft above the roadbed at the ends of the bridge and 10 ft above it in the center of the bridge. The roadbed is 200 ft long. Vertical cables are to be spaced every 20 ft along the bridge. Calculate the lengths of these vertical cables.

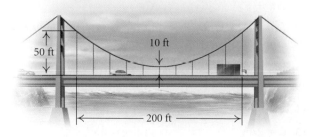

50 ft   10 ft   200 ft

6.2

The Circle and the Ellipse

• *Given an equation of a circle, complete the square, if necessary, and then find the center and the radius and graph the circle.*
• *Given an equation of an ellipse, complete the square, if necessary, and then find the center, the vertices, and the foci and graph the ellipse.*

Circles

We can define a circle geometrically.

> ### Circle
> A **circle** is the set of all points in a plane that are at a fixed distance from a fixed point (the **center**) in the plane.

CIRCLES

REVIEW SECTION 1.1.

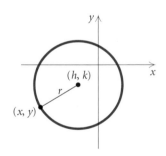

Circles were introduced in Section 1.1. Recall the standard equation of a circle with center (h, k) and radius r.

> ### Standard Equation of a Circle
> The standard equation of a circle with center (h, k) and radius r is
> $$(x - h)^2 + (y - k)^2 = r^2.$$

EXAMPLE 1 For the circle

$$x^2 + y^2 - 16x + 14y + 32 = 0,$$

find the center and the radius. Then graph the circle.

Solution First, we complete the square twice:

$$x^2 + y^2 - 16x + 14y + 32 = 0$$
$$x^2 - 16x \quad + y^2 + 14y \quad = -32$$
$$x^2 - 16x + 64 + y^2 + 14y + 49 = -32 + 64 + 49$$

$\left[\frac{1}{2}(-16)\right]^2 = (-8)^2 = 64$ and $\left(\frac{1}{2} \cdot 14\right)^2 = 7^2 = 49$; adding 64 and 49 on both sides to complete the square twice on the left side

$$(x - 8)^2 + (y + 7)^2 = 81$$
$$(x - 8)^2 + [y - (-7)]^2 = 9^2. \qquad \textbf{Writing standard form}$$

The center is $(8, -7)$ and the radius is 9. We graph the circle.

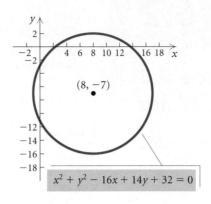

$$x^2 + y^2 - 16x + 14y + 32 = 0$$

Technology Connection

To use a graphing calculator to graph the circle in Example 1, it might be necessary to solve for y first. The original equation can be solved using the quadratic formula, or the standard form of the equation can be solved using the principle of square roots. The second alternative is illustrated here:

$$(x - 8)^2 + (y + 7)^2 = 81$$
$$(y + 7)^2 = 81 - (x - 8)^2$$
$$y + 7 = \pm\sqrt{81 - (x - 8)^2} \qquad \text{Using the principle of square roots}$$
$$y = -7 \pm \sqrt{81 - (x - 8)^2}.$$

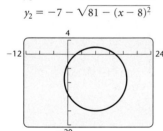

$(x - 8)^2 + (y + 7)^2 = 81$

$y_1 = -7 + \sqrt{81 - (x - 8)^2}$,
$y_2 = -7 - \sqrt{81 - (x - 8)^2}$

Then we graph

$$y_1 = -7 + \sqrt{81 - (x - 8)^2}$$

and

$$y_2 = -7 - \sqrt{81 - (x - 8)^2}$$

in a squared viewing window.

Some graphing calculators have a DRAW feature that provides a quick way to graph a circle when the center and the radius are known. This feature is described on p. 63.

Ellipses

We have studied two conic sections, the parabola and the circle. Now we turn our attention to a third, the *ellipse*.

Ellipse

An **ellipse** is the set of all points in a plane, the sum of whose distances from two fixed points (the **foci**) is constant. The **center** of an ellipse is the midpoint of the segment between the foci.

We can draw an ellipse by first placing two thumbtacks in a piece of cardboard. These are the foci (singular, *focus*). We then attach a piece of string to the tacks. Its length is the constant sum of the distances $d_1 + d_2$ from the foci to any point on the ellipse. Next, we trace a curve with a pen held tight against the string. The figure traced is an ellipse.

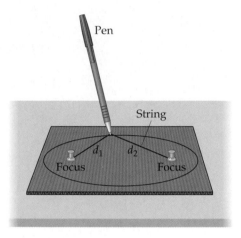

Let's first consider the ellipse shown below with center at the origin. The points F_1 and F_2 are the foci. The segment $\overline{A'A}$ is the **major axis,** and the points A' and A are the **vertices.** The segment $\overline{B'B}$ is the **minor axis,** and the points B' and B are the **y-intercepts.** Note that the major axis of an ellipse is longer than the minor axis.

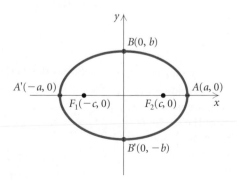

Standard Equation of an Ellipse with Center at the Origin

Major Axis Horizontal

$$\frac{x^2}{a^2} + \frac{y^2}{b^2} = 1, \ a > b > 0$$

Vertices: $(-a, 0), (a, 0)$

y-intercepts:
$(0, -b), (0, b)$

Foci: $(-c, 0), (c, 0)$,
where $c^2 = a^2 - b^2$

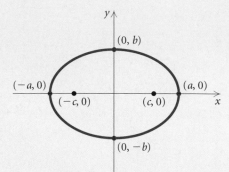

Major Axis Vertical

$$\frac{x^2}{b^2} + \frac{y^2}{a^2} = 1, \ a > b > 0$$

Vertices: $(0, -a), (0, a)$

x-intercepts:
$(-b, 0), (b, 0)$

Foci: $(0, -c), (0, c)$,
where $c^2 = a^2 - b^2$

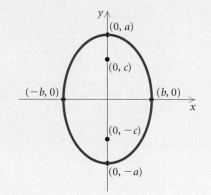

EXAMPLE 2 Find the standard equation of the ellipse with vertices $(-5, 0)$ and $(5, 0)$ and foci $(-3, 0)$ and $(3, 0)$. Then graph the ellipse.

Solution Since the foci are on the x-axis and the origin is the midpoint of the segment between them, the major axis is horizontal and $(0, 0)$ is the center of the ellipse. Thus the equation is of the form

$$\frac{x^2}{a^2} + \frac{y^2}{b^2} = 1.$$

Since the vertices are $(-5, 0)$ and $(5, 0)$ and the foci are $(-3, 0)$ and $(3, 0)$, we know that $a = 5$ and $c = 3$. These values can be used to find b^2:

$$c^2 = a^2 - b^2$$
$$3^2 = 5^2 - b^2$$
$$9 = 25 - b^2$$
$$b^2 = 16.$$

Study Tip

If you are finding it difficult to master a particular topic or concept, talk about it with a classmate. Verbalizing your questions about the material might help clarify it for you. If your classmate is also finding the material difficult, it is possible that the majority of the students in your class are confused, and you can ask your instructor to explain the concept again.

Thus the equation of the ellipse is

$$\frac{x^2}{5^2} + \frac{y^2}{4^2} = 1, \quad \text{or} \quad \frac{x^2}{25} + \frac{y^2}{16} = 1.$$

To graph the ellipse, we plot the vertices $(-5, 0)$ and $(5, 0)$. Since $b^2 = 16$, we know that $b = 4$ and the y-intercepts are $(0, -4)$ and $(0, 4)$. We plot these points as well and connect the four points we have plotted with a smooth curve.

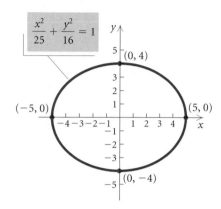

Technology Connection

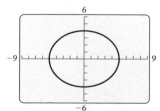

To draw the graph in Example 2 using a graphing calculator, it might be necessary to solve for y first:

$$y = \pm\sqrt{\frac{400 - 16x^2}{25}}.$$

Then we graph

$$y_1 = -\sqrt{\frac{400 - 16x^2}{25}} \quad \text{and} \quad y_2 = \sqrt{\frac{400 - 16x^2}{25}}$$

or

$$y_1 = -\sqrt{\frac{400 - 16x^2}{25}} \quad \text{and} \quad y_2 = -y_1$$

in a squared viewing window.

EXAMPLE 3 For the ellipse

$$9x^2 + 4y^2 = 36,$$

find the vertices and the foci. Then draw the graph.

Solution   We first find standard form:

$$9x^2 + 4y^2 = 36$$

$$\frac{9x^2}{36} + \frac{4y^2}{36} = \frac{36}{36} \qquad \text{Dividing by 36 on both sides to get 1 on the right side}$$

$$\frac{x^2}{4} + \frac{y^2}{9} = 1$$

$$\frac{x^2}{2^2} + \frac{y^2}{3^2} = 1. \qquad \text{Writing standard form}$$

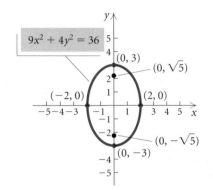

Thus, $a = 3$ and $b = 2$. The major axis is vertical, so the vertices are $(0, -3)$ and $(0, 3)$. Since we know that $c^2 = a^2 - b^2$, we have $c^2 = 3^2 - 2^2 = 5$, so $c = \sqrt{5}$ and the foci are $\left(0, -\sqrt{5}\right)$ and $\left(0, \sqrt{5}\right)$.

To graph the ellipse, we plot the vertices. Note also that since $b = 2$, the x-intercepts are $(-2, 0)$ and $(2, 0)$. We plot these points as well and connect the four points we have plotted with a smooth curve.   ▬

If the center of an ellipse is not at the origin but at some point (h, k), then we can think of an ellipse with center at the origin being translated horizontally $|h|$ units and vertically $|k|$ units.

Standard Equation of an Ellipse with Center at (h, k)

Major Axis Horizontal

$$\frac{(x-h)^2}{a^2} + \frac{(y-k)^2}{b^2} = 1, \; a > b > 0$$

Vertices: $(h - a, k), (h + a, k)$

Length of minor axis: $2b$

Foci: $(h - c, k), (h + c, k)$, where $c^2 = a^2 - b^2$

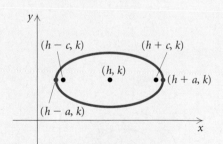

Major Axis Vertical

$$\frac{(x-h)^2}{b^2} + \frac{(y-k)^2}{a^2} = 1, \; a > b > 0$$

Vertices: $(h, k - a), (h, k + a)$

Length of minor axis: $2b$

Foci: $(h, k - c), (h, k + c)$, where $c^2 = a^2 - b^2$

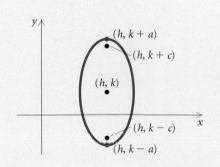

EXAMPLE 4 For the ellipse

$$4x^2 + y^2 + 24x - 2y + 21 = 0,$$

find the center, the vertices, and the foci. Then draw the graph.

Solution First, we complete the square twice to get standard form:

$$4x^2 + y^2 + 24x - 2y + 21 = 0$$

$$4(x^2 + 6x \quad) + (y^2 - 2y \quad) = -21$$

$$4(x^2 + 6x + 9) + (y^2 - 2y + 1) = -21 + 4 \cdot 9 + 1$$

> **Completing the square twice by adding $4 \cdot 9$ and 1 on both sides**

$$4(x + 3)^2 + (y - 1)^2 = 16$$

$$\frac{1}{16}[4(x + 3)^2 + (y - 1)^2] = \frac{1}{16} \cdot 16$$

$$\frac{(x + 3)^2}{4} + \frac{(y - 1)^2}{16} = 1$$

$$\frac{[x - (-3)]^2}{2^2} + \frac{(y - 1)^2}{4^2} = 1.$$

> **Writing standard form:** $\dfrac{(x - h)^2}{b^2} + \dfrac{(y - k)^2}{a^2} = 1$

The center is $(-3, 1)$. Note that $a = 4$ and $b = 2$. The major axis is vertical, so the vertices are 4 units above and below the center:

$$(-3, 1 + 4) \text{ and } (-3, 1 - 4), \quad \text{or} \quad (-3, 5) \text{ and } (-3, -3).$$

We know that $c^2 = a^2 - b^2$, so $c^2 = 4^2 - 2^2 = 12$ and $c = \sqrt{12}$, or $2\sqrt{3}$. Then the foci are $2\sqrt{3}$ units above and below the center:

$$\left(-3, 1 + 2\sqrt{3}\right) \quad \text{and} \quad \left(-3, 1 - 2\sqrt{3}\right).$$

To graph the ellipse, we plot the vertices. Note also that since $b = 2$, two other points on the graph are the endpoints of the minor axis, 2 units right and left of the center:

$$(-3 + 2, 1) \quad \text{and} \quad (-3 - 2, 1),$$

or

$$(-1, 1) \quad \text{and} \quad (-5, 1).$$

We plot these points as well and connect the four points with a smooth curve.

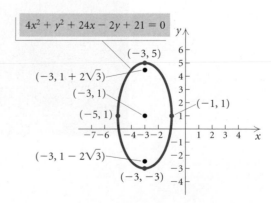

Applications

One exciting application of an ellipse is a medical device called a *lithotripter*. This machine uses underwater shock waves to pulverize kidney stones. The waves originate at one focus of an ellipse and are reflected to the kidney stone, which is positioned at the other focus. Recovery time following the use of this technique is much shorter than with conventional surgery and the mortality rate is far lower.

A room with an ellipsoidal ceiling is known as a *whispering gallery*. In such a room, a word whispered at one focus can be clearly heard at the other. Whispering galleries are found in the rotunda of the Capitol Building in Washington, D.C., and in the Mormon Tabernacle in Salt Lake City.

Ellipses have many other applications. Planets travel around the sun in elliptical orbits with the sun at one focus, for example, and satellites travel around the earth in elliptical orbits as well.

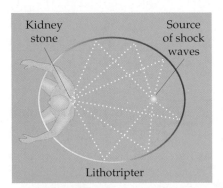

6.2

Exercise Set

In Exercises 1–6, match the equation with one of the graphs (a)–(f), which follow.

a)
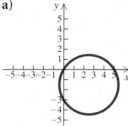

b)
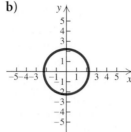

c)
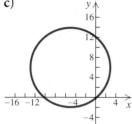

d)
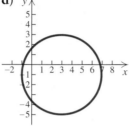

e)
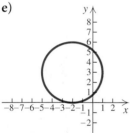

f)
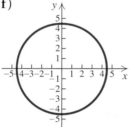

1. $x^2 + y^2 = 5$

2. $y^2 = 20 - x^2$

3. $x^2 + y^2 - 6x + 2y = 6$

4. $x^2 + y^2 + 10x - 12y = 3$

5. $x^2 + y^2 - 5x + 3y = 0$

6. $x^2 + 4x - 2 = 6y - y^2 - 6$

Find the center and the radius of the circle with the given equation. Then draw the graph.

7. $x^2 + y^2 - 14x + 4y = 11$

8. $x^2 + y^2 + 2x - 6y = -6$

9. $x^2 + y^2 + 6x - 2y = 6$

10. $x^2 + y^2 - 4x + 2y = 4$

11. $x^2 + y^2 + 4x - 6y - 12 = 0$

12. $x^2 + y^2 - 8x - 2y - 19 = 0$

13. $x^2 + y^2 - 6x - 8y + 16 = 0$

14. $x^2 + y^2 - 2x + 6y + 1 = 0$

15. $x^2 + y^2 + 6x - 10y = 0$

16. $x^2 + y^2 - 7x - 2y = 0$

17. $x^2 + y^2 - 9x = 7 - 4y$

18. $y^2 - 6y - 1 = 8x - x^2 + 3$

In Exercises 19–22, match the equation with one of the graphs (a)–(d), which follow.

a)
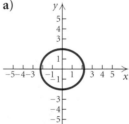

b)
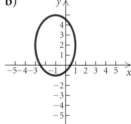

c)
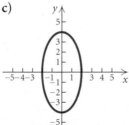

d)
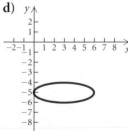

19. $16x^2 + 4y^2 = 64$

20. $4x^2 + 5y^2 = 20$

21. $x^2 + 9y^2 - 6x + 90y = -225$

22. $9x^2 + 4y^2 + 18x - 16y = 11$

Find the vertices and the foci of the ellipse with the given equation. Then draw the graph.

23. $\dfrac{x^2}{4} + \dfrac{y^2}{1} = 1$

24. $\dfrac{x^2}{25} + \dfrac{y^2}{36} = 1$

25. $16x^2 + 9y^2 = 144$

26. $9x^2 + 4y^2 = 36$

27. $2x^2 + 3y^2 = 6$

28. $5x^2 + 7y^2 = 35$

29. $4x^2 + 9y^2 = 1$

30. $25x^2 + 16y^2 = 1$

Find an equation of an ellipse satisfying the given conditions.

31. Vertices: $(-7, 0)$ and $(7, 0)$;
foci: $(-3, 0)$ and $(3, 0)$

32. Vertices: $(0, -6)$ and $(0, 6)$;
foci: $(0, -4)$ and $(0, 4)$

33. Vertices: $(0, -8)$ and $(0, 8)$;
length of minor axis: 10

34. Vertices: $(-5, 0)$ and $(5, 0)$;
length of minor axis: 6

35. Foci: $(-2, 0)$ and $(2, 0)$;
length of major axis: 6

36. Foci: $(0, -3)$ and $(0, 3)$;
length of major axis: 10

Find the center, the vertices, and the foci of the ellipse. Then draw the graph.

37. $\dfrac{(x-1)^2}{9} + \dfrac{(y-2)^2}{4} = 1$

38. $\dfrac{(x-1)^2}{1} + \dfrac{(y-2)^2}{4} = 1$

39. $\dfrac{(x+3)^2}{25} + \dfrac{(y-5)^2}{36} = 1$

40. $\dfrac{(x-2)^2}{16} + \dfrac{(y+3)^2}{25} = 1$

41. $3(x+2)^2 + 4(y-1)^2 = 192$

42. $4(x-5)^2 + 3(y-4)^2 = 48$

43. $4x^2 + 9y^2 - 16x + 18y - 11 = 0$

44. $x^2 + 2y^2 - 10x + 8y + 29 = 0$

45. $4x^2 + y^2 - 8x - 2y + 1 = 0$

46. $9x^2 + 4y^2 + 54x - 8y + 49 = 0$

*The **eccentricity** of an ellipse is defined as $e = c/a$. For an ellipse, $0 < c < a$, so $0 < e < 1$. When e is close to 0, an ellipse appears to be nearly circular. When e is close to 1, an ellipse is very flat.*

47. Observe the shapes of the ellipses in Examples 2 and 4. Which ellipse has the smaller eccentricity? Confirm your answer by computing the eccentricity of each ellipse.

48. Which ellipse below has the smaller eccentricity? (Assume that the coordinate systems have the same scale.)

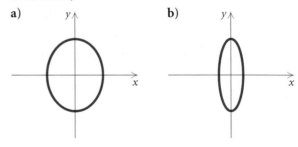

a) **b)**

49. Find an equation of an ellipse with vertices $(0, -4)$ and $(0, 4)$ and $e = \frac{1}{4}$.

50. Find an equation of an ellipse with vertices $(-3, 0)$ and $(3, 0)$ and $e = \frac{7}{10}$.

51. *Bridge Supports.* The bridge support shown in the figure below is the top half of an ellipse. Assuming that a coordinate system is superimposed on the drawing in such a way that the center of the ellipse is at point Q, find an equation of the ellipse.

52. *The Ellipse.* In Washington, D.C., there is a large grassy area south of the White House known as the Ellipse. It is actually an ellipse with major axis of length 1048 ft and minor axis of length 898 ft. Assuming that a coordinate system is superimposed on the area in such a way that the center is at the origin and the major and minor axes are on the *x*- and *y*-axes of the coordinate system, respectively, find an equation of the ellipse.

53. *The Earth's Orbit.* The maximum distance of the earth from the sun is 9.3×10^7 mi. The minimum distance is 9.1×10^7 mi. The sun is at one focus of the elliptical orbit. Find the distance from the sun to the other focus.

54. *Carpentry.* A carpenter is cutting a 3-ft by 4-ft elliptical sign from a 3-ft by 4-ft piece of plywood. The ellipse will be drawn using a string attached to the board at the foci of the ellipse.

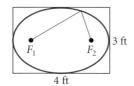

a) How far from the ends of the board should the string be attached?
b) How long should the string be?

Technology Connection

Use a graphing calculator to find the center and the vertices of each of the following.

55. $4x^2 + 9y^2 - 16.025x + 18.0927y - 11.346 = 0$

56. $9x^2 + 4y^2 + 54.063x - 8.016y + 49.872 = 0$

Collaborative Discussion and Writing

57. Explain why function notation is not used in this section.

58. Is the center of an ellipse part of the graph of the ellipse? Why or why not?

Skill Maintenance

In each of Exercises 59–66, fill in the blank with the correct term. Some of the given choices will not be used.

piecewise function
linear equation
factor
remainder
solution
zero
x-intercept
y-intercept
parabola
circle
ellipse
midpoint
distance
one real-number solution
two different real-number solutions
two different imaginary-number solutions

59. The _____ between two points (x_1, y_1) and (x_2, y_2) is given by $\left(\dfrac{x_1 + x_2}{2}, \dfrac{y_1 + y_2}{2} \right)$.

60. An input c of a function f is a(n) _____ of the function if $f(c) = 0$.

61. A(n) _____ of the graph of an equation is a point $(0, b)$.

62. For a quadratic equation $ax^2 + bx + c = 0$, if $b^2 - 4ac > 0$, the equation has _____.

63. Given a polynomial $f(x)$, then $f(c)$ is the _____ that would be obtained by dividing $f(x)$ by $x - c$.

64. A(n) _____ is the set of all points in a plane the sum of whose distances from two fixed points is constant.

65. A(n) _____ is the set of all points in a plane equidistant from a fixed line and a fixed point not on the line.

66. A(n) _____ is the set of all points in a plane that are at a fixed distance from a fixed point in the plane.

Synthesis

Find an equation of an ellipse satisfying the given conditions.

67. Vertices: $(3, -4), (3, 6)$;
endpoints of minor axis: $(1, 1), (5, 1)$

68. Vertices: $(-1, -1), (-1, 5)$;
endpoints of minor axis: $(-3, 2), (1, 2)$

69. Vertices: $(-3, 0)$ and $(3, 0)$;
passing through $\left(2, \frac{22}{3}\right)$

70. Center: $(-2, 3)$; major axis vertical;
length of major axis: 4;
length of minor axis: 1

71. *Bridge Arch.* A bridge with a semielliptical arch spans a river as shown below. What is the clearance 6 ft from the riverbank?

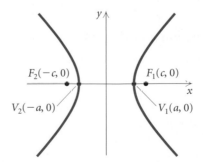

14 ft

← 50 ft →

6.3

The Hyperbola

• Given an equation of a hyperbola, complete the square, if necessary, and then find the center, the vertices, and the foci and graph the hyperbola.

The last type of conic section that we will study is the *hyperbola*.

Hyperbola

A **hyperbola** is the set of all points in a plane for which the absolute value of the difference of the distances from two fixed points (the **foci**) is constant. The midpoint of the segment between the foci is the **center** of the hyperbola.

Standard Equations of Hyperbolas

We first consider the equation of a hyperbola with center at the origin. In the figure below, F_1 and F_2 are the foci. The segment $\overline{V_2 V_1}$ is the **transverse axis** and the points V_2 and V_1 are the **vertices.**

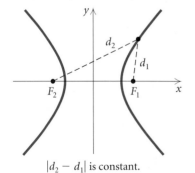

$|d_2 - d_1|$ is constant.

$F_2(-c, 0)$ $F_1(c, 0)$

$V_2(-a, 0)$ $V_1(a, 0)$

Standard Equation of a Hyperbola with Center at the Origin

Transverse Axis Horizontal

$$\frac{x^2}{a^2} - \frac{y^2}{b^2} = 1$$

Vertices: $(-a, 0), (a, 0)$

Foci: $(-c, 0), (c, 0)$, where $c^2 = a^2 + b^2$

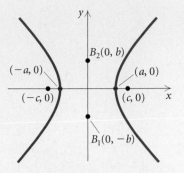

Transverse Axis Vertical

$$\frac{y^2}{a^2} - \frac{x^2}{b^2} = 1$$

Vertices: $(0, -a), (0, a)$

Foci: $(0, -c), (0, c)$, where $c^2 = a^2 + b^2$

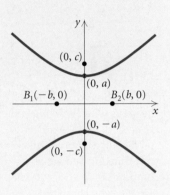

The segment $\overline{B_1 B_2}$ is the **conjugate axis** of the hyperbola.

To graph a hyperbola with a horizontal transverse axis, it is helpful to begin by graphing the lines $y = -(b/a)x$ and $y = (b/a)x$. These are the **asymptotes** of the hyperbola. For a hyperbola with a vertical transverse axis, the asymptotes are $y = -(a/b)x$ and $y = (a/b)x$. As $|x|$ gets larger and larger, the graph of the hyperbola gets closer and closer to the asymptotes.

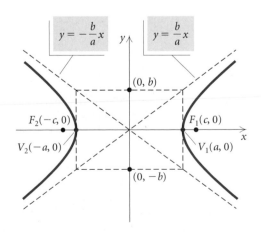

EXAMPLE 1 Find an equation of the hyperbola with vertices $(0, -4)$ and $(0, 4)$ and foci $(0, -6)$ and $(0, 6)$.

Solution We know that $a = 4$ and $c = 6$. We find b^2:

$$c^2 = a^2 + b^2$$
$$6^2 = 4^2 + b^2$$
$$36 = 16 + b^2$$
$$20 = b^2.$$

Since the vertices and the foci are on the y-axis, we know that the transverse axis is vertical. We can now write the equation of the hyperbola:

$$\frac{y^2}{a^2} - \frac{x^2}{b^2} = 1$$
$$\frac{y^2}{16} - \frac{x^2}{20} = 1.$$

Study Tip

Take the time to include all the steps when working your homework problems. Doing so will help you organize your thinking and avoid computational errors. It will also give you complete, step-by-step solutions of the exercises that can be used to study for an exam.

EXAMPLE 2 For the hyperbola given by

$$9x^2 - 16y^2 = 144,$$

find the vertices, the foci, and the asymptotes. Then graph the hyperbola.

Solution First, we find standard form:

$$9x^2 - 16y^2 = 144$$
$$\frac{1}{144}(9x^2 - 16y^2) = \frac{1}{144} \cdot 144 \qquad \text{Multiplying by } \tfrac{1}{144} \text{ to get 1 on the right side}$$
$$\frac{x^2}{16} - \frac{y^2}{9} = 1$$
$$\frac{x^2}{4^2} - \frac{y^2}{3^2} = 1. \qquad \text{Writing standard form}$$

The hyperbola has a horizontal transverse axis, so the vertices are $(-a, 0)$ and $(a, 0)$, or $(-4, 0)$ and $(4, 0)$. From the standard form of the equation, we know that $a^2 = 4^2$, or 16, and $b^2 = 3^2$, or 9. We find the foci:

$$c^2 = a^2 + b^2$$
$$c^2 = 16 + 9$$
$$c^2 = 25$$
$$c = 5.$$

Thus the foci are $(-5, 0)$ and $(5, 0)$.

Next, we find the asymptotes:

$$y = -\frac{b}{a}x = -\frac{3}{4}x \quad \text{and} \quad y = \frac{b}{a}x = \frac{3}{4}x.$$

To draw the graph, we sketch the asymptotes first. This is easily done by drawing the rectangle with horizontal sides passing through $(0, 3)$ and $(0, -3)$ and vertical sides through $(4, 0)$ and $(-4, 0)$. Then we draw and extend the diagonals of this rectangle. The two extended diagonals are the asymptotes of the hyperbola. Next, we plot the vertices and draw the branches of the hyperbola outward from the vertices toward the asymptotes.

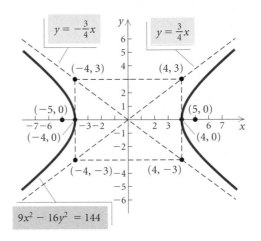

Technology Connection

To graph the hyperbola in Example 2 on a graphing calculator, it might be necessary to solve for y first and then graph the top and bottom halves of the hyperbola in the same squared viewing window.

$$9x^2 - 16y^2 = 144$$

$$y_1 = \sqrt{\frac{9x^2 - 144}{16}}, \quad y_2 = -\sqrt{\frac{9x^2 - 144}{16}}$$

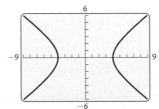

If a hyperbola with center at the origin is translated horizontally $|h|$ units and vertically $|k|$ units, the center is at the point (h, k).

Standard Equation of a Hyperbola with Center (h, k)

Transverse Axis Horizontal

$$\frac{(x - h)^2}{a^2} - \frac{(y - k)^2}{b^2} = 1$$

Vertices: $(h - a, k), (h + a, k)$

Asymptotes: $y - k = \dfrac{b}{a}(x - h), y - k = -\dfrac{b}{a}(x - h)$

Foci: $(h - c, k), (h + c, k)$, where $c^2 = a^2 + b^2$

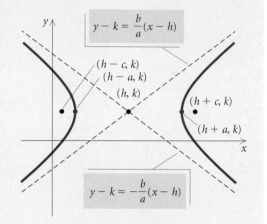

Transverse Axis Vertical

$$\frac{(y - k)^2}{a^2} - \frac{(x - h)^2}{b^2} = 1$$

Vertices: $(h, k - a), (h, k + a)$

Asymptotes: $y - k = \dfrac{a}{b}(x - h), y - k = -\dfrac{a}{b}(x - h)$

Foci: $(h, k - c), (h, k + c)$, where $c^2 = a^2 + b^2$

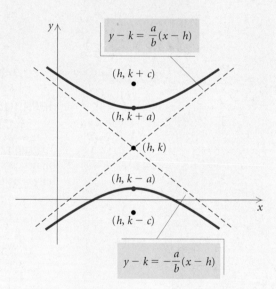

EXAMPLE 3 For the hyperbola given by

$$4y^2 - x^2 + 24y + 4x + 28 = 0,$$

find the center, the vertices, and the foci. Then draw the graph.

Solution   First, we complete the square to get standard form:

$$4y^2 - x^2 + 24y + 4x + 28 = 0$$

$$4(y^2 + 6y \qquad) - (x^2 - 4x \qquad) = -28$$

$$4(y^2 + 6y + 9 - 9) - (x^2 - 4x + 4 - 4) = -28$$

$$4(y^2 + 6y + 9) + 4(-9) - (x^2 - 4x + 4) - (-4) = -28$$

$$4(y^2 + 6y + 9) - 36 - (x^2 - 4x + 4) + 4 = -28$$

$$4(y^2 + 6y + 9) - (x^2 - 4x + 4) = -28 + 36 - 4$$

$$4(y + 3)^2 - (x - 2)^2 = 4$$

$$\frac{(y + 3)^2}{1} - \frac{(x - 2)^2}{4} = 1 \qquad \textbf{Dividing by 4}$$

$$\frac{[y - (-3)]^2}{1^2} - \frac{(x - 2)^2}{2^2} = 1. \qquad \textbf{Standard form}$$

The center is $(2, -3)$. Note that $a = 1$ and $b = 2$. The transverse axis is vertical, so the vertices are 1 unit below and above the center:

$$(2, -3 - 1) \text{ and } (2, -3 + 1), \quad \text{or} \quad (2, -4) \text{ and } (2, -2).$$

We know that $c^2 = a^2 + b^2$, so $c^2 = 1^2 + 2^2 = 1 + 4 = 5$ and $c = \sqrt{5}$. Thus the foci are $\sqrt{5}$ units below and above the center:

$$\left(2, -3 - \sqrt{5}\right) \quad \text{and} \quad \left(2, -3 + \sqrt{5}\right).$$

The asymptotes are

$$y - (-3) = \frac{1}{2}(x - 2) \quad \text{and} \quad y - (-3) = -\frac{1}{2}(x - 2),$$

or

$$y + 3 = \frac{1}{2}(x - 2) \quad \text{and} \quad y + 3 = -\frac{1}{2}(x - 2).$$

We sketch the asymptotes, plot the vertices, and draw the graph.

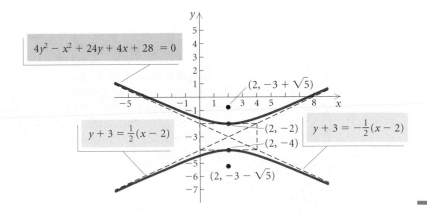

CONNECTING THE CONCEPTS

CLASSIFYING EQUATIONS OF CONIC SECTIONS

EQUATION	TYPE OF CONIC SECTION	GRAPH
$x - 4 + 4y = y^2$	Only one variable is squared, so this cannot be a circle, an ellipse, or a hyperbola. Find an equivalent equation: $$x = (y - 2)^2.$$ This is an equation of a parabola.	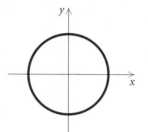
$3x^2 + 3y^2 = 75$	Both variables are squared, so this cannot be a parabola. The squared terms are added, so this cannot be a hyperbola. Divide by 3 on both sides to find an equivalent equation: $$x^2 + y^2 = 25.$$ This is an equation of a circle.	
$y^2 = 16 - 4x^2$	Both variables are squared, so this cannot be a parabola. Add $4x^2$ on both sides to find an equivalent equation: $4x^2 + y^2 = 16$. The squared terms are added, so this cannot be a hyperbola. The coefficients of x^2 and y^2 are not the same, so this is not a circle. Divide by 16 on both sides to find an equivalent equation: $$\frac{x^2}{4} + \frac{y^2}{16} = 1.$$ This is an equation of an ellipse.	
$x^2 = 4y^2 + 36$	Both variables are squared, so this cannot be a parabola. Subtract $4y^2$ on both sides to find an equivalent equation: $x^2 - 4y^2 = 36$. The squared terms are not added, so this cannot be a circle or an ellipse. Divide by 36 on both sides to find an equivalent equation: $$\frac{x^2}{36} - \frac{y^2}{9} = 1.$$ This is an equation of a hyperbola.	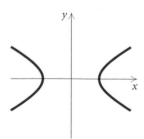

Applications

Some comets travel in hyperbolic paths with the sun at one focus. Such comets pass by the sun only one time, unlike those with elliptical orbits, which reappear at intervals. A cross section of an amphitheater might be one branch of a hyperbola. A cross section of a nuclear cooling tower might also be a hyperbola.

One other application of hyperbolas is in the long-range navigation system LORAN. This system uses transmitting stations in three locations to send out simultaneous signals to a ship or aircraft. The difference in the arrival times of the signals from one pair of transmitters is recorded on the ship or aircraft. This difference is also recorded for signals from another pair of transmitters. For each pair, a computation is performed to determine the difference in the distances from each member of the pair to the ship or aircraft. If each pair of differences is kept constant, two hyperbolas can be drawn. Each has one of the pairs of transmitters as foci, and the ship or aircraft lies on the intersection of two of their branches.

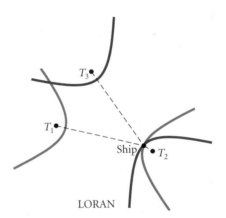

LORAN

6.3

Exercise Set

In Exercises 1–6, match the equation with one of the graphs (a)–(f), which follow.

a)
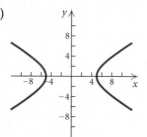

b)
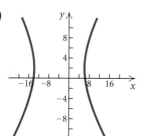

c)
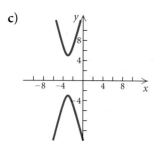

d)
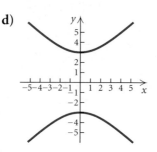

e)
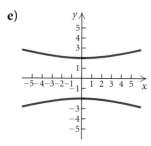

f)

1. $\dfrac{x^2}{25} - \dfrac{y^2}{9} = 1$

2. $\dfrac{y^2}{4} - \dfrac{x^2}{36} = 1$

3. $\dfrac{(y-1)^2}{16} - \dfrac{(x+3)^2}{1} = 1$

4. $\dfrac{(x+4)^2}{100} - \dfrac{(y-2)^2}{81} = 1$

5. $25x^2 - 16y^2 = 400$

6. $y^2 - x^2 = 9$

Find an equation of a hyperbola satisfying the given conditions.

7. Vertices at $(0, 3)$ and $(0, -3)$; foci at $(0, 5)$ and $(0, -5)$

8. Vertices at $(1, 0)$ and $(-1, 0)$; foci at $(2, 0)$ and $(-2, 0)$

9. Asymptotes $y = \frac{3}{2}x$, $y = -\frac{3}{2}x$; one vertex $(2, 0)$

10. Asymptotes $y = \frac{5}{4}x$, $y = -\frac{5}{4}x$; one vertex $(0, 3)$

Find the center, the vertices, the foci, and the asymptotes. Then draw the graph.

11. $\dfrac{x^2}{4} - \dfrac{y^2}{4} = 1$

12. $\dfrac{x^2}{1} - \dfrac{y^2}{9} = 1$

13. $\dfrac{(x-2)^2}{9} - \dfrac{(y+5)^2}{1} = 1$

14. $\dfrac{(x-5)^2}{16} - \dfrac{(y+2)^2}{9} = 1$

15. $\dfrac{(y+3)^2}{4} - \dfrac{(x+1)^2}{16} = 1$

16. $\dfrac{(y+4)^2}{25} - \dfrac{(x+2)^2}{16} = 1$

17. $x^2 - 4y^2 = 4$ 18. $4x^2 - y^2 = 16$

19. $9y^2 - x^2 = 81$ 20. $y^2 - 4x^2 = 4$

21. $x^2 - y^2 = 2$ 22. $x^2 - y^2 = 3$

23. $y^2 - x^2 = \frac{1}{4}$ 24. $y^2 - x^2 = \frac{1}{9}$

Find the center, the vertices, the foci, and the asymptotes of the hyperbola. Then draw the graph.

25. $x^2 - y^2 - 2x - 4y - 4 = 0$

26. $4x^2 - y^2 + 8x - 4y - 4 = 0$

27. $36x^2 - y^2 - 24x + 6y - 41 = 0$

28. $9x^2 - 4y^2 + 54x + 8y + 41 = 0$

29. $9y^2 - 4x^2 - 18y + 24x - 63 = 0$

30. $x^2 - 25y^2 + 6x - 50y = 41$

31. $x^2 - y^2 - 2x - 4y = 4$

32. $9y^2 - 4x^2 - 54y - 8x + 41 = 0$

33. $y^2 - x^2 - 6x - 8y - 29 = 0$

34. $x^2 - y^2 = 8x - 2y - 13$

*The **eccentricity** of a hyperbola is defined as $e = c/a$. For a hyperbola, $c > a > 0$, so $e > 1$. When e is close to 1, a hyperbola appears to be very narrow. As the eccentricity increases, the hyperbola becomes "wider."*

35. Observe the shapes of the hyperbolas in Examples 2 and 3. Which hyperbola has the larger eccentricity? Confirm your answer by computing the eccentricity of each hyperbola.

36. Which hyperbola below has the larger eccentricity? (Assume that the coordinate systems have the same scale.)

a)

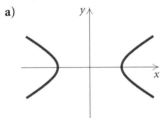

b)

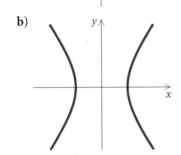

37. Find an equation of a hyperbola with vertices $(3, 7)$ and $(-3, 7)$ and $e = \frac{5}{3}$.

38. Find an equation of a hyperbola with vertices $(-1, 3)$ and $(-1, 7)$ and $e = 4$.

39. *Hyperbolic Mirror.* Certain telescopes contain both a parabolic mirror and a hyperbolic mirror. In the telescope shown in the figure below, the

parabola and the hyperbola share focus F_1, which is 14 m above the vertex of the parabola. The hyperbola's second focus F_2 is 2 m above the parabola's vertex. The vertex of the hyperbolic mirror is 1 m below F_1. Position a coordinate system with the origin at the center of the hyperbola and with the foci on the y-axis. Then find the equation of the hyperbola.

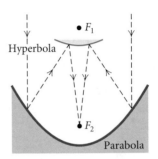

40. *Nuclear Cooling Tower.* A cross section of a nuclear cooling tower is a hyperbola with equation

$$\frac{x^2}{90^2} - \frac{y^2}{130^2} = 1.$$

The tower is 450 ft tall and the distance from the top of the tower to the center of the hyperbola is half the distance from the base of the tower to the center of the hyperbola. Find the diameter of the top and the base of the tower.

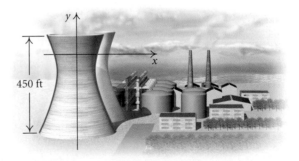

Technology Connection

Use a graphing calculator to find the center, the vertices, and the asymptotes.

41. $5x^2 - 3.5y^2 + 14.6x - 6.7y + 3.4 = 0$

42. $x^2 - y^2 - 2.046x - 4.088y - 4.228 = 0$

Collaborative Discussion and Writing

43. How does the graph of a parabola differ from the graph of one branch of a hyperbola?

44. Are the asymptotes of a hyperbola part of the graph of the hyperbola? Why or why not?

Skill Maintenance

In Exercises 45–48, given the function:

a) *Determine whether it is one-to-one.*
b) *If it is one-to-one, find a formula for the inverse.*

45. $f(x) = 2x - 3$

46. $f(x) = x^3 + 2$

47. $f(x) = \dfrac{5}{x - 1}$

48. $f(x) = \sqrt{x + 4}$

Solve.

49. $x + y = 5,$
$\quad x - y = 7$

50. $3x - 2y = 5,$
$\quad 5x + 2y = 3$

51. $2x - 3y = 7,$
$\quad 3x + 5y = 1$

52. $3x + 2y = -1,$
$\quad 2x + 3y = 6$

Synthesis

Find an equation of a hyperbola satisfying the given conditions.

53. Vertices at $(3, -8)$ and $(3, -2)$;
asymptotes $y = 3x - 14,\ y = -3x + 4$

54. Vertices at $(-9, 4)$ and $(-5, 4)$;
asymptotes $y = 3x + 25,\ y = -3x - 17$

55. *Navigation.*   Two radio transmitters positioned 300 mi apart along the shore send simultaneous signals to a ship that is 200 mi offshore, sailing parallel to the shoreline. The signal from transmitter S reaches the ship 200 microseconds later than the signal from transmitter T. The signals travel at a speed of 186,000 miles per second, or 0.186 mile per microsecond. Find the equation of the hyperbola with foci S and T on which the ship is located. (*Hint*: For any point on the hyperbola, the absolute value of the difference of its distances from the foci is $2a$.)

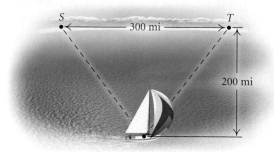

6.4

Nonlinear Systems of Equations

• *Solve a nonlinear system of equations.*
• *Use nonlinear systems of equations to solve applied problems.*

The systems of equations that we have studied so far have been composed of linear equations. Now we consider systems of two equations in two variables in which at least one equation is not linear.

Nonlinear Systems of Equations

The graphs of the equations in a nonlinear system of equations can have no point of intersection or one or more points of intersection. The coordinates of each point of intersection represent a solution of the system of equations. When no point of intersection exists, the system of equations has no real-number solution.

Solutions of nonlinear systems of equations can be found using the substitution or elimination method. The substitution method is preferable for a system consisting of one linear and one nonlinear equation. The elimination method is preferable in most, but not all, cases when both equations are nonlinear.

EXAMPLE 1 Solve the following system of equations:

$$x^2 + y^2 = 25, \qquad (1) \qquad \textbf{The graph is a circle.}$$
$$3x - 4y = 0. \qquad (2) \qquad \textbf{The graph is a line.}$$

Algebraic Solution

We use the substitution method. First, we solve equation (2) for x:

$$x = \tfrac{4}{3}y. \qquad (3) \qquad \textbf{We could have solved for } y \textbf{ instead.}$$

Next, we substitute $\tfrac{4}{3}y$ for x in equation (1) and solve for y:

$$\left(\tfrac{4}{3}y\right)^2 + y^2 = 25$$
$$\tfrac{16}{9}y^2 + y^2 = 25$$
$$\tfrac{25}{9}y^2 = 25$$
$$y^2 = 9 \qquad \textbf{Multiplying by } \tfrac{9}{25}$$
$$y = \pm 3.$$

Now we substitute these numbers for y in equation (3) and solve for x:

$$x = \tfrac{4}{3}(3) = 4, \qquad \textbf{The pair } (4,3) \textbf{ appears to be a solution.}$$
$$x = \tfrac{4}{3}(-3) = -4. \qquad \textbf{The pair } (-4,-3) \textbf{ appears to be a solution.}$$

CHECK: For $(4,3)$:

$$
\begin{array}{c|c}
x^2 + y^2 = 25 & 3x - 4y = 0 \\ \hline
4^2 + 3^2 \ ? \ 25 & 3(4) - 4(3) \ ? \ 0 \\
16 + 9 & 12 - 12 \\
25 \ \big| \ 25 \quad \text{TRUE} & 0 \ \big| \ 0 \quad \text{TRUE}
\end{array}
$$

For $(-4,-3)$:

$$
\begin{array}{c|c}
x^2 + y^2 = 25 & 3x - 4y = 0 \\ \hline
(-4)^2 + (-3)^2 \ ? \ 25 & 3(-4) - 4(-3) \ ? \ 0 \\
16 + 9 & -12 + 12 \\
25 \ \big| \ 25 \quad \text{TRUE} & 0 \ \big| \ 0 \quad \text{TRUE}
\end{array}
$$

The pairs $(4,3)$ and $(-4,-3)$ check, so they are the solutions.

Visualizing the Solution

The ordered pairs corresponding to the points of intersection of the graphs of the equations are the solutions of the system of equations.

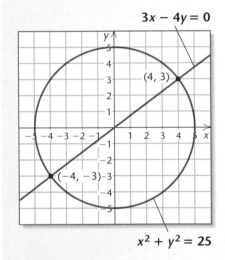

$3x - 4y = 0$

$x^2 + y^2 = 25$

We see that the solutions are $(4,3)$ and $(-4,-3)$.

Technology Connection

To solve the system of equations in Example 1 on a graphing calculator, we graph both equations in the same viewing window. Note that there are two points of intersection. We can find their coordinates using the INTERSECT feature.

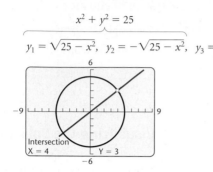
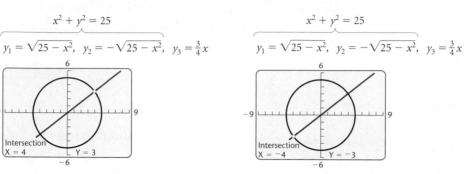

The solutions are $(4, 3)$ and $(-4, -3)$.

The graph can also be used to check the solutions that we found algebraically.

In the solution in Example 1, suppose that to find x, we had substituted 3 and -3 in equation (1) rather than equation (3). If $y = 3$, $y^2 = 9$, and if $y = -3$, $y^2 = 9$, so both substitutions can be performed at the same time:

$$x^2 + y^2 = 25 \qquad (1)$$
$$x^2 + (\pm 3)^2 = 25$$
$$x^2 + 9 = 25$$
$$x^2 = 16$$
$$x = \pm 4.$$

Thus, if $y = 3$, $x = 4$ or $x = -4$, and if $y = -3$, $x = 4$ or $x = -4$. The possible solutions are $(4, 3)$, $(-4, 3)$, $(4, -3)$, and $(-4, -3)$. A check reveals that $(4, -3)$ and $(-4, 3)$ are not solutions of equation (2). Since a circle and a line can intersect in at most two points, it is clear that there can be at most two real-number solutions.

EXAMPLE 2 Solve the following system of equations:

$$x + y = 5, \quad (1) \qquad \text{The graph is a line.}$$
$$y = 3 - x^2. \quad (2) \qquad \text{The graph is a parabola.}$$

Algebraic Solution

We use the substitution method, substituting $3 - x^2$ for y in equation (1):

$$x + 3 - x^2 = 5$$
$$-x^2 + x - 2 = 0 \qquad \text{Subtracting 5 and rearranging}$$
$$x^2 - x + 2 = 0. \qquad \text{Multiplying by } -1$$

Next, we use the quadratic formula:

$$x = \frac{-b \pm \sqrt{b^2 - 4ac}}{2a} = \frac{-(-1) \pm \sqrt{(-1)^2 - 4(1)(2)}}{2(1)}$$
$$= \frac{1 \pm \sqrt{1 - 8}}{2} = \frac{1 \pm \sqrt{-7}}{2} = \frac{1 \pm i\sqrt{7}}{2} = \frac{1}{2} \pm \frac{\sqrt{7}}{2}i.$$

Now, we substitute these values for x in equation (1) and solve for y:

$$\frac{1}{2} + \frac{\sqrt{7}}{2}i + y = 5$$
$$y = 5 - \frac{1}{2} - \frac{\sqrt{7}}{2}i = \frac{9}{2} - \frac{\sqrt{7}}{2}i$$

and

$$\frac{1}{2} - \frac{\sqrt{7}}{2}i + y = 5$$
$$y = 5 - \frac{1}{2} + \frac{\sqrt{7}}{2}i = \frac{9}{2} + \frac{\sqrt{7}}{2}i.$$

The solutions are

$$\left(\frac{1}{2} + \frac{\sqrt{7}}{2}i, \frac{9}{2} - \frac{\sqrt{7}}{2}i \right) \quad \text{and} \quad \left(\frac{1}{2} - \frac{\sqrt{7}}{2}i, \frac{9}{2} + \frac{\sqrt{7}}{2}i \right).$$

There are no real-number solutions.

Visualizing the Solution

We graph the equations, as shown below.

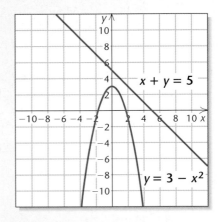

Note that there are no points of intersection. This indicates that there are no real-number solutions.

EXAMPLE 3 Solve the following system of equations:

$$2x^2 + 5y^2 = 39, \quad (1) \qquad \text{The graph is an ellipse.}$$
$$3x^2 - y^2 = -1. \quad (2) \qquad \text{The graph is a hyperbola.}$$

Algebraic Solution

We use the elimination method. First, we multiply equation (2) by 5 and add to eliminate the y^2-term:

$$
\begin{array}{ll}
2x^2 + 5y^2 = 39 & (1) \\
\underline{15x^2 - 5y^2 = -5} & \text{Multiplying (2) by 5} \\
17x^2 = 34 & \text{Adding} \\
x^2 = 2 & \\
x = \pm\sqrt{2}. &
\end{array}
$$

If $x = \sqrt{2}$, $x^2 = 2$, and if $x = -\sqrt{2}$, $x^2 = 2$. Thus substituting $\sqrt{2}$ or $-\sqrt{2}$ for x in equation (2) gives us

$$
\begin{aligned}
3\left(\pm\sqrt{2}\right)^2 - y^2 &= -1 \\
3 \cdot 2 - y^2 &= -1 \\
6 - y^2 &= -1 \\
-y^2 &= -7 \\
y^2 &= 7 \\
y &= \pm\sqrt{7}.
\end{aligned}
$$

Thus, for $x = \sqrt{2}$, we have $y = \sqrt{7}$ or $y = -\sqrt{7}$, and for $x = -\sqrt{2}$, we have $y = \sqrt{7}$ or $y = -\sqrt{7}$. The possible solutions are $\left(\sqrt{2}, \sqrt{7}\right)$, $\left(\sqrt{2}, -\sqrt{7}\right)$, $\left(-\sqrt{2}, \sqrt{7}\right)$, and $\left(-\sqrt{2}, -\sqrt{7}\right)$. All four pairs check, so they are the solutions.

Visualizing the Solution

We graph the equations and observe that there are four points of intersection.

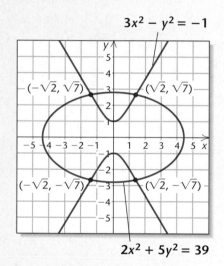

The coordinates of the points of intersection are $\left(\sqrt{2}, \sqrt{7}\right)$, $\left(\sqrt{2}, -\sqrt{7}\right)$, $\left(-\sqrt{2}, \sqrt{7}\right)$, and $\left(-\sqrt{2}, -\sqrt{7}\right)$. These are the solutions of the system of equations.

Technology Connection

To solve the system of equations in Example 3 graphically on a graphing calculator, we first graph both equations in the same viewing window. There are four points of intersection. We can use the INTERSECT feature to find their coordinates.

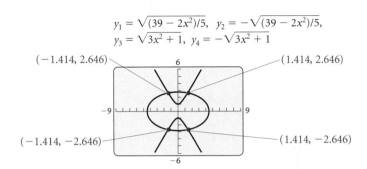

$$y_1 = \sqrt{(39 - 2x^2)/5}, \quad y_2 = -\sqrt{(39 - 2x^2)/5},$$
$$y_3 = \sqrt{3x^2 + 1}, \quad y_4 = -\sqrt{3x^2 + 1}$$

$(-1.414, 2.646)$ $(1.414, 2.646)$

$(-1.414, -2.646)$ $(1.414, -2.646)$

Note that the algebraic method yields exact solutions, whereas the graphical method yields decimal approximations of the solutions on most graphing calculators.

The solutions are approximately $(1.414, 2.646)$, $(1.414, -2.646)$, $(-1.414, 2.646)$, and $(-1.414, -2.646)$.

EXAMPLE 4 Solve the following system of equations:

$$x^2 - 3y^2 = 6, \qquad (1)$$
$$xy = 3. \qquad (2)$$

Algebraic Solution

We use the substitution method. First, we solve equation (2) for y:

$$xy = 3 \qquad (2)$$

$$y = \frac{3}{x}. \qquad (3) \qquad \textbf{Dividing by } x$$

Next, we substitute $3/x$ for y in equation (1) and solve for x:

$$x^2 - 3\left(\frac{3}{x}\right)^2 = 6$$

$$x^2 - 3 \cdot \frac{9}{x^2} = 6$$

$$x^2 - \frac{27}{x^2} = 6$$

$$x^4 - 27 = 6x^2 \qquad \textbf{Multiplying by } x^2$$

$$x^4 - 6x^2 - 27 = 0$$

$$u^2 - 6u - 27 = 0 \qquad \textbf{Letting } u = x^2$$

$$(u - 9)(u + 3) = 0 \qquad \textbf{Factoring}$$

$$u = 9 \quad or \quad u = -3 \qquad \textbf{Principle of zero products}$$

$$x^2 = 9 \quad or \quad x^2 = -3 \qquad \textbf{Substituting } x^2 \textbf{ for } u$$

$$x = \pm 3 \quad or \quad x = \pm i\sqrt{3}.$$

Since $y = 3/x$,

when $x = 3$, $\qquad y = \dfrac{3}{3} = 1;$

when $x = -3$, $\qquad y = \dfrac{3}{-3} = -1;$

when $x = i\sqrt{3}$, $\quad y = \dfrac{3}{i\sqrt{3}} = \dfrac{3}{i\sqrt{3}} \cdot \dfrac{-i\sqrt{3}}{-i\sqrt{3}} = -i\sqrt{3};$

when $x = -i\sqrt{3}$, $\quad y = \dfrac{3}{-i\sqrt{3}} = \dfrac{3}{-i\sqrt{3}} \cdot \dfrac{i\sqrt{3}}{i\sqrt{3}} = i\sqrt{3}.$

The pairs $(3, 1)$, $(-3, -1)$, $\left(i\sqrt{3}, -i\sqrt{3}\right)$, and $\left(-i\sqrt{3}, i\sqrt{3}\right)$ check, so they are the solutions.

Visualizing the Solution

The coordinates of the points of intersection of the graphs of the equations give us the real-number solutions of the system of equations. These graphs do not show us the imaginary-number solutions.

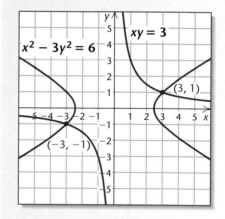

Modeling and Problem Solving

EXAMPLE 5 *Dimensions of a Piece of Land.* For a student recreation building at Southport Community College, an architect wants to lay out a rectangular piece of land that has a perimeter of 204 m and an area of 2565 m^2. Find the dimensions of the piece of land.

Solution

1. **Familiarize.** We make a drawing and label it, letting $l =$ the length of the piece of land, in meters, and $w =$ the width, in meters.

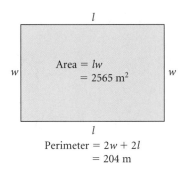

2. **Translate.** We now have the following:

   Perimeter: $2w + 2l = 204$,   (1)
   Area: $lw = 2565$.   (2)

3. **Carry out.** We solve the system of equations:

   $$2w + 2l = 204,$$
   $$lw = 2565.$$

   Solving the second equation for l gives us $l = 2565/w$. We then substitute $2565/w$ for l in equation (1) and solve for w:

   $$2w + 2\left(\frac{2565}{w}\right) = 204$$

   $$2w^2 + 2(2565) = 204w \qquad \textbf{Multiplying by } w$$

   $$2w^2 - 204w + 2(2565) = 0$$

   $$w^2 - 102w + 2565 = 0 \qquad \textbf{Multiplying by } \tfrac{1}{2}$$

   $$(w - 57)(w - 45) = 0$$

   $$w = 57 \quad or \quad w = 45. \qquad \textbf{Principle of zero products}$$

   If $w = 57$, then $l = 2565/w = 2565/57 = 45$. If $w = 45$, then $l = 2565/w = 2565/45 = 57$. Since length is generally considered to be longer than width, we have the solution $l = 57$ and $w = 45$, or $(57, 45)$.

4. **Check.** If $l = 57$ and $w = 45$, the perimeter is $2 \cdot 45 + 2 \cdot 57$, or 204. The area is $57 \cdot 45$, or 2565. The numbers check.

5. **State.** The length of the piece of land is 57 m and the width is 45 m.

Study Tip

Prepare for each chapter test by rereading the text, reviewing your homework, reading the important properties and formulas in the Chapter Summary and Review, and doing the review exercises. Then take the chapter test at the end of the chapter.

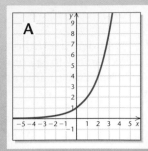

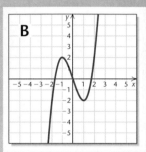

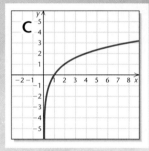

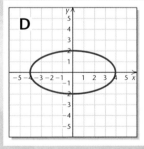

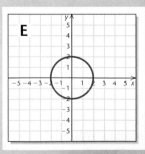

Visualizing the Graph

Match the equation with its graph.

1. $y = x^3 - 3x$

2. $y = x^2 + 2x - 3$

3. $y = \dfrac{x - 1}{x^2 - x - 2}$

4. $y = -3x + 2$

5. $x + y = 3,$
$2x + 5y = 3$

6. $9x^2 - 4y^2 = 36,$
$x^2 + y^2 = 9$

7. $5x^2 + 5y^2 = 20$

8. $4x^2 + 16y^2 = 64$

9. $y = \log_2 x$

10. $y = 2^x$

Answers on page A-46

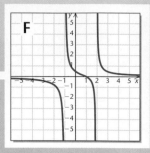

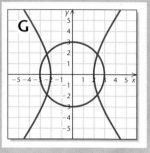

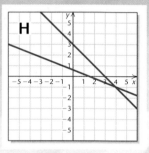

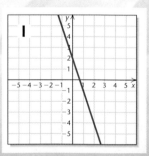

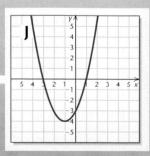

6.4

Exercise Set

In Exercises 1–6, match the system of equations with one of the graphs (a)–(f), which follow.

a)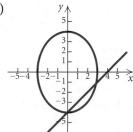

b)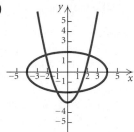

c)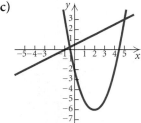

d)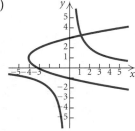

e)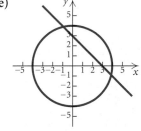

f)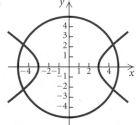

1. $x^2 + y^2 = 16,$
$\quad x + y = 3$

2. $16x^2 + 9y^2 = 144,$
$\quad x - y = 4$

3. $y = x^2 - 4x - 2,$
$\quad 2y - x = 1$

4. $4x^2 - 9y^2 = 36,$
$\quad x^2 + y^2 = 25$

5. $y = x^2 - 3,$
$\quad x^2 + 4y^2 = 16$

6. $y^2 - 2y = x + 3,$
$\quad xy = 4$

Solve.

7. $x^2 + y^2 = 25,$
$\quad y - x = 1$

8. $x^2 + y^2 = 100,$
$\quad y - x = 2$

9. $4x^2 + 9y^2 = 36,$
$\quad 3y + 2x = 6$

10. $9x^2 + 4y^2 = 36,$
$\quad 3x + 2y = 6$

11. $x^2 + y^2 = 25,$
$\quad y^2 = x + 5$

12. $y = x^2,$
$\quad x = y^2$

13. $x^2 + y^2 = 9,$
$\quad x^2 - y^2 = 9$

14. $y^2 - 4x^2 = 4,$
$\quad 4x^2 + y^2 = 4$

15. $y^2 - x^2 = 9,$
$\quad 2x - 3 = y$

16. $x + y = -6,$
$\quad xy = -7$

17. $y^2 = x + 3,$
$\quad 2y = x + 4$

18. $y = x^2,$
$\quad 3x = y + 2$

19. $x^2 + y^2 = 25,$
$\quad xy = 12$

20. $x^2 - y^2 = 16,$
$\quad x + y^2 = 4$

21. $x^2 + y^2 = 4,$
$\quad 16x^2 + 9y^2 = 144$

22. $x^2 + y^2 = 25,$
$\quad 25x^2 + 16y^2 = 400$

23. $x^2 + 4y^2 = 25,$
$\quad x + 2y = 7$

24. $y^2 - x^2 = 16,$
$\quad 2x - y = 1$

25. $x^2 - xy + 3y^2 = 27,$
$\quad x - y = 2$

26. $2y^2 + xy + x^2 = 7,$
$\quad x - 2y = 5$

27. $x^2 + y^2 = 16,$
$\quad y^2 - 2x^2 = 10$

28. $x^2 + y^2 = 14,$
$\quad x^2 - y^2 = 4$

29. $x^2 + y^2 = 5,$
$\quad xy = 2$

30. $x^2 + y^2 = 20,$
$\quad xy = 8$

31. $3x + y = 7,$
$\quad 4x^2 + 5y = 56$

32. $2y^2 + xy = 5,$
$\quad 4y + x = 7$

33. $a + b = 7,$
$\quad ab = 4$

34. $p + q = -4,$
$\quad pq = -5$

35. $x^2 + y^2 = 13,$
$\quad xy = 6$

36. $x^2 + 4y^2 = 20,$
$\quad xy = 4$

37. $x^2 + y^2 + 6y + 5 = 0,$
$\quad x^2 + y^2 - 2x - 8 = 0$

38. $2xy + 3y^2 = 7,$
$\quad 3xy - 2y^2 = 4$

39. $2a + b = 1,$
$\quad b = 4 - a^2$

40. $4x^2 + 9y^2 = 36,$
$\quad x + 3y = 3$

41. $a^2 + b^2 = 89,$
$\quad a - b = 3$

42. $xy = 4,$
$\quad x + y = 5$

43. $xy - y^2 = 2,$
$\quad 2xy - 3y^2 = 0$

44. $4a^2 - 25b^2 = 0,$
$\quad 2a^2 - 10b^2 = 3b + 4$

45. $m^2 - 3mn + n^2 + 1 = 0,$
$\quad 3m^2 - mn + 3n^2 = 13$

46. $ab - b^2 = -4,$
$\quad ab - 2b^2 = -6$

47. $x^2 + y^2 = 5$,
$x - y = 8$

48. $4x^2 + 9y^2 = 36$,
$y - x = 8$

49. $a^2 + b^2 = 14$,
$ab = 3\sqrt{5}$

50. $x^2 + xy = 5$,
$2x^2 + xy = 2$

51. $x^2 + y^2 = 25$,
$9x^2 + 4y^2 = 36$

52. $x^2 + y^2 = 1$,
$9x^2 - 16y^2 = 144$

53. $5y^2 - x^2 = 1$,
$xy = 2$

54. $x^2 - 7y^2 = 6$,
$xy = 1$

55. *Picture Frame Dimensions.* Frank's Frame Shop is building a frame for a rectangular oil painting with a perimeter of 28 cm and a diagonal of 10 cm. Find the dimensions of the painting.

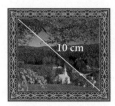

56. *Landscaping.* Green Leaf Landscaping is planting a rectangular wildflower garden with a perimeter of 6 m and a diagonal of $\sqrt{5}$ m. Find the dimensions of the garden.

57. *Graphic Design.* Marcia Graham, owner of Graham's Graphics, is designing an advertising brochure for the Art League's spring show. Each page of the brochure is rectangular with an area of 20 in^2 and a perimeter of 18 in. Find the dimensions of the brochure.

58. *Sign Dimensions.* Peden's Advertising is building a rectangular sign with an area of 2 yd^2 and a perimeter of 6 yd. Find the dimensions of the sign.

59. *Banner Design.* A rectangular banner with an area of $\sqrt{3}$ m^2 is being designed to advertise an exhibit at the Davis Gallery. The length of a diagonal is 2 m. Find the dimensions of the banner.

60. *Carpentry.* A carpenter wants to make a rectangular tabletop with an area of $\sqrt{2}$ m^2 and a diagonal of $\sqrt{3}$ m. Find the dimensions of the tabletop.

61. *Fencing.* It will take 210 yd of fencing to enclose a rectangular dog run. The area of the run is 2250 yd^2. What are the dimensions of the run?

62. *Office Dimensions.* The diagonal of the floor of a rectangular office cubicle is 1 ft longer than the length of the cubicle and 3 ft longer than twice the width. Find the dimensions of the cubicle.

63. *Seed Test Plots.* The Burton Seed Company has two square test plots. The sum of their areas is 832 ft^2 and the difference of their areas is 320 ft^2. Find the length of a side of each plot.

64. *Investment.* Jenna made an investment for 1 yr that earned $7.50 simple interest. If the principal had been $25 more and the interest rate 1% less, the interest would have been the same. Find the principal and the interest rate.

Technology Connection

Solve using a graphing calculator.

65. $y - \ln x = 2$,
$y = x^2$

66. $y = \ln(x + 4)$,
$x^2 + y^2 = 6$

67. $e^x - y = 1$,
$3x + y = 4$

68. $y - e^{-x} = 1$,
$y = 2x + 5$

69. $y = e^x$,
$x - y = -2$

70. $y = e^{-x}$,
$x + y = 3$

71. $x^2 + y^2 = 19{,}380{,}510.36$,
$27{,}942.25x - 6.125y = 0$

72. $2x + 2y = 1660$,
$xy = 35{,}325$

73. $14.5x^2 - 13.5y^2 - 64.5 = 0$,
$5.5x - 6.3y - 12.3 = 0$

74. $13.5xy + 15.6 = 0$,
$5.6x - 6.7y - 42.3 = 0$

75. $0.319x^2 + 2688.7y^2 = 56{,}548$,
$0.306x^2 - 2688.7y^2 = 43{,}452$

76. $18.465x^2 + 788.723y^2 = 6408$,
$106.535x^2 - 788.723y^2 = 2692$

Collaborative Discussion and Writing

77. Which conic sections, if any, have equations that can be expressed using function notation? Explain why you answered as you did.

78. Write a problem that can be translated to a nonlinear system of equations, and ask a classmate to solve it. Devise the problem so that the solution is "The dimensions of the rectangle are 6 ft by 8 ft."

Skill Maintenance

Solve.

79. $2^{3x} = 64$

80. $5^x = 27$

81. $\log_3 x = 4$

82. $\log(x - 3) + \log x = 1$

Synthesis

83. Find an equation of the circle that passes through the points $(2, 4)$ and $(3, 3)$ and whose center is on the line $3x - y = 3$.

84. Find an equation of the circle that passes through the points $(-2, 3)$ and $(-4, 1)$ and whose center is on the line $5x + 8y = -2$.

85. Find an equation of an ellipse centered at the origin that passes through the points $\left(1, \sqrt{3}/2\right)$ and $\left(\sqrt{3}, 1/2\right)$.

86. Find an equation of a hyperbola of the type

$$\frac{x^2}{b^2} - \frac{y^2}{a^2} = 1$$

that passes through the points $\left(-3, -3\sqrt{5}/2\right)$ and $(-3/2, 0)$.

87. Find an equation of the circle that passes through the points $(4, 6)$, $(-6, 2)$, and $(1, -3)$.

88. Find an equation of the circle that passes through the points $(2, 3)$, $(4, 5)$, and $(0, -3)$.

89. Show that a hyperbola does not intersect its asymptotes. That is, solve the system of equations

$$\frac{x^2}{a^2} - \frac{y^2}{b^2} = 1,$$

$$y = \frac{b}{a}x \ \left(\text{or } y = -\frac{b}{a}x\right).$$

90. *Numerical Relationship.* Find two numbers whose product is 2 and the sum of whose reciprocals is $\frac{33}{8}$.

91. *Numerical Relationship.* The square of a number exceeds twice the square of another number by $\frac{1}{8}$. The sum of their squares is $\frac{5}{16}$. Find the numbers.

92. *Box Dimensions.* Four squares with sides 5 in. long are cut from the corners of a rectangular metal sheet that has an area of 340 in². The edges are bent up to form an open box with a volume of 350 in³. Find the dimensions of the box.

93. *Numerical Relationship.* The sum of two numbers is 1, and their product is 1. Find the sum of their cubes. There is a method to solve this problem that is easier than solving a nonlinear system of equations. Can you discover it?

94. Solve for x and y:

$$x^2 - y^2 = a^2 - b^2,$$
$$x - y = a - b.$$

Solve.

95. $x^3 + y^3 = 72,$
$x + y = 6$

96. $a + b = \dfrac{5}{6},$
$\dfrac{a}{b} + \dfrac{b}{a} = \dfrac{13}{6}$

97. $p^2 + q^2 = 13,$
$\dfrac{1}{pq} = -\dfrac{1}{6}$

98. $x^2 + y^2 = 4,$
$(x - 1)^2 + y^2 = 4$

99. $5^{x+y} = 100,$
$3^{2x-y} = 1000$

100. $e^x - e^{x+y} = 0,$
$e^y - e^{x-y} = 0$

Chapter 6 Summary and Review

Important Properties and Formulas

Standard Equation of a Parabola with Vertex at the Origin

The standard equation of a parabola with vertex $(0,0)$ and directrix $y = -p$ is

$$x^2 = 4py.$$

The focus is $(0, p)$ and the y-axis is the axis of symmetry.

The standard equation of a parabola with vertex $(0,0)$ and directrix $x = -p$ is

$$y^2 = 4px.$$

The focus is $(p, 0)$ and the x-axis is the axis of symmetry.

Standard Equation of a Parabola with Vertex (h,k) and Vertical Axis of Symmetry

The standard equation of a parabola with vertex (h, k) and vertical axis of symmetry is

$$(x - h)^2 = 4p(y - k),$$

where the vertex is (h, k), the focus is $(h, k + p)$, and the directrix is $y = k \quad p$.

Standard Equation of a Parabola with Vertex (h,k) and Horizontal Axis of Symmetry

The standard equation of a parabola with vertex (h, k) and horizontal axis of symmetry is

$$(y - k)^2 = 4p(x - h),$$

where the vertex is (h, k), the focus is $(h + p, k)$, and the directrix is $x = h - p$.

Standard Equation of a Circle

The standard equation of a circle with center (h, k) and radius r is

$$(x - h)^2 + (y - k)^2 = r^2.$$

Standard Equation of an Ellipse with Center at the Origin

Major axis horizontal

$$\frac{x^2}{a^2} + \frac{y^2}{b^2} = 1, \quad a > b > 0$$

Vertices: $(-a, 0), (a, 0)$

y-intercepts: $(0, -b), (0, b)$

Foci: $(-c, 0), (c, 0)$, where $c^2 = a^2 - b^2$

(continued)

Major axis vertical

$$\frac{x^2}{b^2} + \frac{y^2}{a^2} = 1, \ a > b > 0$$

Vertices: $(0, -a), (0, a)$

x-intercepts: $(-b, 0), (b, 0)$

Foci: $(0, -c), (0, c)$, where $c^2 = a^2 - b^2$

Standard Equation of an Ellipse with Center at (h, k)

Major axis horizontal

$$\frac{(x - h)^2}{a^2} + \frac{(y - k)^2}{b^2} = 1, \ a > b > 0$$

Vertices: $(h - a, k), (h + a, k)$

Length of minor axis: $2b$

Foci: $(h - c, k), (h + c, k)$, where $c^2 = a^2 - b^2$

Major axis vertical

$$\frac{(x - h)^2}{b^2} + \frac{(y - k)^2}{a^2} = 1, \ a > b > 0$$

Vertices: $(h, k - a), (h, k + a)$

Length of minor axis: $2b$

Foci: $(h, k - c), (h, k + c)$, where $c^2 = a^2 - b^2$

Standard Equation of a Hyperbola with Center at the Origin

Transverse axis horizontal

$$\frac{x^2}{a^2} - \frac{y^2}{b^2} = 1$$

Vertices: $(-a, 0), (a, 0)$

Asymptotes: $y = -\frac{b}{a}x, y = \frac{b}{a}x$

Foci: $(-c, 0), (c, 0)$, where $c^2 = a^2 + b^2$

Transverse axis vertical

$$\frac{y^2}{a^2} - \frac{x^2}{b^2} = 1$$

Vertices: $(0, -a), (0, a)$

Asymptotes: $y = -\frac{a}{b}x, y = \frac{a}{b}x$

Foci: $(0, -c), (0, c)$, where $c^2 = a^2 + b^2$

Standard Equation of a Hyperbola with Center at (h, k)

Transverse axis horizontal

$$\frac{(x - h)^2}{a^2} - \frac{(y - k)^2}{b^2} = 1$$

Vertices: $(h - a, k), (h + a, k)$

Asymptotes: $y - k = \frac{b}{a}(x - h),$

$$y - k = -\frac{b}{a}(x - h)$$

Foci: $(h - c, k), (h + c, k)$, where $c^2 = a^2 + b^2$

Transverse axis vertical

$$\frac{(y - k)^2}{a^2} - \frac{(x - h)^2}{b^2} = 1$$

Vertices: $(h, k - a), (h, k + a)$

Asymptotes: $y - k = \frac{a}{b}(x - h),$

$$y - k = -\frac{a}{b}(x - h)$$

Foci: $(h, k - c), (h, k + c)$, where $c^2 = a^2 + b^2$

Review Exercises

In Exercises 1–8, match the equation with one of the graphs (a)–(h), which follow.

a)

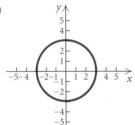

b)

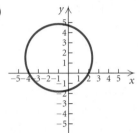

c)

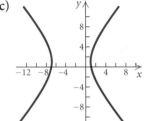

d)

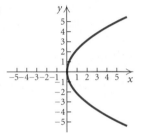

e)

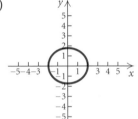

f)

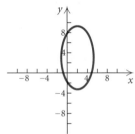

g)

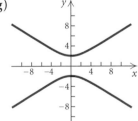

h)

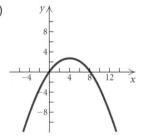

1. $y^2 = 5x$

2. $y^2 = 9 - x^2$

3. $3x^2 + 4y^2 = 12$

4. $9y^2 - 4x^2 = 36$

5. $x^2 + y^2 + 2x - 3y = 8$

6. $4x^2 + y^2 - 16x - 6y = 15$

7. $x^2 - 8x + 6y = 0$

8. $\dfrac{(x + 3)^2}{16} - \dfrac{(y - 1)^2}{25} = 1$

9. Find an equation of the parabola with directrix $y = \frac{3}{2}$ and focus $\left(0, -\frac{3}{2}\right)$.

10. Find the focus, the vertex, and the directrix of the parabola given by
$$y^2 = -12x.$$

11. Find the vertex, the focus, and the directrix of the parabola given by
$$x^2 + 10x + 2y + 9 = 0.$$

12. Find the center, the vertices, and the foci of the ellipse given by
$$16x^2 + 25y^2 - 64x + 50y - 311 = 0.$$
Then draw the graph.

13. Find an equation of the ellipse having vertices $(0, -4)$ and $(0, 4)$ with minor axis of length 6.

14. Find the center, the vertices, the foci, and the asymptotes of the hyperbola given by
$$x^2 - 2y^2 + 4x + y - \tfrac{1}{8} = 0.$$

15. *Spotlight.* A spotlight has a parabolic cross section that is 2 ft wide at the opening and 1.5 ft deep at the vertex. How far from the vertex is the focus?

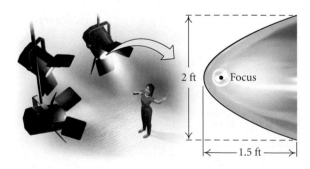

Solve.

16. $x^2 - 16y = 0,$
$x^2 - y^2 = 64$

17. $4x^2 + 4y^2 = 65,$
$6x^2 - 4y^2 = 25$

18. $x^2 - y^2 = 33,$
$x + y = 11$

19. $x^2 - 2x + 2y^2 = 8,$
$2x + y = 6$

20. $x^2 - y = 3,$
$\quad 2x - y = 3$

21. $x^2 + y^2 = 25,$
$\quad x^2 - y^2 = 7$

22. $x^2 - y^2 = 3,$
$\quad y = x^2 - 3$

23. $x^2 + y^2 = 18,$
$\quad 2x + y = 3$

24. $x^2 + y^2 = 100,$
$\quad 2x^2 - 3y^2 = -120$

25. $x^2 + 2y^2 = 12,$
$\quad xy = 4$

26. *Numerical Relationship.* The sum of two numbers is 11 and the sum of their squares is 65. Find the numbers.

27. *Dimensions of a Rectangle.* A rectangle has a perimeter of 38 m and an area of 84 m². What are the dimensions of the rectangle?

28. *Numerical Relationship.* Find two positive integers whose sum is 12 and the sum of whose reciprocals is $\frac{3}{8}$.

29. *Perimeter.* The perimeter of a square is 12 cm more than the perimeter of another square. The area of the first square exceeds the area of the other by 39 cm². Find the perimeter of each square.

30. *Radius of a Circle.* The sum of the areas of two circles is 130π ft². The difference of the areas is 112π ft². Find the radius of each circle.

Collaborative Discussion and Writing

31. What would you say to a classmate who tells you that it is always possible to visualize all of the solutions of a nonlinear system of equations?

32. Is a circle a special type of ellipse? Why or why not?

Synthesis

33. Find an equation of the ellipse containing the point $\left(-1/2, 3\sqrt{3}/2\right)$ and with vertices $(0, -3)$ and $(0, 3)$.

34. Find two numbers whose product is 4 and the sum of whose reciprocals is $\frac{65}{56}$.

35. Find an equation of the circle that passes through the points $(10, 7)$, $(-6, 7)$, and $(-8, 1)$.

36. *Navigation.* Two radio transmitters positioned 400 mi apart along the shore send simultaneous signals to a ship that is 250 mi offshore, sailing parallel to the shoreline. The signal from transmitter A reaches the ship 300 microseconds before the signal from transmitter B. The signals travel at a speed of 186,000 miles per second, or 0.186 mile per microsecond. Find the equation of the hyperbola with foci A and B on which the ship is located. (*Hint:* For any point on the hyperbola, the absolute value of the difference of its distances from the foci is $2a$.)

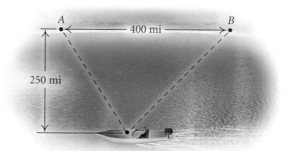

Chapter 6 Test

In Exercises 1–4, match the equation with one of the graphs (a)–(d), which follow.

a)

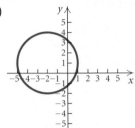

b)

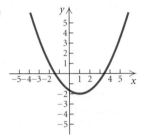

c)

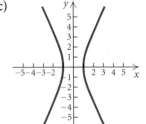

d)

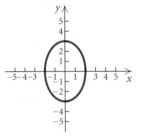

1. $4x^2 - y^2 = 4$

2. $x^2 - 2x - 3y = 5$

3. $x^2 + 4x + y^2 - 2y - 4 = 0$

4. $9x^2 + 4y^2 = 36$

Find the vertex, the focus, and the directrix of the parabola. Then draw the graph.

5. $x^2 = 12y$

6. $y^2 + 2y - 8x - 7 = 0$

7. Find an equation of the parabola with focus $(0, 2)$ and directrix $y = -2$.

8. Find the center and the radius of the circle given by $x^2 + y^2 + 2x - 6y - 15 = 0$. Then draw the graph.

Find the center, the vertices, and the foci of the ellipse. Then draw the graph.

9. $9x^2 + 16y^2 - 144$

10. $\dfrac{(x + 1)^2}{4} + \dfrac{(y - 2)^2}{9} = 1$

11. Find an equation of the ellipse having vertices $(0, -5)$ and $(0, 5)$ and with minor axis of length 4.

Find the center, the vertices, the foci, and the asymptotes of the hyperbola. Then draw the graph.

12. $4x^2 - y^2 = 4$

13. $\dfrac{(y - 2)^2}{4} - \dfrac{(x + 1)^2}{9} = 1$

14. Find the asymptotes of the hyperbola given by $2y^2 - x^2 = 18$.

15. *Satellite Dish.*   A satellite dish has a parabolic cross section that is 18 in. wide at the opening and 6 in. deep at the vertex. How far from the vertex is the focus?

Solve.

16. $2x^2 - 3y^2 = -10$,
    $x^2 + 2y^2 = 9$

17. $x^2 + y^2 = 13$,
    $x + y = 1$

18. $x + y = 5$,
    $xy = 6$

19. *Landscaping.*   Leisurescape is planting a rectangular flower garden with a perimeter of 18 ft and a diagonal of $\sqrt{41}$ ft. Find the dimensions of the garden.

20. *Fencing.*   It will take 210 ft of fencing to enclose a rectangular playground with an area of 2700 ft^2. Find the dimensions of the playground.

Synthesis

21. Find an equation of the circle for which the endpoints of a diameter are $(1, 1)$ and $(5, -3)$.

Sequences, Series, and Combinatorics

7

A P P L I C A T I O N

When a parachutist jumps from an airplane, the distances, in feet, that the parachutist falls in each successive second before pulling the ripcord to release the parachute are as follows:

$$16, 48, 80, 112, 144, \ldots .$$

What is the total distance fallen after 10 sec?

This problem appears as Exercise 44 in Section 7.2.

549

7.1

Sequences and Series

- *Find terms of sequences given the nth term.*
- *Look for a pattern in a sequence and try to determine a general term.*
- *Convert between sigma notation and other notation for a series.*
- *Construct the terms of a recursively defined sequence.*

In this section, we discuss sets or lists of numbers, considered in order, and their sums.

Sequences

Suppose that $1000 is invested at 6%, compounded annually. The amounts to which the account will grow after 1 yr, 2 yr, 3 yr, 4 yr, and so on, form the following sequence of numbers:

$$\begin{array}{cccc} (1) & (2) & (3) & (4) \\ \downarrow & \downarrow & \downarrow & \downarrow \\ \$1060.00, & \$1123.60, & \$1191.02, & \$1262.48, \dots \end{array}$$

We can think of this as a function that pairs 1 with $1060.00, 2 with $1123.60, 3 with $1191.02, and so on. A **sequence** is thus a *function*, where the domain is a set of consecutive positive integers beginning with 1.

If we continue to compute the amounts of money in the account forever, we obtain an **infinite sequence** with function values

$$\$1060.00, \$1123.60, \$1191.02, \$1262.48, \$1338.23, \$1418.52, \dots .$$

The dots "..." at the end indicate that the sequence goes on without stopping. If we stop after a certain number of years, we obtain a **finite sequence:**

$$\$1060.00, \$1123.60, \$1191.02, \$1262.48.$$

Sequences

An **infinite sequence** is a function having for its domain the set of positive integers, $\{1, 2, 3, 4, 5, \dots\}$.

A **finite sequence** is a function having for its domain a set of positive integers, $\{1, 2, 3, 4, 5, \dots, n\}$, for some positive integer n.

Consider the sequence given by the formula

$$a(n) = 2^n, \quad \text{or} \quad a_n = 2^n.$$

Some of the function values, also known as the **terms** of the sequence, follow:

$$a_1 = 2^1 = 2,$$
$$a_2 = 2^2 = 4,$$
$$a_3 = 2^3 = 8,$$
$$a_4 = 2^4 = 16,$$
$$a_5 = 2^5 = 32.$$

The first term of the sequence is denoted as a_1, the fifth term as a_5, and the nth term, or **general term,** as a_n. This sequence can also be denoted as

$$2, 4, 8, \ldots, \quad \text{or as} \quad 2, 4, 8, \ldots, 2^n, \ldots .$$

EXAMPLE 1   Find the first 4 terms and the 23rd term of the sequence whose general term is given by $a_n = (-1)^n n^2$.

Solution   We have $a_n = (-1)^n n^2$, so

$$a_1 = (-1)^1 \cdot 1^2 = -1,$$
$$a_2 = (-1)^2 \cdot 2^2 = 4,$$
$$a_3 = (-1)^3 \cdot 3^2 = -9,$$
$$a_4 = (-1)^4 \cdot 4^2 = 16,$$
$$a_{23} = (-1)^{23} \cdot 23^2 = -529.$$

Note in Example 1 that the power $(-1)^n$ causes the signs of the terms to alternate between positive and negative, depending on whether n is even or odd. This kind of sequence is called an **alternating sequence.**

Technology Connection

We can use a graphing calculator to find the desired terms of the sequence in Example 1. We enter $y_1 = (-1)^x x^2$. We then set up a table in ASK mode and enter 1, 2, 3, 4, and 23 as values for x.

We can also use the SEQ feature to find the terms of a sequence. Suppose, for example, that we want to find the first 5 terms of the sequence whose general term is given by $a_n = n/(n + 1)$. We select SEQ from the LIST OPS menu and enter the general term, the variable, and the numbers of the first and last terms desired. The calculator will write the terms horizontally as a list. The list can also be written in fraction notation.

X	Y₁	
1	−1	
2	4	
3	−9	
4	16	
23	−529	

X =

```
seq(X/(X+1),X,1,5)▶Frac
{1/2  2/3  3/4  4/...
```

```
seq(X/(X+1),X,1,5)▶Frac
.../3  3/4  4/5  5/6}
```

We use the $\boxed{\triangleright}$ key to view the two items that do not initially appear on the screen. The first 5 terms of the sequence are $1/2, 2/3, 3/4, 4/5,$ and $5/6$.

Study Tip

Refer to the *Graphing Calculator Manual* that accompanies this text to find the keystrokes for finding the terms of a sequence.

We can graph a sequence just as we graph other functions. Consider the function given by $f(x) = x + 1$ and the sequence whose general term is given by $a_n = n + 1$. The graph of $f(x) = x + 1$ is shown on the left below. If we remove all the points from this graph except those whose first coordinates are positive integers, we have the graph of the sequence $a_n = n + 1$ as shown on the right below.

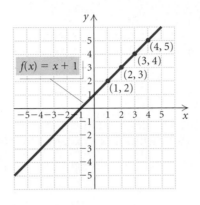
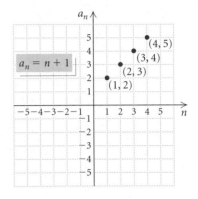

Finding the General Term

When only the first few terms of a sequence are known, we do not know for sure what the general term is, but we might be able to make a prediction by looking for a pattern.

EXAMPLE 2 For each of the following sequences, predict the general term.

a) $1, \sqrt{2}, \sqrt{3}, 2, \ldots$ **b)** $-1, 3, -9, 27, -81, \ldots$

c) $2, 4, 8, \ldots$

Solution

a) These are square roots of consecutive integers, so the general term might be $\sqrt{n}$.

b) These are powers of 3 with alternating signs, so the general term might be $(-1)^n 3^{n-1}$.

c) If we see the pattern of powers of 2, we will see 16 as the next term and guess 2^n for the general term. Then the sequence could be written with more terms as

$$2, 4, 8, 16, 32, 64, 128, \ldots .$$

If we see that we can get the second term by adding 2, the third term by adding 4, and the next term by adding 6, and so on, we will see 14 as the next term. A general term for the sequence is $n^2 - n + 2$, and the sequence can be written with more terms as

$$2, 4, 8, 14, 22, 32, 44, 58, \ldots .$$ —

Example 2(c) illustrates that, in fact, you can never be certain about the general term when only a few terms are given. The fewer the given terms, the greater the uncertainty.

Sums and Series

> ### *Series*
>
> Given the infinite sequence
>
> $$a_1, a_2, a_3, a_4, \ldots, a_n, \ldots,$$
>
> the sum of the terms
>
> $$a_1 + a_2 + a_3 + \cdots + a_n + \cdots$$
>
> is called an **infinite series.** A **partial sum** is the sum of the first n terms:
>
> $$a_1 + a_2 + a_3 + \cdots + a_n.$$
>
> A partial sum is also called a **finite series,** or **nth partial sum,** and is denoted S_n.

EXAMPLE 3   For the sequence $-2, 4, -6, 8, -10, 12, -14, \ldots$, find each of the following.

a) S_1 **b)** S_4 **c)** S_5

Solution

a) $S_1 = -2$

b) $S_4 = -2 + 4 + (-6) + 8 = 4$

c) $S_5 = -2 + 4 + (-6) + 8 + (-10) = -6$

Technology
    Connection

We can use a graphing calculator to find partial sums of a sequence when a formula for the general term is known. Suppose, for example, that we want to find S_1, S_2, S_3, and S_4 for the sequence whose general term is given by $a_n = n^2 - 3$. We can use the CUMSUM feature from the LIST OPS menu. The calculator will write the partial sums as a list. (Note that the calculator can be set in either FUNCTION mode or SEQUENCE mode. Here we show SEQUENCE mode.)

```
cumSum(seq(n²−3,n,1,4))
                {−2 −1 5 18}
```

We have $S_1 = -2$, $S_2 = -1$, $S_3 = 5$, and $S_4 = 18$.

Sigma Notation

The Greek letter Σ (sigma) can be used to simplify notation when the general term of a sequence is a formula. For example, the sum of the first four terms of the sequence 3, 5, 7, 9,..., $2k + 1$,... can be named as follows, using what is called **sigma notation,** or **summation notation:**

$$\sum_{k=1}^{4} (2k + 1).$$

This is read "the sum as k goes from 1 to 4 of $2k + 1$." The letter k is called the **index of summation.** The index of summation might start at a number other than 1, and letters other than k can be used.

EXAMPLE 4   Find and evaluate each of the following sums.

a) $\displaystyle\sum_{k=1}^{5} k^3$ b) $\displaystyle\sum_{k=0}^{4} (-1)^k 5^k$ c) $\displaystyle\sum_{i=8}^{11} \left(2 + \frac{1}{i}\right)$

Solution

a) We replace k with 1, 2, 3, 4, and 5. Then we add the results.

$$\sum_{k=1}^{5} k^3 = 1^3 + 2^3 + 3^3 + 4^3 + 5^3$$
$$= 1 + 8 + 27 + 64 + 125$$
$$= 225$$

b) $\displaystyle\sum_{k=0}^{4} (-1)^k 5^k = (-1)^0 5^0 + (-1)^1 5^1 + (-1)^2 5^2 + (-1)^3 5^3 + (-1)^4 5^4$

$$= 1 - 5 + 25 - 125 + 625 = 521$$

c) $\displaystyle\sum_{i=8}^{11} \left(2 + \frac{1}{i}\right) = \left(2 + \frac{1}{8}\right) + \left(2 + \frac{1}{9}\right) + \left(2 + \frac{1}{10}\right) + \left(2 + \frac{1}{11}\right)$

$$= 8\frac{1691}{3960}$$

Technology Connection

We can combine the SUM and SEQ features on a graphing calculator to add the terms of a sequence. We find the sum in Example 4(a) as shown below.

sum(seq($n^{3,}n$,1,5))
                    225

EXAMPLE 5   Write sigma notation for each sum.

a) $1 + 2 + 4 + 8 + 16 + 32 + 64$

b) $-2 + 4 - 6 + 8 - 10$

c) $x + \dfrac{x^2}{2} + \dfrac{x^3}{3} + \dfrac{x^4}{4} + \cdots$

Solution

a) $1 + 2 + 4 + 8 + 16 + 32 + 64$

This is the sum of powers of 2, beginning with 2^0, or 1, and ending with 2^6, or 64. Sigma notation is $\sum_{k=0}^{6} 2^k$.

b) $-2 + 4 - 6 + 8 - 10$

Disregarding the alternating signs, we see that this is the sum of the first 5 even integers. Note that $2k$ is a formula for the kth positive even integer, and $(-1)^k = -1$ when k is odd and $(-1)^k = 1$ when k is even. Thus the general term is $(-1)^k(2k)$. The sum begins with $k = 1$ and ends with $k = 5$, so sigma notation is $\sum_{k=1}^{5} (-1)^k(2k)$.

c) $x + \dfrac{x^2}{2} + \dfrac{x^3}{3} + \dfrac{x^4}{4} + \cdots$

The general term is x^k/k, beginning with $k = 1$. This is also an infinite series. We use the symbol ∞ for infinity and write the series using sigma notation: $\sum_{k=1}^{\infty} (x^k/k)$. ▬

Recursive Definitions

A sequence may be defined **recursively** or by using a **recursion formula.** Such a definition lists the first term, or the first few terms, and then describes how to determine the remaining terms from the given terms.

EXAMPLE 6   Find the first 5 terms of the sequence defined by

$$a_1 = 5, \qquad a_{n+1} = 2a_n - 3, \quad \text{for } n \geq 1.$$

Solution

$$a_1 = 5,$$

$$a_2 = 2a_1 - 3 = 2 \cdot 5 - 3 = 7,$$

$$a_3 = 2a_2 - 3 = 2 \cdot 7 - 3 = 11,$$

$$a_4 = 2a_3 - 3 = 2 \cdot 11 - 3 = 19,$$

$$a_5 = 2a_4 - 3 = 2 \cdot 19 - 3 = 35.$$ ▬

Technology
Connection

Many graphing calculators have the capability to work with recursively defined sequences when they are set in SEQUENCE mode. In Example 6, for instance, the function could be entered as $u(n) = 2 * u(n - 1) - 3$ with $u(n\text{Min}) = 5$. We can read the terms of the sequence from a table.

```
Plot1  Plot2  Plot3
nMin=1
\u(n)■2*u(n−1)−3

u(nMin)■{5}
\v(n)=
 v(nMin)=
\w(n)=
```

n	$u(n)$	
1	5	
2	7	
3	11	
4	19	
5	35	
6	67	
7	131	
$n = 1$		

7.1

Exercise Set

In each of the following, the nth term of a sequence is given. Find the first 4 terms, a_{10}, and a_{15}.

1. $a_n = 4n - 1$

2. $a_n = (n - 1)(n - 2)(n - 3)$

3. $a_n = \dfrac{n}{n - 1}, \ n \geq 2$

4. $a_n = n^2 - 1, \ n \geq 3$

5. $a_n = \dfrac{n^2 - 1}{n^2 + 1}$

6. $a_n = \left(-\dfrac{1}{2}\right)^{n-1}$

7. $a_n = (-1)^n n^2$

8. $a_n = (-1)^{n-1}(3n - 5)$

9. $a_n = 5 + \dfrac{(-2)^{n+1}}{2^n}$

10. $a_n = \dfrac{2n - 1}{n^2 + 2n}$

Find the indicated term of the given sequence.

11. $a_n = 5n - 6; \ a_8$

12. $a_n = (3n - 4)(2n + 5); \ a_7$

13. $a_n = (2n + 3)^2; \ a_6$

14. $a_n = (-1)^{n-1}(4.6n - 18.3); \ a_{12}$

15. $a_n = 5n^2(4n - 100); \ a_{11}$

16. $a_n = \left(1 + \dfrac{1}{n}\right)^2; \ a_{80}$

17. $a_n = \ln e^n; \ a_{67}$

18. $a_n = 2 - \dfrac{1000}{n}; \ a_{100}$

Predict the general term, or nth term, a_n, of the sequence. Answers may vary.

19. $2, 4, 6, 8, 10, \ldots$

20. $3, 9, 27, 81, 243, \ldots$

21. $-2, 6, -18, 54, \ldots$

22. $-2, 3, 8, 13, 18, \ldots$

23. $\dfrac{2}{3}, \dfrac{3}{4}, \dfrac{4}{5}, \dfrac{5}{6}, \dfrac{6}{7}, \ldots$

24. $\sqrt{2}, 2, \sqrt{6}, 2\sqrt{2}, \sqrt{10}, \ldots$

25. $1 \cdot 2, 2 \cdot 3, 3 \cdot 4, 4 \cdot 5, \ldots$

26. $-1, -4, -7, -10, -13, \ldots$

27. $0, \log 10, \log 100, \log 1000, \ldots$

28. $\ln e^2, \ln e^3, \ln e^4, \ln e^5, \ldots$

Find the indicated partial sums for the sequence.

29. $1, 2, 3, 4, 5, 6, 7, \ldots; \ S_3 \text{ and } S_7$

30. $1, -3, 5, -7, 9, -11, \ldots; \ S_2 \text{ and } S_5$

31. $2, 4, 6, 8, \ldots; \ S_4 \text{ and } S_5$

32. $1, \dfrac{1}{4}, \dfrac{1}{9}, \dfrac{1}{16}, \dfrac{1}{25}, \ldots; \ S_1 \text{ and } S_5$

Find and evaluate the sum.

33. $\displaystyle\sum_{k=1}^{5} \dfrac{1}{2k}$

34. $\displaystyle\sum_{i=1}^{6} \dfrac{1}{2i + 1}$

35. $\displaystyle\sum_{i=0}^{6} 2^i$

36. $\displaystyle\sum_{k=4}^{7} \sqrt{2k - 1}$

37. $\displaystyle\sum_{k=7}^{10} \ln k$

38. $\displaystyle\sum_{k=1}^{4} \pi k$

39. $\displaystyle\sum_{k=1}^{8} \dfrac{k}{k + 1}$

40. $\displaystyle\sum_{i=1}^{5} \dfrac{i - 1}{i + 3}$

41. $\displaystyle\sum_{i=1}^{5} (-1)^i$

42. $\displaystyle\sum_{k=0}^{5} (-1)^{k+1}$

43. $\displaystyle\sum_{k=1}^{8} (-1)^{k+1} 3k$

44. $\displaystyle\sum_{k=0}^{7} (-1)^k 4^{k+1}$

45. $\displaystyle\sum_{k=0}^{6} \dfrac{2}{k^2 + 1}$

46. $\displaystyle\sum_{i=1}^{10} i(i + 1)$

47. $\displaystyle\sum_{k=0}^{5} (k^2 - 2k + 3)$

48. $\displaystyle\sum_{k=1}^{10} \dfrac{1}{k(k + 1)}$

49. $\displaystyle\sum_{i=0}^{10} \dfrac{2^i}{2^i + 1}$

50. $\displaystyle\sum_{k=0}^{3} (-2)^{2k}$

Write sigma notation.

51. $5 + 10 + 15 + 20 + 25 + \cdots$

52. $7 + 14 + 21 + 28 + 35 + \cdots$

53. $2 - 4 + 8 - 16 + 32 - 64$

54. $3 + 6 + 9 + 12 + 15$

55. $-\dfrac{1}{2} + \dfrac{2}{3} - \dfrac{3}{4} + \dfrac{4}{5} - \dfrac{5}{6} + \dfrac{6}{7}$

56. $\dfrac{1}{1^2} + \dfrac{1}{2^2} + \dfrac{1}{3^2} + \dfrac{1}{4^2} + \dfrac{1}{5^2}$

57. $4 - 9 + 16 - 25 + \cdots + (-1)^n n^2$

58. $9 - 16 + 25 + \cdots + (-1)^{n+1} n^2$

59. $\dfrac{1}{1 \cdot 2} + \dfrac{1}{2 \cdot 3} + \dfrac{1}{3 \cdot 4} + \dfrac{1}{4 \cdot 5} + \cdots$

60. $\dfrac{1}{1 \cdot 2^2} + \dfrac{1}{2 \cdot 3^2} + \dfrac{1}{3 \cdot 4^2} + \dfrac{1}{4 \cdot 5^2} + \cdots$

Find the first 4 terms of the recursively defined sequence.

61. $a_1 = 4, \quad a_{n+1} = 1 + \dfrac{1}{a_n}$

62. $a_1 = 256, \quad a_{n+1} = \sqrt{a_n}$

63. $a_1 = 6561, \quad a_{n+1} = (-1)^n \sqrt{a_n}$

64. $a_1 = e^Q, \quad a_{n+1} = \ln a_n$

65. $a_1 = 2, a_2 = 3, \quad a_{n+1} = a_n + a_{n-1}$

66. $a_1 = -10, a_2 = 8, \quad a_{n+1} = a_n - a_{n-1}$

67. *Compound Interest.* Suppose that $1000 is invested at 6.2%, compounded annually. The value of the investment after n years is given by the sequence model

$$a_n = \$1000(1.062)^n, \quad n = 1, 2, 3, \ldots .$$

a) Find the first 10 terms of the sequence.
b) Find the value of the investment after 20 yr.

68. *Salvage Value.* The value of an office machine is $5200. Its salvage value each year is 75% of its value the year before. Give a sequence that lists the salvage value of the machine for each year of a 10-yr period.

69. *Bacteria Growth.* Suppose a single cell of bacteria divides into two every 15 min. Suppose that the same rate of division is maintained for 4 hr. Give a sequence that lists the number of cells after successive 15-min periods.

70. *Salary Sequence.* Torrey is paid $8.30 per hour for working at Red Freight Limited. Each year he receives a $0.30 hourly raise. Give a sequence that lists Torrey's hourly salary over a 10-yr period.

71. *Fibonacci Sequence: Rabbit Population Growth.* One of the most famous recursively defined sequences is the *Fibonacci sequence.* In 1202, the Italian mathematician Leonardo Fibonacci (also called Leonardo da Pisa) proposed the following model for rabbit population growth. Start with one pair of rabbits, one female and one male. These rabbits mature and produce a new pair, again one female and one male. The first pair lives until each pair can reproduce, and then each produces a new pair. This pattern continues. The population of mature rabbits can be modeled by the following recursively defined sequence:

$$a_1 = 1, \quad a_2 = 1, \quad a_{n+1} = a_n + a_{n-1}, \quad \text{for } n \ge 2,$$

where a_n is the total number of pairs of rabbits after $n - 2$ reproductions. Find the first 7 terms of the Fibonacci sequence.

Technology Connection

Use a graphing calculator to construct a table of values for the first 10 terms of the sequence.

72. $a_n = \sqrt{n + 1} - \sqrt{n}$

73. $a_n = \left(1 + \dfrac{1}{n}\right)^n$

74. $a_1 = 2$, $a_{n+1} = \dfrac{1}{2}\left(a_n + \dfrac{2}{a_n}\right)$

75. $a_1 = 2$, $a_{n+1} = \sqrt{1 + \sqrt{a_n}}$

76. *Median Household Income by Age.* The following table contains data relating median household income by age.

AGE, n	MEDIAN HOUSEHOLD INCOME, a_n
19	$27,711
29	44,477
39	53,243
49	58,217
59	44,993
69	23,047

Source: U.S. Bureau of the Census

a) Use a graphing calculator to fit a quadratic sequence regression function
$$a_n = an^2 + bn + c$$
to the data.

b) Find the median household income of people whose age is 15 yr, 25 yr, 40 yr, 60 yr, and 75 yr. Round to the nearest dollar.

77. *Driver Fatalities by Age.* The number of licensed drivers per 100,000 who died in motor vehicle accidents at age n can be estimated by the sequence model
$$a_n = 0.018n^2 - 1.7n + 48.7, \quad n = 15, 16, 17, \ldots, 90.$$
a) Find the first 10 terms of the sequence.
b) Find the number of fatalities per 100,000 of those drivers whose age is 16 yr, 23 yr, 50 yr, 75 yr, and 85 yr.

Collaborative Discussion and Writing

78. a) Find the first few terms of the sequence $a_n = n^2 - n + 41$ and describe the pattern you observe.
b) Does the pattern you found in part (a) hold for all choices of n? Why or why not?

79. The Fibonacci sequence has intrigued mathematicians for centuries. In fact, a journal called the *Fibonacci Quarterly* is devoted to publishing the results pertaining to such sequences. Do some research on the connection of the Fibonacci sequence to the idea of the "Golden Section."

Skill Maintenance

Solve.

80. $3x - 2y = 3$,
$2x + 3y = -11$

81. *Individual Income Taxes.* A total of $1943.5 billion in individual income taxes was collected by the federal government in 2001 and 2002. The amount collected in 2002 was $45.1 billion less than the amount collected in 2001. (*Source:* U.S. Office of Management and Budget) Find the amount collected in each year.

Find the center and the radius of the circle with the given equation.

82. $x^2 + y^2 - 6x + 4y = 3$

83. $x^2 + y^2 + 5x - 8y = 2$

Synthesis

Find the first 5 terms of the sequence, and then find S_5.

84. $a_n = \dfrac{1}{2^n} \log 1000^n$

85. $a_n = i^n$, $i = \sqrt{-1}$

86. $a_n = \ln(1 \cdot 2 \cdot 3 \cdots n)$

For each sequence, find a formula for S_n.

87. $a_n = \ln n$

88. $a_n = \dfrac{1}{n} - \dfrac{1}{n+1}$

7.2

Arithmetic Sequences and Series

- *For any arithmetic sequence, find the nth term when n is given and n when the nth term is given, and given two terms, find the common difference and construct the sequence.*
- *Find the sum of the first n terms of an arithmetic sequence.*

A sequence in which each term after the first is found by adding the same number to the preceding term is an **arithmetic sequence.**

Arithmetic Sequences

The sequence 2, 5, 8, 11, 14, 17,... is arithmetic because adding 3 to any term produces the next term. In other words, the difference between any term and the preceding one is 3. Arithmetic sequences are also called *arithmetic progressions.*

> ### *Arithmetic Sequence*
>
> A sequence is **arithmetic** if there exists a number d, called the **common difference,** such that $a_{n+1} = a_n + d$ for any integer $n \geq 1$.

EXAMPLE 1 For each of the following arithmetic sequences, identify the first term, a_1, and the common difference, d.

a) 4, 9, 14, 19, 24,...

b) 34, 27, 20, 13, 6, -1, -8,...

c) 2, $2\frac{1}{2}$, 3, $3\frac{1}{2}$, 4, $4\frac{1}{2}$,...

Solution The first term, a_1, is the first term listed. To find the common difference, d, we choose any term beyond the first and subtract the preceding term from it.

SEQUENCE	FIRST TERM, a_1	COMMON DIFFERENCE, d
a) 4, 9, 14, 19, 24,...	4	5 $\quad (9 - 4 = 5)$
b) 34, 27, 20, 13, 6, -1, -8,...	34	-7 $\quad (27 - 34 = -7)$
c) 2, $2\frac{1}{2}$, 3, $3\frac{1}{2}$, 4, $4\frac{1}{2}$,...	2	$\frac{1}{2}$ $\quad \left(2\frac{1}{2} - 2 = \frac{1}{2}\right)$

We obtained the common difference by subtracting a_1 from a_2. Had we subtracted a_2 from a_3 or a_3 from a_4, we would have obtained the same values for d. Thus we can check by adding d to each term in a sequence to see if we progress correctly to the next term.

CHECK:

a) $4 + 5 = 9, \quad 9 + 5 = 14, \quad 14 + 5 = 19, \quad 19 + 5 = 24$

b) $34 + (-7) = 27, \quad 27 + (-7) = 20, \quad 20 + (-7) = 13,$
   $13 + (-7) = 6, \quad 6 + (-7) = -1, \quad -1 + (-7) = -8$

c) $2 + \frac{1}{2} = 2\frac{1}{2}, \quad 2\frac{1}{2} + \frac{1}{2} = 3, \quad 3 + \frac{1}{2} = 3\frac{1}{2}, \quad 3\frac{1}{2} + \frac{1}{2} = 4,$
   $4 + \frac{1}{2} = 4\frac{1}{2}$

To find a formula for the general, or nth, term of any arithmetic sequence, we denote the common difference by d, write out the first few terms, and look for a pattern:

$a_1,$

$a_2 = a_1 + d,$

$a_3 = a_2 + d = (a_1 + d) + d = a_1 + 2d$ **Substituting for a_2**

$a_4 = a_3 + d = (a_1 + 2d) + d = a_1 + 3d$ **Substituting for a_3**

Note that the coefficient of d in each case is 1 less than the subscript.

Generalizing, we obtain the following formula.

nth Term of an Arithmetic Sequence

The **nth term** of an arithmetic sequence is given by
$a_n = a_1 + (n - 1)d$, for any integer $n \geq 1$.

EXAMPLE 2 Find the 14th term of the arithmetic sequence 4, 7, 10, 13, ...

Solution We first note that $a_1 = 4$, $d = 7 - 4$, or 3, and $n = 14$. Then using the formula for the nth term, we obtain

$$a_n = a_1 + (n - 1)d$$
$$a_{14} = 4 + (14 - 1) \cdot 3 \quad \text{**Substituting**}$$
$$= 4 + 13 \cdot 3 = 4 + 39$$
$$= 43.$$

The 14th term is 43.

EXAMPLE 3 In the sequence of Example 2, which term is 301? That is, find n if $a_n = 301$.

Solution We substitute 301 for a_n, 4 for a_1, and 3 for d in the formula for the nth term and solve for n:

$$a_n = a_1 + (n - 1)d$$
$$301 = 4 + (n - 1) \cdot 3 \qquad \text{Substituting}$$
$$301 = 4 + 3n - 3$$
$$301 = 3n + 1$$
$$300 = 3n$$
$$100 = n.$$

Solving for n

The term 301 is the 100th term of the sequence. ▬

Given two terms and their places in an arithmetic sequence, we can construct the sequence.

EXAMPLE 4 The 3rd term of an arithmetic sequence is 8, and the 16th term is 47. Find a_1 and d and construct the sequence.

Solution We know that $a_3 = 8$ and $a_{16} = 47$. Thus we would have to add d 13 times to get from 8 to 47. That is,

$$8 + 13d = 47. \qquad a_3 \text{ and } a_{16} \text{ are } 16 - 3, \text{ or } 13, \text{ terms apart.}$$

Solving $8 + 13d = 47$, we obtain

$$13d = 39$$
$$d = 3.$$

Since $a_3 = 8$, we subtract d twice to get a_1. Thus,

$$a_1 = 8 - 2 \cdot 3 = 2. \qquad a_1 \text{ and } a_3 \text{ are } 3 - 1, \text{ or } 2, \text{ terms apart.}$$

The sequence is 2, 5, 8, 11,.... Note that we could also subtract d 15 times from a_{16} in order to find a_1. ▬

In general, d should be subtracted $n - 1$ times from a_n in order to find a_1.

Sum of the First n Terms of an Arithmetic Sequence

Consider the arithmetic sequence

$$3, 5, 7, 9, \ldots.$$

When we add the first 4 terms of the sequence, we get S_4, which is

$$3 + 5 + 7 + 9, \quad \text{or} \quad 24.$$

This sum is called an **arithmetic series.** To find a formula for the sum of the first n terms, S_n, of an arithmetic sequence, we first denote an arithmetic sequence, as follows:

> **This term is two terms back from the last. If you add d to this term, the result is the next-to last term, $a_n - d$.**

$$a_1, \quad (a_1 + d), \quad (a_1 + 2d), \quad \ldots, \quad \overbrace{(a_n - 2d)}, \quad \underbrace{(a_n - d)}, \quad a_n.$$

> **This is the next-to-last term. If you add d to this term, the result is a_n.**

Then S_n is given by

$$S_n = a_1 + (a_1 + d) + (a_1 + 2d) + \cdots + (a_n - 2d)$$
$$+ (a_n - d) + a_n. \tag{1}$$

Reversing the order of the addition gives us

$$S_n = a_n + (a_n - d) + (a_n - 2d) + \cdots + (a_1 + 2d)$$
$$+ (a_1 + d) + a_1. \tag{2}$$

If we add corresponding terms of each side of equations (1) and (2), we get

$$2S_n = [a_1 + a_n] + [(a_1 + d) + (a_n - d)] + [(a_1 + 2d) + (a_n - 2d)]$$
$$+ \cdots + [(a_n - 2d) + (a_1 + 2d)]$$
$$+ [(a_n - d) + (a_1 + d)] + [a_n + a_1].$$

In the expression for $2S_n$, there are n expressions in square brackets. Each of these expressions is equivalent to $a_1 + a_n$. Thus the expression for $2S_n$ can be written in simplified form as

$$2S_n = [a_1 + a_n] + [a_1 + a_n] + [a_1 + a_n] + \cdots + [a_n + a_1]$$
$$+ [a_n + a_1] + [a_n + a_1].$$

Since $a_1 + a_n$ is being added n times, it follows that

$$2S_n = n(a_1 + a_n),$$

from which we get the following formula.

Sum of the First n Terms

The sum of the first n terms of an arithmetic sequence is given by

$$S_n = \frac{n}{2}(a_1 + a_n).$$

EXAMPLE 5   Find the sum of the first 100 natural numbers.

Solution   The sum is

$$1 + 2 + 3 + \cdots + 99 + 100.$$

This is the sum of the first 100 terms of the arithmetic sequence for which

$$a_1 = 1, \qquad a_n = 100, \quad \text{and} \quad n = 100.$$

Thus substituting into the formula

$$S_n = \frac{n}{2}(a_1 + a_n),$$

we get

$$S_{100} = \frac{100}{2}(1 + 100) = 50(101) = 5050.$$

The sum of the first 100 natural numbers is 5050.   ▬

EXAMPLE 6   Find the sum of the first 15 terms of the arithmetic sequence 4, 7, 10, 13, ...

Solution   Note that $a_1 = 4$, $d = 3$, and $n = 15$. Before using the formula

$$S_n = \frac{n}{2}(a_1 + a_n),$$

we find the last term, a_{15}:

$$a_{15} = 4 + (15 - 1)3 \qquad \textbf{Substituting into the formula } a_n = a_1 + (n-1)d$$
$$= 4 + 14 \cdot 3 = 46.$$

Thus,

$$S_{15} = \frac{15}{2}(4 + 46) = \frac{15}{2}(50) = 375.$$

The sum of the first 15 terms is 375.   ▬

EXAMPLE 7   Find the sum: $\displaystyle\sum_{k=1}^{130}(4k + 5)$.

Solution   It is helpful to first write out a few terms:

$$9 + 13 + 17 + \cdots.$$

It appears that this is an arithmetic series coming from an arithmetic sequence with $a_1 = 9$, $d = 4$, and $n = 130$. Before using the formula

$$S_n = \frac{n}{2}(a_1 + a_n),$$

we find the last term, a_{130}:

$$a_{130} = 4 \cdot 130 + 5 \qquad \text{The } k\text{th term is } 4k + 5.$$
$$= 520 + 5$$
$$= 525.$$

Thus,

$$S_{130} = \frac{130}{2}(9 + 525) \qquad \text{Substituting into } S_n = \frac{n}{2}(a_1 + a_n)$$
$$= 34{,}710.$$

Applications

The translation of some applications and problem-solving situations may involve arithmetic sequences or series. We consider some examples.

EXAMPLE 8 *Hourly Wages.* Gloria accepts a job, starting with an hourly wage of $14.25, and is promised a raise of 15¢ per hour every 2 months for 5 yr. At the end of 5 yr, what will Gloria's hourly wage be?

Solution It helps to first write down the hourly wage for several 2-month time periods:

Beginning: $14.25,
After two months: $14.40,
After four months: $14.55,
and so on.

What appears is a sequence of numbers: 14.25, 14.40, 14.55,.... This sequence is arithmetic, because adding 0.15 each time gives us the next term.

We want to find the last term of an arithmetic sequence, so we use the formula $a_n = a_1 + (n - 1)d$. We know that $a_1 = 14.25$ and $d = 0.15$, but what is n? That is, how many terms are in the sequence? Each year there are 12/2, or 6 raises, since Gloria gets a raise every 2 months. There are 5 yr, so the total number of raises will be $5 \cdot 6$, or 30. Thus there will be 31 terms: the original wage and 30 increased rates.

Substituting in the formula $a_n = a_1 + (n - 1)d$ gives us

$$a_{31} = 14.25 + (31 - 1) \cdot 0.15$$
$$= 18.75.$$

Thus, at the end of 5 yr, Gloria's hourly wage will be $18.75.

The calculations in Example 8 could be done in a number of ways. There is often a variety of ways in which a problem can be solved. In this chapter, we will concentrate on the use of sequences and series and their related formulas.

EXAMPLE 9 *Total in a Stack.* A stack of telephone poles has 30 poles in the bottom row. There are 29 poles in the second row, 28 in the next row, and so on. How many poles are in the stack if there are 5 poles in the top row?

Solution A picture will help in this case. The following figure shows the ends of the poles and the way in which they stack.

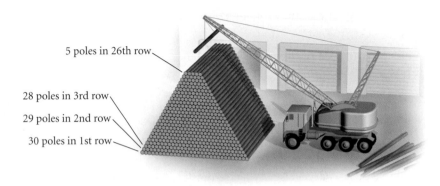

5 poles in 26th row

28 poles in 3rd row

29 poles in 2nd row

30 poles in 1st row

Since the number of poles goes from 30 in a row up to 5 in the top row, there must be 26 rows. We want the sum

$$30 + 29 + 28 + \cdots + 5.$$

Thus we have an arithmetic series. We use the formula

$$S_n = \frac{n}{2}(a_1 + a_n),$$

with $n = 26$, $a_1 = 30$, and $a_{26} = 5$.
    Substituting, we get

$$S_{26} = \frac{26}{2}(30 + 5) = 455.$$

There are 455 poles in the stack.

7.2

Exercise Set

Find the first term and the common difference.

1. 3, 8, 13, 18, …

2. $1.08, $1.16, $1.24, $1.32, …

3. 9, 5, 1, −3, …

4. −8, −5, −2, 1, 4, …

5. $\frac{3}{2}, \frac{9}{4}, 3, \frac{15}{4}, \ldots$

6. $\frac{3}{5}, \frac{1}{10}, -\frac{2}{5}, \ldots$

7. $316, $313, $310, $307, …

8. Find the 11th term of the arithmetic sequence
$0.07, 0.12, 0.17, \ldots$.

9. Find the 12th term of the arithmetic sequence
$2, 6, 10, \ldots$.

10. Find the 17th term of the arithmetic sequence
$7, 4, 1, \ldots$.

11. Find the 14th term of the arithmetic sequence
$3, \frac{7}{3}, \frac{5}{3}, \ldots$.

12. Find the 13th term of the arithmetic sequence
$\$1200, \$964.32, \$728.64, \ldots$.

13. Find the 10th term of the arithmetic sequence
$\$2345.78, \$2967.54, \$3589.30, \ldots$.

14. In the sequence of Exercise 9, what term is the
number 106?

15. In the sequence of Exercise 8, what term is the
number 1.67?

16. In the sequence of Exercise 10, what term is -296?

17. In the sequence of Exercise 11, what term is -27?

18. Find a_{20} when $a_1 = 14$ and $d = -3$.

19. Find a_1 when $d = 4$ and $a_8 = 33$.

20. Find d when $a_1 = 8$ and $a_{11} = 26$.

21. Find n when $a_1 = 25$, $d = -14$, and $a_n = -507$.

22. In an arithmetic sequence, $a_{17} = -40$ and
$a_{28} = -73$. Find a_1 and d. Write the first 5 terms of
the sequence.

23. In an arithmetic sequence, $a_{17} = \frac{25}{3}$ and $a_{32} = \frac{95}{6}$.
Find a_1 and d. Write the first 5 terms of the
sequence.

24. Find the sum of the first 14 terms of the series
$11 + 7 + 3 + \cdots$.

25. Find the sum of the first 20 terms of the series
$5 + 8 + 11 + 14 + \cdots$.

26. Find the sum of the first 300 natural numbers.

27. Find the sum of the first 400 even natural numbers.

28. Find the sum of the odd numbers 1 to 199, inclusive.

29. Find the sum of the multiples of 7 from 7 to 98,
inclusive.

30. Find the sum of all multiples of 4 that are between
14 and 523.

31. If an arithmetic series has $a_1 = 2$, $d = 5$, and
$n = 20$, what is S_n?

32. If an arithmetic series has $a_1 = 7$, $d = -3$, and
$n = 32$, what is S_n?

Find the sum.

33. $\displaystyle\sum_{k=1}^{40} (2k + 3)$

34. $\displaystyle\sum_{k=5}^{20} 8k$

35. $\displaystyle\sum_{k=0}^{19} \frac{k - 3}{4}$

36. $\displaystyle\sum_{k=2}^{50} (2000 - 3k)$

37. $\displaystyle\sum_{k=12}^{57} \frac{7 - 4k}{13}$

38. $\displaystyle\sum_{k=101}^{200} (1.14k - 2.8) - \sum_{k=1}^{5} \left(\frac{k + 4}{10}\right)$

39. *Pole Stacking.* How many poles will be in a stack of
telephone poles if there are 50 in the first layer, 49 in
the second, and so on, with 6 in the top layer?

40. *Investment Return.* Max is an investment
counselor. He sets up an investment situation for a
client that will return $5000 the first year, $6125 the
second year, $7250 the third year, and so on, for
25 yr. How much is received from the investment
altogether?

41. *Garden Plantings.* A gardener is making a planting
in the shape of a trapezoid. It will have 35 plants in
the front row, 31 in the second row, 27 in the third
row, and so on. If the pattern is consistent, how
many plants will there be in the last row? How
many plants are there altogether?

42. *Band Formation.* A formation of a marching band
has 10 marchers in the front row, 12 in the second
row, 14 in the third row, and so on, for 8 rows. How
many marchers are in the last row? How many
marchers are there altogether?

43. *Total Savings.* If 10¢ is saved on October 1, 20¢ is
saved on October 2, 30¢ on October 3, and so on,
how much is saved during the 31 days of October?

44. *Parachutist Free Fall.* When a parachutist jumps
from an airplane, the distances, in feet, that the
parachutist falls in each successive second before
pulling the ripcord to release the parachute are
as follows:

$$16, 48, 80, 112, 144, \ldots .$$

Is this sequence arithmetic? What is the common difference? What is the total distance fallen after 10 sec?

45. *Theater Seating.*   Theaters are often built with more seats per row as the rows move toward the back. Suppose that the first balcony of a theater has 28 seats in the first row, 32 in the second, 36 in the third, and so on, for 20 rows. How many seats are in the first balcony altogether?

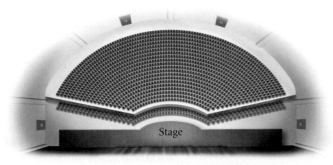

46. *Small Group Interaction.*   In a social science study, Stephan found the following data regarding an interaction measurement r_n for groups of size n.

n	r_n	n	r_n
3	0.5908	7	0.6596
4	0.6080	8	0.6768
5	0.6252	9	0.6940
6	0.6424	10	0.7112

Source: American Sociological Review, 17 (1952)

Is this sequence arithmetic? What is the common difference?

47. *Raw Material Production.*   In an industrial situation, it took 3 units of raw materials to produce 1 unit of a product. The raw material needs thus formed the sequence

$$3, 6, 9, \ldots, 3n, \ldots.$$

Is this sequence arithmetic? What is the common difference?

Collaborative Discussion and Writing

48. The sum of the first n terms of an arithmetic sequence can be given by

$$S_n = \frac{n}{2}[2a_1 + (n-1)d].$$

Compare this formula to

$$S_n = \frac{n}{2}(a_1 + a_n).$$

Discuss the reasons for the use of one formula over the other.

49. It is said that as a young child, the mathematician Karl F. Gauss (1777–1855) was able to compute the sum $1 + 2 + 3 + \cdots + 100$ very quickly in his head to the amazement of a teacher. Explain how Gauss might have done this had he possessed some knowledge of arithmetic sequences and series. Then give a formula for the sum of the first n natural numbers.

Skill Maintenance

Solve.

50. $7x - 2y = 4,$
    $x + 3y = 17$

51. $2x + y + 3z = 12,$
    $x - 3y + 2z = 11,$
    $5x + 2y - 4z = -4$

52. Find the vertices and the foci of the ellipse with the equation $9x^2 + 16y^2 = 144$.

53. Find an equation of the ellipse with vertices $(0, -5)$ and $(0, 5)$ and minor axis of length 4.

Synthesis

54. Find three numbers in an arithmetic sequence such that the sum of the first and third is 10 and the product of the first and second is 15.

55. Find a formula for the sum of the first n odd natural numbers:
$$1 + 3 + 5 + \cdots + (2n - 1).$$

56. Find the first 10 terms of the arithmetic sequence for which
$$a_1 = \$8760 \quad \text{and} \quad d = -\$798.23.$$
Then find the sum of the first 10 terms.

57. Find the first term and the common difference for the arithmetic sequence for which
$$a_2 = 40 - 3q \quad \text{and} \quad a_4 = 10p + q.$$

58. The zeros of this polynomial function form an arithmetic sequence. Find them.
$$f(x) = x^4 + 4x^3 - 84x^2 - 176x + 640$$

*If p, m, and q form an arithmetic sequence, it can be shown that $m = (p + q)/2$. (See Exercise 65.) The number m is the **arithmetic mean**, or **average**, of p and q. Given two numbers p and q, if we find k other numbers $m_1, m_2, \ldots, m_k$ such that*
$$p, m_1, m_2, \ldots, m_k, q$$
forms an arithmetic sequence, we say that we have "inserted k arithmetic means between p and q."

59. Insert three arithmetic means between 4 and 12.

60. Insert three arithmetic means between -3 and 5.

61. Insert four arithmetic means between 4 and 13.

62. Insert ten arithmetic means between 27 and 300.

63. Insert enough arithmetic means between 1 and 50 so that the sum of the resulting series will be 459.

64. *Straight-Line Depreciation.* A company buys an office machine for $5200 on January 1 of a given year. The machine is expected to last for 8 yr, at the end of which time its **trade-in value,** or **salvage value,** will be $1100. If the company's accountant figures the decline in value to be the same each year, then its **book values,** or **salvage values,** after t years, $0 \le t \le 8$, form an arithmetic sequence given by
$$a_t = C - t\left(\frac{C - S}{N}\right),$$
where C is the original cost of the item ($5200), N is the number of years of expected life (8), and S is the salvage value ($1100).

a) Find the formula for a_t for the straight-line depreciation of the office machine.

b) Find the salvage value after 0 yr, 1 yr, 2 yr, 3 yr, 4 yr, 7 yr, and 8 yr.

65. Prove that if p, m, and q form an arithmetic sequence, then
$$m = \frac{p + q}{2}.$$

7.3
Geometric Sequences and Series

- *Identify the common ratio of a geometric sequence, and find a given term and the sum of the first n terms.*
- *Find the sum of an infinite geometric series, if it exists.*

A sequence in which each term after the first is found by multiplying the preceding term by the same number is a **geometric sequence.**

Geometric Sequences

Consider the sequence:
$$2, \ 6, \ 18, \ 54, \ 162, \ldots.$$

Note that multiplying each term by 3 produces the next term. We call the number 3 the **common ratio** because it can be found by dividing any term by the preceding term. A geometric sequence is also called a *geometric progression.*

> ## Geometric Sequence
>
> A sequence is **geometric** if there is a number r, called the **common ratio,** such that
>
> $$\frac{a_{n+1}}{a_n} = r, \quad \text{or} \quad a_{n+1} = a_n r, \quad \text{for any integer } n \geq 1.$$

EXAMPLE 1 For each of the following geometric sequences, identify the common ratio.

a) 3, 6, 12, 24, 48,...

b) $1, -\dfrac{1}{2}, \dfrac{1}{4}, -\dfrac{1}{8}, \ldots$

c) \$5200, \$3900, \$2925, \$2193.75,...

d) \$1000, \$1060, \$1123.60,...

Solution

SEQUENCE	COMMON RATIO
a) 3, 6, 12, 24, 48,...	$2 \quad \left(\frac{6}{3} = 2, \frac{12}{6} = 2, \text{and so on}\right)$
b) $1, -\dfrac{1}{2}, \dfrac{1}{4}, -\dfrac{1}{8}, \ldots$	$-\dfrac{1}{2} \quad \left(\dfrac{-\frac{1}{2}}{1} = -\dfrac{1}{2}, \dfrac{\frac{1}{4}}{-\frac{1}{2}} = -\dfrac{1}{2}, \text{and so on}\right)$
c) \$5200, \$3900, \$2925, \$2193.75,...	$0.75 \quad \left(\dfrac{\$3900}{\$5200} = 0.75, \dfrac{\$2925}{\$3900} = 0.75, \text{and so on}\right)$
d) \$1000, \$1060, \$1123.60,...	$1.06 \quad \left(\dfrac{\$1060}{\$1000} = 1.06, \dfrac{\$1123.60}{\$1060} = 1.06, \text{and so on}\right)$

We now find a formula for the general, or nth, term of a geometric sequence. Let a_1 be the first term and r the common ratio. The first few terms are as follows:

$$a_1,$$
$$a_2 = a_1 r,$$
$$a_3 = a_2 r = (a_1 r)r = a_1 r^2, \qquad \text{Substituting } a_1 r \text{ for } a_2$$
$$a_4 = a_3 r = (a_1 r^2)r = a_1 r^3. \qquad \text{Substituting } a_1 r^2 \text{ for } a_3$$

Note that the exponent is 1 less than the subscript.

Generalizing, we obtain the following.

nth Term of a Geometric Sequence

The **nth term** of a geometric sequence is given by

$$a_n = a_1 r^{n-1}, \quad \text{for any integer } n \geq 1.$$

EXAMPLE 2 Find the 7th term of the geometric sequence 4, 20, 100, ...

Solution We first note that

$$a_1 = 4 \quad \text{and} \quad n = 7.$$

To find the common ratio, we can divide any term (other than the first) by the preceding term. Since the second term is 20 and the first is 4, we get

$$r = \frac{20}{4}, \quad \text{or} \quad 5.$$

Then using the formula $a_n = a_1 r^{n-1}$, we have

$$a_7 = 4 \cdot 5^{7-1} = 4 \cdot 5^6 = 4 \cdot 15{,}625 = 62{,}500.$$

Thus the 7th term is 62,500. ▬

EXAMPLE 3 Find the 10th term of the geometric sequence 64, −32, 16, −8, ...

Solution We first note that

$$a_1 = 64, \qquad n = 10, \quad \text{and} \quad r = \frac{-32}{64}, \text{or} -\frac{1}{2}.$$

Then using the formula $a_n = a_1 r^{n-1}$, we have

$$a_{10} = 64 \cdot \left(-\frac{1}{2}\right)^{10-1} = 64 \cdot \left(-\frac{1}{2}\right)^9 = 2^6 \cdot \left(-\frac{1}{2^9}\right) = -\frac{1}{2^3} = -\frac{1}{8}.$$

Thus the 10th term is $-\frac{1}{8}$. ▬

Sum of the First *n* Terms of a Geometric Sequence

Next, we develop a formula for the sum S_n of the first n terms of a geometric sequence:

$$a_1, \ a_1 r, \ a_1 r^2, \ a_1 r^3, \ldots, \ a_1 r^{n-1}, \ldots .$$

The associated **geometric series** is given by

$$S_n = a_1 + a_1 r + a_1 r^2 + a_1 r^3 + \cdots + a_1 r^{n-1}. \tag{1}$$

We want to find a formula for this sum. If we multiply on both sides of equation (1) by r, we have

$$rS_n = a_1r + a_1r^2 + a_1r^3 + a_1r^4 + \cdots + a_1r^n. \qquad (2)$$

Subtracting equation (2) from equation (1), we see that the differences of the red terms are 0, leaving

$$S_n - rS_n = a_1 - a_1r^n,$$

or

$$S_n(1 - r) = a_1(1 - r^n). \qquad \textbf{Factoring}$$

Dividing on both sides by $1 - r$ gives us the following formula.

Sum of the First n Terms

The sum of the first n terms of a geometric sequence is given by

$$S_n = \frac{a_1(1 - r^n)}{1 - r}, \quad \text{for any } r \neq 1.$$

EXAMPLE 4 Find the sum of the first 7 terms of the geometric sequence $3, 15, 75, 375, \ldots$.

Solution We first note that

$$a_1 = 3, \qquad n = 7, \quad \text{and} \quad r = \frac{15}{3}, \text{ or } 5.$$

Then using the formula

$$S_n = \frac{a_1(1 - r^n)}{1 - r},$$

we have

$$S_7 = \frac{3(1 - 5^7)}{1 - 5}$$

$$= \frac{3(1 - 78{,}125)}{-4}$$

$$= 58{,}593.$$

Thus the sum of the first 7 terms is 58,593. ▬

EXAMPLE 5 Find the sum: $\displaystyle\sum_{k=1}^{11} (0.3)^k$.

Solution This is a geometric series with $a_1 = 0.3$, $r = 0.3$, and $n = 11$. Thus,

$$S_{11} = \frac{0.3(1 - 0.3^{11})}{1 - 0.3}$$

$$\approx 0.42857.$$ ▬

Infinite Geometric Series

The sum of the terms of an infinite geometric sequence is an **infinite geometric series.** For some geometric sequences, S_n gets close to a specific number as n gets large. For example, consider the infinite series

$$\frac{1}{2} + \frac{1}{4} + \frac{1}{8} + \frac{1}{16} + \cdots + \frac{1}{2^n} + \cdots.$$

We can visualize S_n by considering the area of a square. For S_1, we shade half the square. For S_2, we shade half the square plus half the remaining part, or $\frac{1}{4}$. For S_3, we shade the parts shaded in S_2 plus half the remaining part. Again we see that the values of S_n will continue to get close to 1 (shading the complete square).

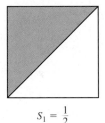 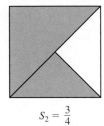 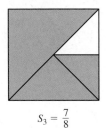 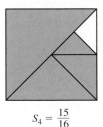

$$S_1 = \frac{1}{2} \qquad S_2 = \frac{3}{4} \qquad S_3 = \frac{7}{8} \qquad S_4 = \frac{15}{16}$$

We examine some partial sums. Note that each of the partial sums is less than 1, but S_n gets very close to 1 as n gets large.

n	S_n
1	0.5
5	0.96875
10	0.9990234375
20	0.9999990463
30	0.9999999991

We say that 1 is the **limit** of S_n and also that 1 is the **sum of the infinite geometric sequence.** The sum of an infinite geometric sequence is denoted S_∞. In this case, $S_\infty = 1$.

Some infinite sequences do not have sums. Consider the infinite geometric series

$$2 + 4 + 8 + 16 + \cdots + 2^n + \cdots.$$

We again examine some partial sums. Note that as n gets large, S_n gets large without bound. This sequence does not have a sum.

n	S_n
1	2
5	62
10	2,046
20	2,097,150
30	2,147,483,646

It can be shown (but we will not do so here) that the sum of the terms of an infinite geometric series exists if and only if $|r| < 1$ (that is, the absolute value of the common ratio is less than 1).

To find a formula for the sum of an infinite geometric series, we first consider the sum of the first n terms:

$$S_n = \frac{a_1(1 - r^n)}{1 - r} = \frac{a_1 - a_1 r^n}{1 - r}. \qquad \textbf{Using the distributive law}$$

For $|r| < 1$, values of r^n get close to 0 as n gets large. As r^n gets close to 0, so does $a_1 r^n$. Thus, S_n gets close to $a_1/(1 - r)$.

Limit or Sum of an Infinite Geometric Series

When $|r| < 1$, the limit or sum of an infinite geometric series is given by

$$S_\infty = \frac{a_1}{1 - r}.$$

EXAMPLE 6 Determine whether each of the following infinite geometric series has a limit. If a limit exists, find it.

a) $1 + 3 + 9 + 27 + \cdots$

b) $-2 + 1 - \frac{1}{2} + \frac{1}{4} - \frac{1}{8} + \cdots$

Solution

a) Here $r = 3$, so $|r| = |3| = 3$. Since $|r| > 1$, the series *does not* have a limit.

b) Here $r = -\frac{1}{2}$, so $|r| = \left|-\frac{1}{2}\right| = \frac{1}{2}$. Since $|r| < 1$, the series *does* have a limit. We find the limit:

$$S_\infty = \frac{a_1}{1 - r} = \frac{-2}{1 - \left(-\frac{1}{2}\right)} = \frac{-2}{\frac{3}{2}} = -\frac{4}{3}.$$

EXAMPLE 7 Find fraction notation for $0.78787878\ldots$, or $0.\overline{78}$.

Solution We can express this as

$$0.78 + 0.0078 + 0.000078 + \cdots.$$

Then we see that this is an infinite geometric series, where $a_1 = 0.78$ and $r = 0.01$. Since $|r| < 1$, this series has a limit:

$$S_\infty = \frac{a_1}{1 - r} = \frac{0.78}{1 - 0.01} = \frac{0.78}{0.99} = \frac{78}{99}, \quad \text{or} \quad \frac{26}{33}.$$

Thus fraction notation for $0.78787878\ldots$ is $\frac{26}{33}$. You can check this on your calculator.

Applications

The translation of some applications and problem-solving situations may involve geometric sequences or series. Examples 9 and 10 in particular show applications in business and economics.

EXAMPLE 8 *A Daily Doubling Salary.* Suppose someone offered you a job for the month of September (30 days) under the following conditions. You will be paid $0.01 for the first day, $0.02 for the second, $0.04 for the third, and so on, doubling your previous day's salary each day. How much would you earn? (Would you take the job? Make a conjecture before reading further.)

Solution You earn $0.01 the first day, $0.01(2)$ the second day, $0.01(2)(2)$ the third day, and so on. The amount earned is the geometric series

$$\$0.01 + \$0.01(2) + \$0.01(2^2) + \$0.01(2^3) + \cdots + \$0.01(2^{29}),$$

where $a_1 = \$0.01$, $r = 2$, and $n = 30$. Using the formula

$$S_n = \frac{a_1(1 - r^n)}{1 - r},$$

we have

$$S_{30} = \frac{\$0.01(1 - 2^{30})}{1 - 2} = \$10{,}737{,}418.23.$$

The pay exceeds $10.7 million for the month. ▬

EXAMPLE 9 *The Amount of an Annuity.* An **annuity** is a sequence of equal payments, made at equal time intervals, that earn interest. Fixed deposits in a savings account are an example of an annuity. Suppose that to save money to buy a car, Andrea deposits $1000 at the *end* of each of 5 yr in an account that pays 8% interest, compounded annually. The total amount in the account at the end of 5 yr is called the **amount of the annuity.** Find that amount.

Solution The following time diagram can help visualize the problem. Note that no deposit is made until the end of the first year.

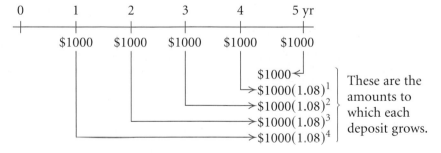

The amount of the annuity is the geometric series

$$\$1000 + \$1000(1.08)^1 + \$1000(1.08)^2 + \$1000(1.08)^3 + \$1000(1.08)^4,$$

where $a_1 = \$1000$, $n = 5$, and $r = 1.08$. Using the formula

$$S_n = \frac{a_1(1 - r^n)}{1 - r},$$

we have

$$S_5 = \frac{\$1000(1 - 1.08^5)}{1 - 1.08} \approx \$5866.60.$$

The amount of the annuity is $5866.60.

EXAMPLE 10 *The Economic Multiplier.*   The finals of the NCAA men's basketball tournament have a significant effect on the economy of the host city. Suppose that 45,000 people visit the city and spend $600 each while there. Then assume that 80% of that money is spent again in the city, and then 80% of that money is spent again, and so on. This is known as the *economic multiplier effect.* Find the total effect on the economy.

Solution   The initial effect is $45{,}000 \cdot \$600$, or $27,000,000. The total effect is given by the infinite series

$$\$27{,}000{,}000 + \$27{,}000{,}000(0.80) + \$27{,}000{,}000(0.80^2) + \cdots.$$

Since $|r| = |0.80| = 0.80 < 1$, the series has a sum. Using the formula for the sum of an infinite geometric series, we have

$$S_\infty = \frac{a_1}{1 - r} = \frac{27{,}000{,}000}{1 - 0.80} = \$135{,}000{,}000.$$

Thus the total effect on the economy of the spending from the tournament is $135,000,000.

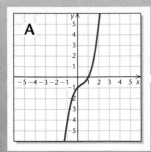

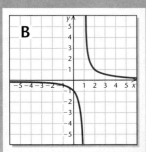

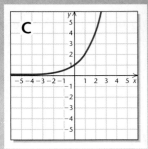

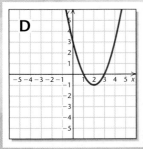

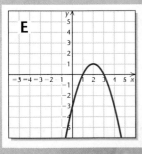

Visualizing the Graph

Match the equation with its graph.

1. $(x - 1)^2 + (y + 2)^2 = 9$

2. $y = x^3 - x^2 + x - 1$

3. $f(x) = 2^x$

4. $f(x) = x$

5. $a_n = n$

6. $y = \log(x + 3)$

7. $f(x) = -(x - 2)^2 + 1$

8. $f(x) = (x - 2)^2 - 1$

9. $y = \dfrac{1}{x - 1}$

10. $y = -3x + 4$

Answers on page A-49

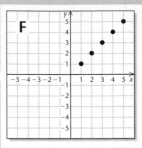

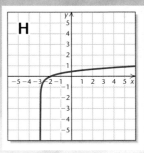

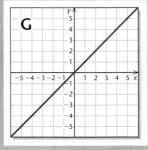

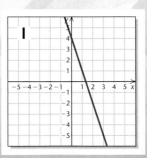

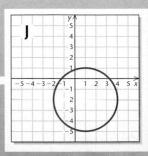

7.3

Exercise Set

Find the common ratio.

1. $2, 4, 8, 16, \ldots$

2. $18, -6, 2, -\frac{2}{3}, \ldots$

3. $-1, 1, -1, 1, \ldots$

4. $-8, -0.8, -0.08, -0.008, \ldots$

5. $\frac{2}{3}, -\frac{4}{3}, \frac{8}{3}, -\frac{16}{3}, \ldots$

6. $75, 15, 3, \frac{3}{5}, \ldots$

7. $6.275, 0.6275, 0.06275, \ldots$

8. $\dfrac{1}{x}, \dfrac{1}{x^2}, \dfrac{1}{x^3}, \ldots$

9. $5, \dfrac{5a}{2}, \dfrac{5a^2}{4}, \dfrac{5a^3}{8}, \ldots$

10. $\$780, \$858, \$943.80, \$1038.18, \ldots$

Find the indicated term.

11. $2, 4, 8, 16, \ldots;$ the 7th term

12. $2, -10, 50, -250, \ldots;$ the 9th term

13. $2, 2\sqrt{3}, 6, \ldots;$ the 9th term

14. $1, -1, 1, -1, \ldots;$ the 57th term

15. $\frac{7}{625}, -\frac{7}{25}, \ldots;$ the 23rd term

16. $\$1000, \$1060, \$1123.60, \ldots;$ the 5th term

Find the nth, or general, term.

17. $1, 3, 9, \ldots$

18. $25, 5, 1, \ldots$

19. $1, -1, 1, -1, \ldots$

20. $-2, 4, -8, \ldots$

21. $\dfrac{1}{x}, \dfrac{1}{x^2}, \dfrac{1}{x^3}, \ldots$

22. $5, \dfrac{5a}{2}, \dfrac{5a^2}{4}, \dfrac{5a^3}{8}, \ldots$

23. Find the sum of the first 7 terms of the geometric series
$$6 + 12 + 24 + \cdots.$$

24. Find the sum of the first 10 terms of the geometric series
$$16 - 8 + 4 - \cdots.$$

25. Find the sum of the first 9 terms of the geometric series
$$\tfrac{1}{18} - \tfrac{1}{6} + \tfrac{1}{2} - \cdots.$$

26. Find the sum of the geometric series
$$-8 + 4 + (-2) + \cdots + \left(-\tfrac{1}{32}\right).$$

Find the sum, if it exists.

27. $4 + 2 + 1 + \cdots$

28. $7 + 3 + \frac{9}{7} + \cdots$

29. $25 + 20 + 16 + \cdots$

30. $100 - 10 + 1 - \frac{1}{10} + \cdots$

31. $8 + 40 + 200 + \cdots$

32. $-6 + 3 - \frac{3}{2} + \frac{3}{4} - \cdots$

33. $0.6 + 0.06 + 0.006 + \cdots$

34. $\displaystyle\sum_{k=0}^{10} 3^k$

35. $\displaystyle\sum_{k=1}^{11} 15\left(\frac{2}{3}\right)^k$

36. $\displaystyle\sum_{k=0}^{50} 200(1.08)^k$

37. $\displaystyle\sum_{k=1}^{\infty} \left(\frac{1}{2}\right)^{k-1}$

38. $\displaystyle\sum_{k=1}^{\infty} 2^k$

39. $\displaystyle\sum_{k=1}^{\infty} 12.5^k$

40. $\displaystyle\sum_{k=1}^{\infty} 400(1.0625)^k$

41. $\displaystyle\sum_{k=1}^{\infty} \$500(1.11)^{-k}$

42. $\displaystyle\sum_{k=1}^{\infty} \$1000(1.06)^{-k}$

43. $\displaystyle\sum_{k=1}^{\infty} 16(0.1)^{k-1}$

44. $\displaystyle\sum_{k=1}^{\infty} \frac{8}{3}\left(\frac{1}{2}\right)^{k-1}$

Find fraction notation.

45. $0.131313\ldots,$ or $0.\overline{13}$

46. $0.2222\ldots,$ or $0.\overline{2}$

47. $8.9\overline{999}$

48. $6.161\overline{616}$

49. $3.4125\overline{125}$

50. $12.7809\overline{809}$

51. *Bouncing Ping-Pong Ball.*   A ping-pong ball is dropped from a height of 16 ft and always rebounds $\frac{1}{4}$ of the distance fallen.

a) How high does it rebound the 6th time?

b) Find the total sum of the rebound heights of the ball.

52. *Daily Doubling Salary.* Suppose someone offered you a job for the month of February (28 days) under the following conditions. You will be paid $0.01 the 1st day, $0.02 the 2nd, $0.04 the 3rd, and so on, doubling your previous day's salary each day. How much would you earn altogether?

53. *Bungee Jumping.* A bungee jumper always rebounds 60% of the distance fallen. A bungee jump is made using a cord that stretches to 200 ft.

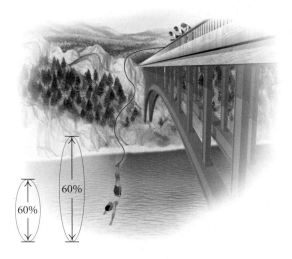

a) After jumping and then rebounding 9 times, how far has a bungee jumper traveled upward (the total rebound distance)?

b) About how far will a jumper have traveled upward (bounced) before coming to rest?

54. *Population Growth.* Hadleytown has a present population of 100,000, and the population is increasing by 3% each year.

a) What will the population be in 15 yr?

b) How long will it take for the population to double?

55. *Amount of an Annuity.* To create a college fund, a parent makes a sequence of 18 yearly deposits of $1000 each in a savings account on which interest is compounded annually at 3.2%. Find the amount of the annuity.

56. *Amount of an Annuity.* A sequence of yearly payments of P dollars is invested at the end of each

of N years at interest rate i, compounded annually. The total amount in the account, or the amount of the annuity, is V.

a) Show that

$$V = \frac{P[(1 + i)^N - 1]}{i}.$$

b) Suppose that interest is compounded n times per year and deposits are made every compounding period. Show that the formula for V is then given by

$$V = \frac{P\left[\left(1 + \dfrac{i}{n}\right)^{nN} - 1\right]}{i/n}.$$

57. *Loan Repayment.* A family borrows $120,000. The loan is to be repaid in 13 yr at 12% interest, compounded annually. How much will be repaid at the end of 13 yr?

58. *Doubling the Thickness of Paper.* A piece of paper is 0.01 in. thick. It is cut and stacked repeatedly in such a way that its thickness is doubled each time for 20 times. How thick is the result?

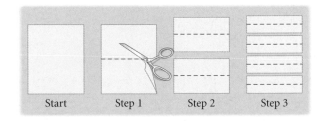

Start   Step 1   Step 2   Step 3

59. *The Economic Multiplier.* Suppose the government is making a $13,000,000,000 expenditure for educational improvement. If 85% of this is spent again, and so on, what is the total effect on the economy?

60. *Advertising Effect.* Great Grains Cereal Company is about to market a new low-fat cereal in a city of 5,000,000 people. They plan an advertising campaign that they think will induce 30% of the people to buy the product. They estimate that if those people like the product, they will induce 30% · 30% · 5,000,000 more to buy the product, and those will induce 30% · 30% · 30% · 5,000,000, and so on. In all, how many people will buy the product as a result of the advertising campaign? What percentage of the population is this?

Technology Connection

61. Use a graphing calculator to find the sum in Exercise 35.

62. Use a graphing calculator to find the sum in Exercise 34.

Collaborative Discussion and Writing

63. Write a problem for a classmate to solve. Devise the problem so that a geometric series is involved and the solution is "The total amount in the bank is $900(1.08)^{40}$, or about \$19,552."

64. The infinite series
$$S_\infty = 2 + \frac{1}{2} + \frac{1}{2 \cdot 3} + \frac{1}{2 \cdot 3 \cdot 4} + \frac{1}{2 \cdot 3 \cdot 4 \cdot 5} + \cdots$$
is not geometric, but it does have a sum. Consider S_1, S_2, S_3, S_4, S_5, and S_6. Construct a table and a graph of the sequence. Expand the sequence of sums, if needed. Make a conjecture about the value of S_∞ and explain your reasoning.

Skill Maintenance

For each pair of functions, find $(f \circ g)(x)$ and $(g \circ f)(x)$.

65. $f(x) = x^2, \ g(x) = 4x + 5$

66. $f(x) = x - 1, \ g(x) = x^2 + x + 3$

Solve.

67. $5^x = 35$

68. $\log_2 x = -4$

Synthesis

69. Prove that
$$\sqrt{3} - \sqrt{2}, \quad 4 - \sqrt{6}, \quad \text{and} \quad 6\sqrt{3} - 2\sqrt{2}$$
form a geometric sequence.

70. Consider the sequence
$$4, \ 20.4, \ 104.04, \ 531.6444, \ldots.$$
What is the error in using $a_{277} = 4(5.1)^{276}$ to find the 277th term?

71. Consider the sequence
$$x + 3, \ x + 7, \ 4x - 2, \ldots.$$
a) If the sequence is arithmetic, find x and then determine each of the 3 terms and the 4th term.
b) If the sequence is geometric, find x and then determine each of the 3 terms and the 4th term.

72. Find the sum of the first n terms of
$$1 + x + x^2 + \cdots.$$

73. Find the sum of the first n terms of
$$x^2 - x^3 + x^4 - x^5 + \cdots.$$

In Exercises 74 and 75, assume that $a_1, a_2, a_3, \ldots$ is a geometric sequence.

74. Prove that $a_1^2, a_2^2, a_3^2, \ldots$, is a geometric sequence.

75. Prove that $\ln a_1, \ln a_2, \ln a_3, \ldots$, is an arithmetic sequence.

76. Prove that $5^{a_1}, 5^{a_2}, 5^{a_3}, \ldots$, is a geometric sequence, if $a_1, a_2, a_3, \ldots$, is an arithmetic sequence.

77. The sides of a square are 16 cm long. A second square is inscribed by joining the midpoints of the sides, successively. In the second square, we repeat the process, inscribing a third square. If this process is continued indefinitely, what is the sum of all the areas of all the squares? (*Hint:* Use an infinite geometric series.)

7.4

Mathematical Induction

- *List the statements of an infinite sequence that is defined by a formula.*
- *Do proofs by mathematical induction.*

In this section we learn to prove a sequence of mathematical statements using a procedure called *mathematical induction*.

Sequences of Statements

Infinite sequences of statements occur often in mathematics. In an infinite sequence of statements, there is a statement for each natural number. For example, consider the sequence of statements represented by the following:

"For each x between 0 and 1, $0 < x^n < 1$."

Let's think of this as $S(n)$, or S_n. Substituting natural numbers for n gives a sequence of statements. We list a few of them.

Statement 1, S_1: For x between 0 and 1, $0 < x^1 < 1$.
Statement 2, S_2: For x between 0 and 1, $0 < x^2 < 1$.
Statement 3, S_3: For x between 0 and 1, $0 < x^3 < 1$.
Statement 4, S_4: For x between 0 and 1, $0 < x^4 < 1$.

In this context, the symbols S_1, S_2, S_3, and so on, do not represent sums.

EXAMPLE 1 List the first four statements in the sequence that can be obtained from each of the following.

a) $\log n < n$
b) $1 + 3 + 5 + \cdots + (2n - 1) = n^2$

Solution

a) This time, S_n is "$\log n < n$."

S_1: $\log 1 < 1$
S_2: $\log 2 < 2$
S_3: $\log 3 < 3$
S_4: $\log 4 < 4$

b) This time, S_n is "$1 + 3 + 5 + \cdots + (2n - 1) = n^2$."

S_1: $1 = 1^2$
S_2: $1 + 3 = 2^2$
S_3: $1 + 3 + 5 = 3^2$
S_4: $1 + 3 + 5 + 7 = 4^2$

Proving Infinite Sequences of Statements

We now develop a method of proof, called **mathematical induction,** which we can use to try to prove that all statements in an infinite sequence of statements are true. The statements usually have the form:

"For all natural numbers n, S_n,"

where S_n is some mathematical sentence such as those of the preceding examples. Of course, we cannot prove each statement of an infinite sequence individually. Instead, we try to show that "whenever S_k holds, then S_{k+1} must hold." We abbreviate this as $S_k \rightarrow S_{k+1}$. (This is also read "*If* S_k, *then* S_{k+1}," or "S_k *implies* S_{k+1}.") Suppose that we could somehow establish that this holds for all natural numbers k. Then we would have the following:

$S_1 \rightarrow S_2$ meaning "if S_1 is true, then S_2 is true";
$S_2 \rightarrow S_3$ meaning "if S_2 is true, then S_3 is true";
$S_3 \rightarrow S_4$ meaning "if S_3 is true, then S_4 is true";
and so on, indefinitely.

Even knowing that $S_k \rightarrow S_{k+1}$, we would still not be certain whether there is *any* k for which S_k is true. All we would know is that "if S_k is true, then S_{k+1} is true." Suppose now that S_k is true for some k, say, $k = 1$. We then must have the following.

S_1 is true.   **We have verified, or proved, this.**
$S_1 \rightarrow S_2$ **This means that whenever S_1 is true, S_2 is true.**
Therefore, S_2 is true.

$S_2 \rightarrow S_3$ **This means that whenever S_2 is true, S_3 is true.**
Therefore, S_3 is true.
and so on.

We conclude that S_n is true for all natural numbers n.

This leads us to the principle of mathematical induction, which we use to prove the types of statements considered here.

The Principle of Mathematical Induction

We can prove an infinite sequence of statements S_n by showing the following.

(1) *Basis step.* S_1 is true.

(2) *Induction step.* For all natural numbers k, $S_k \rightarrow S_{k+1}$.

Mathematical induction is analogous to lining up a sequence of dominoes. The induction step tells us that if any one domino is knocked over, then the one next to it will be hit and knocked over. The basis step tells us that the first domino can indeed be knocked over. Note that in order for all dominoes to fall, *both* conditions must be satisfied.

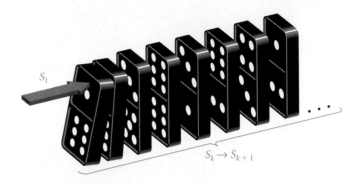

When you are learning to do proofs by mathematical induction, it is helpful to first write out S_n, S_1, S_k, and S_{k+1}. This helps to identify what is to be assumed and what is to be deduced.

EXAMPLE 2   Prove: For every natural number n,
$$1 + 3 + 5 + \cdots + (2n - 1) = n^2.$$

PROOF   We first list S_n, S_1, S_k, and S_{k+1}.

S_n: $1 + 3 + 5 + \cdots + (2n - 1) = n^2$
S_1: $1 = 1^2$
S_k: $1 + 3 + 5 + \cdots + (2k - 1) = k^2$
S_{k+1}: $1 + 3 + 5 + \cdots + (2k - 1) + [2(k + 1) - 1] = (k + 1)^2$

(1) *Basis step.* S_1, as listed, is true since $1 = 1^2$, or $1 = 1$.

(2) *Induction step.* We let k be any natural number. We assume S_k to be true and try to show that it implies that S_{k+1} is true. Now S_k is
$$1 + 3 + 5 + \cdots + (2k - 1) = k^2.$$

Starting with the left side of S_{k+1} and substituting k^2 for $1 + 3 + 5 + \cdots + (2k - 1)$, we have

$$\underbrace{1 + 3 + \cdots + (2k - 1)} + [2(k + 1) - 1]$$

$$\begin{aligned}
&= k^2 + [2(k + 1) - 1] \\
&= k^2 + 2k + 2 - 1 \\
&= k^2 + 2k + 1 = (k + 1)^2.
\end{aligned}$$

We have derived S_{k+1} from S_k. Thus we have shown that for all natural numbers k, $S_k \rightarrow S_{k+1}$. This completes the induction step. It and the basis step tell us that the proof is complete.

EXAMPLE 3   Prove: For every natural number n,

$$\frac{1}{2} + \frac{1}{4} + \frac{1}{8} + \cdots + \frac{1}{2^n} = \frac{2^n - 1}{2^n}.$$

PROOF   We first list S_n, S_1, S_k, and S_{k+1}.

S_n: $\dfrac{1}{2} + \dfrac{1}{4} + \dfrac{1}{8} + \cdots + \dfrac{1}{2^n} = \dfrac{2^n - 1}{2^n}$

S_1: $\dfrac{1}{2^1} = \dfrac{2^1 - 1}{2^1}$

S_k: $\dfrac{1}{2} + \dfrac{1}{4} + \dfrac{1}{8} + \cdots + \dfrac{1}{2^k} = \dfrac{2^k - 1}{2^k}$

S_{k+1}: $\dfrac{1}{2} + \dfrac{1}{4} + \dfrac{1}{8} + \cdots + \dfrac{1}{2^k} + \dfrac{1}{2^{k+1}} = \dfrac{2^{k+1} - 1}{2^{k+1}}$

(1) *Basis step.* We show S_1 to be true as follows:

$$\frac{2^1 - 1}{2^1} = \frac{2 - 1}{2} = \frac{1}{2}.$$

(2) *Induction step.* We let k be any natural number. We assume S_k to be true and try to show that it implies that S_{k+1} is true. Now S_k is

$$\frac{1}{2} + \frac{1}{4} + \frac{1}{8} + \cdots + \frac{1}{2^k} = \frac{2^k - 1}{2^k}.$$

Starting with the left side of S_{k+1} and substituting

$$\frac{2^k - 1}{2^k} \quad \text{for} \quad \frac{1}{2} + \frac{1}{4} + \cdots + \frac{1}{2^k},$$

we have

$$\underbrace{\frac{1}{2} + \frac{1}{4} + \frac{1}{8} + \cdots + \frac{1}{2^k}} + \frac{1}{2^{k+1}}$$

$$\downarrow$$

$$= \frac{2^k - 1}{2^k} + \frac{1}{2^{k+1}} = \frac{2^k - 1}{2^k} \cdot \frac{2}{2} + \frac{1}{2^{k+1}} = \frac{(2^k - 1) \cdot 2 + 1}{2^{k+1}}$$

$$= \frac{2^{k+1} - 2 + 1}{2^{k+1}} = \frac{2^{k+1} - 1}{2^{k+1}}.$$

We have derived S_{k+1} from S_k. Thus we have shown that for all natural numbers k, $S_k \to S_{k+1}$. This completes the induction step. It and the basis step tell us that the proof is complete.

EXAMPLE 4 Prove: For every natural number n, $n < 2^n$.

PROOF We first list S_n, S_1, S_k, and S_{k+1}.

$$S_n: \quad n < 2^n$$
$$S_1: \quad 1 < 2^1$$
$$S_k: \quad k < 2^k$$
$$S_{k+1}: \quad k + 1 < 2^{k+1}$$

(1) *Basis step.* S_1, as listed, is true since $2^1 = 2$ and $1 < 2$.

(2) *Induction step.* We let k be any natural number. We assume S_k to be true and try to show that it implies that S_{k+1} is true. Now

$k < 2^k$	**This is S_k.**
$2k < 2 \cdot 2^k$	**Multiplying by 2 on both sides**
$2k < 2^{k+1}$	**Adding exponents on the right**
$k + k < 2^{k+1}.$	**Rewriting $2k$ as $k + k$**

Since k is any natural number, we know that $1 \leq k$. Thus,

$$k + 1 \leq k + k. \qquad \text{**Adding k on both sides of $1 \leq k$**}$$

Putting the results $k + 1 \leq k + k$ and $k + k < 2^{k+1}$ together gives us

$$k + 1 < 2^{k+1}. \qquad \text{**This is S_{k+1}.**}$$

We have derived S_{k+1} from S_k. Thus we have shown that for all natural numbers k, $S_k \rightarrow S_{k+1}$. This completes the induction step. It and the basis step tell us that the proof is complete. ▬

7.4

Exercise Set

List the first five statements in the sequence that can be obtained from each of the following. Determine whether each of the statements is true or false.

1. $n^2 < n^3$

2. $n^2 - n + 41$ is prime. Find a value for n for which the statement is false.

3. A polygon of n sides has $[n(n-3)]/2$ diagonals.

4. The sum of the angles of a polygon of n sides is $(n-2) \cdot 180°$.

Use mathematical induction to prove each of the following.

5. $2 + 4 + 6 + \cdots + 2n = n(n+1)$

6. $4 + 8 + 12 + \cdots + 4n = 2n(n+1)$

7. $1 + 5 + 9 + \cdots + (4n - 3) = n(2n - 1)$

8. $3 + 6 + 9 + \cdots + 3n = \dfrac{3n(n+1)}{2}$

9. $2 + 4 + 8 + \cdots + 2^n = 2(2^n - 1)$

10. $2 \leq 2^n$

11. $n < n + 1$

12. $3^n < 3^{n+1}$

13. $2n \leq 2^n$

14. $\dfrac{1}{1 \cdot 2} + \dfrac{1}{2 \cdot 3} + \cdots + \dfrac{1}{n(n+1)} = \dfrac{n}{n+1}$

15. $\dfrac{1}{1 \cdot 2 \cdot 3} + \dfrac{1}{2 \cdot 3 \cdot 4} + \dfrac{1}{3 \cdot 4 \cdot 5} + \cdots$

$+ \dfrac{1}{n(n+1)(n+2)} = \dfrac{n(n+3)}{4(n+1)(n+2)}$

16. If x is any real number greater than 1, then for any natural number n, $x \le x^n$.

The following formulas can be used to find sums of powers of natural numbers. Use mathematical induction to prove each formula.

17. $1 + 2 + 3 + \cdots + n = \dfrac{n(n+1)}{2}$

18. $1^2 + 2^2 + 3^2 + \cdots + n^2 = \dfrac{n(n+1)(2n+1)}{6}$

19. $1^3 + 2^3 + 3^3 + \cdots + n^3 = \dfrac{n^2(n+1)^2}{4}$

20. $1^4 + 2^4 + 3^4 + \cdots + n^4$
$= \dfrac{n(n+1)(2n+1)(3n^2+3n-1)}{30}$

21. $1^5 + 2^5 + 3^5 + \cdots + n^5$
$= \dfrac{n^2(n+1)^2(2n^2+2n-1)}{12}$

Use mathematical induction to prove each of the following.

22. $\sum\limits_{i=1}^{n} (3i - 1) = \dfrac{n(3n+1)}{2}$

23. $\sum\limits_{i=1}^{n} i(i+1) = \dfrac{n(n+1)(n+2)}{3}$

24. $\left(1 + \dfrac{1}{1}\right)\left(1 + \dfrac{1}{2}\right)\left(1 + \dfrac{1}{3}\right) \cdots \left(1 + \dfrac{1}{n}\right) = n + 1$

25. The sum of n terms of an arithmetic sequence:
$a_1 + (a_1 + d) + (a_1 + 2d) + \cdots + [a_1 + (n-1)d]$
$= \dfrac{n}{2}[2a_1 + (n-1)d]$

Collaborative Discussion and Writing

26. Write an explanation of the idea behind mathematical induction for a fellow student.

27. Find two statements not considered in this section that are not true for all natural numbers. Then try to find where a proof by mathematical induction fails.

Skill Maintenance

Solve.

28. $2x - 3y = 1,$
$3x - 4y = 3$

29. $x + y + z = 3,$
$2x - 3y - 2z = 5,$
$3x + 2y + 2z = 8$

30. *e-Commerce.* ebooks.com ran a one-day promotion offering a hardback title for $24.95 and a paperback title for $9.95. A total of 80 books were sold and $1546 was taken in. How many of each type of book were sold?

31. *Investment.* Martin received $104 in simple interest one year from three investments. Part is invested at 1.5%, part at 2%, and part at 3%. The amount invested at 2% is twice the amount invested at 1.5%. There is $400 more invested at 3% than at 2%. Find the amount invested at each rate.

Synthesis

Use mathematical induction to prove each of the following.

32. The sum of n terms of a geometric sequence:
$a_1 + a_1 r + a_1 r^2 + \cdots + a_1 r^{n-1} = \dfrac{a_1 - a_1 r^n}{1 - r}$

33. $x + y$ is a factor of $x^{2n} - y^{2n}$.

Prove each of the following using mathematical induction. Do the basis step for $n = 2$.

34. For every natural number $n \ge 2$,
$2n + 1 < 3^n.$

35. For every natural number $n \ge 2$,
$\log_a(b_1 b_2 \cdots b_n)$
$= \log_a b_1 + \log_a b_2 + \cdots + \log_a b_n.$

Prove each of the following for any complex numbers z_1, $z_2, \ldots, z_n$, where $i^2 = -1$ and $\overline{z}$ is the conjugate of z (see Section 2.2).

36. $\overline{z^n} = \overline{z}^n$

37. $\overline{z_1 + z_2 + \cdots + z_n} = \overline{z_1} + \overline{z_2} + \cdots + \overline{z_n}$

38. $\overline{z_1 z_2 \cdots z_n} = \overline{z_1} \cdot \overline{z_2} \cdots \overline{z_n}$

39. i^n is either 1, -1, i, or $-i$.

For any integers a and b, b is a factor of a if there exists an integer c such that a = bc. Prove each of the following for any natural number n.

40. 2 is a factor of $n^2 + n$.

41. 3 is a factor of $n^3 + 2n$.

42. *The Tower of Hanoi Problem.* There are three pegs on a board. On one peg are n disks, each smaller than the one on which it rests. The problem is to move this pile of disks to another peg. The final order must be the same, but you can move only one disk at a time and can never place a larger disk on a smaller one.

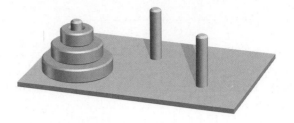

a) What is the *smallest* number of moves needed to move 3 disks? 4 disks? 2 disks? 1 disk?

b) Conjecture a formula for the *smallest* number of moves needed to move n disks. Prove it by mathematical induction.

7.5

Combinatorics: Permutations

- *Evaluate factorial and permutation notation and solve related applied problems.*

In order to study probability, it is first necessary to learn about **combinatorics,** the theory of counting.

Permutations

In this section, we will consider the part of combinatorics called *permutations.*

> The study of permutations involves *order* and *arrangements.*

EXAMPLE 1 How many 3-letter code symbols can be formed with the letters A, B, C *without* repetition (that is, using each letter only once)?

Solution Consider placing the letters in these boxes.

We can select any of the 3 letters for the first letter in the symbol. Once this letter has been selected, the second must be selected from the 2 remaining letters. After this, the third letter is already determined, since only 1 possibility is left. That is, we can place any of the 3 letters in the first box, either of the remaining 2 letters in the second box, and the only remaining letter in the third box. The possibilities can be arrived at using a **tree diagram,** as shown below.

TREE DIAGRAM   OUTCOMES

$$
\begin{array}{cccc}
 & & B \rule[0.5ex]{1em}{0.4pt} C & \text{ABC} \\
A \!\!< & \\
 & & C \rule[0.5ex]{1em}{0.4pt} B & \text{ACB} \\
 & & A \rule[0.5ex]{1em}{0.4pt} C & \text{BAC} \\
B \!\!< & \\
 & & C \rule[0.5ex]{1em}{0.4pt} A & \text{BCA} \\
 & & A \rule[0.5ex]{1em}{0.4pt} B & \text{CAB} \\
C \!\!< & \\
 & & B \rule[0.5ex]{1em}{0.4pt} A & \text{CBA}
\end{array}
$$

Each outcome represents one permutation of the letters A, B, C.

We see that there are 6 possibilities. The set of all the possibilities is

{ABC, ACB, BAC, BCA, CAB, CBA}.

Suppose that we perform an experiment such as selecting letters (as in the preceding example), flipping a coin, or drawing a card. The results are called **outcomes.** An **event** is a set of outcomes. The following principle pertains to the counting of actions that occur together, or are combined to form an event.

The Fundamental Counting Principle

Given a combined action, or *event*, in which the first action can be performed in n_1 ways, the second action can be performed in n_2 ways, and so on, the total number of ways in which the combined action can be performed is the product

$$n_1 \cdot n_2 \cdot n_3 \cdot \cdots \cdot n_k.$$

Thus, in Example 1, there are 3 choices for the first letter, 2 for the second letter, and 1 for the third letter, making a total of $3 \cdot 2 \cdot 1$, or 6 possibilities.

EXAMPLE 2   How many 3-letter code symbols can be formed with the letters A, B, C, D, and E *with* repetition (that is, allowing letters to be repeated)?

Solution   Since repetition is allowed, there are 5 choices for the first letter, 5 choices for the second, and 5 for the third. Thus, by the fundamental counting principle, there are $5 \cdot 5 \cdot 5$, or 125 code symbols.

Permutation

A **permutation** of a set of n objects is an ordered arrangement of all n objects.

Consider, for example, a set of 4 objects

$$\{A, B, C, D\}.$$

To find the number of ordered arrangements of the set, we select a first letter: There are 4 choices. Then we select a second letter: There are 3 choices. Then we select a third letter: There are 2 choices. Finally, there is 1 choice for the last selection. Thus, by the fundamental counting principle, there are $4 \cdot 3 \cdot 2 \cdot 1$, or 24, permutations of a set of 4 objects.

We can find a formula for the total number of permutations of all objects in a set of n objects. We have n choices for the first selection, $n - 1$ choices for the second, $n - 2$ for the third, and so on. For the nth selection, there is only 1 choice.

The Total Number of Permutations of n Objects

The total number of permutations of n objects, denoted $_nP_n$, is given by

$$_nP_n = n(n - 1)(n - 2) \cdots 3 \cdot 2 \cdot 1.$$

EXAMPLE 3 Find each of the following.

a) $_4P_4$ **b)** $_7P_7$

Solution

Start with 4.

a) $_4P_4 = \underbrace{4 \cdot 3 \cdot 2 \cdot 1}_{\text{4 factors}} = 24$

b) $_7P_7 = 7 \cdot 6 \cdot 5 \cdot 4 \cdot 3 \cdot 2 \cdot 1 = 5040$ ▬

EXAMPLE 4 In how many different ways can 9 packages be placed in 9 mailboxes, one package in a box?

Solution We have

$$_9P_9 = 9 \cdot 8 \cdot 7 \cdot 6 \cdot 5 \cdot 4 \cdot 3 \cdot 2 \cdot 1 = 362{,}880. \qquad ▬$$

Factorial Notation

We will use products such as $7 \cdot 6 \cdot 5 \cdot 4 \cdot 3 \cdot 2 \cdot 1$ so often that it is convenient to adopt a notation for them. For the product

$$7 \cdot 6 \cdot 5 \cdot 4 \cdot 3 \cdot 2 \cdot 1,$$

we write 7!, read "7 factorial."

We now define factorial notation for natural numbers and for 0.

> **Factorial Notation**
>
> For any natural number n,
>
> $$n! = n(n-1)(n-2) \cdots 3 \cdot 2 \cdot 1.$$
>
> For the number 0,
>
> $$0! = 1.$$

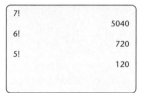
We define 0! as 1 so that certain formulas can be stated concisely and with a consistent pattern.

Here are some examples.

$$7! = 7 \cdot 6 \cdot 5 \cdot 4 \cdot 3 \cdot 2 \cdot 1 = 5040$$
$$6! = 6 \cdot 5 \cdot 4 \cdot 3 \cdot 2 \cdot 1 = 720$$
$$5! = 5 \cdot 4 \cdot 3 \cdot 2 \cdot 1 = 120$$
$$4! = 4 \cdot 3 \cdot 2 \cdot 1 = 24$$
$$3! = 3 \cdot 2 \cdot 1 = 6$$
$$2! = 2 \cdot 1 = 2$$
$$1! = 1 = 1$$
$$0! = 1 = 1$$

We now see that the following statement is true.

$$\boxed{{}_nP_n = n!}$$

We will often need to manipulate factorial notation. For example, note that

$$8! = 8 \cdot 7 \cdot 6 \cdot 5 \cdot 4 \cdot 3 \cdot 2 \cdot 1$$
$$= 8 \cdot (7 \cdot 6 \cdot 5 \cdot 4 \cdot 3 \cdot 2 \cdot 1) = 8 \cdot 7!.$$

Generalizing, we get the following.

> For any natural number n, $n! = n(n-1)!$.

By using this result repeatedly, we can further manipulate factorial notation.

EXAMPLE 5   Rewrite 7! with a factor of 5!.

Solution   We have

$$7! = 7 \cdot 6! = 7 \cdot 6 \cdot 5!.$$

In general, we have the following.

For any natural numbers k and n, with $k < n$,

$$n! = \underbrace{n(n-1)(n-2)\cdots[n-(k-1)]}_{k\text{ factors}} \cdot \underbrace{(n-k)!}_{n-k\text{ factors}}$$

Permutations of n Objects Taken k at a Time

Consider a set of 5 objects

$$\{A, B, C, D, E\}.$$

How many ordered arrangements can be formed using 3 objects without repetition? Examples of such an arrangement are EBA, CAB, and BCD. There are 5 choices for the first object, 4 choices for the second, and 3 choices for the third. By the fundamental counting principle, there are

$$5 \cdot 4 \cdot 3,$$

or

60 *permutations* of a set of 5 objects taken 3 at a time.

Note that

$$5 \cdot 4 \cdot 3 = \frac{5 \cdot 4 \cdot 3 \cdot 2 \cdot 1}{2 \cdot 1}, \quad \text{or} \quad \frac{5!}{2!}.$$

> ### *Permutation of n Objects Taken k at a Time*
>
> A **permutation** of a set of n objects taken k at a time is an ordered arrangement of k objects taken from the set.

Consider a set of n objects and the selection of an ordered arrangement of k of them. There would be n choices for the first object. Then there would remain $n - 1$ choices for the second, $n - 2$ choices for the third, and so on. We make k choices in all, so there are k factors in the product. By the fundamental counting principle, the total number of permutations is

$$\underbrace{n(n-1)(n-2)\cdots[n-(k-1)]}_{k\text{ factors}}.$$

We can express this in another way by multiplying by 1, as follows:

$$n(n-1)(n-2)\cdots[n-(k-1)] \cdot \frac{(n-k)!}{(n-k)!}$$

$$= \frac{n(n-1)(n-2)\cdots[n-(k-1)](n-k)!}{(n-k)!}$$

$$= \frac{n!}{(n-k)!}.$$

This gives us the following.

> ### The Number of Permutations of n Objects Taken k at a Time
>
> The number of permutations of a set of n objects taken k at a time, denoted $_nP_k$, is given by
>
> $$_nP_k = \underbrace{n(n-1)(n-2)\cdots[n-(k-1)]}_{k\text{ factors}} \qquad (1)$$
>
> $$= \frac{n!}{(n-k)!}. \qquad (2)$$

EXAMPLE 6 Compute $_8P_4$ using both forms of the formula.

Solution Using form (1), we have

The 8 tells where to start.

$$_8P_4 = 8 \cdot 7 \cdot 6 \cdot 5 = 1680.$$

The 4 tells how many factors.

Using form (2), we have

$$_8P_4 = \frac{8!}{(8-4)!}$$

$$= \frac{8!}{4!}$$

$$= \frac{8 \cdot 7 \cdot 6 \cdot 5 \cdot 4!}{4!} = \frac{8 \cdot 7 \cdot 6 \cdot 5 \cdot \cancel{4!}}{\cancel{4!}}$$

$$= 8 \cdot 7 \cdot 6 \cdot 5 = 1680.$$

EXAMPLE 7 *Flags of Nations.* The flags of many nations consist of three vertical stripes. For example, the flag of Ireland, shown here, has its first stripe green, second white, and third orange.

Suppose that the following 9 colors are available:

{black, yellow, red, blue, white, gold, orange, pink, purple}.

How many different flags of 3 colors can be made without repetition of colors in a flag? This assumes that the order in which the stripes appear is considered.

Solution We are determining the number of permutations of 9 objects taken 3 at a time. There is no repetition of colors. Using form (1), we get

$$_9P_3 = 9 \cdot 8 \cdot 7 = 504.$$

EXAMPLE 8 *Batting Orders.* A baseball manager arranges the batting order as follows: The 4 infielders will bat first. Then the 3 outfielders, the catcher, and the pitcher will follow, not necessarily in that order. How many different batting orders are possible?

Solution The infielders can bat in $_4P_4$ different ways, the rest in $_5P_5$ different ways. Then by the fundamental counting principle, we have

$$_4P_4 \cdot {_5P_5} = 4! \cdot 5!, \quad \text{or} \quad 2880 \text{ possible batting orders.} \quad \blacksquare$$

If we allow repetition, a situation like the following can occur.

EXAMPLE 9 How many 5-letter code symbols can be formed with the letters A, B, C, and D if we allow a letter to occur more than once?

Solution We can select each of the 5 letters in 4 ways. That is, we can select the first letter in 4 ways, the second in 4 ways, and so on. Thus there are 4^5, or 1024 arrangements. $\quad \blacksquare$

> The number of distinct arrangements of n objects taken k at a time, allowing repetition, is n^k.

Permutations of Sets with Nondistinguishable Objects

Consider a set of 7 marbles, 4 of which are blue and 3 of which are red. Although the marbles are all different, when they are lined up, one red marble will look just like any other red marble. In this sense, we say that the red marbles are nondistinguishable and, similarly, the blue marbles are nondistinguishable.

We know that there are 7! permutations of this set. Many of them will look alike, however. We develop a formula for finding the number of distinguishable permutations.

Consider a set of n objects in which n_1 are of one kind, n_2 are of a second kind, ..., and n_k are of a kth kind. The total number of permutations of the set is $n!$, but this includes many that are nondistinguishable. Let N be the total number of distinguishable permutations. For each of these N permutations, there are $n_1!$ actual, nondistinguishable permutations, obtained by permuting the objects of the first kind. For each of these $N \cdot n_1!$ permutations, there are $n_2!$ nondistinguishable permutations, obtained by permuting the objects of the second kind, and so on.

By the fundamental counting principle, the total number of permutations, including those that are nondistinguishable, is

$$N \cdot n_1! \cdot n_2! \cdot \cdots \cdot n_k!.$$

Then we have $N \cdot n_1! \cdot n_2! \cdot \cdots \cdot n_k! = n!$. Solving for N, we obtain

$$N = \frac{n!}{n_1! \cdot n_2! \cdot \cdots \cdot n_k!}.$$

Now, to finish our problem with the marbles, we have

$$N = \frac{7!}{4! \, 3!}$$

$$= \frac{7 \cdot 6 \cdot 5 \cdot 4!}{4! \cdot 3 \cdot 2 \cdot 1} = \frac{7 \cdot \cancel{3} \cdot \cancel{2} \cdot 5 \cdot \cancel{4!}}{\cancel{4!} \cdot \cancel{3} \cdot \cancel{2} \cdot 1}$$

$$= \frac{7 \cdot 5}{1}, \quad \text{or} \quad 35$$

distinguishable permutations of the marbles.

In general:

For a set of n objects in which n_1 are of one kind, n_2 are of another kind, ..., and n_k are of a kth kind, the number of distinguishable permutations is

$$\frac{n!}{n_1! \cdot n_2! \cdot \cdots \cdot n_k!}.$$

EXAMPLE 10 In how many distinguishable ways can the letters of the word CINCINNATI be arranged?

Solution There are 2 C's, 3 I's, 3 N's, 1 A, and 1 T for a total of 10 letters. Thus,

$$N = \frac{10!}{2! \cdot 3! \cdot 3! \cdot 1! \cdot 1!}, \quad \text{or} \quad 50{,}400.$$

The letters of the word CINCINNATI can be arranged in 50,400 distinguishable ways.

7.5

Exercise Set

Evaluate.

1. $_6P_6$ **2.** $_4P_3$

3. $_{10}P_7$ **4.** $_{10}P_3$

5. $5!$ **6.** $7!$

7. $0!$ **8.** $1!$

9. $\dfrac{9!}{5!}$ **10.** $\dfrac{9!}{4!}$

11. $(8-3)!$ **12.** $(8-5)!$

13. $\dfrac{10!}{7!\,3!}$ **14.** $\dfrac{7!}{(7-2)!}$

15. $_8P_0$ **16.** $_{13}P_1$

17. $_{52}P_4$ **18.** $_{52}P_5$

19. $_nP_3$ **20.** $_nP_2$

21. $_nP_1$ **22.** $_nP_0$

In each of Exercises 23–41, give your answer using permutation notation, factorial notation, or other operations. Then evaluate.

How many permutations are there of the letters in each of the following words, if all the letters are used without repetition?

23. MARVIN   **24.** JUDY

25. UNDERMOST   **26.** COMBINES

27. How many permutations are there of the letters of the word UNDERMOST if the letters are taken 4 at a time?

28. How many permutations are there of the letters of the word COMBINES if the letters are taken 5 at a time?

29. How many 5-digit numbers can be formed using the digits 2, 4, 6, 8, and 9 without repetition? with repetition?

30. In how many ways can 7 athletes be arranged in a straight line?

31. How many distinguishable code symbols can be formed from the letters of the word BUSINESS? BIOLOGY? MATHEMATICS?

32. A professor is going to grade her 24 students on a curve. She will give 3 A's, 5 B's, 9 C's, 4 D's, and 3 F's. In how many ways can she do this?

33. *Phone Numbers.* How many 7-digit phone numbers can be formed with the digits 0, 1, 2, 3, 4, 5, 6, 7, 8, and 9, assuming that the first number cannot be 0 or 1? Accordingly, how many telephone numbers can there be within a given area code, before the area needs to be split with a new area code?

34. *Program Planning.* A program is planned to have 5 rock numbers and 4 speeches. In how many ways can this be done if a rock number and a speech are to alternate and a rock number is to come first?

35. Suppose the expression $a^2b^3c^4$ is rewritten without exponents. In how many ways can this be done?

36. *Coin Arrangements.* A penny, a nickel, a dime, and a quarter are arranged in a straight line.

a) Considering just the coins, in how many ways can they be lined up?

b) Considering the coins and heads and tails, in how many ways can they be lined up?

37. How many code symbols can be formed using 5 out of 6 letters of A, B, C, D, E, F if the letters:

a) are not repeated?
b) can be repeated?
c) are not repeated but must begin with D?
d) are not repeated but must begin with DE?

38. *License Plates.* A state forms its license plates by first listing a number that corresponds to the county in which the owner of the car resides. (The names of the counties are alphabetized and the number is its location in that order.) Then the plate lists a letter of the alphabet, and this is followed by a number from 1 to 9999. How many such plates are possible if there are 80 counties?

39. *Zip Codes.* A U.S. postal zip code is a five-digit number.

a) How many zip codes are possible if any of the digits 0 to 9 can be used?

b) If each post office has its own zip code, how many possible post offices can there be?

40. *Zip-Plus-4 Codes.* A zip-plus-4 postal code uses a 9-digit number like 75247-5456. How many 9-digit zip-plus-4 postal codes are possible?

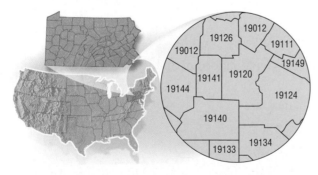

41. *Social Security Numbers.* A social security number is a 9-digit number like 243-47-0825.

a) How many different social security numbers can there be?

b) There are about 285 million people in the United States. Can each person have a unique social security number?

Collaborative Discussion and Writing

42. How "long" is 15!? You own 15 different books and decide to make up all the possible arrangements of the books on a shelf. About how long, in years, would it take you if you make one arrangement per second? Write out the reasoning you used for this problem in the form of a paragraph.

43. *Circular Arrangements.* In how many ways can the numbers on a clock face be arranged? See if you can derive a formula for the number of distinct circular arrangements of n objects. Explain your reasoning.

Skill Maintenance

Find the zero(s) of the function.

44. $f(x) = 4x - 9$

45. $f(x) = x^2 + x - 6$

46. $f(x) = 2x^2 - 3x - 1$

47. $f(x) = x^3 - 4x^2 - 7x + 10$

Synthesis

Solve for n.

48. $_nP_5 = 7 \cdot {_nP_4}$ **49.** $_nP_4 = 8 \cdot {_{n-1}P_3}$

50. $_nP_5 = 9 \cdot {_{n-1}P_4}$ **51.** $_nP_4 = 8 \cdot {_nP_3}$

52. Show that $n! = n(n-1)(n-2)(n-3)!$.

53. *Single-Elimination Tournaments.* In a single-elimination sports tournament consisting of n teams, a team is eliminated when it loses one game. How many games are required to complete the tournament?

54. *Double-Elimination Tournaments.* In a double-elimination softball tournament consisting of n teams, a team is eliminated when it loses two games. At most, how many games are required to complete the tournament?

7.6

Combinatorics: Combinations

• *Evaluate combination notation and solve related applied problems.*

We now consider counting techniques in which order is not considered.

Combinations

We sometimes make a selection from a set *without regard to order*. Such a selection is called a *combination*. If you play cards, for example, you know that in most situations the *order* in which you hold cards is not important. That is,

The hand   is "equivalent" to these hands.

Each hand contains the same combination of three cards.

EXAMPLE 1 Find all the combinations of 3 letters taken from the set of 5 letters {A, B, C, D, E}.

Solution The combinations are

$$
\begin{array}{ll}
\{A, B, C\}, & \{A, B, D\}, \\
\{A, B, E\}, & \{A, C, D\}, \\
\{A, C, E\}, & \{A, D, E\}, \\
\{B, C, D\}, & \{B, C, E\}, \\
\{B, D, E\}, & \{C, D, E\}.
\end{array}
$$

There are 10 combinations of the 5 letters taken 3 at a time. ▬

When we find all the combinations from a set of 5 objects taken 3 at a time, we are finding all the 3-element subsets. When a set is named, the order of the elements is *not* considered. Thus,

$$\{A, C, B\} \quad \text{names the same set as} \quad \{A, B, C\}.$$

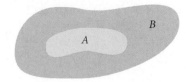

Subset

Set A is a subset of set B, denoted $A \subseteq B$, if every element of A is an element of B.

The elements of a subset are not ordered. When thinking of *combinations*, do *not* think about order!

Combination

A **combination** containing k objects is a subset containing k objects.

We want to develop a formula for computing the number of combinations of n objects taken k at a time without actually listing the combinations or subsets.

Combination Notation

The number of combinations of n objects taken k at a time is denoted $_nC_k$.

We call $_nC_k$ **combination notation.** We want to derive a general formula for $_nC_k$ for any $k \leq n$. First, it is true that $_nC_n = 1$, because a set with n objects has only 1 subset with n objects, the set itself. Second, $_nC_1 = n$, because a set with n objects has n subsets with 1 element each. Finally, $_nC_0 = 1$, because a set with n objects has only one subset with 0 elements, namely, the empty set $\varnothing$. To consider other possibilities, let's return to Example 1 and compare the number of combinations with the number of permutations.

COMBINATIONS						
$\{A, B, C\} \longrightarrow$	ABC	BCA	CAB	CBA	BAC	ACB
$\{A, B, D\} \longrightarrow$	ABD	BDA	DAB	DBA	BAD	ADB
$\{A, B, E\} \longrightarrow$	ABE	BEA	EAB	EBA	BAE	AEB
$\{A, C, D\} \longrightarrow$	ACD	CDA	DAC	DCA	CAD	ADC
$\{A, C, E\} \longrightarrow$	ACE	CEA	EAC	ECA	CAE	AEC
$\{A, D, E\} \longrightarrow$	ADE	DEA	EAD	EDA	DAE	AED
$\{B, C, D\} \longrightarrow$	BCD	CDB	DBC	DCB	CBD	BDC
$\{B, C, E\} \longrightarrow$	BCE	CEB	EBC	ECB	CBE	BEC
$\{B, D, E\} \longrightarrow$	BDE	DEB	EBD	EDB	DBE	BED
$\{C, D, E\} \longrightarrow$	CDE	DEC	ECD	EDC	DCE	CED

PERMUTATIONS

$_5C_3$ of these

$3! \cdot {}_5C_3$ of these

Note that each combination of 3 objects yields 6, or 3!, permutations.

$$3! \cdot {}_5C_3 = 60 = {}_5P_3 = 5 \cdot 4 \cdot 3,$$

so

$${}_5C_3 = \frac{{}_5P_3}{3!} = \frac{5 \cdot 4 \cdot 3}{3 \cdot 2 \cdot 1} = 10.$$

In general, the number of combinations of n objects taken k at a time, ${}_nC_k$, times the number of permutations of these objects, $k!$, must equal the number of permutations of n objects taken k at a time:

$$k! \cdot {}_nC_k = {}_nP_k$$

$$\begin{aligned}
{}_nC_k &= \frac{{}_nP_k}{k!} \\[6pt]
&= \frac{1}{k!} \cdot {}_nP_k \\[6pt]
&= \frac{1}{k!} \cdot \frac{n!}{(n-k)!} = \frac{n!}{k!\,(n-k)!}.
\end{aligned}$$

Combinations of n Objects Taken k at a Time

The total number of combinations of n objects taken k at a time, denoted ${}_nC_k$, is given by

$${}_nC_k = \frac{n!}{k!\,(n-k)!}, \tag{1}$$

or

$${}_nC_k = \frac{{}_nP_k}{k!} = \frac{n(n-1)(n-2)\cdots[n-(k-1)]}{k!}. \tag{2}$$

Another kind of notation for ${}_nC_k$ is **binomial coefficient notation.** The reason for such terminology will be seen later.

Binomial Coefficient Notation

$$\binom{n}{k} = {}_nC_k$$

You should be able to use either notation and either form of the formula.

EXAMPLE 2   Evaluate $\begin{pmatrix} 7 \\ 5 \end{pmatrix}$, using forms (1) and (2).

Solution

a) By form (1),

$$\begin{pmatrix} 7 \\ 5 \end{pmatrix} = \frac{7!}{5!\,(7-5)!} = \frac{7!}{5!\,2!}$$

$$= \frac{7 \cdot 6 \cdot 5 \cdot 4 \cdot 3 \cdot 2 \cdot 1}{5 \cdot 4 \cdot 3 \cdot 2 \cdot 1 \cdot 2 \cdot 1} = \frac{7 \cdot 6}{2 \cdot 1} = 21.$$

b) By form (2),

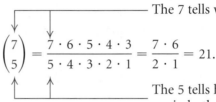

The 7 tells where to start.

$$\begin{pmatrix} 7 \\ 5 \end{pmatrix} = \frac{7 \cdot 6 \cdot 5 \cdot 4 \cdot 3}{5 \cdot 4 \cdot 3 \cdot 2 \cdot 1} = \frac{7 \cdot 6}{2 \cdot 1} = 21.$$

The 5 tells how many factors there are in both the numerator and the denominator and where to start the denominator.

Be sure to keep in mind that $\begin{pmatrix} n \\ k \end{pmatrix}$ does not mean $n \div k$, or n/k.

EXAMPLE 3   Evaluate $\begin{pmatrix} n \\ 0 \end{pmatrix}$ and $\begin{pmatrix} n \\ 2 \end{pmatrix}$.

Solution   We use form (1) for the first expression and form (2) for the second. Then

$$\begin{pmatrix} n \\ 0 \end{pmatrix} = \frac{n!}{0!\,(n-0)!} = \frac{n!}{1 \cdot n!} = 1,$$

using form (1), and

$$\begin{pmatrix} n \\ 2 \end{pmatrix} = \frac{n(n-1)}{2!} = \frac{n(n-1)}{2}, \quad \text{or} \quad \frac{n^2 - n}{2},$$

using form (2).

Note that

$$\begin{pmatrix} 7 \\ 2 \end{pmatrix} = \frac{7 \cdot 6}{2 \cdot 1} = 21,$$

so that using the result of Example 2 gives us

$$\begin{pmatrix} 7 \\ 5 \end{pmatrix} = \begin{pmatrix} 7 \\ 2 \end{pmatrix}.$$

Technology Connection

We can do computations like the one in Example 2 using the $_nC_r$ operation from the MATH PRB (probability) menu on a graphing calculator.

```
7 nCr 5
                21
```

This says that the number of 5-element subsets of a set of 7 objects is the same as the number of 2-element subsets of a set of 7 objects. When 5 elements are chosen from a set, one also chooses *not* to include 2 elements. To see this, consider the set $\{A, B, C, D, E, F, G\}$:

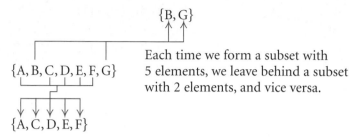

Each time we form a subset with 5 elements, we leave behind a subset with 2 elements, and vice versa.

In general, we have the following. This result provides an alternative way to compute combinations.

Subsets of Size *k* and of Size *n − k*

$$\binom{n}{k} = \binom{n}{n-k} \quad \text{and} \quad {}_nC_k = {}_nC_{n-k}$$

The number of subsets of size k of a set with n objects is the same as the number of subsets of size $n - k$. The number of combinations of n objects taken k at a time is the same as the number of combinations of n objects taken $n - k$ at a time.

We now solve problems involving combinations.

EXAMPLE 4 *Michigan Lottery.* Run by the state of Michigan, WINFALL is a twice-weekly lottery game with a jackpot of at least $2 million. For a purchase price of $1, a player can pick any 6 numbers from 1 through 49. If the numbers match those drawn by the state, the player wins. (*Source:* Michigan Lottery)

a) How many 6-number combinations are there?

b) Suppose that it takes 10 min to pick your numbers and buy a game ticket. How many tickets can you buy in 4 days?

c) How many people would you have to hire to buy tickets with all the possible combinations and ensure that you win?

Solution

a) No order is implied here. You pick any 6 numbers from 1 to 49. Thus the number of combinations is

$$\begin{aligned}
{}_{49}C_6 = \binom{49}{6} &= \frac{49!}{6!\,(49-6)!} = \frac{49!}{6!\,43!} \\
&= \frac{49 \cdot 48 \cdot 47 \cdot 46 \cdot 45 \cdot 44}{6 \cdot 5 \cdot 4 \cdot 3 \cdot 2 \cdot 1} \\
&= 13{,}983{,}816.
\end{aligned}$$

b) First we find the number of minutes in 4 days:

$$4 \text{ days} = 4 \text{ days} \cdot \frac{24 \text{ hr}}{1 \text{ day}} \cdot \frac{60 \text{ min}}{1 \text{ hr}} = 5760 \text{ min.}$$

Thus you could buy 5760/10, or 576 tickets in 4 days.

c) You would need to hire 13,983,816/576, or about 24,278 people to buy tickets with all the possible combinations and ensure a win. (This presumes lottery tickets can be bought 24 hours a day.) ▬

EXAMPLE 5 How many committees can be formed from a group of 5 governors and 7 senators if each committee consists of 3 governors and 4 senators?

Solution The 3 governors can be selected in $_5C_3$ ways and the 4 senators can be selected in $_7C_4$ ways. If we use the fundamental counting principle, it follows that the number of possible committees is

$$_5C_3 \cdot _7C_4 = \frac{5!}{3!\,2!} \cdot \frac{7!}{4!\,3!}$$
$$= \frac{5 \cdot 4 \cdot 3!}{3! \cdot 2 \cdot 1} \cdot \frac{7 \cdot 6 \cdot 5 \cdot 4!}{4! \cdot 3 \cdot 2 \cdot 1}$$
$$= \frac{5 \cdot 2 \cdot 2 \cdot 3!}{3! \cdot 2 \cdot 1} \cdot \frac{7 \cdot 3 \cdot 2 \cdot 5 \cdot 4!}{4! \cdot 3 \cdot 2 \cdot 1}$$
$$= 10 \cdot 35$$
$$= 350.$$ ▬

CONNECTING THE CONCEPTS

PERMUTATIONS AND COMBINATIONS

PERMUTATIONS	COMBINATIONS
Permutations involve order and arrangements of objects.	Combinations do not involve order or arrangements of objects.
Given 5 books, we can arrange 3 of them on a shelf in $_5P_3$, or 60 ways.	Given 5 books, we can select 3 of them in $_5C_3$, or 10 ways.
Placing the books in different orders produces different arrangements.	The order in which the books are chosen does not matter.

7.6

Exercise Set

Evaluate.

1. $_{13}C_2$

2. $_9C_6$

3. $\begin{pmatrix} 13 \\ 11 \end{pmatrix}$

4. $\begin{pmatrix} 9 \\ 3 \end{pmatrix}$

5. $\begin{pmatrix} 7 \\ 1 \end{pmatrix}$

6. $\begin{pmatrix} 8 \\ 8 \end{pmatrix}$

7. $\dfrac{_5P_3}{3!}$

8. $\dfrac{_{10}P_5}{5!}$

9. $\begin{pmatrix} 6 \\ 0 \end{pmatrix}$

10. $\begin{pmatrix} 6 \\ 1 \end{pmatrix}$

11. $\begin{pmatrix} 6 \\ 2 \end{pmatrix}$

12. $\begin{pmatrix} 6 \\ 3 \end{pmatrix}$

13. $\begin{pmatrix} 7 \\ 0 \end{pmatrix} + \begin{pmatrix} 7 \\ 1 \end{pmatrix} + \begin{pmatrix} 7 \\ 2 \end{pmatrix} + \begin{pmatrix} 7 \\ 3 \end{pmatrix} + \begin{pmatrix} 7 \\ 4 \end{pmatrix} + \begin{pmatrix} 7 \\ 5 \end{pmatrix}$
$+ \begin{pmatrix} 7 \\ 6 \end{pmatrix} + \begin{pmatrix} 7 \\ 7 \end{pmatrix}$

14. $\begin{pmatrix} 6 \\ 0 \end{pmatrix} + \begin{pmatrix} 6 \\ 1 \end{pmatrix} + \begin{pmatrix} 6 \\ 2 \end{pmatrix} + \begin{pmatrix} 6 \\ 3 \end{pmatrix} + \begin{pmatrix} 6 \\ 4 \end{pmatrix} + \begin{pmatrix} 6 \\ 5 \end{pmatrix} + \begin{pmatrix} 6 \\ 6 \end{pmatrix}$

15. $_{52}C_4$

16. $_{52}C_5$

17. $\begin{pmatrix} 27 \\ 11 \end{pmatrix}$

18. $\begin{pmatrix} 37 \\ 8 \end{pmatrix}$

19. $\begin{pmatrix} n \\ 1 \end{pmatrix}$

20. $\begin{pmatrix} n \\ 3 \end{pmatrix}$

21. $\begin{pmatrix} m \\ m \end{pmatrix}$

22. $\begin{pmatrix} t \\ 4 \end{pmatrix}$

In each of the following exercises, give an expression for the answer using permutation notation, combination notation, factorial notation, or other operations. Then evaluate.

23. *Fraternity Officers.* There are 23 students in a fraternity. How many sets of 4 officers can be selected?

24. *League Games.* How many games can be played in a 9-team sports league if each team plays all other teams once? twice?

25. *Test Options.* On a test, a student is to select 10 out of 13 questions. In how many ways can this be done?

26. *Test Options.* Of the first 10 questions on a test, a student must answer 7. Of the second 5 questions, the student must answer 3. In how many ways can this be done?

27. *Lines and Triangles from Points.* How many lines are determined by 8 points, no 3 of which are collinear? How many triangles are determined by the same points?

28. *Senate Committees.* Suppose the Senate of the United States consists of 58 Republicans and 42 Democrats. How many committees can be formed consisting of 6 Republicans and 4 Democrats?

29. *Poker Hands.* How many 5-card poker hands are possible with a 52-card deck?

30. *Bridge Hands.* How many 13-card bridge hands are possible with a 52-card deck?

31. *Baskin-Robbins Ice Cream.* Burt Baskin and Irv Robbins began making ice cream in 1945. Initially they developed 31 flavors—one for each day of the month. (*Source*: Baskin-Robbins)

a) How many 2-dip cones are possible using the 31 original flavors if order of flavors is to be considered and no flavor is repeated?

b) How many 2-dip cones are possible if order is to be considered and a flavor can be repeated?

c) How many 2-dip cones are possible if order is not considered and no flavor is repeated?

Collaborative Discussion and Writing

32. Explain why a "combination" lock should really be called a "permutation" lock.

33. Give an explanation that you might use with a fellow student to explain that

$$\binom{n}{k} = \binom{n}{n-k}.$$

Skill Maintenance

Solve.

34. $3x - 7 = 5x + 10$ **35.** $2x^2 - x = 3$

36. $x^2 + 5x + 1 = 0$ **37.** $x^3 + 3x^2 - 10x = 24$

Synthesis

38. *Full House.*   A full house in poker consists of a pair (two of a kind) and three of a kind. How many full houses are there that consist of 3 aces and 2 queens? (See Section 7.8 for a description of a 52-card deck.)

39. *Flush.*   A flush in poker consists of a 5-card hand with all cards of the same suit. How many 5-card hands (flushes) are there that consist of all diamonds?

40. There are n points on a circle. How many quadrilaterals can be inscribed with these points as vertices?

41. *League Games.*   How many games are played in a league with n teams if each team plays each other team once? twice?

Solve for n.

42. $\binom{n+1}{3} = 2 \cdot \binom{n}{2}$ **43.** $\binom{n}{n-2} = 6$

44. $\binom{n}{3} = 2 \cdot \binom{n-1}{2}$ **45.** $\binom{n+2}{4} = 6 \cdot \binom{n}{2}$

46. How many line segments are determined by the n vertices of an n-agon? Of these, how many are diagonals? Use mathematical induction to prove the result for the diagonals.

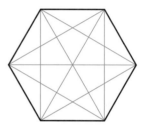

47. Prove that

$$\binom{n}{k-1} + \binom{n}{k} = \binom{n+1}{k}$$

for any natural numbers n and k, $k \leq n$.

7.7

The Binomial Theorem

- *Expand a power of a binomial using Pascal's triangle or factorial notation.*
- *Find a specific term of a binomial expansion.*
- *Find the total number of subsets of a set of n objects.*

In this section, we consider ways of expanding a binomial $(a + b)^n$.

Binomial Expansions Using Pascal's Triangle

Consider the following expanded powers of $(a + b)^n$, where $a + b$ is any binomial and n is a whole number. Look for patterns.

$$(a + b)^0 = 1$$
$$(a + b)^1 = a + b$$
$$(a + b)^2 = a^2 + 2ab + b^2$$
$$(a + b)^3 = a^3 + 3a^2b + 3ab^2 + b^3$$
$$(a + b)^4 = a^4 + 4a^3b + 6a^2b^2 + 4ab^3 + b^4$$
$$(a + b)^5 = a^5 + 5a^4b + 10a^3b^2 + 10a^2b^3 + 5ab^4 + b^5$$

Each expansion is a polynomial. There are some patterns to be noted.

1. There is one more term than the power of the exponent, n. That is, there are $n + 1$ terms in the expansion of $(a + b)^n$.

2. In each term, the sum of the exponents is n, the power to which the binomial is raised.

3. The exponents of a start with n, the power of the binomial, and decrease to 0. The last term has no factor of a. The first term has no factor of b, so powers of b start with 0 and increase to n.

4. The coefficients start at 1 and increase through certain values about "half"-way and then decrease through these same values back to 1.

Let's explore the coefficients further. Suppose that we want to find an expansion of $(a + b)^6$. The patterns we noted above indicate that there are 7 terms in the expansion:

$$a^6 + c_1a^5b + c_2a^4b^2 + c_3a^3b^3 + c_4a^2b^4 + c_5ab^5 + b^6.$$

How can we determine the value of each coefficient, c_i? We can do so in two different ways. The first method involves writing the coefficients in a triangular array, as follows. This is known as **Pascal's triangle:**

$$
\begin{array}{llccccccc}
(a + b)^0: & & & & & 1 & & & & \\
(a + b)^1: & & & & 1 & & 1 & & & \\
(a + b)^2: & & & 1 & & 2 & & 1 & & \\
(a + b)^3: & & 1 & & 3 & & 3 & & 1 & \\
(a + b)^4: & 1 & & 4 & & 6 & & 4 & & 1 \\
(a + b)^5: 1 & & 5 & & 10 & & 10 & & 5 & & 1
\end{array}
$$

There are many patterns in the triangle. Find as many as you can.

Perhaps you discovered a way to write the next row of numbers, given the numbers in the row above it. There are always 1's on the outside. Each remaining number is the sum of the two numbers above it. Let's try to find an expansion for $(a + b)^6$ by adding another row using the patterns we have discovered:

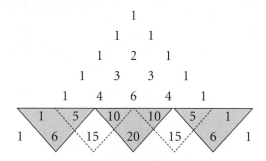

We see that in the last row

the 1st and last numbers are **1**;

the 2nd number is $1 + 5$, or **6**;

the 3rd number is $5 + 10$, or **15**;

the 4th number is $10 + 10$, or **20**;

the 5th number is $10 + 5$, or **15**; and

the 6th number is $5 + 1$, or **6**.

Thus the expansion for $(a + b)^6$ is

$$(a + b)^6 = 1a^6 + 6a^5b + 15a^4b^2 + 20a^3b^3 + 15a^2b^4 + 6ab^5 + 1b^6.$$

To find an expansion for $(a + b)^8$, we complete two more rows of Pascal's triangle:

$$
\begin{array}{ccccccccccccccccc}
 & & & & & & & & 1 & & & & & & & & \\
 & & & & & & & 1 & & 1 & & & & & & & \\
 & & & & & & 1 & & 2 & & 1 & & & & & & \\
 & & & & & 1 & & 3 & & 3 & & 1 & & & & & \\
 & & & & 1 & & 4 & & 6 & & 4 & & 1 & & & & \\
 & & & 1 & & 5 & & 10 & & 10 & & 5 & & 1 & & & \\
 & & 1 & & 6 & & 15 & & 20 & & 15 & & 6 & & 1 & & \\
 & 1 & & 7 & & 21 & & 35 & & 35 & & 21 & & 7 & & 1 & \\
1 & & 8 & & 28 & & 56 & & 70 & & 56 & & 28 & & 8 & & 1 \\
\end{array}
$$

Thus the expansion of $(a + b)^8$ is

$$(a + b)^8 = a^8 + 8a^7b + 28a^6b^2 + 56a^5b^3 + 70a^4b^4 + 56a^3b^5$$
$$+ 28a^2b^6 + 8ab^7 + b^8.$$

We can generalize our results as follows.

The Binomial Theorem Using Pascal's Triangle

For any binomial $a + b$ and any natural number n,

$$(a + b)^n = c_0 a^n b^0 + c_1 a^{n-1} b^1 + c_2 a^{n-2} b^2 + \cdots$$
$$+ c_{n-1} a^1 b^{n-1} + c_n a^0 b^n,$$

where the numbers $c_0, c_1, c_2, \ldots, c_{n-1}, c_n$ are from the $(n + 1)$st row of Pascal's triangle.

EXAMPLE 1 Expand: $(u - v)^5$.

Solution We have $(a + b)^n$, where $a = u$, $b = -v$, and $n = 5$. We use the 6th row of Pascal's triangle:

$$1 \quad 5 \quad 10 \quad 10 \quad 5 \quad 1$$

Then we have

$$(u - v)^5 = [u + (-v)]^5$$
$$= 1(u)^5 + 5(u)^4(-v)^1 + 10(u)^3(-v)^2 + 10(u)^2(-v)^3$$
$$+ 5(u)(-v)^4 + 1(-v)^5$$
$$= u^5 - 5u^4 v + 10u^3 v^2 - 10u^2 v^3 + 5uv^4 - v^5.$$

Note that the signs of the terms alternate between $+$ and $-$. When the power of $-v$ is odd, the sign is $-$. ▬

EXAMPLE 2 Expand: $\left(2t + \dfrac{3}{t}\right)^4$.

Solution We have $(a + b)^n$, where $a = 2t$, $b = 3/t$, and $n = 4$. We use the 5th row of Pascal's triangle:

$$1 \quad 4 \quad 6 \quad 4 \quad 1$$

Then we have

$$\left(2t + \frac{3}{t}\right)^4 = 1(2t)^4 + 4(2t)^3\left(\frac{3}{t}\right)^1 + 6(2t)^2\left(\frac{3}{t}\right)^2 + 4(2t)^1\left(\frac{3}{t}\right)^3 + 1\left(\frac{3}{t}\right)^4$$
$$= 1(16t^4) + 4(8t^3)\left(\frac{3}{t}\right) + 6(4t^2)\left(\frac{9}{t^2}\right) + 4(2t)\left(\frac{27}{t^3}\right) + 1\left(\frac{81}{t^4}\right)$$
$$= 16t^4 + 96t^2 + 216 + 216t^{-2} + 81t^{-4}.$$
▬

Binomial Expansion Using Factorial Notation

Suppose that we want to find the expansion of $(a + b)^{11}$. The disadvantage in using Pascal's triangle is that we must compute all the preceding rows of the triangle to obtain the row needed for the expansion. The following method avoids this. It also enables us to find a specific term—say, the 8th term—without computing all the other terms of the expansion. This method is useful in such courses as finite mathematics, calculus, and statistics, and it uses the *binomial coefficient notation* $\binom{n}{k}$ developed in Section 7.6.

We can restate the binomial theorem as follows.

The Binomial Theorem Using Factorial Notation

For any binomial $a + b$ and any natural number n,

$$(a + b)^n = \binom{n}{0}a^n b^0 + \binom{n}{1}a^{n-1}b^1 + \binom{n}{2}a^{n-2}b^2 + \cdots$$

$$+ \binom{n}{n-1}a^1 b^{n-1} + \binom{n}{n}a^0 b^n$$

$$= \sum_{k=0}^{n} \binom{n}{k}a^{n-k}b^k.$$

The binomial theorem can be proved by mathematical induction, but we will not do so here. This form shows why $\binom{n}{k}$ is called a *binomial coefficient*.

EXAMPLE 3 Expand: $(x^2 - 2y)^5$.

Solution We have $(a + b)^n$, where $a = x^2$, $b = -2y$, and $n = 5$. Then using the binomial theorem, we have

$$(x^2 - 2y)^5 = \binom{5}{0}(x^2)^5 + \binom{5}{1}(x^2)^4(-2y) + \binom{5}{2}(x^2)^3(-2y)^2$$

$$+ \binom{5}{3}(x^2)^2(-2y)^3 + \binom{5}{4}x^2(-2y)^4 + \binom{5}{5}(-2y)^5$$

$$= \frac{5!}{0!\,5!}x^{10} + \frac{5!}{1!\,4!}x^8(-2y) + \frac{5!}{2!\,3!}x^6(4y^2) + \frac{5!}{3!\,2!}x^4(-8y^3)$$

$$+ \frac{5!}{4!\,1!}x^2(16y^4) + \frac{5!}{5!\,0!}(-32y^5)$$

$$= 1 \cdot x^{10} + 5x^8(-2y) + 10x^6(4y^2) + 10x^4(-8y^3)$$

$$+ 5x^2(16y^4) + 1 \cdot (-32y^5)$$

$$= x^{10} - 10x^8 y + 40x^6 y^2 - 80x^4 y^3 + 80x^2 y^4 - 32y^5.$$

EXAMPLE 4   Expand: $\left(\dfrac{2}{x} + 3\sqrt{x}\right)^4$.

Solution   We have $(a + b)^n$, where $a = 2/x$, $b = 3\sqrt{x}$, and $n = 4$. Then using the binomial theorem, we have

$$\left(\frac{2}{x} + 3\sqrt{x}\right)^4 = \binom{4}{0}\left(\frac{2}{x}\right)^4 + \binom{4}{1}\left(\frac{2}{x}\right)^3(3\sqrt{x}) + \binom{4}{2}\left(\frac{2}{x}\right)^2(3\sqrt{x})^2$$

$$+ \binom{4}{3}\left(\frac{2}{x}\right)(3\sqrt{x})^3 + \binom{4}{4}(3\sqrt{x})^4$$

$$= \frac{4!}{0!\,4!}\left(\frac{16}{x^4}\right) + \frac{4!}{1!\,3!}\left(\frac{8}{x^3}\right)(3x^{1/2})$$

$$+ \frac{4!}{2!\,2!}\left(\frac{4}{x^2}\right)(9x) + \frac{4!}{3!\,1!}\left(\frac{2}{x}\right)(27x^{3/2})$$

$$+ \frac{4!}{4!\,0!}(81x^2)$$

$$= \frac{16}{x^4} + \frac{96}{x^{5/2}} + \frac{216}{x} + 216x^{1/2} + 81x^2.$$

Finding a Specific Term

Suppose that we want to determine only a particular term of an expansion. The method we have developed will allow us to find such a term without computing all the rows of Pascal's triangle or all the preceding coefficients.

Note that in the binomial theorem, $\binom{n}{0}a^n b^0$ gives us the 1st term, $\binom{n}{1}a^{n-1}b^1$ gives us the 2nd term, $\binom{n}{2}a^{n-2}b^2$ gives us the 3rd term, and so on. This can be generalized as follows.

Finding the (k + 1)st Term

The $(k + 1)$st term of $(a + b)^n$ is $\binom{n}{k}a^{n-k}b^k$.

EXAMPLE 5   Find the 5th term in the expansion of $(2x - 5y)^6$.

Solution   First, we note that $5 = 4 + 1$. Thus, $k = 4$, $a = 2x$, $b = -5y$, and $n = 6$. Then the 5th term of the expansion is

$$\binom{6}{4}(2x)^{6-4}(-5y)^4, \quad \text{or} \quad \frac{6!}{4!\,2!}(2x)^2(-5y)^4, \quad \text{or} \quad 37{,}500x^2y^4.$$

EXAMPLE 6   Find the 8th term in the expansion of $(3x - 2)^{10}$.

Solution   First, we note that $8 = 7 + 1$. Thus, $k = 7$, $a = 3x$, $b = -2$, and $n = 10$. Then the 8th term of the expansion is

$$\binom{10}{7}(3x)^{10-7}(-2)^7, \quad \text{or} \quad \frac{10!}{7!\,3!}(3x)^3(-2)^7, \quad \text{or} \quad -414{,}720x^3.$$

Total Number of Subsets

Suppose that a set has n objects. The number of subsets containing k elements is $\binom{n}{k}$ by a result of Section 7.6. The total number of subsets of a set is the number of subsets with 0 elements, plus the number of subsets with 1 element, plus the number of subsets with 2 elements, and so on. The total number of subsets of a set with n elements is

$$\binom{n}{0} + \binom{n}{1} + \binom{n}{2} + \cdots + \binom{n}{n}.$$

Now consider the expansion of $(1 + 1)^n$:

$$(1 + 1)^n = \binom{n}{0} \cdot 1^n + \binom{n}{1} \cdot 1^{n-1} \cdot 1^1 + \binom{n}{2} \cdot 1^{n-2} \cdot 1^2$$

$$+ \cdots + \binom{n}{n} \cdot 1^n$$

$$= \binom{n}{0} + \binom{n}{1} + \binom{n}{2} + \cdots + \binom{n}{n}.$$

Thus the total number of subsets is $(1 + 1)^n$, or 2^n. We have proved the following.

> ### Total Number of Subsets
> The total number of subsets of a set with n elements is 2^n.

EXAMPLE 7   The set $\{A, B, C, D, E\}$ has how many subsets?

Solution   The set has 5 elements, so the number of subsets is 2^5, or 32.

EXAMPLE 8   Wendy's, a national restaurant chain, offers the following toppings for its hamburgers:

$$\{catsup, mustard, mayonnaise, tomato,$$

$$lettuce, onions, pickle, relish, cheese\}.$$

How many different kinds of hamburgers can Wendy's serve, excluding size of hamburger or number of patties?

Solution   The toppings on each hamburger are the elements of a subset of the set of all possible toppings, the empty set being a plain hamburger. The total number of possible hamburgers is

$$\binom{9}{0} + \binom{9}{1} + \binom{9}{2} + \cdots + \binom{9}{9} = 2^9 = 512.$$

Thus Wendy's serves hamburgers in 512 different ways.   ▬

7.7

Exercise Set

Expand.

1. $(x + 5)^4$

2. $(x - 1)^4$

3. $(x - 3)^5$

4. $(x + 2)^9$

5. $(x - y)^5$

6. $(x + y)^8$

7. $(5x + 4y)^6$

8. $(2x - 3y)^5$

9. $\left(2t + \dfrac{1}{t}\right)^7$

10. $\left(3y - \dfrac{1}{y}\right)^4$

11. $(x^2 - 1)^5$

12. $(1 + 2q^3)^8$

13. $\left(\sqrt{5} + t\right)^6$

14. $\left(x - \sqrt{2}\right)^6$

15. $\left(a - \dfrac{2}{a}\right)^9$

16. $(1 + 3)^n$

17. $\left(\sqrt{2} + 1\right)^6 - \left(\sqrt{2} - 1\right)^6$

18. $\left(1 - \sqrt{2}\right)^4 + \left(1 + \sqrt{2}\right)^4$

19. $(x^{-2} + x^2)^4$

20. $\left(\dfrac{1}{\sqrt{x}} - \sqrt{x}\right)^6$

Find the indicated term of the binomial expansion.

21. 3rd; $(a + b)^7$

22. 6th; $(x + y)^8$

23. 6th; $(x - y)^{10}$

24. 5th; $(p - 2q)^9$

25. 12th; $(a - 2)^{14}$

26. 11th; $(x - 3)^{12}$

27. 5th; $\left(2x^3 - \sqrt{y}\right)^8$

28. 4th; $\left(\dfrac{1}{b^2} + \dfrac{b}{3}\right)^7$

29. Middle; $(2u - 3v^2)^{10}$

30. Middle two; $\left(\sqrt{x} + \sqrt{3}\right)^5$

Determine the number of subsets of each of the following.

31. A set of 7 elements

32. A set of 6 members

33. The set of letters of the Greek alphabet, which contains 24 letters

34. The set of letters of the English alphabet, which contains 26 letters

35. What is the degree of $(x^5 + 3)^4$?

36. What is the degree of $(2 - 5x^3)^7$?

Expand each of the following, where $i^2 = -1$.

37. $(3 + i)^5$

38. $(1 + i)^6$

39. $\left(\sqrt{2} - i\right)^4$

40. $\left(\dfrac{\sqrt{3}}{2} - \dfrac{1}{2}i\right)^{11}$

41. Find a formula for $(a - b)^n$. Use sigma notation.

42. Expand and simplify:

$$\frac{(x + h)^{13} - x^{13}}{h}.$$

43. Expand and simplify:

$$\frac{(x + h)^n - x^n}{h}.$$

Use sigma notation.

Collaborative Discussion and Writing

44. Discuss the advantages and disadvantages of each method of finding a binomial expansion. Give examples of when you might use one method rather than the other.

45. Blaise Pascal (1623–1662) was a French scientist and philosopher who founded the modern theory of probability. Do some research on Pascal and see if you can find out how he discovered his famous "triangle of numbers."

Skill Maintenance

Given that $f(x) = x^2 + 1$ and $g(x) = 2x - 3$, find each of the following.

46. $(f + g)(x)$

47. $(fg)(x)$

48. $(f \circ g)(x)$

49. $(g \circ f)(x)$

Synthesis

Solve for x.

50. $\displaystyle\sum_{k=0}^{8} \binom{8}{k} x^{8-k} 3^k = 0$

51. $\displaystyle\sum_{k=0}^{4} \binom{4}{k} 5^{4-k} x^k = 64$

52. $\displaystyle\sum_{k=0}^{5} \binom{5}{k} (-1)^k x^{5-k} 3^k = 32$

53. $\displaystyle\sum_{k=0}^{4} \binom{4}{k} (-1)^k x^{4-k} 6^k = 81$

54. Find the term of
$$\left(\frac{3x^2}{2} - \frac{1}{3x} \right)^{12}$$
that does not contain x.

55. Find the middle term of $(x^2 - 6y^{3/2})^6$.

56. Find the ratio of the 4th term of
$$\left(p^2 - \frac{1}{2} p \sqrt[3]{q} \right)^5$$
to the 3rd term.

57. Find the term of
$$\left(\sqrt[3]{x} - \frac{1}{\sqrt{x}} \right)^7$$
containing $1/x^{1/6}$.

58. *Money Combinations.* A money clip contains one each of the following bills: $1, $2, $5, $10, $20, $50, and $100. How many different sums of money can be formed using the bills?

Find the sum.

59. $_{100}C_0 + {}_{100}C_1 + \cdots + {}_{100}C_{100}$

60. $_nC_0 + {}_nC_1 + \cdots + {}_nC_n$

Simplify.

61. $\displaystyle\sum_{k=0}^{23} \binom{23}{k} (\log_a x)^{23-k} (\log_a t)^k$

62. $\displaystyle\sum_{k=0}^{15} \binom{15}{k} i^{30-2k}$

63. Use mathematical induction and the property
$$\binom{n}{r-1} + \binom{n}{r} = \binom{n+1}{r}$$
to prove the binomial theorem.

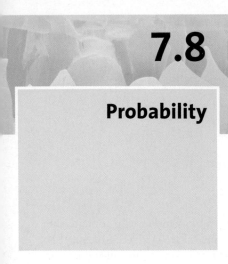

7.8

Probability

• *Compute the probability of a simple event.*

When a coin is tossed, we can reason that the chance, or likelihood, that it will fall heads is 1 out of 2, or the **probability** that it will fall heads is $\frac{1}{2}$. Of course, this does not mean that if a coin is tossed 10 times it will necessarily fall heads 5 times. If the coin is a "fair coin" and it is tossed a great many times, however, it will fall heads very nearly half of the time. Here we give an introduction to two kinds of probability, **experimental** and **theoretical**.

Experimental and Theoretical Probability

If we toss a coin a great number of times—say, 1000—and count the number of times it falls heads, we can determine the probability that it will fall heads. If it falls heads 503 times, we would calculate the probability of its falling heads to be

$$\frac{503}{1000}, \quad \text{or} \quad 0.503.$$

This is an **experimental** determination of probability. Such a determination of probability is discovered by the observation and study of data and is quite common and very useful. Here, for example, are some probabilities that have been determined *experimentally*:

1. The probability that a woman will get breast cancer in her lifetime is $\frac{1}{11}$.
2. If you kiss someone who has a cold, the probability of your catching a cold is 0.07.
3. A person who has just been released from prison has an 80% probability of returning.

If we consider a coin and reason that it is just as likely to fall heads as tails, we would calculate the probability that it will fall heads to be $\frac{1}{2}$. This is a **theoretical** determination of probability. Here are some other probabilities that have been determined *theoretically*, using mathematics:

1. If there are 30 people in a room, the probability that two of them have the same birthday (excluding year) is 0.706.
2. While on a trip, you meet someone and, after a period of conversation, discover that you have a common acquaintance. The typical reaction, "It's a small world!", is actually not appropriate, because the probability of such an occurrence is quite high—just over 22%.

In summary, experimental probabilities are determined by making observations and gathering data. Theoretical probabilities are determined by reasoning mathematically. Examples of experimental and theoretical probability like those above, especially those we do not expect, lead us to see the value of a study of probability. You might ask, "What is the *true* probability?" In fact, there is none. Experimentally, we can determine probabilities within certain limits. These may or may not agree with the probabilities that we obtain theoretically. There are situations in which it is much easier to determine one of these types of probabilities than the other. For example, it would be quite difficult to arrive at the probability of catching a cold using theoretical probability.

Computing Experimental Probabilities

We first consider experimental determination of probability. The basic principle we use in computing such probabilities is as follows.

Principle P (Experimental)

Given an experiment in which n observations are made, if a situation, or event, E occurs m times out of n observations, then we say that the *experimental probability* of the event, $P(E)$, is given by

$$P(E) = \frac{m}{n}.$$

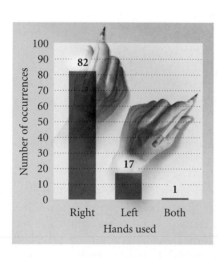

EXAMPLE 1 *Sociological Survey.* The authors of this text conducted an experimental survey to determine the number of people who are left-handed, right-handed, or both. The results are shown in the graph at left.

a) Determine the probability that a person is right-handed.

b) Determine the probability that a person is left-handed.

c) Determine the probability that a person is ambidextrous (uses both hands with equal ability).

d) There are 120 bowlers in most tournaments held by the Professional Bowlers Association. On the basis of the data in this experiment, how many of the bowlers would you expect to be left-handed?

Solution

a) The number of people who are right-handed is 82, the number who are left-handed is 17, and the number who are ambidextrous is 1. The total number of observations is $82 + 17 + 1$, or 100. Thus the probability that a person is right-handed is P, where

$$P = \frac{82}{100}, \quad \text{or} \quad 0.82, \quad \text{or} \quad 82\%.$$

b) The probability that a person is left-handed is P, where

$$P = \frac{17}{100}, \quad \text{or} \quad 0.17, \quad \text{or} \quad 17\%.$$

c) The probability that a person is ambidextrous is P, where

$$P = \frac{1}{100}, \quad \text{or} \quad 0.01, \quad \text{or } 1\%.$$

d) There are 120 bowlers, and from part (b) we can expect 17% to be left-handed. Since

$$17\% \text{ of } 120 = 0.17 \cdot 120 = 20.4,$$

we can expect that about 20 of the bowlers will be left-handed. ▬

EXAMPLE 2 *Quality Control.* It is very important for a manufacturer to maintain the quality of its products. In fact, companies hire quality control inspectors to ensure this process. The goal is to produce as few defective products as possible. But since a company is producing thousands of products every day, it cannot afford to check every product to see if it is defective. To find out what percentage of its products are defective, the company checks a smaller sample.

The U.S. Department of Agriculture requires that 80% of the seeds that a company produces must sprout. To determine the quality of the seeds it produces, a company takes 500 seeds from those it has produced and plants them. It finds that 417 of the seeds sprout.

a) What is the probability that a seed will sprout?

b) Did the seeds meet government standards?

Solution

a) We know that 500 seeds were planted and 417 sprouted. The probability of a seed sprouting is P, where

$$P = \frac{417}{500} = 0.834, \quad \text{or} \quad 83.4\%.$$

b) Since the percentage of seeds that sprouted exceeded the 80% requirement, the seeds meet government standards. ▬

EXAMPLE 3 *Television Ratings.* There are an estimated 105,500,000 households with televisions in the United States. Each week, viewing information is collected and reported. In one recent week, 7,815,000 households tuned in to the popular comedy series "Everybody Loves Raymond" on CBS and 8,302,000 households tuned in to the popular drama series "Law and Order" on NBC (*Source*: Nielsen Media Research). What is the probability that a television household tuned in to "Everybody Loves Raymond" during the given week? to "Law and Order"?

Solution The probability that a television household was tuned in to "Everybody Loves Raymond" is P, where

$$P = \frac{7,815,000}{105,500,000} \approx 0.074 \approx 7.4\%.$$

The probability that a television household was tuned in to "Law and Order" is P, where

$$P = \frac{8,302,000}{105,500,000} \approx 0.079 \approx 7.9\%.$$

These percentages are called *ratings*. ▬

Theoretical Probability

Suppose that we perform an experiment such as flipping a coin, throwing a dart, drawing a card from a deck, or checking an item off an assembly line for quality. Each possible result of such an experiment is called an **outcome.** The set of all possible outcomes is called the **sample space.** An **event** is a set of outcomes, that is, a subset of the sample space.

EXAMPLE 4 *Dart Throwing.* Consider this dartboard. Assume that the experiment is "throwing a dart" and that the dart hits the board. Find each of the following.

a) The outcomes

b) The sample space

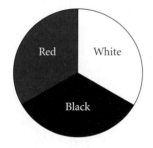

Solution

a) The outcomes are *hitting black* (B), *hitting red* (R), and *hitting white* (W).

b) The sample space is {*hitting black, hitting red, hitting white*}, which can be simply stated as {B, R, W}. ▬

EXAMPLE 5 *Die Rolling.* A die (pl., dice) is a cube, with six faces, each containing a number of dots from 1 to 6 on each side.

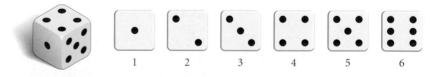

Suppose a die is rolled. Find each of the following.

a) The outcomes

b) The sample space

Solution

a) The outcomes are 1, 2, 3, 4, 5, 6.

b) The sample space is {1, 2, 3, 4, 5, 6}. ▬

We denote the probability that an event E occurs as $P(E)$. For example, "a coin falling heads" may be denoted H. Then $P(H)$ represents the probability of the coin falling heads. When all the outcomes of an experiment have the same probability of occurring, we say that they are *equally likely*. To see the distinction between events that are equally likely and those that are not, consider the dartboards shown below.

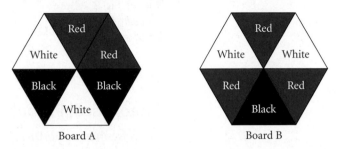

Board A   Board B

For board A, the events *hitting black*, *hitting red*, and *hitting white* are equally likely, because the black, red, and white areas are the same. However, for board B the areas are not the same so these events are not equally likely.

Principle P (Theoretical)

If an event E can occur m ways out of n possible equally likely outcomes of a sample space S, then the **theoretical probability** of the event, $P(E)$, is given by

$$P(E) = \frac{m}{n}.$$

EXAMPLE 6 What is the probability of rolling a 3 on a die?

Solution On a fair die, there are 6 equally likely outcomes and there is 1 way to roll a 3. By Principle P, $P(3) = \frac{1}{6}$. ▬

EXAMPLE 7 What is the probability of rolling an even number on a die?

Solution The event is rolling an *even* number. It can occur 3 ways (getting 2, 4, or 6). The number of equally likely outcomes is 6. By Principle P, $P(\text{even}) = \frac{3}{6}$, or $\frac{1}{2}$. ▬

We now use a number of examples related to a standard bridge deck of 52 cards. Such a deck is made up as shown in the following figure.

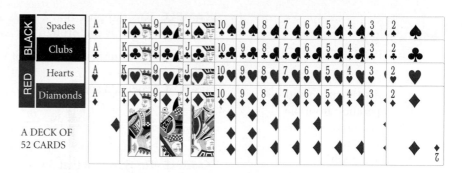

A DECK OF
52 CARDS

EXAMPLE 8 What is the probability of drawing an ace from a well-shuffled deck of cards?

Solution There are 52 outcomes (the number of cards in the deck), they are equally likely (from a well-shuffled deck), and there are 4 ways to obtain an ace, so by Principle *P*, we have

$$P(\text{drawing an ace}) = \frac{4}{52}, \quad \text{or} \quad \frac{1}{13}.$$

EXAMPLE 9 Suppose that we select, without looking, one marble from a bag containing 3 red marbles and 4 green marbles. What is the probability of selecting a red marble?

Solution There are 7 equally likely ways of selecting any marble, and since the number of ways of getting a red marble is 3, we have

$$P(\text{selecting a red marble}) = \frac{3}{7}.$$

The following are some results that follow from Principle *P*.

Probability Properties

a) If an event *E* cannot occur, then $P(E) = 0$.

b) If an event *E* is certain to occur, then $P(E) = 1$.

c) The probability that an event *E* will occur is a number from 0 to 1: $0 \leq P(E) \leq 1$.

For example, in coin tossing, the event that a coin will land on its edge has probability 0. The event that a coin falls either heads or tails has probability 1.

In the following examples, we use the combinatorics that we studied in Sections 7.5 and 7.6 to calculate theoretical probabilities.

EXAMPLE 10 Suppose that 2 cards are drawn from a well-shuffled deck of 52 cards. What is the probability that both of them are spades?

Solution The number of ways n of drawing 2 cards from a well-shuffled deck of 52 cards is $_{52}C_2$. Since 13 of the 52 cards are spades, the number of ways m of drawing 2 spades is $_{13}C_2$. Thus,

$$P(\text{getting 2 spades}) = \frac{m}{n} = \frac{_{13}C_2}{_{52}C_2} = \frac{78}{1326} = \frac{1}{17}.$$

EXAMPLE 11 Suppose that 3 people are selected at random from a group that consists of 6 men and 4 women. What is the probability that 1 man and 2 women are selected?

Solution The number of ways of selecting 3 people from a group of 10 is $_{10}C_3$. One man can be selected in $_6C_1$ ways, and 2 women can be selected in $_4C_2$ ways. By the fundamental counting principle, the number of ways of selecting 1 man and 2 women is $_6C_1 \cdot {_4C_2}$. Thus the probability that 1 man and 2 women are selected is

$$P = \frac{_6C_1 \cdot {_4C_2}}{_{10}C_3} = \frac{3}{10}.$$

EXAMPLE 12 *Rolling Two Dice.* What is the probability of getting a total of 8 on a roll of a pair of dice?

Solution On each die, there are 6 possible outcomes. The outcomes are paired so there are $6 \cdot 6$, or 36, possible ways in which the two can fall. (Assuming that the dice are different—say, one red and one blue—can help in visualizing this.)

The pairs that total 8 are as shown in the figure above. There are 5 possible ways of getting a total of 8, so the probability is $\frac{5}{36}$.

7.8

Exercise Set

1. *Select a Number.*   In a survey conducted by the authors, 100 people were polled and asked to select a number from 1 to 5. The results are shown in the following table.

NUMBER OF CHOICES	1	2	3	4	5
NUMBER WHO CHOSE THAT NUMBER	18	24	23	23	12

a) What is the probability that the number chosen is 1? 2? 3? 4? 5?

b) What general conclusion might be made from the results of the experiment?

2. *Mason Dots®.*   Made by the Tootsie Industries of Chicago, Illinois, Mason Dots® is a gumdrop candy. A box was opened by the authors and was found to contain the following number of gumdrops:

Orange	9
Lemon	8
Strawberry	7
Grape	6
Lime	5
Cherry	4

If we take one gumdrop out of the box, what is the probability of getting a lemon? lime? orange? grape? strawberry? licorice?

3. *Junk Mail.*   In experimental studies, the U.S. Postal Service has found that the probability that a piece of advertising is opened and read is 78%. A business sends out 15,000 pieces of advertising. How many of these can the company expect to be opened and read?

4. *Linguistics.*   An experiment was conducted by the authors to determine the relative occurrence of various letters of the English alphabet. The front page of a newspaper was considered. In all, there were 9136 letters. The number of occurrences of each letter of the alphabet is listed in the following table.

LETTER	NUMBER OF OCCURRENCES	PROBABILITY
A	853	853/9136 ≈ 9.3%
B	136	
C	273	
D	286	
E	1229	
F	173	
G	190	
H	399	
I	539	
J	21	
K	57	
L	417	
M	231	
N	597	
O	705	
P	238	
Q	4	
R	609	
S	745	
T	789	
U	240	
V	113	
W	127	
X	20	
Y	124	
Z	21	21/9136 ≈ 0.2%

a) Complete the table of probabilities with the percentage, to the nearest tenth of a percent, of the occurrence of each letter.

b) What is the probability of a vowel occurring?

c) What is the probability of a consonant occurring?

5. "Wheel of Fortune®."   The results of the experiment in Exercise 4 can be quite useful to a person playing the popular television game show "Wheel of Fortune." Players guess letters in order to spell out a phrase, a person, or a thing.

a) What 5 consonants have the greatest probability of occurring?

b) What vowel has the greatest probability of occurring?

c) The winner of the main part of the show plays for a grand prize and at one time was allowed to guess 5 consonants and a vowel in order to discover the secret wording. The 5 consonants R, S, T, L, N, and the vowel E seemed to be chosen most often. Do the results in parts (a) and (b) support such a choice?

6. *Card Drawing.* Suppose we draw a card from a well-shuffled deck of 52 cards.

a) How many equally likely outcomes are there?

What is the probability of drawing each of the following?

b) A queen   **c)** A heart

d) A 7   **e)** A red card

f) A 9 or a king   **g)** A black ace

7. *Marbles.* Suppose we select, without looking, one marble from a bag containing 4 red marbles and 10 green marbles. What is the probability of selecting each of the following?

a) A red marble

b) A green marble

c) A purple marble

d) A red or a green marble

8. *Production Unit.* The sales force of a business consists of 10 men and 10 women. A production unit of 4 people is set up at random. What is the probability that 2 men and 2 women are chosen?

9. *Coin Drawing.* A sack contains 7 dimes, 5 nickels, and 10 quarters. Eight coins are drawn at random. What is the probability of getting 4 dimes, 3 nickels, and 1 quarter?

10. *Michigan Lottery.* Run by the state of Michigan, WINFALL is a twice-weekly lottery game with a jackpot of at least $2 million. For a purchase price of $1, a player can pick any 6 numbers from 1 through 49. If the numbers match those drawn by the state, the player wins. (*Source:* Michigan Lottery) Kalisha buys 1 game ticket. What is her probability of winning?

Five-Card Poker Hands. Suppose that 5 cards are drawn from a deck of 52 cards. What is the probability of drawing each of the following?

11. 3 sevens and 2 kings

12. 5 aces

13. 5 spades

14. 4 aces and 1 five

15. *Tossing Three Coins.* Three coins are flipped. An outcome might be HTH.

a) Find the sample space.

What is the probability of getting each of the following?

b) Exactly one head

c) At most two tails

d) At least one head

e) Exactly two tails

Roulette. An American roulette wheel contains 38 slots numbered 00, 0, 1, 2, 3, . . . , 35, 36. Eighteen of the slots numbered 1–36 are colored red and 18 are colored black. The 00 and 0 slots are considered to be uncolored. The wheel is spun, and a ball is rolled around the rim until it falls into a slot. What is the probability that the ball falls in each of the following?

16. A red slot   **17.** A black slot

18. The 00 slot   **19.** The 0 slot

20. Either the 00 or the 0 slot

21. A red or a black slot

22. The number 24

23. An odd-numbered slot

24. *Dartboard.*   The figure below shows a dartboard. A dart is thrown and hits the board. Find the probabilities

$$P(\text{red}), \quad P(\text{green}), \quad P(\text{blue}), \quad P(\text{yellow}).$$

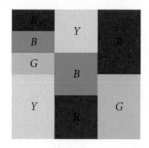

Technology Connection

25. *Random-Number Generator.*   Many graphing calculators have a **random-number generator.** This feature produces a random number in the interval $[0, 1]$. (Consult your user's manual.) We can use such a feature to simulate coin flipping. A number r such that $0 \leq r \leq 0.5$ would indicate heads, H. A number r such that $0.5 < r \leq 1.0$ would indicate tails, T. Use a random-number generator 100 times.

a) What is the experimental probability of getting heads?

b) What is the experimental probability of getting tails?

Collaborative Discussion and Writing

26. *Random Best-Selling Novels.*   Sir Arthur Stanley Eddington, an astronomer, once wrote in a satirical essay that if a monkey were left alone long enough with a typewriter and typed randomly, any great novel could be replicated. What is the probability that the following passage could have been written by a monkey? Ignore capital letters and punctuation and consider only letters and spaces.

"*It was the best of times, it was the worst of times,...*" (Charles Dickens, 1859). Explain your answer.

27. Find at least one use of probability in today's newspaper. Make a report.

Skill Maintenance

In each of Exercises 28–35, fill in the blank with the correct term. Some of the given choices will be used more than once. Others will not be used.

range
domain
function
an inverse function
a composite function
direct variation
inverse variation
factor
solution
zero
y-intercept
one-to-one
rational
permutation
combination
arithmetic sequence
geometric sequence

28. A _____ of a function is an input for which the output is 0.

29. A function is _____ if different inputs have different outputs.

30. A _____ is a correspondence between a first set, called the _____ , and a second set, called the _____ , such that each member of the _____ corresponds to exactly one member of the _____ .

31. The first coordinate of an x-intercept of a function is a _____ of the function.

32. A selection made from a set without regard to order is a _____ .

33. If we have a function $f(x) = k/x$, where k is a positive constant, we have _____ .

34. For a polynomial function $f(x)$, if $f(c) = 0$, then $x - c$ is a _____ of the polynomial.

35. We have $\dfrac{a_{n+1}}{a_n} = r$, for any integer $n \geq 1$, in a _____ .

Synthesis

Five-Card Poker Hands. *Suppose that 5 cards are drawn from a deck of 52 cards. For the following exercises, give both a reasoned expression and an answer.*

36. *Royal Flush.* A *royal flush* consists of a 5-card hand with A-K-Q-J-10 of the same suit.

a) How many royal flushes are there?

b) What is the probability of getting a royal flush?

37. *Straight Flush.* A *straight flush* consists of 5 cards in sequence in the same suit, but excludes royal flushes. An ace can be used low, before a two, or high, following a king.

a) How many straight flushes are there?

b) What is the probability of getting a straight flush?

38. *Four of a Kind.* A *four-of-a-kind* is a 5-card hand in which 4 of the cards are of the same denomination, such as J-J-J-J-6, 7-7-7-7-A, or 2-2-2-2-5.

a) How many four-of-a-kind hands are there?

b) What is the probability of getting four of a kind?

39. *Full House.* A *full house* consists of a pair and 3 of a kind, such as Q-Q-Q-4-4.

a) How many full houses are there?

b) What is the probability of getting a full house?

40. *Three of a Kind.* A *three-of-a-kind* is a 5-card hand in which exactly 3 of the cards are of the same denomination and the other 2 are *not*, such as Q-Q-Q-10-7.

a) How many three-of-a-kind hands are there?

b) What is the probability of getting three of a kind?

41. *Flush.* An ordinary *flush* is a 5-card hand in which all the cards are of the same suit, but not all in sequence (not a straight or royal flush).

a) How many flushes are there?

b) What is the probability of getting a flush?

42. *Two Pairs.* A hand with *two pairs* is a hand like Q-Q-3-3-A.

a) How many are there?

b) What is the probability of getting two pairs?

43. *Straight.* An ordinary *straight* is any 5 cards in sequence, but not of the same suit—for example, 4 of spades, 5 of hearts, 6 of diamonds, 7 of hearts, and 8 of clubs.

a) How many straights are there?

b) What is the probability of getting a straight?

Chapter 7 Summary and Review

Important Properties and Formulas

Arithmetic Sequences and Series

General term: $a_{n+1} = a_n + d$
$a_n = a_1 + (n - 1)d$

Common difference: d

Sum of the first n terms: $S_n = \dfrac{n}{2}(a_1 + a_n)$

Geometric Sequences and Series

General term: $a_{n+1} = a_n r$

$a_n = a_1 r^{n-1}$

Common ratio: r

Sum of the first n terms: $S_n = \dfrac{a_1(1 - r^n)}{1 - r}$

Sum of an infinite geometric series:

$S_\infty = \dfrac{a_1}{1 - r}, \quad |r| < 1$

The Principle of Mathematical Induction

(1) *Basis step*: Prove S_1 is true.
(2) *Induction step*: Prove for all numbers k,
$S_k \rightarrow S_{k+1}$.

The Fundamental Counting Principle

The total number of ways in which k actions can be performed together is $n_1 \cdot n_2 \cdot n_3 \cdots n_k$.

Permutations of n Objects Taken n at a Time

$_nP_n = n! = n(n - 1)(n - 2) \cdots 3 \cdot 2 \cdot 1$

Permutations of n Objects Taken k at a Time

$_nP_k = \underbrace{n(n - 1)(n - 2) \cdots [n - (k - 1)]}_{k \text{ factors}}$

$= \dfrac{n!}{(n - k)!}$

Permutations of Sets with Some Nondistinguishable Objects

$\dfrac{n!}{n_1! \cdot n_2! \cdot \cdots \cdot n_k!}$

Combinations of n Objects Taken k at a Time

$_nC_k = \dbinom{n}{k} = \dfrac{_nP_k}{k!} = \dfrac{n!}{k!\,(n - k)!}$

$= \dfrac{n(n - 1)(n - 2) \cdots [n - (k - 1)]}{k!}$

The Binomial Theorem

$(a + b)^n = \displaystyle\sum_{k=0}^{n} \binom{n}{k} a^{n-k} b^k$

The $(k + 1)$st Term of Binomial Expansion

The $(k + 1)$st term of $(a + b)^n$ is $\dbinom{n}{k} a^{n-k} b^k$.

Total Number of Subsets

The total number of subsets of a set with n elements is 2^n.

Probability Principle P (Experimental)

$P(E) = \dfrac{m}{n}$, where an event E occurs m times out of n observations.

Probability Principle P (Theoretical)

$P(E) = \dfrac{m}{n}$, where an event E can occur m ways out of n possible equally likely outcomes.

Review Exercises

1. Find the first 4 terms, a_{11}, and a_{23}:

$$a_n = (-1)^n \left(\frac{n^2}{n^4 + 1} \right).$$

2. Predict the general, or nth, term. Answers may vary.

$$2, -5, 10, -17, 26, \ldots$$

3. Find and evaluate:

$$\sum_{k=1}^{4} \frac{(-1)^{k+1} 3^k}{3^k - 1}.$$

4. Write sigma notation:

$$0 + 3 + 8 + 15 + 24 + 35 + 48.$$

5. Find the 10th term of the arithmetic sequence

$$\frac{3}{4}, \frac{13}{12}, \frac{17}{12}, \ldots.$$

6. Find the 6th term of the arithmetic sequence

$$a - b, a, a + b, \ldots.$$

7. Find the sum of the first 18 terms of the arithmetic sequence

$$4, 7, 10, \ldots.$$

8. Find the sum of the first 200 natural numbers.

9. The 1st term in an arithmetic sequence is 5, and the 17th term is 53. Find the 3rd term.

10. The common difference in an arithmetic sequence is 3. The 10th term is 23. Find the first term.

11. For a geometric sequence, $a_1 = -2$, $r = 2$, and $a_n = -64$. Find n and S_n.

12. For a geometric sequence, $r = \frac{1}{2}$, $n = 5$, and $S_n = \frac{31}{2}$. Find a_1 and a_n.

Find the sum of each infinite geometric series, if it exists.

13. $25 + 27.5 + 30.25 + 33.275 + \cdots$

14. $0.27 + 0.0027 + 0.000027 + \cdots$

15. $\frac{1}{2} - \frac{1}{6} + \frac{1}{18} - \cdots$

16. Find fraction notation for $2.\overline{13}$.

17. Insert four arithmetic means between 5 and 9.

18. *Bouncing Golfball.* A golfball is dropped from a height of 30 ft to the pavement. It always rebounds three fourths of the distance that it drops. How far (up and down) will the ball have traveled when it hits the pavement for the 6th time?

19. *The Amount of an Annuity.* To create a college fund, a parent makes a sequence of 18 yearly deposits of $2000 each in a savings account on which interest is compounded annually at 2.8%. Find the amount of the annuity.

20. *Total Gift.* You receive 10¢ on the first day of the year, 12¢ on the 2nd day, 14¢ on the 3rd day, and so on.

a) How much will you receive on the 365th day?
b) What is the sum of all these 365 gifts?

21. *The Economic Multiplier.* Suppose the government is making a $24,000,000,000 expenditure for travel to Mars. If 73% of this amount is spent again, and so on, what is the total effect on the economy?

Use mathematical induction to prove each of the following.

22. For every natural number n,

$$1 + 4 + 7 + \cdots + (3n - 2) = \frac{n(3n - 1)}{2}.$$

23. For every natural number n,

$$1 + 3 + 3^2 + \cdots + 3^{n-1} = \frac{3^n - 1}{2}.$$

24. For every natural number $n \geq 2$,

$$\left(1 - \frac{1}{2} \right) \left(1 - \frac{1}{3} \right) \cdots \left(1 - \frac{1}{n} \right) = \frac{1}{n}.$$

25. *Book Arrangements.* In how many ways can 6 books be arranged on a shelf?

26. *Flag Displays.* If 9 different signal flags are available, how many different displays are possible using 4 flags in a row?

27. *Prize Choices.* The winner of a contest can choose any 8 of 15 prizes. How many different sets of prizes can be chosen?

28. *Fraternity–Sorority Names.* The Greek alphabet contains 24 letters. How many fraternity or sorority names can be formed using 3 different letters?

29. *Letter Arrangements.* In how many distinguishable ways can the letters of the word TENNESSEE be arranged?

30. *Floor Plans.* A manufacturer of houses has 1 floor plan but achieves variety by having 3 different roofs, 4 different ways of attaching the garage, and 3 different types of entrances. Find the number of different houses that can be produced.

31. *Code Symbols.* How many code symbols can be formed using 5 out of 6 of the letters of G, H, I, J, K, L if the letters:

a) cannot be repeated?
b) can be repeated?
c) cannot be repeated but must begin with K?
d) cannot be repeated but must end with IGH?

32. Determine the number of subsets of a set containing 8 members.

Expand.

33. $(m + n)^7$

34. $\left(x - \sqrt{2}\right)^5$

35. $(x^2 - 3y)^4$

36. $\left(a + \dfrac{1}{a}\right)^8$

37. $(1 + 5i)^6$, where $i^2 = -1$

38. Find the 4th term of $(a + x)^{12}$.

39. Find the 12th term of $(2a - b)^{18}$. Do not multiply out the factorials.

40. *Rolling Dice.* What is the probability of getting a 10 on a roll of a pair of dice? on a roll of 1 die?

41. *Drawing a Card.* From a deck of 52 cards, 1 card is drawn at random. What is the probability that it is a club?

42. *Drawing Three Cards.* From a deck of 52 cards, 3 are drawn at random without replacement. What is the probability that 2 are aces and 1 is a king?

43. *Election Poll.* Before an election, a poll was conducted to see which candidate was favored. Three people were running for a particular office. During the polling, 86 favored candidate A, 97 favored B, and 23 favored C. Assuming that the poll is a valid indicator of the election, what is the probability that the election will be won by A? B? C?

Technology Connection

44. Use a graphing calculator to construct a table of values for the first 10 terms of this sequence.
$$a_1 = 0.3, \quad a_{k+1} = 5a_k + 1$$

45. *Women in the Civilian Labor Force.* The following table shows the number of women in the civilian labor force in the United States in various years.

YEAR	WOMEN IN THE CIVILIAN LABOR FORCE (IN MILLIONS)
1970	3.1543
1980	4.5487
1990	5.6829
2001	6.6071

Source: U.S. Bureau of Labor Statistics

a) Find a linear sequence function $a_n = an + b$ that models the data. Let n represent the number of years after 1970.
b) Use the sequence found in part (a) to estimate the number of women in the civilian labor force in 2010.

Collaborative Discussion and Writing

46. Write an exercise for a classmate to solve. Design it so that the solution is $_9C_4$.

47. *Chain Business Deals.* Chain letters have been outlawed by the U.S. government. Nevertheless, "chain" business deals still exist and they can be fraudulent. Suppose that a salesperson is charged with the task of hiring 4 new salespersons. Each of them gives half of his or her profits to the person who hires them. Each of these people hires 4 new salespersons. Each of these gives half of his or her profits to the person who hired them. Half of these profits then go back to the original hiring person. Explain the lure of this business to someone who has managed several sequences of hirings. Explain the fallacy of such a business as well. Keep in mind that there are about 285 million people in the United States.

Synthesis

48. Explain why the following cannot be proved by mathematical induction: For every natural number n,

a) $3 + 5 + \cdots + (2n + 1) = (n + 1)^2$.
b) $1 + 3 + \cdots + (2n - 1) = n^2 + 3$.

49. Suppose that $a_1, a_2, \ldots, a_n$ and $b_1, b_2, \ldots, b_n$ are geometric sequences. Prove that $c_1, c_2, \ldots, c_n$ is a geometric sequence, where $c_n = a_n b_n$.

50. Suppose that $a_1, a_2, \ldots, a_n$ is an arithmetic sequence. Is $b_1, b_2, \ldots, b_n$ an arithmetic sequence if:

a) $b_n = |a_n|$?   b) $b_n = a_n + 8$?

c) $b_n = 7a_n$?   d) $b_n = \dfrac{1}{a_n}$?

e) $b_n = \log a_n$?   f) $b_n = a_n^3$?

51. The zeros of this polynomial function form an arithmetic sequence. Find them.
$$f(x) = x^4 - 4x^3 - 4x^2 + 16x$$

52. Write the first 3 terms of the infinite geometric series with $r = -\frac{1}{3}$ and $S_\infty = \frac{3}{8}$.

53. Simplify:
$$\sum_{k=0}^{10} (-1)^k \binom{10}{k} (\log x)^{10-k} (\log y)^k.$$

Solve for n.

54. $\binom{n}{6} = 3 \cdot \binom{n-1}{5}$ **55.** $\binom{n}{n-1} = 36$

56. Solve for a:
$$\sum_{k=0}^{5} \binom{5}{k} 9^{5-k} a^k = 0.$$

Chapter 7 Test

1. For the sequence whose nth term is $a_n = (-1)^n (2n + 1)$, find a_{21}.

2. Find the first 5 terms of the sequence with general term
$$a_n = \frac{n + 1}{n + 2}.$$

3. Find and evaluate:
$$\sum_{k=1}^{4} (k^2 + 1).$$

Write sigma notation.

4. $4 + 8 + 12 + 16 + 20 + 24$

5. $2 + 4 + 8 + 16 + 32 + \cdots$

6. Find the first 4 terms of the recursively defined sequence
$$a_1 = 3, \quad a_{n+1} = 2 + \frac{1}{a_n}.$$

7. Find the 15th term of the arithmetic sequence 2, 5, 8, ...

8. The 1st term of an arithmetic sequence is 8 and the 21st term is 108. Find the 7th term.

9. Find the sum of the first 20 terms of the series $17 + 13 + 9 + \cdots$.

10. Find the sum: $\displaystyle\sum_{k=1}^{25} (2k + 1)$.

11. Find the 11th term of the geometric sequence $10, -5, \frac{5}{2}, -\frac{5}{4}, \ldots$.

12. For a geometric sequence, $r = 0.2$ and $S_4 = 1248$. Find a_1.

Find the sum, if it exists.

13. $\displaystyle\sum_{k=1}^{8} 2^k$

14. $18 + 6 + 2 + \cdots$

15. Find fraction notation for $0.\overline{56}$.

16. *Salvage Value.* The value of an office machine is $10,000. Its salvage value each year is 80% of its value the year before. Give a sequence that lists the salvage value of the machine for each year of a 6-yr period.

17. *Hourly Wage.* Tamika accepts a job, starting with an hourly wage of $8.50, and is promised a raise of 25¢ per hour every three months for 4 yr. What will Tamika's hourly wage be at the end of the 4-yr period?

18. *Amount of an Annuity.* To create a college fund, a parent makes a sequence of 18 equal yearly deposits of $2500 in a savings account on which interest is compounded annually at 5.6%. Find the amount of the annuity.

19. Use mathematical induction to prove that, for every natural number n,

$$2 + 5 + 8 + \cdots + (3n - 1) = \frac{n(3n + 1)}{2}.$$

Evaluate.

20. $_{15}P_6$

21. $_{21}C_{10}$

22. $\dbinom{n}{4}$

23. How many 4-digit numbers can be formed using the digits 1, 3, 5, 6, 7, and 9 without repetition?

24. How many code symbols can be formed using 4 of the 6 letters A, B, C, X, Y, Z if the letters:
   a) can be repeated?
   b) are not repeated but must begin with Z?

25. *Scuba Club Officers.* The Bay Woods Scuba Club has 28 members. How many sets of 4 officers can be selected from this group?

26. *Test Options.* On a test with 20 questions, a student must answer 8 of the first 12 questions and 4 of the last 8. In how many ways can this be done?

27. Expand: $(x + 1)^5$.

28. Find the 5th term of the binomial expansion $(x - y)^7$.

29. Determine the number of subsets of a set containing 9 members.

30. *Marbles.* Suppose we select, without looking, one marble from a bag containing 6 red marbles and 8 blue marbles. What is the probability of selecting a blue marble?

31. *Drawing Coins.* Ethan has 6 pennies, 5 dimes, and 4 quarters in his pocket. Six coins are drawn at random. What is the probability of getting 1 penny, 2 dimes, and 3 quarters?

Synthesis

32. Solve for n: $_nP_7 = 9 \cdot {_nP_6}$.

Appendix

Basic Concepts from Geometry

- *Classify an angle as right, straight, acute, or obtuse.*
- *Identify complementary and supplementary angles and find the measure of a complement or a supplement of a given angle.*
- *Find the lengths of sides of similar triangles.*
- *Given the lengths of any two sides of a right triangle, find the length of the third side.*

In this appendix, we present a review of some basic concepts from geometry. A summary of formulas from geometry is found near the back of the book.

Classifying Angles

The following are ways in which we classify angles.

Types of Angles

Right angle: An angle whose measure is 90°.

Straight angle: An angle whose measure is 180°.

Acute angle: An angle whose measure is greater than 0° and less than 90°.

Obtuse angle: An angle whose measure is greater than 90° and less than 180°.

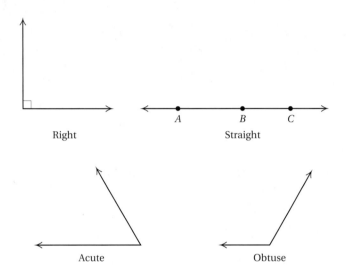

EXAMPLE 1 Classify the angle as right, straight, acute, or obtuse.

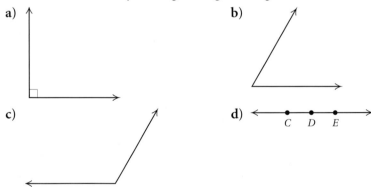

Solution

a) We observe that the measure of this angle is 90°, so it is a right angle. (We could also use a protractor to measure the angle.)

b) We observe that the measure of this angle is greater than 0° and less than 90°. It is an acute angle.

c) We observe that the measure of this angle is greater than 90° and less than 180°. It is an obtuse angle.

d) We observe that the measure of this angle is 180°, so it is a straight angle.

Complementary and Supplementary Angles

We can describe the relationship between certain pairs of angles on the basis of the sum of their measures.

> Two angles are **complementary** if the sum of their measures is 90°. Each angle is called a **complement** of the other.

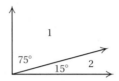

∠1 and ∠2 above are **complementary** angles.

$$m\angle 1 + m\angle 2 = 90°$$
$$75° \ \ + \ \ 15° \ \ = 90°$$

If two angles are complementary, each is an acute angle. When complementary angles are adjacent to each other, they form a right angle.

EXAMPLE 2 Identify each pair of complementary angles.

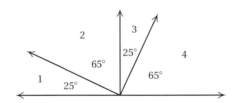

Solution We look for pairs of angles for which the sum of the measures is 90°. They are

$\qquad$ ∠1 and ∠2,

$\qquad$ ∠1 and ∠4,

$\qquad$ ∠2 and ∠3,

$\qquad$ ∠3 and ∠4.

EXAMPLE 3 Find the measure of a complement of an angle of 39°.

Solution

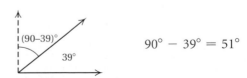

$$90° - 39° = 51°$$

The measure of a complement is 51°.

Next, consider ∠1 and ∠2 as shown below. Because the sum of their measures is 180°, ∠1 and ∠2 are said to be **supplementary.** Note that when supplementary angles are adjacent, they form a straight angle.

$$m\angle 1 + m\angle 2 = 180°;$$
$$30° + 150° = 180°$$

> Two angles are **supplementary** if the sum of their measures is 180°. Each angle is called a **supplement** of the other.

EXAMPLE 4   Identify each pair of supplementary angles.

Solution   We look for pairs of angles for which the sum of the measures is 180°. They are

∠1 and ∠2,
∠1 and ∠4,
∠2 and ∠3,
∠3 and ∠4.

EXAMPLE 5   Find the measure of a supplement of an angle of 112°.

Solution

$$180° - 112° = 68°$$

The measure of a supplement is 68°.

Similar Triangles

Similar figures have the same shape but are not necessarily the same size.

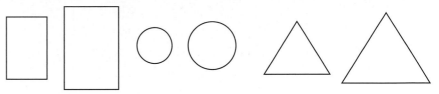

Similar figures

EXAMPLE 6 Which pairs of triangles appear to be similar?

a)

b)

c)

d)

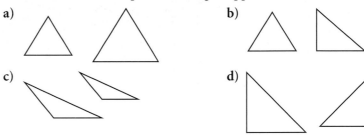

Solution Pairs (a), (c), and (d) appear to be similar because they appear to have the same shape. ▬

Similar triangles have corresponding sides and angles.

EXAMPLE 7 $\triangle ABC$ and $\triangle DEF$ are similar. Name their corresponding sides and angles.

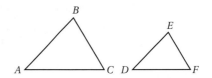

Solution

$\overline{AB} \leftrightarrow \overline{DE}$	$\angle A \leftrightarrow \angle D$	**The symbol $\leftrightarrow$ means "corresponds to."**
$\overline{AC} \leftrightarrow \overline{DF}$	$\angle B \leftrightarrow \angle E$	
$\overline{BC} \leftrightarrow \overline{EF}$	$\angle C \leftrightarrow \angle F$	▬

Two triangles are **similar** if and only if their vertices can be matched so that the corresponding angles have the same measure and the lengths of corresponding sides are proportional.

To say that $\triangle ABC$ and $\triangle DEF$ are similar, we write "$\triangle ABC \sim \triangle DEF$." We will agree that this symbol also tells us the way in which the vertices are matched.

$$\triangle ABC \sim \triangle DEF$$

Thus, $\triangle ABC \sim \triangle DEF$ means that

$$\angle A \leftrightarrow \angle D$$
$$\angle B \leftrightarrow \angle E \quad \text{and} \quad \frac{AB}{DE} = \frac{AC}{DF} = \frac{BC}{EF}.$$
$$\angle C \leftrightarrow \angle F$$

EXAMPLE 8 Suppose that $\triangle PQR \sim \triangle STV$. Name the corresponding angles. Which sides are proportional?

Solution

$$\angle P \leftrightarrow \angle S$$
$$\angle Q \leftrightarrow \angle T \quad \text{and} \quad \frac{PQ}{ST} = \frac{PR}{SV} = \frac{QR}{TV}.$$
$$\angle R \leftrightarrow \angle V$$

EXAMPLE 9 These triangles are similar. Which sides are proportional?

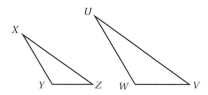

Solution It appears that if we match X with U, Y with W, and Z with V, the corresponding angles have the same measure. Thus,

$$\frac{XY}{UW} = \frac{XZ}{UV} = \frac{YZ}{WV}.$$

Proportions and Similar Triangles

We can find lengths of sides in similar triangles.

EXAMPLE 10 If $\triangle RAE \sim \triangle GQL$, find QL and GL.

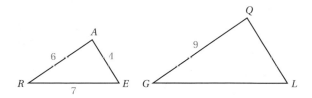

Solution Since $\triangle RAE \sim \triangle GQL$, the corresponding sides are proportional. Thus,

$$\frac{6}{9} = \frac{4}{QL}$$

$6(QL) = 4 \cdot 9$ **Multiplying by 9(QL) on both sides**

$6(QL) = 36$

$QL = 6$ **Dividing by 6 on both sides**

and

$$\frac{6}{9} = \frac{7}{GL}$$

$6(GL) = 7 \cdot 9$

$6(GL) = 63$

$GL = 10\frac{1}{2}.$

Similar triangles and proportions can often be used to find lengths that would ordinarily be difficult to measure. For example, we could find the height of a flagpole without climbing it or the distance across a river without crossing it.

EXAMPLE 11 *Height of a Flagpole.* How high is a flagpole that casts a 56-ft shadow at the same time that a 6-ft man casts a 5-ft shadow?

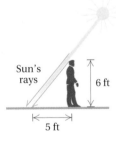
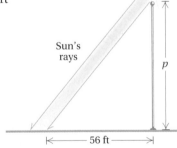

Solution If we use the sun's rays to represent the third side of the triangle in our drawing of the situation, we see that we have similar triangles. Let p = the height of the flagpole. Then the ratio of 6 to p is the same as the ratio of 5 to 56. Thus we have the proportion

$$\text{Height of man} \longrightarrow \frac{6}{p} = \frac{5}{56} \longleftarrow \text{Length of man's shadow} \atop \longleftarrow \text{Length of pole's shadow}$$

$6 \cdot 56 = 5 \cdot p$ **Multiplying by 56p on both sides**

$\dfrac{6 \cdot 56}{5} = p$ **Dividing by 5 on both sides**

$67.2 = p$ **Simplifying**

The height of the flagpole is 67.2 ft.

EXAMPLE 12 *F-106 Blueprint.* A blueprint for an F-106 Delta Dart fighter plane is a scale drawing. Each wing of the plane has a triangular shape. Find the length of side *a* of the wing.

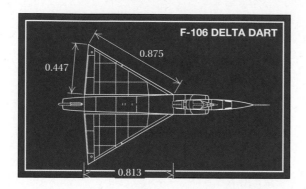

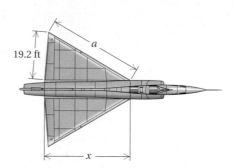

Solution We let *a* = the length of the wing. Thus we have the proportion

Length on the blueprint ⟶ $\dfrac{0.447}{19.2} = \dfrac{0.875}{a}$ ⟵ Length on the blueprint
Length of the wing ⟶ ⟵ Length of the wing

$$0.447 \cdot a = 0.875 \cdot 19.2 \qquad \textbf{Multiplying by 19.2}a$$

$$a = \frac{0.875 \cdot 19.2}{0.447} \qquad \textbf{Dividing by 0.447}$$

$$a \approx 37.6 \text{ ft}$$

The length of side *a* of the wing is about 37.6 ft.

Right Triangles

A **right triangle** is a triangle with a 90° angle, as shown in the figure below. The small square in the corner indicates the 90° angle.

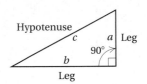

In a right triangle, the longest side is called the **hypotenuse.** It is the side opposite the right angle. The other two sides are called **legs.** We generally use the letters a and b for the lengths of the legs and c for the length of the hypotenuse. They are related as follows.

The Pythagorean Theorem

In any right triangle, if a and b are the lengths of the legs and c is the length of the hypotenuse, then

$$a^2 + b^2 = c^2.$$

The equation $a^2 + b^2 = c^2$ is called the **Pythagorean equation.**

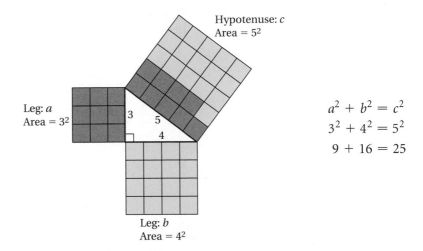

$$a^2 + b^2 = c^2$$
$$3^2 + 4^2 = 5^2$$
$$9 + 16 = 25$$

The Pythagorean theorem is named after the ancient Greek mathematician Pythagoras (569?–500? B.C.). It is uncertain who actually proved this result the first time. A proof can be found in most geometry books.

If we know the lengths of any two sides of a right triangle, we can find the length of the third side.

EXAMPLE 13 Find the length of the hypotenuse of this right triangle. Give an exact answer and an approximation to three decimal places.

Solution

$$4^2 + 5^2 = c^2$$ **Substituting in the Pythagorean equation**

$$16 + 25 = c^2$$

$$41 = c^2$$

$$c = \sqrt{41}$$

$$c \approx 6.403$$ **Using a calculator**

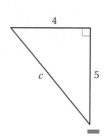

EXAMPLE 14 Find the length of the leg of this right triangle. Give an exact answer and an approximation to three decimal places.

Solution

$$10^2 + b^2 = 12^2$$ **Substituting in the Pythagorean equation**

$$100 + b^2 = 144$$

$$b^2 = 144 - 100$$

$$b^2 = 44$$

$$b = \sqrt{44}$$

$$b \approx 6.633$$ **Using a calculator**

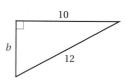

EXAMPLE 15 Find the length of the leg of this right triangle. Give an exact answer and an approximation to three decimal places.

Solution

$$1^2 + b^2 = \left(\sqrt{7}\right)^2$$ **Substituting in the Pythagorean equation**

$$1 + b^2 = 7$$

$$b^2 = 7 - 1 = 6$$

$$b = \sqrt{6}$$

$$b \approx 2.449$$ **Using a calculator**

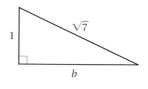

EXAMPLE 16 Find the length of the leg of this right triangle. Give an exact answer and an approximation to three decimal places.

Solution

$$a^2 + 10^2 = 15^2$$

$$a^2 + 100 = 225$$

$$a^2 = 225 - 100$$

$$a^2 = 125$$

$$a = \sqrt{125}$$

$$a \approx 11.180$$ **Using a calculator**

An Application

EXAMPLE 17 *Dimensions of a Softball Diamond.* A slow-pitch softball diamond is actually a square 65 ft on a side. How far is it from home plate to second base? Give an exact answer and an approximation to three decimal places. (This can be helpful information when lining up the bases.)

Solution

a) We first make a drawing. We note that the first and second base lines, together with a line from home plate to second base, form a right triangle. We label the unknown distance d.

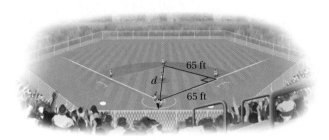

b) We know that $65^2 + 65^2 = d^2$. We solve this equation:

$$4225 + 4225 = d^2$$
$$8450 = d^2.$$

Exact answer: $\quad \sqrt{8450}$ ft $= d$

Approximation: $\quad 91.924$ ft $\approx d$

Exercise Set

Classify the angle as right, straight, acute, or obtuse.

1.

2.

3.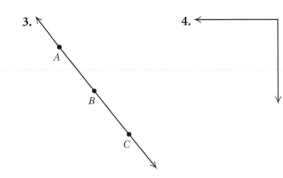

4.

5.

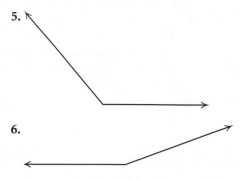

6.

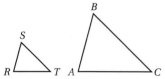

Find the measure of a complement of an angle with the given measure.

7. 11° **8.** 83° **9.** 67° **10.** 5°

11. 58° **12.** 32° **13.** 29° **14.** 54°

Find the measure of a supplement of an angle with the given measure.

15. 3° **16.** 54° **17.** 139° **18.** 13°

19. 85° **20.** 129° **21.** 102° **22.** 45°

For each pair of similar triangles, name the corresponding angles and sides.

23.

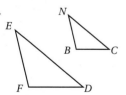

24.

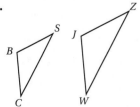

25.

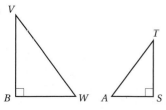

26.

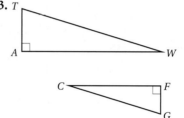

For each pair of similar triangles, name the angles with the same measure and name the proportional sides.

27. △ABC ~ △RST **28.** △PQR ~ △STV

29. △MES ~ △CLF **30.** △SMH ~ △WLK

Name the proportional sides in these similar triangles.

31.

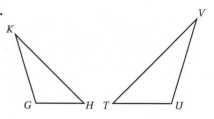

32.

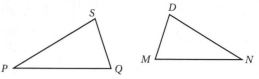

33.

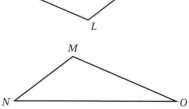

34.

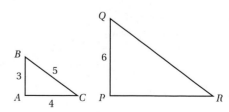

35. If △ABC ~ △PQR, find QR and PR.

36. If $\triangle MAC \sim \triangle GET$, find AM and GT.

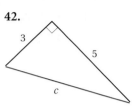

37. How high is a tree that casts a 27-ft shadow at the same time that a 4-ft fence post casts a 3-ft shadow?

38. How high is a flagpole that casts a 42-ft shadow at the same time that a $5\frac{1}{2}$-ft woman casts a 7-ft shadow?

39. Find the distance across the river. Assume that the ratio of d to 25 ft is the same as the ratio of 40 ft to 10 ft.

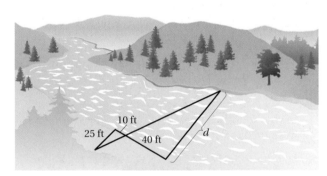

40. To measure the height of a hill, a string is drawn tight from level ground to the top of the hill. A 3-ft yardstick is placed under the string, touching it at point P, a distance of 5 ft from point G, where the string touches the ground. The string is then detached and found to be 120 ft long. How high is the hill?

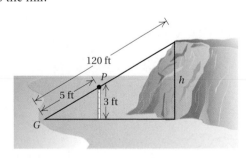

Find the length of the third side of the right triangle. Give an exact answer and an approximation to three decimal places.

41.

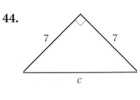

42.

43.

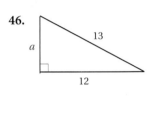

44.

45.

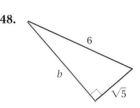

46.

47.

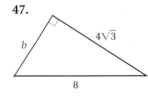

48.

In a right triangle, find the length of the side not given. Give an exact answer and an approximation to three decimal places.

49. $a = 10$, $b = 24$

50. $a = 5$, $b = 12$

51. $a = 9$, $c = 15$

52. $a = 18$, $c = 30$

53. $b = 1$, $c = \sqrt{5}$

54. $b = 1$, $c = \sqrt{2}$

55. $a = 1$, $c = \sqrt{3}$

56. $a = \sqrt{3}$, $b = \sqrt{5}$

57. $c = 10$, $b = 5\sqrt{3}$

58. $a = 5$, $b = 5$

59. $a = \sqrt{2}$, $b = \sqrt{7}$

60. $c = \sqrt{7}$, $a = \sqrt{2}$

Solve. Give an exact answer and an approximation to three decimal places.

61. *Airport Distance.* An airplane is flying at an altitude of 4100 ft. The slanted distance directly to the airport is 15,100 ft. How far is the airplane horizontally from the airport?

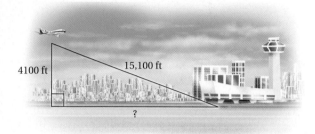

4100 ft

15,100 ft

?

62. *Surveying Distance.* A surveyor had poles located at points *P*, *Q*, and *R*. The distances that the surveyor was able to measure are marked on the drawing. What is the approximate distance from *P* to *R*?

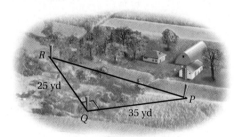

R

25 yd

Q

35 yd

P

63. *Cordless Telephones.* Becky's new cordless telephone has clear reception up to 300 ft from its base. Her phone is located near a window in her apartment, 180 ft above ground level. How far into her backyard can Becky use her phone?

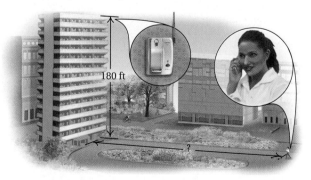

180 ft

?

64. *Rope Course.* An outdoor rope course consists of a cable that slopes downward from a height of 37 ft to a resting place 30 ft above the ground. The trees that the cable connects are 24 ft apart. How long is the cable?

37 ft

30 ft

24 ft

65. *Diagonal of a Square.* Find the length of a diagonal of a square whose sides are 3 cm long.

66. *Ladder Height.* A 10-m ladder is leaning against a building. The bottom of the ladder is 5 m from the building. How high is the top of the ladder?

67. *Guy Wire.* How long is a guy wire reaching from the top of a 12-ft pole to a point on the ground 8 ft from the base of the pole?

68. *Diagonal of a Soccer Field.* The largest regulation soccer field is 100 yd wide and 130 yd long. Find the length of a diagonal of such a field.

Answers

Chapter R

Exercise Set R.1

1. $\sqrt[3]{8}, 0, 9, \sqrt{25}$
3. $\sqrt{7}, 5.242242224\ldots, -\sqrt{14}, \sqrt[5]{5}, \sqrt[3]{4}$
5. $-12, 5.\overline{3}, -\frac{7}{3}, \sqrt[3]{8}, 0, -1.96, 9, 4\frac{2}{3}, \sqrt{25}, \frac{5}{7}$
7. $5.\overline{3}, -\frac{7}{3}, -1.96, 4\frac{2}{3}, \frac{5}{7}$ **9.** $-12, 0$
11. $[-3, 3]$;

$\begin{array}{c}\longleftrightarrow\!\!+\!\!+\!\![\!\!+\!\!+\!\!+\!\!+\!\!+\!\!+\!\!]\!\!+\!\!+\!\!\longrightarrow \\ {\scriptstyle -5\,-4\,-3\,-2\,-1\ \ 0\ \ 1\ \ 2\ \ 3\ \ 4\ \ 5}\end{array}$

13. $[-4, -1)$;

$\begin{array}{c}\longleftrightarrow\!\!+\![\!\!+\!\!+\!\!+\!\!)\!\!+\!\!+\!\!+\!\!+\!\!+\!\!+\!\!\longrightarrow \\ {\scriptstyle -4\qquad -1\ \ 0}\end{array}$

15. $(-\infty, -2]$;

$\begin{array}{c}\longleftrightarrow\!\!+\!\!+\!\!+\!\!]\!\!+\!\!+\!\!+\!\!+\!\!+\!\!+\!\!+\!\!+\!\!\longrightarrow \\ {\scriptstyle -2\qquad 0}\end{array}$

17. $(3.8, \infty)$;

$\begin{array}{c}\longleftrightarrow\!\!+\!\!+\!\!+\!\!+\!\!+\!\!+\!\!+\!\!(\!\!+\!\!\longrightarrow \\ {\scriptstyle 0\qquad\quad 3.8}\end{array}$

19. $(7, \infty)$;

$\begin{array}{c}\longleftrightarrow\!\!+\!\!+\!\!+\!\!+\!\!+\!\!+\!\!+\!\!+\!\!+\!\!(\!\!+\!\!\longrightarrow \\ {\scriptstyle 0\qquad\qquad 7}\end{array}$ **21.** $(0, 5)$

23. $[-9, -4)$ **25.** $[x, x + h]$ **27.** (p, ∞) **29.** True
31. False **33.** True **35.** False **37.** False **39.** True
41. True **43.** True **45.** False **47.** Commutative
property of multiplication
49. Multiplicative identity property
51. Associative property of multiplication
53. Commutative property of multiplication
55. Commutative property of addition
57. Multiplicative inverse property **59.** 7.1 **61.** 347
63. $\sqrt{97}$ **65.** 0 **67.** $\frac{5}{4}$ **69.** 11 **71.** 6 **73.** 5.4
75. $\frac{21}{8}$ **77.** 7 **79.** Discussion and Writing
81. Answers may vary; $0.124124412444\ldots$
83. Answers may vary; -0.00999 **85.**

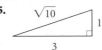

Exercise Set R.2

1. 1 **3.** x^9 **5.** 5^2, or 25 **7.** 1 **9.** y^{-4}, or $\dfrac{1}{y^4}$

11. 7^{-1}, or $\dfrac{1}{7}$ **13.** $6x^5$ **15.** $-15a^{-12}$, or $-\dfrac{15}{a^{12}}$

17. $15a^{-1}b^5$, or $\dfrac{15b^5}{a}$ **19.** $-42x^{-1}y^{-4}$, or $-\dfrac{42}{xy^4}$

21. $72x^5$ **23.** $-200n^5$ **25.** b^3 **27.** x^{-21}, or $\dfrac{1}{x^{21}}$

29. x^3y^{-3}, or $\dfrac{x^3}{y^3}$ **31.** $8xy^{-5}$, or $\dfrac{8x}{y^5}$ **33.** $8a^3b^6$

35. $-32x^{15}$ **37.** $\dfrac{c^2d^4}{25}$ **39.** $432m^{-8}$, or $\dfrac{432}{m^8}$

41. $\dfrac{8x^{-9}y^{21}}{z^{-3}}$, or $\dfrac{8y^{21}z^3}{x^9}$ **43.** $2^{-5}a^{-20}b^{25}c^{-10}$, or $\dfrac{b^{25}}{32a^{20}c^{10}}$

45. 4.05×10^5 **47.** 3.9×10^{-7} **49.** 2.346×10^{11}
51. 1.04×10^{-3} **53.** 1.6×10^{-5} **55.** 0.000083
57. 20,700,000 **59.** 34,960,000,000 **61.** 0.0000000541
63. 231,900,000 **65.** 1.395×10^3 **67.** 2.21×10^{-10}
69. 8×10^{-14} **71.** 2.5×10^5 **73.** $\$1.19 \times 10^7$
75. 1.332×10^{14} disintegrations **77.** 2 **79.** 2048 **81.** 5
83. \$2883.67 **85.** \$8763.54 **87.** Discussion and Writing
89. \$170,797.30 **91.** \$309.79 **93.** x^{8t} **95.** t^{8x}
97. $9x^{2a}y^{2b}$

Exercise Set R.3

1. $-5y^4, 3y^3, 7y^2, -y, -4; 4$ **3.** $3a^4b, -7a^3b^3, 5ab, -2; 6$
5. $3x^2y - 5xy^2 + 7xy + 2$ **7.** $3x + 2y - 2z - 3$
9. $-2x^2 + 6x - 2$ **11.** $x^4 - 3x^3 - 4x^2 + 9x - 3$
13. $2a^4 - 2a^3b - a^2b + 4ab^2 - 3b^3$ **15.** $x^2 + 2x - 15$
17. $x^2 + 10x + 24$ **19.** $2a^2 + 13a + 15$

21. $4x^2 + 8xy + 3y^2$ **23.** $y^2 + 10y + 25$
25. $x^2 - 8x + 16$ **27.** $25x^2 - 30x + 9$
29. $4x^2 + 12xy + 9y^2$ **31.** $4x^4 - 12x^2y + 9y^2$
33. $a^2 - 9$ **35.** $4x^2 - 25$ **37.** $9x^2 - 4y^2$
39. $4x^2 + 12xy + 9y^2 - 16$ **41.** $x^4 - 1$
43. Discussion and Writing **45.** $a^{2n} - b^{2n}$
47. $a^{2n} + 2a^nb^n + b^{2n}$ **49.** $x^6 - 1$ **51.** $x^{a^2-b^2}$
53. $a^2 + b^2 + c^2 + 2ab + 2ac + 2bc$

Exercise Set R.4

1. $2(x - 5)$ **3.** $3x^2(x^2 - 3)$ **5.** $4(a^2 - 3a + 4)$
7. $(b - 2)(a + c)$ **9.** $(x + 3)(x^2 + 6)$
11. $(y - 1)(y^2 + 3)$ **13.** $12(2x - 3)(x^2 + 3)$
15. $(a - 3)(a^2 - 2)$ **17.** $(x - 1)(x^2 - 5)$
19. $(p + 2)(p + 4)$ **21.** $(x + 2)(x + 6)$
23. $(t + 3)(t + 5)$ **25.** $(x + 3y)(x - 9y)$
27. $2(n - 12)(n + 2)$ **29.** $(y^2 + 3)(y^2 - 7)$
31. $(2n - 7)(n + 8)$ **33.** $(3x + 2)(4x + 1)$
35. $(4x + 3)(x + 3)$ **37.** $(2y - 3)(y + 2)$
39. $(3a - 4b)(2a - 7b)$ **41.** $4(3a - 4)(a + 1)$
43. $(m + 2)(m - 2)$ **45.** $(3x + 5)(3x - 5)$
47. $6(x + y)(x - y)$ **49.** $4x(y^2 + z)(y^2 - z)$
51. $7p(q^2 + y^2)(q + y)(q - y)$ **53.** $(y - 3)^2$
55. $(2z + 3)^2$ **57.** $(1 - 4x)^2$ **59.** $a(a + 12)^2$
61. $4(p - q)^2$ **63.** $(x + 2)(x^2 - 2x + 4)$
65. $(m - 1)(m^2 + m + 1)$ **67.** $2(y - 4)(y^2 + 4y + 16)$
69. $3a^2(a - 2)(a^2 + 2a + 4)$ **71.** $(t^2 + 1)(t^4 - t^2 + 1)$
73. $3ab(6a - 5b)$ **75.** $(x - 4)(x^2 + 5)$
77. $8(x + 2)(x - 2)$ **79.** Prime **81.** $(m + 3n)(m - 3n)$
83. $(x + 4)(x + 5)$ **85.** $(y - 5)(y - 1)$
87. $(2a + 1)(a + 4)$ **89.** $(3x - 1)(2x + 3)$
91. $(y - 9)^2$ **93.** $(3z - 4)^2$ **95.** $(xy - 7)^2$
97. $4a(x + 7)(x - 2)$ **99.** $3(z - 2)(z^2 + 2z + 4)$
101. $2ab(2a^2 + 3b^2)(4a^4 - 6a^2b^2 + 9b^4)$
103. $(y - 3)(y + 2)(y - 2)$ **105.** $(x - 1)(x^2 + 1)$
107. $5(m^2 + 2)(m^2 - 2)$ **109.** $2(x + 3)(x + 2)(x - 2)$
111. $(2c - d)^2$ **113.** $(m^3 + 10)(m^3 - 2)$
115. $p(1 - 4p)(1 + 4p + 16p^2)$
117. Discussion and Writing **119.** $(y^2 + 12)(y^2 - 7)$
121. $\left(y + \frac{4}{7}\right)\left(y - \frac{2}{7}\right)$ **123.** $\left(x + \frac{3}{2}\right)^2$ **125.** $\left(x - \frac{1}{2}\right)^2$
127. $h(3x^2 + 3xh + h^2)$ **129.** $(y + 4)(y - 7)$
131. $(x^n + 8)(x^n - 3)$ **133.** $(x + a)(x + b)$
135. $(5y^m + x^n - 1)(5y^m - x^n + 1)$
137. $y(y - 1)^2(y - 2)$

Exercise Set R.5

1. $\{x \mid x \text{ is a real number}\}$
3. $\{x \mid x \text{ is a real number } and \ x \neq 0 \ and \ x \neq 1\}$
5. $\{x \mid x \text{ is a real number } and \ x \neq -5 \ and \ x \neq 1\}$
7. $\{x \mid x \text{ is a real number } and \ x \neq -2 \ and \ x \neq 2 \ and \ x \neq -5\}$
9. $\frac{1}{x - y}$ **11.** $\frac{(x + 5)(2x + 3)}{7x}$ **13.** $\frac{a + 2}{a - 5}$ **15.** $m + n$

17. $\frac{3(x - 4)}{2(x + 4)}$ **19.** $\frac{1}{x + y}$ **21.** $\frac{x - y - z}{x + y + z}$ **23.** $\frac{3}{x}$ **25.** 1
27. $\frac{7}{8z}$ **29.** $\frac{3x - 4}{(x + 2)(x - 2)}$ **31.** $\frac{-y + 10}{(y + 4)(y - 5)}$
33. $\frac{4x - 8y}{(x + y)(x - y)}$ **35.** $\frac{y - 2}{y - 1}$ **37.** $\frac{x + y}{2x - 3y}$
39. $\frac{3x - 4}{(x - 2)(x - 1)}$ **41.** $\frac{5a^2 + 10ab - 4b^2}{(a + b)(a - b)}$
43. $\frac{11x^2 - 18x + 8}{(2 + x)(2 - x)^2}$, or $\frac{11x^2 - 18x + 8}{(x + 2)(x - 2)^2}$ **45.** 0
47. $\frac{x + y}{x}$ **49.** $x - y$ **51.** $\frac{c^2 - 2c + 4}{c}$ **53.** $\frac{xy}{x - y}$
55. $\frac{a^2 - 1}{a^2 + 1}$ **57.** $\frac{3(x - 1)^2(x + 2)}{(x - 3)(x + 3)(-x + 10)}$ **59.** $\frac{1 + a}{1 - a}$
61. $\frac{b + a}{b - a}$ **63.** Discussion and Writing **65.** $2x + h$
67. $3x^2 + 3xh + h^2$ **69.** x^5 **71.** $\frac{(n + 1)(n + 2)(n + 3)}{2 \cdot 3}$
73. $\frac{x^3 + 2x^2 + 11x + 20}{2(x + 1)(2 + x)}$

Exercise Set R.6

1. 11 **3.** $4|y|$ **5.** $|b + 1|$ **7.** $-3x$ **9.** $3x^2$ **11.** 2
13. $6\sqrt{5}$ **15.** $6\sqrt{2}$ **17.** $3\sqrt[3]{2}$ **19.** $8\sqrt{2}|c|d^2$
21. $2|x||y|\sqrt[4]{3x^2}$ **23.** $|x - 2|$ **25.** $10\sqrt{3}$ **27.** $6\sqrt{11}$
29. $2x^2y\sqrt{6}$ **31.** $3x\sqrt[3]{4y}$ **33.** $2(x + 4)\sqrt[3]{(x + 4)^2}$
35. $\frac{m^2n^4}{2}$ **37.** 2 **39.** $\frac{1}{2x}$ **41.** $\frac{4a\sqrt[3]{a}}{3b}$ **43.** $\frac{x\sqrt{7x}}{6y^3}$
45. $51\sqrt{2}$ **47.** $4\sqrt{5}$ **49.** $-2x\sqrt{2} - 12\sqrt{5x}$ **51.** 1
53. $-9 - 5\sqrt{15}$ **55.** $4 + 2\sqrt{3}$ **57.** $11 - 2\sqrt{30}$
59. About 13,709.5 ft **61.** (a) $h = \frac{a}{2}\sqrt{3}$; (b) $A = \frac{a^2}{4}\sqrt{3}$
63. 8 **65.** $\frac{\sqrt{6}}{3}$ **67.** $\frac{\sqrt[3]{10}}{2}$ **69.** $\frac{2\sqrt[3]{6}}{3}$ **71.** $\frac{9 - 3\sqrt{5}}{2}$
73. $-\frac{\sqrt{6}}{6}$ **75.** $\frac{6\sqrt{m} + 6\sqrt{n}}{m - n}$ **77.** $\frac{6}{5\sqrt{3}}$ **79.** $\frac{7}{\sqrt[3]{98}}$
81. $\frac{11}{\sqrt{33}}$ **83.** $\frac{76}{27 + 3\sqrt{5} - 9\sqrt{3} - \sqrt{15}}$
85. $\frac{a - b}{3a\sqrt{a} - 3a\sqrt{b}}$ **87.** $\sqrt[4]{x^3}$ **89.** 8 **91.** $\frac{1}{5}$
93. $\frac{a\sqrt[4]{a}}{\sqrt[4]{b^3}}$, or $a\sqrt[4]{\frac{a}{b^3}}$ **95.** $mn^2\sqrt[3]{m^2n}$ **97.** $13^{5/4}$ **99.** $20^{2/3}$
101. $11^{1/6}$ **103.** $5^{5/6}$ **105.** 4 **107.** $8a^2$ **109.** $\frac{x^3}{3b^{-2}}$, or
$\frac{x^3b^2}{3}$ **111.** $x\sqrt[3]{y}$ **113.** $n\sqrt[3]{mn^2}$ **115.** $a\sqrt[3]{a^5} + a^2\sqrt[12]{a}$
117. $\sqrt[6]{288}$ **119.** $\sqrt[12]{x^{11}y^7}$ **121.** $a\sqrt[6]{a^5}$
123. $(a + x)\sqrt[7]{(a + x)^{11}}$ **125.** Discussion and Writing
127. $\frac{(2 + x^2)\sqrt{1 + x^2}}{1 + x^2}$ **129.** $a^{a/2}$

Review Exercises: Chapter R

1. [R.1] $12, -3, -1, -19, 31, 0$ **2.** [R.1] $12, 31$
3. [R.1] $-43.89, 12, -3, -\frac{1}{5}, -1, -\frac{4}{3}, 7\frac{2}{3}, -19, 31, 0$
4. [R.1] All of them **5.** [R.1] $\sqrt{7}, \sqrt[3]{10}$ **6.** [R.1] $12, 31, 0$
7. [R.1] $[-3, 5)$ **8.** [R.1] 3.5 **9.** [R.1] 16 **10.** [R.1] 10
11. [R.2] 117 **12.** [R.2] -10 **13.** [R.2] $3,261,000$
14. [R.2] 0.00041 **15.** [R.2] 1.432×10^{-2}
16. [R.2] 4.321×10^4 **17.** [R.2] 7.8125×10^{-22}
18. [R.2] 5.46×10^{-32} **19.** [R.2] $-14a^{-2}b^7$, or $\dfrac{-14b^7}{a^2}$
20. [R.2] $6x^9 y^{-6} z^6$, or $\dfrac{6x^9 z^6}{y^6}$ **21.** [R.6] 3 **22.** [R.6] -2
23. [R.5] $\dfrac{b}{a}$ **24.** [R.5] $\dfrac{x+y}{xy}$ **25.** [R.6] -4
26. [R.6] $25x^4 - 10\sqrt{2}x^2 + 2$ **27.** [R.6] $13\sqrt{5}$
28. [R.3] $x^3 + t^3$ **29.** [R.3] $10a^2 - 7ab - 12b^2$
30. [R.3] $8xy^4 - 9xy^2 + 4x^2 + 2y - 7$
31. [R.4] $(x + 2)(x^2 - 3)$
32. [R.4] $3a(2a + 3b^2)(2a - 3b^2)$ **33.** [R.4] $(x + 12)^2$
34. [R.4] $x(9x - 1)(x + 4)$
35. [R.4] $(2x - 1)(4x^2 + 2x + 1)$
36. [R.4] $(3x^2 + 5y^2)(9x^4 - 15x^2 y^2 + 25y^4)$
37. [R.4] $6(x + 2)(x^2 - 2x + 4)$
38. [R.4] $(x - 1)(2x + 3)(2x - 3)$
39. [R.4] $(3x - 5)^2$ **40.** [R.4] $3(6x^2 - x + 2)$
41. [R.4] $(ab - 3)(ab + 2)$ **42.** [R.5] 3
43. [R.5] $\dfrac{x - 5}{(x + 5)(x + 3)}$ **44.** [R.6] $y^3 \sqrt[6]{y}$
45. [R.6] $\sqrt[3]{(a + b)^2}$ **46.** [R.6] $b\sqrt[5]{b^2}$ **47.** [R.6] $\dfrac{m^4 n^2}{3}$
48. [R.6] $\dfrac{x - 2\sqrt{xy} + y}{x - y}$ **49.** [R.6] About 18.8 ft
50. Discussion and Writing [R.2] Anya is probably not following the rules for order of operations. She is subtracting 6 from 15 first, then dividing the difference by 3, and finally multiplying the quotient by 4. The correct answer is 7.
51. Discussion and Writing [R.2] When the number 4 is raised to an integer power greater than 1, the last digit of the result is 4 or 6. Since the calculator returns $4.398046511 \times 10^{12}$, or $4,398,046,511,000$, we can conclude that this result is an approximation. **52.** [R.2] $553.67 **53.** [R.2] $606.92
54. [R.2] $942.54 **55.** [R.2] $857.57
56. [R.3] $x^{2n} + 6x^n - 40$ **57.** [R.3] $t^{2a} + 2 + t^{-2a}$
58. [R.3] $y^{2b} - z^{2c}$ **59.** [R.3] $a^{3n} - 3a^{2n}b^n + 3a^n b^{2n} - b^{3n}$
60. [R.4] $(y^n + 8)^2$ **61.** [R.4] $(x^t - 7)(x^t + 4)$
62. [R.4] $m^{3n}(m^n - 1)(m^{2n} + m^n + 1)$

Test: Chapter R

1. [R.1] **(a)** $-8, 0, 36$; **(b)** $-8, \frac{11}{3}, 0, -5.49, 36, 10\frac{1}{6}$;
(c) $\frac{11}{3}, -5.49, 10\frac{1}{6}$; **(d)** $-8, 0$ **2.** [R.1] $\frac{14}{5}$
3. [R.1] 19.4 **4.** [R.1] $1.2|x||y|$

5. [R.1] $(-3, 6]$;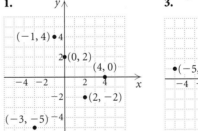
6. [R.1] 12 **7.** [R.2] -5 **8.** [R.2] 3.67×10^{-5}
9. [R.2] $4,510,000$ **10.** [R.2] 7.5×10^6 **11.** [R.2] x^{-3}, or $\dfrac{1}{x^3}$ **12.** [R.2] $72y^{14}$ **13.** [R.2] $-15a^4 b^{-1}$, or $-\dfrac{15a^4}{b}$
14. [R.3] $3x^4 - 5x^3 + x^2 + 5x$ **15.** [R.3] $2x^2 + x - 15$
16. [R.3] $4y^2 - 4y + 1$ **17.** [R.5] $\dfrac{x - y}{xy}$ **18.** [R.6] $3\sqrt{6}$
19. [R.6] $2\sqrt[3]{5}$ **20.** [R.6] $21\sqrt{3}$ **21.** [R.6] $6\sqrt{5}$
22. [R.6] $4 + \sqrt{3}$ **23.** [R.4] $(y + 3)(y - 6)$
24. [R.4] $x(x + 5)^2$ **25.** [R.4] $(2n - 3)(n + 4)$
26. [R.4] $2(2x + 3)(2x - 3)$
27. [R.4] $(m - 2)(m^2 + 2m + 4)$ **28.** [R.5] $\dfrac{x - 5}{x - 2}$
29. [R.5] $\dfrac{x + 3}{(x + 1)(x + 5)}$ **30.** [R.6] $\dfrac{35 + 5\sqrt{3}}{46}$
31. [R.6] $\sqrt[7]{t^5}$ **32.** [R.6] $7^{3/5}$ **33.** [R.6] 13 ft
34. [R.3] $x^2 - 2xy + y^2 - 2x + 2y + 1$

Chapter 1

Visualizing the Graph

1. H **2.** B **3.** D **4.** A **5.** G **6.** I **7.** C
8. J **9.** F **10.** E

Exercise Set 1.1

1.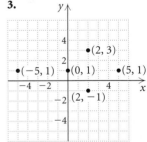

3.

5. (1998, $14 billion), (1999, $19 billion), (2000, $25 billion), (2001, $32 billion), (2002, $38 billion) **7.** Yes; no
9. Yes; no **11.** No; yes **13.** No; yes
15.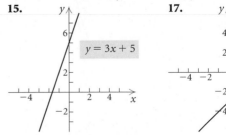

17.

19.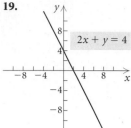

$2x + y = 4$

21.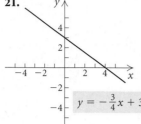

$y = -\frac{3}{4}x + 3$

35.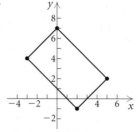

$y = -x^2 + 2x + 3$

37. $\sqrt{10}$, 3.162

23.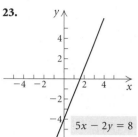

$5x - 2y = 8$

25.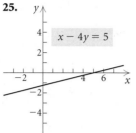

$x - 4y = 5$

39. $\sqrt{45}$, 6.708   **41.** $\sqrt{128.05}$, 11.316   **43.** 3
45. $\sqrt{14 + 6\sqrt{2}}$, 4.742   **47.** $\sqrt{a^2 + b^2}$ **49.** 6.5
51. Yes   **53.** No   **55.** $(-4, -6)$ **57.** $(4.95, -4.95)$
59. $\left(-6, \frac{13}{2}\right)$ **61.** $\left(-\frac{5}{12}, \frac{13}{40}\right)$ **63.** $\left(2\sqrt{3}, \frac{3}{2}\right)$

27.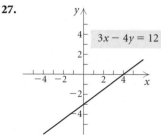

$3x - 4y = 12$

65.

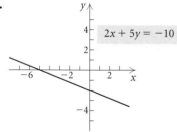

$\left(-\frac{1}{2}, \frac{3}{2}\right)$, $\left(\frac{7}{2}, \frac{1}{2}\right)$, $\left(\frac{5}{2}, \frac{9}{2}\right)$, $\left(-\frac{3}{2}, \frac{11}{2}\right)$; no

67. $\left(\dfrac{\sqrt{7} + \sqrt{2}}{2}, -\dfrac{1}{2}\right)$ **69.** $(x - 2)^2 + (y - 3)^2 = \dfrac{25}{9}$

71. $(x + 1)^2 + (y - 4)^2 = 25$
73. $(x - 2)^2 + (y - 1)^2 = 169$
75. $(x + 2)^2 + (y - 3)^2 = 4$

29.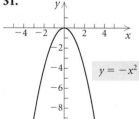

$2x + 5y = -10$

77. $(0, 0)$; 2;

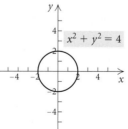

$x^2 + y^2 = 4$

79. $(0, 3)$; 4;

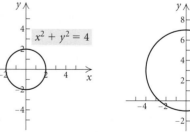

$x^2 + (y - 3)^2 = 16$

31.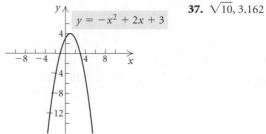

$y = -x^2$

33.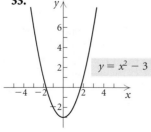

$y = x^2 - 3$

81. $(1, 5)$; 6;

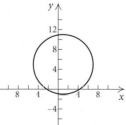

$(x - 1)^2 + (y - 5)^2 = 36$

83. $(-4, -5)$; 3;

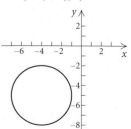

$(x + 4)^2 + (y + 5)^2 = 9$

85. $(3,1)$; 4;

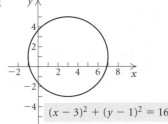

$(x-3)^2 + (y-1)^2 = 16$

87. $(-1,-1)$; 3;

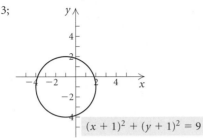

$(x+1)^2 + (y+1)^2 = 9$

89. $(x+2)^2 + (y-1)^2 = 3^2$
91. $(x-5)^2 + (y+5)^2 = 15^2$
93.

$4x + y = 7$

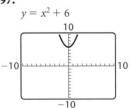

95.

$y = \frac{1}{3}x + 2$

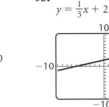

97.

$y = x^2 + 6$

99.

$y = 2 - x^2$

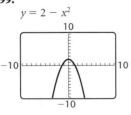

101. Standard window   **103.** $[-1, 1, -0.3, 0.3]$
105. Square the window; for example, $[-12, 9, -4, 10]$.
107.–111. Left to the student   **113.** Discussion and Writing
115. Third   **117.** $\sqrt{h^2 + h + 2a - 2\sqrt{a^2 + ah}}$,
$\left(\dfrac{2a + h}{2}, \dfrac{\sqrt{a} + \sqrt{a+h}}{2}\right)$
119. $(x-2)^2 + (y+7)^2 = 36$ **121.** $(0, 4)$
123. $a_1 \approx 2.7$ ft, $a_2 \approx 37.3$ ft   **125.** Yes   **127.** Yes
129. Let $P_1 = (x_1, y_1)$, $P_2 = (x_2, y_2)$, and
$M = \left(\dfrac{x_1 + x_2}{2}, \dfrac{y_1 + y_2}{2}\right)$. Let $d(AB)$ denote the distance
from point A to point B.

(a) $d(P_1M) = \sqrt{\left(\dfrac{x_1 + x_2}{2} - x_1\right)^2 + \left(\dfrac{y_1 + y_2}{2} - y_1\right)^2}$

$\qquad = \dfrac{1}{2}\sqrt{(x_2 - x_1)^2 + (y_2 - y_1)^2};$

$d(P_2M) = \sqrt{\left(\dfrac{x_1 + x_2}{2} - x_2\right)^2 + \left(\dfrac{y_1 + y_2}{2} - y_2\right)^2}$

$\qquad = \dfrac{1}{2}\sqrt{(x_1 - x_2)^2 + (y_1 - y_2)^2}$

$\qquad = \dfrac{1}{2}\sqrt{(x_2 - x_1)^2 + (y_2 - y_1)^2} = d(P_1M).$

(b) $d(P_1M) + d(P_2M) = \dfrac{1}{2}\sqrt{(x_2 - x_1)^2 + (y_2 - y_1)^2}$

$\qquad\qquad + \dfrac{1}{2}\sqrt{(x_2 - x_1)^2 + (y_2 - y_1)^2}$

$\qquad = \sqrt{(x_2 - x_1)^2 + (y_2 - y_1)^2}$

$\qquad = d(P_1P_2).$

Exercise Set 1.2

1. Yes   **3.** Yes   **5.** No   **7.** Yes   **9.** Yes   **11.** Yes
13. No   **15.** Function; domain: $\{2, 3, 4\}$; range: $\{10, 15, 20\}$
17. Not a function; domain: $\{-7, -2, 0\}$; range: $\{3, 1, 4, 7\}$
19. Function; domain: $\{-2, 0, 2, 4, -3\}$; range: $\{1\}$
21. $h(1) = -2$; $h(3) = 2$; $h(4) = 1$
23. $s(-4) = 3$; $s(-2) = 0$; $s(0) = -3$
25. $f(-1) = 2$; $f(0) = 0$; $f(1) = -2$
27. **(a)** 1; **(b)** 6; **(c)** 22; **(d)** $3x^2 + 2x + 1$; **(e)** $3t^2 - 4t + 2$
29. **(a)** 8; **(b)** -8; **(c)** $-x^3$; **(d)** $27y^3$; **(e)** $8 + 12h + 6h^2 + h^3$
31. **(a)** $\dfrac{1}{8}$; **(b)** 0; **(c)** does not exist; **(d)** $\dfrac{81}{53}$, or approximately

1.5283; **(e)** $\dfrac{x + h - 4}{x + h + 3}$ **33.** 0; does not exist; does not exist

as a real number; $\dfrac{1}{\sqrt{3}}$, or $\dfrac{\sqrt{3}}{3}$

35. All real numbers, or $(-\infty, \infty)$
37. $\{x \mid x \neq 0\}$, or $(-\infty, 0) \cup (0, \infty)$
39. $\{x \mid x \neq 2\}$, or $(-\infty, 2) \cup (2, \infty)$
41. $\{x \mid x \neq -1 \text{ and } x \neq 5\}$, or $(-\infty, -1) \cup (-1, 5) \cup (5, \infty)$
43. $\{x \mid x \leq 8\}$, or $(-\infty, 8]$ **45.** All real numbers, or $(-\infty, \infty)$
47. Domain: all real numbers; range: $[0, \infty)$
49. Domain: $[-3, 3]$; range: $[0, 3]$
51. Domain: all real numbers; range: all real numbers
53. Domain: $(-\infty, 7]$; range: $[0, \infty)$
55. Domain: all real numbers; range: $(-\infty, 3]$
57. No   **59.** Yes   **61.** Yes   **63.** No
65. Domain $[0, 5]$; range: $[0, 3]$
67. Domain: $[-2\pi, 2\pi]$; range: $[-1, 1]$
69. 645 m; 0 m   **71.** 0.4 acre; 20.4 acres; 50.6 acres;
416.9 acres; 1033.6 acres   **73.** $g(-2.1) \approx -21.8$;
$g(5.08) \approx -130.4$; $g(10.003) \approx -468.3$
75. Discussion and Writing

77. [1.1] $(0, -7)$, no; $(8, 11)$, yes
78. [1.1] $\left(\frac{4}{5}, -2\right)$, yes; $\left(\frac{11}{5}, \frac{1}{10}\right)$, yes
79. [1.1]   **80.** [1.1]

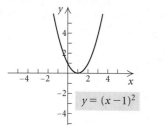
$y = (x-1)^2$

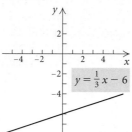
$y = \frac{1}{3}x - 6$

81. [1.1]

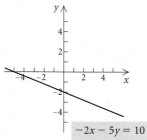
$-2x - 5y = 10$

82. [1.1]

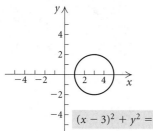
$(x - 3)^2 + y^2 = 4$

83. $f(x) = x, g(x) = x + 1$ **85.**

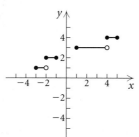

87. -7 **89.** (a) $f(x) = -2x + 1$; (b) $f(x) = 1$;
(c) $f(x) = 2x - 1$

Exercise Set 1.3

1. (a) Yes; (b) yes; (c) yes   **3.** (a) Yes; (b) no; (c) no
5. $-\frac{3}{5}$ **7.** 0   **9.** $\frac{1}{5}$ **11.** $-\frac{5}{3}$ **13.** 0.3   **15.** 0
17. $-\frac{6}{5}$ **19.** $-\frac{1}{3}$ **21.** Not defined

23.

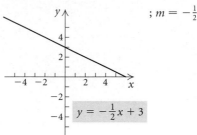
$y = -\frac{1}{2}x + 3$; $m = -\frac{1}{2}$

25.

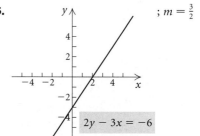
$2y - 3x = -6$; $m = \frac{3}{2}$

27.

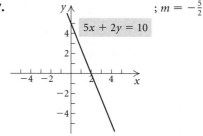
$5x + 2y = 10$; $m = -\frac{5}{2}$

29.

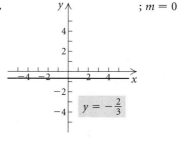
$y = -\frac{2}{3}$; $m = 0$

31. The average rate of change over the 10-year period was $566.50 per year.   **33.** The average rate of change over the 10-year period was about 1363 adoptions per year.
35. $\frac{1}{6}$ km per minute   **37.** (a) $W(h) = 4h - 130$;
(b) 118 lb; (c) $\{h \mid h > 32.5\}$, or $(32.5, \infty)$
39. (a) 115 ft, 75 ft, 135 ft, 179 ft; (b) Below $-57.5°$, stopping distance is negative; above $32°$, ice doesn't form.
41. (a) $\frac{11}{10}$. For each mile per hour faster that the car travels, it takes $\frac{11}{10}$ ft longer to stop; (b) 6, 11.5, 22.5, 55.5, 72;
(c) $\{r \mid r > 0\}$, or $(0, \infty)$. If r is allowed to be 0, the function says that a stopped car has a reaction distance of $\frac{1}{2}$ ft.
43. $C(t) = 60 + 29t$; $C(6) = \$234$
45. $C(x) = 800 + 3x$; $C(75) = \$1025$

47. Left to the student   **49.** Discussion and Writing
50. [1.2] 10   **51.** [1.2] 40   **52.** [1.2] $a^2 + 3a$
53. [1.2] $a^2 + 2ah + h^2 - 3a - 3h$ **55.** $-\dfrac{d}{10c}$ **57.** 0
59. $2a + h$ **61.** False   **63.** False   **65.** $f(x) = x + b$

Visualizing the Graph

1. E   **2.** D   **3.** A   **4.** J   **5.** C   **6.** F   **7.** H
8. G   **9.** B   **10.** I

Exercise Set 1.4

1. $\frac{3}{5}$; $(0, -7)$ **3.** Slope is not defined; there is no y-intercept.
5. $-\frac{1}{2}$; $(0, 5)$ **7.** $-\frac{3}{2}$; $(0, 5)$ **9.** 0; $(0, -6)$
11. $\frac{4}{5}$; $\left(0, \frac{8}{5}\right)$ **13.** 4; $(0, -2)$; $y = 4x - 2$
15. -1, $(0, 0)$; $y = -x$ **17.** 0, $(0, -3)$; $y = -3$
19.

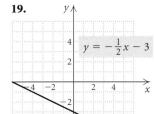

21.

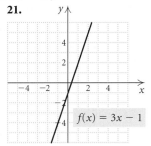

23.

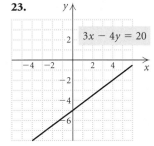

25.

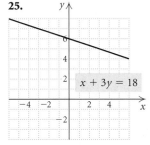

27. $y = \frac{2}{9}x + 4$ **29.** $y = -4x - 7$ **31.** $y = -4.2x + \frac{3}{4}$
33. $y = \frac{2}{9}x + \frac{19}{3}$ **35.** $y = 3x - 5$ **37.** $y = -\frac{3}{5}x - \frac{17}{5}$
39. $y = -3x + 2$ **41.** $y = -\frac{1}{2}x + \frac{7}{2}$ **43.** $y = \frac{2}{3}x - 6$
45. Horizontal: $y = -3$; vertical: $x = 0$
47. Horizontal: $y = -1$; vertical: $x = \frac{2}{11}$ **49.** Perpendicular
51. Neither parallel nor perpendicular   **53.** Parallel
55. Perpendicular   **57.** $y = \frac{2}{7}x + \frac{29}{7}$; $y = -\frac{7}{2}x + \frac{31}{2}$
59. $y = -0.3x - 2.1$; $y = \frac{10}{3}x + \frac{70}{3}$
61. $y = -\frac{3}{4}x + \frac{1}{4}$; $y = \frac{4}{3}x - 6$ **63.** $x = 3$; $y = -3$
65. **(a)** Model 1, using $(0, 7.8)$ and $(20, 6.4)$:
$y = -0.07x + 7.8$; model 2, using $(10, 7.3)$ and $(31, 4.9)$:
$y = -\frac{8}{70}x + \frac{591}{70}$; **(b)** model 1: about 5.4 days; model 2: about
4.4 days; **(c)** model 2   **67.** Using $(1, 10,424)$ and $(3, 11,717)$:
$y = 646.5x + 9777.5$; in 2004–2005: \$14,303; in 2006–2007:
\$15,596; in 2010–2011: \$18,182   **69.** Using $(3, 77.5)$ and
$(5, 82.7)$: $y = 2.6x + 69.7$; in 2004: 95.7 billion; in 2008:

106.1 billion; in 2012: 116.5 billion
71. **(a)** $M = 0.2H + 156$; **(b)** 164, 169, 171, 173; **(c)** $r = 1$;
the regression line fits the data perfectly and should be a good
predictor.   **73.** **(a)** $y = 0.02x + 1.77$, where x is the number
of years after 1999; **(b)** 2005: \$1.89 billion; 2010: \$1.99 billion;
(c) $r \approx 0.3162$. The line does not fit the data well.
75. **(a)** $y = -0.093641074x + 8.028026378$; **(b)** 4.8 days; this
value is only 0.4 days more than the value found with model 2;
(c) $r \approx -0.9786$. The line fits the data well.
77. Discussion and Writing   **78.** [1.3] Not defined
79. [1.3] -1 **80.** [1.1] $x^2 + (y - 3)^2 = 6.25$
81. [1.1] $(x + 7)^2 + (y + 1)^2 = \frac{81}{25}$ **83.** -7.75

Exercise Set 1.5

1. **(a)** $(-5, 1)$; **(b)** $(3, 5)$; **(c)** $(1, 3)$
3. **(a)** $(-3, -1), (3, 5)$; **(b)** $(1, 3)$; **(c)** $(-5, -3)$
5. **(a)** $(-\infty, -8), (-3, -2)$; **(b)** $(-8, -6)$;
(c) $(-6, -3), (-2, \infty)$
7. Domain: $[-5, 5]$; range: $[-3, 3]$
9. Domain: $[-5, -1] \cup [1, 5]$; range: $[-4, 6]$
11. Domain: $(-\infty, \infty)$; range: $(-\infty, 3]$
13. Relative maximum: 3.25 at $x = 2.5$; increasing: $(-\infty, 2.5)$;
decreasing: $(2.5, \infty)$
15. Relative maximum: 2.370 at $x = -0.667$; relative
minimum: 0 at $x = 2$; increasing: $(-\infty, -0.667), (2, \infty)$;
decreasing: $(-0.667, 2)$
17. Increasing: $(0, \infty)$; decreasing: $(-\infty, 0)$; relative minimum:
0 at $x = 0$
19. Increasing: $(-\infty, 0)$; decreasing: $(0, \infty)$; relative maximum:
5 at $x = 0$
21. Increasing: $(3, \infty)$; decreasing: $(-\infty, 3)$; relative minimum:
1 at $x = 3$
23. $A(x) = 30x - x^2$ **25.** $d(t) = \sqrt{(120t)^2 + (400)^2}$
27. $A(w) = 10w - \dfrac{w^2}{2}$ **29.** $d(s) = \dfrac{14}{s}$
31. **(a)** $A(x) = x(30 - x)$, or $30x - x^2$; **(b)** $\{x \mid 0 < x < 30\}$;
(c) 15 ft by 15 ft
33. **(a)** $V(x) = x(12 - 2x)(12 - 2x)$, or $4x(6 - x)^2$;
(b) $\{x \mid 0 < x < 6\}$; **(c)** 8 cm by 8 cm by 2 cm
35. $g(-4) = 0$; $g(0) = 4$; $g(1) = 5$; $g(3) = 5$
37. $h(-5) = 1$; $h(0) = 1$; $h(1) = 3$; $h(4) = 6$
39.   **41.**

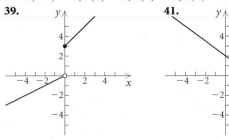

43.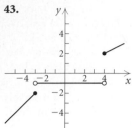

45.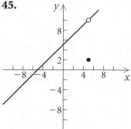

47.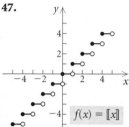

$f(x) = [\![x]\!]$

49.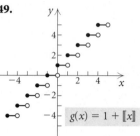

$g(x) = 1 + [\![x]\!]$

51. Domain: $(-\infty, \infty)$; range: $(-\infty, 0) \cup [4, \infty)$
53. Domain: $(-\infty, \infty)$; range: $[-1, \infty)$
55. Domain: $(-\infty, \infty)$; range: $\{y \mid y \leq -2 \text{ or } y = -1 \text{ or } y > 2\}$
57. Domain: $(-\infty, \infty)$; range: $\{-5, -2, 4\}$

$$f(x) = \begin{cases} -2, & \text{for } x < 2, \\ -5, & \text{for } x = 2, \\ 4, & \text{for } x > 2 \end{cases}$$

59. Domain: $(-\infty, \infty)$; range: $(-\infty, -1] \cup [2, \infty)$;

$$g(x) = \begin{cases} x, & \text{for } x \leq -1, \\ 2, & \text{for } -1 < x < 2, \\ x, & \text{for } x \geq 2 \end{cases}$$

61. Domain: $[-5, 3]$; range: $(-3, 5)$;

$$h(x) = \begin{cases} x + 8, & \text{for } -5 \leq x < -3, \\ 3, & \text{for } -3 \leq x \leq 1, \\ 3x - 6, & \text{for } 1 < x \leq 3 \end{cases}$$

63. (a) $y = -0.1x^2 + 1.2x + 98.6$ **(b)** 6 days after the illness began, the temperature is 102.2°F.

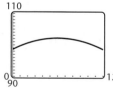

65. Increasing: $(-1, 1)$; decreasing: $(-\infty, -1), (1, \infty)$
67. Increasing: $(-1.414, 1.414)$; decreasing: $(-2, -1.414)$, $(1.414, 2)$ **69.** (a) $A(x) = x\sqrt{256 - x^2}$;
(b) $\{x \mid 0 < x < 16\}$; **(c)** $y = x\sqrt{256 - x^2}$

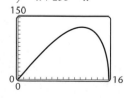

(d) 11.314 ft by 11.314 ft **71.** Discussion and Writing
73. [1.2] Function; domain; range; domain; exactly one; range
74. [1.1] Midpoint formula **75.** [1.1] x-intercept
76. [1.3] Constant; identity
77. $\{x \mid -5 \leq x < -4 \text{ or } 5 \leq x < 6\}$
79. (a) $h(r) = \dfrac{30 - 5r}{3}$; **(b)** $V(r) = \pi r^2 \left(\dfrac{30 - 5r}{3}\right)$;

(c) $V(h) = \pi h \left(\dfrac{30 - 3h}{5}\right)^2$

Exercise Set 1.6

1. 33 **3.** -1 **5.** Does not exist **7.** 0 **9.** 1
11. Does not exist **13.** 0 **15.** 5
17. (a) Domain of $f, g, f + g, f - g, fg,$ and ff: $(-\infty, \infty)$;
domain of f/g: $\left(-\infty, \frac{3}{5}\right) \cup \left(\frac{3}{5}, \infty\right)$;
domain of g/f: $\left(-\infty, -\frac{3}{2}\right) \cup \left(-\frac{3}{2}, \infty\right)$;
(b) $(f + g)(x) = -3x + 6$; $(f - g)(x) = 7x$;
$(fg)(x) = -10x^2 - 9x + 9$; $(ff)(x) = 4x^2 + 12x + 9$;
$(f/g)(x) = \dfrac{2x + 3}{3 - 5x}$; $(g/f)(x) = \dfrac{3 - 5x}{2x + 3}$
19. (a) Domain of f: $(-\infty, \infty)$; domain of g: $[-4, \infty)$;
domain of $f + g, f - g,$ and fg: $[-4, \infty)$;
domain of ff: $(-\infty, \infty)$; domain of f/g: $(-4, \infty)$;
domain of g/f: $[-4, 3) \cup (3, \infty)$;
(b) $(f + g)(x) = x - 3 + \sqrt{x + 4}$;
$(f - g)(x) = x - 3 - \sqrt{x + 4}$; $(fg)(x) = (x - 3)\sqrt{x + 4}$;
$(ff)(x) = x^2 - 6x + 9$; $(f/g)(x) = \dfrac{x - 3}{\sqrt{x + 4}}$;
$(g/f)(x) = \dfrac{\sqrt{x + 4}}{x - 3}$
21. (a) Domain of $f, g, f + g, f - g, fg,$ and ff: $(-\infty, \infty)$;
domain of f/g: $(-\infty, 0) \cup (0, \infty)$;
domain of g/f: $\left(-\infty, \frac{1}{2}\right) \cup \left(\frac{1}{2}, \infty\right)$
(b) $(f + g)(x) = -2x^2 + 2x - 1$;
$(f - g)(x) = 2x^2 + 2x - 1$; $(fg)(x) = -4x^3 + 2x^2$;
$(ff)(x) = 4x^2 - 4x + 1$; $(f/g)(x) = \dfrac{2x - 1}{-2x^2}$;
$(g/f)(x) = \dfrac{-2x^2}{2x - 1}$
23. (a) Domain of f: $[3, \infty)$; domain of g: $[-3, \infty)$;
domain of $f + g, f - g, fg,$ and ff: $[3, \infty)$;
domain of f/g: $[3, \infty)$; domain of g/f: $(3, \infty)$;
(b) $(f + g)(x) = \sqrt{x - 3} + \sqrt{x + 3}$;
$(f - g)(x) = \sqrt{x - 3} - \sqrt{x + 3}$; $(fg)(x) = \sqrt{x^2 - 9}$;
$(ff)(x) = |x - 3|$; $(f/g)(x) = \dfrac{\sqrt{x - 3}}{\sqrt{x + 3}}$; $(g/f)(x) = \dfrac{\sqrt{x + 3}}{\sqrt{x - 3}}$

25. (a) Domain of f, g, $f + g$, $f - g$, fg, and ff: $(-\infty, \infty)$;
domain of f/g: $(-\infty, 0) \cup (0, \infty)$;
domain of g/f: $(-\infty, -1) \cup (-1, \infty)$;
(b) $(f + g)(x) = x + 1 + |x|$; $(f - g)(x) = x + 1 - |x|$;
$(fg)(x) = (x + 1)|x|$; $(ff)(x) = x^2 + 2x + 1$;
$(f/g)(x) = \dfrac{x + 1}{|x|}$; $(g/f)(x) = \dfrac{|x|}{x + 1}$

27. (a) Domain of f, g, $f + g$, $f - g$, fg, and ff: $(-\infty, \infty)$;
domain of f/g: $(-\infty, -3) \cup \left(-3, \frac{1}{2}\right) \cup \left(\frac{1}{2}, \infty\right)$;
domain of g/f: $(-\infty, 0) \cup (0, \infty)$;
(b) $(f + g)(x) = x^3 + 2x^2 + 5x - 3$;
$(f - g)(x) = x^3 - 2x^2 - 5x + 3$;
$(fg)(x) = 2x^5 + 5x^4 - 3x^3$; $(ff)(x) = x^6$;
$(f/g)(x) = \dfrac{x^3}{2x^2 + 5x - 3}$; $(g/f)(x) = \dfrac{2x^2 + 5x - 3}{x^3}$

29. (a) Domain of f: $(-\infty, -1) \cup (-1, \infty)$;
domain of g: $(-\infty, 6) \cup (6, \infty)$; domain of $f + g$, $f - g$, and
fg: $(-\infty, -1) \cup (-1, 6) \cup (6, \infty)$;
domain of ff: $(-\infty, -1) \cup (-1, \infty)$;
domain of f/g and g/f: $(-\infty, -1) \cup (-1, 6) \cup (6, \infty)$;
(b) $(f + g)(x) = \dfrac{4}{x + 1} + \dfrac{1}{6 - x}$;

$(f - g)(x) = \dfrac{4}{x + 1} - \dfrac{1}{6 - x}$; $(fg)(x) = \dfrac{4}{(x + 1)(6 - x)}$;

$(ff)(x) = \dfrac{16}{(x + 1)^2}$; $(f/g)(x) = \dfrac{4(6 - x)}{x + 1}$;

$(g/f)(x) = \dfrac{x + 1}{4(6 - x)}$

31. (a) Domain of f: $(-\infty, 0) \cup (0, \infty)$;
domain of g: $(-\infty, \infty)$; domain of $f + g$, $f - g$, fg, and ff:
$(-\infty, 0) \cup (0, \infty)$; domain of f/g: $(-\infty, 0) \cup (0, 3) \cup (3, \infty)$;
domain of g/f: $(-\infty, 0) \cup (0, \infty)$;
(b) $(f + g)(x) = \dfrac{1}{x} + x - 3$; $(f - g)(x) = \dfrac{1}{x} - x + 3$;

$(fg)(x) = 1 - \dfrac{3}{x}$; $(ff)(x) = \dfrac{1}{x^2}$; $(f/g)(x) = \dfrac{1}{x(x - 3)}$;

$(g/f)(x) = x(x - 3)$

33. Domain of F: $[0, 9]$; domain of G: $[3, 10]$; domain of
$F + G$: $[3, 9]$ **35.** $[3, 6) \cup (6, 8) \cup (8, 9]$

37.

39. (a) $P(x) = -0.4x^2 + 57x - 13$; **(b)** $R(100) = 2000$;
$C(100) = 313$; $P(100) = 1687$ **41.** $2x + h$ **43.** 3

45. $6x + 3h - 2$ **47.** $\dfrac{5|x + h| - 5|x|}{h}$

49. $3x^2 + 3xh + h^2$ **51.** $\dfrac{7}{(x + h + 3)(x + 3)}$ **53.** -8

55. 64   **57.** 218   **59.** -80
61. $(f \circ g)(x) = (g \circ f)(x) = x$;
domain of $f \circ g$ and $g \circ f$: $(-\infty, \infty)$

63. $(f \circ g)(x) = \dfrac{4x}{x - 5}$; $(g \circ f)(x) = \dfrac{1 - 5x}{4}$;

domain of $f \circ g$: $(-\infty, 0) \cup (0, 5) \cup (5, \infty)$;
domain of $g \circ f$: $\left(-\infty, \frac{1}{5}\right) \cup \left(\frac{1}{5}, \infty\right)$
65. $(f \circ g)(x) = (g \circ f)(x) = x$;
domain of $f \circ g$ and $g \circ f$: $(-\infty, \infty)$
67. $(f \circ g)(x) = 2\sqrt{x} + 1$; $(g \circ f)(x) = \sqrt{2x + 1}$;
domain of $f \circ g$: $[0, \infty)$; domain of $g \circ f$: $\left[-\frac{1}{2}, \infty\right)$
69. $(f \circ g)(x) = 20$; $(g \circ f)(x) = 0.05$; domain of $f \circ g$ and
$g \circ f$: $(-\infty, \infty)$
71. $(f \circ g)(x) = |x|$; $(g \circ f)(x) = x$;
domain of $f \circ g$: $(-\infty, \infty)$; domain of $g \circ f$: $[-5, \infty)$
73. $(f \circ g)(x) = 5 - x$; $(g \circ f)(x) = \sqrt{1 - x^2}$;
domain of $f \circ g$: $(-\infty, 3]$; domain of $g \circ f$: $[-1, 1]$
75. $(f \circ g)(x) = (g \circ f)(x) = x$;
domain of $f \circ g$: $(-\infty, -1) \cup (-1, \infty)$;
domain of $g \circ f$: $(-\infty, 0) \cup (0, \infty)$
77. $(f \circ g)(x) = x^3 - 2x^2 - 4x + 6$;
$(g \circ f)(x) = x^3 - 5x^2 + 3x + 8$;
domain of $f \circ g$ and $g \circ f$: $(-\infty, \infty)$
79. $f(x) = x^5$; $g(x) = 4 + 3x$

81. $f(x) = \dfrac{1}{x}$; $g(x) = (x - 2)^4$

83. $f(x) = \dfrac{x - 1}{x + 1}$; $g(x) = x^3$

85. $f(x) = x^6$; $g(x) = \dfrac{2 + x^3}{2 - x^3}$

87. $f(x) = \sqrt{x}$; $g(x) = \dfrac{x - 5}{x + 2}$

89. $f(x) = x^3 - 5x^2 + 3x - 1$; $g(x) = x + 2$
91. $f(x) = 2(x - 20)$ **93.** Left to the student
95. Discussion and Writing   **97.** [1.4] (c)
98. [1.4] None   **99.** [1.3] (b), (d), (f), and (h)
100. [1.3] (b)   **101.** [1.4] (a)   **102.** [1.4] (c) and (g)
103. [1.4] (c) and (g)   **104.** [1.4] (a) and (f)

105. $f(x) = 2x + 5$, $g(x) = \dfrac{x - 5}{2}$; answers may vary

107. $(-\infty, -1) \cup (-1, 1) \cup \left(1, \frac{7}{3}\right) \cup \left(\frac{7}{3}, 3\right) \cup (3, \infty)$

Visualizing the Graph

1. C   **2.** B   **3.** A   **4.** E   **5.** G   **6.** D   **7.** H
8. I   **9.** F

Exercise Set 1.7

1. *x*-axis, no; *y*-axis, yes; origin, no
3. *x*-axis, yes; *y*-axis, no; origin, no
5. *x*-axis, no; *y*-axis, no; origin, yes
7. *x*-axis, no; *y*-axis, no; origin, yes
9. *x*-axis, yes; *y*-axis, yes; origin, yes
11. *x*-axis, no; *y*-axis, yes; origin, no
13. *x*-axis, yes; *y*-axis, yes; origin, yes
15. *x*-axis, no; *y*-axis, no; origin, no
17. *x*-axis, no; *y*-axis, no; origin, yes
19. *x*-axis: $(-5, -6)$; *y*-axis: $(5, 6)$; origin: $(5, -6)$
21. *x*-axis: $(-10, 7)$; *y*-axis: $(10, -7)$; origin: $(10, 7)$
23. *x*-axis: $(0, 4)$; *y*-axis: $(0, -4)$; origin: $(0, 4)$
25. Even **27.** Odd **29.** Neither **31.** Odd **33.** Even
35. Odd **37.** Even **39.** Even
41. Start with the graph of $f(x) = x^2$. Shift it right 3 units.

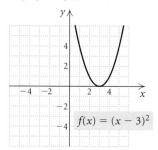

43. Start with the graph of $g(x) = x$. Shift it down 3 units.

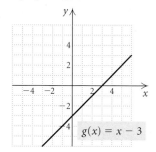

45. Start with the graph of $h(x) = \sqrt{x}$. Reflect it across the *x*-axis.

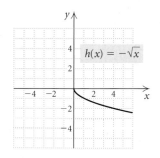

47. Start with the graph of $h(x) = \dfrac{1}{x}$. Shift it up 4 units.

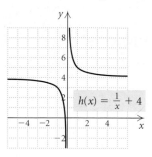

49. Start with the graph of $h(x) = x$. Stretch it vertically by multiplying each *y*-coordinate by 3. Then reflect it across the *x*-axis and shift it up 3 units.

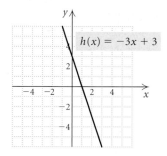

51. Start with the graph of $h(x) = |x|$. Shrink it vertically by multiplying each *y*-coordinate by $\frac{1}{2}$. Then shift it down 2 units.

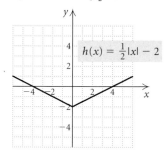

53. Start with the graph of $g(x) = x^3$. Shift it right 2 units. Then reflect it across the *x*-axis.

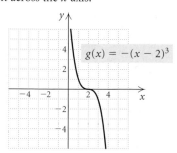

55. Start with the graph of $g(x) = x^2$. Shift it left 1 unit. Then shift it down 1 unit.

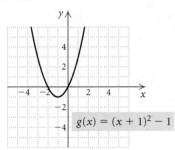

$g(x) = (x + 1)^2 - 1$

57. Start with the graph of $g(x) = x^3$. Shrink it vertically by multiplying each y-coordinate by $\frac{1}{3}$. Then shift it up 2 units.

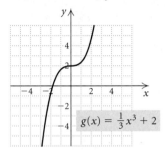

$g(x) = \frac{1}{3}x^3 + 2$

59. Start with the graph of $f(x) = \sqrt{x}$. Shift it left 2 units.

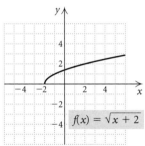

$f(x) = \sqrt{x + 2}$

61. Start with the graph of $f(x) = \sqrt[3]{x}$. Shift it down 2 units.

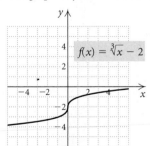

$f(x) = \sqrt[3]{x} - 2$

63. Start with the graph of $f(x) = |x|$. Shrink it horizontally by multiplying each x-coordinate by $\frac{1}{3}$ (or dividing each x-coordinate by 3).

65. Start with the graph of $h(x) = \frac{1}{x}$. Stretch it vertically by multiplying each y-coordinate by 2.

67. Start with the graph of $g(x) = \sqrt{x}$. Stretch it vertically by multiplying each y-coordinate by 3. Then shift it down 5 units.

69. Start with the graph of $f(x) = |x|$. Stretch it horizontally by multiplying each x-coordinate by 3. Then shift it down 4 units.

71. Start with the graph of $g(x) = x^2$. Shift it right 5 units, shrink it vertically by multiplying each y-coordinate by $\frac{1}{4}$, and reflect it across the x-axis.

73. Start with the graph of $g(x) = 1/x$. Shift it left 3 units, then up 2 units.

75. Start with the graph of $h(x) = x^2$. Shift it right 3 units. Then reflect it across the x-axis and shift it up 5 units.

77. $(-12, 2)$ **79.** $(12, 4)$ **81.** $(-12, 2)$ **83.** $(-12, 16)$

85. $f(x) = -(x - 8)^2$ **87.** $f(x) = |x + 7| + 2$

89. $f(x) = \frac{1}{2x} - 3$ **91.** $f(x) = -(x - 3)^2 + 4$

93. $f(x) = \sqrt{-(x + 2)} - 1$

95.

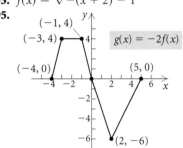

97.

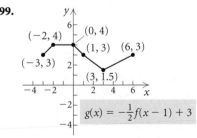

99.

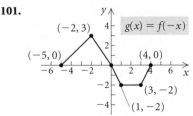

101.

103.

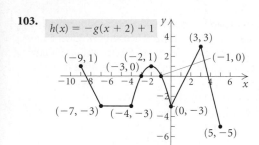

105.

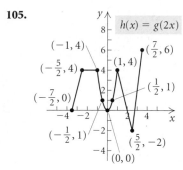

107. (f) **109.** (f) **111.** (d) **113.** (c)
115. $f(-x) = 2(-x)^4 - 35(-x)^3 + 3(-x) - 5 = 2x^4 + 35x^3 - 3x - 5 = g(x)$
117. $g(x) = x^3 - 3x^2 + 2$
119. $k(x) = (x + 1)^3 - 3(x + 1)^2$
121. Discussion and Writing **123.** Discussion and Writing
125. [1.2] **(a)** 38; **(b)** 38; **(c)** $5a^2 - 7$; **(d)** $5a^2 - 7$
126. [1.2] **(a)** 22; **(b)** -22; **(c)** $4a^3 - 5a$; **(d)** $-4a^3 + 5a$
127. [1.3] $y = -\frac{1}{8}x + \frac{7}{8}$
128. [1.3] Slope is $\frac{2}{9}$; y-intercept is $\left(0, \frac{1}{9}\right)$.
129.   **131.**

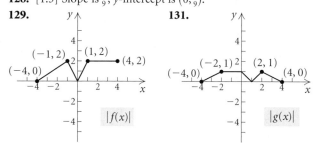

133. Odd **135.** x-axis, yes; y-axis, no; origin, no
137. x-axis, yes; y-axis, no; origin, no **139.** 5 **141.** True
143. $E(-x) = \dfrac{f(-x) + f(-(-x))}{2} = \dfrac{f(-x) + f(x)}{2} = E(x)$

145. **(a)** $E(x) + O(x) = \dfrac{f(x) + f(-x)}{2} + \dfrac{f(x) - f(-x)}{2} = \dfrac{2f(x)}{2} = f(x)$; **(b)** $f(x) = \dfrac{-22x^2 + \sqrt{x} + \sqrt{-x} - 20}{2} + \dfrac{8x^3 + \sqrt{x} - \sqrt{-x}}{2}$

Review Exercises: Chapter 1

1. [1.1] Yes; no **2.** [1.1] Yes; no
3. [1.1] **4.** [1.1]

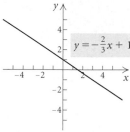
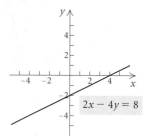

5. [1.1]

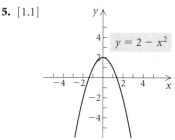

6. [1.1] $\sqrt{34} \approx 5.831$ **7.** [1.1] $\left(\frac{1}{2}, \frac{11}{2}\right)$
8. [1.1] $(x + 2)^2 + (y - 6)^2 = 13$
9. [1.1] Center: $(-1, 3)$; radius: 4
10. [1.1] Center: $(3, -5)$; radius: 1
11. [1.1] $(x - 2)^2 + (y - 4)^2 = 26$
12. [1.2] Not a function; domain: $\{3, 5, 7\}$; range: $\{1, 3, 5, 7\}$
13. [1.2] Function; domain: $\{-2, 0, 1, 2, 7\}$;
range: $\{-7, -4, -2, 2, 7\}$ **14.** [1.2] No **15.** [1.2] Yes
16. [1.2] No **17.** [1.2] Yes **18.** [1.2] $f(2) = -1$;
$f(-3) = -1$; $f(0) = -1$ **19.** [1.2] All real numbers
20. [1.2] $\{x \mid x \neq 0\}$ **21.** [1.2] $\{x \mid x \neq 5 \text{ and } x \neq 1\}$
22. [1.2] $\{x \mid x \neq -4 \text{ and } x \neq 4\}$
23. [1.2] Domain: $[-4, 4]$; range: $[0, 4]$
24. [1.2] Domain: $(-\infty, \infty)$; range: $[0, \infty)$
25. [1.2] Domain: $(-\infty, \infty)$; range: $(-\infty, \infty)$
26. [1.2] Domain: $(-\infty, \infty)$; range: $[0, \infty)$
27. [1.2] **(a)** -3; **(b)** 9; **(c)** $a^2 - 3a - 1$; **(d)** $2x + h - 1$
28. [1.3] **(a)** Yes; **(b)** no; **(c)** no, strictly speaking, but data
might be modeled by a linear regression function.
29. [1.3] **(a)** Yes; **(b)** yes; **(c)** yes **30.** [1.4] $\frac{5}{3}$ **31.** [1.3] 0
32. [1.3] Not defined **33.** [1.3] $m = -2$; y-intercept: $(0, -7)$
34. [1.3] The cost of a formal wedding rose \$590.75 per year
from 1995 to 2003. **35.** [1.3]

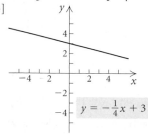

36. [1.4] $y = -\frac{2}{3}x - 4$ **37.** [1.4] $y = 3x + 5$
38. [1.4] $y = \frac{1}{3}x - \frac{1}{3}$ **39.** [1.4] Parallel
40. [1.4] Neither   **41.** [1.4] Perpendicular
42. [1.4] $y = -\frac{2}{3}x - \frac{1}{3}$ **43.** [1.4] $y = \frac{3}{2}x - \frac{5}{2}$
44. [1.3] $C(t) = 25 + 20t$; \$145
45. [1.3] **(a)** 70°C, 220°C, 10,020°C;   **(b)** $[0, 5600]$
46. [1.4] Using $(10, 16.5)$ and $(50, 57.5)$, $y = 1.025x + 6.25$;
67.75 million. Answers may vary.
47. [1.5]   **48.** [1.5]

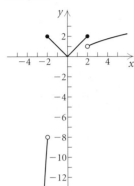

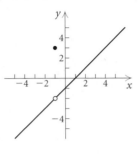

49. [1.5]   **50.** [1.5]

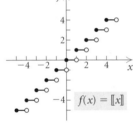

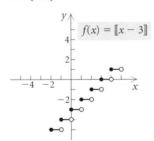

51. [1.5] $f(-1) = 1$; $f(5) = 2$; $f(-2) = 2$; $f(-3) = 27$
52. [1.5] $A(x) = 2x\sqrt{4 - x^2}$

53. [1.5] **(a)** $A(x) = x^2 + \dfrac{432}{x}$;   **(b)** $(0, \infty)$;

(c) $x = 6$ in.; height $= 3$ in.
54. [1.6] -33 **55.** [1.6] 0   **56.** [1.6] Does not exist
57. [1.6] **(a)** Domain of f, $\{x \mid x \neq 0\}$; domain of g, all real
numbers; domain of $f + g$, $f - g$, and fg, $\{x \mid x \neq 0\}$;

domain of f/g, $\left\{x \mid x \neq 0 \text{ and } x \neq \dfrac{3}{2}\right\}$;

(b) $(f + g)(x) = \dfrac{4}{x^2} + 3 - 2x$; $(f - g)(x) = \dfrac{4}{x^2} - 3 + 2x$;

$(fg)(x) = \dfrac{12}{x^2} - \dfrac{8}{x}$; $(f/g)(x) = \dfrac{4}{x^2(3 - 2x)}$

58. [1.6] **(a)** Domain of f, g, $f + g$, $f - g$, and fg, all real

numbers; domain of f/g, $\left\{x \mid x \neq \dfrac{1}{2}\right\}$;

(b) $(f + g)(x) = 3x^2 + 6x - 1$; $(f - g)(x) = 3x^2 + 2x + 1$;

$(fg)(x) = 6x^3 + 5x^2 - 4x$; $(f/g)(x) = \dfrac{3x^2 + 4x}{2x - 1}$

59. [1.6] $P(x) = -0.5x^2 + 105x - 6$ **60.** [1.6] $-2x - h$
61. [1.6] **(a)** Domain of $f \circ g$, $\left\{x \mid x \neq \dfrac{3}{2}\right\}$; domain of $g \circ f$,

$\{x \mid x \neq 0\}$;   **(b)** $(f \circ g)(x) = \dfrac{4}{(3 - 2x)^2}$; $(g \circ f)(x) = 3 - \dfrac{8}{x^2}$

62. [1.6] **(a)** Domain of $f \circ g$ and $g \circ f$, all real numbers;
(b) $(f \circ g)(x) = 12x^2 - 4x - 1$; $(g \circ f)(x) = 6x^2 + 8x - 1$
63. [1.6] $f(x) = \sqrt{x}$, $g(x) = 5x + 2$. Answers may vary.
64. [1.6] $f(x) = 4x^2 + 9$, $g(x) = 5x - 1$. Answers may vary.
65. [1.7] x-axis, yes; y-axis, yes; origin, yes
66. [1.7] x-axis, yes; y-axis, yes; origin, yes
67. [1.7] x-axis, no; y-axis, no; origin, no
68. [1.7] x-axis, no; y-axis, yes; origin, no
69. [1.7] x-axis, no; y-axis, no; origin, yes
70. [1.7] x-axis, no; y-axis, yes; origin, no
71. [1.7] Even   **72.** [1.7] Even   **73.** [1.7] Odd
74. [1.7] Even   **75.** [1.7] Even   **76.** [1.7] Neither
77. [1.7] Odd   **78.** [1.7] Even   **79.** [1.7] Even
80. [1.7] Odd   **81.** [1.7] $f(x) = (x + 3)^2$
82. [1.7] $f(x) = -\sqrt{x - 3} + 4$ **83.** [1.7] $f(x) = 2|x - 3|$
84. [1.7]   **85.** [1.7]

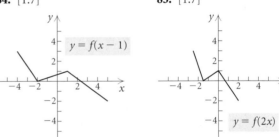

86. [1.7]   **87.** [1.7]

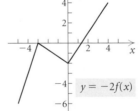
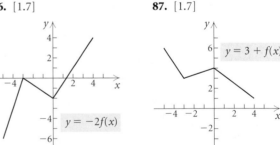

88. [1.4] **(a)** $y = 0.9352625415x + 9.67817685$; 65.8 million;
(b) $r = 0.9946$, a good fit
89. Discussion and Writing [1.2], [1.7] **(a)** $4x^3 - 2x + 9$;
(b) $4x^3 + 24x^2 + 46x + 35$; **(c)** $4x^3 - 2x + 42$. **(a)** Adds 2
to each function value; **(b)** adds 2 to each input before finding
a function value; **(c)** adds the output for 2 to the output for x
90. Discussion and Writing [1.7] In the graph of $y = f(cx)$,
the constant c stretches or shrinks the graph of $y = f(x)$
horizontally. The constant c in $y = cf(x)$ stretches or shrinks
the graph of $y = f(x)$ vertically. For $y = f(cx)$, the
x-coordinates of $y = f(x)$ are divided by c; for $y = cf(x)$, the
y-coordinates of $y = f(x)$ are multiplied by c.

91. Discussion and Writing [1.7] **(a)** To draw the graph of y_2 from the graph of y_1, reflect across the x-axis the portions of the graph for which the y-coordinates are negative. **(b)** To draw the graph of y_2 from the graph of y_1, draw the portion of the graph of y_1 to the right of the y-axis; then draw its reflection across the y-axis.
92. [1.2] $\{x \mid x < 0\}$
93. [1.2] $\{x \mid x \neq -3 \ and \ x \neq 0 \ and \ x \neq 3\}$
94. [1.6], [1.7] Let $f(x)$ and $g(x)$ be odd functions. Then by definition, $f(-x) = -f(x)$, or $f(x) = -f(-x)$, and $g(-x) = -g(x)$, or $g(x) = -g(-x)$. Thus, $(f + g)(x) = f(x) + g(x) = -f(-x) + [-g(-x)] = -[f(-x) + g(-x)] = -(f + g)(-x)$ and $f + g$ is odd.
95. [1.7] Reflect the graph of $y = f(x)$ across the x-axis and then across the y-axis.

Test: Chapter 1

1. [1.1]

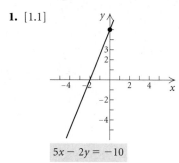

$5x - 2y = -10$

2. [1.1] $\sqrt{45} \approx 6.708$

3. [1.1] $\left(-3, \frac{9}{2}\right)$ **4.** [1.1] $(x + 1)^2 + (y - 2)^2 = 5$
5. [1.1] Center: $(-4, 5)$; radius: 6
6. [1.2] **(a)** Yes; **(b)** $\{-4, 3, 1, 0\}$; **(c)** $\{7, 0, 5\}$
7. [1.2] **(a)** 8; **(b)** $2a^2 + 7a + 11$
8. [1.2] **(a)**

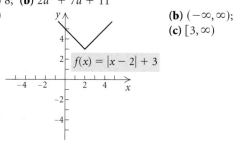

$f(x) = |x - 2| + 3$

(b) $(-\infty, \infty)$;
(c) $[3, \infty)$

9. [1.2] $\{x \mid x \neq 4\}$, or $(-\infty, 4) \cup (4, \infty)$
10. [1.2] $(-\infty, \infty)$
11. [1.2] $\{x \mid -5 \leq x \leq 5\}$, or $[-5, 5]$
12. [1.2] **(a)** No; **(b)** yes   **13.** [1.3] Not defined
14. [1.3] $-\frac{11}{6}$ **15.** [1.3] 0
16. [1.3] Debit-card transactions increased approximately $16.6 billion per year from 1990 to 1999.
17. [1.4] Slope: $\frac{3}{2}$; y-intercept: $\left(0, \frac{5}{2}\right)$
18. [1.4] $y = -\frac{5}{8}x - 5$ **19.** [1.4] $y - 4 = -\frac{3}{4}(x - (-5))$, or $y - (-2) = -\frac{3}{4}(x - 3)$, or $y = -\frac{3}{4}x + \frac{1}{4}$
20. [1.4] $y - 3 = -\frac{1}{2}(x + 1)$, or $y = -\frac{1}{2}x + \frac{5}{2}$

21. [1.4] Perpendicular   **22.** [1.4] Using $(0, 23.3)$ and $(2, 25.1)$, $y = 0.9x + 23.3$; 30.5 gal
23. [1.5]

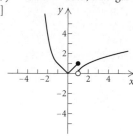

24. [1.5] $f\left(-\frac{7}{8}\right) = \frac{7}{8}$; $f(5) = 2$; $f(-4) = 16$
25. [1.5] $(f - g)(-1) = 6$
26. [1.6] **(a)** $(-\infty, \infty)$; **(b)** $[3, \infty)$; **(c)** $(f - g)(x) = x^2 - \sqrt{x - 3}$; **(d)** $(fg)(x) = x^2\sqrt{x - 3}$; **(e)** $(3, \infty)$
27. [1.6] $2x + h$ **28.** [1.6] $(f \circ g)(x) = \sqrt{x^2 - 4}$; $(g \circ f)(x) = x - 4$
29. [1.6] Domain of $(f \circ g)(x)$: $(-\infty, -2) \cup (2, \infty)$; domain of $(g \circ f)(x) = [5, \infty)$
30. [1.7] x-axis: no; y-axis: yes; origin: no   **31.** [1.7] Odd
32. [1.7] $f(x) = (x - 2)^2 - 1$
33. [1.7] $f(x) = (x + 2)^2 - 3$
34. [1.7]

$y = -\frac{1}{2}f(x)$

35. [1.7] $(-1, 1)$

Chapter 2

Exercise Set 2.1

1. 4   **3.** $-\frac{3}{4}$ **5.** -9 **7.** $\frac{11}{5}$ **9.** 8   **11.** -4 **13.** 6
15. -1 **17.** $\frac{4}{5}$ **19.** $-\frac{3}{2}$ **21.** $-\frac{2}{3}$ **23.** $\frac{1}{2}$
25. About $3708   **27.** About 65.3%   **29.** 29.2 million homes   **31.** 2080 calories   **33.** $1300   **35.** $9800
37. 12 mi   **39.** 26°, 130°, 24°   **41.** Length: 93 m; width: 68 m   **43.** Length: 100 yd; width: 65 yd
45. 67.5 lb   **47.** Freight: 66 mph; passenger: 80 mph
49. 4.5 hr   **51.** 1.25 hr   **53.** $2400 at 3%; $2600 at 4%
55. I: about 1724 violations; II: about 265 violations
57. AOL: 52.5 million users; Microsoft: 29.4 million users
59. 660 yr   **61.** -5 **63.** 18   **65.** 16   **67.** -12
69. 6   **71.** 20   **73.** 6   **75.** 15   **77.** **(a)** $(4, 0)$; **(b)** 4
79. **(a)** $(-2, 0)$; **(b)** -2 **81.** **(a)** $(-4, 0)$; **(b)** -4
83. $b = \dfrac{2A}{h}$ **85.** $w = \dfrac{P - 2l}{2}$ **87.** $h = \dfrac{2A}{b_1 + b_2}$

89. $\pi = \dfrac{3V}{4r^3}$ **91.** $C = \dfrac{5}{9}(F - 32)$ **93.** $A = \dfrac{C - By}{x}$

95. $h = \dfrac{p - l - 2w}{2}$ **97.** $y = \dfrac{2x - 6}{3}$ **99.** $b = \dfrac{a}{1 + cd}$

101. $x = \dfrac{z}{y - y^2}$ **103.** Left to the student

105. Discussion and Writing   **107.** [1.4] $y = -\frac{3}{4}x + \frac{13}{4}$

108. [1.4] $y = -\frac{3}{4}x + \frac{1}{4}$

109. [1.6] All real numbers, or $(-\infty, \infty)$

110. [1.6] $(-\infty, -2) \cup (-2, \infty)$ **111.** [1.6] $-x - 7$

112. [1.6] -9 **113.** Yes   **115.** No   **117.** $-\frac{2}{3}$

119. No; the 6-oz cup costs about 6.4% more per ounce.

121. 7.5 mi

Exercise Set 2.2

1. $2 + 11i$ **3.** $5 - 12i$ **5.** $4 + 8i$ **7.** $-4 - 2i$
9. $5 + 9i$ **11.** $5 + 4i$ **13.** $5 + 7i$ **15.** $11 - 5i$
17. $-1 + 5i$ **19.** $2 - 12i$ **21.** $35 + 14i$
23. $6 + 16i$ **25.** $13 - i$ **27.** $-11 + 16i$
29. $-10 + 11i$ **31.** $-31 - 34i$ **33.** $-14 + 23i$
35. 41 **37.** 13 **39.** 74 **41.** $12 + 16i$
43. $-45 - 28i$ **45.** $-8 - 6i$ **47.** $2i$ **49.** $-7 + 24i$
51. $\frac{15}{146} + \frac{33}{146}i$ **53.** $\frac{10}{13} - \frac{15}{13}i$ **55.** $-\frac{14}{13} + \frac{5}{13}i$
57. $\frac{11}{25} - \frac{27}{25}i$ **59.** $\dfrac{-4\sqrt{3} + 10}{41} + \dfrac{5\sqrt{3} + 8}{41}i$
61. $-\frac{1}{2} + \frac{1}{2}i$ **63.** $-\frac{1}{2} - \frac{13}{2}i$ **65.** $-i$ **67.** $-i$
69. 1 **71.** i **73.** 625 **75.** Left to the student
77. Discussion and Writing   **79.** [1.4] $y = -2x + 1$
80. [1.6] All real numbers, or $(-\infty, \infty)$
81. [1.6] $\left(-\infty, -\frac{5}{3}\right) \cup \left(-\frac{5}{3}, \infty\right)$
82. [1.6] $x^2 - 3x - 1$ **83.** [1.6] $\frac{8}{11}$
84. [1.6] $2x + h - 3$ **85.** True   **87.** True   **89.** $a^2 + b^2$

Exercise Set 2.3

1. $\frac{2}{3}, \frac{3}{2}$ **3.** $-2, 10$ **5.** $-1, \frac{2}{3}$ **7.** $-\sqrt{3}, \sqrt{3}$
9. $-\sqrt{7}, \sqrt{7}$ **11.** $-\sqrt{2}i, \sqrt{2}i$ **13.** $-\sqrt{17}, \sqrt{17}$
15. $0, 3$ **17.** $-\frac{1}{3}, 0, 2$ **19.** $-1, -\frac{1}{7}, 1$
21. (a) $(-4, 0), (2, 0)$; (b) $-4, 2$
23. (a) $(-1, 0), (3, 0)$; (b) $-1, 3$
25. (a) $(-2, 0), (2, 0)$; (b) $-2, 2$ **27.** $-7, 1$
29. $4 \pm \sqrt{7}$ **31.** $-4 \pm 3i$ **33.** $-2, \frac{1}{3}$ **35.** $-3, 5$
37. $-1, \frac{2}{5}$ **39.** $\dfrac{5 \pm \sqrt{7}}{3}$ **41.** $-\dfrac{1}{2} \pm \dfrac{\sqrt{7}}{2}i$
43. $\dfrac{4 \pm \sqrt{31}}{5}$ **45.** $\dfrac{5}{6} \pm \dfrac{\sqrt{23}}{6}i$ **47.** $4 \pm \sqrt{11}$
49. $\dfrac{-1 \pm \sqrt{61}}{6}$ **51.** $\dfrac{5 \pm \sqrt{17}}{4}$ **53.** $-\dfrac{1}{5} \pm \dfrac{3}{5}i$
55. 144; two real   **57.** -7; two imaginary
59. 49; two real   **61.** $-5, -1$ **63.** $\dfrac{3 \pm \sqrt{21}}{2}$

65. $\dfrac{5 \pm \sqrt{21}}{2}$ **67.** $-1 \pm \sqrt{6}$ **69.**
71. $\dfrac{1 \pm \sqrt{13}}{6}$ **73.** $\dfrac{1 \pm \sqrt{6}}{5}$ **75.** $\dfrac{-3}{}$
77. $\pm 1, \pm\sqrt{2}$ **79.** $\pm\sqrt{2}, \pm\sqrt{5}i$ **81.**   **83.** 16
85. $-8, 64$ **87.** $1, 16$ **89.** $\frac{5}{2}, 3$ **91.** $-\frac{3}{2}, -1, \frac{1}{2}, 1$
93. About 11.5 sec   **95.** 1985, 2002
97. Length: 4 ft; width: 3 ft   **99.** 4 and 9; -9 and -4
101. 2 cm   **103.** Length: 8 ft; width: 6 ft   **105.** Linear
107. Quadratic   **109.** Linear   **111.** $2, 6$ **113.** $0.143, 6$
115. $-0.151, 1.651$ **117.** $-0.637, 3.137$
119. $-1.535, 0.869$ **121.** $-0.347, 1.181$
123. Discussion and Writing
125. [1.2] 551,453 associate's degrees
126. [1.2] 660,605 associate's degrees
127. [1.7] x-axis: yes; y-axis: yes; origin: yes
128. [1.7] x-axis: no; y-axis: yes; origin: no
129. [1.7] Odd   **130.** [1.7] Neither   **131.** (a) 2; (b) $\frac{11}{2}$
133. (a) 2; (b) $1 - i$ **135.** 1 **137.** $-\sqrt{7}, -\frac{3}{2}, 0, \frac{1}{3}, \sqrt{7}$
139. $\dfrac{-1 \pm \sqrt{1 + 4\sqrt{2}}}{2}$ **141.** $3 \pm \sqrt{5}$ **143.** 19
145. $-2 \pm \sqrt{2}, \dfrac{1}{2} \pm \dfrac{\sqrt{7}}{2}i$ **147.** $t = \dfrac{-v_0 \pm \sqrt{v_0^2 - 2ax_0}}{a}$

Visualizing the Graph

1. C   **2.** B   **3.** A   **4.** J   **5.** F   **6.** D   **7.** I
8. G   **9.** H   **10.** E

Exercise Set 2.4

1. (a) $\left(-\frac{1}{2}, -\frac{9}{4}\right)$; (b) $x = -\frac{1}{2}$; (c) minimum: $-\frac{9}{4}$
3. (a) $(4, -4)$; (b) $x = 4$; (c) minimum: -4;
(d)

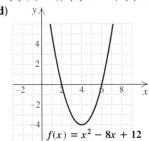

$f(x) = x^2 - 8x + 12$

5. (a) $\left(\frac{7}{2}, -\frac{1}{4}\right)$; (b) $x = \frac{7}{2}$; (c) minimum: $-\frac{1}{4}$;
(d)

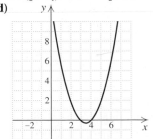

$f(x) = x^2 - 7x + 12$

(a) $(-2, 1)$; (b) $x = -2$; (c) minimum: 1;
(d)

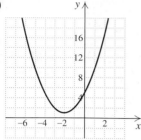

$$f(x) = x^2 + 4x + 5$$

9. (a) $(-3, 12)$; (b) $x = -3$; (c) maximum: 12;
(d)

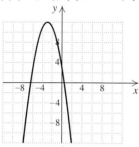

$$f(x) = -x^2 - 6x + 3$$

11. (a) $\left(-\frac{3}{2}, \frac{7}{2}\right)$; (b) $x = -\frac{3}{2}$; (c) minimum: $\frac{7}{2}$;
(d)

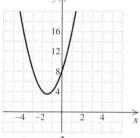

$$g(x) = 2x^2 + 6x + 8$$

13. (a) $\left(\frac{1}{2}, \frac{3}{2}\right)$; (b) $x = \frac{1}{2}$; (c) maximum: $\frac{3}{2}$;
(d)

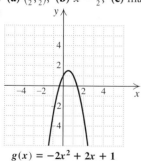

$$g(x) = -2x^2 + 2x + 1$$

15. (f)　**17.** (b)　**19.** (h)　**21.** (c)
23. (a) $(3, -4)$; (b) minimum: -4; (c) $[-4, \infty)$;
(d) increasing: $(3, \infty)$; decreasing: $(-\infty, 3)$
25. (a) $(-1, -18)$; (b) minimum: -18; (c) $[-18, \infty)$;
(d) increasing: $(-1, \infty)$; decreasing: $(-\infty, -1)$

27. (a) $\left(5, \frac{9}{2}\right)$; (b) maximum: $\frac{9}{2}$; (c) $\left(-\infty, \frac{9}{2}\right]$;
(d) increasing: $(-\infty, 5)$; decreasing: $(5, \infty)$
29. (a) $(-1, 2)$; (b) minimum: 2; (c) $[2, \infty)$;
(d) increasing: $(-1, \infty)$; decreasing: $(-\infty, -1)$
31. (a) $\left(-\frac{3}{2}, 18\right)$; (b) maximum: 18; (c) $\left(-\infty, 18\right]$;
(d) increasing: $\left(-\infty, -\frac{3}{2}\right)$; decreasing: $\left(-\frac{3}{2}, \infty\right)$
33. 0.625 sec; 12.25 ft　**35.** 3.75 sec; 305 ft　**37.** 4.5 in.
39. Base: 10 cm; height: 10 cm　**41.** 350　**43.** $797; 40
45. 350.6 ft　**47.** 4800 yd^2　**49.** Left to the student
51. Left to the student　**53.** Discussion and Writing
55. Discussion and Writing　**56.** [1.6] 3
57. [1.6] $4x + 2h - 1$
58. [1.7]

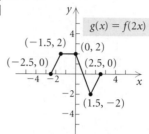

59. [1.7]　　　　　　　　　　　　　　　**61.** -236.25

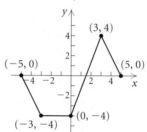

$$g(x) = -2f(x)$$

63.　$y = (|x| - 5)^2 - 3$

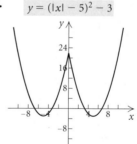

Exercise Set 2.5

1. $\frac{20}{9}$　**3.** 286　**5.** 6　**7.** No solution
9. No solution　**11.** No solution
13. $\{x \mid x \text{ is a real number } and \ x \neq 0 \text{ and } x \neq 6\}$
15. No solution　**17.** No solution　**19.** $-\frac{69}{14}$　**21.** $-\frac{37}{18}$
23. 2　**25.** $\frac{5}{3}$　**27.** $\frac{9}{2}$　**29.** 3　**31.** -4　**33.** -5
35. $\pm\sqrt{2}$　**37.** No solution　**39.** 6　**41.** -1　**43.** $\frac{35}{2}$
45. -98　**47.** -6　**49.** 5　**51.** 7　**53.** 2　**55.** $-1, 2$

57. 7   **59.** 7   **61.** No solution   **63.** 1   **65.** 3, 7
67. 5   **69.** -1 **71.** -8 **73.** 81   **75.** $-7, 7$
77. No solution   **79.** $-3, 5$ **81.** $-\frac{1}{3}, \frac{1}{3}$ **83.** 0
85. $-1, -\frac{1}{3}$ **87.** $-24, 44$ **89.** $-2, 4$ **91.** $-13, 7$
93. $-\frac{4}{3}, \frac{2}{3}$ **95.** $-\frac{3}{4}, \frac{9}{4}$ **97.** $-13, 1$
99. $T_1 = \dfrac{P_1 V_1 T_2}{P_2 V_2}$ **101.** $R_2 = \dfrac{R R_1}{R_1 - R}$ **103.** $p = \dfrac{Fm}{m - F}$
105. Left to the student   **107.** Discussion and Writing
108. [2.1] 3   **109.** [2.1] 7.5   **110.** [2.1] 16.4%
111. [2.1] 419,000   **113.** $3 \pm 2\sqrt{2}$ **115.** -1

Exercise Set 2.6

1. $\{x \mid x > 3\}$, or $(3, \infty)$;
3. $\left\{x \mid x \geq -\frac{5}{12}\right\}$, or $\left[-\frac{5}{12}, \infty\right)$;
5. $\left\{y \mid y \geq \frac{22}{13}\right\}$, or $\left[\frac{22}{13}, \infty\right)$;
7. $\left\{x \mid x \leq \frac{15}{34}\right\}$, or $\left(-\infty, \frac{15}{34}\right]$;
9. $\{x \mid x < 1\}$, or $(-\infty, 1)$;
11. $[-3, 3)$;
13. $[8, 10]$;
15. $[-7, -1]$;
17. $\left(-\frac{3}{2}, 2\right)$;
19. $(1, 5]$;
21. $\left(-\frac{11}{3}, \frac{13}{3}\right)$;
23. $(-\infty, -2] \cup (1, \infty)$;
25. $\left(-\infty, -\frac{7}{2}\right] \cup \left[\frac{1}{2}, \infty\right)$;
27. $(-\infty, 9.6) \cup (10.4, \infty)$;
29. $\left(-\infty, -\frac{57}{4}\right] \cup \left[-\frac{55}{4}, \infty\right)$;
31. $(-7, 7)$;
33. $(-\infty, -4.5) \cup [4.5, \infty)$;
35. $(-17, 1)$;
37. $(-\infty, -17] \cup [1, \infty)$;

39. $\left(-\frac{1}{4}, \frac{3}{4}\right)$;
41. $\left(-\frac{1}{3}, \frac{1}{3}\right)$;
43. $[-6, 3]$;
45. $(-\infty, 4.9) \cup (5.1, \infty)$;
47. $\left[-\frac{1}{2}, \frac{7}{2}\right]$;
49. $\left[-\frac{7}{3}, 1\right]$;
51. $(-\infty, -8) \cup (7, \infty)$;
53. No solution   **55.** More than 4 yr after 2002
57. Less than 4 hr   **59.** $5000   **61.** More than 20 checks
63. Sales greater than $18,000   **65.** Left to the student
67. Discussion and Writing   **69.** [1.1] y-intercept
70. [1.1] Distance formula   **71.** [1.2] Relation
72. [1.2] Function   **73.** [1.3] Horizontal lines
74. [1.4] Parallel   **75.** [1.5] Decreasing
76. [1.7] Symmetric with respect to the y-axis
77. $\left(-\frac{1}{4}, \frac{5}{9}\right]$ **79.** $\left(-\infty, \frac{1}{2}\right)$ **81.** No solution
83. $\left(-\infty, -\frac{8}{3}\right) \cup (-2, \infty)$

Review Exercises: Chapter 2

1. [2.1] $\frac{3}{2}$ **2.** [2.1] -6 **3.** [2.1] -1 **4.** [2.1] -21
5. [2.3] $-\frac{5}{2}, \frac{1}{3}$ **6.** [2.3] $-5, 1$ **7.** [2.3] $-2, \frac{4}{3}$
8. [2.3] $-\sqrt{3}, \sqrt{3}$ **9.** [2.3] $-\sqrt{10}, \sqrt{10}$ **10.** [2.1] 3
11. [2.1] 4   **12.** [2.1] 0.2, or $\frac{1}{5}$ **13.** [2.1] 4   **14.** [2.3] 1
15. [2.3] $-5, 3$ **16.** [2.3] $\dfrac{1 \pm \sqrt{41}}{4}$ **17.** [2.3] $\dfrac{-1 \pm \sqrt{10}}{3}$
18. [2.5] $\frac{27}{7}$ **19.** [2.5] $-\frac{1}{2}, \frac{9}{4}$ **20.** [2.5] 0, 3   **21.** [2.5] 5
22. [2.5] 1, 7   **23.** [2.5] $-8, 1$
24. [2.6] $\left[-\frac{4}{3}, \frac{4}{3}\right]$;
25. [2.6] $\left(\frac{2}{5}, 2\right]$;
26. [2.6] $(-\infty, \infty)$;
27. [2.6] $\left(-\infty, -\frac{5}{3}\right] \cup [1, \infty)$;
28. [2.6] $\left(-\frac{2}{3}, 1\right)$;
29. [2.6] $(-\infty, -6] \cup [-2, \infty)$;
30. [2.1] $h = \dfrac{V}{lw}$ **31.** [2.1] $s = \dfrac{M - n}{0.3}$ **32.** [2.5] $h = \dfrac{v^2}{2g}$

33. [2.5] $t = \dfrac{ab}{a + b}$ **34.** [2.2] $-2\sqrt{10}i$

35. [2.2] $-4\sqrt{15}$ **36.** [2.2] $-\frac{7}{8}$ **37.** [2.2] $-18 - 26i$

38. [2.2] $\frac{11}{10} + \frac{3}{10}i$ **39.** [2.2] $1 - 4i$ **40.** [2.2] $2 - i$

41. [2.2] $-i$ **42.** [2.2] 3^{28}

43. [2.3] $x^2 - 3x + \frac{9}{4} = 18 + \frac{9}{4}; \left(x - \frac{3}{2}\right)^2 = \frac{81}{4}; x = \frac{3}{2} \pm \frac{9}{2};$
$-3, 6$ **44.** [2.3] $x^2 - 4x = 2; x^2 - 4x + 4 = 2 + 4;$
$(x - 2)^2 = 6; x = 2 \pm \sqrt{6}; 2 - \sqrt{6}, 2 + \sqrt{6}$

45. [2.3] $-4, \frac{2}{3}$ **46.** [2.3] $1 - 3i, 1 + 3i$

47. [2.3] $-3, 6$ **48.** [2.3] 1 **49.** [2.3] $\pm\sqrt{\dfrac{3 \pm \sqrt{5}}{2}}$

50. [2.3] $-\sqrt{3}, 0, \sqrt{3}$ **51.** [2.3] $-2, -\frac{2}{3}, 3$

52. [2.3] $-5, -2, 2$ **53.** [2.4] **(a)** $\left(\frac{3}{8}, -\frac{7}{16}\right);$ **(b)** $x = \frac{3}{8};$
(c) maximum: $-\frac{7}{16};$ **(d)** $\left(-\infty, -\frac{7}{16}\right];$
(e)

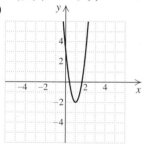

$f(x) = -4x^2 + 3x - 1$

54. [2.4] **(a)** $(1, -2);$ **(b)** $x = 1;$ **(c)** minimum: $-2;$
(d) $[-2, \infty);$ **(e)**

$f(x) = 5x^2 - 10x + 3$

55. [2.4] (d) **56.** [2.4] (c) **57.** [2.4] (b) **58.** [2.4] (a)

59. [2.3] 30 ft, 40 ft **60.** [2.5] 6 mph **61.** [2.3] 80 km/h

62. [2.3] $35 - 5\sqrt{33}$ ft, or about 6.3 ft

63. [2.4] 6 ft by 6 ft

64. [2.3] $\dfrac{15 - \sqrt{115}}{2}$ cm, or about 2.1 cm

65. [2.6] Years after 2004

66. [2.6] Fahrenheit temperatures less than 113°

67. Left to the student **68.** Left to the student

69. Left to the student **70.** Left to the student

71. Discussion and Writing [2.1] If an equation contains no fractions, using the addition principle before using the multiplication principle eliminates the need to add or subtract fractions.

72. Discussion and Writing [2.4] You can conclude that $|a_1| = |a_2|$ since these constants determine how wide the parabolas are. Nothing can be concluded about the h's and the k's.

73. [2.5] 256 **74.** [2.5] $-7, 9$ **75.** [2.5] $4 \pm \sqrt[4]{243}$, or $0.052, 7.948$ **76.** [2.3] -1 **77.** [2.3] $-\frac{1}{4}, 2$

78. [2.4] ± 6 **79.** [2.3] 9%

Test: Chapter 2

1. [2.1] -1 **2.** [2.1] -5 **3.** [2.1] $\frac{21}{11}$ **4.** [2.3] $\frac{1}{2}, -5$

5. [2.3] $-\sqrt{6}, \sqrt{6}$ **6.** [2.3] $-2i, 2i$ **7.** [2.3] $-1, 3$

8. [2.3] $\dfrac{5 \pm \sqrt{13}}{2}$ **9.** [2.3] $\dfrac{3}{4} \pm \dfrac{\sqrt{23}}{4}i$

10. [2.5] 16 **11.** [2.5] $-1, \frac{13}{6}$ **12.** [2.5] 5

13. [2.5] 5 **14.** [2.5] $-\frac{1}{2}, 2$

15. [2.6] $(-5, 3);$

16. [2.6] $(-\infty, 2] \cup [4, \infty);$

17. [2.6] $[-7, 1];$

18. [2.6] $(-\infty, -7) \cup (-3, \infty);$

19. [2.1] $h = \dfrac{3V}{2\pi r^2}$ **20.** [2.5] $n = \dfrac{R^2}{3p}$

21. [2.3] $x^2 + 4x = 1; x^2 + 4x + 4 = 1 + 4; (x + 2)^2 = 5;$
$x = -2 \pm \sqrt{5}; -2 - \sqrt{5}, -2 + \sqrt{5}$

22. [2.1] Length: 60 m; width: 45 m **23.** [2.5] 3 km/h

24. [2.1] $1.80 **25.** [2.2] $\sqrt{43}i$ **26.** [2.2] $-5i$

27. [2.2] $3 - 5i$ **28.** [2.2] $10 + 5i$ **29.** [2.2] $\frac{1}{10} - \frac{1}{5}i$

30. [2.2] i **31.** [2.1] -3 **32.** [2.3] $-\frac{1}{4}, 3$

33. [2.3] $\dfrac{1 \pm \sqrt{57}}{4}$ **34.** [2.4] **(a)** $(1, 9);$ **(b)** $x = 1;$
(c) maximum: 9; **(d)** $(-\infty, 9];$
(e)

$f(x) = -x^2 + 2x + 8$

35. [2.4] 15 ft by 30 ft **36.** [2.6] More than 6 hr
37. [2.4] $-\frac{4}{9}$

Chapter 3

Visualizing the Graph

1. H   **2.** D   **3.** J   **4.** B   **5.** A   **6.** C   **7.** I
8. E   **9.** G   **10.** F

Exercise Set 3.1

1. Cubic; $\frac{1}{2}x^3$; $\frac{1}{2}$; 3   **3.** Linear; $0.9x$; 0.9; 1
5. Quartic; $305x^4$; 305; 4   **7.** Quartic; x^4; 1; 4
9. Cubic; $4x^3$; 4; 3   **11.** (d)   **13.** (b)   **15.** (c)   **17.** (a)
19. (d)   **21.** (f)   **23.** (b)   **25.** Yes; no; no
27. No; yes; yes   **29.** -3, multiplicity 2; 1, multiplicity 1
31. 4, multiplicity 3; -6, multiplicity 1
33. ± 3, each has multiplicity 3
35. 0, multiplicity 3; 1, multiplicity 2; -4, multiplicity 1
37. 3, multiplicity 2; -4, multiplicity 3; 0, multiplicity 4
39. $\pm\sqrt{3}$, ± 1, each has multiplicity 1
41. $-3, -1, 1$, each has multiplicity 1
43. $\pm 2, \frac{1}{2}$, each has multiplicity 1

45.

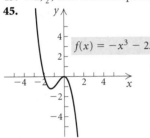

$f(x) = -x^3 - 2x^2$

47.

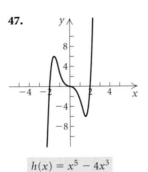

$h(x) = x^5 - 4x^3$

49.

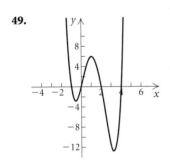

$h(x) = x(x - 4)(x + 1)(x - 2)$

51.

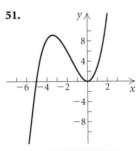

$f(x) = \frac{1}{2}x^3 + \frac{5}{2}x^2$

53.

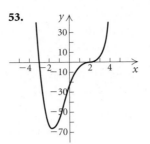

$g(x) = (x - 2)^3(x + 3)$

55.

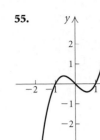

$f(x) = x^3 - x$

57.

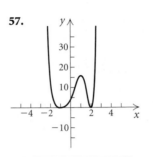

$f(x) = (x - 2)^2(x + 1)^4$

59.

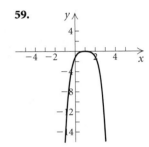

$g(x) = -(x - 1)^4$

61.

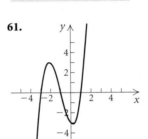

$h(x) = x^3 + 3x^2 - x - 3$

63.

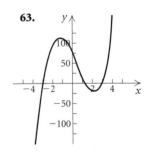

$f(x) = 6x^3 - 8x^2 - 54x + 72$

65. $f(-5) = -18$ and $f(-4) = 7$. By the intermediate value theorem, since $f(-5)$ and $f(-4)$ have opposite signs, then $f(x)$ has a zero between -5 and -4.   **67.** $f(-3) = 22$ and $f(-2) = 5$. Both $f(-3)$ and $f(-2)$ are positive. We cannot use the intermediate value theorem to determine if there is a zero between -3 and -2.   **69.** $f(2) = 2$ and $f(3) = 57$. Both $f(2)$ and $f(3)$ are positive. We cannot use the intermediate value theorem to determine if there is a zero between 2 and 3.   **71.** $f(4) = -12$ and $f(5) = 4$. By the intermediate value theorem, since $f(4)$ and $f(5)$ have opposite signs, then $f(x)$ has a zero between 4 and 5.

73. 689,000; 535,000 **75.** 0.866 sec **77.** $3240
79. 150; 103; 269; 207 **81.** (a) $4\frac{1}{2}$%; (b) 10%
83. Relative maximum: 1.506 at $x = -0.632$, relative
minimum: 0.494 at $x = 0.632$; $(-\infty, \infty)$
85. Relative minimum: -3.8 at $x = 0$, no relative maxima;
$[-3.8, \infty)$ **87.** Relative maximum: 11.012 at $x = 1.258$,
relative minimum: -8.183 at $x = -1.116$; $(-\infty, \infty)$
89. (a) Linear: $y = 3.730909091x + 83.72727273$;
quadratic: $y = 0.025011655x^2 + 1.22974359x + 121.2447552$;
cubic: $y = -0.0011822067x^3 + 0.2023426573x^2 - 5.532478632x + 163.8041958$;
quartic: $y = -0.00002634033x^4 + 0.0040858586x^3 - 0.1269114219x^2 + 1.052602953x + 144.8391608$, where x is
the number of years after 1900;
(b) cubic; 445, 430; answers may vary
91. (a) Linear: $y = 378.7575758x + 3701.090909$;
quadratic: $y = 43.10227273x^2 - 9.162878788x + 4218.318182$, where x is the number of years after 1992;
(b) quadratic; $11,383 billion; answers may vary
93. Discussion and Writing **95.** [1.1] 5 **96.** [1.1] $6\sqrt{2}$
97. [1.1] Center: $(3, -5)$; radius 7
98. [1.1] Center: $(-4, 3)$; radius $2\sqrt{2}$
99. [2.6] $\{y | y \geq 3\}$, or $[3, \infty)$ **100.** [2.6] $\{x | x > \frac{5}{3}\}$, or $(\frac{5}{3}, \infty)$
101. [2.6] $\{x | x \leq -13 \ or \ x \geq 1\}$, or $(-\infty, -13] \cup [1, \infty)$
102. [2.6] $\{x | -\frac{11}{12} \leq x \leq \frac{5}{12}\}$, or $[-\frac{11}{12}, \frac{5}{12}]$ **103.** 7%

Exercise Set 3.2

1. (a) No; (b) yes; (c) no **3.** (a) Yes; (b) no; (c) yes
5. $P(x) = (x + 2)(x^2 - 2x + 4) - 16$
7. $P(x) = (x + 9)(x^2 - 3x + 2) + 0$
9. $P(x) = (x + 2)(x^3 - 2x^2 + 2x - 4) + 11$
11. $Q(x) = 2x^3 + x^2 - 3x + 10, R(x) = -42$
13. $Q(x) = x^2 - 4x + 8, R(x) = -24$
15. $Q(x) = 3x^2 - 4x + 8, R(x) = -18$
17. $Q(x) = x^4 + 3x^3 + 10x^2 + 30x + 89, R(x) = 267$
19. $Q(x) = x^3 + x^2 + x + 1, R(x) = 0$
21. $Q(x) = 2x^3 + x^2 + \frac{7}{2}x + \frac{7}{4}, R(x) = -\frac{1}{8}$
23. 0; -60; 0 **25.** 10; 80; 998 **27.** 5,935,988; -772
29. 0; 0; 65; $1 - 12\sqrt{2}$ **31.** Yes; no **33.** Yes; yes
35. No; yes **37.** No; no
39. $f(x) = (x - 1)(x + 2)(x + 3)$; 1, -2, -3
41. $f(x) = (x - 2)(x - 5)(x + 1)$; 2, 5, -1
43. $f(x) = (x - 2)(x - 3)(x + 4)$; 2, 3, -4
45. $f(x) = (x - 3)^3(x + 2)$; 3, -2
47. $f(x) = (x - 1)(x - 2)(x - 3)(x + 5)$; 1, 2, 3, -5

49.

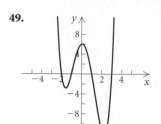

$f(x) = x^4 - x^3 - 7x^2 + x + 6$

51.

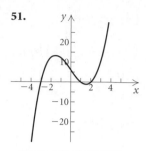

$f(x) = x^3 - 7x + 6$

53.

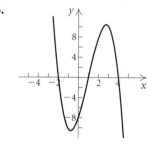

$f(x) = -x^3 + 3x^2 + 6x - 8$

55. Left to the student

57. Discussion and Writing **58.** [2.1] -5
59. [2.3] $\dfrac{5 \pm i\sqrt{71}}{4}$ **60.** [2.3] $-1, \dfrac{3}{7}$ **61.** [2.3] $-5, 0$
62. [1.2] 10 **63.** [2.3] $-3, -2$
64. [1.4] $f(x) = \frac{56}{15}x + 27.2$; $68.3 billion, $86.9 billion,
$105.6 billion **65.** [2.3] $b = 15$ in., $h = 15$ in.
67. (a) $x + 4, x + 3, x - 2, x - 5$;
(b) $P(x) = (x + 4)(x + 3)(x - 2)(x - 5)$; (c) yes; two
examples are $f(x) = c \cdot P(x)$ for any nonzero constant c;
and $g(x) = (x - a)P(x)$; (d) no **69.** $\frac{14}{3}$ **71.** $-1 \pm \sqrt{7}$
73. Answers can vary. One possibility is $P(x) = x^{15} - x^{14}$.
75. $x - 3 + i, R \, 6 - 3i$

Exercise Set 3.3

1. $f(x) = x^3 - 6x^2 - x + 30$
3. $f(x) = x^3 + 3x^2 + 4x + 12$
5. $f(x) = x^3 - 3x^2 - 2x + 6$ **7.** $f(x) = x^3 - 6x - 4$
9. $f(x) = x^3 + 2x^2 + 29x + 148$
11. $f(x) = x^3 - \frac{5}{3}x^2 - \frac{2}{3}x$
13. $f(x) = x^5 + 2x^4 - 2x^2 - x$
15. $f(x) = x^4 + 3x^3 + 3x^2 + x$ **17.** $-\sqrt{3}$
19. $i, 2 + \sqrt{5}$ **21.** $-3i$ **23.** $-4 + 3i, 2 + \sqrt{3}$

25. $-\sqrt{5}, 4i$ **27.** $2 + i$ **29.** $-3 - 4i, 4 + \sqrt{5}$
31. $4 + i$ **33.** $f(x) = x^3 - 4x^2 + 6x - 4$
35. $f(x) = x^2 + 16$ **37.** $f(x) = x^3 - 5x^2 + 16x - 80$
39. $f(x) = x^4 - 2x^3 - 3x^2 - 10x - 10$
41. $f(x) = x^4 + 4x^2 - 45$ **43.** $-\sqrt{2}, \sqrt{2}$ **45.** $i, 2, 3$
47. $1 + 2i, 1 - 2i$ **49.** ± 1 **51.** $\pm 1, \pm \frac{1}{2}, \pm 2, \pm 4, \pm 8$
53. $\pm 1, \pm 2, \pm \frac{1}{3}, \pm \frac{1}{5}, \pm \frac{2}{3}, \pm \frac{2}{5}, \pm \frac{1}{15}, \pm \frac{2}{15}$
55. **(a)** Rational: -3; other: $\pm \sqrt{2}$;
(b) $f(x) = (x + 3)\left(x + \sqrt{2}\right)\left(x - \sqrt{2}\right)$
57. **(a)** Rational: $-2, 1$; other: none;
(b) $f(x) = (x + 2)(x - 1)^2$
59. **(a)** Rational: -1; other: $3 \pm 2\sqrt{2}i$;
(b) $f(x) = (x + 1)\left(x - 3 - 2\sqrt{2}i\right)\left(x - 3 + 2\sqrt{2}i\right)$
61. **(a)** Rational: $-\frac{1}{5}, 1$; other: $\pm 2i$;
(b) $f(x) = (5x + 1)(x - 1)(x + 2i)(x - 2i)$
63. **(a)** Rational: $-2, -1$; other: $3 \pm \sqrt{13}$;
(b) $f(x) = (x + 2)(x + 1)\left(x - 3 - \sqrt{13}\right)\left(x - 3 + \sqrt{13}\right)$
65. **(a)** Rational: 2; other: $1 \pm \sqrt{3}$;
(b) $f(x) = (x - 2)\left(x - 1 - \sqrt{3}\right)\left(x - 1 + \sqrt{3}\right)$
67. **(a)** Rational: -2; other: $1 \pm \sqrt{3}i$;
(b) $f(x) = (x + 2)\left(x - 1 - \sqrt{3}i\right)\left(x - 1 + \sqrt{3}i\right)$
69. **(a)** Rational: $\frac{1}{2}$; other: $\frac{1 \pm \sqrt{5}}{2}$;
(b) $f(x) = \frac{1}{3}\left(x - \frac{1}{2}\right)\left(x - \frac{1 + \sqrt{5}}{2}\right)\left(x - \frac{1 - \sqrt{5}}{2}\right)$
71. No rational zeros **73.** No rational zeros
75. No rational zeros **77.** $-2, 1, 2$ **79.** 3 or 1; 0
81. 0; 3 or 1 **83.** 2 or 0; 2 or 0 **85.** 1; 1 **87.** 1; 0
89. 2 or 0; 2 or 0 **91.** 3 or 1; 1 **93.** 1; 1
95.

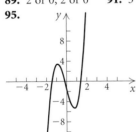

$f(x) = 4x^3 + x^2 - 8x - 2$

97.

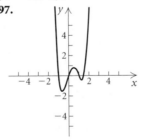

$f(x) = 2x^4 - 3x^3 - 2x^2 + 3x$

99. Left to the student **101.** Discussion and Writing
102. [2.4] **(a)** $(1, -4)$; **(b)** $x = 1$; **(c)** minimum: -4 at $x = 1$
103. [2.4] **(a)** $(4, -6)$; **(b)** $x = 4$; **(c)** minimum: -6 at $x = 4$
104. [2.3] $-3, 11$ **105.** [2.1] 10
106. [2.4] Quadratic; $-x^2$; -1; 2; as $x \to \infty$, $f(x) \to -\infty$, and as $x \to -\infty$, $f(x) \to -\infty$
107. [3.1] Cubic; $-x^3$; -1; 3; as $x \to \infty$, $g(x) \to -\infty$, and as $x \to -\infty$, $g(x) \to \infty$
108. [1.3] Linear; x; 1; 1; as $x \to \infty$, $h(x) \to \infty$, and as $x \to -\infty$, $h(x) \to -\infty$
109. [1.3] Constant; $-\frac{4}{9}$; $-\frac{4}{9}$; zero degree; for all x, $f(x) = -\frac{4}{9}$

110. [3.1] Cubic; x^3; 1; 3; as $x \to \infty$, $h(x) \to \infty$, and as $x \to -\infty$, $h(x) \to -\infty$
111. [3.1] Quartic; x^4; 1; 4; as $x \to \infty$, $g(x) \to \infty$, and as $x \to -\infty$, $g(x) \to \infty$
113. $Q(x) = x^3 + x^2y + xy^2 + y^3$, $R(x) = 0$
115. **(a)** $-1, \frac{1}{2}, 3$; **(b)** $0, \frac{3}{2}, 4$; **(c)** $-3, -\frac{3}{2}, 1$; **(d)** $-\frac{1}{2}, \frac{1}{4}, \frac{3}{2}$
117. $-8, -\frac{3}{2}, 4, 7, 15$

Visualizing the Graph

1. A **2.** C **3.** D **4.** H **5.** G **6.** F **7.** B
8. I **9.** J **10.** E

Exercise Set 3.4

1. (d); $x = 2, x = -2, y = 0$ **3.** (e); $x = 2, x = -2, y = 0$
5. (c); $x = 2, x = -2, y = 8x$ **7.** $x = 0$ **9.** $x = 2$
11. $x = 4, x = -6$ **13.** $x = \frac{3}{2}, x = -1$ **15.** $y = \frac{3}{4}$
17. $y = 0$ **19.** No horizontal asymptote **21.** $y = x + 1$
23. $y = x$ **25.** $y = x - 3$
27. Domain: $(-\infty, 0) \cup (0, \infty)$; no x-intercepts, no y-intercepts;

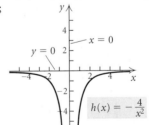

29. Domain: $(-\infty, 0) \cup (0, \infty)$; no x-intercepts, no y-intercepts;

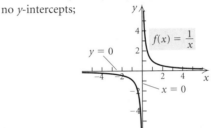

31. Domain: $(-\infty, -1) \cup (-1, \infty)$; x-intercepts: $(1, 0)$ and $(3, 0)$, y-intercept: $(0, 3)$;

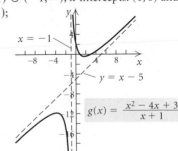

33. Domain: $(-\infty, -3) \cup (-3, \infty)$; no x-intercepts, y-intercept: $\left(0, \frac{1}{3}\right)$;

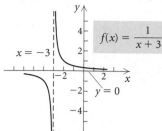

35. Domain: $(-\infty, 5) \cup (5, \infty)$; no x-intercepts, y-intercept: $\left(0, \frac{2}{5}\right)$;

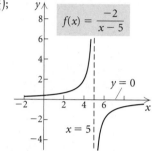

37. Domain: $(-\infty, 0) \cup (0, \infty)$; x-intercept: $\left(-\frac{1}{2}, 0\right)$, no y-intercept;

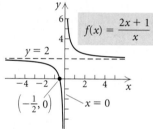

39. Domain: $(-\infty, 2) \cup (2, \infty)$; no x-intercepts, y-intercept: $\left(0, \frac{1}{4}\right)$;

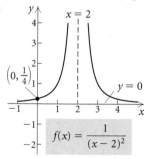

41. Domain: $(-\infty, 0) \cup (0, \infty)$; no x-intercepts, no y-intercept;

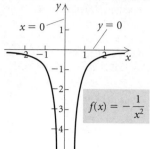

43. Domain: $(-\infty, \infty)$; no x-intercepts, y-intercept: $\left(0, \frac{1}{3}\right)$;

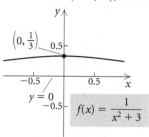

45. Domain: $(-\infty, 2) \cup (2, \infty)$; x-intercept: $(-2, 0)$, y-intercept: $(0, 2)$;

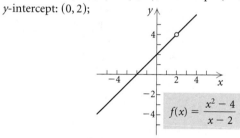

47. Domain: $(-\infty, -2) \cup (-2, \infty)$; x-intercept: $(1, 0)$, y-intercept: $\left(0, -\frac{1}{2}\right)$;

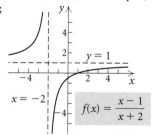

49. Domain: $\left(-\infty, -\frac{1}{2}\right) \cup \left(-\frac{1}{2}, 3\right) \cup (3, \infty)$;
x-intercept: $(-3, 0)$, y-intercept: $(0, -1)$;

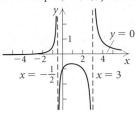

$$f(x) = \frac{x + 3}{2x^2 - 5x - 3}$$

51. Domain: $(-\infty, -1) \cup (-1, \infty)$; x-intercepts: $(-3, 0)$ and $(3, 0)$, y-intercept: $(0, -9)$;

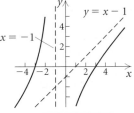

$$f(x) = \frac{x^2 - 9}{x + 1}$$

53. Domain: $(-\infty, \infty)$; x-intercepts: $(-2, 0)$ and $(1, 0)$, y-intercept: $(0, -2)$;

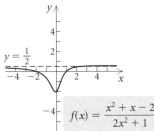

$$f(x) = \frac{x^2 + x - 2}{2x^2 + 1}$$

55. Domain: $(-\infty, 1) \cup (1, \infty)$; x-intercept: $\left(-\frac{2}{3}, 0\right)$, y-intercept: $(0, 2)$;

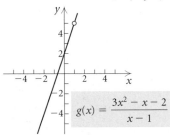

$$g(x) = \frac{3x^2 - x - 2}{x - 1}$$

57. Domain: $(-\infty, -1) \cup (-1, 3) \cup (3, \infty)$; x-intercept: $(1, 0)$, y-intercept: $\left(0, \frac{1}{3}\right)$;

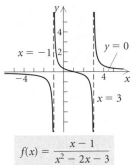

$$f(x) = \frac{x - 1}{x^2 - 2x - 3}$$

59. Domain: $(-\infty, -1) \cup (-1, \infty)$; x-intercept: $(3, 0)$, y-intercept: $(0, -3)$;

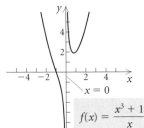

$$f(x) = \frac{x - 3}{(x + 1)^3}$$

61. Domain: $(-\infty, 0) \cup (0, \infty)$; x-intercept: $(-1, 0)$, no y-intercept;

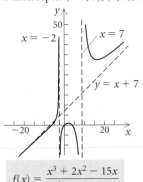

$$f(x) = \frac{x^3 + 1}{x}$$

63. Domain: $(-\infty, -2) \cup (-2, 7) \cup (7, \infty)$; x-intercepts: $(-5, 0)$, $(0, 0)$, and $(3, 0)$, y-intercept: $(0, 0)$;

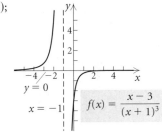

$$f(x) = \frac{x^3 + 2x^2 - 15x}{x^2 - 5x - 14}$$

65. Domain: $(-\infty, \infty)$; x-intercept: $(0,0)$, y-intercept: $(0,0)$;

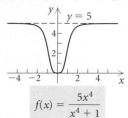

$$f(x) = \frac{5x^4}{x^4 + 1}$$

67. Domain: $(-\infty, -1) \cup (-1, 2) \cup (2, \infty)$; x-intercept: $(0,0)$, y-intercept: $(0,0)$;

$$f(x) = \frac{x^2}{x^2 - x - 2}$$

$x = -1$ $x = 2$

69. $f(x) = \dfrac{1}{x^2 - x - 20}$ **71.** $f(x) = \dfrac{3x^2 + 12x + 12}{2x^2 - 2x - 40}$

73. (a) $N(t) \to 0.16$ as $t \to \infty$; **(b)** The medication never completely disappears from the body; a trace amount remains.
75. (a) $P(0) = 0$; $P(1) = 45{,}455$; $P(3) = 55{,}556$; $P(8) = 29{,}197$; **(b)** $P(t) \to 0$ as $t \to \infty$; **(c)** In time, no one lives in Lordsburg. **77.** $58{,}926$ at $t \approx 2.12$ months
79. Discussion and Writing **81.** [1.3] Slope
82. [1.4] Slope–intercept equation
83. [1.4] Point–slope equation
84. [1.2] Domain, range, domain, range
85. [1.7] $f(-x) = -f(x)$ **86.** [1.1] x-intercept
87. [1.1] Midpoint formula **88.** [1.3] Vertical lines
89. $y = x^3 + 4$ **91.** $x = -3$ $x = 3$

$y = 2$

$x = -1$

$$f(x) = \frac{2x^3 + x^2 - 8x - 4}{x^3 + x^2 - 9x - 9}$$

93. $(-\infty, -3) \cup (7, \infty)$

Exercise Set 3.5

1. $\{-5, 3\}$ **3.** $[-5, 3]$ **5.** $(-\infty, -5] \cup [3, \infty)$
7. $(-\infty, -3) \cup (0, 3)$ **9.** $(-3, 0) \cup (3, \infty)$

11. $(-\infty, -5) \cup (-3, 2)$ **13.** $(-2, 0] \cup (2, \infty)$
15. $(-4, 1)$ **17.** $(-\infty, -2] \cup [4, \infty)$
19. $(-\infty, -2) \cup (1, \infty)$ **21.** $(-\infty, -5) \cup (5, \infty)$
23. $(-\infty, -2] \cup [2, \infty)$ **25.** $(-\infty, 3) \cup (3, \infty)$
27. $\varnothing$ **29.** $\left(-\infty, -\frac{5}{4}\right) \cup [0, 3]$ **31.** $[-3, -1] \cup [1, \infty)$
33. $(-\infty, -2) \cup (1, 3)$ **35.** $\left[-\sqrt{2}, -1\right] \cup \left[\sqrt{2}, \infty\right)$
37. $(-\infty, -1] \cup \left[\frac{3}{2}, 2\right]$ **39.** $(-\infty, 5]$ **41.** $(-4, \infty)$
43. $\left(-\frac{5}{2}, \infty\right)$ **45.** $\left(-3, -\frac{1}{5}\right] \cup (1, \infty)$
47. $(-\infty, -3) \cup \left[\dfrac{5 - \sqrt{105}}{10}, -\dfrac{1}{3}\right) \cup \left[\dfrac{5 + \sqrt{105}}{10}, \infty\right)$
49. $\left(2, \frac{7}{2}\right]$ **51.** $\left(1 - \sqrt{2}, 0\right) \cup \left(1 + \sqrt{2}, \infty\right)$
53. $(-\infty, -3) \cup (1, 3) \cup \left[\frac{11}{3}, \infty\right)$ **55.** $(-\infty, \infty)$
57. $\left(-3, \dfrac{1 - \sqrt{61}}{6}\right) \cup \left(-\dfrac{1}{2}, 0\right) \cup \left(\dfrac{1 + \sqrt{61}}{6}, \infty\right)$
59. $(-1, 0) \cup \left(\frac{2}{7}, \frac{7}{2}\right)$
61. $\left[-6 - \sqrt{33}, -5\right) \cup \left[-6 + \sqrt{33}, 1\right) \cup (5, \infty)$
63. $(0.408, 2.449)$ **65. (a)** $(10, 200)$; **(b)** $(0, 10) \cup (200, \infty)$
67. $\{n \,|\, 9 \leq n \leq 23\}$ **69.** Left to the student
71. Discussion and Writing **72.** [1.1] $x^2 + (y + 3)^2 = \frac{49}{16}$
73. [1.1] $(x + 2)^2 + (y - 4)^2 = 9$
74. [2.4] **(a)** $(5, -23)$; **(b)** minimum: -23 when $x = 5$;
(c) $[-23, \infty)$ **75.** [2.4] **(a)** $\left(\frac{3}{4}, -\frac{55}{8}\right)$; **(b)** maximum: $-\frac{55}{8}$
when $x = \frac{3}{4}$; **(c)** $\left(-\infty, -\frac{55}{8}\right]$ **77.** $(-\infty, \infty)$
79. $\left[-\sqrt{5}, \sqrt{5}\right]$ **81.** $\left[-\frac{3}{2}, \frac{3}{2}\right]$ **83.** $\left(-\infty, -\frac{1}{4}\right) \cup \left(\frac{1}{2}, \infty\right)$
85. $(-4, -2) \cup (-1, 1)$
87. $x^2 + x - 12 < 0$; answers may vary

Exercise Set 3.6

1. 4.5; $y = 4.5x$ **3.** 36; $y = \dfrac{36}{x}$ **5.** 4; $y = 4x$
7. 4; $y = \dfrac{4}{x}$ **9.** $\dfrac{3}{8}$; $y - \dfrac{3}{8}x$ **11.** 0.54; $y = \dfrac{0.54}{x}$
13. 3.5 hr **15.** 90 g **17.** About 686 kg **19.** 40 lb
21. $66\frac{2}{3}$ cm **23.** 1.92 ft **25.** $y = \dfrac{0.0015}{x^2}$
27. $y = 15x^2$ **29.** $y = xz$ **31.** $y = \frac{3}{10}xz^2$
33. $y = \dfrac{xz}{5wp}$ **35.** 2.5 m **37.** 36 mph
39. About 75 earned runs **41.** Discussion and Writing
43. [1.5]

y

44. [1.7] x-axis, no; y-axis, yes; origin, no
45. [1.7] x-axis, yes; y-axis, no; origin, no

46. [1.7] x-axis, no; y-axis, no; origin, yes

47. $2.32; $2.80   **49.** $\dfrac{\pi}{4}$

Review Exercises: Chapter 3

1. [3.1] Quartic, $0.45x^4$, 0.45, 4
2. [3.1] Constant, -25, -25, 0
3. [3.1] Linear, $-0.5x$, -0.5, 1
4. [3.1] Cubic, $\frac{1}{3}x^3$, $\frac{1}{3}$, 3
5. [3.1] As $x \to \infty$, $f(x) \to -\infty$, and as $x \to -\infty$, $f(x) \to -\infty$.
6. [3.1] As $x \to \infty$, $f(x) \to \infty$, and as $x \to -\infty$, $f(x) \to -\infty$.
7. [3.1] $\frac{2}{3}$, multiplicity 1; -2, multiplicity 3; 5, multiplicity 2
8. [3.1] ± 1, ± 5, each has multiplicity 1
9. [3.1] ± 3, -4, each has multiplicity 1
10. [3.1]

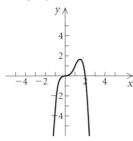

$$f(x) = -x^4 + 2x^3$$

11. [3.1]

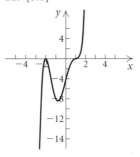

$$g(x) = (x - 1)^3(x + 2)^2$$

12. [3.1]

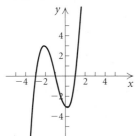

$$h(x) = x^3 + 3x^2 - x - 3$$

13. [3.2]

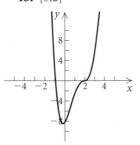

$$f(x) = x^4 - 5x^3 + 6x^2 + 4x - 8$$

14. [3.3]

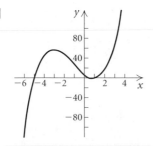

$$g(x) = 2x^3 + 7x^2 - 14x + 5$$

15. [3.1] $f(1) = -4$ and $f(2) = 3$. Since $f(1)$ and $f(2)$ have opposite signs, $f(x)$ has a zero between 1 and 2.
16. [3.1] $f(0) = 2$ and $f(1) = -0.5$. Since $f(0)$ and $f(1)$ have opposite signs, $f(x)$ has a zero between 0 and 1.
17. [3.1] **(a)** 4%; **(b)** 5%
18. [3.2] $Q(x) = 6x^2 + 16x + 52$, $R(x) = 155$; $P(x) = (x - 3)(6x^2 + 16x + 52) + 155$
19. [3.2] $Q(x) = x^3 - 3x^2 + 3x - 2$, $R(x) = 7$; $P(x) = (x + 1)(x^3 - 3x^2 + 3x - 2) + 7$
20. [3.2] $x^2 + 7x + 22$, R 120
21. [3.2] $x^3 + x^2 + x + 1$, R 0
22. [3.2] $x^4 - x^3 + x^2 - x - 1$, R 1   **23.** [3.2] 120
24. [3.2] 0   **25.** [3.2] $-141{,}220$ **26.** [3.2] Yes, no
27. [3.2] No, yes   **28.** [3.2] Yes, no   **29.** [3.2] No, yes
30. [3.2] $f(x) = (x - 1)^2(x + 4)$; -4, 1
31. [3.2] $f(x) = (x - 2)(x + 3)^2$; -3, 2
32. [3.2] $f(x) = (x - 2)^2(x - 5)(x + 5)$; 2, 5, -5
33. [3.2] $f(x) = (x - 1)(x + 1)\left(x - \sqrt{2}\right)\left(x + \sqrt{2}\right)$; $-1, 1, -\sqrt{2}, \sqrt{2}$
34. [3.3] $f(x) = x^3 + 3x^2 - 6x - 8$
35. [3.3] $f(x) = x^3 + x^2 - 4x + 6$
36. [3.3] $f(x) = x^3 - \frac{5}{2}x^2 + \frac{1}{2}$, or $2x^3 - 5x^2 + 1$
37. [3.3] $f(x) = x^4 + \frac{29}{2}x^3 + \frac{135}{2}x^2 + \frac{175}{2}x - \frac{125}{2}$, or $2x^4 + 29x^3 + 135x^2 + 175x - 125$
38. [3.3] $f(x) = x^5 + 4x^4 - 3x^3 - 18x^2$
39. [3.3] $-\sqrt{5}, -i$ **40.** [3.3] $1 - \sqrt{3}, \sqrt{3}$ **41.** [3.3] $\sqrt{2}$
42. [3.3] $f(x) = x^2 - 11$
43. [3.3] $f(x) = x^3 - 6x^2 + x - 6$
44. [3.3] $f(x) = x^4 - 5x^3 + 4x^2 + 2x - 8$
45. [3.3] $f(x) = x^4 - x^2 - 20$
46. [3.3] $f(x) = x^3 + \frac{8}{3}x^2 - x$
47. [3.3] $\pm\frac{1}{4}, \pm\frac{1}{2}, \pm\frac{3}{4}, \pm 1, \pm\frac{3}{2}, \pm 2, \pm 3, \pm 4, \pm 6, \pm 12$
48. [3.3] $\pm\frac{1}{3}, \pm 1$
49. [3.3] $\pm 1, \pm 2, \pm 3, \pm 4, \pm 6, \pm 8, \pm 12, \pm 24$
50. [3.3] **(a)** Rational: $0, -2, \frac{1}{3}, 3$; other: none; **(b)** $f(x) = x(3x - 1)(x + 2)^2(x - 3)$
51. [3.3] **(a)** Rational: 2; other: $\pm\sqrt{3}$; **(b)** $f(x) = (x - 2)\left(x + \sqrt{3}\right)\left(x - \sqrt{3}\right)$
52. [3.3] **(a)** Rational: -1, 1; other: $3 \pm i$; **(b)** $f(x) = (x + 1)(x - 1)(x - 3 - i)(x - 3 + i)$
53. [3.3] **(a)** Rational: -5; other: $1 \pm \sqrt{2}$; **(b)** $f(x) = (x + 5)\left(x - 1 - \sqrt{2}\right)\left(x - 1 + \sqrt{2}\right)$
54. [3.3] **(a)** Rational: $\frac{2}{3}$, 1; other: none; **(b)** $f(x) = (3x - 2)(x - 1)^2$
55. [3.3] **(a)** Rational: 2; other: $1 \pm \sqrt{5}$; **(b)** $f(x) = (x - 2)^3\left(x - 1 + \sqrt{5}\right)\left(x - 1 - \sqrt{5}\right)$
56. [3.3] **(a)** Rational: $-4, 0, 3, 4$; other: none; **(b)** $f(x) = x^2(x + 4)^2(x - 3)(x - 4)$
57. [3.3] **(a)** Rational: $\frac{5}{2}$, 1; other: none; **(b)** $f(x) = (2x - 5)(x - 1)^4$
58. [3.3] 3 or 1; 0   **59.** [3.3] 4 or 2 or 0; 2 or 0
60. [3.3] 3 or 1; 0

61. [3.4] Domain: $(-\infty, -2) \cup (-2, \infty)$;
x-intercepts: $(-\sqrt{5}, 0)$ and $(\sqrt{5}, 0)$, y-intercept: $\left(0, -\frac{5}{2}\right)$

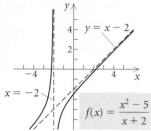

$$y = x - 2$$
$$x = -2$$
$$f(x) = \frac{x^2 - 5}{x + 2}$$

62. [3.4] Domain: $(-\infty, 2) \cup (2, \infty)$; x-intercepts: none, y-intercept: $\left(0, \frac{5}{4}\right)$

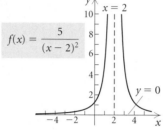

$$f(x) = \frac{5}{(x-2)^2}$$
$$x = 2$$
$$y = 0$$

63. [3.4] Domain: $(-\infty, -4) \cup (-4, 5) \cup (5, \infty)$;
x-intercepts: $(-3, 0)$ and $(2, 0)$, y-intercept: $\left(0, \frac{3}{10}\right)$

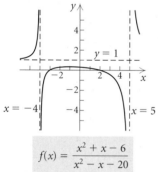

$$y = 1$$
$$x = -4$$
$$x = 5$$
$$f(x) = \frac{x^2 + x - 6}{x^2 - x - 20}$$

64. [3.4] Domain: $(-\infty, -3) \cup (-3, 5) \cup (5, \infty)$;
x-intercept: $(2, 0)$, y-intercept: $\left(0, \frac{2}{15}\right)$

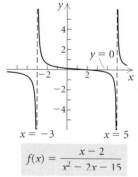

$$y = 0$$
$$x = -3$$
$$x = 5$$
$$f(x) = \frac{x - 2}{x^2 - 2x - 15}$$

65. [3.4] $f(x) = \dfrac{1}{x^2 - x - 6}$ **66.** [3.4] $f(x) = \dfrac{4x^2 + 12x}{x^2 - x - 6}$

67. [3.4] **(a)** $N(t) \to 0.0875$ as $t \to \infty$; **(b)** The medication never completely disappears from the body; a trace amount remains. **68.** [3.5] $(-3, 3)$ **69.** [3.5] $\left(-\infty, -\frac{1}{2}\right) \cup (2, \infty)$
70. [3.5] $[-4, 1] \cup [2, \infty)$ **71.** [3.5] $\left(-\infty, -\frac{14}{3}\right) \cup (-3, \infty)$
72. [3.1], [3.5] **(a)** $t = 7$; **(b)** $(2, 3)$

73. [3.5] $\left[\dfrac{5 - \sqrt{15}}{2}, \dfrac{5 + \sqrt{15}}{2}\right]$ **74.** [3.6] $y = 4x$

75. [3.6] $y = \dfrac{2500}{x}$ **76.** [3.6] 20 min **77.** [3.6] 500 watts

78. [3.6] About 78 **79.** [3.6] $y = \dfrac{48}{x^2}$ **80.** [3.6] $y = \dfrac{xz^2}{10w}$

81. [3.1] **(a)** $-2.637, 1.137$; **(b)** relative maximum: 7.125 at $x = -0.75$; **(c)** none; **(d)** domain: all real numbers; range: $(-\infty, 7.125]$
82. [3.1] **(a)** $-3, -1.414, 1.414$; **(b)** relative maximum: 2.303 at $x = -2.291$; **(c)** relative minimum: -6.303 at $x = 0.291$; **(d)** domain: all real numbers; range: all real numbers
83. [3.1] **(a)** $0, 1, 2$; **(b)** relative maximum: 0.202 at $x = 0.610$; **(c)** relative minima: 0 at $x = 0, -0.620$ at $x = 1.640$; **(d)** domain: all real numbers; range: $[-0.620, \infty)$
84. [3.1] **(a)** Linear: $f(x) = 0.5408695652x - 30.30434783$; quadratic: $f(x) = 0.0030322581x^2 - 0.5764516129x + 57.53225806$; cubic: $f(x) = 0.0000247619x^3 - 0.0112857143x^2 + 2.002380952x - 82.14285714$; **(b)** the cubic function; **(c)** 298, 498
85. Discussion and Writing [3.1], [3.4] A polynomial function is a function that can be defined by a polynomial expression. A rational function is a function that can be defined as a quotient of two polynomials.
86. Discussion and Writing [3.4] Vertical asymptotes occur at any x-values that make the denominator zero. The graph of a rational function does not cross any vertical asymptotes. Horizontal asymptotes occur when the degree of the numerator is less than or equal to the degree of the denominator. Oblique asymptotes occur when the degree of the numerator is 1 greater than the degree of the denominator. Graphs of rational functions may cross horizontal or oblique asymptotes.
87. [3.1] 9% **88.** [3.5] $\left(-\infty, -1 - \sqrt{6}\right] \cup \left[-1 + \sqrt{6}, \infty\right)$
89. [3.5] $\left(-\infty, -\frac{1}{2}\right) \cup \left(\frac{1}{2}, \infty\right)$
90. [3.3] $\{1 + i, 1 - i, i, -i\}$ **91.** [3.5] $(-\infty, 2)$
92. [3.3] $(x - 1)\left(x + \dfrac{1}{2} - \dfrac{\sqrt{3}}{2}i\right)\left(x + \dfrac{1}{2} + \dfrac{\sqrt{3}}{2}i\right)$
93. [3.2] 7 **94.** [3.2] -4 **95.** [3.5] $(-\infty, -5] \cup [2, \infty)$
96. [3.5] $(-\infty, 1.1] \cup [2, \infty)$ **97.** [3.5] $\left(-1, \frac{3}{7}\right)$

Test: Chapter 3

1. [3.1] Quartic, $-x^4, -1, 4$ **2.** [3.1] Linear, $-4.7x, -4.7, 1$
3. [3.1] $0, \frac{5}{3}$, each has multiplicity 1; 3, multiplicity 2; -1, multiplicity 3

4. [3.1]

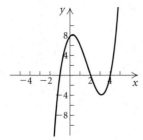

$$f(x) = x^3 - 5x^2 + 2x + 8$$

5. [3.1]

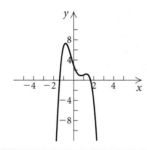

$$f(x) = -2x^4 + 3x^3 + 3x^2 - 6x + 3$$

6. [3.1] $f(0) = 3$ and $f(2) = -17$. Since $f(0)$ and $f(2)$ have opposite signs, $f(x)$ has a zero between 0 and 2.

7. [3.1] $g(-2) = 5$ and $g(-1) = 1$. Both $g(-2)$ and $g(-1)$ are positive. We cannot use the intermediate value theorem to determine if there is a zero between -2 and -1.

8. [3.1] 3388; 5379; 3514

9. [3.2] $Q(x) = x^3 + 4x^2 + 4x + 6$, $R(x) = 1$; $P(x) = (x - 1)(x^3 + 4x^2 + 4x + 6) + 1$

10. [3.2] $3x^2 + 15x + 63$, R 322   **11.** [3.2] -115

12. [3.2] Yes   **13.** [3.2] $f(x) = x^4 - 27x^2 - 54x$

14. [3.3] $-\sqrt{3}, 2 + i$

15. [3.3] $f(x) = x^3 + 10x^2 + 9x + 90$

16. [3.3] $f(x) = x^5 - 2x^4 - x^3 + 6x^2 - 6x$

17. [3.3] $\pm 1, \pm 2, \pm 3, \pm 4, \pm 6, \pm 12, \pm \frac{1}{2}, \pm \frac{3}{2}$

18. [3.3] $\pm \frac{1}{10}, \pm \frac{1}{5}, \pm \frac{1}{2}, \pm 1, \pm \frac{5}{2}, \pm 5$

19. [3.3] **(a)** Rational: -1; other: $\pm \sqrt{5}$; **(b)** $f(x) = (x + 1)(x - \sqrt{5})(x + \sqrt{5})$

20. [3.3] **(a)** Rational: $-\frac{1}{2}, 1, 2, 3$; other: none; **(b)** $g(x) = (2x + 1)(x - 1)(x - 2)(x - 3)$

21. [3.3] **(a)** Rational: -4; other: $\pm 2i$; **(b)** $h(x) = (x - 2i)(x + 2i)(x + 4)$

22. [3.3] **(a)** Rational: $\frac{2}{3}, 1$; other: none; **(b)** $f(x) = (x - 1)^3(3x - 2)$

23. [3.3] 2 or 0; 2 or 0

24. [3.4] Domain: $(-\infty, 3) \cup (3, \infty)$; x-intercepts: none, y-intercept: $\left(0, \frac{2}{9}\right)$;

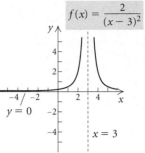

$$f(x) = \frac{2}{(x - 3)^2}$$

25. [3.4] Domain: $(-\infty, -1) \cup (-1, 4) \cup (4, \infty)$; x-intercept: $(-3, 0)$, y-intercept: $\left(0, -\frac{3}{4}\right)$;

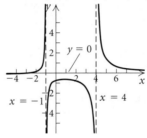

$$f(x) = \frac{x + 3}{x^2 - 3x - 4}$$

26. [3.4] Answers may vary; $f(x) = \dfrac{x + 4}{x^2 - x - 2}$

27. [3.5] $\left(-\infty, -\frac{1}{2}\right) \cup (3, \infty)$ **28.** [3.5] $(-\infty, 4) \cup \left[\frac{13}{2}, \infty\right)$

29. [3.4] **(a)** 6 sec; **(b)** $(1, 3)$ **30.** [3.6] $y = \dfrac{30}{x}$

31. [3.6] 50 ft   **32.** [3.6] $y = \dfrac{50xz^2}{w}$

33. [3.1] $(-\infty, -4] \cup [3, \infty)$

Chapter 4

Exercise Set 4.1

1. $\{(8, 7), (8, -2), (-4, 3), (-8, 8)\}$ **3.** $\{(-1, -1), (4, -3)\}$

5. $x = 4y - 5$ **7.** $y^3x = -5$ **9.** $y = x^2 - 2x$

11.

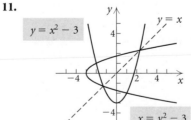

13.

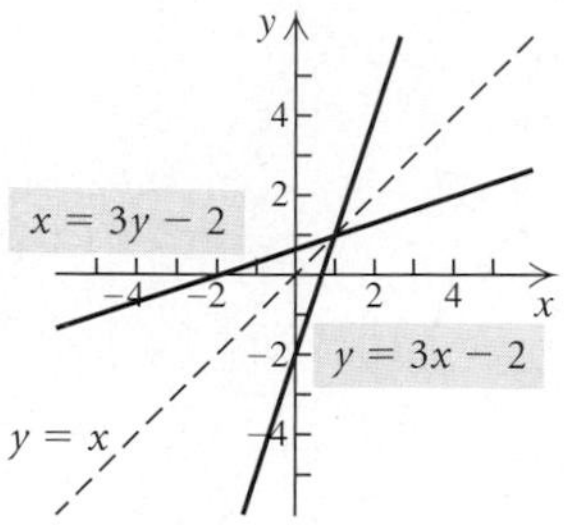

15.

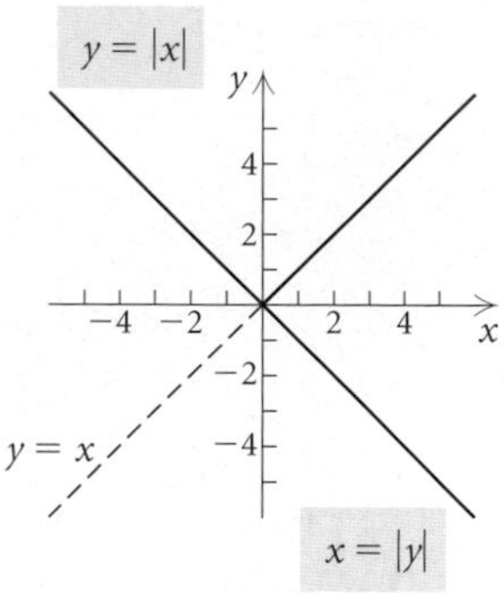

17. Assume $f(a) = f(b)$ for any numbers a and b in the domain of f. Since $f(a) = \frac{1}{3}a - 6$ and $f(b) = \frac{1}{3}b - 6$, we have

$\frac{1}{3}a - 6 = \frac{1}{3}b - 6$

$\frac{1}{3}a = \frac{1}{3}b$ Adding 6

$a = b.$ Multiplying by 3

Thus, if $f(a) = f(b)$, then $a = b$ and f is one-to-one.

19. Assume $f(a) = f(b)$ for any numbers a and b in the domain of f. Since $f(a) = a^3 + \frac{1}{2}$ and $f(b) = b^3 + \frac{1}{2}$, we have

$a^3 + \frac{1}{2} = b^3 + \frac{1}{2}$

$a^3 = b^3$ Subtracting $\frac{1}{2}$

$a = b.$ Taking the cube root

Thus, if $f(a) = f(b)$, then $a = b$ and f is one-to-one.

21. Find two numbers a and b for which $a \neq b$ and $g(a) = g(b)$. Two such numbers are -2 and 2, because $g(-2) = g(2) = -3$. Thus, g is not one-to-one.

23. Find two numbers a and b for which $a \neq b$ and $g(a) = g(b)$. Two such numbers are -1 and 1, because $g(-1) = g(1) = 0$. Thus, g is not one-to-one.

25. Yes **27.** No **29.** No **31.** Yes **33.** Yes
35. No **37.** No **39.** Yes **41.** No **43.** No

45. **(a)** One-to-one; **(b)** $f^{-1}(x) = x - 4$

47. **(a)** One-to-one; **(b)** $f^{-1}(x) = \dfrac{x + 1}{2}$

49. **(a)** One-to-one; **(b)** $f^{-1}(x) = \dfrac{4}{x} - 7$

51. **(a)** One-to-one; **(b)** $f^{-1}(x) = \dfrac{3x + 4}{x - 1}$

53. **(a)** One-to-one; **(b)** $f^{-1}(x) = \sqrt[3]{x + 1}$

55. **(a)** Not one-to-one; **(b)** does not have an inverse that is a function

57. **(a)** One-to-one; **(b)** $f^{-1}(x) = \sqrt{\dfrac{x + 2}{5}}$

59. **(a)** One-to-one; **(b)** $f^{-1}(x) = x^2 - 1, x \geq 0$

61. $\frac{1}{3}x$ **63.** $-x$ **65.** $x^3 + 5$

67.

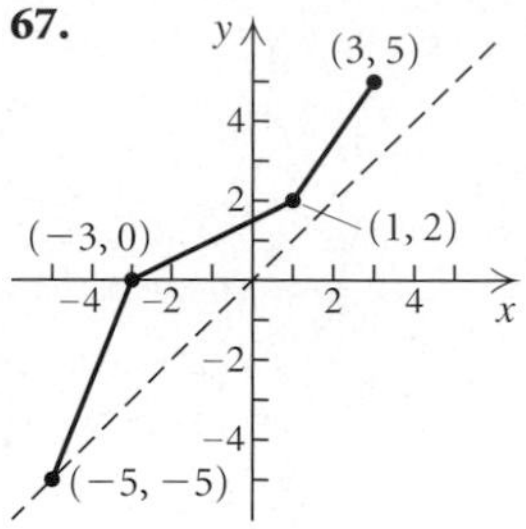

69.

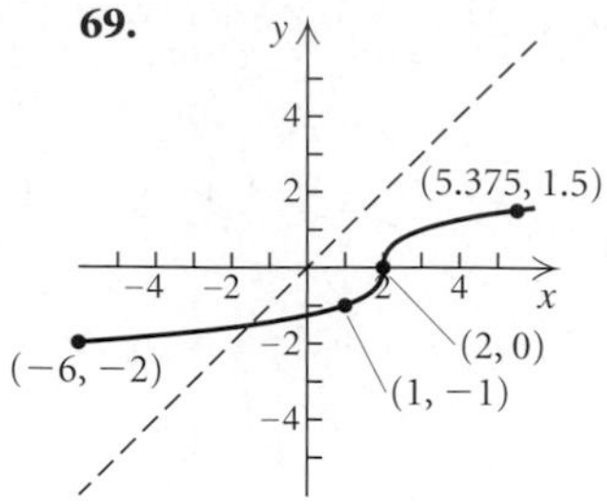

71.

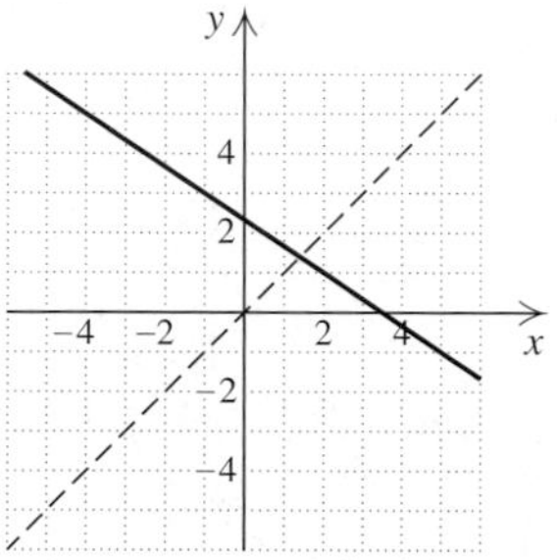

73. $f^{-1}(f(x)) = f^{-1}\left(\frac{7}{8}x\right) = \frac{8}{7} \cdot \frac{7}{8}x = x;$
$f(f^{-1}(x)) = f\left(\frac{8}{7}x\right) = \frac{7}{8} \cdot \frac{8}{7}x = x$

75. $f^{-1}(f(x)) = f^{-1}\left(\dfrac{1 - x}{x}\right) = \dfrac{1}{\dfrac{1 - x}{x} + 1} = \dfrac{1}{\dfrac{1 - x + x}{x}} = \dfrac{1}{\dfrac{1}{x}} = 1 \cdot \dfrac{x}{1} = x;\ f(f^{-1}(x)) = f\left(\dfrac{1}{x + 1}\right) = \dfrac{1 - \dfrac{1}{x + 1}}{\dfrac{1}{x + 1}} = \dfrac{\dfrac{x + 1 - 1}{x + 1}}{\dfrac{1}{x + 1}} = \dfrac{x}{x + 1} \cdot \dfrac{x + 1}{1} = x$

77. $f^{-1}(f(x)) = f^{-1}\left(\dfrac{2}{5}x + 1\right) = \dfrac{5\left(\dfrac{2}{5}x + 1\right) - 5}{2} = \dfrac{2x + 5 - 5}{2} = \dfrac{2x}{2} = x;\ f(f^{-1}(x)) = f\left(\dfrac{5x - 5}{2}\right) = \dfrac{2}{5}\left(\dfrac{5x - 5}{2}\right) + 1 = x - 1 + 1 = x$

79. $f^{-1}(x) = \frac{1}{5}x + \frac{3}{5}$; domain of f and f^{-1}: $(-\infty, \infty)$; range of f and f^{-1}: $(-\infty, \infty)$;

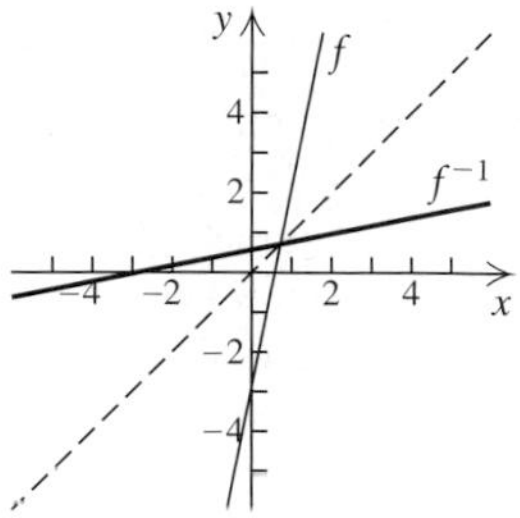

81. $f^{-1}(x) = \dfrac{2}{x}$; domain of f and f^{-1}: $(-\infty, 0) \cup (0, \infty)$; range of f and f^{-1}: $(-\infty, 0) \cup (0, \infty)$;

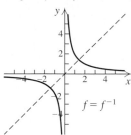

$f = f^{-1}$

83. $f^{-1}(x) = \sqrt[3]{3x + 6}$; domain of f and f^{-1}: $(-\infty, \infty)$; range of f and f^{-1}: $(-\infty, \infty)$

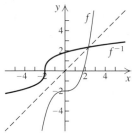

85. $f^{-1}(x) = \dfrac{3x + 1}{x - 1}$; domain of f: $(-\infty, 3) \cup (3, \infty)$; range of f: $(-\infty, 1) \cup (1, \infty)$; domain of f^{-1}: $(-\infty, 1) \cup (1, \infty)$; range of f^{-1}: $(-\infty, 3) \cup (3, \infty)$;

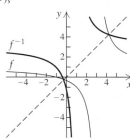

87. 5; a **89.** **(a)** $36, 40, 44, 52, 60$; **(b)** $g^{-1}(x) = \dfrac{x}{2} - 12$; **(c)** $6, 8, 10, 14, 18$

91.
$$y_1 = 0.8x + 1.7,$$
$$y_2 = \dfrac{x - 1.7}{0.8}$$
Domain and range of both f and f^{-1}: all real numbers

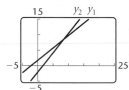

93.
$$y_1 = \tfrac{1}{2}x - 4,$$
$$y_2 = 2x + 8$$
Domain and range of both f and f^{-1}: all real numbers

95.
$$y_1 = \sqrt{x - 3},$$
$$y_2 = x^2 + 3, \ x \geqslant 0$$
Domain of f: $[3, \infty)$, range of f: $[0, \infty)$; domain of f^{-1}: $[0, \infty)$, range of f^{-1}: $[3, \infty)$

97. $y_1 = x^2 - 4, \ x \geqslant 0$; $y_2 = \sqrt{4 + x}$ Domain of f: $[0, \infty)$, range of f: $[-4, \infty)$; domain of f^{-1}: $[-4, \infty)$, range of f^{-1}: $[0, \infty)$

99. $y_1 = (3x - 9)^3$, $y_2 = \dfrac{\sqrt[3]{x} + 9}{3}$ Domain and range of both f and f^{-1}: all real numbers

101. **(a)** $0.5, 11.5, 22.5, 55.5, 72$; **(b)** $D^{-1}(r) = \dfrac{10r - 5}{11}$; the speed, in miles per hour, that the car is traveling when the reaction distance is r feet; **(c)** $y_1 = \dfrac{11x + 5}{10}$, $y_2 = \dfrac{10x - 5}{11}$

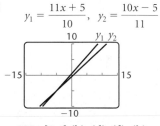

103. Discussion and Writing   **105.** $[2.4]$ (b), (d), (f), (h)
106. $[2.4]$ (a), (c), (e), (g)   **107.** $[2.4]$ (a)   **108.** $[2.4]$ (d)
109. $[2.4]$ (f)   **110.** $[2.4]$ (a), (b), (c), (d)
111. $f(x) = x^2 - 3$, for inputs $x \geq 0$; $f^{-1}(x) = \sqrt{x + 3}$, for inputs $x \geq -3$ **113.** Answers may vary. $f(x) = 3/x$, $f(x) = 1 - x$, $f(x) = x$ **115.** If the graph of $y = f^{-1}(x)$ is symmetric with respect to $y = x$, then $f^{-1}(x) = f(x)$.

Exercise Set 4.2

1. 54.5982   **3.** 0.0856   **5.** (f)   **7.** (e)   **9.** (a)

11.

$f(x) = 3^x$

13.

$f(x) = 6^x$

15.

$f(x) = \left(\frac{1}{4}\right)^x$

17.

$y = -2^x$

19.

$f(x) = -0.25^x + 4$

21.

$f(x) = 1 + e^{-x}$

23.

$y = \frac{1}{4}e^x$

25.

$f(x) = 1 - e^{-x}$

27. Shift the graph of $y = 2^x$ left 1 unit.

$f(x) = 2^{x+1}$

29. Shift the graph of $y = 2^x$ down 3 units.

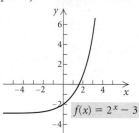

$f(x) = 2^x - 3$

31. Reflect the graph of $y = 3^x$ across the y-axis, then across the x-axis, and then shift it up 4 units.

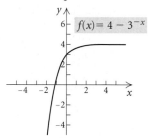

$f(x) = 4 - 3^{-x}$

33. Shift the graph of $y = \left(\frac{3}{2}\right)^x$ right 1 unit.

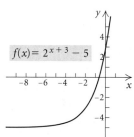

$f(x) = \left(\frac{3}{2}\right)^{x-1}$

35. Shift the graph of $y = 2^x$ left 3 units, and then down 5 units.

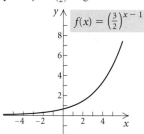

$f(x) = 2^{x+3} - 5$

37. Shrink the graph of $y = e^x$ horizontally.

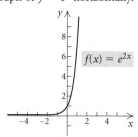

$f(x) = e^{2x}$

39. Shift the graph of $y = e^x$ left 1 unit and reflect it across the y-axis.

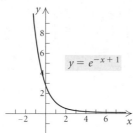

$y = e^{-x+1}$

41. Reflect the graph of $y = e^x$ across the y-axis, then across the x-axis, then shift it up 1 unit, and then stretch it vertically.

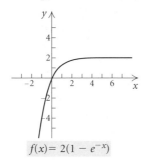

$f(x) = 2(1 - e^{-x})$

43. (a) $A(t) = 82{,}000(1.01125)^{4t}$; (b) $82{,}000, \$89{,}677.22$, $\$102{,}561.54, \$128{,}278.90$ **45.** $\$4930.86$ **47.** $\$3247.30$
49. $\$153{,}610.15$ **51.** $\$76{,}305.59$ **53.** $\$26{,}086.69$
55. (a) 350,000; 233,333; 69,136; 6070;
(b)

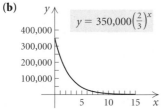

$y = 350{,}000\left(\frac{2}{3}\right)^x$

57. 23.3%; 31.3%; 41.9%
59. 2003: $E = \$845.5$ million, $I = \$3807.2$ million;
2005: $E = \$1258.5$ million, $I = \$7676.8$ million;
2008: $E = \$2285.2$ million, $I = \$21{,}980.9$ million
61. $\$1800; \$1440; \$1152; \$589.82; \$193.27$
63. About 71.5 billion ft^3, 78.2 billion ft^3
65. About 63%   **67.** (c)   **69.** (a)   **71.** (1)   **73.** (g)
75. (i)   **77.** (k)   **79.** (m)   **81.** Discussion and Writing
83. Discussion and Writing   **85.** [2.2] $31 - 22i$
86. [2.2] $\frac{1}{2} - \frac{1}{2}i$ **87.** [2.3] $\left(-\frac{1}{2}, 0\right), (7, 0); -\frac{1}{2}, 7$
88. [3.3] $(1, 0); 1$ **89.** [3.1] $(-1, 0), (0, 0), (1, 0); -1, 0, 1$
90. [3.1] $(-4, 0), (0, 0), (3, 0); -4, 0, 3$ **91.** [3.1] $-8, 0, 2$
92. [3.1] $\dfrac{5 \pm \sqrt{97}}{6}$ **93.** $\pi^7; 70^{80}$

Visualizing the Graph

1. J   **2.** F   **3.** H   **4.** B   **5.** E   **6.** A   **7.** C
8. I   **9.** D   **10.** G

Exercise Set 4.3

1.

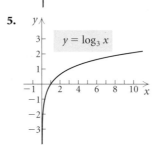

$x = 3^y$

3.

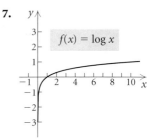

$x = \left(\frac{1}{2}\right)^y$

5.

$y = \log_3 x$

7.

$f(x) = \log x$

9. 4   **11.** 3   **13.** -3 **15.** -2 **17.** 0   **19.** 1
21. 4   **23.** $\frac{1}{4}$ **25.** -7 **27.** $\frac{1}{2}$ **29.** $\frac{3}{4}$ **31.** 0   **33.** $\frac{1}{2}$
35. $\log_{10} 1000 = 3$ **37.** $\log_8 2 = \frac{1}{3}$ **39.** $\log_e t = 3$
41. $\log_e 7.3891 = 2$ **43.** $\log_p 3 = k$ **45.** $5^1 = 5$
47. $10^{-2} = 0.01$ **49.** $e^{3.4012} = 30$ **51.** $a^{-x} = M$
53. $a^x = T^3$ **55.** 0.4771   **57.** 2.7259   **59.** -0.2441
61. Does not exist   **63.** 0.6931   **65.** 6.6962
67. Does not exist   **69.** 3.3219   **71.** -0.2614
73. 0.7384   **75.** 2.2619   **77.** 0.5880
79.

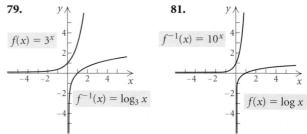

$f(x) = 3^x$
$f^{-1}(x) = \log_3 x$

81.

$f^{-1}(x) = 10^x$
$f(x) = \log x$

83. Shift the graph of $y = \log_2 x$ left 3 units. Domain: $(-3, \infty)$; vertical asymptote: $x = -3$;

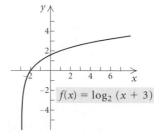

$f(x) = \log_2 (x + 3)$

85. Shift the graph of $y = \log_3 x$ down 1 unit. Domain: $(0,\infty)$; vertical asymptote: $x = 0$;

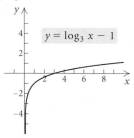

$y = \log_3 x - 1$

87. Stretch the graph of $y = \ln x$ vertically. Domain: $(0,\infty)$; vertical asymptote: $x = 0$;

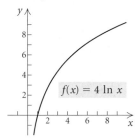

$f(x) = 4 \ln x$

89. Reflect the graph of $y = \ln x$ across the x-axis and shift it up 2 units. Domain: $(0,\infty)$; vertical asymptote: $x = 0$;

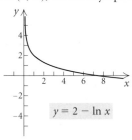

$y = 2 - \ln x$

91. (a) 2.5 ft/sec; (b) 2.3 ft/sec; (c) 2.1 ft/sec; (d) 1.8 ft/sec; (e) 2.3 ft/sec; (f) 2.2 ft/sec; (g) 3.4 ft/sec; (h) 3.0 ft/sec
93. (a) 78%; (b) 67.5%, 57%
95. (a) 10^{-7}; (b) 4.0×10^{-6}; (c) 6.3×10^{-4}; (d) 1.6×10^{-5}
97. (a) 34 decibels; (b) 64 decibels; (c) 60 decibels; (d) 90 decibels **99.** Discussion and Writing
101. [1.4] $m = 0$; y-intercept: $(0,6)$
102. [1.4] $m = \frac{3}{10}$; y-intercept: $\left(0, -\frac{7}{5}\right)$
103. [1.4] $m = 2$; y-intercept: $\left(0, -\frac{3}{13}\right)$
104. [1.4] Slope is not defined; no y-intercept
105. [3.2] -280 **106.** [3.2] -4
107. [3.3] $f(x) = x^3 - 7x$
108. [3.3] $f(x) = x^3 - x^2 + 16x - 16$ **109.** 3
111. $(0,\infty)$ **113.** $(-\infty,0) \cup (0,\infty)$ **115.** $\left(-\frac{5}{2}, -2\right)$
117. (d) **119.** (b)

Exercise Set 4.4

1. $\log_3 81 + \log_3 27$ **3.** $\log_5 5 + \log_5 125$
5. $\log_t 8 + \log_t Y$ **7.** $\ln x + \ln y$ **9.** $3 \log_b t$

11. $8 \log y$ **13.** $-6 \log_c K$ **15.** $\frac{1}{3} \ln 4$
17. $\log_t M - \log_t 8$ **19.** $\log x - \log y$ **21.** $\ln r - \ln s$
23. $\log_a 6 + \log_a x + 5 \log_a y + 4 \log_a z$
25. $2 \log_b p + 5 \log_b q - 4 \log_b m - 9$
27. $\ln 2 - \ln 3 - 3 \ln x - \ln y$
29. $\frac{3}{2} \log r + \frac{1}{2} \log t$ **31.** $3 \log_a x - \frac{5}{2} \log_a p - 4 \log_a q$
33. $2 \log_a m + 3 \log_a n - \frac{3}{4} - \frac{5}{4} \log_a b$ **35.** $\log_a 150$
37. $\log 100 = 2$ **39.** $\log m^3 \sqrt{n}$
41. $\log_a x^{-5/2} y^4$, or $\log_a \dfrac{y^4}{x^{5/2}}$ **43.** $\ln x$ **45.** $\ln (x - 2)$
47. $\log \dfrac{x - 7}{x - 2}$ **49.** $\ln \dfrac{x}{(x^2 - 25)^3}$ **51.** $\ln \dfrac{2^{11/5} x^9}{y^8}$
53. -0.74 **55.** 1.991 **57.** 0.356 **59.** 4.827
61. -1.792 **63.** 0.099 **65.** 3 **67.** $|x - 4|$ **69.** $4x$
71. w **73.** $8t$ **75.** $\frac{1}{2}$ **77.** Discussion and Writing
79. [3.1] Quartic **80.** [4.2] Exponential
81. [1.4] Linear (constant) **82.** [4.2] Exponential
83. [3.4] Rational **84.** [4.3] Logarithmic
85. [3.1] Cubic **86.** [3.4] Rational **87.** [1.4] Linear
88. [2.4] Quadratic **89.** 4 **91.** $\log_a (x^3 - y^3)$
93. $\frac{1}{2} \log_a (x - y) - \frac{1}{2} \log_a (x + y)$ **95.** 7 **97.** True
99. True **101.** True **103.** -2 **105.** 3
107. $e^{-xy} = \dfrac{a}{b}$

109. $\log_a \left(\dfrac{x + \sqrt{x^2 - 5}}{5} \cdot \dfrac{x - \sqrt{x^2 - 5}}{x - \sqrt{x^2 - 5}} \right)$

$= \log_a \dfrac{5}{5\left(x - \sqrt{x^2 - 5}\right)}$

$= -\log_a \left(x - \sqrt{x^2 - 5}\right)$

Exercise Set 4.5

1. 4 **3.** $\frac{3}{2}$ **5.** 5.044 **7.** $\frac{5}{2}$ **9.** $-3, \frac{1}{2}$ **11.** 0.959
13. 0 **15.** 6.908 **17.** 84.191 **19.** -1.710 **21.** 2.844
23. $-1.567, 1.567$ **25.** 0.347 **27.** 625 **29.** 0.0001
31. e **33.** $\frac{22}{3}$ **35.** 10 **37.** 4 **39.** $\frac{1}{63}$ **41.** 2
43. 5 **45.** $\frac{21}{8}$ **47.** $\frac{8}{7}$ **49.** 0.367 **51.** 0.621
53. -1.532 **55.** 7.062 **57.** 2.444 **59.** $(4.093, 0.786)$
61. $(7.586, 6.684)$ **63.** Discussion and Writing
64. [2.4] (a) $(3, 1)$; (b) $x = 3$; (c) maximum: 1 when $x = 3$
65. [2.4] (a) $(0, -6)$; (b) $x = 0$; (c) minimum: -6 when $x = 0$ **66.** [2.4] (a) $(2, 4)$; (b) $x = 2$; (c) minimum: 4 when $x = 2$ **67.** [2.4] (a) $(-1, -5)$; (b) $x = -1$; (c) maximum: -5 when $x = -1$ **69.** 10
71. $1, e^4$ or $1, 54.598$ **73.** $\frac{1}{3}, 27$ **75.** $1, e^2$ or $1, 7.389$
77. $0, 0.431$ **79.** $-9, 9$ **81.** e^{-2}, e^2 or $0.135, 7.389$
83. $\frac{7}{4}$ **85.** 5 **87.** $a = \frac{2}{3} b$ **89.** 88

Exercise Set 4.6

1. (a) $P(t) = 6.2e^{0.012t}$; (b) 6.5 billion, 6.9 billion; (c) in 21.2 yr; (d) 57.8 yr
3. (a) 27.7 yr; (b) 0.1% per year; (c) 21.0 yr; (d) 346.6 yr;

(e) 1.5% per year; (f) 1.9% per year; (g) 77.0 yr;
(h) 0.7% per year; (i) 63.0 yr; (j) 231.0 yr
5. About 640 yr   7. (a) $P(t) = 10,000e^{0.054t}$;
(b) $10,555; $11,140; $13,100; $17,160; (c) 12.8 yr
9. About 5135 yr   11. (a) 23.1% per minute; (b) 3.15% per year; (c) 7.2 days; (d) 11 yr; (e) 2.8% per year; (f) 0.015% per year; (g) 0.003% per year
13. (a) $P(t) = 3395e^{-0.175t}$; (b) $103; $51; (c) about 2010
15. (a) 167; (b) 500; 1758; 3007; 3449; 3495; (c) as $t \to \infty$, $N(t) \to 3500$; the number approaches 3500 but never actually reaches it.   17. 46.7°F   19. 59.6°F
21. (a) $y = 4.195491964(1.025306189)^x$, or
$y = 4.195491964e^{0.024991289x}$, where x is the number of years after 2000; since $r \approx 0.9954$, the function is a good fit.
(b)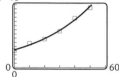
(c) 4.8 million, 7.8 million, 51.1 million
23. (a)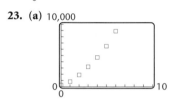
(b) linear: $y = 1429.214286x - 530.0714286$, $r^2 = 0.9641$;
quadratic: $y = 158.0952381x^2 + 480.6428571x + 260.4047619$, $r^2 = 0.9995$; exponential:
$y = 445.8787388(1.736315606)^x$, $r^2 = 0.9361$; according to r^2, the quadratic has the best fit;
(c)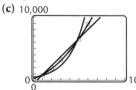
(d) linear: $19,479 million, or $19.479 billion;
quadratic: $37,976 million, or $37.976 billion;
exponential: $1,009,295 million, or $1009.295 billion; the revenue from the quadratic model seems most realistic. Answers may vary.
25. (a) $y = 21.13877717(1.038467095)^x$, where x is the number of years after 1980; (b) 54.3%; 65.6%; (c) in 2012
27. Discussion and Writing   29. [2.6] Multiplication principle for inequalities   30. [4.4] Product rule
31. [2.3] Principle of zero products   32. [2.3] Principle of square roots   33. [4.4] Power rule
34. [2.1] Multiplication principle for equations
35. $14,182.70   37. $166.16
39. $t = -\dfrac{L}{R}\left[\ln\left(1 - \dfrac{iR}{V}\right)\right]$ 41. Linear

Review Exercises: Chapter 4

1. [4.1] $\{(-2.7, 1.3), (-3, 8), (3, -5), (-3, 6), (-5, 7)\}$
2. [4.1] (a) $x = -2y + 3$; (b) $x = 3y^2 + 2y - 1$;
(c) $0.8y^3 - 5.4x^2 = 3y$ 3. [4.1] No   4. [4.1] No
5. [4.1] Yes   6. [4.1] Yes   7. [4.1] (a) Yes;
(b) $f^{-1}(x) = \dfrac{-x + 2}{3}$ 8. [4.1] (a) Yes; (b) $f^{-1}(x) = \dfrac{x + 2}{x - 1}$
9. [4.1] (a) Yes; (b) $f^{-1}(x) = x^2 + 6, x \geq 0$
10. [4.1] (a) Yes; (b) $f^{-1}(x) = \sqrt[3]{x + 8}$ 11. [4.1] (a) No
12. [4.1] (a) Yes; (b) $f^{-1}(x) = \ln x$
13. [4.1] $f^{-1}(f(x)) = f^{-1}(6x - 5) = \dfrac{6x - 5 + 5}{6} = \dfrac{6x}{6} = x$;
$f(f^{-1}(x)) = f\left(\dfrac{x + 5}{6}\right) = 6\left(\dfrac{x + 5}{6}\right) - 5 = x + 5 - 5 = x$
14. [4.1] $f^{-1}(f(x)) = f^{-1}\left(\dfrac{x + 1}{x}\right) = \dfrac{1}{\dfrac{x + 1}{x} - 1} =$
$\dfrac{1}{\dfrac{x + 1 - x}{x}} = \dfrac{1}{\dfrac{1}{x}} = x$; $f(f^{-1}(x)) = f\left(\dfrac{1}{x - 1}\right) =$
$\dfrac{\dfrac{1}{x - 1} + 1}{\dfrac{1}{x - 1}} = \dfrac{\dfrac{1 + x - 1}{x - 1}}{\dfrac{1}{x - 1}} = \dfrac{x}{x - 1} \cdot \dfrac{x - 1}{1} = x$
15. [4.1] $f^{-1}(x) = \dfrac{2 - x}{5}$; domain of f and f^{-1}: $(-\infty, \infty)$;
range of f and f^{-1}: $(-\infty, \infty)$;
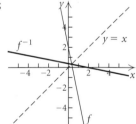
16. [4.1] $f^{-1}(x) = \dfrac{-2x - 3}{x - 1}$;
domain of f: $(-\infty, -2) \cup (-2, \infty)$;
range of f: $(-\infty, 1) \cup (1, \infty)$;
domain of f^{-1}: $(-\infty, 1) \cup (1, \infty)$;
range of f^{-1}: $(-\infty, -2) \cup (-2, \infty)$
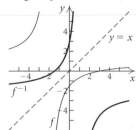

17. [4.1] 657   **18.** [4.1] a
19. [4.2]

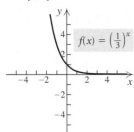

20. [4.2]

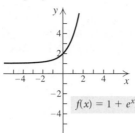

21. [4.2]

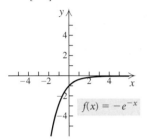

22. [4.3]

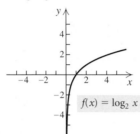

23. [4.3]

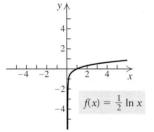

24. [4.3]

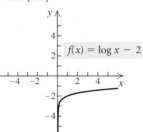

25. [4.2] (c)   **26.** [4.3] (a)   **27.** [4.3] (b)   **28.** [4.2] (f)
29. [4.2] (e)   **30.** [4.3] (d)   **31.** [4.3] 3   **32.** [4.3] 5
33. [4.3] 1   **34.** [4.3] 0   **35.** [4.3] $\frac{1}{4}$ **36.** [4.3] $\frac{1}{2}$
37. [4.3] 0   **38.** [4.3] 1   **39.** [4.3] $\frac{1}{3}$ **40.** [4.3] -2
41. [4.3] $4^2 = x$ **42.** [4.3] $a^k = Q$
43. [4.3] $\log_4 \frac{1}{64} = -3$ **44.** [4.3] $\ln 80 = x$
45. [4.3] 1.0414   **46.** [4.3] -0.6308 **47.** [4.3] 1.0986
48. [4.3] -3.6119 **49.** [4.3] Does not exist
50. [4.3] Does not exist   **51.** [4.3] 1.9746
52. [4.3] 0.5283   **53.** [4.4] $\log_b \dfrac{x^3\sqrt{z}}{y^4}$
54. [4.4] $\ln(x^2 - 4)$ **55.** [4.4] $\frac{1}{4}\ln w + \frac{1}{2}\ln r$
56. [4.4] $\frac{2}{3}\log M - \frac{1}{3}\log N$ **57.** [4.4] 0.477
58. [4.4] 1.699   **59.** [4.4] -0.699 **60.** [4.4] 0.233
61. [4.4] $-5k$ **62.** [4.4] $-6t$ **63.** [4.5] 16   **64.** [4.5] $\frac{1}{5}$
65. [4.5] 4.382   **66.** [4.5] 2   **67.** [4.5] $\frac{1}{7}$ **68.** [4.5] 5
69. [4.5] 4   **70.** [4.5] 9   **71.** [4.5] 1   **72.** [4.5] 3.912
73. [4.2] **(a)** $A(t) = 16,000(1.0105)^{4t}$; **(b)** \$16,000,
\$20,588.51, \$26,415.77, \$33,941.80
74. [4.2] 59.6%; 71.5%   **75.** [4.6] 8.1 yr   **76.** [4.6] 2.3%

77. [4.6] About 2623 yr   **78.** [4.3] 5.6   **79.** [4.3] 6.3
80. [4.3] 30 decibels   **81.** [4.3] **(a)** 1.9 ft/sec; **(b)** 8,553,143
82. [4.6] **(a)** $k = 0.304$; **(b)** $P(t) = 7e^{0.304t}$, where t is the
number of years since 1967; **(c)** 727.9 billion; 3.328 trillion;
(d) 2006
83. [4.6] **(a)** $P(t) = 174.5e^{0.008t}$, where t is the number of
years since 2001; **(b)** 178.7 million; 190.6 million; **(c)** in 17 yr;
(d) 86.6 yr
84. [4.6] **(a)** $y = 11.96557466(1.00877703)^x$;
(b) $y = 11.96557466(1.00877703)^x$ **(c)** 44, 255, 394

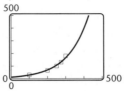

85. [4.1] No   **86.** [4.2], [4.3] **(a)** $y = 5e^{-x}\ln x$

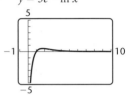

(b) relative maximum: 0.486 at $x = 1.763$; no relative
minimum   **87.** Discussion and Writing [4.4] By the product
rule, $\log_2 x + \log_2 5 = \log_2 5x$, not $\log_2(x + 5)$. Also,
substituting various numbers for x shows that both sides of
the inequality are indeed unequal. You could also graph each
side and show that the graphs do not coincide.
88. Discussion and Writing [4.1] The inverse of a function
$f(x)$ is written $f^{-1}(x)$, whereas $[f(x)]^{-1}$ means $\dfrac{1}{f(x)}$.
89. [4.5] $\frac{1}{64}$, 64   **90.** [4.5] 1   **91.** [4.5] 16
92. [4.3] $(1, \infty)$

Test: Chapter 4

1. [4.1] $\{(5, -2), (3, 4), (-1, 0), (-3, -6)\}$
2. [4.1] No   **3.** [4.1] Yes
4. [4.1] **(a)** Yes; **(b)** $f^{-1}(x) = \sqrt[3]{x - 1}$
5. [4.1] **(a)** Yes; **(b)** $f^{-1}(x) = 1 - x$
6. [4.1] **(a)** Yes; **(b)** $f^{-1}(x) = \dfrac{2x}{1 + x}$ **7.** [4.1] **(a)** No

8. [4.1] $f^{-1}(f(x)) = f^{-1}(-4x + 3) = \dfrac{3 - (-4x + 3)}{4} =$
$\dfrac{4x}{4} = x$; $f(f^{-1}(x)) = f\left(\dfrac{3 - x}{4}\right) = -4\left(\dfrac{3 - x}{4}\right) + 3 =$
$-3 + x + 3 = x$
9. [4.1] $f^{-1}(x) = \dfrac{4x + 1}{x}$; domain of f: $(-\infty, 4) \cup (4, \infty)$;
range of f: $(-\infty, 0) \cup (0, \infty)$;
domain of f^{-1}: $(-\infty, 0) \cup (0, \infty)$;

range of f^{-1}: $(-\infty, 4) \cup (4, \infty)$;

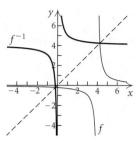

10. [4.2] **11.** [4.3]

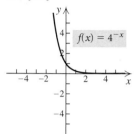 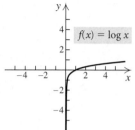

12. [4.2] **13.** [4.3]

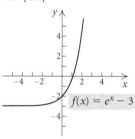 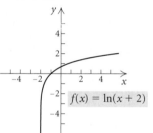

14. [4.3] -5 **15.** [4.3] 1 **16.** [4.3] 0 **17.** [4.3] $\frac{1}{5}$
18. [4.3] $x = e^4$ **19.** [4.3] $x = \log_3 5.4$ **20.** [4.3] 2.7726
21. [4.3] -0.5331 **22.** [4.3] 1.2851 **23.** [4.4] $\log_a \dfrac{x^2 \sqrt{z}}{y}$
24. [4.4] $\frac{2}{5} \ln x + \frac{1}{5} \ln y$ **25.** [4.4] 0.656 **26.** [4.4] $-4t$
27. [4.5] $\frac{1}{2}$ **28.** [4.5] 1 **29.** [4.5] 1 **30.** [4.5] 4.174
31. [4.6] 0.0154 **32.** [4.6] **(a)** 4.5%;
(b) $P(t) = 1000e^{0.045t}$; **(c)** \$1433.33; **(d)** 15.4 yr
33. [4.5] $\frac{27}{8}$

Chapter 5

Visualizing the Graph

1. C **2.** G **3.** D **4.** J **5.** A **6.** F **7.** I
8. B **9.** H **10.** E

Exercise Set 5.1

1. (c) **3.** (f) **5.** (b) **7.** $(-1, 3)$ **9.** $(-1, 1)$
11. No solution **13.** $(-2, 4)$ **15.** Infinitely many

solutions; $\left(x, \dfrac{x-1}{2}\right)$, or $(2y+1, y)$ **17.** $(5, 4)$
19. $(1, -3)$ **21.** $(2, -2)$ **23.** $\left(\frac{39}{11}, -\frac{1}{11}\right)$ **25.** $(1, -1)$
27. $\left(\frac{1}{2}, \frac{3}{4}\right)$ **29.** $(1, 3)$; consistent, independent
31. $(-4, -2)$; consistent, independent
33. $(4y + 2, y)$ or $\left(x, \frac{1}{4}x - \frac{1}{2}\right)$; consistent, dependent
35. $(1, 1)$; consistent, independent
37. $(-3, 0)$; consistent, independent
39. $(10, 8)$; consistent, independent
41. Ice: 200 tons; hot dogs: 4 tons
43. Adult: \$9.50; child: \$4
45. Free rentals: 10; popcorn: 38 **47.** $(15, \$100)$
49. 140 **51.** 6000 **53.** 1.5 servings of spaghetti,
2 servings of lettuce **55.** Boat: 20 km/h; stream: 3 km/h
57. \$6000 at 7%, \$9000 at 9% **59.** 6 lb of French roast, 4 lb
of Kenyan **61.** \$21,428.57 **63.** Left to the student
65. **(a)** $b(x) = 0.1714285714x + 63.48571429$;
$c(x) = 1.007142857x + 47.92857143$;
(b) about 19 yr after 1995 **67.** Discussion and Writing
69. [2.1] About 120 million DVDs **70.** [2.1] \$303 million
71. [2.3] $-2, 6$ **72.** [1.2] 15 **73.** [2.3] $-1, 5$
74. [2.3] 1, 3 **75.** 4 km **77.** First train: 36 km/h;
second train: 54 km/h **79.** $A = \frac{1}{10}, B = -\frac{7}{10}$
81. City: 386.4 mi; highway: 234.6 mi

Exercise Set 5.2

1. $(3, -2, 1)$ **3.** $(-3, 2, 1)$ **5.** $\left(2, \frac{1}{2}, -2\right)$
7. No solution **9.** $\left(\dfrac{11y + 19}{5}, y, \dfrac{9y + 11}{5}\right)$ **11.** $\left(\frac{1}{2}, \frac{2}{3}, -\frac{5}{6}\right)$
13. $(-1, 4, 3)$ **15.** $(1, -2, 4, -1)$ **17.** Under 10 lb: 60;
10 lb up to 15 lb: 70; 15 lb or more: 20 **19.** $1\frac{1}{4}$ servings of
beef, 1 baked potato, $\frac{3}{4}$ serving of strawberries
21. 3%: \$1300; 4%: \$900; 6%: \$2800
23. Orange juice: \$1; bagel: \$1.25; coffee: \$0.75
25. Private automobile: 1850 billion passenger-miles;
bus: 35 billion passenger-miles;
railroad: 14 billion passenger-miles
27. Par-3: 4; par-4: 10; par-5: 4
29. **(a)** $f(x) = \frac{57}{55}x^2 - \frac{692}{55}x + 237$; **(b)** about 346 thousand
31. **(a)** $f(x) = \frac{49}{150}x^2 - \frac{149}{30}x + 594$; **(b)** about 643 lb
33. **(a)** $f(x) = 0.1793514633x^2 - 11.15232206x + 468.8939939$; **(b)** about 836 newspapers,
about 897 newspapers **35.** Discussion and Writing
37. [1.4] Perpendicular **38.** [3.1] The leading-term test
39. [1.2] A vertical line **40.** [4.1] A one-to-one function
41. [3.4] A rational function **42.** [3.6] Inverse variation
43. [3.4] A vertical asymptote
44. [3.4] A horizontal asymptote
45. $\left(-1, \frac{1}{5}, -\frac{1}{2}\right)$ **47.** 180° **49.** $3x + 4y + 2z = 12$
51. $y = -4x^3 + 5x^2 - 3x + 1$
53. Adults: 5; students: 1; children: 94

Exercise Set 5.3

1. 3×2 **3.** 1×4 **5.** 3×3 **7.** $\begin{bmatrix} 2 & -1 & | & 7 \\ 1 & 4 & | & -5 \end{bmatrix}$

9. $\begin{bmatrix} 1 & -2 & 3 & | & 12 \\ 2 & 0 & -4 & | & 8 \\ 0 & 3 & 1 & | & 7 \end{bmatrix}$

11. $3x - 5y = 1,$
$\quad x + 4y = -2$

13. $2x + y - 4z = 12,$
$\quad 3x \quad\;\; + 5z = -1,$
$\quad x - y + z = 2$

15. $\left(\frac{3}{2}, \frac{5}{2}\right)$ **17.** $\left(-\frac{63}{29}, -\frac{114}{29}\right)$ **19.** $\left(-1, \frac{5}{2}\right)$
21. $(0, 3)$ **23.** No solution **25.** $(3y - 2, y)$
27. $(-1, 2, -2)$ **29.** $\left(\frac{3}{2}, -4, 3\right)$ **31.** $(-1, 6, 3)$
33. $\left(\frac{1}{2}z + \frac{1}{2}, -\frac{1}{2}z - \frac{1}{2}, z\right)$ **35.** $(r - 2, -2r + 3, r)$
37. No solution **39.** $(1, -3, -2, -1)$ **41.** 1:00 A.M.
43. $8000 at 8%; $12,000 at 10%; $10,000 at 12%
45. Discussion and Writing **47.** [4.2] Exponential
48. [1.3] Linear **49.** [3.4] Rational **50.** [3.1] Quartic
51. [4.3] Logarithmic **52.** [3.1] Cubic **53.** [1.3] Linear
54. [2.3] Quadratic **55.** $y = 3x^2 + \frac{5}{2}x - \frac{15}{2}$

57. $\begin{bmatrix} 1 & 5 \\ 0 & 1 \end{bmatrix}, \begin{bmatrix} 1 & 0 \\ 0 & 1 \end{bmatrix}$ **59.** $\left(-\frac{4}{3}, -\frac{1}{3}, 1\right)$

61. $\left(-\frac{14}{13}z - 1, \frac{3}{13}z - 2, z\right)$ **63.** $(-3, 3)$

Exercise Set 5.4

1. $x = -3, y = 5$ **3.** $x = -1, y = 1$

5. $\begin{bmatrix} -2 & 7 \\ 6 & 2 \end{bmatrix}$ **7.** $\begin{bmatrix} 1 & 3 \\ 2 & 6 \end{bmatrix}$ **9.** $\begin{bmatrix} 9 & 9 \\ -3 & -3 \end{bmatrix}$

11. $\begin{bmatrix} 11 & 13 \\ 5 & 3 \end{bmatrix}$ **13.** $\begin{bmatrix} -4 & 3 \\ -2 & -4 \end{bmatrix}$ **15.** $\begin{bmatrix} 17 & 9 \\ -2 & 1 \end{bmatrix}$

17. $\begin{bmatrix} 0 & 0 \\ 0 & 0 \end{bmatrix}$ **19.** $\begin{bmatrix} 1 & 2 \\ 4 & 3 \end{bmatrix}$ **21.** $\begin{bmatrix} 1 \\ 40 \end{bmatrix}$

23. $\begin{bmatrix} -10 & 28 \\ 14 & -26 \\ 0 & -6 \end{bmatrix}$ **25.** Not defined **27.** $\begin{bmatrix} 3 & 16 & 3 \\ 0 & -32 & 0 \\ -6 & 4 & 5 \end{bmatrix}$

29. **(a)** $[150 \quad 80 \quad 40]$; **(b)** $[157.5 \quad 84 \quad 42]$;
(c) $[307.5 \quad 164 \quad 82]$, the total budget for each area in June
and July **31.** **(a)** $C = [140 \quad 27 \quad 3 \quad 13 \quad 64]$,
$P = [180 \quad 4 \quad 11 \quad 24 \quad 662]$, $B = [50 \quad 5 \quad 1 \quad 82 \quad 20]$;
(b) $[650 \quad 50 \quad 28 \quad 307 \quad 1448]$, the total nutritional values
of a meal of 1 serving of chicken, 1 cup of potato salad, and
3 broccoli spears

33. **(a)** $\begin{bmatrix} 45.29 & 6.63 & 10.94 & 7.42 & 8.01 \\ 53.78 & 4.95 & 9.83 & 6.16 & 12.56 \\ 47.13 & 8.47 & 12.66 & 8.29 & 9.43 \\ 51.64 & 7.12 & 11.57 & 9.35 & 10.72 \end{bmatrix}$;

(b) $[65 \quad 48 \quad 93 \quad 57]$;
(c) $[12{,}851.86 \quad 1862.1 \quad 3019.81 \quad 2081.9 \quad 2611.56]$;
(d) the total cost, in cents, for each item for the day's meals

35. **(a)** $\begin{bmatrix} 8 & 15 \\ 6 & 10 \\ 4 & 3 \end{bmatrix}$; **(b)** $[3 \quad 1.50 \quad 2]$; **(c)** $[41 \quad 66]$;

(d) the total cost, in dollars, of ingredients for each coffee shop
37. **(a)** $[6 \quad 4.50 \quad 5.20]$; **(b)** $PS = [95.80 \quad 150.60]$

39. $\begin{bmatrix} 2 & -3 \\ 1 & 5 \end{bmatrix} \begin{bmatrix} x \\ y \end{bmatrix} = \begin{bmatrix} 7 \\ -6 \end{bmatrix}$

41. $\begin{bmatrix} 1 & 1 & -2 \\ 3 & -1 & 1 \\ 2 & 5 & -3 \end{bmatrix} \begin{bmatrix} x \\ y \\ z \end{bmatrix} = \begin{bmatrix} 6 \\ 7 \\ 8 \end{bmatrix}$

43. $\begin{bmatrix} 3 & -2 & 4 \\ 2 & 1 & -5 \end{bmatrix} \begin{bmatrix} x \\ y \\ z \end{bmatrix} = \begin{bmatrix} 17 \\ 13 \end{bmatrix}$

45. $\begin{bmatrix} -4 & 1 & -1 & 2 \\ 1 & 2 & -1 & -1 \\ -1 & 1 & 4 & -3 \\ 2 & 3 & 5 & -7 \end{bmatrix} \begin{bmatrix} w \\ x \\ y \\ z \end{bmatrix} = \begin{bmatrix} 12 \\ 0 \\ 1 \\ 9 \end{bmatrix}$

47. Left to the student **49.** Discussion and Writing
51. [2.4] **(a)** $\left(\frac{1}{2}, -\frac{25}{4}\right)$; **(b)** $x = \frac{1}{2}$; **(c)** minimum: $-\frac{25}{4}$;
(d)

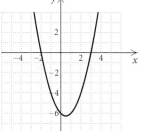

$f(x) = x^2 - x - 6$

52. [2.4] **(a)** $\left(\frac{5}{4}, -\frac{49}{8}\right)$; **(b)** $x = \frac{5}{4}$; **(c)** minimum: $-\frac{49}{8}$;
(d)

$f(x) = 2x^2 - 5x - 3$

53. [2.4] **(a)** $\left(-\frac{3}{2}, \frac{17}{4}\right)$; **(b)** $x = -\frac{3}{2}$; **(c)** maximum: $\frac{17}{4}$;
(d)

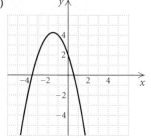

$$f(x) = -x^2 - 3x + 2$$

54. [2.4] **(a)** $\left(\frac{2}{3}, \frac{16}{3}\right)$; **(b)** $x = \frac{2}{3}$; **(c)** maximum: $\frac{16}{3}$;
(d)

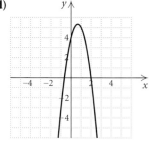

$$f(x) = -3x^2 + 4x + 4$$

55. $(\mathbf{A} + \mathbf{B})(\mathbf{A} - \mathbf{B}) = \begin{bmatrix} -2 & 1 \\ 2 & -1 \end{bmatrix}$; $\mathbf{A}^2 - \mathbf{B}^2 = \begin{bmatrix} 0 & 3 \\ 0 & -3 \end{bmatrix}$

57. $(\mathbf{A} + \mathbf{B})(\mathbf{A} - \mathbf{B}) = \begin{bmatrix} -2 & 1 \\ 2 & -1 \end{bmatrix}$
$$= \mathbf{A}^2 + \mathbf{B}\mathbf{A} - \mathbf{A}\mathbf{B} - \mathbf{B}^2$$

59. $\mathbf{A} + \mathbf{B} =$
$$\begin{bmatrix} a_{11} + b_{11} & a_{12} + b_{12} & a_{13} + b_{13} & \cdots & a_{1n} + b_{1n} \\ a_{21} + b_{21} & a_{22} + b_{22} & a_{23} + b_{23} & \cdots & a_{2n} + b_{2n} \\ a_{31} + b_{31} & a_{32} + b_{32} & a_{33} + b_{33} & \cdots & a_{3n} + b_{3n} \\ \vdots & \vdots & \vdots & & \vdots \\ a_{m1} + b_{m1} & a_{m2} + b_{m2} & a_{m3} + b_{m3} & \cdots & a_{mn} + b_{mn} \end{bmatrix}$$
$$=$$
$$\begin{bmatrix} b_{11} + a_{11} & b_{12} + a_{12} & b_{13} + a_{13} & \cdots & b_{1n} + a_{1n} \\ b_{21} + a_{21} & b_{22} + a_{22} & b_{23} + a_{23} & \cdots & b_{2n} + a_{2n} \\ b_{31} + a_{31} & b_{32} + a_{32} & b_{33} + a_{33} & \cdots & b_{3n} + a_{3n} \\ \vdots & \vdots & \vdots & & \vdots \\ b_{m1} + a_{m1} & b_{m2} + a_{m2} & b_{m3} + a_{m3} & \cdots & b_{mn} + a_{mn} \end{bmatrix}$$
$$= \mathbf{B} + \mathbf{A}$$

61. $(kl)\mathbf{A} = \begin{bmatrix} (kl)a_{11} & (kl)a_{12} & (kl)a_{13} & \cdots & (kl)a_{1n} \\ (kl)a_{21} & (kl)a_{22} & (kl)a_{23} & \cdots & (kl)a_{2n} \\ (kl)a_{31} & (kl)a_{32} & (kl)a_{33} & \cdots & (kl)a_{3n} \\ \vdots & \vdots & \vdots & & \vdots \\ (kl)a_{m1} & (kl)a_{m2} & (kl)a_{m3} & \cdots & (kl)a_{mn} \end{bmatrix}$

$$= \begin{bmatrix} k(la_{11}) & k(la_{12}) & k(la_{13}) & \cdots & k(la_{1n}) \\ k(la_{21}) & k(la_{22}) & k(la_{23}) & \cdots & k(la_{2n}) \\ k(la_{31}) & k(la_{32}) & k(la_{33}) & \cdots & k(la_{3n}) \\ \vdots & \vdots & \vdots & & \vdots \\ k(la_{m1}) & k(la_{m2}) & k(la_{m3}) & \cdots & k(la_{mn}) \end{bmatrix}$$

$$= k \begin{bmatrix} la_{11} & la_{12} & la_{13} & \cdots & la_{1n} \\ la_{21} & la_{22} & la_{23} & \cdots & la_{2n} \\ la_{31} & la_{32} & la_{33} & \cdots & la_{3n} \\ \vdots & \vdots & \vdots & & \vdots \\ la_{m1} & la_{m2} & la_{m3} & \cdots & la_{mn} \end{bmatrix}$$

$$= k(l\mathbf{A})$$

63. $(k + l)\mathbf{A} =$
$$\begin{bmatrix} (k+l)a_{11} & (k+l)a_{12} & (k+l)a_{13} & \cdots & (k+l)a_{1n} \\ (k+l)a_{21} & (k+l)a_{22} & (k+l)a_{23} & \cdots & (k+l)a_{2n} \\ (k+l)a_{31} & (k+l)a_{32} & (k+l)a_{33} & \cdots & (k+l)a_{3n} \\ \vdots & \vdots & \vdots & & \vdots \\ (k+l)a_{m1} & (k+l)a_{m2} & (k+l)a_{m3} & \cdots & (k+l)a_{mn} \end{bmatrix}$$
$$=$$
$$\begin{bmatrix} ka_{11} + la_{11} & ka_{12} + la_{12} & ka_{13} + la_{13} & \cdots & ka_{1n} + la_{1n} \\ ka_{21} + la_{21} & ka_{22} + la_{22} & ka_{23} + la_{23} & \cdots & ka_{2n} + la_{2n} \\ ka_{31} + la_{31} & ka_{32} + la_{32} & ka_{33} + la_{33} & \cdots & ka_{3n} + la_{3n} \\ \vdots & \vdots & \vdots & & \vdots \\ ka_{m1} + la_{m1} & ka_{m2} + la_{m2} & ka_{m3} + la_{m3} & \cdots & ka_{mn} + la_{mn} \end{bmatrix}$$
$$= k\mathbf{A} + l\mathbf{A}$$

Exercise Set 5.5

1. Yes   **3.** No   **5.** $\begin{bmatrix} -3 & 2 \\ 5 & -3 \end{bmatrix}$ **7.** Does not exist

9. $\begin{bmatrix} \frac{2}{5} & -\frac{3}{5} \\ \frac{1}{5} & -\frac{4}{5} \end{bmatrix}$ **11.** $\begin{bmatrix} \frac{3}{8} & -\frac{1}{4} & \frac{1}{8} \\ -\frac{1}{8} & \frac{3}{4} & -\frac{3}{8} \\ -\frac{1}{4} & \frac{1}{2} & \frac{1}{4} \end{bmatrix}$ **13.** Does not exist

15. $\begin{bmatrix} -1 & -1 & -6 \\ 1 & 0 & 2 \\ 0 & 1 & 3 \end{bmatrix}$ **17.** $\begin{bmatrix} 1 & 1 & 2 \\ 1 & 1 & 1 \\ 2 & 3 & 4 \end{bmatrix}$

19. Does not exist   **21.** $\begin{bmatrix} 1 & -2 & 3 & 8 \\ 0 & 1 & -3 & 1 \\ 0 & 0 & 1 & -2 \\ 0 & 0 & 0 & -1 \end{bmatrix}$

23. $\begin{bmatrix} 0.25 & 0.25 & 1.25 & -0.25 \\ 0.5 & 1.25 & 1.75 & -1 \\ -0.25 & -0.25 & -0.75 & 0.75 \\ 0.25 & 0.5 & 0.75 & -0.5 \end{bmatrix}$

25. $(-23, 83)$ **27.** $(-1, 5, 1)$ **29.** $(2, -2)$ **31.** $(0, 2)$

33. $(3, -3, -2)$ **35.** $(-1, 0, 1)$ **37.** $(1, -1, 0, 1)$

39. 50 sausages, 95 hot dogs

41. Topsoil: \$239; mulch: \$179; pea gravel: \$222

43. Left to the student **45.** Left to the student

47. Discussion and Writing **49.** [3.2] -48 **50.** [3.2] 194

51. [2.3] $\dfrac{-1 \pm \sqrt{57}}{4}$ **52.** [2.5] $-3, -2$ **53.** [2.5] 4

54. [2.5] 9 **55.** [3.2] $(x + 2)(x - 1)(x - 4)$

56. [3.2] $(x + 5)(x + 1)(x - 1)(x - 3)$

57. $\mathbf{A}^{-1}$ exists if and only if $x \neq 0$. $\mathbf{A}^{-1} = \begin{bmatrix} \dfrac{1}{x} \end{bmatrix}$

59. $\mathbf{A}^{-1}$ exists if and only if $xyz \neq 0$. $\mathbf{A}^{-1} = \begin{bmatrix} 0 & 0 & \dfrac{1}{z} \\ 0 & \dfrac{1}{y} & 0 \\ \dfrac{1}{x} & 0 & 0 \end{bmatrix}$

Exercise Set 5.6

1. -11 **3.** $x^3 - 4x$ **5.** -109 **7.** $-x^4 + x^2 - 5x$

9. $M_{11} = 6, M_{32} = -9, M_{22} = -29$

11. $A_{11} = 6, A_{32} = 9, A_{22} = -29$

13. -10 **15.** -10 **17.** $M_{41} = -14, M_{33} = 20$

19. $A_{24} = 15, A_{43} = 30$ **21.** 110 **23.** $\left(-\frac{25}{2}, -\frac{11}{2}\right)$

25. $(3, 1)$ **27.** $\left(\frac{1}{2}, -\frac{1}{3}\right)$ **29.** $(1, 1)$ **31.** $\left(\frac{3}{7}, \frac{13}{14}, \frac{33}{14}\right)$

33. $(3, -2, 1)$ **35.** $(1, 3, -2)$ **37.** $\left(\frac{1}{2}, \frac{2}{3}, -\frac{5}{6}\right)$

39. Left to the student **41.** 110 **43.** Left to the student

45. Discussion and Writing **47.** [4.1] $f^{-1}(x) = \dfrac{x - 2}{3}$

48. [4.1] Not one-to-one **49.** [4.1] Not one-to-one

50. [4.1] $f^{-1}(x) = (x - 1)^3$ **51.** [2.2] $5 - 3i$

52. [2.2] $6 - 2i$ **53.** [2.2] $10 - 10i$ **54.** [2.2] $\frac{9}{25} + \frac{13}{25}i$

55. ± 2 **57.** $\left(-\infty, -\sqrt{3}\,\right] \cup \left[\sqrt{3}, \infty\right)$ **59.** -34

61. 4 **63.** Answers may vary. **65.** Answers may vary.

$$\begin{vmatrix} L & -W \\ 2 & 2 \end{vmatrix} \qquad \begin{vmatrix} a & b \\ -b & a \end{vmatrix}$$

67. Answers may vary.

$$\begin{vmatrix} 2\pi r & 2\pi r \\ h & r \end{vmatrix}$$

Exercise Set 5.7

1. (f) **3.** (h) **5.** (g) **7.** (b)

9.

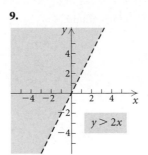

$y > 2x$

11.

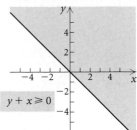

$y + x \geq 0$

13.

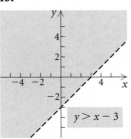

$y > x - 3$

15.

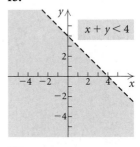

$x + y < 4$

17.

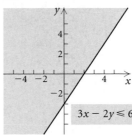

$3x - 2y \leq 6$

19.

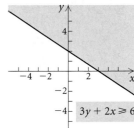

$3y + 2x \geq 6$

21.

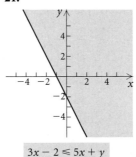

$3x - 2 \leq 5x + y$

23.

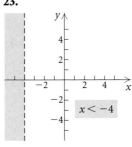

$x < -4$

25.

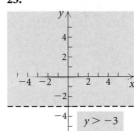

$y > -3$

27.

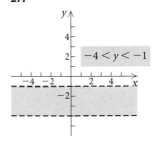

$-4 < y < -1$

29.

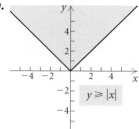

$y \geq |x|$

31. (f) **33.** (a) **35.** (b)

37. $y \leq -x + 4,$
   $y \leq 3x$

39. $x < 2,$
   $y > -1$

41. $y \leq -x + 3,$
   $y \leq x + 1,$
   $x \geq 0,$
   $y \geq 0$

43.

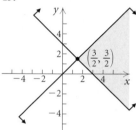

45.

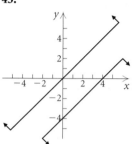

47.

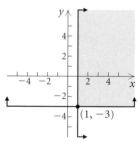

49.

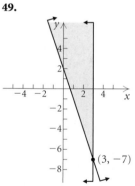

51.

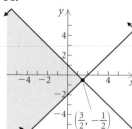

53.

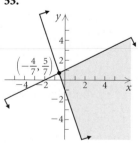

55.

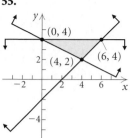

57.

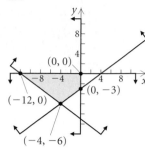

59.

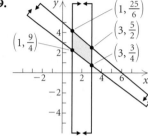

61. Maximum: 179 when $x = 7$ and $y = 0$; minimum: 48 when $x = 0$ and $y = 4$

63. Maximum: 216 when $x = 0$ and $y = 6$; minimum: 0 when $x = 0$ and $y = 0$

65. Maximum income of $18 is achieved when 100 of each type of biscuit is made.

67. Maximum profit of $11,000 is achieved by producing 100 units of lumber and 300 units of plywood.

69. Minimum cost of $36\frac{12}{13}$ is achieved by using $1\frac{11}{13}$ sacks of soybean meal and $1\frac{11}{13}$ sacks of oats.

71. Maximum income of $3110 is achieved when $22,000 is invested in corporate bonds and $18,000 is invested in municipal bonds.

73. Minimum cost of $460 thousand is achieved using 30 P_1's and 10 P_2's.

75. Maximum profit per day of $192 is achieved when 2 knit suits and 4 worsted suits are made.

77. Minimum weekly cost of $19.05 is achieved when 1.5 lb of meat and 3 lb of cheese are used.

79. Maximum total number of 800 is achieved when there are 550 of A and 250 of B.

81. Left to the student **83.** Discussion and Writing

85. [2.6] $\{x \mid -7 \leq x < 2\}$, or $[-7, 2)$

86. [2.6] $\{x \mid x \leq 1 \text{ or } x \geq 5\}$, or $(-\infty, 1] \cup [5, \infty)$

87. [3.5] $\{x \mid -1 \leq x \leq 3\}$, or $[-1, 3]$

88. [3.5] $\{x \mid -3 < x < -2\}$, or $(-3, -2)$

89.

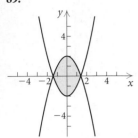

91.

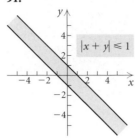

$|x + y| \leq 1$

93.

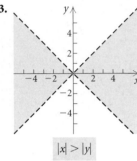

$|x| > |y|$

95. Maximum income of $28,500 is achieved by making 30 less expensive assemblies and 30 more expensive assemblies.

Exercise Set 5.8

1. $\dfrac{2}{x-3} - \dfrac{1}{x+2}$ **3.** $-\dfrac{4}{3x-1} + \dfrac{5}{2x-1}$

5. $-\dfrac{3}{x-2} + \dfrac{2}{x+2} + \dfrac{4}{x+1}$

7. $-\dfrac{3}{(x+2)^2} - \dfrac{1}{x+2} + \dfrac{1}{x-1}$ **9.** $\dfrac{3}{x-1} - \dfrac{4}{2x-1}$

11. $x - 2 - \dfrac{\frac{11}{4}}{(x+1)^2} + \dfrac{\frac{17}{16}}{x+1} - \dfrac{\frac{17}{16}}{x-3}$

13. $\dfrac{3x+5}{x^2+2} - \dfrac{4}{x-1}$ **15.** $-\dfrac{2}{x+2} + \dfrac{10}{(x+2)^2} + \dfrac{3}{2x-1}$

17. $3x + 1 + \dfrac{2}{2x-1} + \dfrac{3}{x+1}$

19. $-\dfrac{1}{x-3} + \dfrac{3x}{x^2+2x-5}$

21. $\dfrac{5}{3x+5} - \dfrac{3}{x+1} + \dfrac{4}{(x+1)^2}$ **23.** $\dfrac{8}{4x-5} + \dfrac{3}{3x+2}$

25. $\dfrac{2x-5}{3x^2+1} - \dfrac{2}{x-2}$ **27.** Discussion and Writing

29. Discussion and Writing **30.** [3.3] $3, \pm i$

31. [3.3] $-2, \dfrac{1 \pm \sqrt{5}}{2}$ **32.** [3.3] $-2, 3, \pm i$

33. [3.3] $-3, -1 \pm \sqrt{2}$

35. $-\dfrac{\frac{1}{2a^2}x}{x^2+a^2} + \dfrac{\frac{1}{4a^2}}{x-a} + \dfrac{\frac{1}{4a^2}}{x+a}$

37. $-\dfrac{3}{25(\ln x + 2)} + \dfrac{3}{25(\ln x - 3)} + \dfrac{7}{5(\ln x - 3)^2}$

Review Exercises: Chapter 5

1. [5.1] (a) **2.** [5.1] (e) **3.** [5.1] (h) **4.** [5.1] (d)
5. [5.7] (b) **6.** [5.7] (g) **7.** [5.7] (c) **8.** [5.7] (f)
9. [5.1] $(-2, -2)$ **10.** [5.1] $(-5, 4)$
11. [5.1] No solution **12.** [5.1] $(y - 2, y)$, or $(x, x + 2)$
13. [5.2] No solution **14.** [5.2] $(0, 0, 0)$
15. [5.2] $(-5, 13, 8, 2)$ **16.** [5.1], [5.2] Consistent: 9, 10, 12, 14, 15; the others are inconsistent.
17. [5.1], [5.2] Dependent: 12; the others are independent.
18. [5.3] $(1, 2)$ **19.** [5.3] $(-3, 4, -2)$

20. [5.3] $\left(\dfrac{z}{2}, -\dfrac{z}{2}, z\right)$ **21.** [5.3] $(-4, 1, -2, 3)$

22. [5.1] 31 nickels, 44 dimes
23. [5.1] $1600 at 3%, $3400 at 3.5%
24. [5.2] 1 bagel, $\frac{1}{2}$ serving cream cheese, 2 bananas
25. [5.2] 74.5, 68.5, 82
26. [5.2] **(a)** $f(x) = -\dfrac{19}{150}x^2 + \dfrac{46}{75}x + \dfrac{37}{5}$; **(b)** 4.2 lb

27. [5.4] $\begin{bmatrix} 0 & -1 & 6 \\ 3 & 1 & -2 \\ -2 & 1 & -2 \end{bmatrix}$ **28.** [5.4] $\begin{bmatrix} -3 & 3 & 0 \\ -6 & -9 & 6 \\ 6 & 0 & -3 \end{bmatrix}$

29. [5.4] $\begin{bmatrix} -1 & 1 & 0 \\ -2 & -3 & 2 \\ 2 & 0 & -1 \end{bmatrix}$ **30.** [5.4] $\begin{bmatrix} -2 & 2 & 6 \\ 1 & -8 & 18 \\ 2 & 1 & -15 \end{bmatrix}$

31. [5.4] Not possible **32.** [5.4] $\begin{bmatrix} 2 & -1 & -6 \\ 1 & 5 & -2 \\ -2 & -1 & 4 \end{bmatrix}$

33. [5.4] $\begin{bmatrix} -13 & 1 & 6 \\ -3 & -7 & 4 \\ 8 & 3 & -5 \end{bmatrix}$ **34.** [5.4] $\begin{bmatrix} -2 & -1 & 18 \\ 5 & -3 & -2 \\ -2 & 3 & -8 \end{bmatrix}$

35. [5.4] **(a)** $\begin{bmatrix} 46.1 & 5.9 & 10.1 & 8.5 & 11.4 \\ 54.6 & 4.6 & 9.6 & 7.6 & 10.6 \\ 48.9 & 5.5 & 12.7 & 9.4 & 9.3 \\ 51.3 & 4.8 & 11.3 & 6.9 & 12.7 \end{bmatrix}$

(b) $\begin{bmatrix} 32 & 19 & 43 & 38 \end{bmatrix}$;
(c) $\begin{bmatrix} 6564.7 & 695.1 & 1481.1 & 1082.8 & 1448.7 \end{bmatrix}$;
(d) the total cost, in cents, for each item for the day's meal

36. [5.5] $\begin{bmatrix} -\frac{1}{2} & 0 \\ \frac{1}{6} & \frac{1}{3} \end{bmatrix}$ **37.** [5.5] $\begin{bmatrix} 0 & 0 & \frac{1}{4} \\ 0 & -\frac{1}{2} & 0 \\ \frac{1}{3} & 0 & 0 \end{bmatrix}$

38. [5.5] $\begin{bmatrix} 1 & 0 & 0 & 0 \\ 0 & \frac{1}{9} & \frac{5}{18} & 0 \\ 0 & -\frac{1}{9} & \frac{2}{9} & 0 \\ 0 & 0 & 0 & 1 \end{bmatrix}$

39. [5.4] $\begin{bmatrix} 3 & -2 & 4 \\ 1 & 5 & -3 \\ 2 & -3 & 7 \end{bmatrix} \begin{bmatrix} x \\ y \\ z \end{bmatrix} = \begin{bmatrix} 13 \\ 7 \\ -8 \end{bmatrix}$ **40.** [5.5] $(-8, 7)$

41. [5.5] $(1, -2, 5)$ **42.** [5.5] $(2, -1, 1, -3)$
43. [5.6] 10   **44.** [5.6] -18 **45.** [5.6] -6
46. [5.6] -1 **47.** [5.6] $(3, -2)$ **48.** [5.6] $(-1, 5)$
49. [5.6] $\left(\frac{3}{2}, \frac{13}{14}, \frac{33}{14}\right)$ **50.** [5.6] $(2, -1, 3)$
51. [5.7]   **52.** [5.7]

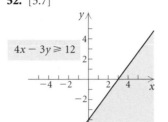

53. [5.7]

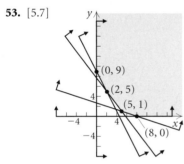

54. [5.7] Minimum $= 52$ at $(2, 4)$; maximum $= 92$ at $(2, 8)$
55. [5.7] Maximum score of 96 when 0 group A questions and 8 group B questions are answered

56. [5.8] $\dfrac{5}{x + 1} - \dfrac{5}{x + 2} - \dfrac{5}{(x + 2)^2}$

57. [5.8] $\dfrac{2}{2x - 3} - \dfrac{5}{x + 4}$

58. Discussion and Writing [5.1] During a holiday season, a caterer sold a total of 55 food trays. She sold 15 more seafood trays than cheese trays. How many of each were sold?
59. Discussion and Writing [5.4] In general, $(\mathbf{AB})^2 \neq \mathbf{A}^2\mathbf{B}^2$. $(\mathbf{AB})^2 = \mathbf{ABAB}$ and $\mathbf{A}^2\mathbf{B}^2 = \mathbf{AABB}$. Since matrix multiplication is not commutative, $\mathbf{BA} \neq \mathbf{AB}$, so $(\mathbf{AB})^2 \neq \mathbf{A}^2\mathbf{B}^2$.
60. [5.2] 4%: \$10,000; 5%: \$12,000; $5\frac{1}{2}$%: \$18,000
61. [5.1] $\left(\frac{5}{18}, \frac{1}{7}\right)$ **62.** [5.2] $\left(1, \frac{1}{2}, \frac{1}{3}\right)$

63. [5.7]   **64.** [5.7]

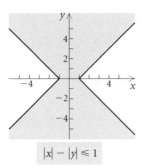

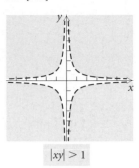

$|x| - |y| \leq 1$ $|xy| > 1$

Test: Chapter 5

1. [5.1] $(-3, 5)$; consistent, independent
2. [5.1] $(x, 2x - 3)$ or $\left(\dfrac{y + 3}{2}, y\right)$; consistent, dependent
3. [5.1] No solution; inconsistent, independent
4. [5.1] $(1, -2)$; consistent, independent
5. [5.2] $(-1, 3, 2)$ **6.** [5.1] Student: 342; nonstudent: 408
7. [5.2] Tricia: 120 orders; Maria: 104 orders; Antonio: 128 orders

8. [5.4] $\begin{bmatrix} -2 & -3 \\ -3 & 4 \end{bmatrix}$ **9.** [5.4] Not defined

10. [5.4] $\begin{bmatrix} -7 & -13 \\ 5 & -1 \end{bmatrix}$ **11.** [5.4] Not defined

12. [5.4] $\begin{bmatrix} 2 & -2 & 6 \\ -4 & 10 & 4 \end{bmatrix}$ **13.** [5.5] $\begin{bmatrix} 0 & -1 \\ -\frac{1}{4} & -\frac{3}{4} \end{bmatrix}$

14. [5.4] **(a)** $\begin{bmatrix} 49 & 10 & 13 \\ 43 & 12 & 11 \\ 51 & 8 & 12 \end{bmatrix}$; **(b)** $[26 \quad 18 \quad 23]$;

(c) $[3221 \quad 660 \quad 812]$; **(d)** the total cost, in cents, for each type of menu item served on the given day

15. [5.4] $\begin{bmatrix} 3 & -4 & 2 \\ 2 & 3 & 1 \\ 1 & -5 & -3 \end{bmatrix} \begin{bmatrix} x \\ y \\ z \end{bmatrix} = \begin{bmatrix} -8 \\ 7 \\ 3 \end{bmatrix}$ **16.** [5.5] $(-2, 1, 1)$

17. [5.6] 61   **18.** [5.6] -33 **19.** [5.6] $\left(-\frac{1}{2}, \frac{3}{4}\right)$
20. [5.7]

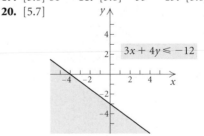

21. [5.7] Maximum: 15 at $(3, 3)$; minimum: 2 at $(1, 0)$
22. [5.7] Pound cakes: 25; carrot cakes: 75
23. [5.8] $-\dfrac{2}{x-1} + \dfrac{5}{x+3}$
24. [5.2] $A = 1, B = -3, C = 2$

Chapter 6

Exercise Set 6.1

1. (f)   **3.** (b)   **5.** (d)
7. $V: (0, 0)$; $F: (0, 5)$; $D: y = -5$

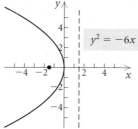

$x^2 = 20y$

9. $V: (0, 0)$; $F: \left(-\frac{3}{2}, 0\right)$; $D: x = \frac{3}{2}$

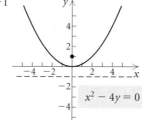

$y^2 = -6x$

11. $V: (0, 0)$; $F: (0, 1)$; $D: y = -1$

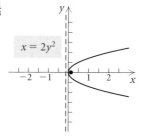

$x^2 - 4y = 0$

13. $V: (0, 0)$; $F: \left(\frac{1}{8}, 0\right)$; $D: x = -\frac{1}{8}$

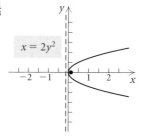

$x = 2y^2$

15. $y^2 = 16x$ **17.** $x^2 = -4\pi y$ **19.** $(y - 2)^2 = 14\left(x + \frac{1}{2}\right)$

21. $V: (-2, 1)$; $F: \left(-2, -\frac{1}{2}\right)$; $D: y = \frac{5}{2}$

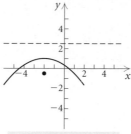

$(x + 2)^2 = -6(y - 1)$

23. $V: (-1, -3)$; $F: \left(-1, -\frac{7}{2}\right)$; $D: y = -\frac{5}{2}$

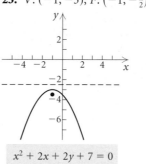

$x^2 + 2x + 2y + 7 = 0$

25. $V: (0, -2)$; $F: \left(0, -1\frac{3}{4}\right)$; $D: y = -2\frac{1}{4}$

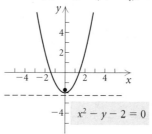

$x^2 - y - 2 = 0$

27. $V: (-2, -1)$; $F: \left(-2, -\frac{3}{4}\right)$; $D: y = -1\frac{1}{4}$

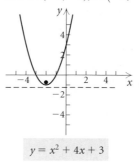

$y = x^2 + 4x + 3$

29. $V: \left(5\frac{3}{4}, \frac{1}{2}\right)$; $F: \left(6, \frac{1}{2}\right)$; $D: x = 5\frac{1}{2}$

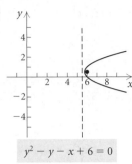

$$y^2 - y - x + 6 = 0$$

31. **(a)** $y^2 = 16x$; **(b)** $3\frac{33}{64}$ ft **33.** $\frac{2}{3}$ ft, or 8 in.
35. $V: (0.867, 0.348)$; $F: (0.867, -0.190)$; $D: y = 0.887$
37. Discussion and Writing **39.** [1.1] (h)
40. [1.1], [1.4] (d) **41.** [1.3] (a), (b), (f), (g)
42. [1.3] (b) **43.** [1.4] (b) **44.** [1.1] (f)
45. [1.4] (a) and (g) **46.** [1.4] (a) and (h); (g) and (h);
(b) and (c) **47.** $(x + 1)^2 = -4(y - 2)$
49. 10 ft, 11.6 ft, 16.4 ft, 24.4 ft, 35.6 ft, 50 ft

Exercise Set 6.2

1. (b) **3.** (d) **5.** (a)
7. $(7, -2)$; 8 **9.** $(-3, 1)$; 4

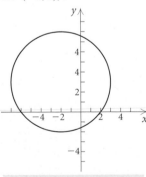

$$x^2 + y^2 - 14x + 4y = 11$$

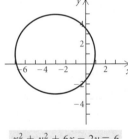

$$x^2 + y^2 + 6x - 2y = 6$$

11. $(-2, 3)$; 5 **13.** $(3, 4)$; 3

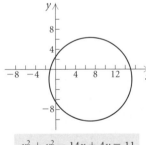

$$x^2 + y^2 + 4x - 6y - 12 = 0$$

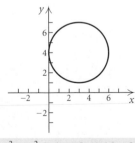

$$x^2 + y^2 - 6x - 8y + 16 = 0$$

15. $(-3, 5)$; $\sqrt{34}$ **17.** $\left(\frac{9}{2}, -2\right)$; $\frac{5\sqrt{5}}{2}$

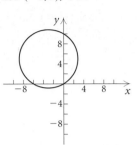

$$x^2 + y^2 + 6x - 10y = 0$$

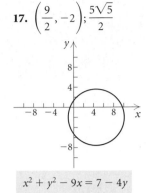

$$x^2 + y^2 - 9x = 7 - 4y$$

19. (c) **21.** (d)
23. $V: (2, 0), (-2, 0)$; **25.** $V: (0, 4), (0, -4)$;
$F: \left(\sqrt{3}, 0\right), \left(-\sqrt{3}, 0\right)$ $F: \left(0, \sqrt{7}\right), \left(0, -\sqrt{7}\right)$

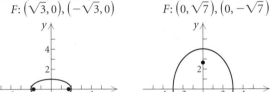

$$\frac{x^2}{4} + \frac{y^2}{1} = 1$$

$$16x^2 + 9y^2 = 144$$

27. $V: \left(-\sqrt{3}, 0\right), \left(\sqrt{3}, 0\right)$;
$F: (-1, 0), (1, 0)$

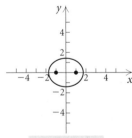

$$2x^2 + 3y^2 = 6$$

29. $V: \left(-\frac{1}{2}, 0\right), \left(\frac{1}{2}, 0\right)$;
$F: \left(-\frac{\sqrt{5}}{6}, 0\right), \left(\frac{\sqrt{5}}{6}, 0\right)$

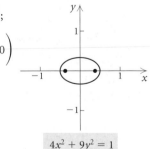

$$4x^2 + 9y^2 = 1$$

31. $\dfrac{x^2}{49} + \dfrac{y^2}{40} = 1$ **33.** $\dfrac{x^2}{25} + \dfrac{y^2}{64} = 1$ **35.** $\dfrac{x^2}{9} + \dfrac{y^2}{5} = 1$

37. $C: (1, 2);\ V: (4, 2), (-2, 2);\ F: \left(1 + \sqrt{5}, 2\right), \left(1 - \sqrt{5}, 2\right)$

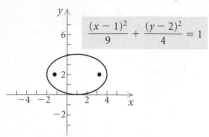

$$\dfrac{(x-1)^2}{9} + \dfrac{(y-2)^2}{4} = 1$$

39. $C: (-3, 5);\ V: (-3, 11), (-3, -1);\ F: \left(-3, 5 + \sqrt{11}\right),$
$\left(-3, 5 - \sqrt{11}\right)$

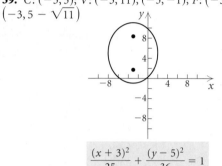

$$\dfrac{(x+3)^2}{25} + \dfrac{(y-5)^2}{36} = 1$$

41. $C: (-2, 1);\ V: (-10, 1), (6, 1);\ F: (-6, 1), (2, 1)$

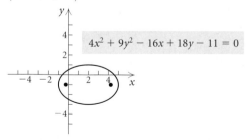

$$3(x+2)^2 + 4(y-1)^2 = 192$$

43. $C: (2, -1);\ V: (-1, -1), (5, -1);\ F: \left(2 + \sqrt{5}, -1\right),$
$\left(2 - \sqrt{5}, -1\right)$

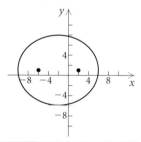

$$4x^2 + 9y^2 - 16x + 18y - 11 = 0$$

45. $C: (1, 1);\ V: (1, 3), (1, -1);\ F: \left(1, 1 + \sqrt{3}\right), \left(1, 1 - \sqrt{3}\right)$

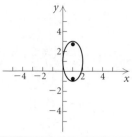

$$4x^2 + y^2 - 8x - 2y + 1 = 0$$

47. Example 2; $\dfrac{3}{5} < \dfrac{\sqrt{12}}{4}$ **49.** $\dfrac{x^2}{15} + \dfrac{y^2}{16} = 1$

51. $\dfrac{x^2}{2500} + \dfrac{y^2}{144} = 1$ **53.** 2×10^6 mi

55. $C: (2.003, -1.005);\ V: (-1.017, -1.005), (5.023, -1.005)$

57. Discussion and Writing **59.** [1.1] Midpoint

60. [2.1] Zero **61.** [1.1] y-intercept

62. [2.3] Two different real-number solutions

63. [3.2] Remainder **64.** [6.2] Ellipse **65.** [6.1] Parabola

66. [6.2] Circle **67.** $\dfrac{(x-3)^2}{4} + \dfrac{(y-1)^2}{25} = 1$

69. $\dfrac{x^2}{9} + \dfrac{y^2}{484/5} = 1$ **71.** About 9.1 ft

Exercise Set 6.3

1. (b) **3.** (c) **5.** (a) **7.** $\dfrac{y^2}{9} - \dfrac{x^2}{16} = 1$ **9.** $\dfrac{x^2}{4} - \dfrac{y^2}{9} = 1$

11. $C: (0, 0);\ V: (2, 0), (-2, 0);\ F: \left(2\sqrt{2}, 0\right), \left(-2\sqrt{2}, 0\right);$
$A: y = x,\ y = -x$

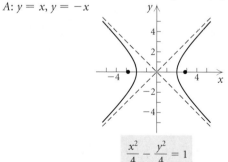

$$\dfrac{x^2}{4} - \dfrac{y^2}{4} = 1$$

13. $C: (2, -5)$; $V: (-1, -5), (5, -5)$; $F: \left(2 - \sqrt{10}, -5\right)$, $\left(2 + \sqrt{10}, -5\right)$; $A: y = -\dfrac{x}{3} - \dfrac{13}{3}, y = \dfrac{x}{3} - \dfrac{17}{3}$

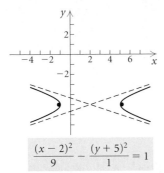

$$\frac{(x-2)^2}{9} - \frac{(y+5)^2}{1} = 1$$

15. $C: (-1, -3)$; $V: (-1, -1), (-1, -5)$; $F: \left(-1, -3 + 2\sqrt{5}\right), \left(-1, -3 - 2\sqrt{5}\right)$; $A: y = \frac{1}{2}x - \frac{5}{2}, y = -\frac{1}{2}x - \frac{7}{2}$

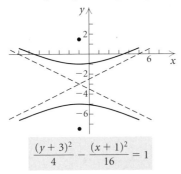

$$\frac{(y+3)^2}{4} - \frac{(x+1)^2}{16} = 1$$

17. $C: (0, 0)$; $V: (-2, 0), (2, 0)$; $F: \left(-\sqrt{5}, 0\right), \left(\sqrt{5}, 0\right)$; $A: y = -\frac{1}{2}x, y = \frac{1}{2}x$

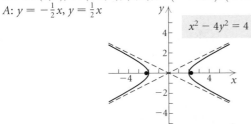

$$x^2 - 4y^2 = 4$$

19. $C: (0, 0)$; $V: (0, -3), (0, 3)$; $F: \left(0, -3\sqrt{10}\right), \left(0, 3\sqrt{10}\right)$; $A: y = \frac{1}{3}x, y = -\frac{1}{3}x$

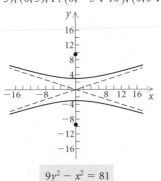

$$9y^2 - x^2 = 81$$

21. $C: (0, 0)$; $V: \left(-\sqrt{2}, 0\right), \left(\sqrt{2}, 0\right)$; $F: (-2, 0), (2, 0)$; $A: y = x, y = -x$

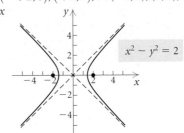

$$x^2 - y^2 = 2$$

23. $C: (0, 0)$; $V: \left(0, -\dfrac{1}{2}\right), \left(0, \dfrac{1}{2}\right)$; $F: \left(0, -\dfrac{\sqrt{2}}{2}\right)$, $\left(0, \dfrac{\sqrt{2}}{2}\right)$; $A: y = x, y = -x$

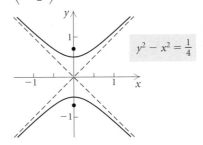

$$y^2 - x^2 = \frac{1}{4}$$

25. $C: (1, -2)$; $V: (0, -2), (2, -2)$; $F: \left(1 - \sqrt{2}, -2\right)$, $\left(1 + \sqrt{2}, -2\right)$; $A: y = -x - 1, y = x - 3$

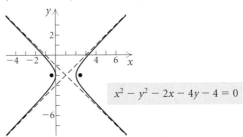

$$x^2 - y^2 - 2x - 4y - 4 = 0$$

27. $C: \left(\frac{1}{3}, 3\right)$; $V: \left(-\frac{2}{3}, 3\right), \left(\frac{4}{3}, 3\right)$; $F: \left(\frac{1}{3} - \sqrt{37}, 3\right)$, $\left(\frac{1}{3} + \sqrt{37}, 3\right)$; $A: y = 6x + 1, y = -6x + 5$

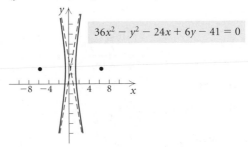

$$36x^2 - y^2 - 24x + 6y - 41 = 0$$

29. C: $(3, 1)$; V: $(3, 3)$, $(3, -1)$; F: $\left(3, 1 + \sqrt{13}\right)$, $\left(3, 1 - \sqrt{13}\right)$; A: $y = \frac{2}{3}x - 1$, $y = -\frac{2}{3}x + 3$

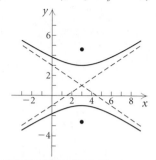

$$9y^2 - 4x^2 - 18y + 24x - 63 = 0$$

31. C: $(1, -2)$; V: $(2, -2)$, $(0, -2)$; F: $\left(1 + \sqrt{2}, -2\right)$, $\left(1 - \sqrt{2}, -2\right)$; A: $y = x - 3$, $y = -x - 1$

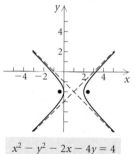

$$x^2 - y^2 - 2x - 4y = 4$$

33. C: $(-3, 4)$; V: $(-3, 10)$, $(-3, -2)$; F: $\left(-3, 4 + 6\sqrt{2}\right)$, $\left(-3, 4 - 6\sqrt{2}\right)$; A: $y = x + 7$, $y = -x + 1$

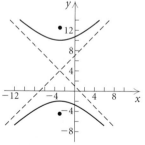

$$y^2 - x^2 - 6x - 8y - 29 = 0$$

35. Example 3; $\dfrac{\sqrt{5}}{1} > \dfrac{5}{4}$ **37.** $\dfrac{x^2}{9} - \dfrac{(y - 7)^2}{16} = 1$

39. $\dfrac{y^2}{25} - \dfrac{x^2}{11} = 1$ **41.** C: $(-1.460, -0.957)$;

V: $(-2.360, -0.957)$, $(-0.560, -0.957)$;

A: $y = -1.429x - 3.043$, $y = 1.429x + 1.129$

43. Discussion and Writing

45. [4.1] **(a)** Yes; **(b)** $f^{-1}(x) = \dfrac{x + 3}{2}$

46. [4.1] **(a)** Yes; **(b)** $f^{-1}(x) = \sqrt[3]{x - 2}$

47. [4.1] **(a)** Yes; **(b)** $f^{-1}(x) = \dfrac{5}{x} + 1$, or $\dfrac{5 + x}{x}$

48. [4.1] **(a)** Yes; **(b)** $f^{-1}(x) = x^2 - 4$, $x \geq 0$

49. [5.1], [5.3], [5.5], [5.6] $(6, -1)$

50. [5.1], [5.3], [5.5], [5.6] $(1, -1)$

51. [5.1], [5.3], [5.5], [5.6] $(2, -1)$

52. [5.1], [5.3], [5.5], [5.6] $(-3, 4)$

53. $\dfrac{(y + 5)^2}{9} - (x - 3)^2 = 1$

55. $\dfrac{x^2}{345.96} - \dfrac{y^2}{22,154.04} = 1$

Visualizing the Graph

1. B   **2.** J   **3.** F   **4.** I   **5.** H   **6.** G   **7.** E
8. D   **9.** C   **10.** A

Exercise Set 6.4

1. (e)   **3.** (c)   **5.** (b)   **7.** $(-4, -3)$, $(3, 4)$

9. $(0, 2)$, $(3, 0)$ **11.** $(-5, 0)$, $(4, 3)$, $(4, -3)$

13. $(3, 0)$, $(-3, 0)$ **15.** $(0, -3)$, $(4, 5)$ **17.** $(-2, 1)$

19. $(3, 4)$, $(-3, -4)$, $(4, 3)$, $(-4, -3)$

21. $\left(\dfrac{6\sqrt{21}}{7}, \dfrac{4i\sqrt{35}}{7}\right)$, $\left(\dfrac{6\sqrt{21}}{7}, -\dfrac{4i\sqrt{35}}{7}\right)$, $\left(-\dfrac{6\sqrt{21}}{7}, \dfrac{4i\sqrt{35}}{7}\right)$, $\left(-\dfrac{6\sqrt{21}}{7}, -\dfrac{4i\sqrt{35}}{7}\right)$

23. $(3, 2)$, $\left(4, \frac{3}{2}\right)$

25. $\left(\dfrac{5 + \sqrt{70}}{3}, \dfrac{-1 + \sqrt{70}}{3}\right)$, $\left(\dfrac{5 - \sqrt{70}}{3}, \dfrac{-1 - \sqrt{70}}{3}\right)$

27. $\left(\sqrt{2}, \sqrt{14}\right)$, $\left(-\sqrt{2}, \sqrt{14}\right)$, $\left(\sqrt{2}, -\sqrt{14}\right)$, $\left(-\sqrt{2}, -\sqrt{14}\right)$

29. $(1, 2)$, $(-1, -2)$, $(2, 1)$, $(-2, -1)$

31. $\left(\dfrac{15 + \sqrt{561}}{8}, \dfrac{11 - 3\sqrt{561}}{8}\right)$, $\left(\dfrac{15 - \sqrt{561}}{8}, \dfrac{11 + 3\sqrt{561}}{8}\right)$

33. $\left(\dfrac{7 - \sqrt{33}}{2}, \dfrac{7 + \sqrt{33}}{2}\right)$, $\left(\dfrac{7 + \sqrt{33}}{2}, \dfrac{7 - \sqrt{33}}{2}\right)$

35. $(3, 2)$, $(-3, -2)$, $(2, 3)$, $(-2, -3)$

37. $\left(\dfrac{5 - 9\sqrt{15}}{20}, \dfrac{-45 + 3\sqrt{15}}{20}\right)$, $\left(\dfrac{5 + 9\sqrt{15}}{20}, \dfrac{-45 - 3\sqrt{15}}{20}\right)$

39. $(3, -5)$, $(-1, 3)$ **41.** $(8, 5)$, $(-5, -8)$ **43.** $(3, 2)$, $(-3, -2)$ **45.** $(2, 1)$, $(-2, -1)$, $(1, 2)$, $(-1, -2)$

47. $\left(4 + \dfrac{3i\sqrt{6}}{2}, -4 + \dfrac{3i\sqrt{6}}{2}\right)$, $\left(4 - \dfrac{3i\sqrt{6}}{2}, -4 - \dfrac{3i\sqrt{6}}{2}\right)$

49. $\left(3, \sqrt{5}\right)$, $\left(-3, -\sqrt{5}\right)$, $\left(\sqrt{5}, 3\right)$, $\left(-\sqrt{5}, -3\right)$

51. $\left(\dfrac{8\sqrt{5}}{5}i, \dfrac{3\sqrt{105}}{5}\right), \left(\dfrac{8\sqrt{5}}{5}i, -\dfrac{3\sqrt{105}}{5}\right),$
$\left(-\dfrac{8\sqrt{5}}{5}i, \dfrac{3\sqrt{105}}{5}\right), \left(-\dfrac{8\sqrt{5}}{5}i, -\dfrac{3\sqrt{105}}{5}\right)$

53. $(2,1), (-2,-1), \left(-i\sqrt{5}, \dfrac{2i\sqrt{5}}{5}\right), \left(i\sqrt{5}, -\dfrac{2i\sqrt{5}}{5}\right)$

55. 6 cm by 8 cm   **57.** 4 in. by 5 in.
59. Length: $\sqrt{3}$ m; width: 1 m   **61.** 30 yd by 75 yd
63. 16 ft, 24 ft   **65.** $(1.564, 2.448), (0.138, 0.019)$
67. $(0.871, 1.388)$ **69.** $(1.146, 3.146), (-1.841, 0.159)$
71. $(0.965, 4402.33), (-0.965, -4402.33)$
73. $(2.112, -0.109), (-13.041, -13.337)$
75. $(400, 1.431), (-400, 1.431), (400, -1.431), (-400, -1.431)$
77. Discussion and Writing   **79.** [4.5] 2   **80.** [4.5] 2.048
81. [4.5] 81   **82.** [4.5] 5   **83.** $(x-2)^2 + (y-3)^2 = 1$

85. $\dfrac{x^2}{4} + y^2 = 1$ **87.** $\left(x + \dfrac{5}{13}\right)^2 + \left(y - \dfrac{32}{13}\right)^2 = \dfrac{5365}{169}$

89. There is no number x such that $\dfrac{x^2}{a^2} - \dfrac{\left(\dfrac{b}{a}x\right)^2}{b^2} = 1$,

because the left side simplifies to $\dfrac{x^2}{a^2} - \dfrac{x^2}{a^2}$, which is 0.

91. $\left(\dfrac{1}{2}, \dfrac{1}{4}\right), \left(\dfrac{1}{2}, -\dfrac{1}{4}\right), \left(-\dfrac{1}{2}, \dfrac{1}{4}\right), \left(-\dfrac{1}{2}, -\dfrac{1}{4}\right)$
93. Factor: $x^3 + y^3 = (x+y)(x^2 - xy + y^2)$. We know
that $x + y = 1$, so $(x+y)^2 = x^2 + 2xy + y^2 = 1$, or
$x^2 + y^2 = 1 - 2xy$. We also know that $xy = 1$, so
$x^2 + y^2 = 1 - 2 \cdot 1 = -1$. Then
$x^3 + y^3 = 1 \cdot (-1 - 1) = -2$.   **95.** $(2,4), (4,2)$
97. $(3,-2), (-3,2), (2,-3), (-2,3)$

99. $\left(\dfrac{2\log 3 + 3\log 5}{3(\log 3 \cdot \log 5)}, \dfrac{4\log 3 - 3\log 5}{3(\log 3 \cdot \log 5)}\right)$

Review Exercises: Chapter 6

1. [6.1] (d)   **2.** [6.2] (a)   **3.** [6.2] (e)   **4.** [6.3] (g)
5. [6.2] (b)   **6.** [6.2] (f)   **7.** [6.1] (h)   **8.** [6.3] (c)
9. [6.1] $x^2 = -6y$ **10.** [6.1] $F: (-3,0); V: (0,0);$
$D: x = 3$ **11.** [6.1] $V: (-5,8); F: \left(-5, \dfrac{15}{2}\right); D: y = \dfrac{17}{2}$
12. [6.2] $C: (2,-1); V: (-3,-1), (7,-1); F: (-1,-1),$
$(5,-1);$

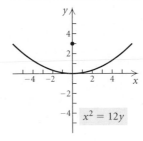

$16x^2 + 25y^2 - 64x + 50y - 311 = 0$

13. [6.2] $\dfrac{x^2}{9} + \dfrac{y^2}{16} = 1$

14. [6.3] $C: \left(-2, \dfrac{1}{4}\right); V: \left(0, \dfrac{1}{4}\right), \left(-4, \dfrac{1}{4}\right);$
$F: \left(-2 + \sqrt{6}, \dfrac{1}{4}\right), \left(-2 - \sqrt{6}, \dfrac{1}{4}\right);$
$A: y - \dfrac{1}{4} = \dfrac{\sqrt{2}}{2}(x+2), y - \dfrac{1}{4} = -\dfrac{\sqrt{2}}{2}(x+2)$
15. [6.1] 0.167 ft   **16.** [6.4] $\left(-8\sqrt{2}, 8\right), \left(8\sqrt{2}, 8\right)$

17. [6.4] $\left(3, \dfrac{\sqrt{29}}{2}\right), \left(-3, \dfrac{\sqrt{29}}{2}\right), \left(3, -\dfrac{\sqrt{29}}{2}\right),$
$\left(-3, -\dfrac{\sqrt{29}}{2}\right)$ **18.** [6.4] $(7,4)$

19. [6.4] $(2,2), \left(\dfrac{32}{9}, -\dfrac{10}{9}\right)$ **20.** [6.4] $(0,-3), (2,1)$
21. [6.4] $(4,3), (4,-3), (-4,3), (-4,-3)$
22. [6.4] $\left(-\sqrt{3}, 0\right), \left(\sqrt{3}, 0\right), (-2,1), (2,1)$
23. [6.4] $\left(-\dfrac{3}{5}, \dfrac{21}{5}\right), (3,-3)$
24. [6.4] $(6,8), (6,-8), (-6,8), (-6,-8)$
25. [6.4] $(2,2), (-2,-2), \left(2\sqrt{2}, \sqrt{2}\right), \left(-2\sqrt{2}, -\sqrt{2}\right)$
26. [6.4] 7, 4   **27.** [6.4] 7 m by 12 m   **28.** [6.4] 4, 8
29. [6.4] 32 cm, 20 cm   **30.** [6.4] 11 ft, 3 ft
31. Discussion and Writing [6.4] Although we can always
visualize the real-number solutions, we cannot visualize the
imaginary-number solutions.
32. Discussion and Writing [6.2] The equation of a circle can
be written as

$$\dfrac{(x-h)^2}{a^2} + \dfrac{(y-k)^2}{b^2} = 1,$$

where $a = b = r$, the radius of the circle. In an ellipse, $a > b$,
so a circle is not a special type of ellipse.

33. [6.2] $x^2 + \dfrac{y^2}{9} = 1$ **34.** [6.4] $\dfrac{8}{7}, \dfrac{7}{2}$

35. [6.2], [6.4] $(x-2)^2 + (y-1)^2 = 100$

36. [6.3] $\dfrac{x^2}{778.41} - \dfrac{y^2}{39{,}221.59} = 1$

Test: Chapter 6

1. [6.3] (c)   **2.** [6.1] (b)   **3.** [6.2] (a)   **4.** [6.2] (d)
5. [6.1] $V: (0,0); F: (0,3); D: y = -3$

$x^2 = 12y$

6. [6.1] $V: (-1, -1)$; $F: (1, -1)$; $D: x = -3$

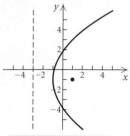

$$y^2 + 2y - 8x - 7 = 0$$

7. [6.1] $x^2 = 8y$

8. [6.2] Center: $(-1, 3)$; radius: 5

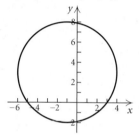

$$x^2 + y^2 + 2x - 6y - 15 = 0$$

9. [6.2] $C: (0, 0)$; $V: (-4, 0), (4, 0)$; $F: \left(-\sqrt{7}, 0\right), \left(\sqrt{7}, 0\right)$

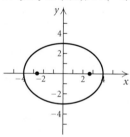

$$9x^2 + 16y^2 = 144$$

10. [6.2] $C: (-1, 2)$; $V: (-1, -1), (-1, 5)$; $F: \left(-1, 2 - \sqrt{5}\right), \left(-1, 2 + \sqrt{5}\right)$

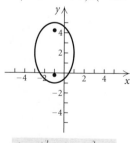

$$\frac{(x + 1)^2}{4} + \frac{(y - 2)^2}{9} = 1$$

11. [6.2] $\dfrac{x^2}{4} + \dfrac{y^2}{25} = 1$

12. [6.3] $C: (0, 0)$; $V: (-1, 0), (1, 0)$; $F: \left(-\sqrt{5}, 0\right), \left(\sqrt{5}, 0\right)$; $A: y = -2x, y = 2x$

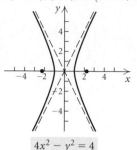

$$4x^2 - y^2 = 4$$

13. [6.3] $C: (-1, 2)$; $V: (-1, 0), (-1, 4)$; $F: \left(-1, 2 - \sqrt{13}\right), \left(-1, 2 + \sqrt{13}\right)$; $A: y = -\frac{2}{3}x + \frac{4}{3}, y = \frac{2}{3}x + \frac{8}{3}$

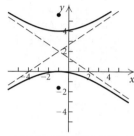

$$\frac{(y - 2)^2}{4} - \frac{(x + 1)^2}{9} = 1$$

14. [6.3] $y = \dfrac{\sqrt{2}}{2}x, y = -\dfrac{\sqrt{2}}{2}x$ **15.** [6.1] $\dfrac{27}{8}$ in.

16. [6.4] $(1, 2), (1, -2), (-1, 2), (-1, -2)$

17. [6.4] $(3, -2), (-2, 3)$ **18.** [6.4] $(2, 3), (3, 2)$

19. [6.4] 5 ft by 4 ft   **20.** [6.4] 60 ft by 45 ft

21. [6.2] $(x - 3)^2 + (y + 1)^2 = 8$

Chapter 7

Exercise Set 7.1

1. $3, 7, 11, 15; 39; 59$ **3.** $2, \frac{3}{2}, \frac{4}{3}, \frac{5}{4}; \frac{10}{9}; \frac{15}{14}$

5. $0, \frac{3}{5}, \frac{4}{5}, \frac{15}{17}; \frac{99}{101}; \frac{112}{113}$ **7.** $-1, 4, -9, 16; 100; -225$

9. $7, 3, 7, 3; 3; 7$ **11.** 34 **13.** 225 **15.** $-33{,}880$

17. 67 **19.** $2n$ **21.** $(-1)^n \cdot 2 \cdot 3^{n-1}$ **23.** $\dfrac{n + 1}{n + 2}$

25. $n(n + 1)$ **27.** $\log 10^{n-1}$, or $n - 1$ **29.** $6; 28$

31. $20; 30$ **33.** $\frac{1}{2} + \frac{1}{4} + \frac{1}{6} + \frac{1}{8} + \frac{1}{10} = \frac{137}{120}$

35. $1 + 2 + 4 + 8 + 16 + 32 + 64 = 127$

37. $\ln 7 + \ln 8 + \ln 9 + \ln 10 = \ln(7 \cdot 8 \cdot 9 \cdot 10) = \ln 5040 \approx 8.5252$

39. $\frac{1}{2} + \frac{2}{3} + \frac{3}{4} + \frac{4}{5} + \frac{5}{6} + \frac{6}{7} + \frac{7}{8} + \frac{8}{9} = \frac{15{,}551}{2520}$

41. $-1 + 1 - 1 + 1 - 1 = -1$

43. $3 - 6 + 9 - 12 + 15 - 18 + 21 - 24 = -12$

45. $2 + 1 + \frac{2}{5} + \frac{1}{5} + \frac{2}{17} + \frac{1}{13} + \frac{2}{37} = \frac{157{,}351}{40{,}885}$

47. $3 + 2 + 3 + 6 + 11 + 18 + 43$

49. $\frac{1}{2} + \frac{2}{3} + \frac{4}{5} + \frac{8}{9} + \frac{16}{17} + \frac{32}{33} + \frac{64}{65} + \frac{128}{129} + \frac{256}{257} + \frac{512}{513} + \frac{1024}{1025} \approx 9.736$

51. $\sum_{k=1}^{\infty} 5k$ **53.** $\sum_{k=1}^{6} (-1)^{k+1}2^k$ **55.** $\sum_{k=1}^{6} (-1)^k \frac{k}{k+1}$

57. $\sum_{k=2}^{n} (-1)^k k^2$ **59.** $\sum_{k=1}^{\infty} \frac{1}{k(k+1)}$ **61.** $4, 1\frac{1}{4}, 1\frac{4}{5}, 1\frac{5}{9}$

63. $6561, -81, 9i, -3\sqrt{i}$ **65.** $2, 3, 5, 8$

67. (a) 1062, 1127.84, 1197.77, 1272.03, 1350.90, 1434.65, 1523.60, 1618.07, 1718.39, 1824.93; **(b)** $3330.35

69. 1, 2, 4, 8, 16, 32, 64, 128, 256, 512, 1024, 2048, 4096, 8192, 16,384, 32,768, 65,536 **71.** 1, 1, 2, 3, 5, 8, 13

73.

n	U_n
1	2
2	2.25
3	2.3704
4	2.4414
5	2.4883
6	2.5216
7	2.5465
8	2.5658
9	2.5812
10	2.5937

75.

n	U_n
1	2
2	1.5538
3	1.4988
4	1.4914
5	1.4904
6	1.4902
7	1.4902
8	1.4902
9	1.4902
10	1.4902

77. (a) 27.25, 26.108, 25.002, 23.932, 22.898, 21.9, 20.938, 20.012, 19.122, 18.268; **(b)** 26, 19, 9, 22, 34

79. Discussion and Writing

80. [5.1], [5.3], [5.5], [5.6] $(-1, -3)$

81. [5.1], [5.3], [5.5], [5.6] 2001: $994.3 billion; 2002: $949.2 billion

82. [6.2] $(3, -2); 4$ **83.** [6.2] $\left(-\frac{5}{2}, 4\right); \frac{\sqrt{97}}{2}$

85. $i, -1, -i, 1, i; i$ **87.** $\ln(1 \cdot 2 \cdot 3 \cdots \cdot n)$

Exercise Set 7.2

1. $a_1 = 3, d = 5$ **3.** $a_1 = 9, d = -4$ **5.** $a_1 = \frac{3}{2}, d = \frac{3}{4}$
7. $a_1 = $316, d = -$3$ **9.** $a_{12} = 46$ **11.** $a_{14} = -\frac{17}{3}$
13. $a_{10} = 7941.62 **15.** 33rd **17.** 46th **19.** $a_1 = 5$
21. $n = 39$ **23.** $a_1 = \frac{1}{3}; d = \frac{1}{2}; \frac{1}{3}, \frac{5}{6}, \frac{4}{3}, \frac{11}{6}, \frac{7}{3}$ **25.** 670
27. 160,400 **29.** 735 **31.** 990 **33.** 1760 **35.** $\frac{65}{2}$
37. $-\frac{6026}{13}$ **39.** 1260 poles **41.** 3 plants; 171 plants
43. 4960¢, or $49.60 **45.** 1320 seats **47.** Yes; 3
49. Discussion and Writing
50. [5.1], [5.3], [5.5], [5.6] $(2, 5)$
51. [5.2], [5.3], [5.5], [5.6] $(2, -1, 3)$
52. [6.2] $(-4, 0), (4, 0); \left(-\sqrt{7}, 0\right), \left(\sqrt{7}, 0\right)$
53. [6.2] $\frac{x^2}{4} + \frac{y^2}{25} = 1$ **55.** n^2
57. $a_1 = 60 - 5p - 5q; d = 5p + 2q - 20$ **59.** 6, 8, 10
61. $5\frac{4}{5}, 7\frac{3}{5}, 9\frac{2}{5}, 11\frac{1}{5}$
63. Insert 16 arithmetic means between 1 and 50 with $d = \frac{49}{17}$.

65.
$$m = p + d$$
$$\underline{m = q - d}$$
$$2m = p + q \quad \textbf{Adding}$$
$$m = \frac{p + q}{2}$$

Visualizing the Graph

1. J **2.** A **3.** C **4.** G **5.** F **6.** H **7.** E
8. D **9.** B **10.** I

Exercise Set 7.3

1. 2 **3.** -1 **5.** -2 **7.** 0.1 **9.** $\frac{a}{2}$ **11.** 128
13. 162 **15.** $7(5)^{40}$ **17.** 3^{n-1} **19.** $(-1)^{n-1}$
21. $\frac{1}{x^n}$ **23.** 762 **25.** $\frac{4921}{18}$ **27.** 8 **29.** 125
31. Does not exist **33.** $\frac{2}{3}$ **35.** $29\frac{38,569}{59,049}$ **37.** 2
39. Does not exist **41.** $4545.\overline{45}$ **43.** $\frac{160}{9}$ **45.** $\frac{13}{99}$
47. 9 **49.** $\frac{34,091}{9990}$ **51. (a)** $\frac{1}{256}$ ft; **(b)** $5\frac{1}{3}$ ft
53. (a) About 297 ft; **(b)** 300 ft **55.** $23,841.50
57. $523,619.17 **59.** $86,666,666,667
61. Left to the student **63.** Discussion and Writing
65. [1.6] $(f \circ g)(x) = 16x^2 + 40x + 25$;
$(g \circ f)(x) = 4x^2 + 5$ **66.** [1.6] $(f \circ g)(x) = x^2 + x + 2$;
$(g \circ f)(x) = x^2 - x + 3$ **67.** [4.5] 2.209 **68.** [4.5] $\frac{1}{16}$
69. $\left(4 - \sqrt{6}\right)/\left(\sqrt{3} - \sqrt{2}\right) = 2\sqrt{3} + \sqrt{2}$,
$\left(6\sqrt{3} - 2\sqrt{2}\right)/\left(4 - \sqrt{6}\right) = 2\sqrt{3} + \sqrt{2}$; there exists a
common ratio, $2\sqrt{3} + \sqrt{2}$; thus the sequence is geometric.
71. (a) $\frac{13}{3}; \frac{22}{3}, \frac{34}{3}, \frac{46}{3}, \frac{58}{3}$; **(b)** $-\frac{11}{3}; -\frac{2}{3}, \frac{10}{3}, -\frac{50}{3}, \frac{250}{3}$ or 5; 8, 12,
18, 27 **73.** $S_n = \frac{x^2(1 - (-x)^n)}{x + 1}$

75. $\frac{a_{n+1}}{a_n} = r$, so $\ln \frac{a_{n+1}}{a_n} = \ln r$. But $\ln \frac{a_{n+1}}{a_n} = \ln a_{n+1} - \ln a_n = \ln r$. Thus, $\ln a_1, \ln a_2, \ldots,$ is an arithmetic sequence
with common difference $\ln r$. **77.** 512 cm^2

Exercise Set 7.4

1. $1^2 < 1^3$, false; $2^2 < 2^3$, true; $3^2 < 3^3$, true; $4^2 < 4^3$, true; $5^2 < 5^3$, true

3. A polygon of 3 sides has $\frac{3(3 - 3)}{2}$ diagonals. True; A
polygon of 4 sides has $\frac{4(4 - 3)}{2}$ diagonals. True; A polygon
of 5 sides has $\frac{5(5 - 3)}{2}$ diagonals. True; A polygon of
6 sides has $\frac{6(6 - 3)}{2}$ diagonals. True; A polygon of 7 sides
has $\frac{7(7 - 3)}{2}$ diagonals. True.

5. S_n: $2 + 4 + 6 + \cdots + 2n = n(n + 1)$
S_1: $2 = 1(1 + 1)$
S_k: $2 + 4 + 6 + \cdots + 2k = k(k + 1)$
S_{k+1}: $2 + 4 + 6 + \cdots + 2k + 2(k + 1)$
        $= (k + 1)(k + 2)$

1. *Basis step*: S_1 true by substitution.
2. *Induction step*: Assume S_k. Deduce S_{k+1}.
   Starting with the left side of S_{k+1}, we have

$$2 + 4 + 6 + \cdots + 2k + 2(k + 1)$$
$$= k(k + 1) + 2(k + 1) \qquad \textbf{By } S_k$$
$$= (k + 1)(k + 2). \qquad \textbf{Distributive law}$$

7. S_n: $1 + 5 + 9 + \cdots + (4n - 3) = n(2n - 1)$
S_1: $1 = 1(2 \cdot 1 - 1)$
S_k: $1 + 5 + 9 + \cdots + (4k - 3) = k(2k - 1)$
S_{k+1}: $1 + 5 + 9 + \cdots + (4k - 3) + [4(k + 1) - 3]$
        $= (k + 1)[2(k + 1) - 1]$
        $= (k + 1)(2k + 1)$

1. *Basis step*: S_1 true by substitution.
2. *Induction step*: Assume S_k. Deduce S_{k+1}.
   Starting with the left side of S_{k+1}, we have

$$1 + 5 + 9 + \cdots + (4k - 3) + [4(k + 1) - 3]$$
$$= k(2k - 1) + \lfloor 4(k + 1) - 3 \rfloor \qquad \textbf{By } S_k$$
$$= 2k^2 - k + 4k + 4 - 3$$
$$= (k + 1)(2k + 1).$$

9. S_n: $2 + 4 + 8 + \cdots + 2^n = 2(2^n - 1)$
S_1: $2 = 2(2 - 1)$
S_k: $2 + 4 + 8 + \cdots + 2^k = 2(2^k - 1)$
S_{k+1}: $2 + 4 + 8 + \cdots + 2^k + 2^{k+1} = 2(2^{k+1} - 1)$

1. *Basis step*: S_1 is true by substitution.
2. *Induction step*: Assume S_k. Deduce S_{k+1}.
   Starting with the left side of S_{k+1}, we have

$$\underbrace{2 + 4 + 8 + \cdots + 2^k}_{} + 2^{k+1}$$
$$= \underbrace{2(2^k - 1)}_{} + 2^{k+1} \qquad \textbf{By } S_k$$
$$= 2^{k+1} - 2 + 2^{k+1}$$
$$= 2 \cdot 2^{k+1} - 2$$
$$= 2(2^{k+1} - 1).$$

11. 1. *Basis step*: Since $1 < 1 + 1$, S_1 is true.
2. *Induction step*: Assume S_k. Deduce S_{k+1}. Now

$$k < k + 1 \qquad \textbf{By } S_k$$
$$k + 1 < k + 1 + 1 \qquad \textbf{Adding 1}$$
$$k + 1 < k + 2.$$

13. 1. *Basis step*: Since $2 = 2$, S_1 is true.
2. *Induction step*: Let k be any natural number. Assume S_k. Deduce S_{k+1}.

$$2k \le 2^k \qquad \textbf{By } S_k$$
$$2 \cdot 2k \le 2 \cdot 2^k \qquad \textbf{Multiplying by 2}$$
$$4k \le 2^{k+1}$$

Since $1 \le k$, $k + 1 \le k + k$, or $k + 1 \le 2k$.
Then $2(k + 1) \le 4k$.
Thus, $2(k + 1) \le 4k \le 2^{k+1}$, so $2(k + 1) \le 2^{k+1}$.

15.

S_n: $\dfrac{1}{1 \cdot 2 \cdot 3} + \dfrac{1}{2 \cdot 3 \cdot 4} + \dfrac{1}{3 \cdot 4 \cdot 5} + \cdots$

   $+ \dfrac{1}{n(n + 1)(n + 2)} = \dfrac{n(n + 3)}{4(n + 1)(n + 2)}$

S_1: $\dfrac{1}{1 \cdot 2 \cdot 3} = \dfrac{1(1 + 3)}{4(1 + 1)(1 + 2)}$

S_k: $\dfrac{1}{1 \cdot 2 \cdot 3} + \dfrac{1}{2 \cdot 3 \cdot 4} + \cdots + \dfrac{1}{k(k + 1)(k + 2)}$

   $= \dfrac{k(k + 3)}{4(k + 1)(k + 2)}$

S_{k+1}: $\dfrac{1}{1 \cdot 2 \cdot 3} + \dfrac{1}{2 \cdot 3 \cdot 4} + \cdots + \dfrac{1}{k(k + 1)(k + 2)}$

   $+ \dfrac{1}{(k + 1)(k + 2)(k + 3)}$

   $= \dfrac{(k + 1)(k + 1 + 3)}{4(k + 1 + 1)(k + 1 + 2)} = \dfrac{(k + 1)(k + 4)}{4(k + 2)(k + 3)}$

1. *Basis step*: Since $\dfrac{1}{1 \cdot 2 \cdot 3} = \dfrac{1}{6}$ and $\dfrac{1(1 + 3)}{4(1 + 1)(1 + 2)} =$

$\dfrac{1 \cdot 4}{4 \cdot 2 \cdot 3} = \dfrac{1}{6}$, S_1 is true.

2. *Induction step*: Assume S_k. Deduce S_{k+1}.

   Add $\dfrac{1}{(k + 1)(k + 2)(k + 3)}$ on both sides of S_k and

   simplify the right side. Only the right side is shown here.

$$\dfrac{k(k + 3)}{4(k + 1)(k + 2)} + \dfrac{1}{(k + 1)(k + 2)(k + 3)}$$
$$= \dfrac{k(k + 3)(k + 3) + 4}{4(k + 1)(k + 2)(k + 3)}$$
$$= \dfrac{k^3 + 6k^2 + 9k + 4}{4(k + 1)(k + 2)(k + 3)}$$
$$= \dfrac{(k + 1)^2(k + 4)}{4(k + 1)(k + 2)(k + 3)}$$
$$= \dfrac{(k + 1)(k + 4)}{4(k + 2)(k + 3)}$$

17.

S_n: $1 + 2 + 3 + \cdots + n = \dfrac{n(n + 1)}{2}$

S_1: $1 = \dfrac{1(1 + 1)}{2}$

S_k: $1 + 2 + 3 + \cdots + k = \dfrac{k(k + 1)}{2}$

S_{k+1}: $1 + 2 + 3 + \cdots + k + (k + 1) = \dfrac{(k + 1)(k + 2)}{2}$

1. *Basis step*: S_1 true by substitution.

2. *Induction step*: Assume S_k. Deduce S_{k+1}.
Starting with the left side of S_{k+1}, we have

$$\underbrace{1 + 2 + 3 + \cdots + k} + (k + 1)$$

$$= \frac{k(k + 1)}{2} + (k + 1) \qquad \textbf{By } S_k$$

$$= \frac{k(k + 1) + 2(k + 1)}{2} \qquad \textbf{Adding}$$

$$= \frac{(k + 1)(k + 2)}{2}. \qquad \textbf{Distributive law}$$

19. 1. *Basis step.* S_1: $\quad 1^3 = \dfrac{1^2(1 + 1)^2}{4} = 1$. True.

2. *Induction step*: Assume S_k. Deduce S_{k+1}.

$$S_k: \quad 1^3 + 2^3 + \cdots + k^3 = \frac{k^2(k + 1)^2}{4}$$

$$1^3 + 2^3 + \cdots + (k + 1)^3 = \frac{k^2(k + 1)^2}{4} + (k + 1)^3 \qquad \textbf{By } S_k$$

$$= \frac{(k + 1)^2}{4}[k^2 + 4(k + 1)]$$

$$= \frac{(k + 1)^2(k + 2)^2}{4}$$

21.

1. *Basis step.* S_1: $\quad 1^5 = \dfrac{1^2(1 + 1)^2(2 \cdot 1^2 + 2 \cdot 1 - 1)}{12}$. True.

2. *Induction step.* Assume S_k:

$$1^5 + 2^5 + \cdots + k^5 = \frac{k^2(k + 1)^2(2k^2 + 2k - 1)}{12}.$$

Then $1^5 + 2^5 + \cdots + k^5 + (k + 1)^5$

$$= \frac{k^2(k + 1)^2(2k^2 + 2k - 1)}{12} + (k + 1)^5$$

$$= \frac{k^2(k + 1)^2(2k^2 + 2k - 1) + 12(k + 1)^5}{12}$$

$$= \frac{(k + 1)^2(2k^4 + 14k^3 + 35k^2 + 36k + 12)}{12}$$

$$= \frac{(k + 1)^2(k + 2)^2(2k^2 + 6k + 3)}{12}$$

$$= \frac{(k + 1)^2(k + 1 + 1)^2(2(k + 1)^2 + 2(k + 1) - 1)}{12}.$$

23.

1. *Basis step.* S_1: $\quad 1(1 + 1) = \dfrac{1(1 + 1)(1 + 2)}{3}$. True.

2. *Induction step.* Assume S_k:

$$1(1 + 1) + 2(2 + 1) + \cdots + k(k + 1)$$
$$= \frac{k(k + 1)(k + 2)}{3}.$$

Then $1(1 + 1) + 2(2 + 1) + \cdots + k(k + 1)$
$\quad + (k + 1)(k + 1 + 1)$

$$= \frac{k(k + 1)(k + 2)}{3} + (k + 1)(k + 2)$$

$$= \frac{(k + 1)(k + 2)(k + 3)}{3}$$

$$= \frac{(k + 1)(k + 1 + 1)(k + 1 + 2)}{3}.$$

25. 1. *Basis step*: Since $\frac{1}{2}[2a_1 + (1 - 1)d] = \frac{1}{2} \cdot 2a_1 = a_1$,
S_1 is true.

2. *Induction step*: Assume S_k. Deduce S_{k+1}. Starting with
the left side of S_{k+1}, we have

$$\underbrace{a_1 + (a_1 + d) + \cdots + [a_1 + (k - 1)d]} + [a_1 + kd]$$

$$= \qquad \frac{k}{2}[2a_1 + (k - 1)d] \qquad + [a_1 + kd]$$
$$\qquad\qquad\qquad\qquad\qquad\qquad\qquad \textbf{By } S_k$$

$$= \frac{k[2a_1 + (k - 1)d]}{2} + \frac{2[a_1 + kd]}{2}$$

$$= \frac{2ka_1 + k(k - 1)d + 2a_1 + 2kd}{2}$$

$$= \frac{2a_1(k + 1) + k(k - 1)d + 2kd}{2}$$

$$= \frac{2a_1(k + 1) + (k - 1 + 2)kd}{2}$$

$$= \frac{2a_1(k + 1) + (k + 1)kd}{2}$$

$$= \frac{k + 1}{2}[2a_1 + kd].$$

27. Discussion and Writing
28. [5.1], [5.3], [5.5], [5.6] $(5, 3)$
29. [5.2], [5.3], [5.5], [5.6] $(2, -3, 4)$
30. [5.1], [5.3], [5.5], [5.6] Hardback: 50; paperback: 30
31. [5.2], [5.3], [5.5], [5.6] $800 at 1.5%, $1600 at 2%,
$2000 at 3%
33.
1. *Basis step.* S_1: $\quad x + y$ is a factor of $x^2 - y^2$. True.
$\qquad\qquad\quad S_2$: $\quad x + y$ is a factor of $x^4 - y^4$. True.
2. *Induction step.* Assume S_{k-1}: $x + y$ is a factor of
$x^{2(k-1)} - y^{2(k-1)}$. Then $x^{2(k-1)} - y^{2(k-1)} = (x + y)Q(x)$ for
some polynomial Q.
Assume S_k: $x + y$ is a factor of $x^{2k} - y^{2k}$. Then
$x^{2k} - y^{2k} = (x + y)P(x)$ for some polynomial P.
$x^{2(k+1)} - y^{2(k+1)}$
$$= (x^{2k} - y^{2k})(x^2 + y^2) - (x^{2(k-1)} - y^{2(k-1)})(x^2y^2)$$
$$= (x + y)P(x)(x^2 + y^2) - (x + y)Q(x)(x^2y^2)$$
$$= (x + y)[P(x)(x^2 + y^2) - Q(x)(x^2y^2)]$$
so $x + y$ is a factor of $x^{2(k+1)} - y^{2(k+1)}$.

35.

S_2: $\log_a (b_1 b_2) = \log_a b_1 + \log_a b_2$

S_k: $\log_a (b_1 b_2 \cdots b_k) = \log_a b_1 + \log_a b_2 + \cdots + \log_a b_k$

S_{k+1}: $\log_a (b_1 b_2 \cdots b_{k+1}) = \log_a b_1 + \log_a b_2 + \cdots + \log_a b_{k+1}$

1. *Basis step*: S_2 is true by the properties of logarithms.
2. *Induction step*: Let k be a natural number $k \geq 2$. Assume S_k. Deduce S_{k+1}.

$\log_a (b_1 b_2 \cdots b_{k+1})$ **Left side of S_{k+1}**
$= \log_a (b_1 b_2 \cdots b_k) + \log_a b_{k+1}$ **By S_2**
$= \log_a b_1 + \log_a b_2 + \cdots + \log_a b_k + \log_a b_{k+1}$

37. S_2: $\overline{z_1 + z_2} = \bar{z}_1 + \bar{z}_2$:

$\overline{(a + bi) + (c + di)} = \overline{(a + c) + (b + d)i}$
$= (a + c) - (b + d)i$
$\overline{(a + bi)} + \overline{(c + di)} = a - bi + c - di$
$= (a + c) - (b + d)i.$

S_k: $\overline{z_1 + z_2 + \cdots + z_k} = \bar{z}_1 + \bar{z}_2 + \cdots + \bar{z}_k.$

$\overline{(z_1 + z_2 + \cdots + z_k) + z_{k+1}}$
$= \overline{(z_1 + z_2 + \cdots + z_k)} + \overline{z_{k+1}}$ **By S_2**
$= \bar{z}_1 + \bar{z}_2 + \cdots + \bar{z}_k + \bar{z}_{k+1}$ **By S_k**

39. S_1: i is either i or -1 or $-i$ or 1.
S_k: i^k is either i or -1 or $-i$ or 1.
$i^{k+1} = i^k \cdot i$ is then $i \cdot i = -1$ or $-1 \cdot i = -i$ or $-i \cdot i = 1$ or $1 \cdot i = i$.

41. S_1:   3 is a factor of $1^3 + 2 \cdot 1$.
S_k: 3 is a factor of $k^3 + 2k$, i.e., $k^3 + 2k = 3 \cdot m$.
S_{k+1}:   3 is a factor of $(k + 1)^3 + 2(k + 1)$.
Consider

$(k + 1)^3 + 2(k + 1) = k^3 + 3k^2 + 5k + 3$
$= (k^3 + 2k) + 3k^2 + 3k + 3$
$= 3m + 3(k^2 + k + 1).$

A multiple of 3

Exercise Set 7.5

1. 720   **3.** 604,800   **5.** 120   **7.** 1   **9.** 3024   **11.** 120
13. 120   **15.** 1   **17.** 6,497,400   **19.** $n(n - 1)(n - 2)$
21. n **23.** $6! = 720$ **25.** $9! = 362,880$ **27.** $_9P_4 = 3024$

29. $_5P_5 = 120; 5^5 = 3125$ **31.** $\dfrac{8!}{3!} = 6720; \dfrac{7!}{2!} = 2520;$

$\dfrac{11!}{2! \, 2! \, 2!} = 4,989,600$ **33.** $8 \cdot 10^6 = 8,000,000$; 8 million

35. $\dfrac{9!}{2! \, 3! \, 4!} = 1260$ **37. (a)** $_6P_5 = 720$; **(b)** $6^5 = 7776$;

(c) $1 \cdot _5P_4 = 120$; **(d)** $1 \cdot 1 \cdot _4P_3 = 24$
39. (a) 10^5, or 100,000; **(b)** 100,000
41. (a) $10^9 = 1,000,000,000$; **(b)** yes
43. Discussion and Writing   **44.** $[2.1] \frac{9}{4}$, or 2.25

45. $[2.3]$ $-3, 2$ **46.** $[2.3]$ $\dfrac{3 \pm \sqrt{17}}{4}$ **47.** $[3.3]$ $-2, 1, 5$

49. 8   **51.** 11   **53.** $n - 1$

Exercise Set 7.6

1. 78   **3.** 78   **5.** 7   **7.** 10   **9.** 1   **11.** 15   **13.** 128
15. 270,725   **17.** 13,037,895   **19.** n **21.** 1
23. $_{23}C_4 = 8855$ **25.** $_{13}C_{10} = 286$

27. $\dbinom{8}{2} = 28; \dbinom{8}{3} = 56$ **29.** $\dbinom{52}{5} = 2,598,960$

31. (a) $_{31}P_2 = 930$; **(b)** $31^2 = 961$; **(c)** $_{31}C_2 = 465$
33. Discussion and Writing   **34.** $[2.1]$ $-\frac{17}{2}$

35. $[2.3]$ $-1, \frac{3}{2}$ **36.** $[2.3]$ $\dfrac{-5 \pm \sqrt{21}}{2}$

37. $[3.3]$ $-4, -2, 3$ **39.** $\dbinom{13}{5} = 1287$ **41.** $\dbinom{n}{2}; 2\dbinom{n}{2}$

43. 4   **45.** 7

47. $\dbinom{n}{k - 1} + \dbinom{n}{k}$

$= \dfrac{n!}{(k - 1)! \, (n - k + 1)!} \cdot \dfrac{k}{k}$

$\quad + \dfrac{n!}{k! \, (n - k)!} \cdot \dfrac{(n - k + 1)}{(n - k + 1)}$

$= \dfrac{n! \, (k + (n - k + 1))}{k! \, (n - k + 1)!}$

$= \dfrac{(n + 1)!}{k! \, (n - k + 1)!} = \dbinom{n + 1}{k}$

Exercise Set 7.7

1. $x^4 + 20x^3 + 150x^2 + 500x + 625$
3. $x^5 - 15x^4 + 90x^3 - 270x^2 + 405x - 243$
5. $x^5 - 5x^4y + 10x^3y^2 - 10x^2y^3 + 5xy^4 - y^5$
7. $15,625x^6 + 75,000x^5y + 150,000x^4y^2 + 160,000x^3y^3 + 96,000x^2y^4 + 30,720xy^5 + 4096y^6$
9. $128t^7 + 448t^5 + 672t^3 + 560t + 280t^{-1} + 84t^{-3} + 14t^{-5} + t^{-7}$
11. $x^{10} - 5x^8 + 10x^6 - 10x^4 + 5x^2 - 1$
13. $125 + 150\sqrt{5}\,t + 375t^2 + 100\sqrt{5}\,t^3 + 75t^4 + 6\sqrt{5}\,t^5 + t^6$
15. $a^9 - 18a^7 + 144a^5 - 672a^3 + 2016a - 4032a^{-1} + 5376a^{-3} - 4608a^{-5} + 2304a^{-7} - 512a^{-9}$
17. $140\sqrt{2}$ **19.** $x^{-8} + 4x^{-4} + 6 + 4x^4 + x^8$
21. $21a^5b^2$ **23.** $-252x^5y^5$ **25.** $-745,472a^3$
27. $1120x^{12}y^2$ **29.** $-1,959,552u^5v^{10}$ **31.** 128
33. 2^{24}, or 16,777,216   **35.** 20   **37.** $-12 + 316i$

39. $-7 - 4\sqrt{2}i$ **41.** $\displaystyle\sum_{k=0}^{n} \dbinom{n}{k}(-1)^k a^{n-k}b^k$

43. $\displaystyle\sum_{k=1}^{n} \dbinom{n}{k}x^{n-k}h^{k-1}$ **45.** Discussion and Writing

46. $[1.6]$ $x^2 + 2x - 2$ **47.** $[1.6]$ $2x^3 - 3x^2 + 2x - 3$
48. $[1.6]$ $4x^2 - 12x + 10$ **49.** $[1.6]$ $2x^2 - 1$
51. $-5 \pm 2\sqrt{2}, -5 \pm 2\sqrt{2}\,i$ **53.** $3, 9, 6 \pm 3i$

55. $-4320x^6y^{9/2}$ **57.** $-\dfrac{35}{x^{1/6}}$ **59.** 2^{100} **61.** $[\log_a (xt)]^{23}$

63. 1. *Basis step*: Since $a + b = (a + b)^1$, S_1 is true.

2. *Induction step*: Let S_k be the statement of the binomial theorem with n replaced by k. Multiply both sides of S_k by $(a + b)$ to obtain

$$(a + b)^{k+1}$$
$$= \left[a^k + \cdots + \binom{k}{r-1}a^{k-(r-1)}b^{r-1} \right.$$
$$\left. + \binom{k}{r}a^{k-r}b^r + \cdots + b^k \right](a + b)$$
$$= a^{k+1} + \cdots + \left[\binom{k}{r-1} + \binom{k}{r} \right]a^{(k+1)-r}b^r$$
$$+ \cdots + b^{k+1}$$
$$= a^{k+1} + \cdots + \binom{k+1}{r}a^{(k+1)-r}b^r + \cdots + b^{k+1}.$$

This proves S_{k+1}, assuming S_k. Hence S_n is true for $n = 1, 2, 3, \ldots$.

Exercise Set 7.8

1. **(a)** 0.18, 0.24, 0.23, 0.23, 0.12; **(b)** Opinions may vary, but it seems that people tend not to pick the first or last numbers.
3. 11,700 pieces   **5.** **(a)** T, S, R, N, L; **(b)** E; **(c)** yes
7. **(a)** $\frac{2}{7}$; **(b)** $\frac{5}{7}$; **(c)** 0 **(d)** 1   **9.** $\frac{350}{31,977}$ **11.** $\frac{1}{108,290}$
13. $\frac{33}{66,640}$ **15.** **(a)** HHH, HHT, HTH, HTT, THH, THT, TTH, TTT; **(b)** $\frac{3}{8}$; **(c)** $\frac{7}{8}$; **(d)** $\frac{7}{8}$; **(e)** $\frac{3}{8}$ **17.** $\frac{9}{19}$ **19.** $\frac{1}{38}$
21. $\frac{18}{19}$ **23.** $\frac{9}{19}$ **25.** Answers will vary.
27. Discussion and Writing   **28.** [2.1] Zero
29. [4.1] One-to-one   **30.** [1.2] Function; domain; range; domain; range   **31.** [2.1] Zero   **32.** [7.6] Combination
33. [3.6] Inverse variation   **34.** [3.2] Factor
35. [7.3] Geometric sequence
37. **(a)** 36; **(b)** $\dfrac{36}{{}_{52}C_5} \approx 1.39 \times 10^{-5}$
39. **(a)** $(13 \cdot {}_4C_3) \cdot (12 \cdot {}_4C_2) = 3744$; **(b)** 0.00144
41. **(a)** $4 \cdot \binom{13}{5} - 4 - 36 = 5108$; **(b)** 0.00197
43. **(a)** $\binom{10}{1}\binom{4}{1}\binom{4}{1}\binom{4}{1}\binom{4}{1}\binom{4}{1} - 4 - 36 = 10{,}200$;
(b) 0.00392

Review Exercises: Chapter 7

1. [7.1] $-\frac{1}{2}, \frac{4}{17}, -\frac{9}{82}, \frac{16}{257}; -\frac{121}{14,642}; -\frac{529}{279,842}$
2. [7.1] $(-1)^{n+1}(n^2 + 1)$ **3.** [7.1] $\frac{3}{2} - \frac{9}{8} + \frac{27}{26} - \frac{81}{80} = \frac{417}{1040}$
4. [7.1] $\displaystyle\sum_{k=1}^{7} (k^2 - 1)$ **5.** [7.2] $\frac{15}{4}$ **6.** [7.2] $a + 4b$
7. [7.2] 531   **8.** [7.2] 20,100   **9.** [7.2] 11   **10.** [7.2] -4
11. [7.3] $n = 6, S_n = -126$ **12.** [7.3] $a_1 = 8, a_5 = \frac{1}{2}$
13. [7.3] Does not exist   **14.** [7.3] $\frac{3}{11}$ **15.** [7.3] $\frac{3}{8}$

16. [7.3] $\frac{241}{99}$ **17.** [7.2] $5\frac{4}{5}, 6\frac{3}{5}, 7\frac{2}{5}, 8\frac{1}{5}$ **18.** [7.3] 167.3 ft
19. [7.3] \$45,993.04   **20.** [7.2] **(a)** \$7.38; **(b)** \$1365.10
21. [7.3] \$88,888,888,889

22. [7.4] S_n: $1 + 4 + 7 + \cdots + (3n - 2) = \dfrac{n(3n - 1)}{2}$

S_1: $1 = \dfrac{1(3 - 1)}{2}$

S_k: $1 + 4 + 7 + \cdots + (3k - 2) = \dfrac{k(3k - 1)}{2}$

S_{k+1}: $1 + 4 + 7 + \cdots + [3(k + 1) - 2]$
$= 1 + 4 + 7 + \cdots + (3k - 2) + (3k + 1)$
$= \dfrac{(k + 1)(3k + 2)}{2}$

1. *Basis step*: $\dfrac{1(3 - 1)}{2} = \dfrac{2}{2} = 1$ is true.

2. *Induction step*: Assume S_k. Add $(3k + 1)$ to both sides.

$1 + 4 + 7 + \cdots + (3k - 2) + (3k + 1)$
$$= \frac{k(3k - 1)}{2} + (3k + 1)$$
$$= \frac{k(3k - 1)}{2} + \frac{2(3k + 1)}{2}$$
$$= \frac{3k^2 - k + 6k + 2}{2}$$
$$= \frac{3k^2 + 5k + 2}{2}$$
$$= \frac{(k + 1)(3k + 2)}{2}$$

23. [7.4] S_n: $1 + 3 + 3^2 + \cdots + 3^{n-1} = \dfrac{3^n - 1}{2}$

S_1: $1 = \dfrac{3^1 - 1}{2}$

S_k: $1 + 3 + 3^2 + \cdots + 3^{k-1} = \dfrac{3^k - 1}{2}$

S_{k+1}: $1 + 3 + 3^2 + \cdots + 3^{(k+1)-1} = \dfrac{3^{k+1} - 1}{2}$

1. *Basis step*: $\dfrac{3^1 - 1}{2} = \dfrac{2}{2} = 1$ is true.

2. *Induction step*: Assume S_k. Add 3^k on both sides.

$1 + 3 + \cdots + 3^{k-1} + 3^k$
$$= \frac{3^k - 1}{2} + 3^k = \frac{3^k - 1}{2} + 3^k \cdot \frac{2}{2}$$
$$= \frac{3 \cdot 3^k - 1}{2} = \frac{3^{k+1} - 1}{2}$$

24. [7.4]

S_n: $\left(1 - \dfrac{1}{2}\right)\left(1 - \dfrac{1}{3}\right) \cdots \left(1 - \dfrac{1}{n}\right) = \dfrac{1}{n}$

S_2: $\left(1 - \dfrac{1}{2}\right) = \dfrac{1}{2}$

S_k: $\left(1 - \dfrac{1}{2}\right)\left(1 - \dfrac{1}{3}\right) \cdots \left(1 - \dfrac{1}{k}\right) = \dfrac{1}{k}$

S_{k+1}: $\left(1 - \dfrac{1}{2}\right)\left(1 - \dfrac{1}{3}\right) \cdots \left(1 - \dfrac{1}{k}\right)\left(1 - \dfrac{1}{k+1}\right)$
$= \dfrac{1}{k+1}$

1. *Basis step*: S_2 is true by substitution.
2. *Induction step*: Assume S_k. Deduce S_{k+1}. Starting with the left side of S_{k+1}, we have

$\underbrace{\left(1 - \dfrac{1}{2}\right)\left(1 - \dfrac{1}{3}\right) \cdots \left(1 - \dfrac{1}{k}\right)}\left(1 - \dfrac{1}{k+1}\right)$

$= \dfrac{1}{k} \cdot \left(1 - \dfrac{1}{k+1}\right)$ **By S_k**

$= \dfrac{1}{k} \cdot \left(\dfrac{k+1-1}{k+1}\right)$

$= \dfrac{1}{k} \cdot \dfrac{k}{k+1}$

$= \dfrac{1}{k+1}.$ **Simplifying**

25. [7.5] $6! = 720$ **26.** [7.5] $9 \cdot 8 \cdot 7 \cdot 6 = 3024$
27. [7.6] $\dbinom{15}{8} = 6435$ **28.** [7.5] $24 \cdot 23 \cdot 22 = 12{,}144$
29. [7.5] $\dfrac{9!}{1! \, 4! \, 2! \, 2!} = 3780$ **30.** [7.5] 36
31. [7.5] **(a)** $_6P_5 = 720$; **(b)** $6^5 = 7776$; **(c)** $_5P_4 = 120$;
(d) $_3P_2 = 6$ **32.** [7.7] 2^8, or 256
33. [7.7] $m^7 + 7m^6n + 21m^5n^2 + 35m^4n^3 + 35m^3n^4 + 21m^2n^5 + 7mn^6 + n^7$
34. [7.7] $x^5 - 5\sqrt{2}\,x^4 + 20x^3 - 20\sqrt{2}\,x^2 + 20x - 4\sqrt{2}$
35. [7.7] $x^8 - 12x^6y + 54x^4y^2 - 108x^2y^3 + 81y^4$
36. [7.7] $a^8 + 8a^6 + 28a^4 + 56a^2 + 70 + 56a^{-2} + 28a^{-4} + 8a^{-6} + a^{-8}$ **37.** [7.7] $-6624 + 16{,}280i$
38. [7.7] $220a^9x^3$ **39.** [7.7] $-\dbinom{18}{11}128a^7b^{11}$
40. [7.8] $\frac{1}{12}$; 0   **41.** [7.8] $\frac{1}{4}$ **42.** [7.8] $\frac{6}{5525}$
43. [7.8] $\frac{86}{206} \approx 0.42$, $\frac{97}{206} \approx 0.47$, $\frac{23}{206} \approx 0.11$

44. [7.1]

n	U_n
1	0.3
2	2.5
3	13.5
4	68.5
5	343.5
6	1718.5
7	8593.5
8	42968.5
9	214843.5
10	1074218.5

45. [7.1] **(a)** $a_n = 0.1112988224n + 3.300942958$;
(b) 7.7529 million
46. Discussion and Writing [7.6] A list of 9 candidates for an office is to be narrowed down to 4 candidates. In how many ways can this be done?
47. Discussion and Writing [7.3] Someone who has managed several sequences of hiring has a considerable income from the sales of the people in the lower levels. However, with a finite population, it will not be long before the salespersons in the lowest level have no one to hire and no one to sell to.
48. [7.4] S_1 fails for both (a) and (b).
49. [7.3] $\dfrac{a_{k+1}}{a_k} = r_1$, $\dfrac{b_{k+1}}{b_k} = r_2$, so $\dfrac{a_{k+1}b_{k+1}}{a_kb_k} = r_1r_2$, a constant
50. [7.2] **(a)** No (unless a_n is all positive or all negative);
(b) yes; **(c)** yes; **(d)** no (unless a_n is constant); **(e)** no (unless a_n is constant); **(f)** no (unless a_n is constant)
51. [7.2] $-2, 0, 2, 4$ **52.** [7.3] $\frac{1}{2}, -\frac{1}{6}, \frac{1}{18}$
53. [7.6] $\left(\log \dfrac{x}{y}\right)^{10}$ **54.** [7.6] 18   **55.** [7.6] 36
56. [7.7] -9

Test: Chapter 7

1. [7.1] -43 **2.** [7.1] $\frac{2}{3}, \frac{3}{4}, \frac{4}{5}, \frac{5}{6}, \frac{6}{7}$
3. [7.1] $2 + 5 + 10 + 17 = 34$ **4.** [7.1] $\displaystyle\sum_{k=1}^{6} 4k$
5. [7.1] $\displaystyle\sum_{k=1}^{\infty} 2^k$ **6.** [7.1] $3, 2\frac{1}{3}, 2\frac{3}{7}, 2\frac{7}{17}$ **7.** [7.2] 44
8. [7.2] 38   **9.** [7.2] -420 **10.** [7.2] 675   **11.** [7.3] $-\frac{5}{512}$
12. [7.3] 1000   **13.** [7.3] 510   **14.** [7.3] 27   **15.** [7.3] $\frac{56}{99}$
16. [7.1] $10{,}000, 8000, 6400, 5120, 4096, 3276.80$
17. [7.2] $12.50   **18.** [7.3] $74{,}399.77

19. [7.4]

S_n: $\quad 2 + 5 + 8 + \cdots + (3n - 1) = \dfrac{n(3n + 1)}{2}$

S_1: $\quad 2 = \dfrac{1(3 \cdot 1 + 1)}{2}$

S_k: $\quad 2 + 5 + 8 + \cdots + (3k - 1) = \dfrac{k(3k + 1)}{2}$

S_{k+1}: $\quad 2 + 5 + 8 + \cdots + (3k - 1) + [3(k + 1) - 1]$
$$= \dfrac{(k + 1)[3(k + 1) + 1]}{2}$$

1. *Basis step*: $\dfrac{1(3 \cdot 1 + 1)}{2} = \dfrac{1 \cdot 4}{2} = 2$, so S_1 is true.

2. *Induction step*:

$2 + 5 + 8 + \cdots + (3k - 1) + [3(k + 1) - 1]$
$$= \dfrac{k(3k + 1)}{2} + [3k + 3 - 1]$$
$$= \dfrac{3k^2}{2} + \dfrac{k}{2} + 3k + 2$$
$$= \dfrac{3k^2}{2} + \dfrac{7k}{2} + 2$$
$$= \dfrac{3k^2 + 7k + 4}{2}$$
$$= \dfrac{(k + 1)(3k + 4)}{2}$$
$$= \dfrac{(k + 1)[3(k + 1) + 1]}{2}$$

20. [7.5] 3,603,600   **21.** [7.6] 352,716

22. [7.6] $\dfrac{n(n - 1)(n - 2)(n - 3)}{24}$ **23.** [7.5] $_6P_4 = 360$

24. [7.5] **(a)** $6^4 = 1296$; **(b)** $_5P_3 = 60$

25. [7.6] $_{28}C_4 = 20{,}475$ **26.** [7.6] $_{12}C_8 \cdot {_8}C_4 = 34{,}650$

27. [7.7] $x^5 + 5x^4 + 10x^3 + 10x^2 + 5x + 1$

28. [7.7] $35x^3y^4$ **29.** [7.7] $2^9 = 512$ **30.** [7.8] $\frac{4}{7}$

31. [7.8] $\frac{48}{1001}$ **32.** [7.5] 15

23. $\angle R \leftrightarrow \angle A, \angle S \leftrightarrow \angle B, \angle T \leftrightarrow \angle C$;
$\overline{RS} \leftrightarrow \overline{AB}, \overline{RT} \leftrightarrow \overline{AC}, \overline{ST} \leftrightarrow \overline{BC}$

24. $\angle D \leftrightarrow \angle C, \angle E \leftrightarrow \angle N, \angle F \leftrightarrow \angle B$;
$\overline{DE} \leftrightarrow \overline{CN}, \overline{EF} \leftrightarrow \overline{NB}, \overline{FD} \leftrightarrow \overline{BC}$

25. $\angle C \leftrightarrow \angle W, \angle B \leftrightarrow \angle J, \angle S \leftrightarrow \angle Z$;
$\overline{CB} \leftrightarrow \overline{WJ}, \overline{CS} \leftrightarrow \overline{WZ}, \overline{BS} \leftrightarrow \overline{JZ}$

26. $\angle B \leftrightarrow \angle S, \angle V \leftrightarrow \angle T, \angle W \leftrightarrow \angle A$;
$\overline{BV} \leftrightarrow \overline{ST}, \overline{VW} \leftrightarrow \overline{TA}, \overline{WB} \leftrightarrow \overline{AS}$

27. $\angle A \leftrightarrow \angle R, \angle B \leftrightarrow \angle S, \angle C \leftrightarrow \angle T; \dfrac{AB}{RS} = \dfrac{AC}{RT} = \dfrac{BC}{ST}$

28. $\angle P \leftrightarrow \angle S, \angle Q \leftrightarrow \angle T, \angle R \leftrightarrow \angle V; \dfrac{PQ}{ST} = \dfrac{PR}{SV} = \dfrac{QR}{TV}$

29. $\angle M \leftrightarrow \angle C, \angle E \leftrightarrow \angle L, \angle S \leftrightarrow \angle F; \dfrac{ME}{CL} = \dfrac{MS}{CF} = \dfrac{ES}{LF}$

30. $\angle S \leftrightarrow \angle W, \angle M \leftrightarrow \angle L, \angle H \leftrightarrow \angle K; \dfrac{SM}{WL} = \dfrac{SH}{WK} = \dfrac{MH}{LK}$

31. $\dfrac{PS}{ND} = \dfrac{SQ}{DM} = \dfrac{PQ}{NM}$ **32.** $\dfrac{GH}{UT} = \dfrac{GK}{UV} = \dfrac{HK}{TV}$

33. $\dfrac{TA}{GF} = \dfrac{TW}{GC} = \dfrac{AW}{FC}$ **34.** $\dfrac{JK}{ON} = \dfrac{JL}{OM} = \dfrac{KL}{NM}$

35. $QR = 10, PR = 8$ **36.** $AM = 24, GT = 25$

37. 36 ft   **38.** 33 ft   **39.** 100 ft   **40.** 72 ft   **41.** 17

42. $\sqrt{34} \approx 5.831$ **43.** $\sqrt{32} \approx 5.657$ **44.** $\sqrt{98} \approx 9.899$

45. 12   **46.** 5   **47.** 4   **48.** $\sqrt{31} \approx 5.568$ **49.** 26

50. 13   **51.** 12   **52.** 24   **53.** 2   **54.** 1

55. $\sqrt{2} \approx 1.414$ **56.** $\sqrt{8} \approx 2.828$ **57.** 5

58. $\sqrt{50} \approx 7.071$ **59.** 3   **60.** $\sqrt{5} \approx 2.236$

61. $\sqrt{211{,}200{,}000}$ ft ≈ 14.533 ft

62. $\sqrt{1850}$ yd ≈ 43.012 yd   **63.** 240 ft   **64.** 25 ft

65. $\sqrt{18}$ cm ≈ 4.243 cm   **66.** $\sqrt{75}$ m ≈ 8.660 m

67. $\sqrt{208}$ ft ≈ 14.422 ft   **68.** $\sqrt{26{,}900}$ yd ≈ 164.012 yd

Appendix

1. Acute   **2.** Acute   **3.** Straight   **4.** Right

5. Obtuse   **6.** Obtuse   **7.** 79°   **8.** 7°   **9.** 23°

10. 85°   **11.** 32°   **12.** 58°   **13.** 61°   **14.** 36°

15. 177°   **16.** 126°   **17.** 41°   **18.** 167°   **19.** 95°

20. 51°   **21.** 78°   **22.** 135°

Photo Credits

Index

Index of Applications

STATISTICS/DEMOGRAPHICS

TRANSPORTATION

Geometry

Plane Geometry

Rectangle
Area: $A = lw$
Perimeter: $P = 2l + 2w$

Square
Area: $A = s^2$
Perimeter: $P = 4s$

Triangle
Area: $A = \frac{1}{2}bh$

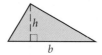

Sum of Angle Measures
$A + B + C = 180°$

Right Triangle
Pythagorean theorem
(equation):
$\quad a^2 + b^2 = c^2$

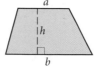

Parallelogram
Area: $A = bh$

Trapezoid
Area: $A = \frac{1}{2}h(a + b)$

Circle
Area: $A = \pi r^2$
Circumference:
$\quad C = \pi d = 2\pi r$

Solid Geometry

Rectangular Solid
Volume: $V = lwh$

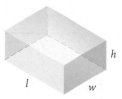

Cube
Volume: $V = s^3$

Right Circular Cylinder
Volume: $V = \pi r^2 h$
Lateral surface area:
$\quad L = 2\pi rh$
Total surface area:
$\quad S = 2\pi rh + 2\pi r^2$

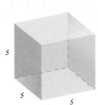

Right Circular Cone
Volume: $V = \frac{1}{3}\pi r^2 h$
Lateral surface area:
$\quad L = \pi rs$
Total surface area:
$\quad S = \pi r^2 + \pi rs$
Slant height:
$\quad s = \sqrt{r^2 + h^2}$

Sphere
Volume: $V - \frac{4}{3}\pi r^3$
Surface area: $S = 4\pi r^2$

Algebra

Properties of Real Numbers

Commutative: $\quad a + b = b + a; \quad ab = ba$

Associative: $\quad a + (b + c) = (a + b) + c;$
$a(bc) = (ab)c$

Additive Identity: $\quad a + 0 = 0 + a = a$

Additive Inverse: $\quad -a + a = a + (-a) = 0$

Multiplicative Identity: $\quad a \cdot 1 = 1 \cdot a = a$

Multiplicative Inverse: $\quad a \cdot \dfrac{1}{a} = 1, \ a \neq 0$

Distributive: $\quad a(b + c) = ab + ac$

Exponents and Radicals

$$a^m \cdot a^n = a^{m+n} \qquad \frac{a^m}{a^n} = a^{m-n}$$

$$(a^m)^n = a^{mn} \qquad (ab)^m = a^m b^m$$

$$\left(\frac{a}{b}\right)^m = \frac{a^m}{b^m} \qquad a^{-n} = \frac{1}{a^n}$$

If n is even, $\sqrt[n]{a^n} = |a|$.

If n is odd, $\sqrt[n]{a^n} = a$.

$\sqrt[n]{a} \cdot \sqrt[n]{b} = \sqrt[n]{ab}, \ a, b \geq 0$

$$\sqrt[n]{\frac{a}{b}} = \frac{\sqrt[n]{a}}{\sqrt[n]{b}}$$

$\sqrt[n]{a^m} = \left(\sqrt[n]{a}\right)^m = a^{m/n}$

Special-Product Formulas

$(a + b)(a - b) = a^2 - b^2$

$(a + b)^2 = a^2 + 2ab + b^2$

$(a - b)^2 = a^2 - 2ab + b^2$

$(a + b)^3 = a^3 + 3a^2b + 3ab^2 + b^3$

$(a - b)^3 = a^3 - 3a^2b + 3ab^2 - b^3$

$$(a + b)^n = \sum_{k=0}^{n} \binom{n}{k} a^{n-k} b^k, \quad \text{where}$$

$$\binom{n}{k} = \frac{n!}{k!\,(n - k)!}$$

$$= \frac{n(n - 1)(n - 2)\cdots[n - (k - 1)]}{k!}$$

Factoring Formulas

$a^2 - b^2 = (a + b)(a - b)$

$a^2 + 2ab + b^2 = (a + b)^2$

$a^2 - 2ab + b^2 = (a - b)^2$

$a^3 + b^3 = (a + b)(a^2 - ab + b^2)$

$a^3 - b^3 = (a - b)(a^2 + ab + b^2)$

Interval Notation

$(a, b) = \{x \mid a < x < b\}$

$[a, b] = \{x \mid a \leq x \leq b\}$

$(a, b] = \{x \mid a < x \leq b\}$

$[a, b) = \{x \mid a \leq x < b\}$

$(-\infty, a) = \{x \mid x < a\}$

$(a, \infty) = \{x \mid x > a\}$

$(-\infty, a] = \{x \mid x \leq a\}$

$[a, \infty) = \{x \mid x \geq a\}$

Absolute Value

$|a| \geq 0$

For $a > 0$,

$$|X| = a \rightarrow X = -a \quad \text{or} \quad X = a,$$

$$|X| < a \rightarrow -a < X < a,$$

$$|X| > a \rightarrow X < -a \quad \text{or} \quad X > a.$$

Equation-Solving Principles

$a = b \rightarrow a + c = b + c$

$a = b \rightarrow ac = bc$

$a = b \rightarrow a^n = b^n$

$ab = 0 \leftrightarrow a = 0 \quad \text{or} \quad b = 0$

$x^2 = k \rightarrow x = \sqrt{k} \quad \text{or} \quad x = -\sqrt{k}$

Inequality-Solving Principles

$a < b \rightarrow a + c < b + c$

$a < b$ and $c > 0 \rightarrow ac < bc$

$a < b$ and $c < 0 \rightarrow ac > bc$

(Algebra continued)

Algebra *(continued)*

The Distance Formula

The distance from (x_1, y_1) to (x_2, y_2) is given by

$$d = \sqrt{(x_2 - x_1)^2 + (y_2 - y_1)^2}.$$

Midpoint Formula

The midpoint of the line segment from (x_1, y_1) to (x_2, y_2) is given by

$$\left(\frac{x_1 + x_2}{2}, \frac{y_1 + y_2}{2} \right).$$

Formulas Involving Lines

The slope of the line containing points (x_1, y_1) and (x_2, y_2) is given by

$$m = \frac{y_2 - y_1}{x_2 - x_1}.$$

Slope–intercept equation: $\quad y = f(x) = mx + b$

Horizontal line: $\qquad\qquad y = b \quad \text{or} \quad f(x) = b$

Vertical line: $\qquad\qquad\quad x = a$

Point–slope equation: $\qquad y - y_1 = m(x - x_1)$

The Quadratic Formula

The solutions of $ax^2 + bx + c = 0, a \neq 0$, are given by

$$x = \frac{-b \pm \sqrt{b^2 - 4ac}}{2a}.$$

Compound Interest Formulas

Compounded n times per year: $\quad A = P\left(1 + \frac{i}{n}\right)^{nt}$

Compounded continuously: $\qquad P(t) = P_0 e^{kt}$

Properties of Exponential and Logarithmic Functions

$\log_a x = y \leftrightarrow x = a^y$ $\qquad\qquad a^x = a^y \leftrightarrow x = y$

$\log_a MN = \log_a M + \log_a N$ $\qquad \log_a M^p = p \log_a M$

$\log_a \dfrac{M}{N} = \log_a M - \log_a N$

$\log_b M = \dfrac{\log_a M}{\log_a b}$

$\log_a a = 1$ $\qquad\qquad\qquad\qquad \log_a 1 = 0$

$\log_a a^x = x$ $\qquad\qquad\qquad\qquad a^{\log_a x} = x$

Conic Sections

Circle: $\qquad (x - h)^2 + (y - k)^2 = r^2$

Ellipse: $\qquad \dfrac{(x - h)^2}{a^2} + \dfrac{(y - k)^2}{b^2} = 1,$

$\qquad\qquad \dfrac{(x - h)^2}{b^2} + \dfrac{(y - k)^2}{a^2} = 1$

Parabola: $\qquad (x - h)^2 = 4p(y - k),$

$\qquad\qquad (y - k)^2 = 4p(x - h)$

Hyperbola: $\qquad \dfrac{(x - h)^2}{a^2} - \dfrac{(y - k)^2}{b^2} = 1,$

$\qquad\qquad \dfrac{(y - k)^2}{a^2} - \dfrac{(x - h)^2}{b^2} = 1$

Arithmetic Sequences and Series

$a_1, a_1 + d, a_1 + 2d, a_1 + 3d, \ldots$

$a_{n+1} = a_n + d \qquad a_n = a_1 + (n - 1)d$

$S_n = \dfrac{n}{2}(a_1 + a_n)$

Geometric Sequences and Series

$a_1, a_1 r, a_1 r^2, a_1 r^3, \ldots$

$a_{n+1} = a_n r \qquad\qquad a_n = a_1 r^{n-1}$

$S_n = \dfrac{a_1(1 - r^n)}{1 - r} \qquad S_\infty = \dfrac{a_1}{1 - r}, \ |r| < 1$